JAGUAR
XJ-S

XJ-S 3.6
XJ-SC 3.6
XJ-S 4.0

SERVICE MANUAL

Jaguar Cars Ltd   **AKM 9063 and Supplement AKM 9063BB**

> # This manual and supplement covers all 6-cylinder powered XJS models - XJS 3.6, XJSC 3.6 and XJS 4.0.
>
> # The supplement covers later vehicles, mainly dealing with those modifications brought about in line with the introduction of the automatic transmission, 4 Litre engine and related systems, ABS brake systems, body modifications and more.

Service Tools

Where performance of an operation requires the use of a service tool the tool number is quoted under the operation heading and is repeated in, or following, the instruction involving its use. A list of all necessary tools is included in Section 99.

References

References to the LH or RH side in the manual are made when viewing from the rear. With the engine and gearbox assembly removed the timing cover end of the engine is referred to as the front. A key to abbreviations and symbols is given in Section 01.

REPAIRS AND REPLACEMENTS

When service parts are required it is essential that only genuine Jaguar or Unipart replacements are used.

Attention is particularly drawn to the following points concerning repairs and the fitting of replacement parts and accessories.

1. Safety features embodied in the vehicle may be impaired if other than genuine parts are fitted. In certain territories, legislation prohibits the fitting of parts not to the vehicle manufacturer's specification.
2. Torque wrench setting figures given in this Service Manual must be strictly adhered to.
3. Locking devices, where specified, must be fitted. If the effiency of a locking device is impaired during removal it must be replaced.
4. Owners purchasing accessories while travelling abroad should ensure that the accessory and its fitted location on the vehicle conform to mandatory requirements existing in their country of origin.
5. The vehicle warranty may be invalidated by the fitting of other than genuine Jaguar or Unipart parts. All Jaguar and Unipart replacements have the full backing of the factory warranty.
6. Jaguar Distributors and Dealers are obliged to supply only genuine service parts.

SPECIFICATION

Purchasers are advised that the specification details set out in this Manual apply to a range of vehicles and not to any one. For the specification of a particular vehicle, purchasers should consult their Distributor or Dealer.

The Manufacturers reserve the right to vary their specifications with or without notice, and at such times and in such manner as they think fit. Major as well as minor changes may be involved in accordance with the Manufacturer's policy of constant product improvement.

Whilst every effort is made to ensure the accuracy of the particulars contained in this Manual, neither the Manufacturer nor the Distributor or Dealer, by whom this Manual is supplied, shall in any circumstances be held liable for any inaccuracy or the consequences thereof.

COPYRIGHT

XJ-S 3.6, XJ-SC 3.6 Service Manual Contents

XJ-S 4.0 Service Manual Supplement Contents

STANDARDIZED ABBREVIATIONS AND SYMBOLS IN THIS MANUAL

Abbreviation or Symbol	Term	Abbreviation or Symbol	Term
A	Ampere	L.H.Thd.	Left-hand thread
A.B.D.C.	After bottom dead centre	l.t.	Low tension (electrical)
a.c.	Alternating current		
A.F.	Across flats (bolt/nut size)	M	Metric (screw thread)
Ah	Ampere hour	m	Metres
A.T.D.C.	After top dead centre	max.	Maximum
Atm	Atmospheres	MES	Miniature Edison Screw
Auto.	Automatic transmission	min.	Minimum
		mm	Millimetres
B.A.	British Association (screw thread)	mmHg	Millimetres of mercury
B.B.D.C.	Before bottom dead centre	m.p.g.	Miles per gallon
B.D.C.	Bottom dead centre	m.p.h.	Miles per hour
b.h.p.	Brake horse-power		
b.m.e.p.	Brake mean effective pressure	N	Newton
B.S.	British Standards	Nm	Newton metres
B.S.F.	British Standard Fine (screw thread)	No.	Numbers
B.S.P.	British Standard Pipe (thread)	Nox	Oxides of nitrogen
B.S.W.	British Standard Whitworth (screw thread)	N.P.T.F.	American Standard Taper Pipe (thread)
B.T.D.C.	Before top dead centre		
		O_2	Oxygen
C	Centigrade (Celsius)	O/D	Overdrive
cm	Centimetres	o.dia.	Outside diameter
cm^2	Square centimetres	oz	Ounces (mass)
cm^3	Cubic centimetres	ozf	Ounces (force)
c/min	Cycles per minute	ozf in	Ounces inch (torque)
CO	Carbon monoxide		
cwt	Hundredweight	para.	Paragraph
		Part no.	Part number
d.c.	Direct current	PAS	Power assisted steering
deg.	Degree (angle or temperature)	pt	Imperial pints
dia.	Diameter		
DIN	Deutsche Industrie Norm (Standard)	r	Radius
		ref.	Reference
E.C.U.	Electronic Control Unit	rev/min	Revolutions per minute
E.G.R.	Exhaust Gas Recirculation	R.H.	Right-hand
		R.H.Stg.	Right-hand steering
F	Fahrenheit		
F.I.	Fuel Injection	S.A.E.	Society of Automotive Engineers
Fig.	Figure (illustration)	S.C.	Single carburetters
ft	Feet	sp. gr.	Specific gravity
ft/min	Feet per minute	Std.	Standard
		s.w.g	Standard wire gauge
g	Grammes (mass)	Synchro	{ Synchronizer
gal	Imperial gallons		{ Synchromesh
gf	Grammes (force)		
		T.C.	Twin carburetters
h.c.	High compression	T.D.C.	Top dead centre
hp	Horse-power	t.p.i.	Threads per inch
h.t.	High tension (electrical)		
		U.N.C.	Unified Coarse (screw thread)
i.dia.	Internal diameter	U.N.F.	Unified Fine (screw thread)
i.f.s.	Independent front suspension	U.K.	United Kingdom
in	Inches	U.S. gal	Gallons (US)
in^2	Square inches	U.S. pt	Pints (US)
in^3	Cubic inches		
inHg	Inches of mercury	V	Volts
kg	Kilogrammes (mass)	W	Watts
kgf/cm^2	Kilogrammes per square centimetre		
kgf m	Kilogramme metres	1st	First
km	Kilometres	2nd	Second
km/h	Kilometres per hour	3rd	Third
kPa	Kilopascals	4th	Fourth
k.p.i.	King pin inclination	5th	Fifth
kV	Kilovolts	°	Degree (angle or temperature)
kW	Kilowatts	∞	Infinity
		/	Minute (angle)
lb	Pounds (mass)		{ Minus (tolerance)
lbf	Pounds (force)	−	{ Negative (electrical)
lbf ft	Pounds feet (torque)	%	Percentage
lbf/ft^2	Pounds per square foot	+	Plus (tolerance)
lbf in	Pounds inches (torque)	+ve	Positive (electrical)
lbf/in^2	Pounds per square inch	−ve	Negative (electrical)
l.c.	Low compression	±	Plus or minus (tolerance)
L.H.	Left-hand	//	Second (angle)
L.H.Stg.	Left-hand steering	Ω	Ohms

4

Engine	See engine tuning and data	Section 05
Final Drive Unit	Type	Hypoid with Powr-Lok differential
	Ratio : Standard	3.54 : 1
	North American market	3.77 : 1
Manual Gearbox	Type	Five speed with baulk-ring synchromesh on all forward gears
	Ratios : First gear	3.573 : 1
	Second gear	2.056 : 1
	Third gear	1.391 : 1
	Fourth gear	1.0 : 1
	Fifth gear	0.76 : 1
	Reverse	3.463 : 1
Cooling System	Water pump : type	Centrifugal
	drive	Belt
	Number of cooling fans	One 11-bladed fan, driven through viscous coupling
	Cooling system and control	Thermostat
	Auxiliary cooling	One fan blowing air through radiator; controlled by a sensor in the radiator
	Thermostat opening temperature	88°C (190°F)
	Filler cap : pressure rating	1.05 kgf/cm² (15 lbf/in²)
	Make	AC Delco
Fuel Injection Equipment	Make and type	Digital, P type Lucas Bosch
Fuel System Pump	Make and type	Electrial, Lucas 73175 roller cell pump with integral relief valve and non-return valve
	Fuel pressure	2.5 bar (36 lbf/in²)
Braking System	Front brakes, make and type	Girling, ventilated discs, bridge type calipers
	Rear brakes, make and type	Girling, damped discs, bridge type calipers incorporating handbrake friction pads
	Handbrake : type	Mechanical, operating on rear discs
	Disc diameter : Front	284 mm (11.18 in)
	Rear	263,5 mm (10.375 in)
	Disc thickness : Front	24,13 mm (0.95 in), 22,86 mm (0.90 in) min
	Rear	12,7 mm (0.50 in), 11,43 mm (0.45 in) min
	Master cylinder bore diameter	23,800 to 23,825 mm (0.937 to 0.939 in)
	Brake operation	Hydraulic
	Hydraulic fluid	Castrol/Girling Universal Brake and Clutch Fluid — exceeding specification SAE J1703/D
	Main brake friction pad material	Ferodo 3401 slotted
	Handbrake friction pad material	Mintex M 68/1
	Servo units refs : RHD cars	Girling 64049669
	LHD cars	Girling 64049668
Front Suspension	Type	Independent coil spring
	Castor angle	$3\frac{1}{2}° \pm \frac{1}{4}°$ positive
	Camber angle	$\frac{1}{2}° \pm \frac{1}{4}°$ negative
	Front wheel alignment	0 to 3.18 mm toe in (0 to $\frac{1}{8}$ in toe in)
	Dampers	Telescopic, gas filled
Rear Suspension	Type	Independent coil springs, co-axial with dampers
	Camber angle	$\frac{3}{4}° \pm \frac{1}{4}°$ negative
	Rear wheel alignment	Parallel $\pm$ 0,08 mm (Parallel $\pm \frac{1}{32}$ in)
	Dampers	Telescopic, gas filled
Power Assisted Steering	Type	Rack and Pinion
	Number of turns lock to lock	2.75
	Turning circle — between walls	12,6 m (41.3 ft)

Electrical Equipment		
Battery	Make and type	RH Stg cars — Lucas CP11
	Voltage	12V
	Number of plates per cell	11
	Capacity at 20 hour rate	66 Ah
Alternator	Make and type: all air conditioned cars	Lucas A133
	Normal voltage	12V
	Cut-in voltage	13.5V at 1500 rev/min
	Earth polarity	Negative
	Maximum output	75 A
	Maximum operating speed	15,000 rpm
	Rotor winding resistance	2.43 at 20°C
	Brush spring pressure	4.7 to 9.8
Starter Motor	Make and type	Lucas M45 pre-engaged
	Lock torque at 940 amps	4.01 kgf m (29 lbf ft)
	Torque at 1000 rev/min at 535 amps	1.80 kgf m (13 lbf ft)
	Light running current	100A at 5000 to 6000 rev/min
Distributor	Make and type	Lucas 45DM6
Windscreen Wiper Motor	Make and type	Lucas 16W
	Light running speed, rack disconnected (after 60 seconds from cold)	Normal: 46 to 52 rpm; high: 60 to 70 rpm
	Light running current (after 60 seconds from cold)	Normal: 1.5A; high: 2.0A

TYRE DATA

Fitted as complete sets only

Tyres

Type	Dunlop 215/70 VR 15 SP or Pirelli P5 215/70	
	Sport Super D7	VR15 Cinturato

Inflation Pressure

	Front	Rear
For speeds above 160 km/hr (100 mph)	2.25 kg/cm²	2.25 kg/cm²
under all	32 lb/in²	32 lb/in²
load conditions	2.20 Bars	2.20 Bars

For maximum comfort in countries where speeds are not in excess of 160 km/hr (100 mph) the above inflation pressures may be reduced by 0.42 kg/cm² (6 lb/in², 0.41 Bars) on front and rear tyres.

Tyre Replacement and Wheel Interchanging

When replacement of tyres is necessary, it is preferable to fit a complete car set. Should either front or rear tyres only show a necessity for replacement, new tyres must be fitted to replace the worn ones. No attempt must be made to interchange tyres from front to rear or vice-versa as tyre wear produces characteristic patterns depending upon their position and if such position is changed after wear has occurred, the performance of the tyre will be adversely affected. It should be remembered that new tyres require to be balanced.

The radial-ply tyres specified above are designed to meet the high-speed performance of which this car is capable.

Only tyres of identical specification as shown under 'TYRE DATA' must be fitted as replacements and, if to different tread pattern, should not be fitted in mixed form.

UNDER NO CIRCUMSTANCES SHOULD CROSS-PLY TYRES BE FITTED.

RECOMMENDED SNOW TYRE

The following information relates to the only snow tyre recommended for Jaguar Cars. The use of snow tyres fitted with studs is not permitted in certain countries.

TYRE DESIGNATION	RECOMMENDED FITMENT	ROAD SPEED AND TYRE PRESSURES		REMARKS
Dunlop Weathermaster 185SR 15 SP M & S (Mud and Slush)	Complete sets only	Up to 137 km/h (85 mph)		1. Snow chains may be fitted to rear wheels only.
		FRONT	**REAR**	
		1,90 kgf/cm²	1,83 kgf/cm²	2. Tyres may be fitted with studs provided maximum speed does not exceed 121 km/h (75 mph)
		27 lbf/in²	26 lbf/in²	
		1.86 bar	1.79 bar	
		From 137 km/h (85 mph) up to a maximum of 161 km/h (100 mph)		3. Inner tubes with the wording 'Weathermaster only' are available and MUST be fitted when using 185SR 15 SP M & S Dunlop Weathermaster tyres
		FRONT	**REAR**	
		2,46 kgf/cm²	2,39 kgf/cm²	
		35 lbf/in²	34 lbf/in²	
		2.41 bar	2.35 bar	

Stopping distance

(a) fully operational maximum load	....	182 ft
(b) fully operational light load	....	163 ft
(c) emergency — rears only	....	387 ft
(d) boost failure	....	258 ft

BULB CHART

	Watts	Lucas Part No.	Unipart No.	Notes
Headlamps — not France or USA — main/dip	60/55	472	GLB 472	Halogen H4 base bulb
— France only — main/dip	60/55	476	GLB 476	Yellow Halogen H4 base bulb
— USA only — outer	37.5/60			Tungsten sealed beam light unit
— USA only — inner	50			Tungsten sealed beam light unit
Headlamp pilot bulb — not USA	4	233	GLB 233	
Front flasher lamp	21	382	GLB 382	Not USA
Front flasher and side lamp	21/5	380	GLB 380	USA only
Stop lamp	21	382	GLB 382	
Tail lamp	5	207	GLB 207	
Rear flasher	21	382	GLB 382	
Reversing lamp	21	273	GLB 273	Festoon bulb
Number plate lamp	6	254	GLB 254	Festoon bulb
Sidemarker	4	233	GLB 233	USA only
Flasher side repeaters — when fitted	4	233	GLB 233	
Rear fog guard lamps	21	382	GLB 382	Not USA
Interior/map lamps	6	254	GLB 254	Festoon bulb
Roof lamp	10	272	GLB 272	Festoon bulb
Boot lamp	5	239	GLB 239	Festoon bulb
Fibre optic light source	6	254	GLB 254	
Instrument illumination	2.2	987	GLB 987	
Warning lights	1.2	286	GLB 286	
Automatic selector illumination	2.2	987	GLB 987	
Cigar lighter illumination	2.2	987	GLB 987	
Door puddle lamp	5	501	GLB 501	

MAIN FUSE BOX — RH Steering

Fuse No.	Protected Circuit	Fuse Capacity	Unipart Number
1.	Cigar lighter	20A	GFS420
2.	Hazard warning, seat belt logic	15A	GFS415
3.	Clock, aerial, caravan, boot lamp	35A	GFS435
4.	Panel instruments, reverse light	10A	GFS410
5.	Direction indicators, stop lamps, auto kick down switch	15A	GFS415
6.	Fog rear guard	10A	GFS410
7.	Panel/cigar lighter/selector illumination	10A	GFS410
8.	Door locks, electric mirrors	3A	GFS43
9.	Wipers	35A	GFS435
10.	Air conditioning motors	10A	GFS410
11.	Air conditioning controls, horn, washers, radiator cooling fan	35A	GFS435
12.	Heated rear screen, heated mirrors	35A	GFS435

MAIN FUSE BOX — LH Steering

Fuse No.	Protected Circuit	Fuse Capacity	Unipart Number
1.	Front fog lights	20A	GFS240
2.	Hazard warning, seat belt logic	15A	GFS415
3.	Clock, aerial, caravan, boot lamp	35A	GFS435
4.	Panel instruments, reverse light	10A	GFS410
5.	Direction indicators, stop lamps, auto kick down switch	15A	GFS415
6.	Fog rear guard	10A	GFS410
7.	Panel/cigar lighter/selector illumination	10A	GFS410
8.	Door locks, electric mirrors	3A	GFS43
9.	Wipers	35A	GFS435
10.	Air conditioning motors	50A	GFS450
11.	Air conditioning controls, horn, washers, radiator cooling fan	35A	GFS435
12.	Heated rear screen, heated mirrors	35A	GFS435

HEADLAMP FUSE BOX

Fuse No.	Protected Circuit	Fuse Capacity	Unipart Number
1.	Radiator auxiliary cooling fan motor relay	25A	GFS425
2.	LH main beam	25A	GFS425
3.	LH dip beam	10A	GFS410
4.	RH main beam	25A	GFS425
5.	RH dip beam	10A	GFS410

AUXILIARY FUSE BOX — RH Steering

Fuse No.	Protected Circuit	Fuse Capacity	Unipart Number
13.	Interior and map lights	10A	GFS410
14.	LH side lights	3A	GFS43
15.	RH side lights	3A	GFS43
16.	Front fog lights	20A	GFS420
17.	Speed control	3A	GFS43

AUXILIARY FUSE BOX — LH Steering

Fuse No.	Protected Circuit	Fuse Capacity	Unipart Number
13.	Interior and map lights	10A	GFS410
14.	LH side lights	3A	GFS43
15.	RH side lights	3A	GFS43
16.	Cigar lighter	20A	GFS420
17.	Speed control	3A	GFS43

This page is intentionally blank

ENGINE TUNING AND DATA

Ignition Timing	Federal	18° BTDC @ 2000 RPM vacuum pipe disconnected at normal running temperature
	European	21° BTDC @ 2000 RPM vacuum pipe disconnected at normal running temperature

Valve Clearances	Inlet and exhaust	0,30 to 0,36 mm (0.012 to 0.014 in)

Spark Plugs	Make/type	Champion RC12 YC
	Gap	0,64 mm (0.025 in)

Ignition Coil	Make/type	Lucas 35 C6
	Primary resistance at 20°C (68°F)	0.9 to 1.1 ohms
	Secondary resistance	3.95 to 9.52K
	Consumption : stationary	5.0 to 6.5 amps
	running	2.5 to 3.0 amps

Distributor	Make/type	Lucas constant energy 45 DM6
	Rotation of rotor (looking down at rotor)	Clockwise
	Pick-up module to timing rotor gap	0,20 to 0,36 mm (0.008 to 0.014 in)
	Firing order	1, 5, 3, 6, 2, 4 (No. 1 cylinder at front)

Spark plug lead resistences	No. 1	5.13K to 12.3K
	2	5.47K to 13.2K
	3	6.11K to 14.69K
	4	7.24K to 17.34K
	5	9.0K to 21.48K
	6	8.61K to 20.56K

DISTRIBUTOR ADVANCE CURVE (CENTRIFUGAL) EUROPEAN VEHICLES

Advance Characteristics (Dynamic)

Engine rpm	Advance (degrees)
6400	$7\frac{1}{4}°$ to $9\frac{1}{2}°$
5000	$7°$ to $9°$
3200	$6°$ to $8°$
1800	$5\frac{1}{4}°$ to $7\frac{1}{4}°$
1250	$3°$ to $6°$
1050	$\frac{1}{2}°$ to $3\frac{1}{2}°$
850	$-1.2°$ to $0.8°$
700	$-1.1°$ to $0.9°$
500	$-\frac{1}{4}°$ to $\frac{1}{4}°$
300	$-0.9°$ to $1.1°$

Vacuum Advance Curve

inHg	Advance (degrees)
24	$5°$ to $7°$
12	$4°$ to $6°$
$8\frac{1}{2}$	$2°$ to $4°$
6	$\frac{1}{2}°$ to $2\frac{1}{2}°$
4	$0°$ to $1°$
2	$0°$

DISTRIBUTOR ADVANCE CURVE (CENTRIFUGAL)
USA FEDERAL VEHICLES

Advance Characteristics (Dynamic)

Engine rpm	Advance (degrees)
6400	$7\frac{3}{4}°$ to 10°
5200	7° to 9°
3500	$5\frac{1}{2}°$ to $7\frac{1}{2}°$
2000	4° to 6°
1300	2° to 4°
1100	$\frac{1}{2}°$ to $2\frac{1}{2}°$
900	−1° to 1°
700	−1.1° to 0.9°
500	$-\frac{1}{4}°$ to $\frac{1}{4}°$
300	−0.9° to 1.1°

Vacuum Advance Curve

inHg	Advance (degrees)
20	5° to 7°
9	4° to 6°
7	2° to 4°
$5\frac{1}{2}$	$\frac{1}{2}°$ to $2\frac{1}{2}°$
4	0° to 1°
2	0°

Fuel Injection Equipment Digital .... Lucas digital P type
Fuel pressure .... 2.5 bar (36 lbf/in²)

Exhaust Emission Exhaust gas analyser reading at engine idle speed 1 to 2% max. CO at 800 rev/min without air injection

Idle Speed 800±50 rev/min.

Compression Pressure European 200 to 220 lb/in² Federal 160 to 175 lb/in²

Differential between Cylinders 10 lb/in² (max) 10 lb/in² (max)

Note: Compressions to be checked with all sparking plugs removed, the throttle wide open, the engine at operating temperature and a cranking speed of 300 rpm (minimum).

ENGINE DATA

General Data
Number of cylinders	6
Bore	91 mm
Stroke	92 mm
Cubic capacity	3590 cc
Compression ratio	9.6:1
Firing order	1, 5, 3, 6, 2, 4 (No. 1 cylinder at front)

Cylinder Block
Material (cylinder block)	Cast Aluminium alloy
Bore diameters after honing:	
Piston grade	
A	91,002 to 90,990 mm
B	91,018 to 91,005 mm
+ 0.010 in	91,272 to 91,259 mm
+ 0.020 in	91,526 to 91,513 mm

Cylinder head
Material	Aluminium alloy
Valve seat angle (inclusive)	89° 30' to 89° 00'

Crankshaft
Material	SE Cast Iron
Number of main bearings	7
Main bearing type	Vandervell VP2C
Journal diameter	76,230 to 76,218 mm
Journal length (over $\frac{3}{32}$ in radii)	
Front	30,48 mm
Centre	36,233 to 36,208 mm
Remainder	30,53 to 30,43 mm
Thrust washer material	VP10 or VP2
Thrust washer diameter	3,888 to 3,898 mm
Thrust washer thickness Std	2,57 to 2,62 mm (0.101 to 0.103 in)
Oversize	2,67 to 2,72 mm (0.105 to 0.107 in)
Permissible end-float	0,10 to 0,25 mm (0.004 to 0.010 in)
Width of main bearing:	
Front intermediate and rear	24,71 to 24,33 mm
Centre	30,48 to 30,10 mm
Balancing	Crankshaft to be balanced to 45 gm/cm ($\frac{5}{8}$ oz/in)
	Unbalance to be corrected by drilling up to 4 holes in each balance weight 9,5 x 29 mm max depth.
Diametrical clearance	0,041 to 0,084 mm
Crankpin: Diameter	52,987 to 52,974 mm
Length	30,193 to 30,142 mm

Connecting Rods
Length between centres	166,42 to 166,32 mm
Big end bearing material	Vandervell VP2C
Bore for big end bearing	56,731 mm
Width of big end bearing	24,77 to 24,38 mm (0.975 to 0.960 in)
Big end diametrical clearance	0,001 to 0,0027 mm
Big end side clearance	0,132 to 0,233 mm
Small end bush material	VP10
Bore for small end bush	27,000 to 26,975 mm
Width of small end bush	30,00 to 29,50 mm
Fitted I/D of small end bush	23,818 to 23,814 mm
Bore diameter of small end bush	23,48 to 23,38 mm
Small end balance weight:	
Minimum dimension after balancing	17 mm
Big end balance weight:	
Minimum dimension after balancing	42 mm

Pistons	Type	HG413 (BS 1490 — 1970 — LM13TF)
		(phosphorous modified)
	Grade	
	Diameter: A	90,972 to 90,960 mm
	B	90,987 to 90,975 mm
	+ 0.010 in	91,241 to 91,229 mm
	+ 0.020 in	91,495 to 91,483 mm

Piston Rings	No. of compression rings	2
	No. of oil control rings	1
	Top compression ring:	
	Gap	0,4 to 0,65 mm
	Width	1,490 to 1,495 mm
	Diameter	91,00 mm
	Depth of chrome	0.06 mm
	Tangential load of 6.62N to 9.93N to close gap to	0,40 to 0,65 mm
	Material	Cast iron HG22C
	Second compression ring:	
	Gap	0,4 to 0,65 mm
	Width	1,990 to 1,975 mm
	Diameter	91,00 mm
	Tangential load of 5.8N to 8.75N to close gap to	0,40 to 0,65 mm
	Material	Cast iron HG10
	Oil control ring:	
	Gap	0,3 to 0,55 mm

Gudgeon Pins	Type	Chamfer locking type
	Length	77,25 to 77,12 mm
	Outside diameter:	23,812 to 23,807 mm
	Inside diameter	14,81 to 14,30 mm
	Permissible offset of centre web	1,0 mm
	Material	EN 32

Camshafts	Number of journals	7
	Journal diameter	26,950 to 29,937 mm
	Diametrical clearance	0,063 to 0,037 mm
	Material	Cast iron to BS 1452 Grade 17

Valves	Inlet valve material	EN 52
	Exhaust valve material	Steel 21—4—N
	Inlet valve head diameter	35,43 to 35,17 mm
	Exhaust valve head diameter	29,90 mm
	Valve Stem diameter:	
	Inlet and exhaust	7,897 to 7,866 mm
	Valve lift	9,525 mm
	Inlet valve clearance	0,031 to 0,036 mm (0.012 to 0.014 in)
	Exhaust valve clearance	0,031 to 0,036 mm (0.012 to 0.014 in)
	Valve stem seal:	
	Material	BLS.RU28/E3 (VITRON)
	Internal diameter	7,90 to 7,85 mm

Valve Guides and Seats	Valve guide material	Brico alloy 2 or BS 1452/12 (Brinell hardness 190 to 260)
	Guide length	48,51 mm (1.910 in)
	Circlip material	18 SWG (0.048 in) zinc coated
Identification	No groove — Standard	12,75 to 12,74 mm (0.5020 to 0.5015 in)
	1 groove — First oversize (Production)	12,80 to 12,79 mm (0.5040 to 0.5035 in)
	2 grooves — Second oversize (Service)	12,88 to 12,87 mm (0.5070 to 0.5065 in)
	3 grooves — Third oversize (Service)	13,00 to 12,99 mm (0.5120 to 0.5115 in)
Valve guide (service)	Bore to be concentric with the outside diameter of the guide within 0.001 in. i.e. total indicator reading of 0.002 in.	
Valve seat	Inlet valve seat outside diameter:	
	Standard	36,404 to 36,388 mm
	Service	36,785 to 36,769 mm
	Interference fit in cylinder head	0,077 mm (0.003 in)
	Exhaust valve seat outside diameter:	
	Standard	32,884 to 32,868 mm
	Service	33,265 to 33,249 mm
	Interference fit in cylinder head	0,077 mm (0.003 in)
	Valve seat angle (inclusive)	89° 30' to 89° 00'
Tappets	Tappet material	Chilled cast iron
	Outside diameter of tappet	33,34 to 33,35 mm (1.3125 to 1.3130 in)
	Diametrical clearance of tappet	0,051 to 0,020 mm (0.002 to 0.0008 in)
Valve spring	Wire diameter	3,76 mm (0.148 in)
	Inside diameter	20,62 ± 0,25 mm (0.812 ± 0.010 in)
	Total No. of coils	6
	No. of working coils	3.75
	Helix	RH
	Rate	35,51 N/mm (202.5 lb/in)
	Natural frequency	603,4 Hz (36,203 C/min)
	Free length	40,13 mm (1.580 in)
	Material	OTEVA 60
Camshaft Sprocket	No. of teeth	30
	Pitch of teeth	9,525 mm (0.375 in)
	Pitch circle diameter	91,135 mm (3.588 in)
	Internal surrations	131
	Pitch circle diameter	66,675 mm (2.625 in)
	Pitch of teeth	1,600 mm (0.063 in)
	o. dia of sprocket	95,25 mm ± 0,050 mm (3.75 in ± 0.002 in)
	Total width	18,6531 mm ($\frac{47}{64}$ in)
	Total width across both teeth	15,341 to 15,570 mm (0.604 to 0.613 in)
	Concentricity of teeth and sprocket not to exceed	0,076 mm (0.003 in)

Crankshaft Sprocket	No. of teeth	21
	Pitch	9,525 mm (0.375 in)
	Pitch circle diameter	63,906 mm (2.516 in)
	Root diameter	57,556 mm (2.266 in)
	Roller diameter	6,35 mm (0.25 in)
	Measured over 6,35 mm (0.25 in) dia. pins.	70,008 to 69,95 mm (2.759 to 2.754 in)
	Tooth profile cut to	BS 228
	Internal diameter	42,875 to 42,857 mm
	Overall diameter	67,87 to 67,77 mm
	Width across 1 tooth	5,33 to 5,10 mm
	Width across all the teeth	38,43 to 38,00 mm
	Width across 2 teeth (timing chain)	15,57 to 15,34 mm
Crankshaft damper oil seal spacer	Inside diameter	42,875 to 42,860 mm (1.6880 to 1.6874 in)
	Outside diameter	52,00 to 51,81 mm (2.047 to 2.040 in)
	Width	17,10 to 16,90 mm
	Outside diameter to be plunge ground	
	All chamfers to be polished and free from burrs	
Intermediate shaft bush	Inside diameter (when fitted)	22,010 to 21,964 mm (0.8665 to 0.8647 in)
	Outside diameter	25,083 to 25,057 mm (0.9875 to 0.9865 in)
	Width	17,25 to 16,25 mm (0.679 to 0.659 in)
	Press fit	0.036 to 0,083 mm (0.0015 to 0.0033 in)
	Shaft size	22 mm (0.8645 to 0.8637 in)
	Running clearance	0,071 to 0,005 mm (0.0028 to 0.0002 in)
	Material	Vandervell No. L10083/3
Intermediate sprocket inner teeth	Number of teeth	28
	Pitch	9,525 mm (0.375 in)
	Pitch circle diameter	85,063 mm (3.349 in)
	Root diameter	78,70 mm (3.009 in)
	Roller diameter	6,35 mm (0.25 in)
	Measured over 6,35 mm (0.25 in) dia. pins	91,41 to 91,29 mm (3.599 to 3.594 in)
	Tooth profile cut to	BS 228
	Width across the depth	15,57 to 15,34 mm (0.613 to 0.604 in)
	Overall diameter	89,20 to 89,10 mm (3.512 to 3.508 in)
Intermediate sprocket outer teeth	Number of teeth	20
	Pitch	9,525 mm (0.375 in)
	Pitch circle diameter	60,884 mm (2.397 in)
	Root diameter	54,53 mm (2.147 in)
	Roller diameter	6,35 mm (0.25 in)
	Measured over 6,35 mm (0.25 in) dia. pins	67,23 to 67,11 mm (2.647 to 2.642 in)
	Tooth profile cut to	BS 228
	Overall diameter	65,07 to 64,97 mm (2.562 to 2.558 in)
	Width across both teeth	15,57 to 15,34 mm (0.613 to 0.604 in)
	Total width of sprocket	48,80 to 48,70 mm (1.921 to 1.917 in)
Primary timing chain	Pitches	80
	Roller diameter	6,35 mm (0.25 in)
	Pitch	9,525 mm (0.375 in)
	Type	Endless duplex chain
Secondary timing chain	Pitches	86
	Roller diameter	6,35 mm (0.25 in)
	Pitch	9,525 mm (0.375 in)
	Type	Endless duplex chain
Oil pump chain	Pitches	52
	Roller diameter	6,35 mm (0.25 in)
	Pitch	9,525 mm (0.375 in)
	Type	Endless simplex chain

| Dampers and tensioners | Backing material | Steel BS1449 CR4 |
| | Damper material | Rubber |

| Hydraulic chain tension housing | Bore diameter | 21,38 to 21,36 mm (0.842 to 0.841 in) |

Hydraulic chain tensioner

	Overall length	45 mm (1.771 in)
	Overall length to bottom of groove in piston	41 mm (1.614 in)
	Diameter	21,349 to 31,336 mm (0.8405 to 0.8400 in)
	Width of groove in piston	15 mm (0.59 in)
	Non return valve primary ball diameter	4,762 mm (0.1875 in)
	Ball free movement	0,74 to 0,45 mm (0.290 to 0.018 in)

| Non return valve secondary | Ball diameter | 4,762 mm (0.1875 in) |
| | Ball free movement | 0,74 to 0,45 mm (0.29 to 0.018 in) |

Oil pressure relief valve	Relief valve:	
	O/D	17,98 to 17,95 mm (0.7079 to 0.7067 in)
	I/D	15,00 to 14,50 mm (0.591 to 0.571 in)
	Overall length	35 mm (1.378 in)
	Working length	30 mm (1.181 in)
	Bore depth	27 mm (1.063 in)
	Material	ENIA
	Cyanide Harden	0,020 to 0,013 mm (0.008 to 0.005 in) deep
	Mandrel: Length	57 mm (2.244 in)
	Diameter	11,11 mm (0.437 in)
	Material	Steel tube to BS970 (CDS2 or ERW1)
	Spring: Wall thickness	1,63 (0.064 in, 16 SWG)
		or 1,22 (0.048, 18 SWG)
	Free length	106 mm (4.173 in)
	Outside diameter	16,34 mm (0.643 in)

Spring Specification	Wire diameter	2,34 mm (0.092 in, 13 SWG)
	Mean diameter coils	14,0 mm (0.551 in)
	Total number of coils	26
	Number of active coils	24
	Helix of coil	Left
	Spring rate	4,49 N/mm (25.65 lb/IN)

| | Length | | Load | | Stress | |
	mm	inches	kg	lbs	N/mm²	lb/in²
Fitted	76,00	2.992	13.74	30.29	472	68,395
Valve opening	72,00	2.835	15.58	34.35	535	77.543
Max opening	62,00	2.441	20.15	44.43	692	100.304
Solid (ref)	60,80	2.392	20,71	45,66	711	103.083

Material
BS. 970/735/A50 (EN.47) Spring steel wire. Ends to be close coiled and ground square to axis, feather edges to be removed to a minimum thickness of 0,5 mm (0.020 in).

Oil Filter   Oil filter: type .... Full flow disposable canister

Canister Specification

Maximum working pressure	100 lb/in²
Relief valve setting	13 to 16 lb/in² (0.90 to 1,1 kgf/cm²)
Canister to withstand	17,60 kgf/cm² (250 lb/in²) pressure
Element filtration area	3555 cm² (551 lb/in²)
Canister diameter	95,70 mm (3.78 in)
Canister length	148,00 to 145,50 mm (5.827 to 5.728 in)
Canister case material thickness	24 SWG
Adaptor thread	1 in 12 UNF — 2B

Oil Pump

Oil pump	Rotor type
Outer rotor outside diameter	69825 to 69,774 mm (2.749 to 2.747 in)
Outer rotor width	27,975 to 27,962 mm (1.1014 to 1.1009 in)
Inner rotor width	27,975 to 27,962 mm (1,1014 to 1.1009 in)
Clearance outer rotor to body	0,1 mm (0.010 in)
Material	Cast iron grade 12
Oil pump body rotor bore	69,951 to 69,926 mm (2.7539 to 2.7529 in)
End float	0,1 mm (0.005 in)
2 horn clearance	0,2 mm (0.010 in)
3 horn clearance	0,2 mm (0.010 in)

TORQUE WRENCH SETTINGS

For the Torque wrench settings, refer to the front of the relevant section.

GENERAL FITTING INSTRUCTIONS

Precautions against damage

Always fit covers to protect the wings before commencing work in the engine compartment. Cover the seats and carpets, wear clean overalls and wash your hands or wear gloves before working inside the car.
Avoid spilling hydraulic fluid or battery acid on paintwork. Wash off with water immediately if this occurs.
Use polythene sheets in the boot to protect carpets.
Always use a recommended service tool, or a satisfactory equivalent, where specified.
Protect temporarily exposed screw threads by replacing nuts or fitting plastic caps.

Safety precautions

Whenever possible use a ramp or pit when working beneath a car, in preference to jacking. Chock the wheels as well as applying the handbrake.
Never rely on a jack alone to support a car. Use axle stands or blocks carefully placed at the jacking points to provide a rigid location.
Ensure that a suitable form of fire extinguisher is conveniently located.
Check that any lifting equipment used has adequate capacity and is fully serviceable.
Inspect power leads of any mains electrical equipment for damage, and check that it is properly earthed.
Disconnect the earth (grounded) terminal of a car battery.
Do not disconnect any pipes in the air conditioning refrigeration system, if fitted, unless trained and instructed to do so. A refrigerant is used which can cause blindness if allowed to contact the eyes.
Ensure that adequate ventilation is provided when volatile de-greasing agents are being used.

CAUTION: Fume extraction equipment must be in operation when trichlorethylene, carbon tetrachloride, methylene chloride, chloroform, or perchlorethylene are used for cleaning purposes.

Do not apply heat in an attempt to free stiff nuts or fittings; as well as causing damage to protective coatings, there is a risk of damage to electronic equipment and brake lines from stray heat.
Do not leave tools, equipment, spilt oil, etc., around or on work area.
Wear protective overalls and use barrier creams when necessary.

Preparation

Before removing a component, clean it and its surrounding area as thoroughly as possible.
Blank off any openings exposed by component removal, using greaseproof paper and masking tape.

Immediately seal fuel, oil or hydraulic lines when separated, using plastic caps or plugs, to prevent loss of fluid and entry of dirt.
Close the open ends of oilways, exposed by component removal, with tapered hardwood plugs or readily visible plastic plugs.
Immediately a component is removed, place it in a suitable container; use a separate container for each component and its associated parts.
Before dismantling a component clean it thoroughly with a recommended cleaning agent; check that the agent is suitable for all materials of component.
Clean the bench and provide marking materials, labels, containers and locking wire before dismantling a component.

Dismantling

Observe scrupulous cleanliness when dismantling components, particularly when brake, fuel or hydraulic system parts are being worked on.
A particle of dirt or a cloth fragment could cause a dangerous malfunction if trapped in these systems.
Blow out all tapped holes, crevices, oilways and fluid passages with an air line. Ensure that any 'O' rings used for sealing are correctly replaced or renewed if disturbed.
Mark mating parts to ensure that they are replaced as dismantled. Whenever possible use marking ink, which avoids possibilities of distortion or initiation of cracks, liable if centrepunch or scriber are used.
Wire together mating parts where necessary to prevent accidental interchange (e.g. roller bearing components).
Wire labels onto all parts which are to be renewed, and to parts requiring further inspection before being passed for reassembly; place these parts in separate containers from those containing parts for rebuild.
Do not discard a part due for renewal until after comparing it with a new part, to ensure that its correct replacement has been obtained.

Inspection — General

Never inspect a component for wear or dimensional check unless it is absolutely clean; a slight smear of grease can conceal an incipient failure. When a component is to be checked dimensionally against figures quoted for it, use correct equipment (surface plates, micrometers, dial gauges, etc.) in serviceable condition. Makeshift checking equipment can be dangerous. Reject a component if its dimensions are outside the limits quoted, or if damage is apparent. A part may, however, be refitted if its critical dimension is exactly limit size, and is otherwise satisfactory.
Use Plastigauge 12 Type PG-1 for checking bearing surface clearances.
Directions for its use, and a scale giving bearing clearances in 0.00025 mm (0.0001 in) steps are provided with it.

Ball and roller bearings

NEVER REPLACE A BALL OR ROLLER BEARING WITHOUT FIRST ENSURING THAT IT IS IN AS-NEW CONDITION.

Remove all traces of lubricant from a bearing under inspection by washing it in petrol or a suitable de-greaser; maintain absolute cleanliness throughout the operations.
Inspect visually for markings of any form on rolling elements, raceways, outer surface of outer rings or inner surface of inner rings. Reject any bearings found to be marked, since any markings in these areas indicates onset of wear.
Holding the inner race between fingers and thumb of one hand, spin the outer race and check that it revolves absolutely smoothly. Repeat, holding the outer race and spinning the inner race.
Rotate the outer ring with a reciprocating motion, while holding the inner ring; feel for any check or obstruction to rotation, and reject the bearing if action is not perfectly smooth.
Lubricate the bearing generously with lubricant appropriate to installation. Inspect shaft and bearing housing for discolouration or other marking suggesting that movement has taken place between bearing and seatings.
If markings are found use Loctite in installation of replacement bearing.
Ensure that the shaft and housing are clean and free from burrs before fitting the bearing.
If one bearing of a pair shows an imperfection it is generally advisable to renew both bearings; an exception could be made only if the faulty bearing had covered a low mileage, and it could be established that damage was confined to it. When fitting bearing to shaft, apply force only to inner ring of bearing, and only to outer ring when fitting into housing (Fig. 1).

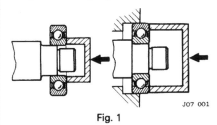

Fig. 1

In the case of grease-lubricated bearings (e.g. hub bearings) fill the space between the bearing and outer seal with a recommended grade of grease before fitting the seal.
Always mark components of separable bearings (e.g. taper-roller bearings) in dismantling, to ensure correct reassembly. Never fit new rollers in a used cup.

Oil seals

Always fit new oil seals when rebuilding an assembly. It is not physically possible to replace a seal exactly as it had bedded down.
Carefully examine the seal before fitting to ensure that it is clean and undamaged.

Smear sealing lips with clean grease, pack dust excluder seals with grease, and heavily grease duplex seals in cavity between sealing lips.

Ensure that seal spring, if provided, is correctly fitted.

Place lip of seal towards fluid to be sealed and slide into position on shaft, using fitting sleeve (Fig. 2) when possible to protect sealing lip from damage by sharp corners, threads or splines. If fitting sleeve is not available, use plastic tube or adhesive tape to prevent damage to sealing lip.

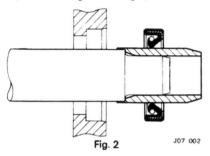

Fig. 2 J07 002

Grease the outside diameter of the seal, place it square to the housing recess and press it into position, using great care and if possible a 'bell piece' (Fig. 3) to ensure that seal is not tilted. (In some cases it may be preferable to fit the seal to the housing before fitting to the shaft.) Never let the weight of an unsupported shaft rest in a seal.

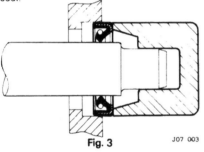

Fig. 3 J07 003

If correct service tool is not available, use a suitable drift approximately 0,4 mm (0.015 in) smaller than the outside diameter of the seal. Use a hammer VERY GENTLY on the drift if a press is not suitable.

Press or drift a seal in to the depth of housing if the housing is shouldered, or flush with the face of the housing where no shoulder is provided.

NOTE: Most cases of failure of leakage or oil seals are due to careless fitting, and resulting damage to both seals and sealing surfaces. Care in fitting is essential if good results are to be obtained.

Joints and joint faces

Always use the correct gaskets where they are specified.

Use jointing compound only when recommended. Otherwise fit joints dry.

When jointing compound is used, apply in a thin uniform film to metal surfaces; take great care to prevent it from entering oilways, pipes or blind tapped holes.

Remove all traces of old jointing materials prior to reassembly. Do not use a tool which could damage joint faces.

Inspect joint faces for scratches or burrs and remove with a fine file or oil-stone; do not allow swarf or dirt to enter tapped holes or enclosed parts. Blow out any pipes, channels or crevices with compressed air, renewing any 'O' rings or seals displaced by air blast.

Flexible hydraulic pipes, hoses

Before removing any brake or power steering hose, clean end fittings and area surrounding them as thoroughly as possible. Obtain appropriate blanking caps before detaching hose end fittings, so that ports can be immediately covered to exclude dirt.

Clean hose externally and blow through with airline. Examine carefully for cracks, separation of plies, security of end fittings and external damage. Reject any hose found faulty.

When refitting hose, ensure that no unnecessary bends are introduced, and that hose is not twisted before or during tightening of union nuts.

Containers for hydraulic fluid must be kept absolutely clean.

Do not store hydraulic fluid in an unsealed container. It will absorb water, and fluid in this condition would be dangerous to use due to a lowering of its boiling point.

Do not allow hydraulic fluid to be contaminated with mineral oil, or use a container which has previously contained mineral oil.

Do not re-use fluid bled from system. Always use clean brake fluid, or a recommended alternative, to clean hydraulic components.

Fit a blanking cap to a hydraulic union and a plug to its socket after removal to prevent ingress of dirt.

Absolute cleanliness must be observed with hydraulic components at all times.

After any work on hydraulic systems, inspect carefully for leaks underneath the car while a second operator applies maximum pressure to the brakes (engine running) and operates the steering.

Metric bolt identification

An ISO metric bolt or screw, made of steel and larger than 6 mm in diameter can be identified by either of the symbols ISO M or M embossed or indented on top of the head (Fig. 4).

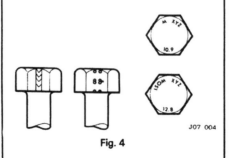

Fig. 4 J07 004

In addition to marks to identify the manufacture, the head is also marked with symbols to indicate the strength grade e.g. 8.8, 10.9, 12.9, or 14.9, where the first figure gives the minimum tensile strength of the bolt material in tens of kgf/mm². Zinc plated ISO metric bolts and nuts are chromate passivated, a greenish-khaki to gold-bronze colour.

Metric nut identification

A nut with an ISO metric thread is marked on one face (1, Fig. 5) or on one of the flats (2, Fig. 5) of the hexagon with the strength grade symbol 8, 12 or 14. Some nuts with a strength 4, 5 or 6 are also marked and some have the metric symbol M on the flat opposite the strength grade marking.

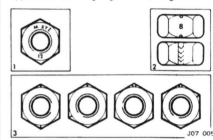

Fig. 5 J07 005

A clock face system (3, Fig. 5) is used as an alternative method of indicating the strength grade. The external chamfers or a face of the nut is marked in a position relative to the appropriate hour mark on a clock face to indicate the strength grade.

A dot is used to locate the 12 o'clock position and a dash to indicate the strength grade. If the grade is above 12, two dots identify the 12 o'clock position.

Hydraulic fittings — Metrification

WARNING: METRIC AND UNIFIED THREADED HYDRAULIC PARTS. ALTHOUGH PIPE CONNECTIONS TO BRAKE SYSTEM UNITS INCORPORATE THREADS OF METRIC FORM, THOSE FOR POWER ASSISTED STEERING ARE OF UNF TYPE. IT IS VITALLY IMPORTANT THAT THESE TWO THREAD FORMS ARE NOT CONFUSED, AND CAREFUL STUDY SHOULD BE MADE OF THE FOLLOWING NOTES.

Metric threads and metric sizes are being introduced into motor vehicle manufacture and some duplication of parts must be expected. Although standardization must in the long run be good, it would be wrong not to give warning of the dangers that exist while UNF and metric threaded hydraulic parts continue together in service.

Fitting UNF pipe nuts into metric ports and vice-versa should not happen, but experience of the change from BSF to UNF indicated that there is no certainty in relying upon the difference in thread size when safety is involved.

To provide permanent identification of metric parts is not easy but recognition has been assisted by the following means:

All metric pipe nuts, hose ends, unions and bleed screws are coloured black.

The hexagon area of pipe nuts is indented with the letter 'M'.

Metric and UNF pipe nuts are slightly different in shape.

NOTE: In Figs. 6 to 9, A indicates the metric type and 'B' the UNF type.

The metric female nut is **always** used with a trumpet flared pipe and the metric male nut is **always** used with a convex flared pipe (Fig. 6).

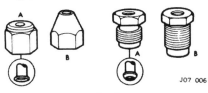

Fig. 6

All metric ports in cylinders and calipers have no counterbores, but unfortunately a few cylinders with UNF threads also have no counterbore. The situation is, all ports with counterbores are UNF, but ports not counterbored are most likely to be metric (Fig. 7).

Fig. 7

The colour of the protective plugs in hydraulic ports indicates the size and the type of the threads, but the function of the plugs is protective and not designed as positive identification. In production it is difficult to use the wrong plug but human error must be taken into account.

The plug colours and thread sizes are:

	UNF
RED	$\frac{3}{8}$ in × 24 UNF
GREEN	$\frac{7}{16}$ in × 20 UNF
YELLOW	$\frac{1}{2}$ in × 20 UNF
PINK	$\frac{5}{8}$ in × 18 UNF

	METRIC
BLACK	10 × 1 mm
GREY	12 × 1 mm
BROWN	14 × 1,5 mm

Hose ends differ slightly between metric and UNF (Fig. 8).

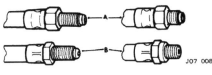

Fig. 8

Gaskets are not used with metric hoses. The UNF hose is sealed on the cylinder or caliper face by a copper gasket but the metric hose seals against the bottom of the port and there is a gap between faces of the hose end and cylinder (Fig. 9).

Fig. 9

Pipe sizes for UNF are $\frac{3}{16}$ in, $\frac{1}{4}$ in, and $\frac{5}{16}$ in outside diameter.
Metric pipe sizes are 4,75 mm, 6 mm and 8 mm.
4,75 mm pipe is exactly the same as $\frac{3}{16}$ in pipe.
6 mm pipe is 0.014 in smaller than $\frac{1}{4}$ in pipe.
8 mm pipe is 0.002 in larger than $\frac{5}{16}$ in pipe.
Convex pipe flares are shaped differently for metric sizes and when making pipes for metric equipment, metric pipe flaring tools must be used.
The greatest danger lies with the confusion of 10 mm and $\frac{3}{8}$ in UNF pipe nuts used for $\frac{3}{16}$ in (or 4,75 mm) pipe. The $\frac{3}{8}$ in UNF pipe nut or hose can be screwed into a 10 mm port but is very slack and easily stripped. The thread engagement is very weak and cannot provide an adequate seal. The opposite condition, a 10 mm nut in a $\frac{3}{8}$ in port, is difficult and unlikely to cause trouble. The 10 mm nut will screw in $1\frac{1}{2}$ or two turns and seize. It has a crossed thread 'feel' and it is impossible to force the nut far enough to seal the pipe. With female pipe nuts the position is of course reversed.
The other combinations are so different that there is no danger of confusion.

Keys and keyways

Remove burrs from edges of keyways with a fine file and clean thoroughly before attempting to refit key.
Clean and inspect key closely; keys are suitable for refitting only if indistinguishable from new, as any indentation may indicate the onset of wear.

Split pins

Fit new split pins throughout when replacing any unit.
Always fit split pins where split pins were originally used. Do not substitute spring washers; there is always a good reason for the use of a split pin.
All split pins should be fitted as shown in Fig. 10 unless otherwise stated.

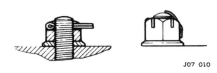

Fig. 10

Tab washers

Fit new tab washers in all places where they are used. Never replace a used tab washer. Ensure that the new tab washer is of the same design as that replaced.

Nuts

When tightening up a slotted or castellated nut **never slacken it back** to insert split pin or locking wire except in those recommended cases where this forms part of an adjustment. If difficulty is experienced, alternative washers or nuts should be selected, or washer thickness reduced.

Where self-locking nuts have been removed it is advisable to replace them with new ones of the same type.

NOTE: Where bearing pre-load is involved nuts should be tightened in accordance with special instructions.

Locking wire

Fit new locking wire of the correct type for all assemblies incorporating it.
Arrange wire so that its tension tends to tighten the bolt heads, or nuts, to which it is fitted.

Screw threads

Both UNF and Metric threads to ISO standards are used. See below for thread identification.
Damaged threads must always be discarded. Cleaning up threads with a die or tap impairs the strength and closeness of fit of the threads and is not recommended.
Always ensure that replacement bolts are at least equal in strength to those replaced.
Do not allow oil, grease or jointing compound to enter blind threaded holes. The hydraulic action on screwing in the bolt or stud could split the housing.
Always tighten a nut or bolt to the recommended torque figure. Damaged or corroded threads can affect the torque reading.
To check or re-tighten a bolt or screw to a specified torque figure, first slacken a quarter of a turn, then re-tighten to the correct figure. Always oil thread lightly before tightening to ensure a free running thread, except in the case of self-locking nuts.

Unified thread identification bolts

A circular recess is stamped in the upper surface of the bolt head (1, Fig. 11).

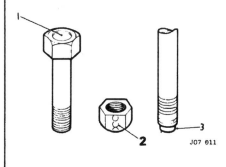

Fig. 11

Nuts

A continuous line of circles is indented on one of the flats of the hexagon, parallel to the axis of the nut (2, Fig. 11)

Studs, brake rods, etc.

The component is reduced to the core diameter for a short length at its extremity (3, Fig. 11).

TOWED RECOVERY

The car may be towed by another vehicle provided the following precautions are taken:

1. The gear box oil is at the correct level.
2. The gear lever is in neutral, and the ignition key is in position '1'.
3. The towing speed should not exceed 48 km/h (30 mph) and the towing distance should be limited to 48 km (30 miles).
4. The registration number of the towing vehicle and an 'ON TOW' sign or warning triangle must be displayed in a prominent position on the rear of the vehicle being towed.

WARNING: WHEN THE ENGINE IS NOT RUNNING THE STEERING WILL NO LONGER BE POWER ASSISTED, AND THE BRAKE SERVO WILL BECOME INEFFECTIVE AFTER A FEW APPLICATIONS OF THE BRAKES. THEREFORE BE PREPARED FOR RELATIVELY HEAVY STEERING, AND THE NEED FOR INCREASED BRAKE PEDAL PRESSURE.

If the distance to be towed will exceed 48 km (30 miles) then the car should be towed with the rear wheels clear of the ground or the propellor shaft disconnected from the final drive input flange.

If the propellor shaft is disconnected then it must be firmly secured away from the final drive flange.

TAPER ROLLER BEARING — FAULT DIAGNOSIS

CONDITION OF BEARING	CAUSE	REMEDY
Fig. 12 Good Bearing J51-049	—	—
Fig. 13 Bent cage J51-050	Improper handling or tool usage	Replace the bearing
Fig. 14 Bent cage J51-051	Improper handling or tool usage	Replace the bearing
Fig. 15 Galling J51-052	Marks on roller ends due to overheating. Lubricant failure or overloading.	Replace the bearing. Check the seals and ensure that the bearing is properly lubricated

CONDITION OF BEARING	CAUSE	REMEDY
Fig. 16 Step wear	Wear on the roller ends caused by fine abrasives	Clean all components and housing. Check the seals and bearings and replace if leaking, rough or noisy.
Fig. 17 Etching	The bearing surfaces are grey or greyish black in colour. With the rollers and track material being etched away usually related to roller spacing	Replace the bearings, check the seals and also ensure there is adequate lubrication.
Fig. 18 Misalignment	Outer track misalignment usually due to a foreign body under the track.	Clean all components and replace the bearing and ensure that the new track is correctly seated.
Fig. 19 Indentations	Surfaces are depressed on the race and the track caused by hard particles of foreign material.	Clean all parts and housings, check the seals and replace if rough or noisy.

CONDITION OF BEARING	CAUSE	REMEDY
J51-057 Fig. 20 Flaking	Flaking of the suface material due to fatigue	Replace the bearing and clean all related components.
J51-058 Fig. 21 Indentations	Surface indentations in track caused by rollers either vibrating or impact loading while the bearing is not rotating	Replace the bearing if rough or noisy.
J51-059 Fig. 22 Cage wear	Wear around the outside diameter of the cage and roller pockets caused by poor lubrication and abrasive material	Replace the bearings and check the conditions of the seals.
J51-060 Fig. 23 Roller wear	Marks on track and rollers caused by fine abrasives.	Clean all components and housings. Check the seal and bearing condition and replace if leaking or noisy

CONDITION OF BEARING	CAUSE	REMEDY
 Fig. 24 Discolouration	Discolouration ranges from black to light brown caused by moisture or incorrect use of lubricants.	Re-use bearings if stains can be removed by light polishing or if no evidence of overheating is apparent. Check the seal and other component part condition.
 Fig. 25 Heat discolouration	Heat discolouration ranges from blue to faint yellow resulting from overload or incorrect lubricant. As excessive heat can cause softening of tracks and rollers, check be drawing a fine file over a softened area. If faultly the file with grab and cut metal, if it remains hard the file will skid over the suface without removing any material.	Replace the bearings and seals if any heat damage is evident.
 Fig. 26 Cracked race	Race cracked due to incorrect fitment to shaft, tipping or poor seating.	Replace the bearing and check the condition of the seals
 Fig. 27 Rotating track and inner race	Removal of material due to slippage. This can be caused by poor fits, incorrect lubrication, overheating, overloading and poor assembly	Replace bearings and clean all related parts, check the fit and ensure the replacement bearings are the correct type. Replace the shaft or housing if damaged.

CONDITION OF BEARING	CAUSE	REMEDY
Fig. 28 Fretting	Corrosion caused by small movement of components with no lubrication.	Replace the bearing, check the seals for leakage and ensure there is adequate lubrication.
Fig. 29 Seizure	Caused by lack of lubrication, excessive loads on the ingress of foreign matter.	Change the bearings. Check the seals for wear and ensure there is adequate lubrication.

LIFTING AND JACKING

Stands

When carrying out any work on the car which requires a wheel to be raised (apart from a simple wheel change) it is essential that the jack is replaced by a stand, located by the jacking spigot, to provide a secure support for the car.

Jacking Points (Fig. 1)

The jack provided in the car's tool kit engages with spigots situated below the body side members, in front of the rear wheels and behind the front wheels.
Always chock wheels as well as applying handbrake when using the jack.

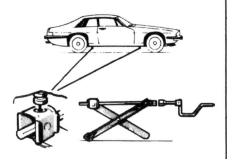

Fig. 1

Workshop Jack

Front — one wheel (Fig. 2)

Jack under the lower spring support pan, using a suitable wooden block on the jack head. Place a stand in position at the adjacent spigot when the wheel is raised.

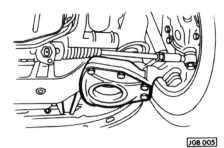

Fig. 2

Rear — one wheel (Fig. 3)

Locate the jack with a wooden block on its head, under the outer fork of the wishbone at the wheel to be raised. Take care to avoid damage to the aluminium alloy hub carrier or to the grease nipple fitted to it. Place a stand under the adjacent jacking spigot when the wheel is raised.

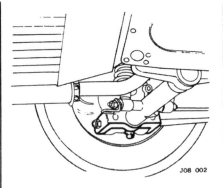

Fig. 3

Front — both wheels (Fig. 4)

Place the jack, with a wooden block on its head, centrally under the front suspension. Place stands under both front jacking spigots when the car is raised.

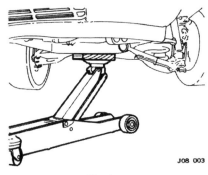

Fig. 4

Rear — both wheels (Fig. 5)

Place a suitable shaped wooden block between the jack head and the plate in the centre of the rear crossmember, ensuring that the jacking load is not applied to the flanges of the plate. Place stands under both rear jacking spigots when the car is raised.

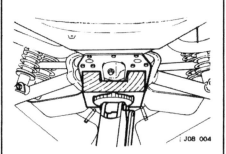

Fig. 5

Lifting

Locate lifting pads at the four jacking spigots.

TOWED RECOVERY

The car may be towed by another vehicle provided the following precautions are taken:

1. The gearbox oil is at the correct level.
2. The gear lever is in Neutral, and the ignition key is in position '1'.
3. The towing speed should not exceed 48 km/h (30 m.p.h.) and the towing distance should be limited to 48 km (30 miles).
4. The registration number of the towing vehicle and an 'ON TOW' sign or warning triangle, must be displayed in a prominent position on the rear of the vehicle being towed.

WARNING: WHEN THE ENGINE IS NOT RUNNING THE STEERING WILL NO LONGER BE POWER ASSISTED, AND THE BRAKE SERVO WILL BECOME INEFFECTIVE AFTER A FEW APPLICATIONS OF THE BRAKES. THEREFORE BE PREPARED FOR RELATIVELY HEAVY STEERING, AND THE NEED FOR INCREASED BRAKE PEDAL PRESSURE.

If the distance to be towed will exceed 48 km (30 miles) then the car should be towed with the rear wheels clear of the ground or the propellor shaft disconnected from the final drive input flange.
If the propellor shaft is disconnected then it must be firmly secured away from the final drive flange.

Engine Oil — Recommended S.A.E. Viscosity Range/Ambient Temperature Scale

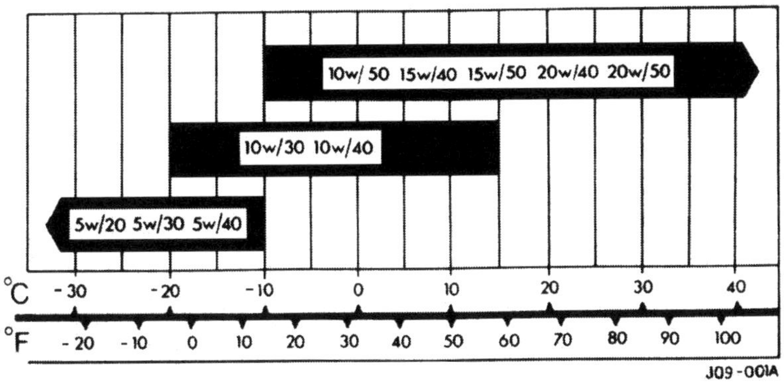

J09-001A

Fig. 1

Component — Model	Temperature Range	Specification	SAE Viscosity Rating	Approved Brands Available in UK for Temperatures Above −10°C (14°F)
Engine Distributor Oil Can	Above −10°C (14°F) −20°C to 10°C (−4°F to 50°F) Below −10°C (14°F)	BLSO OL.02 or MIL-L-2104 B or A.P.I. SE	10W/50, 15W/50, 20W/40, 20W/50 10W/30, 10W/40, 10W/50 5W/20, 5W/30	Unipart Super Multigrade, BP Super Viscostatic, Castrol GTX, Duckhams Q Motor Oil, Esso Uniflow, Fina Super Grade, Mobiloil Super, Shell Super Oil, Texaco Havoline
Manual Gearbox	All All	— MIL-L-2105A, APIGL4	SAE 80	BP gear oil SAE 80EP, Shell spirax 80EP, Castrol Hypoy, Duckhams Hypoid 80, Esso gear oil GX 85W/140, Mobilube HD 80
Powr-Lok Differential — Initial Fill — Refill	All All	Use only approved brands of fluid specially formulated for Powr-Lok	90 90	Shell Spirax Super 90, Shell Spirax Super 90, BP Gear Oil 1453, BP Limslip Gear Oil 90/1, Castrol G722, Castrol Hypoy LS, Duckhams Hypoid 90 DL, Texaco 3450 Gear Oil, Veedol Multigear Limited Slip SAE 90
Power Assisted Steering	All	Type G	—	BP Autran G, Castrol TQF, Duckhams Q-Matic, Esso Glide Type G Fina Purfimatic 33F, Mobil AT210, Shell Donax TF, Texaco Texamatic Type G
Grease Points — All	All	Multipurpose Lithium Grease, N.L.C.I. Consistency No. 2	—	BP Energrease L8, Castrol LM, Duckhams LB10, Esso Multipurpose H, Fina Marson HTL2, Mobilgrease MP, Shell Retinax A, Texaco Marfak

COOLING SYSTEM

Additive   Barr's Leak Inhibitor, 1 sachet per car

Anti-freeze   BP Type HS25 Bluecol, 'U', Union Carbide UT 184 or Unipart Universal
If these are not available, phosphate free anti-freeze conforming to
specification BS3150 or 3152 may be used.
Concentration — UK and RHD export markets only 40% sp.gr 1,065
In North America use Jaguar Part No. ZVW 244101
All other markets   55% sp.gr 1,074

In territories where anti-freeze is unnecessary the cooling system must be filled with a
solution of Marston Corrosion Inhibitor Concentrate SQ36.

Always top-up the cooling system with recommended strength of anti-freeze or Corrosion
Inhibitor, NEVER with water only.

CAPACITIES	Litres	Imperial Pints	US Quarts
Engine refill (including filter)	8,50	15	9
(excluding filter)	7,95	14	5.8
Automatic transmission unit from dry	10,60	18.60	11.2
Drain and refill	3,30	7	4.20
Manual gearbox	1,40	2.50	1.50
Final drive unit	1,50	2.75	1.65
Cooling system including reservoir and air conditioning: Initial fill from dry	11,60	20.50	12.3
Drain and refill	9,90	17.50	10.5
Washer bottle	2,60	4.58	2.75
Washer bottle (headlamp wash/wipe)	5,40	9.50	5.70
Fuel tank	91	20 gal	24 gal

DIMENSIONS	mm	Inches
Wheelbase	2591	102
Track: Front	1482	58.4
Rear	1504	59.2
Overall length: European cars	4764	187.6
USA and Canada	4859	191.3
Overall width	1793	70.6
Overall height	1261	49.6
Turning circle between walls	12,6 m	41ft 4in
kerbs	12,0 m	39ft 4in
Ground clearance: kerb condition	140	5.5

WEIGHTS AND FUEL REQUIREMENTS

UK AND EUROPEAN MODELS	kg	lb
Kerb weight	1660	3652
Front axle weight	873	1921
Rear axle weight	787	1731
Gross vehicle weight	2010	4422
*Gross car weight	3510	7722
Maximum permitted front axle load	980	2156
Maxmimum permitted rear axle load	1070	2354

* Gross car weight is the gross vehicle weight plus maximum trailer weight.

FEDERAL MODELS	kg	lb
Gross vehicle weight rating	2040	4487
Gross axle weight rating — Front	982	2161
Gross axle weight rating — Rear	1057	2326
Front axle weight	906	1993
Rear axle weight	806	1774
Kerb weight	1721	3787
CANADIAN MODELS	kg	lb
Gross vehicle weight	2035	4477
Front axle weight	980	2156
Rear axle weight	1055	2321

ALL MARKETS	kg	lb
Trailer weight maximum — braked	1500	3300
unbraked	750	1650

Maximum permitted luggage compartment load with 5 passengers is 70 kg (154.3 lb).

FUEL REQUIREMENTS

Only cars with 'S' compression ratio engine require 97 octane fuel.
Cars with 'L' compression ratio engines should use 94 octane fuel.
In USA use unleaded fuel with a minimum octane rating of 91 RON.
In the United Kingdom, use '4 Star' fuel.
If, of necessity, the car has to be operated on lower octane fuel, do not use full throttle otherwise detonation may occur with resultant piston trouble.

RECOMMENDED HYDRAULIC FLUID

Braking System

Castrol-Girling Universal Brake and Clutch fluid. This fluid exceeds SAE J1703/D specification.

NOTE: Check all pipes in the brake system at the start and finish of each winter period for possible corrosion due to salt and grit used on the roads.

MAINTENANCE SUMMARY—UK & EUROPE

OPERATION	Interval in Kilometres x 1000 Interval in Miles x 1000	1.6 1	12 7.5	24 15
PASSENGER COMPARTMENT				
Fit protection kit		X	X	X
Check condition and security of seats and seat belts		X	X	X
Check operation of seat belt warning system		X		
Check footbrake operation		X	X	X
Drive on lift, stop engine		X	X	X
Check operation of lamps		X		
Check operation of horns		X		
Check operation of warning indicators		X		
Check operation of windscreen wipers		X		
Check operation of windscreen washers		X		
Check security of handbrake — release fully after checking		X	X	X
Check rear-view mirrors for security and function		X		
Check operation of boot lamp		X		
EXTERIOR				
Open bonnet — fit wing covers		X	X	X
Raise lift to convenient working height with wheels free to rotate		X	X	X
Mark stud to wheel relationship			X	X
Remove front wheels			X	
Remove road wheels — front and rear				X
Check that tyres are the correct size and type		X	X	X
Check tyre tread depth		X	X	X
Check tyres visually for external lumps, bulges and uneven wear		X	X	X
Check tyres visually for external exposure of ply or cord		X	X	X
Check/adjust tyre pressures		X	X	X
Inspect brake pads for wear and discs for condition			X	X
Adjust front hub bearing end-float				X
Grease hubs				X
Check for oil leaks from steering and fluid leaks from suspension system		X	X	X
Check condition and security of steering unit joints and gaiters		X	X	X
Refit road wheels in original position			X	X
Check tightness of road wheel fastenings		X	X	X
UNDERBODY				
Raise lift to convenient height		X	X	X
Drain engine oil		X	X	X
Check/top-up gearbox oil		X	X	X
Change gearbox oil		X		
Grease all points excluding hubs			X	X
Check/top-up rear axle/final drive oil		X	X	X
Check visually hydraulic hoses, pipes and unions for chafing, cracks, leaks and corrosion		X	X	X
Check exhaust system for leakage and security		X	X	X
Lubricate handbrake mechanical linkage and cables		X	X	X
Check condition of handbrake pads				X
Check tightness of propshaft coupling bolts		X		X
Check security of accessible engine mountings		X		
Check condition and security of steering unit, joints and gaiters		X	X	X
Check security and condition of suspension fixings		X	X	X
Check steering rack for oil leaks		X	X	X
Check power steering for leaks, hydraulic pipes and unions for chafing, corrosion and security		X	X	X
Check shock absorbers for fluid leaks		X	X	X
Renew engine oil filter element			X	X
Refit engine drain plug		X	X	X
Check for oil leaks — engine and transmission		X	X	X
Lower lift		X	X	X

MAINTENANCE SUMMARY — UK & EUROPE

OPERATION	Interval in Kilometres x 1000 Interval in Miles x 1000	1.6 1	12 7.5	24 15
ENGINE COMPARTMENT				
Fit exhaust extractor pipe		X	X	X
Fill engine with oil		X	X	X
Lubricate accelerator control linkage and pedal pivot		X		
Renew air cleaner element				X
Check security of accessible engine mountings		X		
Check driving belts, adjust or renew		X		X
Check and adjust spark plugs			X	
Renew spark plugs				X
Check/top-up battery electrolyte		X	X	X
Clean and grease battery connections		X	X	X
Check/top-up clutch fluid reservoir		X	X	X
Check/top-up brake fluid reservoir		X	X	X
Check brake servo hose(s) for security and condition		X	X	X
Check/top-up windscreen washer reservoir		X		
Check cooling and heater system for leaks and hoses for security and condition		X	X	X
Check/top-up cooling system		X		
Renew fuel filter				X
Clean engine breather filter (where applicable)				X
Check crankcase breathing system for leaks, hoses for security and condition		X		X
Check/top-up fluid in power steering reservoir; check security and condition of oil pressure hose at oil filter		X	X	X
Run engine and check for sealing of oil filter; stop engine			X	X
Check/top-up engine oil			X	X
Connect electronic instruments and check data		X		X
Lubricate distributor (not cam wiping pad) — run engine		X		X
Check ignition timing		X	X	X
Check distributor automatic advance		X		X
Check advance increases as vacuum pipe is reconnected		X		X
DOOR AND WINDOW MECHANISMS				
Lubricate all locks, hinges and door check mechanisms (not steering lock)		X		X
Check operation of all door, bonnet and boot locks		X		
Check operation of window controls		X		
Check and if necessary renew windscreen wiper blades			X	X
UNDER BONNET				
Check/adjust engine idle speed, stop engine — disconnect instruments		X		X
Check power steering system for leaks, hydraulic pipes and unions for chafing and corrosion		X	X	X
Check for oil leaks from engine and transmission		X	X	X
Re-check tension if driving belt has been renewed		X		X
Remove wing covers		X	X	X
Fill in details and fix appropriate Unipart underbonnet stickers		X	X	X
Close bonnet		X	X	X
Remove exhaust extractor pipe		X	X	X
SPARE WHEEL				
Remove spare wheel		X	X	X
Check that tyre complies with manufacturer's specification		X	X	X
Check tyre tread depth		X	X	X

MAINTENANCE SUMMARY — UK & EUROPE

OPERATION	Interval in Kilometres x 1000 / Interval in Miles x 1000	1.6 / 1	12 / 7.5	24 / 15
SPARE WHEEL cont.				
Check tyre visually for external exposure of cord or ply		X	X	X
Check tyre visually for external lumps or bulges		X	X	X
Check/adjust tyre pressure		X	X	X
Refit spare wheel		X	X	X
MISCELLANEOUS				
Check/adjust headlamp alignment		X		X
Check/adjust front wheel alignment		X		X
Drive off lift		X	X	X
Carry out road or roller test		X	X	X
Check operation of seat belt inertia mechanism		X	X	X
Ensure cleanliness of controls, door handles, steering wheel, etc		X	X	X
Remove protection kit		X	X	X
Report additional work required		X	X	X

At 18 month intervals
Renew brake fluid

At 48 000 km (30 000 miles) intervals
Change final drive oil
Change coolant ensuring that the correct anti-freeze content is present upon replenishment, i.e. 40% all UK and RHD export markets. 55% all other markets.
Change gearbox oil

At 3 years or 60 000 km (37 500 miles) intervals — whichever is the sooner
Renew all fluid seals in hydraulic system, examine and renew if necessary all flexible hoses
Examine working surfaces of master cylinder and calipers. Renew if necessary

OPTIONAL SERVICES

OPERATION	Interval in Kilometres x 1000 / Interval in Miles x 1000	12 / 7.5	24 / 15
Check operation of lamps			X
Check operation of horns			X
Check operation of warning indicators			X
Check operation of windscreen wipers			X
Check operation of windscreen washers			X
Check operation of window controls		X	X
Check sunroof and controls for correct operation (if fitted)			X
Check operation of headlamp wipe/wash (if fitted)			X
Check rear view mirrors for security and function			X
Check operation of boot lamp			X
Check/top-up windscreen washer reservoir			X
Check/top-up cooling system		X	X
Check operation of all door, bonnet and boot locks		X	X
Check operation of cruise control (if fitted)			X
Lubricate all locks, hinges and door check mechanisms (not steering lock)		X	
Clean aerial mast		X	
Check/adjust headlamp alignment		X	
Check/adjust front wheel alignment		X	

MAINTENANCE SUMMARY — NORTH AMERICAN MARKETS

Service Code Letter		Distance x 1000 in miles and kilometres The period between services should not exceed 12 months.							
A	Km	1.5							
	Miles	1							
B	Km		12		36		60		92
	Miles		7.5		22.5		37.5		52.5
C	Km			24				72	
	Miles			15				45	
D	Km					48			96
	Miles					30			60

OPERATION DESCRIPTION	SERVICE			
	A	B	C	D
LUBRICATION				
Lubricate all grease points		X	X	X
Renew engine oil and engine oil filter	X	X	X	X
Check/top-up brake fluid reservoir	X	X	X	X
Check/top-up gearbox oil	X	X	X	X
Change gearbox oil	X			X
Check battery condition	X	X	X	X
Check/top-up cooling system	X	X	X	X
Check/top-up rear axle oil	X	X	X	X
Check/top-up clutch fluid reservoir	X	X	X	X
Lubricate all locks and hinges (not steering lock)	X	X	X	X
Check/top-up power steering reservoir	X	X	X	X
ENGINE				
Check all driving belts — adjust				X
Renew air cleaner element				X
Check security of engine mountings	X			
Check for oil leaks	X	X	X	X
Renew air pump filter		52.5 only		
IGNITION				
Renew spark plugs				X
Lubricate distributor				X
FUEL AND EXHAUST SYSTEMS				
Check fuel system for leaks, pipes and unions for chafing and corrosion	X	X	X	X
Check exhaust system for leaks and security	X	X	X	X
Renew oxygen sensor				X
Renew fuel filter		52.5 only		
TRANSMISSION, BRAKES, STEERING AND SUSPENSION				
Check condition and security of steering unit, joints and gaiters		X	X	X
Inspect brake pads for wear and discs for condition		X	X	X
Check brake servo hoses for security and condition	X	X	X	X
Check/adjust front wheel alignment	X	X	X	X
Check foot and hand brakes	X			
Check visually brake hydraulic pipes and unions for cracks, chafing, leaks and corrosion	X	X	X	X
Check/adjust front hub bearing end float			X	X
Check tightness of propeller shaft coupling bolts			X	X

OPERATION DESCRIPTION	SERVICE			
	A	B	C	D
WHEELS AND TYRES				
Check tyres for tread depth and visually for external cuts in fabric, exposure of ply or cord structure, lumps or bulges	X	X	X	X
Check that tyres comply with manufacturer's specification	X	X	X	X
Check/adjust tyre pressure, including spare wheel	X	X	X	X
Check tightness of road wheel fastenings	X	X	X	X
ELECTRICAL				
Check/adjust operation of washers and top up reservoir	X	X	X	X
Check function of original equipment, i.e. lamps, horns, wipers and all warning indicators	X	X	X	X
Check wiper blades and arms, renew if necessary		X	X	X
Check/adjust headlamp alignment	X	X	X	X
BODY				
Check operation, security and operation of seats and seat belts	X	X	X	X
Check operation of all door, bonnet and luggage compartment locks	X	X	X	X
Check operation of window controls	X	X	X	X
GENERAL				
Road/roller test and check function of all instrumentation	X	X	X	X
Report additional work required	X	X	X	X

At 18 month intervals
Renew brake fluid

At 48 000 km (30 000 mile) intervals
Change final drive oil
Change coolant ensuring that 55% anti-freeze content is present upon replenishment.
Change gearbox oil.

At 3 years or 60 000 km (37 500 miles) intervals — whichever is the sooner
Renew all fluid seals in hydraulic system, examine and renew if necessary all flexible hoses.
Examine working surfaces of master cylinder and calipers. Renew if necessary.

CONTENTS

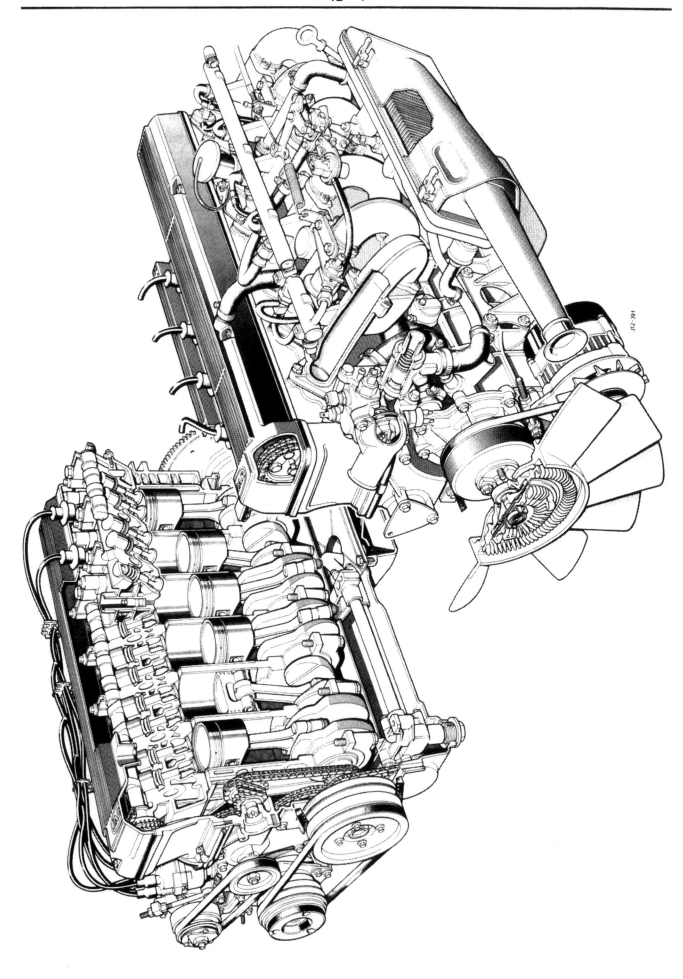

J12·391

3.6 4 VALVE ENGINE

Configuration	In line 6 cylinder
Bore	91 mm
Stroke	92 mm
Capacity	3590 cc
Compression ratio	9.6:1 (Europe)
	8.2:1 (Federal)

The new slant 6 cylinder engine is of an all alloy construction. The cylinders are of an in-line configuration and have a cubic capacity of 3590 cc. The engine is canted over at 15° from the vertical to the RH side of the vehicle ensuring there is ample room for the induction manifolds necessary to provide low speed flexibility and torque output.

ENGINE CONSTRUCTION

The skirted design crankcase is manufactured in cast aluminium alloy with shrink fit dry cast iron cylinder liners.

The crankshaft is manufactured from SG cast iron and is nitro carburise treated to enable a very high quality finish on the bearing surfaces and increase the life of the journals. The crankshaft is supported by seven iron bearing caps having bearings, which are lead bronze on split steel backed shells with a lead indium overlay.

Crankshaft end float is controlled by half thrust washers fitted on each side of the centre main bearing journal.

The connecting rods are manufactured from manganese molybdenum steel, forged in an 'H' section. The small end bushes are lead bronze with steel backing, machined to size after pressing into connecting rod.

The big end bearings are of a lead bronze alloy on split steel backed shells and with lead indium overlay.

The pistons are manufactured from aluminium alloy and have either a raised or dished crown to produce the alternative compression ratios required to suit varying market needs. Each piston has a spring assisted micro land oil control ring situated below a barrel faced chrome plated compression ring and an externally stepped taper faced secondary ring.

The pistons are internally strutted for thermal control and run on hardened steel gudgeon pins offset from the centre line of the piston towards the thrust face.

The cylinder head is cast from aluminium alloy with pent roof shaped combustion chambers with crossflow valve porting.

Running directly in the cylinder head are two cast iron camshafts retained by machined aluminium caps. Each camshaft uses chilled cams to drive two valves per cylinder via chilled cast iron bucket tappets with shim adjustment, control of each of the four valves per cylinder is maintained by single valve springs.

The camshafts are operated by a two stage 'duplex' chain drive from the crankshaft. Each stage is controlled by a hydraulic tensioner operating through a pivoted rubber faced curved tensioner blade.

The first stage incorporates a three point drive via the crankshaft, intermediate shaft and auxiliary shaft.

The intermediate shaft is live and provides a 0.75 x crankspeed drive through the timing cover. This drive access is blanked off.

The 'live' auxiliary shaft is driven at crankshaft speed and is situated on the RH side of the engine (looking from rear). In addition to driving the distributor via a set of 2:1 reduction spiral gears, it provides

external drives for:

(a) The power assisted steering pump at the rear.

(b) The air injection pump drive pulley at the front. On non air injection engines this drive is blanked off.

The second stage is a three point drive via the intermediate shaft and two camshafts. The 2:1 reduction ratio from crankspeed is achieved by the combined ratio of the intermediate and camshaft sprocket sizes.

The oil pump is a rotor type mounted on the underside of the front of the crankcase and driven by a 'simplex' chain from the crankshaft nose. The pump incorporates a built in pressure relief valve which is accessible via a plug at the front of the sump.

Below the line of the crankcase are two windage trays; these prevent oil being sucked up and thrown into the crankcase; this alleviates windage and power losses.

Above the sump oil level is a sump tray and baffle assembly to prevent oil surge.

At the rear of the crankshaft is a new design of lip type oil seal which provides a high degree of oil retention. It also allows the use of higher engine speed and easier serviceability as opposed to the conventional asbestos rope seal.

LUBRICATION SYSTEM

Oil is drawn from the sump via a gauze filter. Pressurised oil, having been regulated by a relief valve, is then fed via internal galleries on the LH side of the cylinder block. An adaptor block is fitted between the cylinder block and the oil filter housing which diverts the full oil flow through the oil cooler prior to filtering. A balance valve is fitted which bypasses the oil cooler circuit when a pressure differential of 10 to 15 lb/in² occurs.

Pressurised and filtered oil is fed into the main oil gallery, the seven main bearings are fed and thence via crankshaft drillings to the big end bearings.

The intermediate shaft, auxiliary shaft and camshaft bearings are pressure lubricated by means of internal drillings directly fed from the front of the main oil gallery.

NOTE: Oil coolers are specified on all models to date, but if at any time a cooler is not required, then the oil filter housing can be bolted directly to the cylinder block through omitting the oil cooler adaptor.

SPS JOINT CONTROL SYSTEM

A feature of the engine is the adoption of the SPS system on the cylinder head bolts. This system ensure that the joints receive maximum clamp load for a given fixing size and type of material by tightening the fixing to its particular yield point. This greatly helps to prevent any premature cylinder head gasket failures.

Apart from the cylinder head the manifolds and the auxiliary housing, the engine is gasketless. The use of a rubber sealing material on all other joints ensures a leak free engine.

NOTE: The SPS system in service requires the nut or bolt to be set to a specified torque initially, the fastening then has to be rotated clockwise through exactly 90°. The accuracy of this cannot be too highly stressed.

It is therefore recommended that a tool is manufactured to aid the turning of the fixing through exactly 90°. A suggested tool is illustrated below.

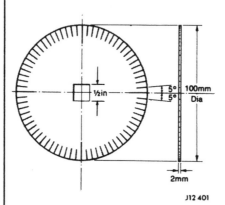

J12 401

CRANKCASE BREATHER

Blowby gases are recycled via the induction system to maintain a crankcase depression and so prevent their escape to the atmosphere.

A dual system is employed by means of:

(a) An oil filler stack pipe which vents into the crankcase.

(b) A baffled vent from the cam box cover, the pipe from which connects the air cleaner casing (clean side) with a feed branch via a water heated restrictor into the plenium chamber.

In this way, a crankcase depression is maintained under both closed and open throttle workload conditions.

CYLINDER HEAD DESIGN

The advantage of four valves per cylinder are: although the valves are smaller in diameter than on a conventional engine combined they have a greater effective area and, also being lighter, the operating gear has less stress applied to it. This design also increases the power at high engine revs and allows an efficient combustion of the fuel.

It also allows the sparking plug to be situated in its ideal central position which also creates efficient combustion and consequently enhances fuel economy.

COOLANT SYSTEM

The engine is liquid cooled by a mixture of water and anti-freeze circulating around the coolant passages. The coolant pump is mounted on the LH side of the cylinder block and is driven from the crankshaft nose by a three point belt drive (which includes the alternator). The pump is a fully assembled bolt on unit. The coolant is fed into the cylinder block at two places via an external delivery pipe. The coolant is drawn from the cylinder head via a self-contained thermostat housing back to the radiator or recirculated according to the thermostat position.

SEALANTS

The sealant used on all gasketless joints on this engine must be the Marston compound supplied by Unipart known as Hylosil 101, a white **amine** cure system rubber. Should this not be available an **amine** cure sealant **must be used.**

Under no circumstances should any acidtoxy cure system be used.

SERVICE TOOLS

Auxiliary shaft oil seal remover	18G 1468
Auxiliary shaft oil seal replacer	18G 1469
Rear main oil seal installer	18G 134-8
Timing chain tensioner restrainer and crankshaft pulley remover	18G 1436
Camshaft timing tool	18G 1433
Jackshaft inner bearing remover (2 part tool)	18G 1434
Piston ring clamp	18G 55A
Valve spring compressor	18G 106A
Valve guide remover/replacer	18G 1432
Front pulley lock	18G 1437
Dummy camshaft caps	18G 1435
Remover/replacer P.A.S. pump drive and flange	18G 1445
Valve seat cutters	MS 204
Pilot	MS 150-8
Handle kit	MS 76

DESCRIPTION	THREAD SIZE	SPANNER SIZE	TIGHTENING TORQUE		
			Nm	kgf/cm	lbf/ft
Adjust bolt to compressor — Flange headed bolt	M8	10 mm	23 to 27	2,38 to 2,8	17 to 20
Adjusting bolt, mounting bracket — Flange headed bolt	M8	10 mm	23 to 27	2,38 to 2,8	17 to 20
Adjust bolt to sleeve — Nut	⅜ UNF	⁹⁄₁₆ in AF	12,2 to 13,6	1,26 to 1,4	9 to 10
Adjust sleeve — Nut	⅜ UNF	⁹⁄₁₆ in AF	12,2 to 13,6	1,26 to 1,4	9 to 10
Adjusting bolt to timing cover — Nut	M8	10 mm	23 to 27	2,38 to 2,8	17 to 20
Alternator pivot bolt — Nut	M8	10 mm	23 to 27	2,38 to 2,8	17 to 20
Air conditioning adaptor to throttle body — Flange headed bolt	M6	8 mm	9,5 to 12,2	0,98 to 1,26	7 to 9
Air conditioning ass to adaptor plate — Flange headed setscrew	M6	8 mm	9,5 to 12,2	0,98 to 1,26	7 to 9
Air pump mounting — Nut	M14	22 mm	30 to 40	3,10 to 4,2	22 to 30
Air pump to filter — Clip	M6	8 mm	2,4	0,24	1.75
Air pump to switching valve — Clip	M6	8 mm	2,4	0,24	1.74
Air temperature sensor	M10	19 mm	49 to 54	5,04 to 5,6	36 to 40
Auxiliary shaft pulley — Nut	M12	19 mm	57 to 64	5,9 to 6,53	42 to 47
Baffle tray — Flange headed setscrew	M8	10 mm	23 to 27	2,38 to 2,8	17 to 20
Base plate to pedestal — Setscrew	M6	8 mm	9,5 to 12,2	0,98 to 1,26	7 to 9
Bearing housing to body — Flange headed setscrew	M8	10 mm	23 to 27	2,38 to 2,8	17 to 20
Bracket to compressor — Flange headed setscrew	M8	10 mm	23 to 27	2,38 to 2,8	17 to 20
Bracket to compressor — Flange headed setscrew	M10	13 mm	49 to 54	5,04 to 5,6	36 to 40
Bracket to cylinder block — Flange headed bolt	M8	10 mm	23 to 27	2,38 to 2,8	17 to 20
Bracket to sump — Flange headed setscrew	M8	10 mm	23 to 27	2,38 to 2,8	17 to 20
Breather pipe to air cleaner — Clip	M6	8 mm	2,4	0,24	1.75
Breather pipe to independent manifold — Clip	M4	6 mm	0,51	0,051	0.38
Button head socket screw	M5	3 mm	1,7	0,17	1.25
By-pass elbow — Flange headed setscrew	M8	10 mm	23 to 27	2,38 to 2,8	17 to 20
By-pass hose — Hose clip	M6	8 mm	2,4	0,24	1.74
Cable abutment bracket — Flange headed setscrew	M6	8 mm	9,5 to 12,2	0,98 to 1,26	7 to 9
Cam cover assembly — Flange headed setscrew	M8	10 mm	23 to 27	2,38 to 2,8	17 to 20
Cam cover to filler pipe — Clip	M6	8 mm	2,4	0,24	1.75
Camshaft coupling assembly — Setscrew	¼ UNF	⁹⁄₁₆ in AF	23 to 27	2,38 to 2,8	17 to 20
Cap to head — Flange headed bolt	M8	10 mm	23 to 27	2,38 to 2,8	17 to 20
Check valve to valve switching — Clip	M6	8 mm	2,4	0,24	1.75
Clamp plate to compressor — Flange headed bolt	M10	13 mm	49 to 54	5,04 to 5,6	36 to 40
Connecting rod nut	⅜ in UNF	½ in AF	52 to 55	5,5 to 5,8	39 to 41
C S inj — Socket head counter-sunk screw	M6	8 mm	9,5 to 12,2	0,98 to 1,26	7 to 9
Cover to adaptor — Taptite screw	M5	8 mm	5,4 to 8,1	0,50 to 0,83	4 to 6
Cover to cylinder head — Screw	M6	8 mm	9,5 to 12,2	0,98 to 1,26	7 to 9
Cover to housing — Flange headed setscrew	M8	10 mm	23 to 27	2,38 to 2,8	17 to 20

DESCRIPTION	THREAD SIZE	SPANNER SIZE	TIGHTENING TORQUE		
			Nm	kgf/cm	lbf/ft
Crank pulley to damper — Flange headed bolt	M8	10 mm	23 to 27	2.38 to 2.8	17 to 20
Crankshaft bolt	$\frac{3}{4}$ in UNF	$1\frac{5}{16}$ in AF	203,4	20,8	150
Cylinder head — Bolt	$\frac{7}{16}$ in UNC	$\frac{11}{16}$ in AF	Initial torque 38 to 40 lbf/ft and 90° angle (see note SPS system 12-6).		
Cylinder head to timing cover — Flange headed bolt	M8	10 mm	23 to 27	2.38 to 2.8	17 to 20
Damper distance piece — Setscrew	M8	10 mm	23 to 27	2.38 to 2.8	17 to 20
Damper/distance piece to cylinder block — Bolt . . .	M8	10 mm	23 to 27	2.38 to 2.8	17 to 20
Damper to cylinder block — Setscrew	M8	10 mm	23 to 27	2.38 to 2.8	17 to 20
Damper to saddle — Setscrew	M8	10 mm	23 to 27	2.38 to 2.8	17 to 20
Distance piece to block — Flange headed bolt . . .	M8	10 mm	23 to 27	2.38 to 2.8	17 to 20
Distributor clamp plate to timing cover — Flange headed setscrew	M6	8 mm	9,5 to 12,2	0,98 to 1,26	7 to 9
Drain plug	M20	30 mm	41 to 47	4,2 to 4,9	30 to 35
End plate to cylinder head — Flange headed setscrew	M8	10 mm	23 to 27	2.38 to 2.8	17 to 20
Engine dipstick — Flange headed setscrew	M6	8 mm	9,5 to 12,2	0,98 to 1,26	7 to 9
Engine dipstick — Taptite screw	$\frac{1}{4}$ UNC	$\frac{3}{8}$ in AF	5,4 to 8,1	0,56 to 0,83	4 to 6
Exhaust manifold to head — Flange headed bolt . .	M10	13 mm	49 tó 54	5,04 to 5,6	36 to 40
Exhaust pressure switch and solenoid vacuum valve to speed control actuator — Flange headed setscrew	M5	10 mm	1,7	0,17	1.25
Extra air valve — Nut	M6	8 mm	9,5 to 12,2	0,98 to 1,26	7 to 9
Fan drive unit to water pump pulley — Nut	M8	10 mm	23 to 27	2.38 to 2.8	17 to 20
Fan to drive unit — Setscrew	$\frac{5}{16}$ in UNC	$\frac{1}{2}$ in AF	13,6	1,4	10
Filler pipe to breather pipe — Clip	M6	8 mm	2,4	0,24	1.75
Filter head to cylinder block — Flange headed setscrew	M8	10 mm	23 to 27	2.38 to 2.8	17 to 20
Filter head to cylinder block — Flange headed bolt .	M8	10 mm	23 to 27	2.38 to 2.8	17 to 20
Filter to air pump — Flange headed setscrew	M8	.10 mm	23 to 27	2.38 to 2.8	17 to 20
Flywheel	M12	17 mm	98	10,08	72
Front pivot to mounting bracket — Flange headed setscrew	M10	13 mm	49 to 54	5,04 to 5,6	36 to 40
Fuel rail — Flange headed setscrew	M6	8 mm	9,5 to 12,2	0,98 to 1,26	7 to 9
Heater feed — Adaptor	$\frac{3}{4}$ BSPF	$\frac{7}{16}$ WHIT	49 to 54	5,04 to 5,6	36 to 40
High tension lead clip to cover — Flange headed setscrew	M6	8 mm	9,5 to 12,2	0,98 to 1,26	7 to 9
Hose to air pipe — Clip	M6	8 mm	2,4	0,24	1.75
Hose to backplate and air pipe — Clip	M6	8 mm	2,4	0,24	1.75
Hose to EA valve — Clip	M6	8 mm	2,4	0,24	1.75
Hose to elbow — Clip	M6	8 mm	2,4	0,24	1.75
Hose to housing — Clip	M6	8 mm	2,4	0,24	1.75
Hose to pipe — Clip	M6	8 mm	2,4	0,24	1.75
Housing to cylinder block — Flange headed setscrew	M8	10 mm	23 to 27	2.38 to 2.8	17 to 20
Housing to cylinder block — Flange headed bolt . .	M8	10 mm	23 to 27	2.38 to 2.8	17 to 20

DESCRIPTION	THREAD SIZE	SPANNER SIZE	TIGHTENING TORQUE		
			Nm	kgf/cm	lbf/ft
Housing to cylinder block — Capscrew	M8	10 mm	23 to 27	2,38 to 2,8	17 to 20
Housing to cylinder head — Flange headed bolt	M8	10 mm	23 to 27	2,38 to 2,8	17 to 20
Housing to cylinder block — Flange headed setscrew	M6	8 mm	9,5 to 12,2	0,98 to 1,26	7 to 9
Housing to cylinder head — Flange headed setscrew	M8	10 mm	23 to 27	2,38 to 2,8	17 to 20
Induction manifold to head — Flange headed bolt	M8	10 mm	23 to 27	2,38 to 2,8	17 to 20
Induction manifold to head — Nut	M8	13 mm	12 to 16	1,26 to 1,66	9 to 12
Inlet elbow — Flange headed setscrew	M6	8 mm	9,5 to 12,2	0,98 to 1,26	7 to 9
Magnetic clutch assembly — Setscrew	M5	8 mm	5,4 to 8,1	0,56 to 0,83	4 to 6
Main bearing caps — Bolt	M12	17mm	136 to 142	13,8 to 14,6	100 to 105
Mounting bracket to block — Flange headed setscrew	M10	13 mm	49 to 54	5,04 to 5,6	36 to 40
Mounting bracket to timing cover — Flange headed setscrew	M8	10 mm	23 to 27	2,38 to 2,8	17 to 20
Mounting bracket to timing cover and block — Flange headed setscrew	M8	10 mm	23 to 27	2,38 to 2,8	17 to 20
Oil drilling blank — Setscrew	M8	6 mm	23 to 27	2,38 to 2,8	17 to 20
Oil filler housing — Flange headed setscrew	M8	10 mm	23 to 27	2,38 to 2,8	17 to 20
Oil filler to support bracket — Flange headed setscrew	M6	8 mm	9,5 to 12,2	0,98 to 1,26	7 to 9
Oil filter can	1 in UNF	—	8,1	0,83	6
Oil filter head — Threaded insert	1 in UNF	—	8,1	0,83	6
Oil gallery blank — Setscrew	$\frac{3}{8}$ in UNC	$\frac{9}{16}$ in AF	23 to 27	2,38 to 2,8	17 to 20
Oil pick-up to carrier — Flange headed setscrew	M6	8 mm	9,5 to 12,2	0,98 to 1,26	7 to 9
Oil pipe carrier to cylinder block — Flange headed bolt	M8	10 mm	23 to 27	2,38 to 2,8	17 to 20
Oil pump to cylinder block — Flange headed bolt	M8	10 mm	23 to 27	2,38 to 2,8	17 to 20
Oil sump to cylinder block — Flange headed bolt	M8	10 mm	23 to 27	2,38 to 2,8	17 to 20
Oil sump to timing cover — Flange headed bolt	M8	10 mm	23 to 27	2,38 to 2,8	17 to 20
Pedestal to body	M6	8 mm	9,5 to 12,2	0,98 to 1,26	7 to 9
Pipe to manifold — Flange headed setscrew	M8	10 mm	23 to 27	2,38 to 2,8	17 to 20
Pivot bracket to cylinder block — Bolt	M8	10 mm	23 to 27	2,38 to 2,8	17 to 20
Pivot plate to cylinder block — Bolt	M8	10 mm	23 to 27	2,38 to 2,8	17 to 20
Plate to adaptor — Taptite screw	M5	8 mm	5,4 to 8,1	0,56 to 0,83	4 to 6
Plate to timing cover — Setscrew	M8	10 mm	23 to 27	2,38 to 2,8	17 to 20
Plate to timing cover — Setscrew	M6	8 mm	9,5 to 12,2	0,98 to 1,26	7 to 9
Pressure regulator to fuel rail — Thin nut	M16	24 mm	10,8	1,12	8
Pump to adaptor — Flange headed setscrew	M10	13 mm	49 to 54	5,04 to 5,6	36 to 40
Pump to engine — Flange headed setscrew	M8	10 mm	23 to 27	2,38 to 2,8	17 to 20
Rail to cylinder block — Flange headed setscrew	M8	10 mm	23 to 27	2,38 to 2,8	17 to 20
RR cover to cylinder head — Flange headed setscrew	M8	10 mm	23 to 27	2,38 to 2,8	17 to 20

DESCRIPTION	THREAD SIZE	SPANNER SIZE	TIGHTENING TORQUE		
			Nm	kgf/cm	lbf/ft
Rear pivot to mounting bracket — Flange headed setscrew	M10	13 mm	49 to 54	5,04 to 5,6	36 to 40
Retainer bracket — Flange headed setscrew	M6	8 mm	9,5 to 12,2	0,98 to 1,26	7 to 9
Relief valve — Access plug	M42	42mm	68 to 75	6,9 to 7,66	50 to 55
Selector lever to shaft — Nut	M10	17 mm	49 to 54	5,04 to 5,6	36 to 40
Sleeve to adjustment bolt — Nut	⅜ UNF	9/16 in AF	12,2 to 13,6	1,26 to 1,4	9 to 10
Sleeve to alternator — Flange headed bolt	M8	10 mm	23 to 27	2,38 to 2,8	17 to 20
Sleeve to pump — Flange headed setscrew	M8	10 mm	23 to 27	2,38 to 2,8	17 to 20
Sprocket to hub — Setscrew	M6	8 mm	9,5 to 12,2	0,98 to 1,26	7 to 9
Starter motor to adaptor — Flange headed bolt	M10	13 mm	49 to 54	5,04 to 5,6	36 to 40
Stiffener plate to adaptor — Flange headed setscrew	M10	13 mm	49 to 54	5,04 to 5,6	36 to 40
SA valve to air pipe hose — Clip	M6	8 mm	2,4	0,24	1.75
Temperature switch (air injection only)	M6	8 mm	9,5 to 12,2	0,98 to 1,26	7 to 9
Thermo time switch	M14	24 mm	15	1,55	11
Throttle body to manifold — Setscrew	M8	10 mm	23 to 27	2,38 to 2,8	17 to 20
Throttle spindle assembly — Nut	5/16 UNF	½ in AF	4,1	0,42	3
Throttle spindle assembly — Setscrew	⅛ WHIT	2 BA	0,17	0,017	0.125
Throttle switch — Setscrew	M4	7 mm	0,45	0,045	0.33
Timing cover to cylinder block — Flange headed bolt	M8	10 mm	23 to 27	2,38 to 2,8	17 to 20
Timing disc to damper — Flange headed setscrew	M6	8 mm	9,5 to 12,2	0,98 to 1,26	7 to 9
Timing indicator — Setscrew	M8	10 mm	23 to 27	2,38 to 2,8	17 to 20
Trans adaptor — Flange headed bolt	M10	13 mm	49 to 54	5,04 to 5,6	36 to 40
Trans adaptor — Flange headed setscrew	M10	13 mm	49 to 54	5,04 to 5,6	36 to 40
Transfer unit to adaptor — Bolt	M10	13 mm	49 to 54	5,04 to 5,6	36 to 40
Water drain cylinder block — Plug	M20	30 mm	75 to 81	7,66 to 8,4	55 to 60
Water pump and timing cover to cylinder block — Flange headed bolt	M8	10 mm	23 to 27	2,38 to 2,8	17 to 20
Water pump to rail — Hose clip	M6	8 mm	2,4	0,24	1 75
Water temperature sensor	M12	19 mm	13,6	1,4	10
Windage tray — Flange headed setscrew	M8	10 mm	23 to 27	2,38 to 2,8	17 to 20

ENGINE FAULT FINDING

SYMPTON	POSSIBLE CAUSE	CHECK	REMEDY
Engine spits back into air box	Excessively weak mixture	Check CO reading of gases leaving the exhaust pipe	Adjust mixture
	Air leaking into induction manifold	Listen for leaks (whistling), spray 'easy start' around suspect area, the engine speed will increase if there is an air leak	Replace gasket or manifold
Engine backfires	Air leakage into/from the exhaust system	Check for leaks or blows in the system	Repair leaks or replace system if necessary
	Leakage past valves and guides	Check HC reading of exhaust gases and also check for crankcase fumes	Remove the cylinder head and overhaul
	Ignition timing retarded	Check ignition timing	Adjust ignition timing
	Mixture too weak	Check CO level from exhaust pipe	Adjust mixture
	Incorrect valve timing	Check camshaft timing	Adjust camshaft timing
	Valves sticking open	Check valve clearances	Adjust valve clearances
		Check for wear or gum in valve guides	Replace guides or decarbonise the cylinder head
		Check for poor seating of valves	Overhaul the cylinder head
Engine fails to idle	Insufficient air supply from idle adjustment screw	Ensure the screw is not wound right in	Wind the screw out until engine idles satsifactorily
	Incorrect ignition timing	Check ignition timing	Adjust ignition timing
	Valve clearances insufficient	Check clearances	Adjust clearances
	Faulty cylinder head gasket	Check for leaks into the oil and water	Replace the gasket
	Exhaust system blocked	Check for restrictions	Remove the restrictions or replace components as necessary
Engine fails to rotate when attempting to start	Battery leads loose or terminals corroded	Check the condition of the leads and terminals	Clean and tighten as necessary
	Battery discharged	Check condition of battery with hydrometer	Charge or replace as necessary
	Starter motor inoperative	If the lights dim when ignition switch is operated, the starter may be jammmed in the starter ring gear	Remove starter motor, free off pinion and refit
		Check for loose and dirty connections to the starter motor	Clean/tighten and/or replace as necessary
Engine rotates but will not fire	Starter motor speed too low	Check battery leads and terminals	Tighten and clean leads as necessary
		Check the state of the battery charge	Charge battery or fit replacement
	Faulty ignition system	Using plug lead pliers, hold the plug lead approximately 16 mm from the cylinder head and crank the engine, check that a good blue spark jumps the gap between the end of the plug lead and the cylinder head	If no spark, a yellow or white spark is evident, check the ignition system
		If the spark is adequate, remove the sparking plugs	Clean and regap the sparking plugs, replace if worn out

SYMPTOM	POSSIBLE CAUSE	CHECK	REMEDY
	Fuel system defect	Remove cold start injector, connect the white and green wires to earth. Crank the engine and check that fuel is emitted from the injector (ensure a suitable receptacle is available to collect the fuel)	If no fuel is emitted, refer to fuel injection section
Overheating	Thermostat stuck closed	Remove thermostat and carry out boiling water check	Change thermostat
	Faulty gauge	Check with fast check equipment	Fig new gauge
	Faulty transmitter	If other equipment in circuit OK fit new transmitter	Fit new transmitter
	Radiator blocked	Check for uneven heat in radiator, i.e. hot and cold spots plus water emission from overflow pipe	Flush or change radiator
	Too high concentration of anti-freeze	Remove the header tank cap and check concentration with a hydrometer	Drain coolant and fill with correct concentration
	Fan belt slack	Check tension	Retension the belt
	Fan belt broken	Open bonnet and check visually	Fit new belt
	Water pump seized	Remove drive belt and turn pulley	Change water pump
	Insufficient coolant	Remove header tank cap and check coolant level visually	Top-up coolant
	Incorrect ignition timing	Check ignition timing	Adjust ignition timing
	Fuel/air mixture too weak	Check CO/HC levels	Adjust mixture
	Incorrect thermostat	Remove thermostat and visually check temperature reading	Change thermostat
	Collapsed hoses	Start engine, run until normal temperature is attained and visually check the hoses for collapsing	Fit new hoses
	Cylinder head gasket leaking	Remove head tank cap, fill tank with water. Run engine until hot, rev engine and return to idle; check for bubbles in water	Change head gasket
	Incorrect valve timing	Check valve timing with Service Tool 18G 1433	Reset valve timing
	Cylinder block waterways restricted/ restricted coolant flow		
	Radiator cap spring weak	Fit cap to pressure testing equipment and check spring release pressure	Fit new cap
At tick over	Viscous fan free wheeling	Run engine and visually check that the large fan is turning	Fit new viscous coupling
No oil pressure	Oil pump drive chain broken/not fitted	Remove sump and check visually	Fit new chain
	Faulty gauge	Check with fast check equipment	Change gauge
	No oil in sump	Dip oil	Fill with oil
	Blocked pick-up pipe strainer	Remove sump and pick-up pipe	Clean pick-up strainer

SYMPTON	POSSIBLE CAUSE	CHECK	REMEDY
Too cold	Thermostat stuck open	Remove thermostat	Fit new thermostat
	Auxiliary fan remains operational	Check relay circuit Check radiator fan thermostat Check diode	Fit new relay Fit new fan thermostat Fit new diode
	Faulty gauge	Check with fast check equipment	Change gauge
	Faulty transmitter	If all other equipment OK, fit new transmitter	Fit new transmitter
	Thermostat missing	Remove thermostat housing	Fit new thermostat
	Wrong temperature rated thermostat	Remove thermostat housing and visually check rate	Fit correct thermostat
Loosing oil (leaking)	Worn front oil seal	Wipe clean, run engine and visually check	Fit new seal
	Worn rear oil seal	Wipe bell housing clean, run engine and visually check	Remove gearbox and fit new seal
	Leaking gaskets	Visual (except head gasket see head gasket blown)	Change gasket or reseal
	Cylinder block cracked	Visual	Change cylinder block
Detonation knock (pinking)	Ignition timing too far advanced	Check ignition timing	Reset ignition timing
	Head gasket blown	Dip engine oil and check for ingress of water. Remove header tank cap, rev engine and check for bubbles in water	Change head gasket
	Thermostat stuck (shut) overheating	Remove thermostat and carry out boiling water test	Fit new thermostat
	Mixture too weak	Check with CO/HC meter	Adjust mixture
	No water in engine/radiator	Remove header tank cap and check water level	Filll with water
	Engine running too hot	——	See overheating fault finding
	Valve timing incorrect	Check valve timing	Adjust valve timing
	No water in engine/radiator	Remove header tank cap and check water level	Fill with water
	Engine running too hot	——	See overheating fault finding
	Valve timing incorrect	Check valve timing	Adjust valve timing
	Incorrect octane fuel	If all other checks OK, this could be the cause	Fill with correct octane fuel
Excessive noise from valve gear	Excessive valve clearance	Check valve clearances	Adjust valve clearances
	Broken valve spring(s)	Remove valves and check springs	Replace as necessary
	Broken valve guide	Remove valves and check guides	Replace as necessary
	Broken valve seat insert	Remove valves and check inserts	Replace as necessary
	Lack of lubrication	Check oil pressure	See insufficient oil pressure fault finding
	Valve clash	Check valve timing	Adjust valve timing

SYMPTON	POSSIBLE CAUSE	CHECK	REMEDY
	Worn camshafts	Valve clearances and lack of lubrication	Adjust valve clearances — see insufficient oil pressure
	Worn camshaft drive chains/ tensioners	Remove front timing cover and check for wear	Replace as necessary
Insufficient oil pressure	Oil requires changing	Dip oil, check colour and viscosity	Change oil
	Main oil gallery seals leaking or gallery blocked	If all other checks OK, this could be a cause	Fit new 'O' rings or clear oil gallery
	Worn crankshaft journals	Listen for rumble or knock	Change crankshaft
	Excessive crankshaft endfloat	Fit dial gauge and measure	Remove sump and fit oversize thrust washers
	Worn main bearing shells	Listen for rumble	Check crankshaft journals for wear and fit new shells
	Worn oil pump	Remove oil pump and check the clearances	Fit new oil pump
	Oil cooler valve stuck shut		
	Oil pressure relief valve sticking open	Remove valve and check for sticking	Fit new valve
	Oil pressure relief valve spring too weak	Remove spring and check spring rates	Fit new spring
	Insufficient oil in sump	Dip oil	Top-up as required
	Faulty gauge	Check gauge with fast check equipment	Fit new gauge
	Engine overheating	———	See overheating fault finding
	Wrong specification oil (too thin)	Dip oil and check viscosity	Change oil and filter
	Water in oil	Dip oil and check if oil is a milky white colour	Change oil and check for blown head gasket
	Cracked oil pump housing	Remove the sump and check visually	Change the pump
	Blocked oil pick-up pipe strainer	Remove sump and check visually	Remove oil pick-up pipe and clean strainer
	Oil pump pipe 'O' rings leaking or missing	Remove sump and pipes and check 'O' ring condition	Fit new 'O' rings
Oil pressure too high	Relief valve stuck shut	Remove valve and check for sticking	Clean or replace the valve
	Wrong pressure relief valve spring	Remove spring and check the rate	Fit new spring
	Incorrect grade engine oil		Drain engine oil and fill with correct oil
	Gauge or transmitter fault	Check gauge with fast check equipment, if OK fit new transmitter	Fit new gauge or transmitter
	Engine temperature too low	Check operation of the thermostat (stuck open) or oil cooler valve stuck open	Fit new thermostat or valve
Loss of power	Burned valves	Check compressions	Remove cylinder head and change valves
	Sticking valves	Check compressions	Remove cylinder head and replace valves/guides or springs

SYMPTON	POSSIBLE CAUSE	CHECK	REMEDY
	Poor engine tune	Check engine tune	Adjust as necessary
	Insufficient valve clearance	Check valve clearance	Adjust as necessary
	Fuel injection fault	Refer to fuel injection section	
	Low compression in cylinders	Check compressions	Rebore/rering as necessary
	Ignition fault	Refer to electrical section	
	Camshaft timing incorrect	Check camshaft timing using Service Tool 18G 1433	Adjust camshaft timing
	Incorrect grade fuel		Drain fuel and refill with correct octane
	Partial seizure of engine	Remove spark plugs and rotate engine	Overhaul engine as necessary
	Worn camshaft	Remove camshaft and check for wear	Replace camshaft
Rough running normal engine speed (less than 6 cylinders)	Sticking valves	Check compressions	Change valves, springs or guides
	Broken valve springs	Check compressions	Change valve springs and check for bent valves
	Piston fault	Check compressions	Change pistons
	Head gasket blown	Check compressions and water level	Change head gasket
	Valve burned out	Check compressions	Change valve
	Valve seat burned out	Check compressions	Cut or change valve seat
	Ignition fault	Refer to electrical section	
	Fuel injection fault	Refer to fuel injection section	
	Air leaking into inlet manifold	Run engine and listen for whistling spray easy start around suspect area. If engine speed increases, confirmed	Change gasket or manifold
	Blowing exhaust	Run engine and check for leaks	Repair leak or change exhaust
Noisy Chains	Low oil pressure	Take reading from gauge when engine is hot	See insufficient oil pressure
	Tensioners not released	Remove camcover and check tension of chain (top chain). Remove timing cover to check bottom chain tension	Insert 3 mm Allen key and turn tensioner anti-clockwise, compress tensioner to release bottom chain
	Chains worn	Check visually/remove and check for wear	Replace as necessary
	Sprockets worn	Check visually	Replace as necessary
	Tensioner worn	Check visually	Replace as necessary
Rough idle	Valve timing incorrect	Check camshaft timing using Service Tool 18G 1433	Adjust camshaft timing
	Incorrect mixture	Check CO/HC reading for exhaust	Adjust mixture
	Incorrect ignition timing	Check ignition timing	Adjust ignition timing
	Valve clearances insufficient	Check valve clearances	Adjust valve clearances

SYMPTON	POSSIBLE CAUSE	CHECK	REMEDY
	Valve burned out	Check compressions	Change valve
	Ignition fault	Refer to electrical section	
	Fuel injection fault	Refer to fuel injection section	
Burning oil	Worn cylinder bores	Check wear with a comparitor	Rebore cylinders as necessary
	Worn valve guides	Insert valve in guide and check side movement	Change valve guides as necessary
	Worn inlet valve seals	Remove seals and check for splits or wear	Replace in sets
	Worn piston rings	Measure rings in bore	Replace rings in sets and rebore as necessary
	Leaking cylinder head gasket	Check for blue smoke from exhaust	Replace head gasket
	Engine oil wrong specification		Drain oil and refill with the correct viscosity

TIMING CHAINS, DAMPERS, TENSIONERS AND SPROCKETS

Overhaul

Depressurise the fuel system, remove the bonnet and drain the coolant.
Remove the cylinder head assembly.
Drain the engine oil, remove the crankshaft damper (1, Fig. 1), timing cover (2, Fig. 1) and remove the upper timing chain.

Fig. 1

Remove the lower timing chain tensioner (1, Fig. 2), the intermediate sprocket (2, Fig. 2), remove the oil pump drive chain and damper (3, Fig. 2) and remove the lower timing chain (4, Fig. 2).

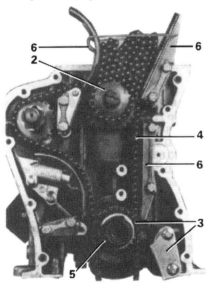

Fig. 2

Remove the crankshaft sprocket (5, Fig. 2) and all remaining dampers and spacers (6, Fig. 2).

Fig. 3

Remove the oil seal (1, Fig. 3) from the power assisted steering end of the distributor drive gear shaft using Service Tool 18G 1468. Remove the shaft securing circlip (2, Fig. 3), the drive gear sprocket and thrust washer. Remove the three distributor drive gear housing bearing set screws (3, Fig. 3), remove the housing (4, Fig. 3) and discard the gasket.

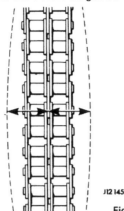

Fig. 4

Inspect all the components for wear, chipped teeth or wear marks on the gears and chains for stretch or picking up. The checking of stretch is demonstrated in Fig. 4, the second check for the chain is to pull it slowly over an outstretched forefinger, the chain should run evenly over the finger, if it picks up or runs jerkily over the finger, the chain should be changed; if any of the components are suspect then they must be renewed.

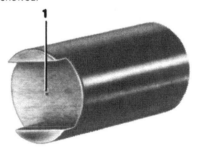

Fig. 5

Check the hydraulic tensioners for wear and scored bores, the spring for tension and also that the oil feed hole (1, Fig. 5) is clear in the end of the piston.
Check the dampers for wear, if any rubber surface has been broken by the chain the component must be replaced.

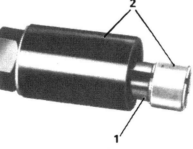

Fig. 6

Check the crankdamper/oil seal spacer for grooving and ensure that the wear does not exceed the limits set.
Locate the legs of service tool (1, Fig. 6)

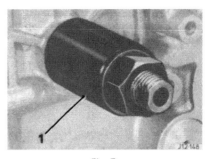

Fig. 7

18G 1434 behind the centre bearing shell in the centre of the cylinder block (1, Fig. 7). Pull the centre peg outward (1, Fig. 8) and tighten the nut (2, Fig. 8) until the bearing and tool (2, Fig. 6) appear from the cylinder block.
All component sizes and tolerances are listed below.

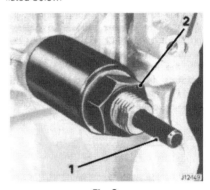

Fig. 8

Remove the remaining bushes from the front cover and distributor drive shaft housing, using Service Tool 18G 1434. Fit new bushes using Service Tool 18G 1434(4) ensuring that the oil feed holes in the bushes are lined up with the holes (1, Fig. 9) in the cylinder block.

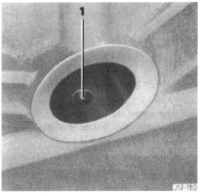

Fig. 9

Apply 'Hylosil' to the timing cover blanking plates and fit the plates to the cover.
Fit a new gasket to the distributor drive gear housing, align and fit the housing to the cylinder block, apply 'Hylomar' to the securing screws, fit and tighten. Clean the drive gear and thrust washer, lubricate the shaft and fit the shaft to the housing, fit the thrust washer and secure with a circlip. Lubricate the oil seal mounting face and the outer edge of the seal and using Service Tool 18G 1469 (1, Fig. 10), fit and seat the oil seal (2, Fig. 10).

NOTE: Under no circumstances must the seal insert be removed prior to fitting the seal.

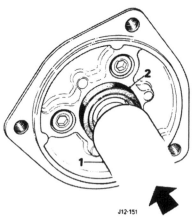

Fig. 10

Assemble the lower pivoting damper (1, Fig. 11) to the spacers, tab washer, bracket and securing bolts (2, Fig. 11), and fit the assembly to the engine.

Refit the lower timing chain (3, Fig. 11), sprockets (4, Fig. 11) and dampers (5, Fig. 11), refit and release the lower timing chain tensioner (6, Fig. 11).

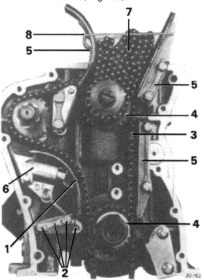

Fig. 11

Refit the upper timing chain (7, Fig. 11), place between the guides and secure with an elastic band (8, Fig. 11).

Fig. 12

Refit the oil pump drive chain and damper. Coat the mating face of the timing cover with 'Hylosil'.
Refit the timing cover (1, Fig. 12), the oil seal/crankshaft spacer and crankshaft damper assembly.
Refit the cylinder head.

Refit the thermostat housing and the air cleaner element.
Refill the engine with oil and coolant, adjust the ignition timing and refit the bonnet.

Camshaft Sprocket

No. of teeth	30
Pitch of teeth	9,525 mm (0.375 in)
Pitch circle diameter	91,135 mm (3.588 in)
Internal surrations	131
Pitch circle diameter	66,675 mm (2.625 in)
Pitch of teeth	1,60 mm (.063 in)
O/D of sprocket	95,25 mm ± 0,050 mm (3.75 in ± .002 in)
Total width	18,653 mm ($\frac{47}{64}$ in)
Total width across both teeth	15,341 to 15,570 mm (0.604 to 0.613 in)
Concentricity of teeth and sprocket not to exceed	0,076 mm (0.003 in)

Crankshaft Sprocket

No. of teeth	21
Pitch	9,525 mm (0.375 in)
Pitch circle diameter	63,906 mm (2.516 in)
Root diameter	57,556 mm (2.266 in)
Roller diameter	6,35 mm (0.25 in)
Measured over Ø 6,35 mm (0.25 in) pins	70,008 to 69,95 mm (2.759 to 2.754 in)
Tooth profile cut to	BS228
Internal diameter	42,875 to 42,857 mm
Overall diameter	67,87 to 67,77 mm
Width across 1 tooth	5,33 to 5,10 mm
Width across all the teeth	38,43 to 38,0 mm
Width across 2 teeth (timing chain)	15,57 to 15,34 mm

Crankshaft damper oil seal spacer

Inside diameter	42,875 to 42,860 mm (1.688 to 1.6874 in)
Outside diameter	52,00 to 51,81 mm (2.047 to 2.040 in)
Width	17,1 to 16,9 mm

Intermediate shaft bush

Inside diameter (when fitted)	22,010 to 21,964 mm (0.8665 to 0.8647 in)
Outside diameter	25,083 to 25,057 mm (0.9875 to 0.9865 in)
Width	17,25 to 16,25 mm (0.679 to 0.659 in)
Press fit	0,036 to 0,083 mm (0.0015 to 0.0033 in)
Shaft size	22 mm (0.8645 to 0.8637 in)
Running clearance	0,071 to 0,005 mm (0.0028 to 0.0002 in)
Material	Vandervell No. L10083/3

Intermediate sprocket inner teeth

Number of teeth	28
Pitch	9,525 mm (0.375 in)
Pitch circle diameter	85,063 mm (3.349 in)
Root diameter	78,7 mm (3.099 in)
Roller diameter	6,35 mm (0.25 in)
Measurement over Ø 6,35 mm (0.25 in) pins	91,41 to 91,29 mm (3.599 to 3.594 in)
Tooth profile cut to	BS228
Width across the depth	15,57 to 15,34 mm (0.613 to 0.604 in)
Overall diameter	89,2 to 89,1 mm (3.512 to 3.508 in)

Intermediate sprocket outer teeth

Number of teeth	20
Pitch	9,525 mm (0.375 in)
Pitch circle diameter	60,884 mm (2.397 in)
Root diameter	54,53 mm (2.147 in)
Roller diameter	6,35 mm (0.25 in)
Measurement over Ø 6,35 mm (0.25 in) pins	67,23 to 67,11 mm (2.647 to 2.642 in)
Tooth profile cut to	BS 228
Overall diameter	65,07 to 64,97 mm (2.562 to 2.558 in)
Width across both teeth	15,57 to 15,34 mm (0.613 to 0.604 in)
Total width of sprocket	48,8 to 48,7 mm (1.921 to 1.917 in)

Primary timing chain
Pitches   80
Roller diameter   6,35 mm (0.25 in)
Pitch   9,525 mm (0.375 in)
Type   Endless duplex chain

Secondary timing chain
Pitches   86
Roller diameter   6,35 mm (0.25 in)
Pitch   9,525 mm (0.375 in)
Type   Endless duplex chain

Oil pump chain
Pitches   52
Roller diameter   6,35 mm (0.25 in)
Pitch   9,525 mm (0.375 in)
Type   Endless simplex chain

Dampers and tensions
Backing material   Steel   BS1449 CR4
Damper material   Rubber

Hydraulic chain tension housing
Bore diameter   21,38 to 21,36 mm (0.842 to 0.841 in)

Hydraulic chain tensioner
Overall length   45 mm (1.771 in)
Overall length to bottom of groove in piston   41 mm (1.614 in)
Diameter   21,349 to 31,336 mm (0.8405 to 0.8400 in)
Width of groove in piston   15 mm (0.59 in)
Non return valve primary ball diameter   4,762 mm (0.1875 in)
Ball free movement   0,74 to 0,45 mm (0.29 to 0.18 in)

Non return valve secondary
Ball diameter   4,762 mm (0.1875 in)
Ball free movement   0,74 to 0,45 mm (0.029 to 0.018 in)

OIL FILTER HEAD GASKET

Renew

Place suitable drain tin under the oil filter. Remove the oil filter cartridge (1, Fig. 13), remove the oil cooler pipe clamp nut and clamp, note the position of, and disconnect the pipes from the oil supply housing, remove and discard the 'O' rings. Remove the oil filter head securing bolts (2, Fig. 13) and remove the inner (3, Fig. 13) and outer (4, Fig. 13) housings. Thoroughly clean the housings, the oil cooler pipes and the cylinder block face.

Apply 'Hylosil' to the outer housing (4, Fig. 13), align the inner housing (3, Fig. 13) with the outer housing, fit the securing bolts to the housing assembly, apply 'Hylosil' to the inner housing, align the assembly to the aperture in the cylinder block and tighten the securing bolts (2, Fig. 13).

Fit new 'O' rings to the oil cooler pipes, lubricate and connect the pipes to the housing, fit the clamp plate, fit a new oil filter (1, Fig. 13), start engine, wait for oil pressure light to extinguish, switch off engine and let stand for 30 seconds; check the engine oil level and top-up as necessary.

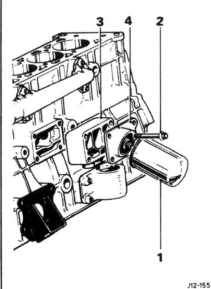

Fig. 13

OIL SUMP

Renew

Remove the air cleaner element, the alternator, the front crossmember and drain the engine oil. Position the steering rack to allow access to the sump securing bolts. Remove the dipstick tube and the adaptor plate to sump securing bolts and the sump securing bolts (1, Fig. 14) leaving two loose. Remove the engine earth strap, the remaining sump bolts and lower the sump (2, Fig. 14).

Apply 'Hylosil' to the new sump and clean the gasket face on the cylinder block (3, Fig. 14). Lift the sump (2, Fig. 14) into position and secure with the bolts (1, Fig. 14). Fit the engine earth strap and adaptor plate. Refit the dipstick tube. Reposition the steering rack under the sump and refit the front crossmember, the alternator and the air cleaner. Refill the engine with oil.

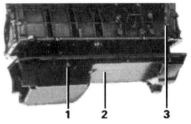

Fig. 14

OIL PRESSURE SWITCH/TRANSMITTERS

Renew

Remove the air cleaner element, disconnect the feed wire (1, Fig. 15, 16) and remove the switch (2, Fig. 15, 16).

Fit the new switch to the cylinder block, reconnect the feed wire and air cleaner element. Start the engine and ensure that either the oil pressure warning light extinguishes or the pressure gauge is operating.

NOTE: Oil pressure transmitter Fig. 16 and oil pressure warning light Fig. 15.

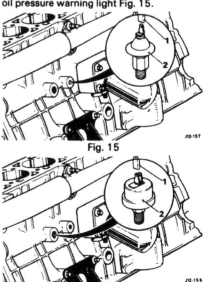

Fig. 15

Fig. 16

OIL FILTER CANISTER

Renew

Place a suitable drain tin under the filter and remove the filter canister (1, Fig. 17). Clean the oil filter mounting face, lubricate the rubber seal on the new filter with clean engine oil. Fit and tighten the filter canister using only the hand pressure and on no account use an oil filter strap designed for removal of canisters. Start the engine, wait for the oil light to extinguish, switch engine off, wait for 30 seconds, check the oil level and top-up as necessary.

NOTE: Normal tightening torque of oil filter is: ⅜ to ½ turn after initial contact.

Fig. 17

Canister Specification

Maximum working pressure	100 lb/in²
Relief valve setting	0,9 to 1,1 kgf/cm² (13 to 16 lb/in²)
Canister to withstand	17.60 kgf/cm² (250 lb/in²) pressure
Element filtration area	3555 cm² (551 in²)
Canister diameter	95,70 mm (3.768 in)
Canister length	148,00 to 145,50 mm (5.827 to 5.728 in)
Canister case material thickness	24 SWG
Adaptor thread	1 in 12UNF — 2B

OIL COOLER

Renew

Remove the horn/bracket assembly, place a drain tray beneath the oil cooler and position a splash guard to prevent oil splashing through the front radiator grille. Slacken and remove the oil cooler pipe unions to drain the oil. Remove the oil cooler securing nuts (1, Fig. 18) and remove the cooler assembly (2, Fig. 18).

Discard the cooler pipe 'O' rings, remove the splash guard and drain tray. Clean the pipe unions.

Fit the new oil cooler (2, Fig. 18) to the mounting rubbers and start but do not tighten the oil cooler securing nuts (1, Fig. 18). Fit new 'O' rings to the cooler pipes, connect the pipes to the cooler and tighten the unions and the cooler mounting nuts.

Refit the horn assembly to the vehicle and top-up the engine with oil. Run the engine and ensure there are no oil leaks.

Fig. 18

OIL FILLER TUBE SEAL AND HOUSING GASKET

Renew

Remove the air cleaner element, the oil filler cap (1, Fig. 19), slacken the oil filler tube to air cleaner backplate hose clips, disconnect and remove the hoses from the air cleaner backplate and manifold, remove the camshaft cover breather pipe. Remove the filler tube upper securing bolt, pull the tube (2, Fig. 19) from its lower housing (3, Fig. 19) and seal assembly (4, Fig. 19) and pull the tube down to ease removal, withdraw the tube and remove the seal from the lower housing. Remove the seal housing to cylinder block securing bolts (5, Fig. 19) and remove the housing, remove the baffle plate (6, Fig. 19), clean the housing, baffle plate, cylinder block mating face and filler tube.

Apply 'Hylosil' to the housing, align the baffle (6, Fig. 19) to the housing, apply 'Hylosil' to the baffle, fit a new filler tube seal (4, Fig. 19) to the housing and align the housing assembly (3, Fig. 19) to the cylinder block. Fit and tighten the housing securing bolts (5, Fig. 19), locate the filler tube (2, Fig. 19) in position from underneath, connect up the camshaft cover breather pipe and seat the filler tube into the lower housing. Fit and tighten the filler tube securing bolt. Connect the hoses to the manifold and air cleaner backplate, tighten all the hose clips, refit the filler cap (1, Fig. 19) and air filter. Check the engine oil level.

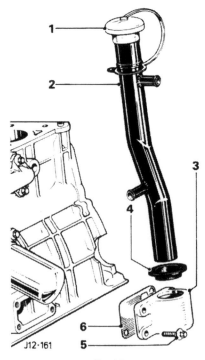

Fig. 19

OIL PUMP

Renew

Remove the air cleaner, the alternator and the front crossmember.

Drain the engine oil and remove the oil sump. Place a suitable drain tray under the oil transfer housing area and remove the transfer housing (1, Fig. 20). Remove the oil

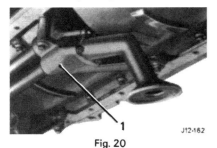

Fig. 20

Fig. 21

pump pick-up pipes (1, Fig. 21), move the lock tabs away from the oil pump sprocket bolts (1, Fig. 22) and remove the bolts (2, Fig. 22) and remove the oil pump drive sprocket (3, Fig. 22), collecting the tab washers (4, Fig. 22) and shims.

Fig. 22

Remove the oil pump securing bolts (2, Fig. 23) and remove the pump (1, Fig. 23).

Fig. 23

Remove the oil pressure relief valve cap (1, Fig. 24), spring (2, Fig. 24), tube (3, Fig. 24) and valve (4, Fig. 24).

Clean the mating face in the cylinder block, the relief valve assembly, sprocket and shims.

Fig. 24

Lubricate the relief valve assembly and assemble into the pump and tighten the relief valve cap. Fit the oil pump to the engine and tighten the securing bolts. Move the drive chain as far rearward as possible, fit a shim pack of 0.015 in thickness to the oil pump, align the bolt holes, fit the drive sprocket and two securing bolts. Using a straight edge (1, Fig. 25), check the alignment of the oil pump drive sprocket (2, Fig. 25) and the crankshaft sprocket (3, Fig. 25). Remove and refit the pump drive

Fig. 25

sprocket as required in order to align the sprockets. When the alignment is satisfactory remove the pump sprocket, fit the chain to the sprocket and the sprocket and shims (4, Fig. 25) to the oil pump, fit the tab washer (4, Fig. 22) and bolts (2, Fig. 22), tighten the bolts and secure with the tab washer.

Fit new 'O' rings to the oil pump pick-up pipes (1, Fig. 26). Lubricate and refit the pipes to the pump. Apply 'Hylosil' to the transfer housing gasket face, fit and seat the housing (1, Fig. 20) and secure with the bolts.

Fig. 26

Refit the engine sump (1, Fig. 27), the crossmember, the alternator, the air cleaner element (Fig. 28) and refill the engine with oil.

Fig. 27

Fig. 28

ENGINE OIL

Drain and refill

Place a suitable tin under the sump rear drain plug, release the plug and collect the oil in the container. Clean the sump plug, remove the drain tin, fit and tighten the sump plug.

Remove the engine filler cap and refill the engine with oil. Refit the oil filler cap. Run the engine to temperature, switch the engine off. Leave for 30 seconds. Remove the dipstick and check the oil level and adjust the oil accordingly.

OIL PUMP

Overhaul

Remove the oil pump and remove the body securing bolts. Remove the pump body, the pump outer rotor, the backplate securing bolts, the backplate and the bearing shell, clean all the component parts of the pump. Check all the clearances and components for undue wear (Fig. 29 & 30).

Clean and lubricate all the component parts, fit the bearing shell to the housing, apply sealant to the backplate gasket face, fit the backplate and securing bolts, fit the outer rotor and refit the oil pump to the engine.

Fig. 29

Fig. 30

Outer rotor O/D	69,825 to 69,774 mm (2.749 to 2.747 in)
Outer rotor width	27,975 to 27,962 mm (1.1014 to 1.1009 in)
Inner rotor width	27,975 to 27,962 mm (1.1014 to 1.1009 in)
Clearance outer rotor to body (Fig. 29)	0,2 mm (0.010 in)
Material	Cast iron grade 12
Oil pump body rotor bore	69,951 to 69,926 mm (2.7539 to 2.7529 in)
End float (Fig. 30)	0,1 mm (0.005 in)
2 horn clearance	0,2 mm (0.010 in)
3 horn clearance	0,2 mm (0.010 in)

CYLINDER HEAD — REAR BLANKING PLATE GASKET

Renew

Drain the coolant and remove the securing bolts (1, Fig. 32), disconnect the injector harness and remove the blanking plate (2, Fig. 32). Clean the blanking plate and using a straight edge check the plate for flatness. Apply 'Hylosil' to the gasket face and carefully align the plate to the cylinder head. Fit the securing bolts and 'P' clip. Reconnect the injector harness and refill the engine with coolant.

CAMSHAFT COVER GASKETS

Renew

Disconnect the spark plug leads, and the breather hose from the camshaft cover. Remove the seven camshaft cover securing screws (1, Fig. 31) and washers. Remove and discard the gaskets and the half moon seals from the rear of the cylinder head.

Clean the camshaft cover and the cylinder head mating face.

Apply sealant to and fit the new half moon seals to the cylinder head.

Fit new plug well seals and camshaft cover gasket (2, Fig. 31) to the camshaft cover (3, Fig. 31), fit and align the cover to the cylinder head and secure with the seven screws (1, Fig. 31) and washers.

Reconnect the breather hose to the cover and tighten the clip.

Reconnect the plug leads.

Fig. 31

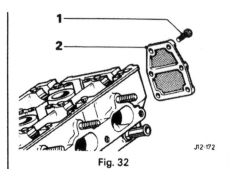

Fig. 32

CRANKSHAFT END FLOAT

Check and adjust

Remove the alternator, the air cleaner element, the front crossmember and drain the engine oil. Remove the engine sump, the oil pump pick-up pipes (1, Fig. 33) and remove the crankshaft windage trays (2, Fig. 33).

Fig. 33

Mount a suitable dial gauge and zero; with the use of suitable levers, measure the crankshaft endfloat; this should be 0.004 to 0.010 in.

Should the endfloat exceed this, remove the centre main bearing cap, remove the thrust washers and replace with suitable oversized thrust washers; these are available in 0.005 in and 0.010 in. These can be added to reduce the endfloat by 0.005 to 0.010 in, 0.015 in and 0.020 in (the oversizes are marked on the steel side of the thrust washer).

NOTE: When refitting the thrust washers, ensure that the bearing face (grooved) and not the steel side contacts the crankshaft.

Recheck the endfloat and if satisfactory, clean the main bearing cap and bearing, lubricate with clean engine oil, refit the cap and bearing assembly, remove the dial gauge and torque up the main bearing cap.

Clean and fit the crankshaft windage trays. Refit the oil pump pick-up pipes with new 'O' rings, coat the engine sump mating face with sealant and refit.

Refit the crossmember, alternator and the air cleaner element.

Refill the engine with new engine oil.

CRANKSHAFT FRONT OIL SEAL

Renew

Remove the air conditioning crankshaft drive pulley and remove the crankshaft damper (1, Fig. 34).

Fig. 34

Remove the oil seal (1, Fig. 35) from the timing cover. Remove the woodruff key and the crankshaft oil seal/crankshaft damper spacer (2, Fig. 35), clean the oil seal face, crankshaft and keyway, key and oil seal/crankshaft damper spacer. Lubricate the oil seal and fit to the front cover and

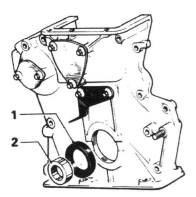

J12·175

Fig. 35

ensure that the seal is flush with the front cover (1, Fig. 36). Inspect the edges of the oil seal/crankshaft damper spacer for nicks, burrs and wear marks and replace if suspect, oil and fit the spacer. Refit the woodruff key and the crankshaft damper. Fit the pulley to the crankshaft.

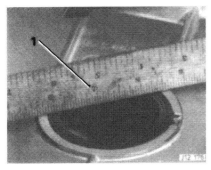

Fig. 36

CRANKSHAFT PULLEY AIR CONDITIONING DRIVE

Renew

Slacken the compressor link arm adjusting nuts and bolts, raise the vehicle on a ramp, slacken the compressor front link bolt and pivot the compressor towards the engine. Remove the compressor drive belt and remove the pulley securing bolts (1, Fig. 37).

J12 177

Fig. 37

Remove the pulley by moving it forwards and to the LH side of the damper.
Fit and align the new pulley to the crankshaft damper and tighten the securing bolts.
Move the compressor away from the engine and fit the drive belt. Tighten the adjusting nut to give the correct belt tension.
Tighten the link arm lock nuts, pivots and bolts.

DISTRIBUTOR DRIVE SHAFT

Renew

Depressurise the fuel system, drain the coolant and remove the air cleaner element. Remove the thermostat housing gasket and

cylinder head gasket. Drain the engine oil. Remove the crankshaft damper, timing cover gaskets, upper timing chain and lower chain tensioner and remove the intermediate sprocket.

Remove the power-assisted steering pump housing to distributor drive shaft housing securing bolts (1, Fig. 38), remove the power-assisted steering pump drive coupling and remove the distributor drive shaft rear oil seal using Service Tool 18G 1468. Using right-angled circlip pliers remove the drive shaft securing circlip, lift the timing chain off the drive shaft gear, remove the drive shaft (2, Fig. 38) and collect the thrust washer (3, Fig. 38).

Clean the distributor drive shaft and bearing housing (4, Fig. 38), thrust washer, drive coupling and pump shaft.

Check for wear and if necessary change the distributor shaft bushes (5, Fig. 38) using Service Tool 18G 1434.

Refit the housing (4, Fig. 38) with new gasket (6, Fig. 38) and secure with the bolts (1, Fig. 38).

Lubricate the drive shaft and refit the shaft into the housing, engaging the chain with the gear as it is fitted. Lubricate and fit the thrust washer (3, Fig. 38) and secure with the circlip.

Lubricate new oil seal and using Service Tool 18G 1469, fit and seat the drive shaft rear oil seal.

NOTE: Under no circumstances must the plastic seal insert be removed prior to fitting the seal.

Fit the drive coupling to the power-assisted steering pump, align the pump shaft with the distributor shaft, connect the pump shaft to the drive shaft and tighten the pump housing securing bolts.
Refit the intermediate sprocket, the lower timing chain tensioner, the upper timing chain, the timing cover crankshaft damper, cylinder head gasket, thermostat housing gasket and the air cleaner element. Refill the cooling system. Refill the engine with oil, start the engine and adjust the ignition timing.

Fig. 38

DISTRIBUTOR DRIVE SHAFT SEAL

Renew

Remove the power steering pump drive coupling and remove the distributor drive shaft oil seal using Service Tool 18G 1468. Clean and lubricate the new seal and mating faces. Using Service Tool 18G 1469 (1, Fig. 39), fit and fully seat the drive shaft seal (2, Fig. 39)

NOTE: Under no circumstances must the plastic seal insert be removed prior to fitting of the seal.

Refit the power steering pump drive coupling.

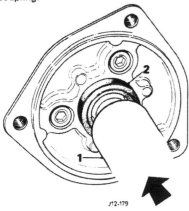

Fig. 39

CRANKSHAFT DAMPER

Renew

NOTE: Under no circumstances must the engine be rotated anti-clockwise (viewed from the front) until Service Tool 18G 1436 is fitted and tensioned.

Remove the air conditioning drive pulley, slacken the alternator drive belt adjusting nut, the link arm adjusting nut, bolt and alternator pivot bolt. Lift the alternator and remove the drive belt, disconnect the plug leads and the HT lead from the ignition coil. Remove the distributor cap and turn the engine to TDC No. 1 cylinder. Disconnect the distributor

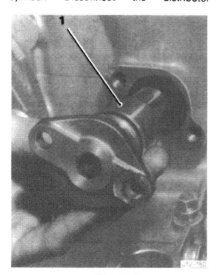

Fig. 40

amplifier block connector and the advance/retard pipe, remove the distributor and place tape over the aperture. Remove the upper chain tensioner valve and using a 3 mm Allen key, retract the tensioner by rotating the key clockwise until the tensioner is in the 'park' position, and remove the tensioner assembly (1, Fig. 40). Fit the Service Tool 18G 1436 (1, Fig. 41) to the cylinder head and tighten the centre bolt (2, Fig. 41) until a pressure of 2 to 4 lbs is exerted on the chain by the tensioner.

Fig. 41

Remove the spoiler/under tray to inner wheel arch cover.

Fit Service Tool 18G 1837 to the front crankshaft damper (1, Fig. 42) and tighten the securing bolts (2, Fig. 42). Wedge the tool against the front crossmember (3, Fig. 42) and remove the damper retaining

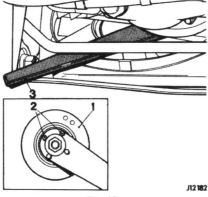

Fig. 42

bolt. Remove Service Tool 18G 1837 from the damper and Service Tool 18G 1836 from the cylinder head, refit the upper

Fig. 43

tensioner assembly with a new gasket and 'O' ring and release by turning the tensioner anti-clockwise using a 3 mm Allen key. Fit the tensioner valve and secure with the clamp.

Fit Service Tool 18G 1436 to the damper (1, Fig. 43) and tighten the securing bolts (2, Fig. 43). Tighten the centre bolt, withdraw the pulley from the crankshaft and remove the tool.

Remove the timing plate securing bolts and remove the timing plate.

Refit the timing plate to the new damper and tighten the securing bolts, fit the damper to the engine, start the damper securing bolt and fit Service Tool 18G 1437. Wedge the tool against the front crossmember, torque up the securing bolt and remove Service Tool 18G 1437.

Reposition the alternator drive belt over the pulleys and refit the air conditioning belt.

Refit the spoiler/under tray. Tension the alternator drive belt and tighten all the securing bolts and nuts. Tension the air conditioning drive belt and tighten the securing bolts.

Fit a new 'O' ring to the distributor, turn the engine to TDC No. 1 cylinder and refit the distributor.

Refit the advance and retard pipe and reconnect the distributor amplifier block connector, refit the distributor cap, leads and HT lead to the ignition coil, start the engine and reset the ignition timing.

OIL COOLER SUPPLY AND RETURN HOSES

Replace

Remove the air cleaner assembly, place a drain tray in position, remove the hose to oil filter head clamp nut, clamp and rubbers. Disconnect and remove the horn assembly. Place a drain tray beneath the oil cooler and cooler union, place a splash guard to prevent oil splashing through the front radiator grille. Slacken the supply and return pipes from the cooler and cooler union nuts and drain the oil. Remove the top rail to fan cowl securing nuts, displace the harness clips, displace the fan cowl from the top rail and remove the receiver/drier. Remove the top rail securing nuts and bolts and displace the earth leads, lift the top rail off and position over the engine. Remove the receiver/drier clamp nuts, bolts and clamps from the receiver/drier. Lift the condenser from its mounting position, disconnect the thermostatic switch feed wires and remove the supply and return hoses from the vehicle. Discard the 'O' rings.

Fit the new hoses in position, lubricate and fit the 'O' rings to the connectors, connect the hoses to the oil cooler and tighten the connections. Connect the horn wires and secure the horn assembly to the front panel. Align and fit the hose clamp rubbers, refit the clamp and secure with the bolt. Reconnect the radiator thermostatic switch feed wires. Refit the condenser to its mounting rubbers, position the clamps over the receiver/drier. Position the earth wires and bolt the top rail into position. Refit the fan cowl over the studs, fit the harness clips to the studs, fit the top rail and secure with the nuts.

Reposition the receiver/drier clamps and secure with the nuts. Connect the supply and return hoses to the filter head.

Refit the clamp plate and secure with nut, refit the air cleaner element and remove the splash guard and drain trays and top-up the engine with oil.

OIL PRESSURE RELIEF VALVE ASSEMBLY

Check and overhaul or renew

Jack up the front of the vehicle and place on two stands, partially drain the engine oil from the front drain plug (1, Fig. 44).

Remove the oil pressure relief valve plug (1, Fig. 45) and collect the valve (2, Fig. 45) spring (3, Fig. 45) and mandrel (4, Fig. 45). Check the valve spring and mandrel for wear, scoring or pitting with the specification below.

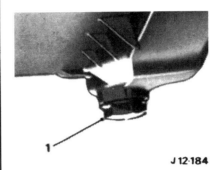

J 12·184

Fig. 44

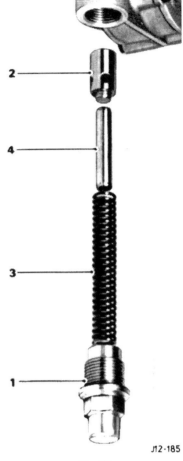

J12·185

Fig. 45

Relief valve:	
O/D	17,98 to 17,95 mm (0.7079 to 0.7067 in)
I/D	15,00 to 14,50 mm (0.591 to 0.571 in)
Overall length	35 mm (1.378 in)
Working length	30 mm (1.181 in)
Bore depth	27 mm (1.063 in)
Material	ENIA
Cyanide Harden	0,020 to 0,013 mm (0.008 to 0.005 in) deep
Mandrel: Length	57 mm (2.244 in)
Diameter	11,11 mm (0.437 in)
Material	Steel tube to BS970 (CDS2 or ERW1)
Spring Wall thickness	1,63 (0.064 in, 16 SWG) or 1,22 (0.048, 18 SWG)
Free length	106 mm (4.173 in)
Outside diameter	16,34 mm (0.643 in)

SPRING SPECIFICATION

Wire diameter	2,34 mm (0.092 in, 13 SWG)
Mean diameter coils	14,0 mm (0.551 in)
Total No. of coils	26
No. Active Coils	24
Helix of coil	Left
Spring rate	4,49 N/mm (25.65 lb/IN)

	Length		Load		Stress	
	mm	inches	kg	lbs	N/mm²	lb/in²
Fitted	76,00	2.992	13,74	30.29	472	68,395
Valve opening	72,00	2.835	15,58	34.35	535	77.543
Max opening	62,00	2.441	20,15	44.43	692	100.304
Solid (ref)	60,80	2.392	20,71	45.66	711	103.083

Material

BS. 970/735/A50 (EN.47) Spring steel wire. Ends to be close coiled and ground square to axis, feather edges to be removed to a minimum thickness of 0.5 mm (0.020 in).

Heat treatment

To be heat stabilised such that when compressed to valve opening position for 24 hours at 150°C, relaxation of load must not exceed 2.5%.

If any of the components are suspect, they should be replaced.

Lubricate and assemble the mandrel, spring and valve, fit and tighten the relief valve plug.

Refit and tighten the sump plug.

Lower the vehicle, open the bonnet and top-up the engine with oil.

ENGINE MOUNTING SET

Renew

Remove the air cleaner element.
Remove the upper engine mounting securing nut (1, Fig. 46) and washer, support the engine with a jack.

NOTE: Ensure that the fan does not come into contact with the fan cowl.

Remove the LH engine mounting bracket from the cylinder block, remove the mounting (2, Fig. 46) and packing from the bracket. Fit new mounting and packing to the bracket, fit the bracket to the cylinder block and tighten the securing bolts.

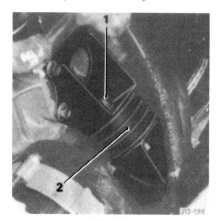

Fig. 46

Take the weight of the engine with a jack.
Remove the RH mounting nuts from the crossmember (1, Fig. 47) and engine (2, Fig. 47), remove the mounting and packer from the bracket (3, Fig. 47), fit a new mounting with the packer to the bracket. Fit the assembly to the engine and secure. Lower the engine, ensure that it is positioned on the mountings without causing any undue strain to the rubber blocks and tighten the securing nuts.

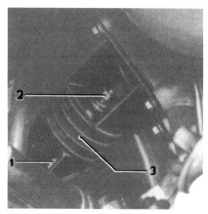

Fig. 47

Fit the engine retaining Service Tool MS 53 across the wing channels (1, Fig. 48), fit the engine support hook, a lifting eye and take the weight of the engine.

Jack up the front of the vehicle and place on stands, remove the intermediate heat shield, remove the rear mounting securing nut (1,

Fig. 48

Fig. 49), take the weight of the gearbox with a jack and remove the rear mounting to body securing bolts (2, Fig. 49). Remove the rear mounting assembly (3, Fig. 49) and collect the spacers, spring and micron insert. Remove the nuts and bolts (4, Fig. 49) securing the mounting rubber to the plate and remove the mounting (5, Fig. 49).

Fit a new mounting to the bracket and secure with the nuts and bolts. Fit a new micron insert, fit the spring and using a jack fit, seat and secure the rear mounting to the body.

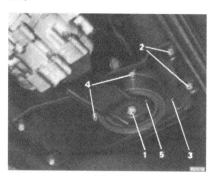

Fig. 49

NOTE: Ensure that the spring is seated in the gearbox spring pan.

Lower the jack and ensure that the gearbox is central over the mounting, finally tighten the mounting securing bolts.
Remove Service Tool MS 53 and the lifting eye, fit and seat the rear mounting centre spacers, tighten and secure with the mounting nut. Refit the intermediate heat shield and lower the vehicle.

OIL PUMP DRIVE CHAIN AND DAMPER

Renew

Remove the engine and the gearbox from the vehicle, detach the gearbox from the engine and fit the engine to a stand.
Remove the distributor and disconnect the leads from the sparking plugs.
Remove the camshaft cover and the upper chain tensioner assembly; discard the 'O' rings and gaskets. Knock back the camshaft top sprocket lock tabs away from the bolts, remove the sprocket securing bolts and collect the four tab washers, remove the camshaft sprockets from the cylinder head.
Position the chain between the upper dampers (1, Fig. 50) and fit an elastic band (2, Fig. 50) across the upper dampers retaining the chain and the pivoting damper

(3, Fig. 50) during the cylinder head removal.

Fig. 50

Remove the cylinder head/camshaft bearing cap securing bolts (1, Fig. 51), remove the cylinder head assembly.

Fig. 51

Remove the flywheel and fit slave bolts to the rear of the crankshaft (1, Fig. 52) and using a suitable bar (2, Fig. 52) to stop the crankshaft rotating, remove the securing bolt and damper assembly.

NOTE: Ensure that the engine does not rotate anti-clockwise (viewed from the front).

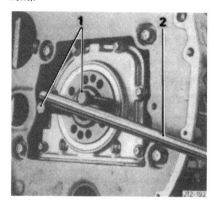

Fig. 52

Turn the engine over in the stand and remove the sump pan securing bolts and sump.
Knock down the oil pump sprocket lock tabs, remove the bolts and tab washers and collect sprocket. Remove the timing cover securing bolts (1, Fig. 53) and timing cover assembly (2, Fig. 53). Remove the oil pump drive chain and damper, remove the crankshaft damper woodruff key and remove the crankshaft damper/oil seal spacer.

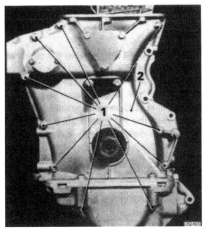

Fig. 53

Remove the front oil seal from the timing cover, clean and inspect the oil pump drive sprocket. Lubricate the oil pump drive chain, fit the chain to the sprocket and the chain/sprocket assembly to the crankshaft sprocket, align the sprocket securing holes, fit and tighten the sprocket securing bolts (1, Fig. 54), and secure with the tab washer

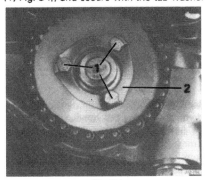

Fig. 54

(2, Fig. 54). Fit a new chain damper and take any slack out of the chain, tighten the securing bolts, fit the front oil seal to the timing cover, apply 'Hylosil' to the timing cover mating face, fit the cover and secure with the fixing bolts. Lubricate the oil seal/crankshaft damper spacer and push onto the crankshaft. Fit and seat the woodruff key to the crankshaft and fit the damper to the crankshaft. Stop the crankshaft from turning using the slave bolts and a suitable bar, fit the crankshaft damper securing bolt and tighten up to the correct torque.

Remove the slave bolts, clean the sump mating faces and coat with 'Hylosil', fit the sump to the engine and torque up the bolts. Turn the engine to TDC, the upper chain should be supported during this operation, and reposition it between the dampers (1, Fig 55).

Fig. 55

Ensure that both the top of the cylinder block and the mating face of the cylinder head are clean and dry and fit a new cylinder head gasket.

Check that both camshafts are on TDC using Service Tool 18G 1433. Fit the cylinder head assembly to the cylinder block; fit the chain damper and pedestal assembly and insert the cylinder head attachment bolts. Tighten the bolts to a torque figure of 38 to 40 lb ft and then turn the bolt clockwise through exactly 90°; carry this out in the sequence illustrated (Fig. 56).

NOTE: To assist the accuracy of the rotation through 90° a suggested tool is illustrated on page 12—6.

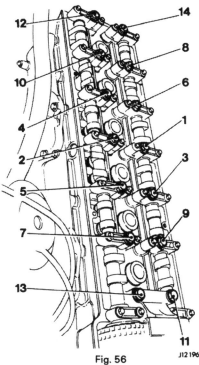

Fig. 56

Lift the timing chain and remove the elastic band, position the chain over the ends of the camshafts. Fit and engage the camshaft drive sprockets to the chain and the camshafts ensuring that all the chain 'slack' is to the pivoting tensioner side.

Strip the upper tensioner and inspect its component parts (1, Fig. 57) for wear or

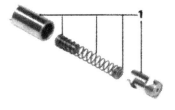

Fig. 57

damage. Check that the ball is free in the non return valve. Fit new 'O' rings to the valves lubricate the rings and insert the assembly into the position housing. Lubricate the snail, spring and mandrel and assemble to the piston. Engage the snail with the guide and rotate clockwise and depress the snail until the peg is engaged in the 'park' position on the snail, fit the assembly into the piston housing. Fit a new

'O' ring to the housing and lubricate, fit a new gasket to the housing and fit the assembly to the cylinder head, ensuring that the tensioner is properly engaged on the pivoting guide. Fit but do not tighten the

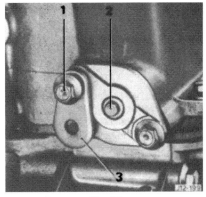

Fig. 58

tensioner securing bolt, fit the valve retainer securing bolt (four threads only) (1, Fig. 58). Fit new 'O' rings to the valve assembly and lubricate. Release the tensioner using a 3 mm Allen key and turn it anticlockwise, fit the valve assembly (2, Fig. 58), engage the retainer plate (3, Fig. 58) and fully tighten the securing bolt.

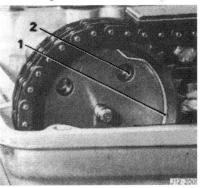

Fig. 59

Remove the inner sprocket securing clip (1, Fig. 59), remove the inner sprocket and align the holes in the inner sprocket with the holes in the camshaft (2, Fig. 59). Fit the tab washers (1, Fig. 60) and securing bolts (2, Fig. 60), but do not fully torque up the bolts,

Fig. 60

refit the inner sprocket securing clip. Repeat for the other camshaft sprocket. Tighten all the sprocket securing bolts (1, Fig. 61), knock up the lock tabs (2, Fig. 61), position the upper chain upper damper (3, Fig. 61) and tighten the securing bolts (4, Fig. 61) and knock up the lock tabs (5, Fig. 61).

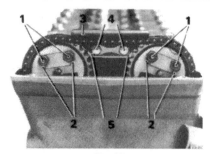

Fig. 61

Fit the plug well seals (1, Fig. 62), the camcover gasket (2, Fig. 62), and the half moon seals (3, Fig. 62) to the cylinder head, fit the camcover (4, Fig. 62) to the engine, fit and tighten the securing screws (5, Fig. 62).

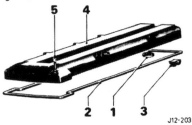

Fig. 62

Check the gap of new sparking plugs and fit to cylinder head, tightening to a torque of 25 lb/ft 3.5 kg/m.

Fit a new 'O' ring to the distributor, lubricate the 'O' ring and gear, turn the rotor arm to the position illustrated in Fig. 123 and fit the distributor in the position shown.

Fit the distributor clamp plate securing bolt, leave the clamp plate securing bolt slightly slack. Fit the distributor cap and tighten the securing screws. Connect the sparking plug leads in the firing order 1, 5, 3, 6, 2, 4.

NOTE: The distributor turns clockwise as viewed from the top.

Refit the water pump and the gearbox, remove the engine from the stand and refit the engine to the vehicle.

VALVE CLEARANCES

Check and adjust

Drain the coolant and remove the camshaft cover.

Rotate the engine as necessary to measure the valve clearances between the heel of the camshaft and the cam followers (1, Fig. 63), this should be 0.012 to 0.014 in.

Fig. 63

Should the clearances be incorrect, remove the camshafts and cam followers. Remove all the shims or only those requiring adjustment, check the size of the existing shims with a micrometer and note the reading. Calculate the size of the shim required, following the examples below.

EXCESSIVE CLEARANCE	INCHES
Size of existing shim	0.100
Plus the actual clearance noted	0.019
	0.119
Less the specified valve clearance	0.013
= required shim size	0.106

INSUFFICIENT CLEARANCE	INCHES
Size of existing shim	0.107
Plus the actual clearance noted	0.010
	0.117
Less the specified valve clearance	0.013
= required shim size	0.104

Refit the cam followers and camshafts, and camshaft cover using new seals and gaskets, refill the cooling system, check and if necessary adjust the ignition timing.

NOTE: See page 12-54 for available shim sizes.

NOTE: To enable the camshafts to be rotated in position, enabling the valve clearances to be checked, without connecting the timing chains and sprockets, the following procedure must be strictly adhered to.

LH and RH refers to the engine being viewed from the front of the cylinder head.
1. Set the crankshaft at 60° BTDC. Dead Centre.
2. Turn the LH camshaft until the No. 3 cylinder valves and No. 6 cylinder valves are 'on the rock' i.e. one set of valves just opening and the other just closing.
3. The RH camshaft can now be rotated and the valve clearances checked.
4. Rotate the RH camshaft to TDC (i.e. groove in the camshaft to the top).
5. Turn the LH camshaft to TDC.
6. Turn the RH camshaft until No. 2 cylinder valves and No. 6 cylinder valves are 'on the rock'.
7. The LH camshaft can now be rotated and the valve clearances checked.
8. Rotate the LH camshaft to TDC.
9. Rotate the RH camshaft to TDC.
10. Rotate the crankshaft to TDC.

NOTE: To set the camshafts at TDC use Service Tool 18G 1433.

CAMSHAFTS

Renew

Drain the coolant and remove the camshaft cover, remove the torquatrol unit securing nuts and remove the torquatrol unit from the studs. Turn the engine over until No. 1 cylinder is at TDC on its firing stroke.

Disconnect the HT lead from the coil and remove the distributor cap, disconnect the amplifier block connector and advance/retard vacuum pipe.

Note the position of the rotor arm relative to the distributor body, and the position of the body relative to the cylinder head; remove the distributor and block off aperture to prevent the ingress of dirt and dust.

Release the lock tabs (1, Fig. 64) on the camshaft sprocket securing bolts and slacken the securing bolts (2, Fig. 64).

Fig. 64

Remove the upper chain tensioner valve clamp bolt (1, Fig. 65), remove the clamp (2, Fig. 65) and valve (3, Fig. 65). Using a 3 mm Allen key wind back the tensioner (turn clockwise) until the snail engages in the 'park' position. Remove the tensioner housing securing bolt and remove the

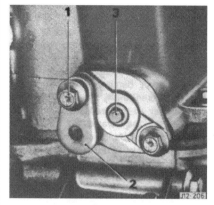

Fig. 65

tensioner assembly (1, Fig. 66). Remove and discard the 'O' ring (2, Fig. 66) and gasket (3, Fig. 66), remove the camshaft sprocket securing bolts, collect the tab washers and remove the sprockets.

Fig. 66

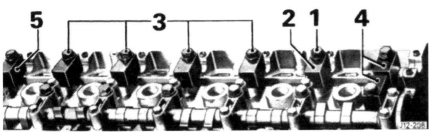

Fig. 67

Remove the cylinder head bolt from the RH No. 2 camshaft cap (1, Fig. 67), remove the remaining two bolts securing the camshaft cap, and remove the cap.

Fit the spacer Service Tool 18G 1435 (2, Fig. 67) to the cylinder head, fit and tighten the cylinder head bolt to specified torque.

Repeat the procedure for cap numbers 3, 4, 5 and 6 (3, Fig. 67).

Fig. 68

Remove the upper chain upper damper pedestal securing bolts (1, Fig. 68) and remove the pedestal assembly (2, Fig. 68) and fit the two spacer Service Tools 18G 1435 (4, Fig. 67).

Remove the head bolt from No. 7 camshaft cap and remove Nos. 7 and 1 camshaft cap securing bolts alternately until the camshaft is free and lift the camshaft from the cylinder head and fit spacer tool 18G 1435 to No. 9 cap bolt and retorque the cylinder head bolts (5, Fig. 67).

Clean new camshaft, oil the journals and install in the cylinder head. Fit the camshaft at approximately TDC.

Start No. 1 camshaft cap securing bolts. Remove No. 7 cylinder head bolt and spacer, refit No. 7 camshaft cap and bolt, but do not tighten. Pull down No. 1 and 7 camshaft caps alternately until the camshaft is fully down. Tighten the cylinder head bolt. Remove No. 4 cylinder head bolt and remove the spacer, lubricate the cap, fit and tighten the cap securing bolts, tighten the cylinder head bolt.

Repeat the above procedure for Nos 2, 3, 5 and 6 caps and using Service Tool 18G 1433 set the camshaft to TDC.

Repeat the procedure for the other camshaft.

Position the chain over the sprockets and align the sprockets to the camshafts, move the blank in the chain to the tensioner side, fit the oil pressure simulator Tool 18G 1436 and tension the chain (tighten the tool to apply a pressure of 2 - 4 lbs).

Remove the clip securing the sprocket inner wheel and align the securing bolt holes, fit the tab washers and securing bolts, knock up the tab washers and refit the securing clip.

Repeat the procedure for the second camshaft.

Turn the engine over and check all the valve clearances.

Should any of the clearances be incorrect, remove the camshafts as previously described, and using a suitable magnet remove the relevant cam followers (1,

Fig. 69

Fig. 69). Remove the shims requiring adjustment; check with a micrometer the size of the existing shims and note the reading. Calculate the size of shim required, following the example below for either excessive clearance or insufficient clearance.

EXCESSIVE CLEARANCE	INCHES
Size of existing shim	0.100
Plus the actual clearance noted	0.019
	0.119
Less the specified valve clearance	0.013
= required shim size	0.106

INSUFFICIENT CLEARANCE	INCHES
Size of existing shim	0.107
Plus the actual clearance noted	0.010
	0.117
Less the specified valve clearance	0.013
= required shim size	0.104

Remove the oil pressure simulator tool 18G 1436. Fit and seat the tensioner assembly to the damper. Fit the top tensioner securing bolt, fit new 'O' rings to the valve and lubricate.

Using a 3 mm Allen key, fully release the chain tensioner (turn anti-clockwise). Fit the valve, position the clamp over the valve and secure with the retaining bolt. Remove the tape from the distributor aperture, fit and lubricate the 'O' ring and drive gear, refit the distributor.

Reconnect the amplifier block connector, refit the distributor cap, plug leads and the HT lead to the coil.

Refit the torquatrol fan unit and secure with the nuts. Refit the gaskets to the camshaft cover, fit the cover and secure with the seven screws.

Refill the engine cooling system, reset the ignition timing and refit the distributor advance/retard pipe.

PISTONS AND CONNECTING RODS ENGINE SET

Replace

Remove the engine from the vehicle and fit to a stand.

Disconnect the plug leads, remove the distributor assembly, remove the camshaft cover. Turn the engine over to TDC firing on No. 1 cylinder, remove the upper chain tensioner assembly (1, Fig. 70).

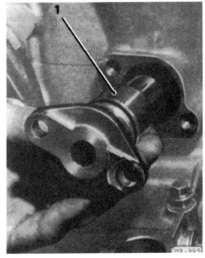

Fig. 70

Remove the camshaft securing bolts and sprockets.

Position the chain between the upper chain dampers, secure the dampers with an elastic band. Remove the cylinder head/camshaft cap securing bolts and remove the cylinder head assembly.

Fig. 71

Remove oil sump securing bolts and sump. Remove the oil pump pick-up pipes and windage trays.

Slacken the big-end connecting rod nuts (1, Fig. 71), in piston and connecting rod pairs, following in sequence (i.e. 1 & 6, 2 & 5, 3 & 4 cylinders) and push the pistons up and through the top of the cylinder block.

Remove the gudgeon pin circlips (1, Fig. 72) from each piston, remove the gudgeon pin (2, Fig. 72) and separate the piston (3, Fig. 72) from the connecting rod (4, Fig. 72).

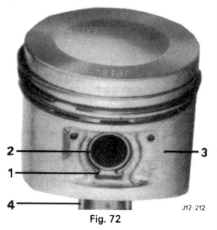

Fig. 72

NOTE: Check the connecting rods for out of balance, twist (Fig. 73) and bend (Fig. 74). If any connecting rod is bent, twisted or out of balance with the other five, then the complete set must be renewed.

Fig. 73

Fig. 74

Insert the new piston rings into the cylinder bores (1, Fig. 75) ensuring they are square in the cylinders and check the gap using a feeler gauge (2, Fig. 75), if the gap is insufficient, then a small flat file can be used on the butting ends of the ring. Ensure that after filing that no burrs remain.

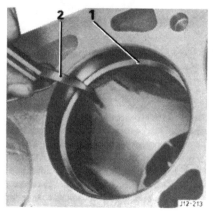

Fig. 75

NOTE: Ensure that the piston is fitted correctly to the connecting rod (Fig. 76).

Fig. 76

Lubricate the small end bush and slide the gudgeon pin (1, Fig. 77) through the piston (2, Fig. 77) and connecting rod and secure with the circlips (3, Fig. 77).

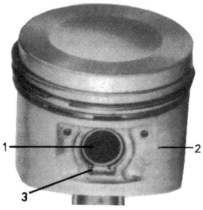

Fig. 77

Fit the rings to the pistons ensuring the gaps are positioned as Fig. 78. Lubricate and compress the rings using Service Tool 18G 55A (1, Fig. 79). Insert the piston skirt into the bore, and using a suitable implement carefully tap the piston into the cylinder bore, ensuring that the connecting rod does not foul either the cylinder block or

Fig. 78

crankshaft. Lubricate and fit the big end bearing shells to the connecting rod and the connecting rod cap. Pull the connecting rod and bearing carefully onto the crankshaft

Fig. 79

and fit the connecting rod cap to the rod. Fit and tighten the connecting rod cap nuts to a torque of 39-41 lbf/ft. Repeat the operation for the remaining five pistons.

NOTE: Ensure the connecting rod caps are fitted as in (1, Fig. 98).

Clean and fit the crankshaft windage trays (1, Fig. 80) and tighten the securing bolts. Clean the oil pump pick-up pipes, fit and lubricate the new 'O' rings, smear the oil transfer housing to cylinder block gasket face with 'Hylosil', fit and seat the assembly (2, Fig. 80) to the cylinder block. Clean the sump, smear the mating face with 'Hylosil', fit and torque up the sump securing bolts.

Fig. 80

Support the top timing chain and turn the engine over to TDC No. 1 cylinder using a suitable dial test indicator (1, Fig. 81) and position the timing chain between the dampers.

Fig. 81

Clean the cylinder head gasket faces, the upper damper and camshaft sprockets. Ensure the camshafts are still at TDC using the Service Tool 18G 1433 (1, Fig. 82).

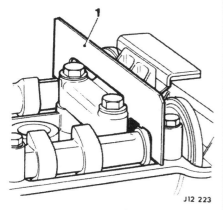

Fig. 82

Fit the cylinder head gasket and cylinder head, fit and tighten the cylinder head securing bolts to a torque of 38-40 lbf/ft and then rotate the bolt clockwise through exactly 90° in the correct sequence (Fig. 56). Tighten the front three bolts.

NOTE: To assist accuracy of rotation through 90° a suggested tool is illustrated on page 12—6.

Fit the top chain damper and pedestal assembly (1, Fig. 83), engage the chain with

Fig. 83

the sprocket teeth and fit the sprockets to the camshafts, rotate the sprockets to ensure any chain slack is to the tensioner side of the engine. Fit Service Tool

18G 1436 to the engine and tighten the securing bolts (1, Fig. 84), tighten the tensioning bolt (2, Fig. 84) to create a pressure on the tensioner of 2 - 4 lb.

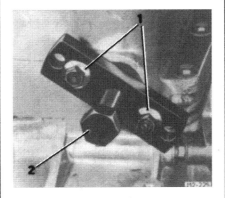

Fig. 84

Remove the inner sprocket clips (1, Fig. 85), lift out the inner serrated portion (2, Fig. 85) and refit, aligning the holes with those in the camshaft (1, Fig. 86). Fit the securing bolts

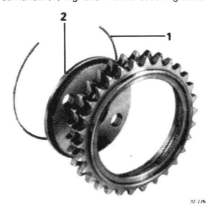

Fig. 85

Fig. 86

(1, Fig. 87) and locktab washers (2, Fig. 87), refit the inner sprocket securing clips (2, Fig. 86), tighten the securing bolts and knock-up the locktabs (3, Fig. 87).

Remove the Service Tool 18G 1436, overhaul the tensioner assembly, lubricate the assembly, fit a new gasket and engage the tensioner with the upper damper. Fit but do not tighten the securing bolt (1, Fig. 88), fit new 'O' rings to the valve (2, Fig. 88), lubricate and release the tensioner using a 3 mm Allen key by turning anticlockwise, fit the valve, secure the valve with the retaining plate (3, Fig. 88), and bolt (4, Fig. 88).

Fig. 87

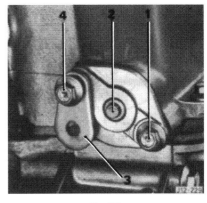

Fig. 88

Clean the camshaft cover and fit new plug well seals, cover gasket and half moon seals, smear with sealant and fit to the cylinder head. Fit and secure the camshaft cover with the seven screws. Refit the distributor.

Refit the distributor cap and connect the spark plug and HT leads. Refit the engine to the vehicle, refill the engine with oil and coolant and reset the ignition timing.

Refit the engine to the vehicle.

CRANKSHAFT

Renewal

Remove the engine assembly and fit to a stand. Remove the cylinder head.
Using Service Tool 18G 1437 (1, Fig. 89) to hold the crankshaft damper, remove the damper securing bolt.

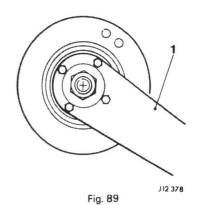

Fig. 89

Remove the oil sump, the oil pump pick-up pipe assemblies (1, Fig. 90), crankshaft windage trays (2, Fig. 90) and oil pump (3, Fig. 90).

Fig. 90

Remove the engine timing cover securing bolts (1, Fig. 91) and remove the cover assembly (2, Fig. 91).

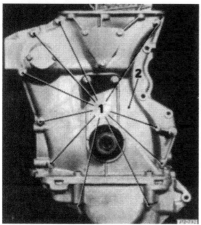

Fig. 91

Slacken the oil pump drive chain damper securing bolts (1, Fig 92) and move the damper (2, Fig 92) clear of the chain

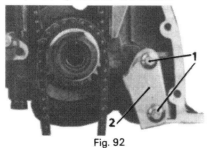

Fig. 92

Remove the oil pump drive chain, remove the elastic band from the upper dampers and remove the upper timing chain.

Remove the lower tensioner securing bolts (1, Fig. 93) and remove the tensioner (2, Fig. 93).

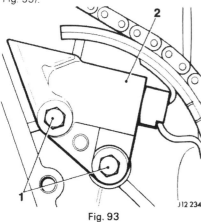

Fig. 93

Release the lower chain fixed damper securing bolt lock tabs (1, Fig. 94), remove the bolts (2, Fig. 94) and collect the damper (3, Fig. 94) and tab washer. Repeat for the remaining fixed and pivot dampers and remove the lower chain and intermediate sprocket.

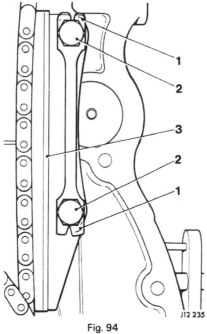

Fig. 94

Remove seal/crankshaft damper spacer, the crank damper woodruff key (1, Fig. 95), oil the crankshaft sprockets (2, Fig. 95) and the inner woodruff key.

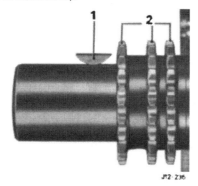

Fig. 95

Remove the rear oil seal housing securing bolts (1, Fig. 96) and remove the housing (2, Fig. 96).

Fig. 96

Remove the connecting rod cap securing nuts (1, Fig. 97), remove the bearing caps in pairs (i.e. 1-6, 2-5, 3-4), turning the crankshaft for access as required. As each

Fig. 97

connecting rod cap is removed ensure that each rod and cap are identified to each other (1, Fig. 98), as this will be necessary during assembly. Also as each cap is removed its relative connecting rod and

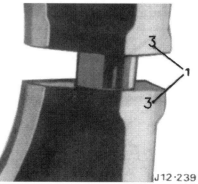

Fig. 98

piston assembly should be pushed up the bore to enable the crankshaft to be rotated to remove the remaining caps. Care should be taken not to push the piston too far up the cylinder bore as this will release the piston rings.

Ensuring that the main bearing caps are marked relative to the cylinder block (1, Fig. 99), remove the main bearing cap bolts, the caps (1, Fig. 100) and carefully lift out the crankshaft (2, Fig. 100). Remove and discard the bearing shells and the thrust washers.

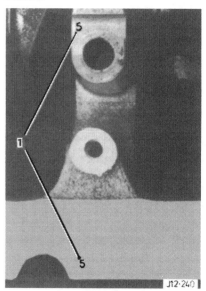

Fig. 99

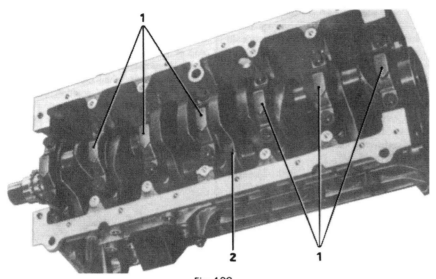

Fig. 100

Clean the bearing caps and the cylinder block main bearing housings.
Fit the new bearing shell halves to the cylinder block (1, Fig. 101) and lubricate with clean engine oil.

Fig. 101

Clean and polish the crankshaft journals, lubricate and carefully assemble the crankshaft into the cylinder block, fit the thrust washers ensuring that the steel side of the washer is mated to the cylinder block (1, Fig. 102).

Fig. 102

Rotate the crankshaft to ensure that it turns freely, fit the main bearing shells into the main bearing caps, lubricate and fit to the cylinder block, start the securing bolts and very carefully tap the main bearing caps to ensure they are seated in the cylinder block, rotate the crankshaft to ensure that it still turns freely, pull down each bearing cap individually and torque up the bolts, rotate the crankshaft between pulling down each bearing cap in order (Fig. 103).

Fig. 103

Check the crankshaft end float (1, Fig. 104), the tolerance is 0.004 to 0.010 in; should the end float exceed this, oversize thrust

washers are available in 0.005 in and 0.010 in and should be fitted accordingly to ensure that the end float is within these tolerances.

Fig. 104

Excessive end float

e.g.

End float measured　　　　　　　(A) 0.023 in

Oversize thrust washers fitted　　(B) 0.000 in
　　　　　　　　　　　　　　　　　(standard)

End float required　　　　　　　　(C) 0.007 in

.. A - C　= Thrust washer size required
　　　　　= 0.016

.. Thrust washer sizes to be fitted:
　　　　　= 0.010 in + 0.005 in

.. End float achieved:
　　　　　= 0.023 – 0.015 in
　　　　　= 0.008 in

Insufficient end float

End float measured　　　　　　　(A) 0.003 in
Oversize thrust washer fitted　　(B) 0.010 in
End float required　　　　　　　　(C) 0.007 in

.. C - A = Thrust washer size required
　　　　　= 0.004 in

.. Thrust washer size to be fitted:
　　　　　= 0.005 in

.. End float achieved is:
　　　　　= 0.003 in + 0.005 in
　　　　　= 0.008 in

DO NOT REMOVE THE PLASTIC 'O' RING PROTECTOR FROM THE SEAL PRIOR TO FITTING TO THE ENGINE

Carefully remove the old rear oil seal from the housing, clean the housing and lubricate the seal mounting face and using tool No. 18G 1293A/1 (1, Fig. 105) and 18G 134 (2, Fig. 105) fit the seal (3, Fig. 105) to the housing (4, Fig. 105). Smear the gasket

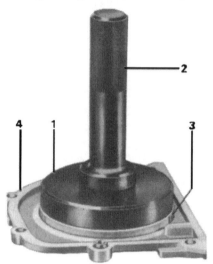

Fig. 105

face with 'Hylosil', locate the plastic protector 'O' ring (1, Fig. 106) onto the end of the crankshaft (2, Fig. 106) and push the rear seal housing (3, Fig. 106) over the crankshaft and up to the rear cylinder block

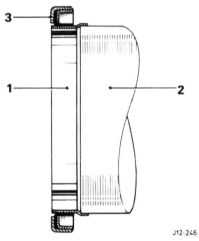

Fig. 106

face. Fit the bolts, check that the sump face and the cylinder block (1, Fig. 107) are flush, and tighten the bolts.

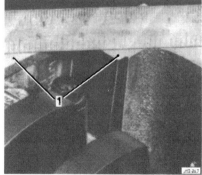

Fig. 107

Fit a new bearing shell to the connecting rod, lubricate and fit it to the crankshaft. Fit a bearing shell to the big end cap, lubricate and fit to the connecting rod. Fit and torque up the nuts. Turn the crankshaft over and ensure that there are no 'tight' spots, i.e. the crankshaft rotates freely. Repeat this procedure for the remaining five cylinders.

Clean and inspect for wear or damage all the timing gears, chains, guides and tensioners, should any be suspect replacement is essential.

Fit and seat the crankshaft sprocket woodruff key and fit the sprocket (1, Fig. 108) to the crankshaft (2, Fig. 108). Lubricate the lower timing chain, fit the chain to the intermediate sprocket, fit the chain to the lower sprocket, and then assemble the intermediate sprocket and chain to the cylinder block.

Should the intermediate sprocket be worn or damaged, the assembly must be replaced.

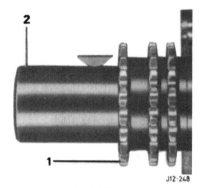

Fig. 108

Fit the spacers (1, Fig. 109) to the lower pivoting damper (2, Fig. 109) and fit the assembly (3, Fig. 109) to the cylinder block, fit a tab washer (4, Fig. 109) and bolts (5, Fig. 109), tighten the bolts and knock the tabs up to secure the bolts.

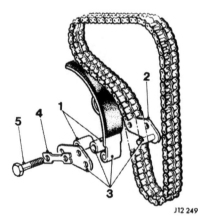

Fig. 109

Clean the lower tensioner assembly, remove the ball valve (1, Fig. 110) and ensure that the bottom of the housing is clean (2, Fig. 110), check that the ball valve is free by shaking it to ensure it rattles. It it does not, replace the valve assembly. If it does, renew and lubricate the 'O' ring (3, Fig. 110), and press the valve back into the housing. The ball free movement is 0.74 to 0.45 mm (0.029 to 0.018 in).

Inspect the housing and hydraulic tensioner (4, Fig. 110) for wear or scoring. Also check that the oil hole (5, Fig. 110) is clear in the end of the piston. Lubricate the component parts. Assemble the tensioner by pressing and twisting the snail clockwise until the pawl locks in the 'park' position. Insert the tensioner assembly into the housing and fit the housing to the cylinder block. Fit and tighten the securing bolts (6, Fig. 110).

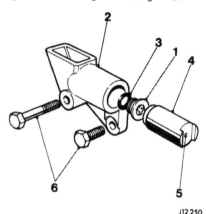

Fig. 110

Fit the lower static damper, fit the tab washers and bolts, tighten the bolts and knock up the tabs.

Release the tensioner by pushing the pivoting damper away from the chain, thus pushing the tensioner back into the housing, releasing the snail, and tensioning the chain.

Fit the upper static damper pedestal securing bolts and tab washer (1, Fig. 111) and the damper (2, Fig. 111). Secure with the tab washer (3, Fig. 111) and tighten the damper securing bolts, finally tighten the damper pedestal securing bolt and knock up the tabs.

Fit a dial gauge to the top of the cylinder block, and turn the engine over until No. 1 on 6 pistons are at TDC. Lubricate the upper chain (4, Fig. 111) and fit it to the intermediate sprocket (5, Fig. 111), fit and engage an elastic band (6, Fig. 111) around the upper dampers to secure the chain.

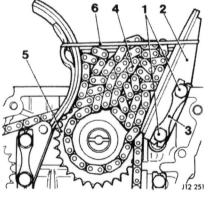

Fig. 111

Lubricate the oil pump drive chain, fit the chain to the crankshaft sprocket and lodge in the correct position.

Remove the crankshaft front oil seal and remove the distributor drive gear shaft seal using Service Tool 18G 1468. Lubricate the new distributor drive shaft seal and fit to the cover using Service Tool 18G 1469 (Fig. 112). Fit the crankshaft front oil seal to the cover ensuring that the front edge of the seal is flush with the timing cover (Fig. 113).

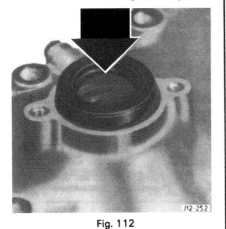

Fig. 112

Fig. 113

Smear the cylinder block timing cover face with "Hylosil", lubricate the bearing shells and carefully fit the timing cover to the engine ensuring that the timing cover/cylinder head mating face is flush with the cylinder block (1, Fig. 114), fit and tighten the securing bolts. Lubricate and fit the oil seal distance piece. Fit and seat the damper woodruff key.

Fig. 114

Remove the oil pump pressure relief valve retaining nut (1, Fig. 115), remove the spring, tube and valve, clean and check for wear or scoring. Remove the pump body securing bolts and remove the outer rotor. Remove the backplate securing bolts, remove the backplate, and remove the

bearing shell from the housing. Clean all dismantled parts and check for obvious wear or scoring.

Fit the outer rotor and check the clearance between the rotor and the oil pump housing. Should any clearance exceed the specified tolerances the assembly, i.e. rotor outer housing assembly, must be replaced. Lubricate the bearing shell and fit to the pump housing. Smear backplate gasket face with sealant and fit the backplate and tighten the securing bolts.

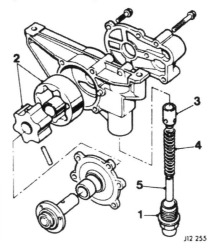

Fig. 115

Lubricate the pump inner and outer rotors (2, Fig. 115), the bearing and the housing. Fit the outer rotor, smear the gasket face with sealant, fit the seal to the pump body and tighten the securing bolts.

Lubricate and assemble the relief valve (3, Fig. 115), spring (4, Fig. 115) and tube (5, Fig. 115) to the pump. Fit and tighten the relief valve cap.

Fit the oil pump to the engine (1, Fig. 116) and tighten the securing bolts (2, Fig. 116).

Fig. 116

Clean oil pump drive sprocket, offer up the sprocket to the oil pump drive flange (1, Fig. 117), place a straight edge between the crankshaft sprocket and the oil pump drive sprocket (2, Fig. 117) and shim out the oil pump drive sprocket until the two sprockets are perfectly in line. Once the necessary shim pack has been selected, fit the shims to the oil pump (3, Fig. 117), locate the sprocket in the drive chain, offer up the sprocket to the pump and align the securing bolt holes, fit the tab washers and the securing bolts, tighten the bolts and knock up the lock tabs, position the oil pump chain damper to take the slack out of the chain and tighten the securing bolts.

Fig. 117

Clean and fit crankshaft windage trays (1, Fig. 118) and tighten the securing bolts.

Clean all oil pick-up pipes and housings. Fit new 'O' rings (1, Fig. 119) to the oil pick-up pipes. Fit the pick-up pipe to the housing and secure with the bolts. Lubricate the 'O' rings on the oil pipes and fit the pipes to the

Fig. 118

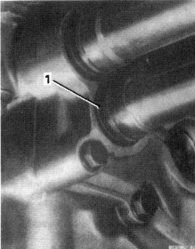

Fig. 119

housing. Smear the housing to the cylinder block gasket face with 'Hylosil', engage the pipes in the rear face of the oil pump, ensure that the 'O' rings enter the housing easily. Fit and tighten the housing securing bolts.

Smear the sump gasket face with 'Hylosil' and fit the sump to the engine and tighten the securing bolts.

Refit the crankshaft damper, fit Service Tool 18G 1437 to the damper and torque up the nut.

Refit the cylinder head, remove the engine from the stand and refit to the vehicle.

CYLINDER HEAD

Overhaul

Depressurise the fuel system, remove the bonnet, drain the coolant, jack up the front of the vehicle and place on two stands.

Remove the front exhaust pipe securing nuts and pull the exhaust clear of the manifold.

Lower the vehicle and support the exhaust.

Disconnect the spark plug leads (1, Fig. 120) and the injector harness block connector. Fit a dummy lifting eye, remove the air cleaner element. Note the position of

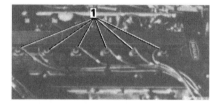

Fig. 120

the air switching valve feed wires prior to disconnection and remove the thermostat housing (1, Fig. 121).

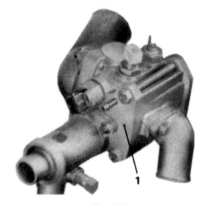

Fig. 121

Disconnect the hose from the pressure regulator and fit blanking plugs to the regulator and the hose.

Disconnect the fuel rail feed hose and fit blanking plugs.

Disconnect the distributor advance and retard pipe, the kickdown cable from the throttle linkage, and the throttle cable from the linkage (1, Fig. 122).

Remove the throttle cable bracket (2, Fig. 122) from the manifold.

Disconnect the brake servo hose, and the air conditioning pipes from the manifold.

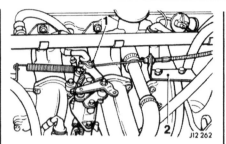

Fig. 122

Slacken the ECU vacuum pipe hose from the manifold.

Remove the engine breather hose from the camshaft cover.

Remove the oil filler tube assembly and dipstick tube assemblies, also remove the heater valve to cylinder head hose.

Remove the camshaft cover and discard all the seals and gaskets.

Remove the fan and torquatrol unit from the drive pulley.

Turn the engine to TDC No. 1 cylinder, disconnect the HT lead from the ignition coil, remove the distributor cap, disconnect the distributor amplifier block connector (1, Fig. 123), remove the vacuum pipe (2, Fig. 123), note the position of the rotor arm (3, Fig. 123) relative to the distributor body and the distributor body to the cylinder head (4, Fig. 123) and remove the distributor assembly from the engine.

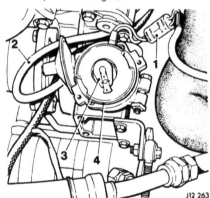

Fig. 123

Remove the camshaft cover.

Release the lock tabs from the bolts on the camshaft sprockets and slacken the bolts.

Slacken but do not remove the upper tensioner non return valve clamp securing bolt, move the clamp to one side and remove the valve from the housing. Using a 3 mm Allen key, rewind the top chain tensioner by engaging the key in the hexagon in the rear of the tensioner and turn clockwise until the tensioner locks in the 'park' position.

Remove the sprocket securing bolts and collect the tab washers, remove the sprockets, lift the chain, pivot the damper inwards and secure it with an elastic band, let the chain rest between the two dampers.

Remove the cylinder head front securing bolts followed by the remainder, remove the upper damper and place to one side and lift the cylinder head off the cylinder block.

Remove and discard the cylinder head gasket, the exhaust sealing rings, the oil

filler tube seal and place the cylinder head on a bench ensuring that the oil filler tube is not trapped.

Remove the upper tensioner securing bolts and remove the tensioner, discard the 'O' ring and gasket.

Remove the sparking plugs and the inlet and exhaust manifolds.

Remove the camshaft cap securing bolts (1, Fig. 124), lift off the caps and remove the camshafts (2, Fig. 124).

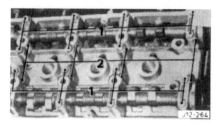

Fig. 124

Using a magnet (1, Fig. 125), lift out the cam followers (2, Fig. 125), remove the

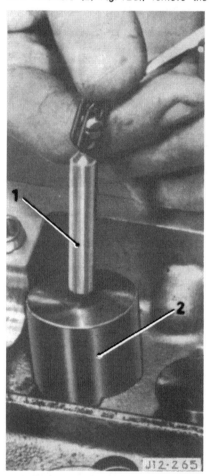

Fig. 125

shims, using Service Tool 18G 106A (1, Fig. 126), compress the valve springs and remove the spring retaining collets (2, Fig. 126). Release the tension on Tool 18G 106A and remove the top valve spring collars (1, Fig. 127), springs (2, Fig. 127) spring seats (3, Fig. 127) and the valves (4, Fig. 127); repeat this operation for the remaining 23 valves.

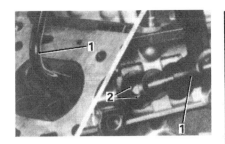

Fig. 126

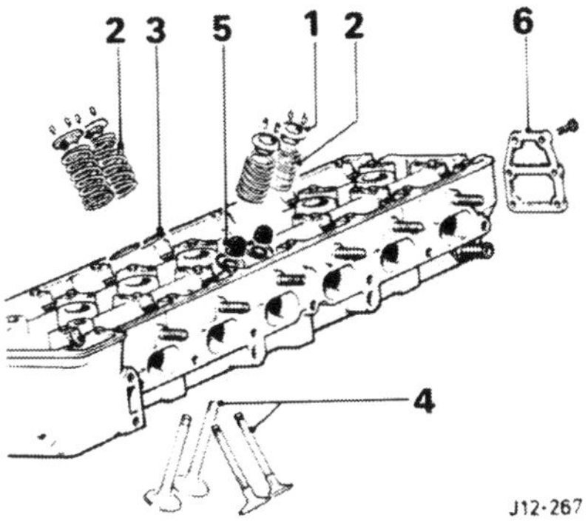

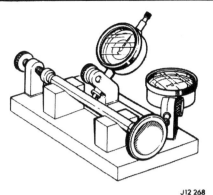

Fig. 128

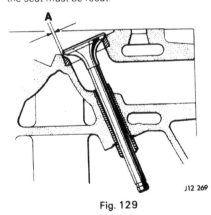

J12·267

Fig. 127

Remove the seals (5, Fig. 127) from the inlet valve guides and remove the cylinder head rear blanking plate (6, Fig. 127).

Clean all component parts and check for wear and also for burning of valves or seats.

Check the cylinder head face for distortion, if any is evident a maximum of 0,021 mm (0.008 in) may be removed by grinding to salvage the cylinder head.

Taking care not to damage the inside surface of the combustion chambers, clean the cylinder head gasket surface and the inlet and exhaust ports. When using scrapers or wire brushes for removing carbon deposits, avoid scratching the valve faces and seats. A soft wire brush is the most suitable implement for this purpose. Clean all carbon and other deposits from the valve guide using a suitable valve guide brush. Thoroughly wash the cylinder head to ensure that all loose carbon is removed and dry off with a high pressure air line.

After cleaning and polishing each valve, examine the stems for straightness and wear, using a suitable tool (Fig. 128), and the faces for burns, pitting and distortion.

Renew valves which are excessively worn, bent or too badly pitted to be salvaged by refacing

J12 268

NOTE: No attempt should be made to clean up a burnt or badly pitted valve face by the extensive 'grinding in' of the valve to the seat.

Lightly lap the valves into the seats with a fine grinding compound. The reseating operation should leave the finished surfaces smooth. Excessive lapping will groove the valve face resulting in a poor seat when hot (see Fig. 138) for valve seat acceptability.

'A' Correctly seated
'B' Undesirable condition
'C' Method of rectification

To test the valves for concentricity with their seats, coat the face of the valve with Engineers' blue or similar, and rotate the valve against the seat. If the valve face is concentric with the valve stem, a mark will be made all round the face.

Should a mark be made on only one side of the face, the face is not concentric with the valve stem. Clean the valve and again coat with Engineers' blue and rotate the valve against the seat to ascertain that the valve guide is concentric with the valve seat, if not the seat must be recut.

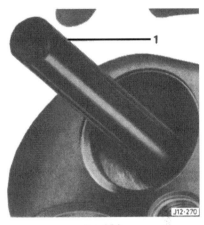

Fig. 129

J12 269

Whenever valves are replaced, the seats must be recut prior to lapping of the valves. Check the valve guide wear by inserting a new valve into the guide to be checked; lift it 3 mm ($\frac{1}{8}$ in) from its seat and rock it sideways. Movement of the valve across its seat must not exceed 0,5 mm (0.020 in) (A, Fig. 129). Should the movement exceed this tolerance, the valve guide must be replaced. This is achieved by using Service Tool 18G 1432 to drift out the old guide (1, Fig. 130). Ensure that the relevant service guide is selected prior to fitting (see chart).

J12·270

Fig. 130

Remove the old valve guide and ream the cylinder head to the relevant dimension — see chart. Coat guide with graphite grease, immerse the cylinder head in boiling water for 30 minutes, and fit the guide to the cylinder head (1, Fig. 131).

NOTE: To fit the guide the interference should not be sufficient to cause the use of excessive force.

Fig. 131

J12·271

IDENTIFICATION FOR OVERSIZE	PART NO.	VALVE GUIDE OUTER DIMENSION	DIMENSION TO BE REAMED TO IN CYLINDER HEAD	APPLICATION
Plain	C29388	0.5020 to 0.5015	—	Production
1 groove	C29389	0.5040 to 0.5035	—	Production
2 grooves	C29390	0.5070 to 0.5065	0.5055 to 0.5048	Service
3 grooves	C29391	0.5120 to 0.5115	0.5105 to 0.5098	Service

Note all dimensions in inches

After fitting a valve guide, the valve seat must be recut using Service Tool MS 204 (1, Fig. 132). Should the insert need

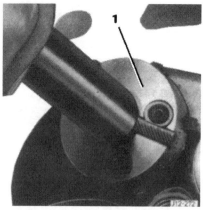

Fig. 132

replacing, the old insert should be bored out until it collapses, the bore should then be measured and the insert ground down until an interference of 0,077 mm 0.003 in) (See 'General Specification Data' for dimensions) exists between the cylinder head and the

Fig. 133

Fig. 134

insert. The cylinder head should then be oven heated at 150° (300°F) for 30 mins and then fit the seal inserts and recut with Service Tool MS 204 when the cylinder head has cooled. Examine the cam followers for wear on the top face, these should be perfectly flat (see Fig. 133), check also for any sign of barrelling on the side faces (1, Fig. 134). Replace all followers

that are worn or suspect. Wash the cylinder head, the valves, spring, collets, cups and followers and air dry.

After the valve springs have been thoroughly washed, they must be examined for fatigue and distortion.

Test the valve springs for pressure, either by comparison with the figures given in the 'General Specification Data' using a recommended valve spring testing machine (Fig. 135), or by comparison with a new valve spring.

To test against a new valve spring, insert both valve springs end to end between the jaws of a vice or under a press with a flat metal plate interposed between the two springs. Apply a load to compress the

Fig. 135

springs partly and measure their comparative lengths (Fig. 136). If the distance 'A' is smaller than 'B' and 'A' is the old spring then it must be replaced.

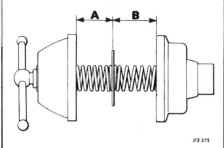

Fig. 136

Distortion is determined by positioning it upright on a surface or table and the squareness of each end is determined by utilising a set square (Fig. 137).

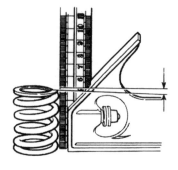

Fig. 137

All valve springs which have diminished in length and/or are not square must be discarded and new replacements fitted.

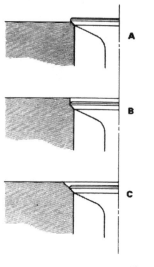

Fig. 138

A Correctly seated
B Undesirable
C Method of rectification

Oil all the valve chain components.

Assemble the valves one at a time, ensuring that the valves are assembled to the valve seat they were lapped to. Fit the spring seat and new valve stem seals, fit the springs and top collar and using Service Tool 18G 106A, compress the spring until the retaining collets can be inserted through the collar, and locate in the valve stem. Before removing the valve spring compressor tool, ensure that the retaining collets (1, Fig. 139) are still securely located in the valve stem (2, Fig. 139). Replace the

Fig. 139

original shim in the top of the collar (1, Fig. 140) ensuring that it is not tipped (1, Fig. 141) or slightly on top of the collar, oil and fit the cam follower (1, Fig. 142).

NOTE: If the cylinder head has been overhauled to the extent of having the valve seats recut, use a shim size 0.010 in smaller than the original.

Repeat this operation for the remaining 23 valves.

Lubricate the camshafts and fit to the cylinder head, ensuring that they are fitted with the slot to the top and that the No. 1 cylinder camshaft lobes are facing each

Fig. 140

Fig. 141

other. Fit the camshaft caps and torque the securing bolts in the sequence (Fig. 143).

Fit slave bolts to the front of the camshafts and measure the clearances between the heel of the camshaft (1, Fig. 144), and the cam followers (2, Fig. 144) turning the camshafts as necessary to measure all the clearances; these should be 0.012 to 0.014 in.

Should any of the clearances be incorrect, remove the camshaft caps, the camshafts and the cam followers using a suitable magnet. Remove all the shims or only those requiring adjustment; check with a micrometer the size of the existing shims and note the reading. Calculate the size of the shim required, following the example below for either excessive clearance or insufficient clearance.

Fig. 142

Fig. 143

Fig. 144

NOTE: See page 12-54 for a available shim sizes.

EXCESSIVE CLEARANCE	INCHES
Size of existing shim	0.100
Plus the actual clearance noted	0.019
	0.119
Less the specified valve clearance	0.013
= required shim size	0.106

INSUFFICIENT CLEARANCE	INCHES
Size of existing shim	0.107
Plus the actual clearance noted	0.010
	0.117
Less the specified valve clearance	0.013
= required shim size	0.104

Fit the shims as determined by the calculations to the appropriate valves and fit the cam followers; refit the camshafts, the camshaft caps and recheck the valve clearances.

NOTE: A final check of the valve clearances should be done once the cylinder head is fitted and torqued to the cylinder block.

Using Tool 18G 1433 set the inlet and exhaust camshafts to TDC (Fig. 145). Remove the camshaft slave bolts.

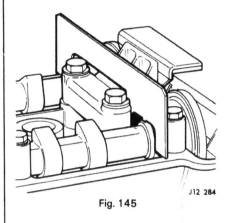

Fig. 145

Clean and check the upper chain upper damper for wear or scoring; should either of these conditions prevail, then the damper must be renewed.
Ensure that both No. 1 and No. 6 pistons are still at TDC and that both the top of the cylinder block and the cylinder head faces are scrupulously clean.
Clean and refit the inlet and exhaust manifolds ensuring that new gaskets are fitted to the exhaust manifolds, fit a new seal to the oil filler tube housing and new fire rings to the exhaust manifolds and fit both lifting eyes.
Fit the cylinder head assembly to the cylinder block, locating it on the dowels.

NOTE: Ensure that the oil filler tube and auxiliary air valve pipe do not foul and that no harnesses or pipes are trapped.

Fit the upper chain damper and pedestal assembly (1, Fig. 146) and fit the cylinder

Fig. 146

head attachment bolts. Tighten the bolts to a torque figure of 38 to 40 lbf ft and then turn the bolt clockwise through exactly 90° carry this out in the sequence illustrated (Fig. 147).

NOTE: To assist the accuracy of the rotation through 90° a suggested tool is illustrated on page 12—6.

Fit and tighten the three front securing bolts.

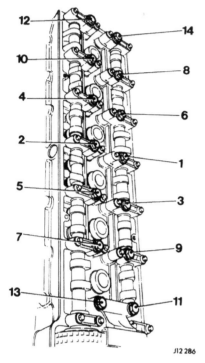

Fig. 147

Lift the timing chain and remove the elastic band, position the chain over the ends of the camshafts. Fit and engage the camshaft drive sprockets to the chain and the camshafts ensuring that all the chain 'slack' is towards the pivoting tensioner.

Fit the oil pressure simulator tool 18G 1436 (1, Fig. 148) to the cylinder head and tighten the bolt (2, Fig. 148) finger tight to exert a pressure of 2 to 4 lb on the chain.

Fig. 148

Remove the inner sprocket securing clip (1, Fig. 149), remove the inner sprocket (2, Fig. 149) and align the holes in the inner sprocket with the holes in the camshaft. Fit the tab washers (3, Fig. 149) and securing bolts (4, Fig. 149) but do not fully torque up the bolts, refit the inner sprocket securing clip. Repeat for the other camshaft sprocket. Tighten all the sprocket securing bolts, knock up the lock tabs, position the upper chain upper damper (5, Fig. 149) and tighten the securing bolts (6, Fig. 149) and knock up the lock tabs (7, Fig. 149).

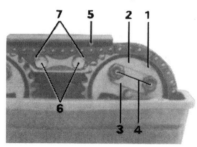

Fig. 149

Remove the oil pressure simulator tool.
Strip the upper tensioner and inspect its component parts for wear or damage. Check that the ball is free in the non return valve and also that the base plate (1, Fig. 150) is adequately secured by the peening (2, Fig. 150). Fit new 'O' rings to the valve (3, Fig. 150) and lubricate the 'O' rings.

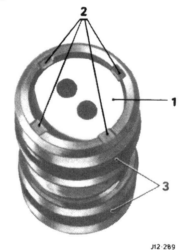

Fig. 150

Lubricate the snail (1, Fig. 151), spring (2, Fig. 151) and spring mandrel (3, Fig. 151) and assemble to the piston (4, Fig. 151).
Engage the snail (1, Fig. 151) with the peg (5, Fig. 151) and rotate the snail (1, Fig. 151) clockwise and depress until the peg (5, Fig. 151) engages in the 'park' position in the snail (1, Fig. 151).

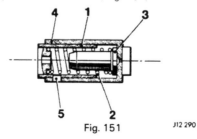

Fig. 151

Fit the assembly (1, Fig. 152) into the piston housing. Fit a new 'O' ring (2, Fig. 152) to the housing and lubricate, fit a new gasket (3, Fig. 152) to the housing and fit the assembly (4, Fig. 152) to the cylinder head, ensuring that the tensioner is engaged on the pivoting guide. Fit but do

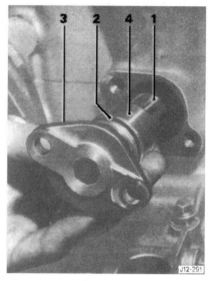

Fig. 152

not tighten the tensioner securing bolt (1, Fig. 153), fit the valve retainer securing bolt (four threads only) (2, Fig. 153). Fully tighten the lower securing bolt, fit new 'O' rings to the valve assembly and lubricate. Release the tensioner using a 3 mm Allen key, fit the non return valve assembly (3, Fig. 153), engage the retainer plate (4, Fig. 153) and fully tighten the securing bolt (2, Fig. 153).

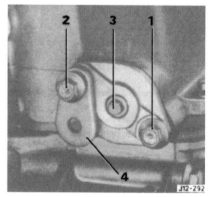

Fig. 153

Remove the baffle plate inside the camshaft cover and clean the camshaft cover and the baffle, refit the baffle to the cover and secure with the bolts. Fit the plug well seals (1, Fig. 154) and the camshaft cover gasket (2, Fig. 154), fit the half moon seals (3, Fig. 154) to the cylinder head and fit the camcover (4, Fig. 154) to the engine, fit and tighten the securing screws (5, Fig. 154).

Check the gap of the new sparking plugs (6, Fig. 154) and fit them to the cylinder head, tightening to a torque of 25 lbf ft

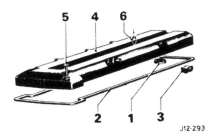

Fig. 154

Fit the new 'O' ring to the distributor, lubricate the 'O' ring and gear, turn the rotor arm and the distributor to the position shown in Fig. 155 (TDC No. 1). Fit the distributor clamp plate securing bolt, leave the clamp plate securing bolt slightly slack.

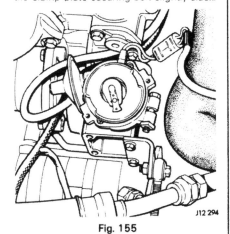

Fig. 155

Fit the distributor cap and tighten the securing screws. Connect the sparking plug leads in the firing order 1, 5, 3, 6, 2, 4.

NOTE: The distributor turns clockwise as viewed from the top.

Refit the torquatrol unit to the drive pulley. Reconnect the heater valve hose to the cylinder head and tighten the clip, reposition the dipstick tube and secure with the 'P' clip to the manifold bracket.

Reposition the oil filler tube (1, Fig. 156) into the lower housing (2, Fig. 156) and secure to the upper bracket. Refit the breather hose (1, Fig. 158) to the camshaft cover and tighten the clip, connect the hose (2, Fig. 158) to the manifold and air cleaner backplate and tighten the clips. Refit

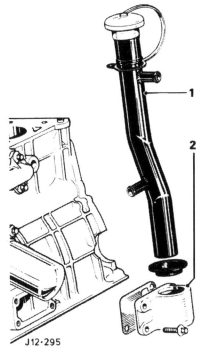

Fig. 156

the distributor advance/retard pipe; the ECU vacuum pipe, the brake servo hose, and the air conditioning pipes to the manifold and tighten the securing clips.

Refit the throttle cable (1, Fig. 157) to the throttle linkage (2, Fig. 157) and secure the throttle cable with the clip provided.

Reconnect the fuel feed hose (3, Fig. 158) to the fuel rail (4, Fig. 158) and tighten the nut; and refit the pressure regulator and hose (5, Fig. 158), secure with the clip.

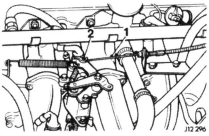

Fig. 157

Identify and reconnect the air switching valve feed wires, reposition the thermostat housing harness and refit the thermostat housing (1, Fig. 159).

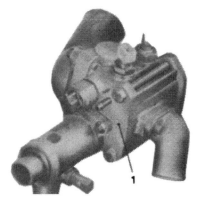

Fig. 159

Reconnect the injector block connectors (1, Fig. 160) and remove the dummy lifting eyes.

Refit the distributor block connector and connect the HT lead to the coil.

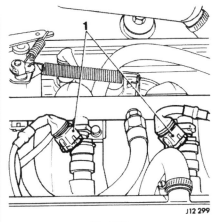

Fig. 160

Raise the front of the vehicle and place on stands, reconnect the exhaust system to the manifolds ensuring that the fire rings are still in position in the manifolds.

Lower the vehicle, refit the air cleaner element (1, Fig. 161), refill the cooling system, check and reset the ignition timing and refit the bonnet.

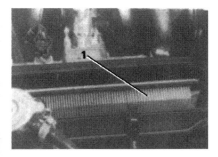

Fig. 161

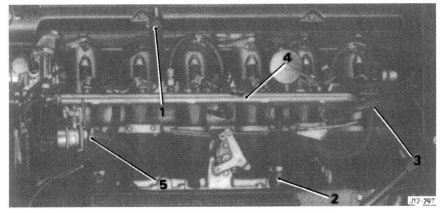

Fig. 158

Fig. 162

ENGINE AND GEARBOX

Renew

Remove the front grille.
Remove the bonnet and support struts from the vehicle.
Drain the coolant.
Disconnect the heater valve hose from the cylinder head and the hose from the heater to water pump return pipe. Remove the securing nuts (1, Fig. 162) for the receiver/dryer clamping brackets and release the receiver/dryer (2, Fig. 162) from the radiator top rail. Disconnect the amplifier harness (3, Fig. 162), the coil and the amplifier from the radiator top rail. Remove the fan cowl securing nuts (4, Fig. 162) and pull the cowl (5, Fig. 162) clear of the radiator. Remove the top rail (6, Fig. 162). Remove the top hose (7, Fig. 162) from the thermostat housing, the bottom hose from the water pump outlet (8, Fig. 162) and the overflow pipe from the radiator (9, Fig. 162). Disconnect the thermostatic switch wires; lift the radiator from the vehicle and remove the fan cowl.
Completely remove the receiver/dryer mounting clamps (10, Fig. 162); remove the radiator auxiliary electric fan complete (11, Fig. 162). Remove the front engine bay tie bars (12, Fig. 162), the horns and brackets, disconnect the engine oil cooler pipes, collect the 'O' rings and remove the oil cooler radiator (13, Fig. 162), position the air conditioning radiator and receiver/dryer as far forward in the engine bay as possible.
Remove the oil cooler pipe securing bracket (14, Fig. 162) and secure the pipes to the engine (1, Fig. 168).
Remove the air conditioning compressor from the engine.
Disconnect the engine electrical 'white' block connectors (15, Fig. 162), disconnect the power steering pump from the drive, and remove the coupling.
Disconnect the high tension lead (16, Fig. 162) from the ignition coil and also the front RH harness plugs (17, Fig. 162).
Remove the water hoses from the expansion tank to the water pump and thermostat housing (18, Fig. 162).
Depressurise the fuel system and disconnect the fuel feed pipe from the fuel rail (19, Fig. 162).
Disconnect the throttle cable from the throttle tower (20, Fig. 162), remove the throttle support bracket from the inlet manifold (21, Fig. 162).
Remove the fuel pipe from the pressure regulator (24, Fig. 162).
Remove the brake servo, the air conditioning, and the ECU vacuum pipes (22, Fig. 162) from the inlet manifold, and disconnect the starter motor electrical feed wires.
Fit two suitable engine lifting eyes.
Using Service Tool MS 53, support the engine by the rear lifting eye (1, Fig. 163).
Raise the front of the vehicle and place on stands. Remove the engine earth strap, slacken and remove the exhaust downpipe.

Fig. 163

Fig. 164

Disconnect the gear lever linkage (1, Fig. 164) and the electrical connections Disconnect the speedometer transducer block connector and remove the intermediate heat shield.
Take the weight of the gearbox with a suitable jack and remove mounting (1, Fig. 165).

Fig. 165

Remove the rear heat shield and propeller shaft, place a plug in the gearbox output to prevent leakage.
Disconnect the clutch slave cylinder from the bell housing.
Lower the vehicle and remove Service Tool MS 53.

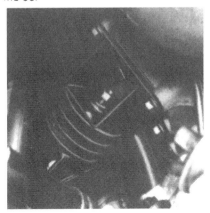
Fig. 166

Attach a suitable engine lifting sling, remove the front engine mounting securing nuts (Figs. 166 & 167) and manoeuvre the engine and gearbox from the vehicle (Fig. 168).

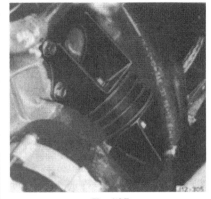
Fig. 167

Fig. 168

Remove the gearbox and all ancillaries from the engine, fit to the replacement engine and refit the gearbox.
Lift the engine and gearbox assembly into the engine bay, align the mountings and gently lower the engine mountings onto the mounting rubbers and secure with the washers and nuts. Remove the hoist and fit Service tool MS 53 (1, Fig. 163).
Raise the front of the vehicle and place on stands.
Take the weight of the gearbox with a suitable jack; lift into position and fit the gearbox rear mounting complete (1, Fig. 165). Remove the blanking plug in the rear of the gearbox, refit the propeller shaft and heat shield. Reconnect all the electrical connections to the gearbox and fit the intermediate heat shield. Reconnect the gear lever (1, Fig. 164), refit the exhaust down pipes and the engine earth strap.
Reconnect the heater return hose and the water valve hose to the cylinder head.
Reconnect the starter and solenoid feed wires (upper spade). Connect the ECU, the brake servo, and air conditioning vacuum pipes to the inlet manifold (22, Fig. 162).
Refit the throttle cable and secure with the clip (20, Fig. 162), refit the throttle support bracket (21, Fig. 162), slacken the locknut (23, Fig. 162) and adjust the cable to obtain the correct tension.
Refit the fuel hoses to the manifold.
Remove the plugs from the pressure regulator and hose and connect the hose and secure with a clip (24, Fig. 162).

Connect the engine harness (15, Fig. 162), the air conditioning compressor and thermal fuse block connectors (17, Fig. 162) and secure the harness with a rachet clip.

Connect the HT lead to the coil (16, Fig. 162), clean the power assisted steering (PAS) pump mating face, fit the coupling to the pump, align the (PAS) pump shaft with the distributor drive shaft, connect the pump assembly and secure with the bolts.

Align the compressor to the mounting bracket and align all mating holes, fit the bolts and spacers and reposition the link arm, refit the compressor drive belt, tension and tighten all the securing bolts. Position the oil cooler pipes, the receiver/drier (10, Fig. 162) and secure the oil cooler pipes with the inner wing bracket (shortest pipe lowest) (14, Fig. 162).

Lubricate and fit new 'O' rings to the oil cooler pipes and fit to the oil cooler radiator. Align the oil cooler (13, Fig. 162) to its mounting studs and secure with the nuts. Refit the horn assembly to the lower panel. Position the condenser and brackets.

Fit the lower crosstube and spacers, repeat for the upper tube (12, Fig. 162). Fit the fan and frame assembly (11, Fig. 162) into position, fit the harness clip to the mounting frame, tighten the screw and securing bolts. Connect the fan motor block connector (3, Fig. 162).

Ensure the condenser is still secured by the mounting brackets and that the pipes run along the inner wing.

Fit the air conditioning vacuum pipe clamps (25, Fig. 162), refit the fan cowl (5, Fig. 162) assembly and engage in the mounting rubbers.

Position the radiator on its mountings and refit the bottom water hose and tighten the clip (8, Fig. 162).

Reconnect the thermostatic switch feed wires and the expansion tank to water pump hose and tighten the clip (18, Fig. 162).

Connect the top hose to the thermostat housing and tighten the clip (18, Fig. 162).

Fit the top rail (6, Fig. 162) to the radiator, condenser and body, position the earth leads to the outer bolt holes (26, Fig. 162) and secure the rail with the bolts.

Reposition the fan cowl over the top rail studs, position the fuse holder over the LH stud (27, Fig. 162) and tighten the centre and LH securing nuts.

Identify and connect the amplifier block connectors (3, Fig. 162), position the harness to the top rail and secure with the harness clip (28, Fig. 162), secure the amplifier harness with ratchet clips.

Position the receiver/drier clamps (1, Fig. 162) over the top rail studs and secure with the nuts and bolts. Fit the bonnet struts, refit the air cleaner element.

Remove the two engine lifting eyes.

Refill the engine with coolant and oil.

Refit the bonnet and close.

CRANKSHAFT REAR OIL SEAL

Renew

Remove the gearbox, bell housing and flywheel, remove the rear oil seal housing securing bolts (1, Fig. 96) and remove the housing (2, Fig. 96).

DO NOT REMOVE THE PLASTIC 'O' RING PROTECTOR FROM THE SEAL PRIOR TO FITTING THE ENGINE.

Carefully remove the old rear oil seal from the housing, clean the housing and lubricate the seal mounting face (1, Fig. 105) and using tool No. 18G 134-8 and 18G 134 (2, Fig. 105) fit the seal (3, Fig. 105), to the housing (4, Fig. 105). Smear the gasket face with 'Hylosil', locate the plastic protector 'O' ring (1, Fig. 106) onto the end of the crankshaft (2, Fig. 106) and push the rear seal housing (3, Fig. 106) over the crankshaft and up to the rear cylinder block face. Fit the bolts, check that the sump face (1, Fig. 106) is flush with the cylinder block (1, Fig. 107) and tighten the bolts.

Refit the flywheel and tighten to the recommended torque, refit the bell housing and the gearbox.

CYLINDER PRESSURES

Check

Run the engine until it reaches its normal operating temperature. Remove all the sparking plugs, open the throttle fully and crank the engine over on the starter motor.

NOTE: The engine speed should not drop below 300rpm. For pressures and differential drop between cylinders see Engine Tuning.

FLYWHEEL

Renew

Remove the gearbox and the bell housing.

Remove the eight securing bolts (1, Fig. 168A) and remove the flywheel (2, Fig. 168A).

Remove the needle roller race (3, Fig. 168A), refit to the replacement flywheel.

Refit the flywheel to the crankshaft and torque up the bolts to the recommended torque figure, refit the bell housing and the gearbox.

Maximum material removal allowance to clean up the clutch face 1mm.

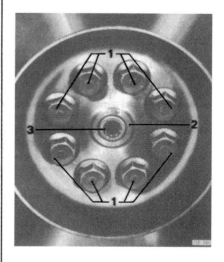

Fig. 168A

ENGINE ASSEMBLY

Overhaul

Remove the engine and gearbox from the vehicle; detach the gearbox from the engine remove all anciliaries and fit the engine to a stand. Remove the distributor and disconnect the plug leads from the sparking plugs.

Remove the camshaft cover and the upper chain tensioner assembly (1, Fig. 169); discard the 'O' rings and gaskets. Move the

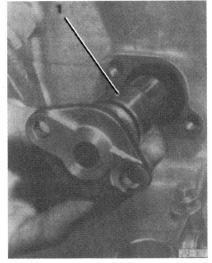

Fig. 169

camshaft top sprocket lock tabs (1, Fig. 170) away from the bolts, remove the sprocket securing bolts (2, Fig. 170) and collect the four tab washers, and remove the camshaft sprockets from the cylinder head.

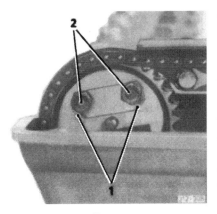

Fig. 170

Position the chain (1, Fig. 171) between the upper dampers and fit an elastic band (2, Fig. 171) across the upper dampers, retaining the chain and the pivoting damper (3, Fig. 171) during the cylinder head removal.

Fig. 171

Remove the cylinder head/camshaft bearing cap securing bolts (1, Fig. 172), remove the cylinder head assembly.

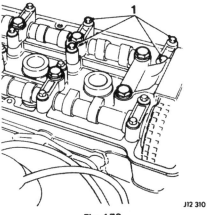

J12 310

Fig. 172

Fit 18G 1437 Service Tool (1, Fig. 173) to the front pulley and remove the damper

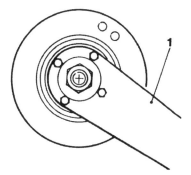

Fig. 173   J12 378

Fig. 174

securing bolt. Using Service Tool 18G 1436 (1, Fig. 174) secure to the damper with two bolts (2, Fig. 174) and remove the damper assembly.

Turn the engine over in the stand and remove the sump pan securing bolts and remove the sump (1, Fig. 175)

Fig. 175

Remove the oil pick-up securing bolts and remove the assembly (1, Fig. 176).

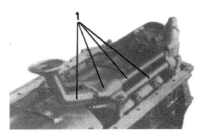

Fig. 176

Remove the crankshaft windage tray securing bolts and remove the trays (1, Fig. 177).

Fig. 177

Knock down the oil pump sprocket lock tabs (1, Fig. 178), remove the bolts (2, Fig. 178) and tabs and collect the sprocket (3, Fig. 178). Note the amount and the thickness of the spacing shims beind the oil pump sprocket. Remove the oil pump securing bolts and the pump assembly.

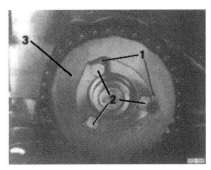

Fig. 178

Remove the timing cover securing bolts (1, Fig. 179) and remove timing cover assembly (2, Fig. 179).

Fig. 179

Remove the oil pump drive chain (1, Fig. 180) and the elastic band (2, Fig. 180) from the upper dampers.

Remove the upper timing chain (3, Fig. 180). Remove the oil pump drive chain damper securing bolts (4, Fig. 180), remove the damper and pedestal (5, Fig. 180).

Remove the lower chain tensioner securing bolts (6, Fig. 180) and remove the tensioner (7, Fig. 180).

Remove the lower tensioner pivoting damper securing bolts (8, Fig. 180), collect the tab washer (9, Fig. 180) and damper assembly (10, Fig. 180).

Remove the spacing tubes from the damper assembly and the lower chain fixed damper securing bolts (11, Fig. 180). Remove the tab washer (12, Fig. 180) and damper (13, Fig. 180).

Remove the upper fixed damper securing bolts (14, Fig. 180) and tab washer (15, Fig. 180), remove the damper (16, Fig. 180). Remove the upper pivoting damper pedestal securing bolts (17,

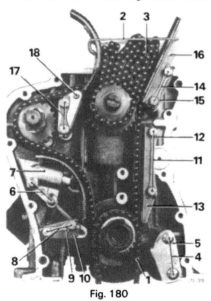

Fig. 180

Fig. 180), the pedestal and damper (18, Fig. 180).

Remove the intermediate sprocket and lower chain.
Remove the crankshaft damper woodruff key (1, Fig. 181) and spacer.
Remove the crankshaft sprocket (2, Fig. 181) and inner woodruff key.

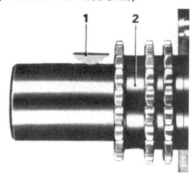

Fig. 181

Remove the oil seal (1, Fig. 182) from the power assisted steering end of the distributor drive gear shaft using Service Tool 18G 1468. Remove the shaft securing circlip (2, Fig. 182), the drive gear sprocket and thrust washer.

Remove the three auxiliary drive gear bearing housing set screws (3, Fig. 182), remove the housing and discard the gasket.

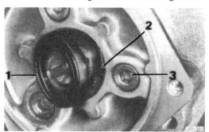

Fig. 182

Remove the rear oil seal housing securing bolts (1, Fig. 183) and remove the housing complete with seal (2, Fig. 183).

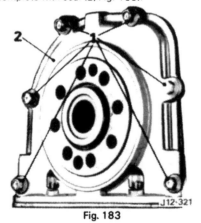

Fig. 183

Turn the crankshaft until Nos. 1 and 6 connecting rod caps are accessible, slacken and remove the big end securing nuts (1, Fig. 184).

Remove the cap (2, Fig. 184) and carefully push the connecting rod (3, Fig. 184) and piston assembly out through the top of the cylinder block, repeat this procedure for Nos. 2 and 5 connecting rods and pistons and 3 and 4. Ensure that upon removal of connecting rods and caps they are not inter-mixed.

Fig. 184

Remove the main bearing cap securing bolts (2, Fig. 185), remove the caps (1, Fig. 185) and then carefully remove the crankshaft (3, Fig. 185), collect the bearing shells and thrust washers.

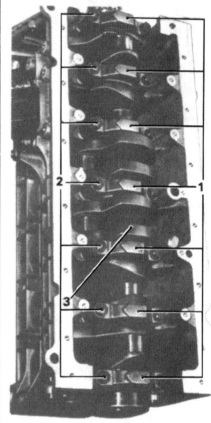

Fig. 185

Fit suitable bolts (1, Fig. 186) to the oil gallery plugs (2, Fig. 186) and pull to remove.

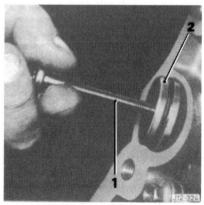

Fig. 186

Clean all the oil galleries, cylinder bores, dowels, distributor shaft mounting, main bearing cap and thrust faces.

Check the bore wear in the cylinder block with a suitable comparitor (Fig. 187). This should be done in at least six positions in the bore (Fig. 188).

NOTE: Maximum bore wear normally occurs towards the top of the bore across its thrust access.

Fig. 189 shows acceptable and unacceptable cylinder bore suface finishes. In all cases except A the cylinder block should be rebored, the sizes of O/S pistons are listed below. In the case of A the cylinder block should be honed using a suitable glaze buster (Fig. 190). Should the bore need to be inspected prior to removal of the cylinder head, a sparking plug may be removed and a suitable borescope may be used (Fig. 191).

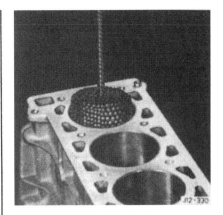
Fig. 190

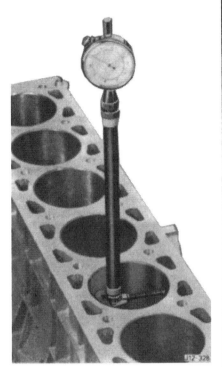
Fig. 187

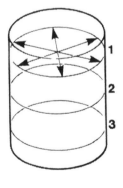
J12 239
Fig. 188

A   B

C   D

E
Fig. 189

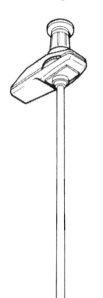
J12 331
Fig. 191

RING GAPS

Top compression	0,4 — 0,65 mm
Second compression	0,4 — 0,65 mm
Oil control	0,3 — 0,55 mm

If new piston rings are being fitted without reboring, deglaze the cylinder bores using a hone or 'glazebuster' without increasing the size of the bore; a deglazed bore will have a cross-hatched finish.

Remove the piston rings and clean the carbon and deposits from the piston crown and skirt. Using a suitable piston groove cleaner, ensure that all deposits are removed from the grooves. Examine the pistons for damage or excessive wear.

If gauge equipment is not available, the piston clearance can be measured using a long feeler gauge. Insert a long suitably sized feeler gauge down the RH side of the cylinder bore, insert the correct piston inverted into the cylinder bore, and position it with the gudgeon pin parallel with the axis of the crankshaft.

Push the piston down the cylinder until it reaches its tightest point in the bore; at this point, withdraw the feeler gauge; a steady resistance should be felt.

If standard sized pistons are being fitted, use pistons of the same grade as the markings on the cylinder block.

Fit the piston rings into the bores and check the clearance with a feeler gauge; should this clearance be insufficient, a small amount of material may be removed from either butting end of the ring using a carburundum stone. Following this operation, ensure that the edges of the ring are chamfered and that no burrs remain before refitting a ring back into a bore.

NOTE: Once these rings are correctly gapped, great care must be taken not to intermix them, and that each piston is numbered to its respective bore.

Fit the bearing shells to the connecting rods and fit each rod in turn to the crankshaft. Torque up the nuts and check the end float between the end face of the connecting rod and the journal shoulder (1, Fig. 192) the tolerance is given in General Specification.

Remove the connecting rods from the crankshaft and retain all parts in related sets.

Check the crankshaft journals for wear and ovality, the tolerances are:

Main	76,231 mm (3.0012 in)
Thrust width	36,233 to 36,208 mm (1.4265 to 1.4255 in)
Big end tolerance	52,987 mm (2.0861 in) 52,974 mm (2.0856 in)
Running clearance	0.0010 to 0.0027 in 0,041 to 0,084 mm (0.0016 to 0.0033 in)

Should the tolerances be outside these dimensions, the crankshaft should be renewed.

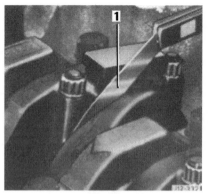

Fig. 192

Prior to fitting the connecting rods to the pistons, check them for twist (Fig. 196) or bend (Fig. 197). If the out of or one is twisted or bent, the rods must be changed as a set.

Fig. 193

Fig. 194

Fit Service Tool No. 18G 1434 to the intermediate sprocket bearing (1, Fig. 195) and remove the bearing.

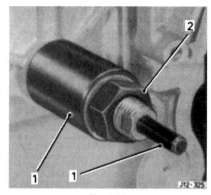

Fig. 195

Refit the new intermediate sprocket bearing (1, Fig. 196), ensuring that the oil feed hole lines up with the drilling (1, Fig. 197) in the cylinder block. Use Service Tool No. 18G 1434 (2, Fig. 196) with the flange towards the cylinder block and tap the end of the tool until the flange rests against the cylinder block.

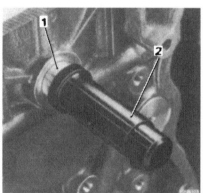

Fig. 196

Fig. 197

Fit new 'O' rings to the oil gallery plugs, lubricate the rings and install the plugs into the cylinder block oil gallery.

Fit the main bearing shells to the cylinder block and lubricate with engine oil (Fig. 198). Clean the crankshaft and inspect the journals for marks and scratches, and if necessary polish them out using lapping tape, and also only move the tape in an anti-clockwise direction against the crankshaft; carefully lower the crankshaft into the cylinder block.

Fig. 198

Fit the thrust washers with the grooved side to the crankshaft (1, Fig. 199), mount a dial test indicator and check the crankshaft end float (1, Fig. 200). This should be 0.004 to 0.010 in. Should the end float exceed these

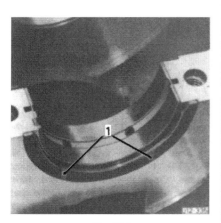

Fig. 199

Fig. 200

figures, then the standard sized thrust washers should be removed and replaced with oversize washers until the crankshaft endfloat is within these tolerances. Lubricate the crankshaft journals with engine oil.

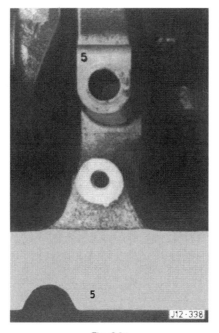

Fig. 201

Fig. 202

Fit the remaining shells to the main bearing caps; lubricate the shells with clean engine oil and fit to the cylinder block. Note the caps are numbered to the cylinder block (Fig. 201), tighten the cap securing bolts in the sequence illustrated (Fig. 202), ensuring that after each cap is torqued down the crankshaft still spins freely with only hand pressure.

DO NOT REMOVE THE PLASTIC 'O' RING PROTECTOR FROM THE SEAL PRIOR TO FITTING TO THE ENGINE.

Carefully remove the old rear oil seal from the housing, clean the housing and lubricate the seal mounting face, and using tool No. 18G 134-8 fit the seal to the housing (1, Fig. 203). Smear the gasket face with

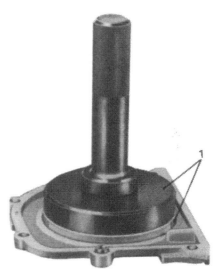

Fig. 203

'Hylosil', locate the plastic protector 'O' ring (1, Fig. 204) onto the end of the crankshaft (2, Fig. 204) and push the rear seal (3, Fig. 204) housing over the crankshaft and up to the rear cylinder block face. Fit the

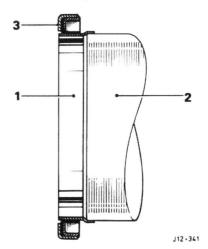

Fig. 204

securing bolts (1, Fig. 205) and by adjusting the housing (2, Fig. 205) ensure that the face is flush with the cylinder block (Fig. 206) and tighten the securing bolts.

Fig. 205

Fig. 206

Fit dummy bolts (1, Fig. 207) to the rear of the crankshaft to assist rotation of the crankshaft.

Fig. 207

Fit the rings to the pistons. Lubricate the small end bushes and fit the rods to the pistons and secure with the circlips (1, Fig. 208), ensuring that the connecting rods are correctly fitted in the piston (Fig. 208).

Fig. 208

Oil the piston rings and the cylinder bores. Position the piston ring gaps as in (Fig. 209) and fit Service Tool 18G 55A to the piston (1, Fig. 210) and compress the rings, insert the piston in the bore with the marking

Fig. 209

'FRONT' facing towards the front of the engine and push the piston down in the bore, ensuring that the connecting rod is the correct side of the crankshaft. Fit the bearing to the connecting rod, oil and fit it to the crankshaft. Fit a bearing to the big end

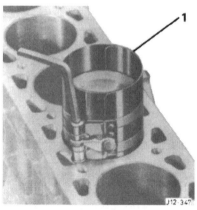

Fig. 210

cap, oil and fit to the connecting rod. Note the cap and rod are numbered to each other, ensure they are fitted as (1, Fig. 211). Fit and torque up the nuts. Turn the crankshaft over and ensure that there are no tight spots. Repeat this procedure for the remaining five cylinders.

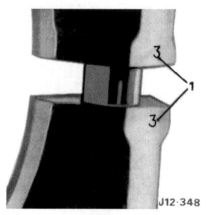

Fig. 211

Using Service Tool 18G 1434 remove the bearings from the timing cover (Fig. 212). Clean the cover and using Service Tool (1, Fig. 213), fit new bearings (2, Fig. 213) to the cover (3, Fig. 213).

Fig. 212

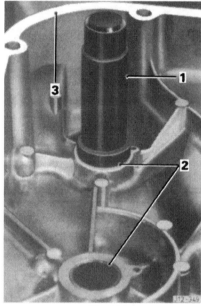

Fig. 213

Fit a new gasket to the auxiliary drive housing (1, Fig. 214), align and fit the housing to the cylinder block, apply 'Hylomar' to the securing screws (2, Fig. 214), fit and tighten the securing screws. Clean the drive gear and thrust washer, lubricate the shaft and fit the shaft (3, Fig. 214) to the housing, fit the thrust washer and secure with a circlip. Lubricate the oil seal mounting face and the outer edge of the seal and using Service Tool 18G 1469 (4, Fig. 214), fit and seat the oil seal (5, Fig. 214).

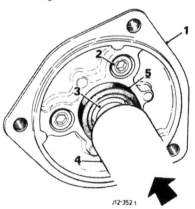

Fig. 214

Clean and inspect for wear or damage all the timing gears, chains, guides and tensioners.

Fit and seat the crankshaft sprocket woodruff key and push the sprocket (1, Fig. 215) onto the crankshaft. Lubricate the lower timing chain, fit the chain to the intermediate sprocket, fit the chain to the lower sprocket (2, Fig. 215), and then assemble the intermediate sprocket (3, Fig. 215) to the cylinder block.

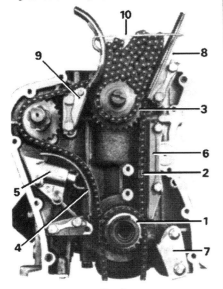

Fig. 215

Fit the spacers to the lower pivoting damper (4, Fig. 215) and fit the assembly to the cylinder block, fit a tab washer and bolts, tighten the bolts and knock the tabs up to secure the bolts.

Clean the lower tensioner assembly, remove the ball valve and ensure that the bottom of the housing is clean, check that the ball valve is free by shaking and ensuring it rattles. If it does not, replace the valve assembly. If it does, renew the 'O' ring, lubricate the ring and press the valve back into the housing.

Inspect the housing and hydraulic tensioner for wear or scoring. Also check that the oil hole is clear in the end of the piston. Lubricate the component parts. Assemble the tensioner by pressing and twisting the snail clockwise until the peg locks in the 'park' position. Insert the tensioner assembly into the housing and fit the housing to the cylinder block (5, Fig. 215). Fit and tighten the securing bolts.

Fit the lower static damper (6, Fig. 215), fit the tab washers and bolts, tighten the bolts and knock up the tabs.

Release the tensioner by pushing the tensioner back into the housing and releasing; this should release the tensioner and so tension the chain.

Fit the oil pump drive chain damper and pedestal (7, Fig. 215). Fit but do not tighten the damper securing bolts. Fit the upper static damper (8, Fig. 215) and pedestal and fit the securing bolts. Fit the upper pivoting damper (9, Fig. 215) and tighten the damper securing bolts, finally tighten the damper pedestal securing bolt and knock up the tabs.

Fit a dial gauge to the top of the cylinder block, and turn the engine over until No. 1 on 6 pistons are at TDC (1, Fig. 216). Lubricate the upper chain and fit it to the intermediate sprocket; fit an elastic band to the upper dampers (10, Fig. 215).

Fig. 216

Lubricate the oil pump drive chain, fit the chain to the crankshaft sprocket and lodge in the correct position.

Remove timing cover blanking plates and pointer. Remove the front oil seal and distributor drive gear shaft seal. Lubricate the oil seals and fit to the cover ensuring that the crankshaft seal is flush with the front cover (1, Fig. 217). Fit the timing pointer, smear the blanking plate surfaces

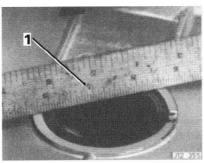

Fig. 217

with 'Hylosil' and fit the blanking plates. Smear the cylinder block with 'Hylosil' lubricate the bearing shells and carefully fit the timing cover to the engine, ensuring the top face is flush with the top of the cylinder block (Fig. 218), fit and tighten the securing bolts. Lubricate the oil seal distance piece. Fit and seat the damper woodruff key.

Fig. 218

Remove the oil pump pressure relief valve retaining nut, remove the spring, tube and valve, clean and check for wear or scoring. Remove the pump body securing bolts and remove the outer rotor. Remove the backplate securing bolts, the backplate, and the bearing shell from the housing. Clean all dismantled parts and check for obvious wear or scoring.

Fit the outer rotor and check the clearance between the rotor and the oil pump housing. Should any clearance exceed the specified tolerances (see General Specification) the assembly, i.e. rotor outer housing assembly, must be replaced.

Lubricate the bearing shell (1, Fig. 219) and fit to the pump housing (2, Fig. 219). Smear backplate gasket face (3, Fig. 219) with sealant and fit the backplate and tighten the securing bolts (4, Fig. 219).

Lubricate the pump inner (5, Fig. 219) and outer rotors (6, Fig. 219), the bearing and the housing. Fit the outer rotor, smear the gasket face with sealant, fit and seat the pump body and tighten the securing bolts.

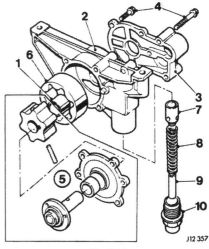

Fig. 219

Lubricate and assemble the relief valve (7, Fig. 219), spring (8, Fig. 219) and tube (9, Fig. 219) to the pump. Fit and tighten the relief valve cap (10, Fig. 219).

Fit the oil pump to the engine and tighten the securing bolts.

Clean oil pump drive sprocket, offer up the sprocket to the oil pump drive flange, place a straight edge (1, Fig. 220) between the crankshaft sprocket and the oil pump drive sprocket and shim out the oil pump drive sprocket until the two sprockets are perfectly in line. Once the necessary shim

Fig. 220

pack has been selected, fit the shims to the oil pump, locate the sprocket (1, Fig 221) in the drive chain (2, Fig 221), offer up the sprocket to the pump and align the securing bolt holes, fit the tab washers (3, Fig 221) and the securing bolts (4, Fig 221), tighten the bolts and knock up the lock tabs, position the oil pump chain damper and tighten the securing bolts.

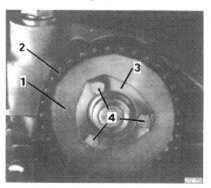

Fig. 221

Clean and fit the crankshaft windage trays (1, Fig. 222) and tighten the securing bolts. Clean all the oil pick-up pipes and housings. Fit new 'O' rings to the oil pick-up pipes. Smear the pick-up gasket face with 'Hylosil' and fit the pick-up pipe to the housing and secure with the bolts. Lubricate the oil pipe 'O' rings and fit the pipes to the housing. Smear the housing (2, Fig. 222) to cylinder block gasket face with 'Hylosil', engage the

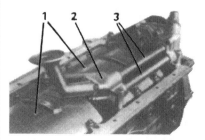

Fig. 222

pipes (3, Fig. 222) in the rear face of the oil pump, ensuring that the 'O' rings (1, Fig. 223) enter the housing easily. Fit and tighten the housing securing bolts.

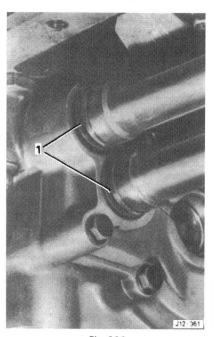

Fig. 223

Remove the sump baffle securing bolts, remove the baffle, clean the sump and the baffle, refit the baffle and secure with the bolts. Smear the sump gasket face with 'Hylosil' and fit the sump (1, Fig. 224) to the engine and tighten the securing bolts.

Fig. 224

Place the cylinder head on suitable wooden blocks.
Remove the sparking plugs, the exhaust manifold securing bolts, the inlet manifolds and collect the gaskets.
Remove the camshaft cap securing bolts, lift off the caps and remove the camshafts.
Using a magnet, lift out the cam followers, remove the shims and using Service Tool 18G 106A (1, Fig 225) compress the valve springs and remove the spring retaining collets (2, Fig 225). Release the tension on Tool 18G 106A and remove the top collar, the springs, the spring seats and the valve, repeat this operation for the remaining 23 valves.

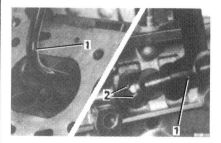

Fig. 225

Remove the seals from the inlet valve guides and remove the cylinder head rear blanking plate.
Clean all component parts and check for wear and also burning of valves or seats.
Taking care not to damage the inside surface of the combustion chambers, clean the cylinder head gasket surface and the inlet and exhaust ports. When using scrapers or wire brushes for removing carbon deposits, avoid scratching the valve faces and seats. A soft wire brush is the most suitable implement for this purpose. Clean all carbon and other deposits from the valve guides using a suitable valve guide brush. Thoroughly wash the cylinder head to ensure that all loose carbon is removed, dry off with a high pressure air line.
After cleaning and polishing each valve, examine the stems for straightness and wear, and the faces for burns, pitting or distortion.
Renew valves which are excessively worn, bent or too badly pitted to be salvaged by refacing.

NOTE: No attempt should be made to clean up a burnt or badly pitted valve face by the extensive 'grinding in' of the valve to the seat.

Lightly tap the valves into the seals with a fine grinding compound. The reseating operation should leave the finished surfaces smooth. Excessive tapping will groove the valve face resulting in a poor seat when hot.

To test the valves for concentricity with their seats, coat the face of the valve with Engineers' blue or similar, and rotate the valve against the seat. If the valve face is concentric with the valve stem, a mark will be made all round the face

Should a mark be made on only one side of the face, the face is not concentric with the valve stem. Should the seat not be concentric the valve seat should be recut. Clean the valve and again coat with Engineers' blue and rotate the valve against the seat to ascertain that the valve guide is concentric with the valve seat.

Whenever valves are replaced, the seats must be recut prior to lapping of the valves.

Check the valve guide wear by inserting a new valve into the guide to be checked; lift it 3 mm ($\frac{1}{8}$ in) from its seat and rock it sideways. Movement of the valve across its seat (A, Fig. 226) must not exceed 0,5 mm (0.020 in) (Fig. 226). Should the movement

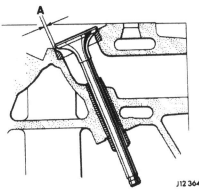

Fig. 226

exceed this tolerance, the valve guide must be replaced. This is achieved by using Service Tool 18G 1432 to drift out of the old guide (1, Fig. 227). Ensure that the relevant Service guide is selected prior to fitting.

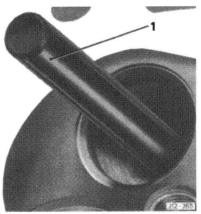

Fig. 227

Ream the cylinder head to the relevant dimension (see chart). Coat the guide with graphite grease and fit to the cylinder head after it has been immersed in boiling water for 30 minutes (1, Fig. 228).

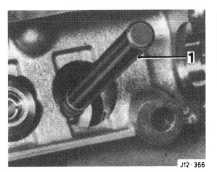

Fig. 228

After fitting a valve guide, the valve seat must be recut using Service Tool MS 204 (1, Fig. 229).

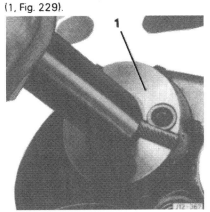

Fig. 229

NOTE: Should the valve seats need replacing see page 12—38.

Examine the cam followers for wear on the top face (Fig. 230) and any sign of barrelling on the side faces (Fig. 231).

Replace all followers that are worn or suspect. Wash the cylinder head, the valves, spring, collets, cups and followers and air dry.

Fig. 230

Fig. 231

After the valve springs have been thoroughly washed, they must be examined for fatigue and distortion.

Fatigue is determined by measuring its free length and distortion by positioning it upright on a surface plate or table and the squareness of each end is determined by utilising a set square (Fig. 232).

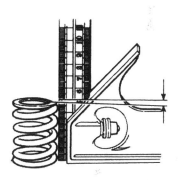

Fig. 232

All valve springs which have diminished in length and/or are not square must be discarded and new replacements fitted.

Test the valve springs for pressure, either by comparison with the figures given in the 'General Specification Data', using a recommended valve spring testing machine (Fig. 233), or by comparison with a new valve spring.

Fig. 233

IDENTIFICATION FOR OVERSIZE	PART NO.	DIMENSION A	DIMENSION TO BE REAMED TO IN CYLINDER HEAD	APPLICATION
Plain	C29388	0.5020 to 0.5015	—	Production
1 groove	C29389	0.5040 to 0.5035	—	Production
2 grooves	C29390	0.5070 to 0.5065	0.5055 to 0.5048	Service
3 grooves	C29391	0.5120 to 0.5115	0.5105 to 0.5098	Service

To test against a new valve spring, insert both valve springs to end between the jaws of a vice or under a press with a flat metal plate interposed between the two springs. Apply a load to compress the springs partly and measure their comparative lengths. If 'B' (Fig. 234) is new spring and distance 'A' is less than B then Spring A must be replaced as it is weak.

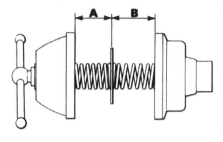

Fig. 234

Oil all the valve train components.
Assemble the valves one at a time, ensuring that the valves are assembled to the valve seat they were lapped to. Fit the spring seat and new valve stem seals, fit the springs and top collar and using Service Tool 18G 106A compress the spring until the retaining collets can be inserted through the collar, and locate in the valve stem. Before removing the valve spring compressor tool, ensure that the retaining collets (1, Fig. 235) are still located in the valve stem (2, Fig. 235). Replace the original shim in

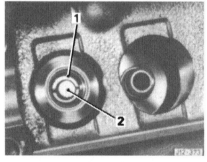

Fig. 235

the top of the collar, ensuring that it is not tipped (1, Fig. 236) or slightly on top of the collar, and fit the cam follower.
Repeat this operation for the remaining 23 valves.

NOTE: If the cylinder head has been overhauled to the extent of having the valve seats recut, use a shim size 0.010 in smaller than the original.

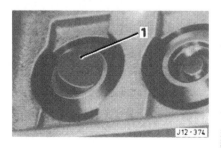

Fig. 236

Lubricate the camshafts and fit to the cylinder head, ensuring that they are fitted with the slot to the top and that the No. 1 cylinder camshaft lobes are facing each other. Fit the camshaft caps and torque up the securing bolts in the sequence (Fig. 237). Fit slave bolts to the camshafts and measure the clearances between the heel of the camshaft and the cam followers (1, Fig. 238) turning the camshafts as necessary to measure all the clearances, valve clearances are 0.012 to 0.014 in.

Fig. 237

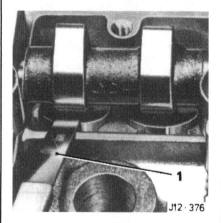

Fig. 238

Should any of the clearances be incorrect, remove the camshaft caps, the camshafts and the cam followers using a suitable magnet. Remove all the shims or only those requiring adjustment; check with a micrometer the size of the existing shims and note the reading. Calculate the size of the shim required following the example below for either excessive clearance or insufficient clearance.

EXCESSIVE CLEARANCE

	INCHES
Size of existing shim	0.100
Plus the actual clearance noted	0.019
	0.119
Less the specified valve clearance	0.013
= required shim size	0.106

INSUFFICIENT CLEARANCE

	INCHES
Size of existing shim	0.107
Plus the actual clearance noted	0.010
	0.117
Less the specified valve clearance	0.013
= required shim size	0.104

SHIM SIZES

Letter	Thickness
A	0.085 in
B	0.086 in
C	0.087 in
D	0.088 in
E	0.089 in
F	0.090 in
G	0.091 in
H	0.092 in
I	0.093 in
J	0.094 in
K	0.095 in
L	0.096 in
M	0.097 in
N	0.098 in
O	0.099 in
P	0.100 in
Q	0.101 in
R	0.102 in
S	0.103 in
T	0.104 in
U	0.105 in
V	0.106 in
W	0.107 in
X	0.108 in

Fit the shims as determined by the calculations to the appropriate valves; refit the camshafts and the camshaft caps and recheck the valve clearances.

NOTE: A final check of the valve clearances should be carried out with the cylinder head fitted to the cylinder block.

Using Tool 18G 1433 set the inlet and exhaust camshafts to TDC (1, Fig. 239).

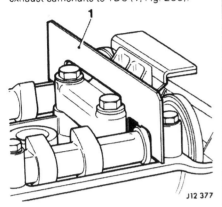

Fig. 239

Remove the camshaft slave bolts and fit the lifting eye to the inlet manifold securing studs; fit and tighten the nuts.

Clean the exhaust manifolds; fit new manifold gaskets, fit the manifolds and the rear lifting eye, fit and tighten the securing nuts.

Clean and check the upper chain upper damper for wear or scoring; should either of these conditions prevail, then the damper must be removed from its pedestal and a new part fitted. Should this be necessary; when refitting the damper to the pedestal, fit but do not tighten the damper to pedestal securing bolts.

Clean the crankshaft damper (1, Fig. 240) and fit the assembly to the crankshaft; fit but do not tighten the damper securing bolt. Fit the front pulley lock, Service Tool No. 18G 1437 to the front damper and tighten the damper nut. Check the front damper run out (Fig. 241) and ensure it is not excessive. Ensure that both No. 1 and No. 6 pistons are still at TDC.

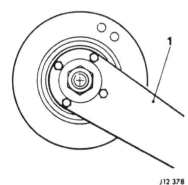

Fig. 240

Remove the camshaft inner sprocket securing clips (1, Fig. 243) and inner sprockets. Clean both the inner (2, Fig. 243) and outer (3, Fig. 243) sprockets and check the outer sprocket for wear and chips on the drive teeth and wear on the inner serrations. Also check the inner sprocket for wear on its outer serrations; replace any suspect items and assemble the sprockets and secure with the circlips. Ensure that both the top of the

Fig. 241

cylinder block and the mating face of the cylinder head are clean and dry and fit a new cylinder head gasket.

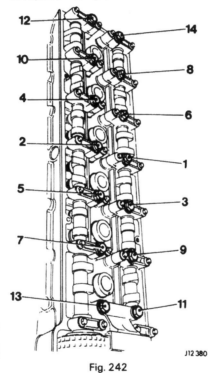

Fig. 242

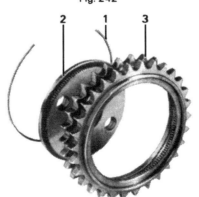

Fig. 243

Fit the cylinder head assembly to the cylinder block; fit the chain damper and pedestal assembly and fit the cylinder head

attachment bolts. Tighten the bolts to a torque figure of 38 to 40 lb ft and then turn the bolt clockwise through exactly 90°; carry this out in the sequence (Fig. 242).

NOTE: To assist the accuracy of the rotation through 90° a suggested tool is illustrated on page 12—6.

Lift the timing chain and remove the elastic band, position the chain over the ends of the camshafts. Fit and engage the camshaft drive sprockets (1, Fig. 244) with the chain and the camshafts ensuring that all the chain 'slack' is towards the pivoting tensioner.

Fig. 244

Strip the upper tensioner and inspect its component parts for wear or damage. Check that the ball is free in the non return valve (1, Fig. 245) and also that the peening securing the end plate is in good condition. Fit new 'O' rings to the valve (2, Fig. 245), fit the assembly into the piston housing.

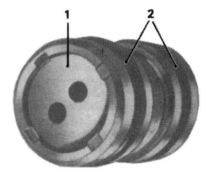

Fig. 245

Lubricate the snail (1, Fig. 246), spring (2, Fig. 246) and spring mandrel (3, Fig. 246) and assemble to the piston (4, Fig. 246), ensure the bleed hole in the end of the piston is clear (1, Fig. 247).

Engage the snail (1, Fig. 246) with the peg and rotate the snail (1, Fig. 246) clockwise and depress until the peg engages in the 'park' position in the snail (1, Fig. 246).

Fig. 246

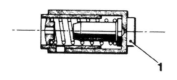

J12 385

Fig. 247

Fit a new 'O' ring (1, Fig. 248) to the housing and lubricate, fit a new gasket (2, Fig. 248) to the housing and fit the assembly (3, Fig. 248) to the cylinder head, ensuring that the tensioner is properly

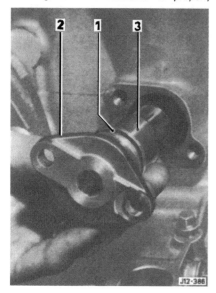

Fig. 248

engaged on the pivoting guide. Fit but do not tighten the tensioner securing bolt (1, Fig. 249), fit the valve retainer securing bolt (2, Fig. 249) (four threads only). Fully tighten the lower securing bolt. Release the tensioner using a 3 mm Allen key, fit the valve assembly (3, Fig. 249), engage the retainer plate (4, Fig. 249) and fully tighten the securing bolt (2, Fig. 249).

Fig. 249

Remove the inner camshaft sprocket securing clip, remove the inner sprocket and align the holes in the inner sprocket with the holes in the camshaft. Fit the tab washers and securing bolts (1, Fig. 250) but do not fully torque up the bolts, refit the inner sprocket securing clip (2, Fig. 250). Repeat for the other camshaft sprocket. Tighten all

the sprocket securing bolts, knock up the lock tabs (3, Fig. 250), position the upper chain upper damper (4, Fig. 250) and tighten the securing bolts and knock-up the lock tabs (5, Fig. 250).

Fig. 250

Remove the baffle plate inside the camshaft cover and clean the camcover, and baffle, refit the baffle to the cover and secure with the bolts. Fit the plug well seals and the cam cover gasket, fit the half moon seals to the cylinder head and fit the camcover to he engine (1, Fig. 251), fit and tighten the securing screws (2, Fig. 251).

Fig. 251

Check gap of new sparking plugs and fit to cylinder head.
Fit a new 'O' ring to the distributor, lubricate the 'O' ring and gear, turn the rotor arm to the position illustrated in (Fig. 252) and fit the distributor in the position shown in (Fig. 252) (TDC No. 1). Fit the distributor clamp plate securing bolt, leave the clamp plate securing bolt, slightly slack. Reconnect the distributor/amplifier block connector (1, Fig. 252). Fit the distributor cap and tighten the securing screws. Connect the sparking plug leads in the firing order 1, 5, 3, 6, 2, 4.

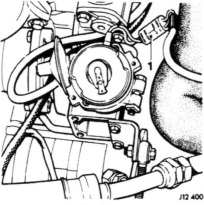

J12 400

Fig. 252

NOTE: The distributor turns clockwise as viewed from the top.

Remove the engine from the stand refit all ancilliaries and refit the engine to the vehicle and reset the ignition timing.

KEY FOR FIG. 253

1. Spark plug
2. Camshaft cover
3. Plug well seal
4. Camshaft cover gasket
5. Half moon sealing plug
6. Camshaft cap
7. Exhaust camshaft
8. Inlet camshaft
9. Spring retaining collar
10. Valve spring
11. Valve stem seal
12. Spring seat
13. Cam follower
14. Shim
15. Collet
16. Cylinder closing plate
17. Cylinder head
18. Cylinder gasket
19. Oil filler tube
20. Exhaust valves
21. Inlet valve
22. Small end bush
23. Connecting rod cap
24. Cylinder block
25. Water pipe
26. Rear oil seal housing
27. Rear oil seal
28. Oil filter housing
29. Oil filter
30. Filler tube oil seal
31. Filler tube mounting
32. Baffle
33. Oil cooler take off housing
34. Crankshaft
35. Engine mounting
36. Thrust washer bearing
37. Main bearing and cap
38. Windage tray
39. Main bearing cap
40. Windage tray
41. Oil pick up
42. Sump baffle
43. Sump
44. Oil drain plug
45. Oil pick up pipes
46. Alternator mounting bracket
47. Washer
48. Inspection plug
49. Shim
50. Oil pump drive sprocket
51. Lock tab plate
52. Oil pump
53. Main bearing and cap
54. Woodruff key
55. Spacer
56. Oil pump chain tensioner
57. Lower static chain damper
58. Lower pivoting chain damper
59. Oil pump drive chain
60. Crankshaft sprocket
61. Lower chain tensioner
62. Timing chain cover
63. Front crankshaft oil seal
64. Spacer
65. Timing disc
66. Crankshaft damper
67. Front pulley
68. Crankshaft front pulley bolt
69. Timing pointer
70. Closing plate
71. Bearing
72. Closing plate
73. Bearing
74. Auxiliary shaft
75. Intermediate timing chain
76. Intermediate sprocket
77. Upper timing chain
78. Upper pivoting chain damper
79. Upper static chain damper
80. Spacer
81. Bearing
82. Sealing plug
83. Connecting rod bearings
84. Connecting rod
85. Piston
86. Gudgeon pin
87. Circlip
88. Piston rings
89. Upper timing chain tensioner
90. 'O' rings
91. Locking plate
92. One way valve
93. Tensioner housing
94. Gasket
95. Upper spacer
96. Chain upper damper
97. Cylinder head securing bolt
98. Spring clip
99. Inner sprocket
100. Outer sprocket
101. Front camshaft cap

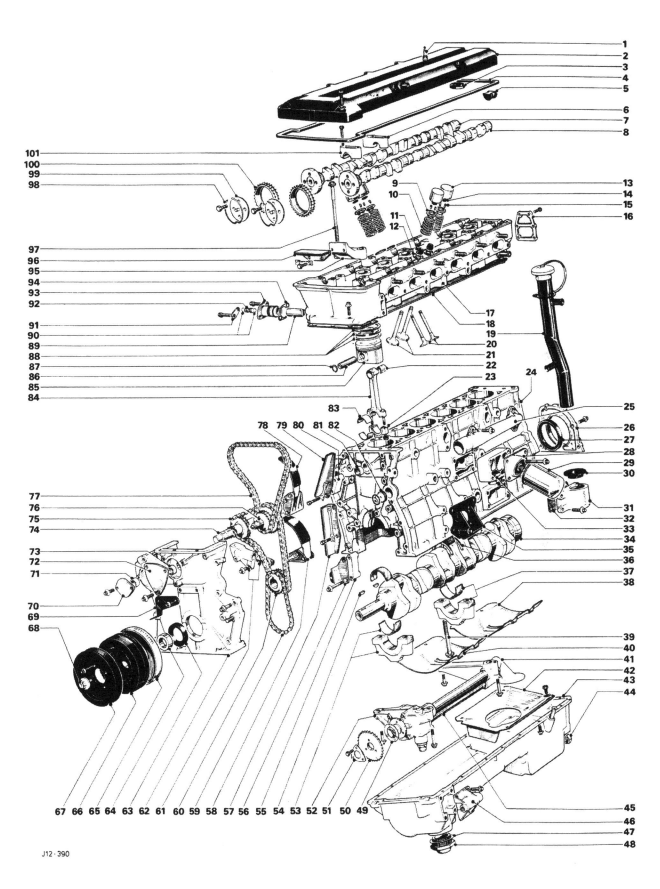

J12·390

Fig. 253

UK SPECIFICATION

CONTENTS

EMISSION CONTROL SYSTEM

Air Injection System

The pump (1, Fig. 1) is a rotary vane positive displacement device, belt driven from the engine crankshaft. The pulley drives the pump via an electrically operated clutch which enables the pump to be disengaged. The pump air intake is cleaned by a separate renewable paper element filter.

The air switching valve (2, Fig. 1) is a normally closed vacuum operated valve, which simultaneously shuts off the secondary air duct when the air pump clutch is disengaged. The vacuum source is controlled by a solenoid vacuum valve (4, Fig. 1) and a thermal switch (5, Fig. 1).

A check valve (3, Fig. 1) fitted in the air delivery pipe protects the air pump from the back flow of harmful exhaust gas in the event that exhaust back pressure exceeds the air pump pressure due to, for example, a pump drive belt failure.

The solenoid vacuum valve (4, Fig. 1) is a three port device incorporating a filter on the atmospheric port. The valve is operated by an electrical solenoid and the atmospheric port is normally open.

The vacuum delay valve (6, Fig. 1) consists of a restrictor which delays the transfer of a vacuum signal, with an umbrella non-return valve.

The thermal switch (5, Fig. 1) contacts open with a rising temperature of 45°C and close with a falling temperature of 38°C. The contacts are actuated by a temperature sensitive bimetallic disc, sensing the engine coolant temperature.

An exhaust back pressure switch (7, Fig. 1) consists of a diaphragm spring operated, normally closed microswitch. When the applied exhaust back pressure exceeds 11" hg, the switch contacts will open.

The throttle microswitch (8, Fig. 1) is in an open circuit position when the throttle is open and the pressure on the back blade and roller is released.

The vacuum switch (9, Fig. 1) consists of a diaphragm spring operated, normally closed microswitch. When the applied vacuum exceeds 20" hg the contacts will open.

OPERATION OF AIR INJECTION

With the engine coolant temperature above 40°C and the throttle open, the electrical supply is broken to the lambda disable relay (10, Fig. 1), thus preventing the operation of feedback inhibit, the solenoid vacuum valve (4, Fig. 1) venting the air switch valve (2, Fig. 1) to atmosphere, and the air pump clutch preventing the operation of the clutch.

With the throttle closed, the throttle microswitch (8, Fig. 1) is also closed supplying battery voltage to the lambda disable relay activating the feedback inhibit and applying full load fuel, the solenoid vacuum valve, thus connecting the air switching panel to vacuum, and also to the air pump clutch to operate the air pump. If the manifold vacuum is more than 20" hg, the vacuum switch (9, Fig. 1) will open or if the exhaust back pressure is greater than 11" hg, the pressure switch (7, Fig. 1) will open, thus interrupting the battery supply to the lambda disable relay, the solenoid vacuum valve and the magnetic clutch. The latter function is required in order to protect the air pump from exhaust gas back flow.

The diode (11, Fig. 1) prevents the energising and closing of the canister purge control valve. It also prevents the idle relay (12, Fig. 1) from being energised opening the relay contacts, preventing the application of full load fuel.

With the engine coolant temperature less than 38°C under all engine operating conditions, the thermal switch contacts are closed, so energising the idle relay, inhibiting the full load fuel, the lambda disable relay, activating the feedback inhibit, the solenoid vacuum valve connecting the air switching valve to vacuum and the air pump clutch connecting the air pump drive. Secondary air is supplied under all operating conditions until the engine coolant temperature exceeds 40°C or if the exhaust back pressure exceeds 11" hg.

A vacuum delay valve (6, Fig. 1) in the vacuum line to the solenoid vacuum valve prevents the air switch valve from interrupting the secondary air during acceleration modes, i.e. when the intake manifold vacuum falls below the minimum operating value of the switching valve.

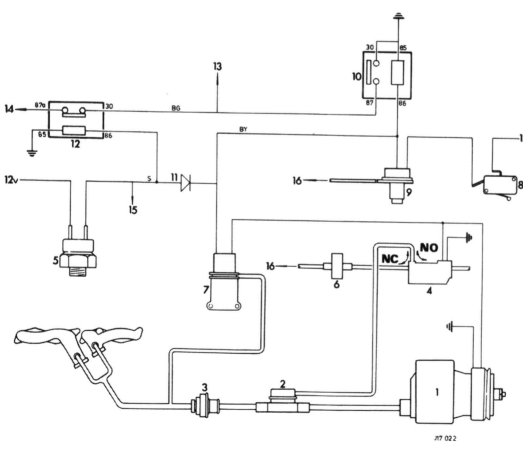

J17 022

Fig. 1

KEY TO DIAGRAM

1. Air pump
2. Air switching valve
3. Check valve
4. Solenoid vacuum valve
5. Thermal switch
6. Vacuum delay valve
7. Exhaust back pressure switch
8. Throttle switch
9. Vacuum switch
10. Lambda disable relay
11. Diode
12. Idle relay
13. Feedback disable relay
14. Pin 3 ECU
15. Canister purge control valve
16. To vacuum source

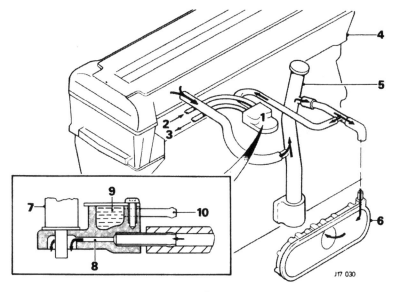

Fig. 2

1. Part throttle control orifice housing
2. From the thermostat housing
3. To the coolant header tank
4. Camshaft cover
5. Oil filler tube
6. Air cleaner
7. Cold start injector
8. Control orifice
9. Coolant
10. Water inlet/outlet

CRANKCASE CONTROL SYSTEM

To ensure that piston blow by gases do not escape from the engine crankcase, a depression is maintained in the crankcase under all operating conditions. This is achieved by scavenging from the camshaft case and crankcase via the oil sump filter tube (1, Fig. 2). The crankcase emissions collected are then fed into the engine intake manifold through a part throttle control orifice (2, Fig. 2) and the air cleaner collector box (3, Fig. 2).

To prevent possible icing up of the control orifice during cold weather the orifice is heated in a water heated housing mounted between the cold start injector and the intake manifold plenum chamber.

EVAPORATIVE EMISSION CONTROL SYSTEM

In order to accommodate up to 10% fuel expansion, the maximum fuel level is limited. This is accomplished by extending the fuel filter inlet tube into the tank connecting into the tube pipe terminating in the tank at the maximum level. When the fuel expands the tank vents through pipes leading down from the LH and RH sides of the tank to the vapour separator (6, Fig. 3) located in the RH rear screen pillar. Any liquid fuel collected in the separator can drain back into the tank. Any excess vapour is directed to the charcoal cannister (1, Fig. 3) by means of a pipe running along the underside of the vehicle.

A pressure relief valve (5, Fig. 3) in the

vapour separator to storage cannister pipe controls the flow of vapour to the cannister.

The cannister (1, Fig. 3) containing activated charcoal located in the LH front wheel arch is used to store hydrocarbon emissions from the fuel tank. Filter pads are fitted above and below the charcoal to prevent the ingress of foreign matter into the charcoal or the passage of charcoal into the purge line. Emissions from the fuel tank enter the top of the cannister and the purging air enters at the bottom of the cannister. The purging passes through the charcoal to the purge outlet at the top of the cannister.

Purging of the cannister is obtained by connecting the purge pipe to a vacuum source and drawing a controlled quantity of air through the charcoal contained in the cannister. The purge depression is obtained from ports located in the throttle housing in close proximity to the intake throttle disc (4, Fig. 3). The maximum purge of air flow rate is controlled by the size of the purge port located in the throttle housing. In order to inhibit cannister purging, a purge control valve (3, Fig. 3) is mounted in the purge line between the charcoal cannister and the purge port. The valve is operated by a

1. Charcoal canister
2. Purge air inlet
3. Purge control valve
4. Purge port
5. Pressure relief valve
6. Vapour separator

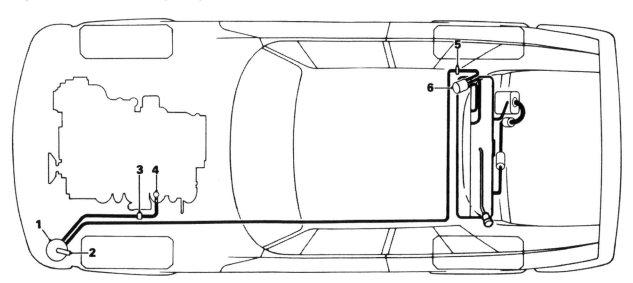

Fig. 3

thermal switch, sensing engine coolant temperature so that when the valve is energised, the purge flow is inhibited. The valve is also energised during engine cranking.

The pressure relief valve (5, Fig. 3) is used to control the transfer of vapour from the vapour separator to the storage cannister. The valve is designed to prevent flow from the tank until a pre-set pressure is exceeded. To allow flow from the cannister to the tank, a vacuum relief is also incorporated.

CLOSED LOOP

In order to make the most of the performance of three-way catalytic convertors, it is necessary to achieve very close control of the engine fuelling level. This is accomplished by using a system in which the oxygen content of the exhaust gas, prior to the entry to the catalyst, is monitored and controlled by trimming of the fuelling level.

The sensor consists of a ceramic probe protected against mechanical influences by a housing which serves for the installation of the probe. The outer part of the ceramic body is located in the exhaust gas stream, while the inner part is in contact with ambient air. The ceramic body is basically zirconium dioxide and its inner and outer surfaces are coated with a thin permeable layer of platinum which acts as an electrode. The ceramic layer becomes conductive to oxygen ions at temperatures of about 300°C (572°F) and higher. If the oxygen content inside the probe differs from that outside an electrical voltage is developed between the two surfaces. Since the oxygen content in the exhaust stream varies with air/fuel ratio, so the voltage of the sensor may be used as a measure of the air/fuel ratio. This voltage is fed to the air/fuel controller unit which continuously adjusts the fuelling to maintain the air/fuel ratio close to the stoichiometric value.

OXYGEN SENSOR SERVICE INTERVAL COUNTER

A warning light is incorporated in the fascia to alert the vehicle operator to the need to renew the oxygen sensor. The warning light is illuminated by the closing of contacts in the service interval counter (Fig. 4). The counter is driven by a small electric motor and the current to drive the motor is obtained from a pulse generator mounted in the vehicle transmission. The warning light is extinguished and the counter reset by fully depressing the reset button until a click is heard. As a bulb check function, the warning light is also illuminated when the starter motor is engaged.

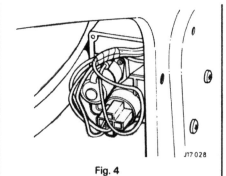

Fig. 4

CATALYTIC CONVERTORS

Catalytic convertors are fitted into the exhaust system to reduce carbon monoxide and hydrocarbon emissions. Twin outlets from the exhaust manifolds form a 'Y' piece in which is the exhaust gas oxygen sensor just above the first catalytic convertor. A second, oval, catalytic convertor is located under the vehicle floor.

The active constituents of the catalytic device are platinum and rhodium. In order for the device to function correctly, it is necessary to control very closely the oxygen concentration in the exhaust gas entering the catalyst. This is achieved by the use of a fuel control system which continuously monitors the oxygen content of the exhaust gas by means of the oxygen sensor and adjusts the fuelling level to obtain the required oxygen content.

Unleaded fuel must be used on catalyst equipped cars, and labels to indicate this are displayed on the instrument panel and below the fuel filler cap. The filler cap is designed to accommodate unleaded fuel pump nozzles only.

The emission control system fitted to this engine is designed to keep emissions within legislated limits, providing the engine is correctly maintained and is in sound mechanical condition.

AIR PUMP

Renew

Disconnect the battery earth lead.

Disconnect the electrical cable block connector from the air pump clutch.

Remove the air cleaner assembly from the air pump.

Disconnect the air outlet hose from the air pump.

Loosen the link arm trunnion bolt (1, Fig. 5), pivot bolt (2, Fig. 5) and adjusting nut (3, Fig. 5).

Loosen the air pump pivot nut and bolt (4, Fig. 5).

Pivot the air pump towards the engine until the drive belt can be removed from the air pump pulley.

Remove the link arm trunnion bolt and the air pump pivot bolt.

Remove the air pump assembly (5, Fig. 5).

Remove the clutch pulley securing nut (6, Fig. 5).

Remove the triangular plate (7, Fig. 5), clutch plate (8, Fig. 5) and spacer (9, Fig. 5).

Remove the pulley housing from the air pump shaft (10, Fig. 5).

Remove the clutch coil retaining screws and remove the coil assembly.

On refitting, ensure the drive belt is adjusted to the correct tension. A load of 2,9 kgf (6.4 lbf) must give the belt a deflection of 5,6 mm (0.22 in) when applied on the longest stretch of the belt.

AIR PUMP CLUTCH

Renew

Disconnect the battery earth lead.

Remove the nut securing the pulley/clutch assembly.

Remove the clutch plates.

Loosen the air pump pivot bolts.

Loosen the adjusting locknut and adjust the air pump towards the engine until the belt can be removed.

Remove the pulley assembly.

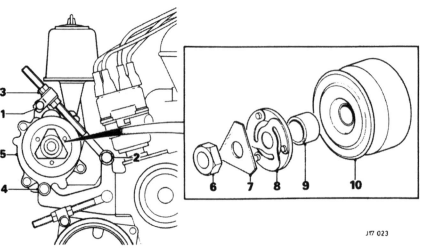

Fig. 5

Remove the trunnion securing bolt and displace the adjusting rod towards the engine.
Remove the three bolts securing the clutch coil.
Disconnect the cable harness block connector and remove the clutch coil assembly.

AIR SWITCHING VALVE

Renew

Remove the inlet and outlet air hoses (1, Fig. 6).
Remove the vacuum pipe (2, Fig. 6) and remove the air switching valve (3, Fig. 6).

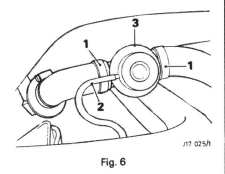

Fig. 6

EXHAUST CHECK VALVE

Renew

Remove the hose to the air switching valve from the check valve (1, Fig. 7).
Unscrew the check valve (2, Fig. 7) from the pipe connected to the manifold.

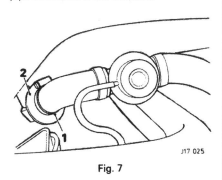

Fig. 7

SOLENOID VACUUM VALVE

Renew

Disconnect the battery earth lead.
Note the position of and disconnect the vacuum pipes (1, Fig. 8).
Disconnect the electrical cables.
Remove the screw and spire nut securing the valve to the speed control actuator bracket (2, Fig. 8).
Remove the valve from the bracket (3, Fig. 8).

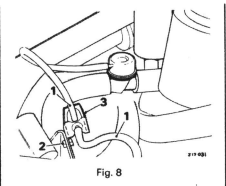

Fig. 8

EXHAUST BACK PRESSURE SWITCH

Renew

Disconnect the battery earth lead.
Remove the exhaust back pressure pipe.
Disconnect the electrical cables.
Remove the securing screw and spire from the speed control actuator bracket.
Remove the back pressure switch.

CANISTER PURGE CONTROL VALVE

Renew

Disconnect the battery earth lead.
Disconnect the electrical cable connectors (1, Fig. 9).
Remove the clip securing the purge hose to the inner wing.
Disconnect the hose from the control valve.
Disconnect and remove the valve assembly from the throttle housing hose (2, Fig. 9).

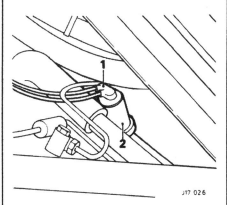

Fig. 9

THERMAL SWITCH

Renew

CAUTION: This operation must only be carried out on a cold or cool engine.

Disconnect the battery earth lead.
Carefully remove the pressure cap from the remote header tank to release any cooling system residual pressure.

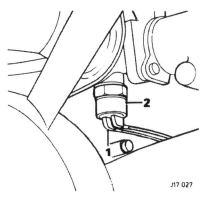

Fig. 10

Disconnect the electrical cables from the lucar connectors (1, Fig. 10).
Drain some of the coolant into a suitable container.
Unscrew and remove the thermal switch (2, Fig. 10).
On refitting, smear the thread of the thermal switch with a suitable sealant.
Refill cooling system with coolant.

EXCESS VACUUM SWITCH

Renew

Disconnect the battery earth lead.
Remove the vacuum hose from the vacuum switch (1, Fig. 11).
Disconnect the electrical cable connectors (2, Fig. 11).
Remove the vacuum switch securing nut and bolt (3, Fig. 11).
Remove the vacuum switch (4, Fig. 11).

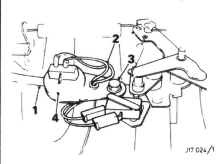

Fig. 11

THROTTLE MICRO SWITCH

Renew

Disconnect the battery earth lead.
Disconnect the electrical cable connectors from the switch (1, Fig. 12).
Remove the switch securing screws.
Remove the switch and plates (2, Fig. 12).
On refitting, connect a battery and test lamp to the switch contacts. Adjust the switch so that the light is on only when the throttle is closed.

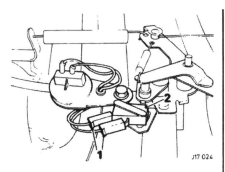

Fig. 12

ADSORPTION CANISTER

Renew

Remove the LH front road wheel.
Remove the inner wheel arch cover securing nuts and bolts.
Remove the cover securing drive fasteners and remove the cover.
Disconnect the canister hoses.
Remove the canister securing clamp nut and bolt.
Open the clamp and remove the canister.

PART THROTTLE CONTROL ORIFICE HOUSING

Renew

CAUTION: This operation must only be carried out on a cold or cool engine.

Carefully remove and refit the pressure cap from the remote header tank.
Remove the breather hose pipe from the housing.
Remove the cold start securing screws and carefully displace the cold start injector from the housing.
Displace the housing above the engine level and disconnect the cooling hoses, which also must be kept above engine level to prevent spillage of coolant.
Remove the orifice housing and discard the gaskets.

IDLE RELAY

Renew

Disconnect the battery earth lead.
Remove the two screws securing the relay cover and remove the cover.
Withdraw the relay cable block connector from the retaining bracket.
Withdraw the relay from the block connector.

LAMBDA (OXYGEN) DISABLE RELAY

Renew

Disconnect the battery earth lead.
Remove the two screws securing the relay cover and remove the cover.
Withdraw the relay cable block connector from the retaining bracket.
Withdraw the relay from the block connector.

SERVICE INTERVAL COUNTER

Renew

Disconnect the battery earth lead.
Remove the trim pad from the LH side of the boot.
Remove the nuts and screws securing the interval counter (1, Fig. 13).
Disconnect the cable harness block connector (2, Fig. 13) and the single cable connector (3, Fig. 13).
Remove the service interval counter (4, Fig. 13).

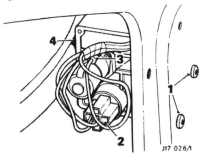

Fig. 13

LAMBDA (OXYGEN) SENSOR

Renew

Disconnect the battery earth lead.
Disconnect the oxygen sensor electrical cable connector.
Unscrew and remove the sensor from the down exhaust down pipe.
On refitting new sensor, smear the sensor with anti-seize compound.

CONTENTS

Operation	Operation No.	Page No.

TEST EQUIPMENT

1. Epitest
2. Epitest adaptor
3. Fuel pressure gauge
4. EFI Throttle pot adjustment gauge
5. EFI Feedback monitor unit
6. Infra red CO meter
7. Multi test meter
8. Idle mixture adjustment key
   Lucas No. 60730551.

GOOD PRACTICE

The following instructions must be strictly observed:

1. Always disconnect the battery before removing any components.
2. Always depressurise the fuel system before disconnecting any fuel pipes.
3. When removing fuelling components always clamp fuel pipes approximately 38 mm (1.5 in) from the unit being removed. Do not overtighten clamp.
4. Ensure that rags are available to absorb any spillage that may occur.
5. When reconnecting electrical components always ensure that good contact is made by the connector before fitting the rubber cover. Always ensure that ground connections are made on to clean bare metal, and are tightly fastened using the correct screws and washers.

WARNING

1. **DO NOT LET THE ENGINE RUN WITHOUT THE BATTERY CONNECTED.**
2. **DO NOT USE A HIGH-SPEED BATTERY CHARGER AS A STARTING AID.**
3. **WHEN USING A HIGH-SPEED CHARGER TO CHARGE THE BATTERY, THE BATTERY MUST BE DISCONNECTED FROM THE REST OF THE VEHICLE'S ELECTRICAL SYSTEM.**
4. **WHEN INSTALLING, ENSURE THAT BATTERY IS CONNECTED WITH CORRECT POLARITY.**
5. **NO BATTERY LARGER THAN 12V MAY BE USED.**
6. **TESTS OR COMPONENT REMOVAL THAT RESULTS IN FUEL VAPOUR BEING PRESENT — IT IS IMPERATIVE THAT ALL PRECAUTIONS ARE TAKEN AGAINST THE RISK OF FIRE AND EXPLOSION.**

FUEL CONTROL SYSTEM

KEY TO DIAGRAM

1. Fuel tank
2. Fuel pump
3. Fuel filter
4. Fuel pressure regulator
5. Extra air valve
6. Cold start injector
7. Injector
8. Air intake temperature sensor
9. Oxygen sensor (Lambda sensor)
10. Throttle potentiometer

11. Coolant temperature sensor
12. Manifold pressure sensor
13. Electronic control unit
14. Thermotime switch
15. Power resistors
16. HT coil
17. Starter and ignition switch
18. Fuel pump relay
19. Main relay
20. Battery

Fuel injection

The system employed is the Lucas electronically controlled, pulsed, port fuel injection system. The system is shown diagramatically on (Fig. 1).

FUEL CONTROL SYSTEM

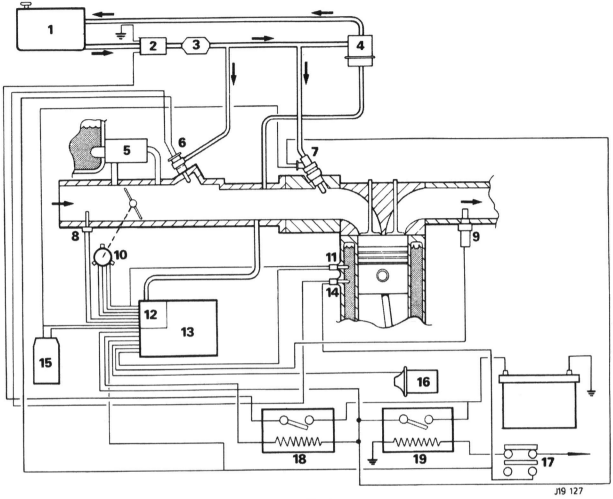

Fig. 1

J19 127

FUEL CONTROL SYSTEM WIRING DIAGRAMS

NAS

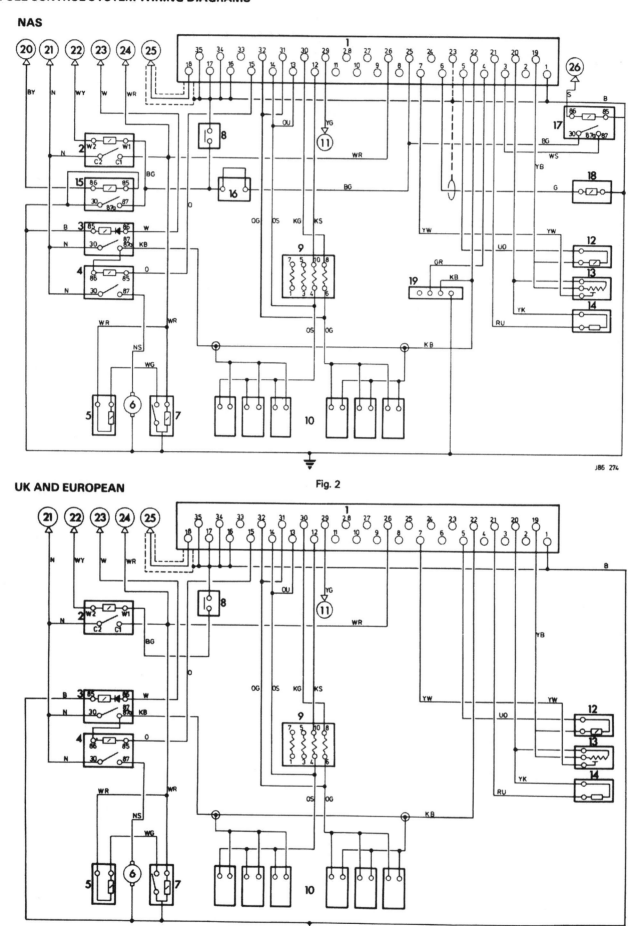

Fig. 2

J86 274

UK AND EUROPEAN

Fig. 3

J86 275

KEY TO DIAGRAMS (Figs. 2 & 3)

Main Diagram Nos

1. Electronic control unit (ECU)	293	
2. Start relay	194	
3. Main relay	312	
4. Pump relay	314	
5. Cold start injector	400	
6. Fuel pump	41	
7. Thermotime switch	298	
8. Start inhibit switch	75	
9. Power resistor	313	
10. Injectors		
11. Trip computer		
12. Coolant temperature sensor	305	
13. Throttle potentiometer	310	

14. Air temperature sensor	297	
15. Lambda disable relay		
16. Disable socket	354	
17. Idle relay		
18. Oxygen (lambda) sensor	316	
19. Feedback monitor socket	353	
20. Air injection system		
21. Positive battery supply		
22. Ignition start switch	38	
23. Inertia switch	250	
24. Starter solenoid	4	
25. Ignition amplifier	261	
26. Thermal switch		

KEY TO LOCATIONS (Fig. 4)

1. Supplementary air valve
2. Auxiliary air valve
3. Thermotime switch
4. Coolant temperature sensor
5. 3-way vacuum valve
6. Cold start injector
7. Lambda (oxygen) sensor
8. Fuel pressure regulator
9. Throttle potentiometer
10. Injector
11. Air temperature sensor

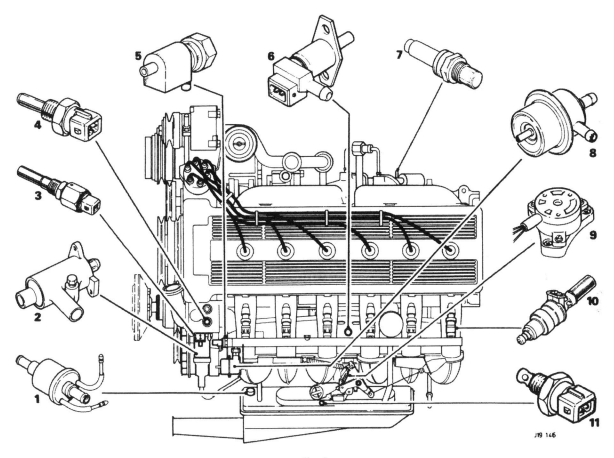

J19 146

Fig. 4

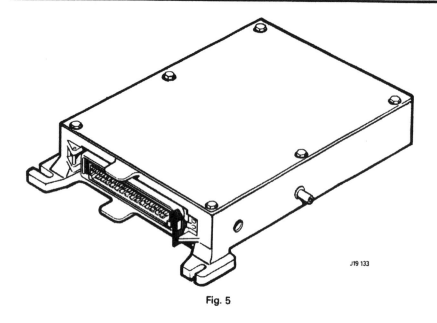

Fig. 5

J19 133

temperature sensor (Fig. 7) is incorporated in the air intake system. The electrical resistance of the device alters with the changes of air temperature. This signal is monitored by the ECU and used to ensure the air/fuel ratio stays correct.

BATTERY VOLTAGE

The electrical system voltages change with the battery state of charge, electrical load etc., and this in turn affects the amount of fuel delivered. The ECU compensates for this by constantly monitoring the system voltage and generating a correction factor to ensure the pulse sent to the injectors is independent of the system voltage.

COOLANT TEMPERATURE SENSOR

The coolant temperature sensor (Fig. 8) located in the coolant thermostat housing controls the warm-up enrichment delivered by the ECU. In addition to the enrichment generated during warm-up, further enrichment is required during engine cranking. The signal to the ECU comes via a connection from the starter relay. When the engine fires and the starter is released, the start enrichment decays slowly to the normal fuelling level as determined by the engine coolant temperature. This over fuelling function is known as the 'after start enrichment'. The 'normal' fuelling level is determined by the engine coolant temperature, such that the fuelling level increases as the temperature decreases.

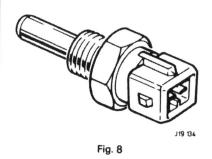

Fig. 8

J19 134

THROTTLE POTENTIOMETER

To ensure the vehicle road performance is satisfactory, with good throttle response, acceleration enrichment is necessary. Signals are provided by the throttle potentiometer (Fig. 9) which is mounted on the throttle spindle and indicates the throttle position to the ECU. When the throttle is opened the fuelling is richened and when closed the fuelling is weakened. When the throttle is opened very quickly, all the injectors are simultaneously energised for one pulse. This ensures that there is enough fuel available at the inlet ports for the air admitted by the sudden opening of the throttle. The duration of this extra pulse is controlled by the engine temperature signal, and is longer with a cold engine.

FUEL METERING

Fuel metering is obtained by controlling the length of time for which the injectors are held open during each engine cycle. The pulse duration is varied by the electronic control unit (ECU) (Fig. 5) according to inputs from the engine and chassis mounted sensors.

The control parameters sensed fall into two groups. The primary consists of intake mainfold absolute pressure and engine speed.

Information on engine speed received by the ECU is derived from the ignition coil negative terminal. The voltage wave form at this point can reach as much as 400 volts when a spark is generated, and it is desirable to suppress this voltage before allowing it into the fuel injection wiring harness and hence to the ECU. If this is not done, interference may occur with other signals. A resistor serves this purpose and it is located inside the ignition amplifier (Fig. 6).

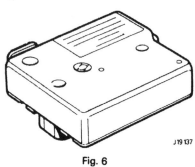

Fig. 6

J19 137

The intake manifold absolute pressure is sensed by a transducer located in the ECU and linked to the intake manifold by a pipe. The transducer is an integral part of the ECU and cannot be replaced without recalibrating the electronic circuits.

The fuelling information for the engine is stored in a preprogrammed memory so that for any combination of manifold absolute pressure and engine speed the memory gives out a number which is proportional to the amount of fuel required by the engine. The injector will be energised for a time proportional to the number computed, plus the constant of proportionality which is varied according to the secondary control parameters.

The secondary control parameters consist of engine coolant temperature, inlet air temperature, throttle movement, and position, closed loop correction and battery voltage.

AIR TEMPERATURE SENSOR

The quantity of fuel supplied per cycle is adjusted according to the air intake temperature, such that it varies approximately with the density of the air aspirated. At low ambient temperatures, the density, and thus the weight of air drawn into the engine is higher, therefore the air/fuel ratio would become leaner. In order to compensate for this, an air intake

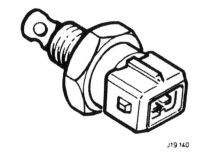

Fig. 7

J19 140

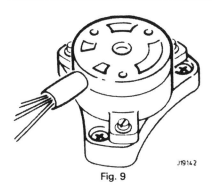

Fig. 9

Lengthening the normal injection pulses is done in proportion to the rate at which the throttle is moved, and it takes a short time to decay when the throttle movement stops. Enrichment in this way is also varied according to the engine temperature.

The fuel cut-off function is controlled by the throttle potentiometer and the conditions under which it occurs are programmed into the electronic control unit memory.

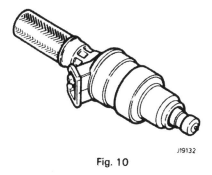

Fig. 10

INJECTORS

The injectors (Fig. 10) consist of a solenoid operated valve. The moveable plunger is rigidly attached to the nozzle needle. In the closed position a helical compression spring holds the nozzle against the valve seat. The electrical pulses from the electronic control unit are passed through the injector solenoid winding creating a magnetic field. As a result the plunger is attracted away from the nozzle seat allowing fuel to enter the inlet port. The injectors are operated in two stages, initially they are operated via a pull-in circuit with current limiting resistors, then when the injectors are open a change to hold on circuit is made via current limiting resistors (Fig. 11) for the remainder of the injector period as determined by the electronic control unit. In this way the heating effect on the output transistors of the electronic control unit is reduced. It also ensures a rapid response from the injectors. To open the injectors at the speeds required by the engine fairly high current is needed. The ECU has an output stage to deliver this current, but to protect the output transistors of the ECU from injector faults and short-circuits a power resistor is wired in series with each three injectors. These resistors will limit fault current to a safe value, thus protecting the ECU. The power resistors (one for each group of injectors) are housed in a single unit secured to the right side of the engine valance by two screws.

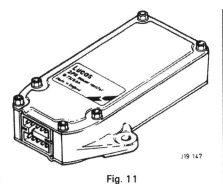

Fig. 11

AUXILIARY AIR VALVE

In order to maintain idling speed during cold start, and warm-up, more air than that supplied by the normal idle bypass is required. This extra air is supplied by the extra air valve (Fig. 12) which is temperature sensitive so that the idle speed can be controlled throughout the warm-up phase. The extra air valve consists of a variable orifice, controlled by an expansion element. It is mounted on the engine cylinder head, where it is responsive to coolant temperatures. By adjusting the profile of the variable orifice according to coolant temperature the engine idling speed can be controlled

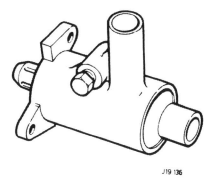

Fig. 12

SUPPLEMENTARY AIR VALVE

An additional throttle bypass device is used to supplement the air necessary to support the engine idle speed when the air conditioning system is in use. This device is a solenoid operated valve (Fig. 13) which supplies supplementary air to the engine.

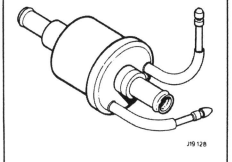

Fig. 13

LAMBDA SENSOR — CARS TO USA

Oxygen sensor

The oxygen sensors (Fig. 14) measure the free oxygen concentration in the exhaust system. Excessive free oxygen over a certain proportion indicates a weak mixture, whereas insufficient free oxygen indicates a rich mixture. A signal is fed to the ECU to compensate for these variations by revising the applied injector pulse width.

A service interval counter in conjunction with a warning lamp indicate when the oxygen sensors must be replaced.

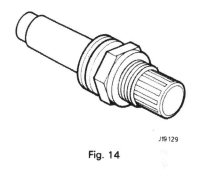

Fig. 14

FUEL SUPPLY SYSTEM

A recirculatory fuel system is used (Fig. 15). The fuel tank (7, Fig. 15) is mounted across the car behind the rear passenger seats. The electric pump (5, Fig. 15) draws fuel from a small sealed sump tank (6, Fig. 15) located at a lower level than the main tank and delivers it to a pressure regulator on the engine (3, Fig. 15). The pressure regulator controls the fuel line pressure so that a constant pressure drop is maintained across the injectors (1 & 2, Fig. 15). This is accomplished by applying the intake manifold pressure to the diaphragm of the pressure regulator. Excess fuel is returned to the main fuel tank. The fuel pump is energised only when the electronic control unit senses the engine cranking signal from the starter relay, or the ignition pulses from the ignition amplifier. This will prevent the engine being flooded with fuel in the event of an injector being held open by a foreign body or control unit fault. The fuel pump is protected by a nylon strainer fitted to the suction line inside the fuel tank. A paper element disposable filter (4, Fig. 15) is fitted in the line between the pump and the pressure regulator to protect the pressure regulator and injectors.

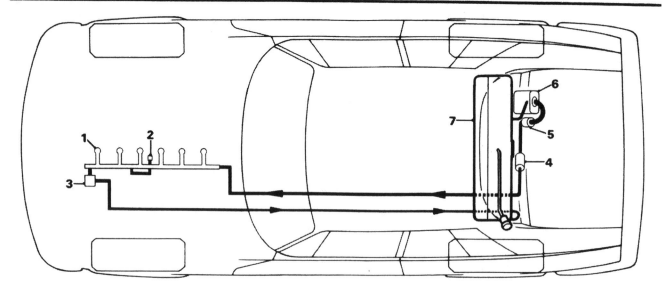

J19 126

Fig. 15

KEY TO FUEL SYSTEM CONFIGURATION
1. Injectors
2. Cold start injector
3. Fuel pressure regulator
4. Fuel filter
5. Fuel pump
6. Sump tank
7. Fuel tank

FUEL PUMP

The fuel pump (Fig. 16) has a roller type cell driven by a permanent magnet electric motor. A small rotor is mounted on the armature shaft and rotates in an eccentric housing. The rotor contains five metal rollers each one housed in a pocket or cutaway on the circumference of the rotor. When the armature rotors centrifugal forces them against the metal housing forming a seal. The trapped fuel is then forced to the pressure scale of the system. The pump is designed as a wet pump which means the motor is full of fuel. However, the motor never contains an ignitable mixture even when the tank empties.

For safety reasons the fuel pump is energised via a relay whose earth connection is complete through the electronic control unit. This ensures that when the ignition is switched on a timing circuit located within the electronic control unit will only allow the pump to operate for approximately one second and then switch the relay earth connection off, thus switching the pump off. The fuel pump relay

will remain switched off until the engine is cranked and a signal is received by the electronic control unit from the starter relay, or when the engine is running and the electronic control unit receives a speed signal from the ignition amplifier. Also, as an additional safety measure an inertia switch is fitted to de-energise the pump and the entire fuel system.

FUEL PRESSURE REGULATOR

The pressure regulator (Fig. 16) is connected to the fuel rail and has an overflow return to the fuel tank. There is also a vacuum pipe connected between the regulator and the inlet manifold via a three-way vacuum valve.

Inside the steel body of the regulator a spring loaded diaphragm is held against the outlet ducts, sealing it off. When the fuel pressure acting on the diaphragm is sufficient to overcome the spring pressure, the diaphragm distorts compressing the spring and allowing the excess fuel to escape through the outlet duct, which is now exposed, causing a reduction in pressure. The reduced fuel pressure allows the diaphragm to return to its original position closing the fuel return outlet. This sequence is repeated as long as the pump is

operating. The continuous movement of the diaphragm which rapidly opens and closes the fuel return valve maintains the fuel rail pressure at 3.1 Bar (43.5 lbf in^2).

It is necessary that the fuel pressure is maintained at a constant amount above manifold pressure which varies with the engine load and it is for this reason that the spring side of the diaphragm is connected to the manifold to keep the pressure constant across the injectors.

The three-way vacuum valve is fitted to the fuel rail to prevent weak fueling after a hot start.

COLD START SYSTEM

In addition to the functions of the Electronic Control Unit which provides extra fuel during starting and the warm-up period, additional fuel is injected into the inlet manifold by a cold start injector (Fig. 18).

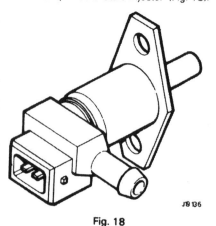

J19 136

Fig. 18

This injector is controlled by a thermotime switch (Fig. 19). The thermotime switch senses the coolant temperature and depending on that temperature makes or breaks the earth circuit of the cold start injector. When the starter switch is operated the cold start injector is energised with its earth circuit completed via the thermotime

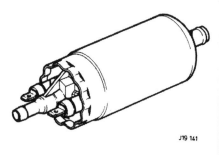

J19 141

Fig. 16

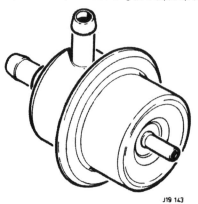

J19 143

Fig. 17

switch, which also limits the time for which the injector is energised to a maximum of 12 seconds under extreme cold conditions.

If the temperature is above the rated value of 35°C (95°F) of the thermotime switch the earth circuit of the cold start injector will be open and the cold start injector will not operate as no start enrichment will be required.

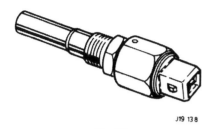

Fig. 19

IGNITION SYSTEM

The ignition system is operated electronically and is known as the Constant Energy System. In this system the distributor (Fig. 21) incorporates a standard automatic advance system, rotor arm, cover, anti-flash shield, reluctor and pick-up assembly. The reluctor is a gear-like component and is mounted on the distributor shaft in the place of the cam. The pick-up consists of a winding around a pole piece attached to a permanent magnet. The pick-up is prewired with two leads terminating in a moulded two-pin block connector.

When the reluctor passes across the pick-up limb the magnetic field strength is intensified creating a voltage in the winding. This voltage signal created by the reluctor and the pick-up assembly is sensed by the amplifier causing the amplifier to switch the current flowing in the primary winding of the HT coil on and off.

The constant energy electronic ignition system employs output current limiting and variable dwell for optimum performance. A long dwell is provided at high speeds for adequate energy storage in the coil and a dwell is provided at low speeds for minimum power dissipation. The output current limiting function of the amplifier maintains the storage energy for spark, and the system open circuit voltage constant over a wide engine speed range. It eliminates the need for ballast resistor whilst ensuring correct current flows at all times even when the engine is cranking. No current flows through the HT coil when the ignition is switched on and the engine is stationary.

The amplifier assembly (Fig. 22) consists of a solid state electronic amplifier module, a zenor diode to protect the amplifier in the event of a current surge, a suppression capacitor, and a moulding containing two resistors.

WARNING: THE AMPLIFIER IS A SEALED UNIT CONTAINING BERYLIA. THIS SUBSTANCE IS EXTREMELY DANGEROUS IF HANDLED. DO NOT ATTEMPT TO OPEN THE AMPLIFIER MODULE.

COLD START CIRCUIT

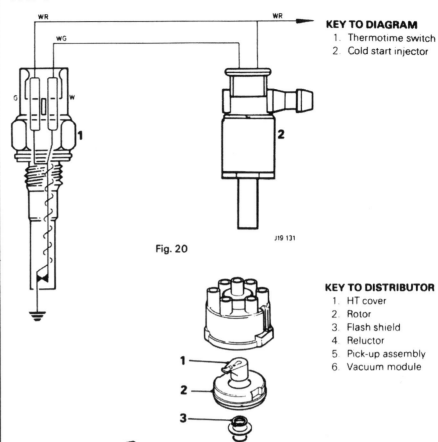

Fig. 20

KEY TO DIAGRAM
1. Thermotime switch
2. Cold start injector

KEY TO DISTRIBUTOR
1. HT cover
2. Rotor
3. Flash shield
4. Reluctor
5. Pick-up assembly
6. Vacuum module

Fig. 21

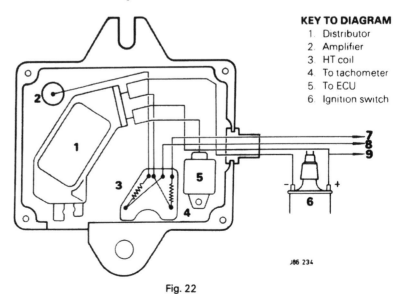

KEY TO DIAGRAM
1. Distributor
2. Amplifier
3. HT coil
4. To tachometer
5. To ECU
6. Ignition switch

Fig. 22

111

RELAYS

Four relays are used in the electrical control system. They are the main, fuel pump, idle speed relay, and the Lambda disable relay. Details of the wiring connections are shown in the circuit diagram.

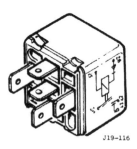

Fig. 23

TEST EQUIPMENT

The Epitest (Fig. 24) and the Epitest Adaptor have been developed to enable quick location of faults to be carried out on the vehicle. It is supplied complete with the necessary multiplugs, fuel pressure gauge and the operating instructions.

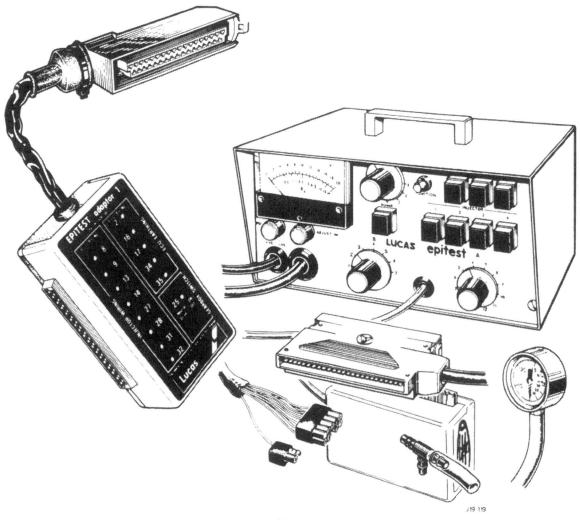

Fig. 24

FAULT FINDING

SYMPTOM	CAUSE	CURE
Engine will not start	1. Low battery or poor connections	1. a) Check battery, recharge. Clean and secure terminals. b) Check for low charge from alternator. c) Check for short circuit.
	2. Start system malfunction	2. Clean and check main starter circuit and connections.
	3. Incorrect or dirty fuel	3. Check grade of fuel. If contamination suspected, drain and flush fuel tank, flush through system, renew fuel filter.
	4. Fuel starvation	4. Check fuel pressure. If not satisfactory, check feed pipes for leaks or blockage. Renew connectors if damaged or deteriorated.
	5. Fuel injection equipment electrical connections	5. Ensure that all connector plugs are securely attached. Pull back rubber boot and ensure that plug is fully home. While replacing boot, press cable towards socket. Ensure ECU multi-pin connector is fully made. Check that all ground connections are clean and tight.
	6. Auxiliary air valve inoperative	6. Remove valve and test.
	7. Cold start system inoperative	7. Check function of cold start system.
	8. Pressure sensor (part of the ECU)	8. Ensure manifold pressure pipe is attached to sensor, and is not twisted, kinked or disconnected anywhere.
	9. Temperature sensors	9. Check sensor for open and short circuit.
	10. HT circuit faults	10. Check for sparking. See electrical book 10.
	11. Power faults	11. Carry out ignition checks. See electrical book 10.
	12. LT switching fault	12. Check pick-up module. See electrical book 10.
	13. Ignition timing incorrect	13. Check and adjust as necessary.
	14. ECU/amplifier	14. As a last resort check by substitution.
Poor or erratic idle	15. Check items 3, 4, 5 and 12 above	15. If trouble persists, proceed with item 18.
	16. Throttle potentiometer	16. Check function of idle and full load switches.
	17. Incorrect idle speed	17. Adjust auxiliary air valve by-pass bleed screw.
	18. Check items 8 and 12 above	18. If trouble still persists, proceed with item 21.
	19. Ignition system deterioration	19. Check ignition wiring for fraying, chafing and security. Inspect distributor cap for cracks and tracking and rotor condition. See electrical book 10.

SYMPTOM	CAUSE	CURE
Poor or erratic idle	20. Spark plug faults	20. Clean, reset and test plugs; renew as necessary.
	21. Check item 14	21. If trouble still persists, proceed with item 24.
	22. Vacuum system faults	22. Check operation of vacuum unit and condition of vacuum pipes. Renew as necessary.
	23. Advance or retard mechanism faults	23. Check operation of advance/retard mechanism. Lubricate or renew as necessary.
	24. Throttle by-pass valves	24. Check and adjust as necessary.
	25. Exhaust system leaking or blocked	25. Check and rectify as necessary.
	26. Incorrect idle mixture	26. Check CO level. Remove the blanking plug from ECU and with special tool No. 60730551 adjust to 1 to 2% max CO at 750 rpm level. Air injection system (where fitted) should be disconnected for this operation.
	27. Poor compressions	27. Check compressions, and rectify as necessary.
	28. Air leaks at inlet manifold	28. Check inlet manifold to cylinder head joint. Remake with new gasket if necessary. Check manifold tappings for leaks.
	29. Check item 6	29. If trouble still persists, proceed with item 32.
	30. Engine oil filler cap loose or leaking	30. Check cap for security. Renew seal if damaged.
	31. Engine breather pipe restrictors missing or blocked	31. Check and clear, or renew as necessary.
	32. Engine breather hoses blocked or leaking	32. Check and clear, or renew as necessary.
	33. Charcoal canister restricted or blocked	33. Inspect, and renew as necessary.
	34. Check items 15, 10 and 9	34. Check in order shown.
Hesitation or flat spot	35. Check items 3, 4 and 5	35. If trouble still persists, proceed with item 38.
	36. Check item 7 with engine cold	36. If trouble still persists, proceed with item 39.
	37. Throttle butterfly	37. Adjust as necessary.
	38. Check item 8	38. If trouble still persists, proceed with item 41.
	39. Brakes	39. Check for binding brakes.
	40. Check items, 12, 19, 20, 13, 22, and 23	40. If trouble still persists, proceed with item 43.
	41. Air cleaner blocked	41. Inspect element, and renew as necessary.
	42. Check items 25, 27, 28, 30, 31, 32, 33, 9 and 14.	42. Check in the order shown.

SYMPTOM	CAUSE	CURE
Excessive fuel consumption	43. Leaking fuel	43. Check fuel system for leaks, rectify and renew connectors as necessary.
	44. Check items 18, 7, 8, 41, 27, 29 and 30.	44. If trouble still persists, proceed with item 47.
	45. Cylinder head gasket leaking	45. Check cylinder head to block joint for signs of leakage. Renew gasket as necessary.
	46. Cooling system blocked or leaking	46. Flush system, check for blockage. Check hoses and connections for security and leaks; renew as necessary. Check functions of thermostats; renew if necessary.
	47. Check items 28, 30, 31, 32, 33, and 14	47. Check in the order shown.
Lack of engine braking or high idle speed	48. Air leaks	48. Any air leak into the manifold will appear as an equivalent throttle opening; correct fuel will then be supplied for that apparent degree of throttle and the engine will run faster. Ensure that all hose and pipe connections are secure. Check all joints for leakage, and remake as necessary.
	49. Throttle sticking	49. Lubricate, check for wear and reset.
	50. Check items 6, 22, 24, 13, 26 and 39	50. If trouble still persists, proceed with item 53.
	51. Throttle spindle leaks	51. Check seals, bearings and spindles for wear; renew as necessary.
	52. Check item 28	52.
Lack of engine power	53. Check items 3, 4, 5 and 7	53. If trouble still persists, proceed with item 56.
	54. Throttle inhibited	54. Check throttle operation, free off and reset as necessary.
	56. Check items 39, 41, 8, 12, 19, 20, 13, 23, 25, 27, 28, 22, 30, 31, 32, 33, 37 and 14	56. Check in order shown.
Engine overheating	57. Check items 46, 45, 13 and 22	57. Check in order shown.
Engine cuts out or stalls	58. Check items 3, 4, 5, 9, 17, 7, 43, 8, 25, 12, 19, 20, 13, 23, 26, 16, 28, 22, 30, 31, 32, 33, 27 and 14	58. Check in order shown.
Engine misfires	59. Check items 3, 4, 5, 7, 8, 12, 19, 20, 13, 23, 41, 25, 26, 27, 28, 22, 30, 31, 32 and 33	59. Check in order shown.

SYMPTOM	CAUSE	CURE
Fuel smells	60. Check items 43, 7 and 32	60. If trouble still persists, proceed with item 61.
	61. Fuel filler cap defective	61. Check seal and cap for deterioration; renew as necessary.
	62. Check items 31, 33, 26, 41 and 14	62. Check in the order shown.
Engine runs on	63. Check items 3, 17, 49, 24, 32, 31, 46, 45, 13, 24, 25 and 15	63. Check in the order shown.
Engine knocking or pinking	64. Check items 3, 13, 23, 22, 46 and 45.	64. Check in the order shown.
Arcing at plugs	65. Check items 19 and 20	65.
Lean running (low CO)	66. Check items 5, 51, 16, 9, 3, 4, 28, 22, 30, 31, 32 and 33	66. Check in the order shown.
Rich running (excess CO)	67. Check items 7, 26, 33 and 14	67. Check in the order shown.
Backfiring in exhaust	68. Check items 3, 4, 5, 41, 25, 28, 22, 46 and 30	65. If trouble persists, proceed with item 71.
	69. Check item 14	69.
Noisy air injection	70. Incorrectly tensioned air pump drive belt	70. Check and adjust drive belt tension; renew belt if necessary.
	71. Check valve faulty or low pump pressure	71. Check that the valve operates. If pump fails to produce enough pressure to lift the valve, check item 70. If satisfactory, renew the pump.
	72. Check valve sticking (Japan and Australia)	72. Check valve operation and hoses for security or blockage. Rectify or renew as necessary.

AIR CLEANER ASSEMBLY

Renew

Disconnect the battery earth lead.
Unclip the air cleaner cover and displace the cover from the element (1, Fig. 25).
Remove the air cleaner element (2, Fig. 25) and the cover.
Disconnect the air temperature sensor lead (3, Fig. 25).
Disconnect the air cleaner to oil filler hose (3, Fig. 25) from the air cleaner.
Remove the four bolts securing the air cleaner to the throttle butterfly housing and displace the air cleaner.
Remove the air cleaner from the air solenoid valve and remove the air cleaner.
Remove the air temperature sensor.

Remove the ten bolts securing the backplate to the housing and remove the back plate.
On refitting, ensure all mating surfaces are clean and smeared with arbrosil sealant.

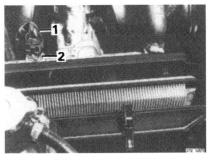

Fig. 25

THROTTLE PEDESTAL HOUSING

Renew

Disconnect the battery earth lead.
Release the throttle cable securing clip and disconnect the throttle cable from the linkage (1, Fig. 26).
Disconnect the throttle linkage return spring (2, Fig. 26).

Fig. 26

Remove the throttle linkage retaining spire clip and remove the linkage assembly from the pedestal (3, Fig. 26). Remove the nylon guide from the pedestal.
Remove the pedestal assembly securing bolts (4, Fig. 26) and displace the pedestal assembly.
Remove the pedestal housing securing bolts and remove the pedestal housing.

THROTTLE CABLE (NAS)

Renew

Disconnect the battery earth lead.
Remove the throttle pedal link retaining split pin, pedal link washer and the pedal link clevis pin.
Remove the nuts and bolts securing the brake reservoir and displace the reservoir for access.
Displace the speed control actuator cable bracket.
Remove the throttle transfer plate securing nuts and bolts, displace the plate for access and discard the gasket.
Remove the split pin and washer securing the cable.
Displace the cable from the transfer anchor pipe.
Remove the cable from the transfer plate.
Release the cable retaining clip and disconnect the cable from the throttle linkage.
Loosen the cable adjusting nuts and remove.
Remove the cable support bracket from the grommet and remove the cable assembly.

THROTTLE CABLE (UK AND EUROPEAN)

Renew

Release the throttle cable retaining clip and disconnect the cable from the throttle linkage.
Remove the cable adjusting nut and displace the cable from the support bracket.
Disconnect the cable from the bulkhead retaining clip.
Remove the cotter pin and the cable retaining collar from the throttle pedal.
Slide the collar forwards to expose the cable end and disconnect the cable from the pedal arm.
Remove the cable from the engine bay.

THROTTLE LINKAGE

Renew

Release the throttle cable securing clip and disconnect the throttle cable from the linkage.
Disconnect the throttle linkage return spring.
Remove the throttle linkage retaining spire clip and remove the linkage assembly from the pedestal.

AUXILIARY AIR VALVE

Renew

CAUTION: The auxiliary air valve must ONLY be removed with a cold or cool engine.

Carefully remove the pressure cap from the remote header tank to release any cooling system residual pressure.
Drain the coolant into a suitable container.
Loosen the auxiliary air valve hose clips and disconnect the hoses from the air valve (1, Fig. 27).
Remove the bolts securing the air valve to the thermostat housing (2, Fig. 27) and remove the auxiliary air valve (3, Fig. 27).
On refitting, clean the thermostat housing and fit new gasket.
Refill with coolant.
Start and run the engine until the normal operating temperature is attained.
Adjust the idle speed adjustment screw on the auxiliary air valve (4, Fig. 27) to give the required idle speed.

Fig. 27

OXYGEN SENSOR (LAMBDA SENSOR)

Renew

Disconnect the battery earth lead.
Disconnect the oxygen sensor cable connector and remove the sensor.
On refitting new sensor, smear the sensor threads with anti-seize compound.

COOLANT TEMPERATURE SENSOR

Test

Disconnect the coolant temperature sensor.
Check the temperature of the coolant.
Connect an ohmmeter between the terminals of the coolant temperature sensor. The ohmmeter should closely approximate to the relevant resistance value given in the table.
Disconnect the ohmmeter.

Temperature (C) Resistance K ohms

−10 ± 1	7	— 11.6
+20 ± 1	2.1	— 2.9
+80 ± 1	0.27	— 0.39

Connect the ohmmeter to the body of the sensor and each terminal in turn. The ohmmeter should register infinity (open circuit). If the ohmmeter indicates a resistance, the sensor is breaking down to earth and should be replaced.

Renew

CAUTION: This operation must ONLY be carried out on a cold or cool engine.

Disconnect the battery earth lead.
Carefully remove the pressure cap from the remote header tank to release any cooling system residual pressure.
Disconnect the coolant temperature sensor electrical connector. (1, Fig. 28).

NOTE: The replacement component should be ready so that the transfer can be made as quickly as possible. Ensure that the sealing washer is located on the replacement sensor and a coat of a suitable sealing compound is smeared on the threads. Unscrew the temperature sensor from the thermostat housing (2, Fig. 28).

On refitting, check the coolant level and top up if necessary.

Fig. 28

THERMOTIME SWITCH

Test

NOTE: Check the coolant temperature with a thermometer and note the reading before carrying out the tests detailed below. Check the rated value of the thermotime switch (stamped on the body).

Disconnect the battery earth lead.
Disconnect the electrical connector from the thermotime switch.
'A' coolant temperature higher than the switch rated value.
Connect the ohmmeter between the terminal 'W' and earth (Fig. 29). The ohmmeter should indicate an open circuit reading. Replace the switch if a low reading (short circuit) is obtained.

'B' coolant temperature lower than the switch rated value.
Connect an ohmmeter between terminal 'W' and earth. A very low resistance reading (closed circuit) should be indicated.
Connect a 12 V supply to the terminal 'G' and earth (Fig. 30) of the thermotime switch.

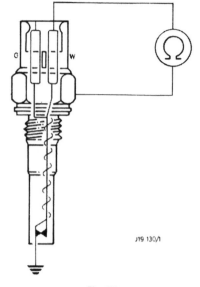

Fig. 29

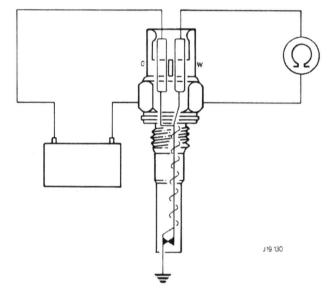

Fig. 30

Using a stop watch, check the time delay of the reading indicated on the ohmmeter change from a low resistance to a high resistance.
The delay period should closely follow the times given in the table:

Coolant Temperature	Delay Period
−20°C	12 seconds
0°C	8 seconds
+35°C	0 seconds

Renew

CAUTION: This operation must ONLY be carried out on a cold or cool engine.

Disconnect the battery earth lead.
Carefully remove the pressure cap from the remote header tank to release any residual pressure in the cooling system.

Fig. 31

Ensure that the new component threads are smeared with a suitable sealing compound and that the sealing washer is fitted correctly.
Disconnect the electrical connector from the thermotime switch (1, Fig 31) and remove the switch (2, Fig 31).
On refitting, check the coolant level and top up if necessary.

AIR TEMPERATURE SENSOR

Test

Disconnect the battery earth lead.
Disconnect the cable connector from the sensor.
Connect a suitable ohmmeter between the terminals of the sensor. Note the resistance indicated on the ohmmeter. The resistance of the sensor is subject to change according to the temperature, and should closely follow the relevant resistance value given in the table below.

Temperature	Resistance K ohms
− 10°C	8.26 — 10.56
+ 20°C	2.28 — 2.72
+ 50°C	0.76 — 0.91

Renew

Disconnect the battery earth lead.
Disconnect the sensor cable connector.
Remove the air temperature sensor from the air cleaner housing.

THROTTLE POTENTIOMETER

Renew

Disconnect the battery earth lead.
Remove the air cleaner element.
Remove the throttle potentiometer securing screws.
Disconnect the multi-pin cable connector.
Remove the throttle potentiometer.
On refitting, do not tighten the securing screws.

Adjust

Connect the throttle potentiometer adjustment gauge to the potentiometer multiplug (1, Fig. 32).
Connect the adjustment gauge crocodile clips to a 12 volt supply and earth (2, Fig. 32).
Move the toggle switch on top of the adjustment gauge to position 'T' (3, Fig. 32).
Adjust by rotating the potentiometer to the right or left until the CORRECT is illuminated (4, Fig. 32).
Tighten the fixing screws.
Refit the air cleaner element.

Test

Remove the air cleaner element.
Disconnect the throttle potentiometer multiplug connector.
Connect an ohmmeter between the yellow and the green leads in the throttle potentiometer multiplug.
The ohmmeter should indicate approximately 5K ohms. Operate the throttle control and the resistance should remain constant.

Remove the ohmmeter leads from the multiplug and reconnect the ohmmeter between the red lead and the green lead of the multiplug. The ohmmeter should indicate approximately 5K ohms.
When the throttle is opened, the resistance should drop to approximately 125 ohms.

PRESSURE REGULATOR

Renew

Remove the luggage compartment RH side panel.
Remove the pump relay from its bracket and remove the relay from the multiplug connector.
Crank the engine to depressurise the fuel system.
Remove the vacuum pipe from pressure regulator (1, Fig. 33).
Clamp the fuel hose and remove the hose from the regulator (2, Fig. 33).
Slacken the pressure regulator securing nut (3, Fig. 33).
Undo the pressure regulator to fuel rail union nut and remove the regulator (4, Fig. 33).

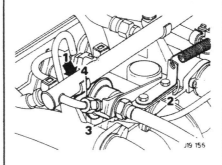

Fig. 33

FUEL INJECTORS
FUEL RAIL

Renew

Depressurise the fuel system.
Disconnect the battery earth lead.
Clamp and disconnect the fuel hose from the pressure regulator.
Clamp and disconnect the fuel hose from the fuel rail.
Disconnect the vacuum pipe from the pressure regulator.
Disconnect the injector electrical block connectors.
Remove the fuel rail securing bolts and lift the fuel rail complete with the injectors from the engine compartment.
Remove the injector retaining clip and remove the injector.

Test

Disconnect the injector electrical block connector.
Connect an ohmmeter between the injector terminals and the ohmmeter should indicate a reading of approximately 3 ohms.
Connect the ohmmeter to each of the injector terminals in turn and to the body of the injector. The ohmmeter should indicate an open circuit.

COLD START INJECTOR

Test

WARNING: THIS TEST RESULTS IN FUEL VAPOUR BEING PRESENT IN THE ENGINE COMPARTMENT. IT IS THEREFORE IMPERATIVE THAT ALL DUE PRECAUTIONS ARE TAKEN AGAINST FIRE AND EXPLOSION.

Remove the bolts securing the cold start injector and place the injector over a suitable container to collect sprayed fuel.
Disconnect the block connector from the thermotime switch.
Connect a jumper lead from the white and green lead in the block connector to earth.
Disconnect the leads from the negative terminal of the ignition coil.
Crank the engine one or two revolutions. The injector should spray while the engine cranks.
If the injector does not spray fuel, check the white and red lead to the start relay. If satisfactory, check the white and green lead to the thermotime switch block connector; if satisfactory, replace the injector.

POWER RESISTORS

Renew

Disconnect the battery earth lead.
Remove the bolts securing the power resistor.

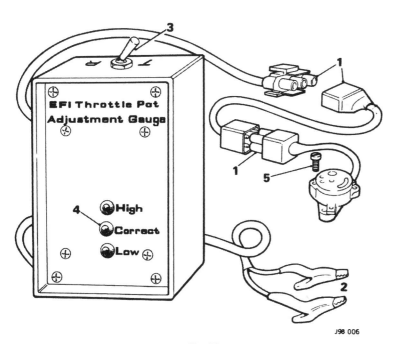

Fig. 32

Remove the multi-pin block connector from beneath the resistor and remove the resistor.

On refitting, ensure the connector is firmly connected.

Test

Connect an ohmmeter between terminals 10 and 4 of the power resistor. The ohmmeter should indicate a reading of approximately 6 ohms. The same reading should be indicated with the ohmmeter connected between terminals 8 and 6 of the power resistor.

NOTE: The power resistors are fitted to protect the ECU against the high current required to operate the injectors. The resistors will limit the current to a safe value, thus protecting the ECU during the pull-in stage. It is most important to check the resistors before renewing the ECU.

ELECTRICAL CONTROL UNIT (ECU)

Renew

Disconnect the battery earth lead.
Remove the RH luggage compartment trim pad.
Withdraw the harness plug from the ECU.
Remove the two securing bolts and washers.
Withdraw the ECU clear of the mounting bracket.
Disconnect the vacuum signal hose and remove the ECU.

Refitting

Ensure the connectors are secure and good contacts are made.
Check the idle CO level using a suitable infra-red CO meter.
When using the CO meter, do not use an exhaust extractor fan.

Fig. 34

OXYGEN (LAMBDA) SENSOR

Test

Check the ignition timing and idle speed.
Remove the blanking plug from the ECU to expose the idling fuel setting adjuster.
Run the engine until it reaches operating temperature (at least 8 minutes from cold or 2 minutes if hot).

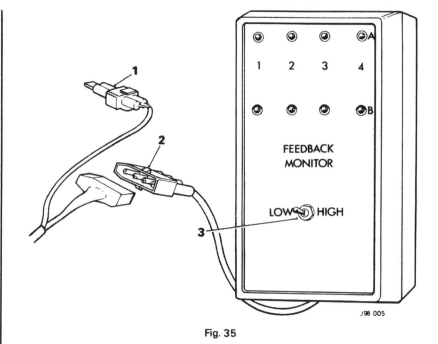

Fig. 35

Disconnect the Lambda diable plug from the harness socket (1, Fig. 35), otherwise the Lambda sensors will not function whilst in neutral or park.
Connect EFI Feedback monitor to the Fuel Setting Diagnostic socket (2, Fig. 35).
Position switch to LOW (3, Fig. 35).
Slowly turn the idling Fuel Setting Adjuster in the ECU with Special Tool No. 607 30551 until LAMP 2 in ROW A or B is lit and LAMP 2 or 3 in other ROW is lit.
Fit a new blanking plug to cover the idling fuel setting adjuster.
Disconnect the pressure regulator vacuum pipe and temporarily seal off the vacuum take off on the manifold pipe.
The Feedback Monitor unit indicators should move towards 'RICH' i.e. LAMP 2 to LAMP 1 and LAMP 3 to LAMP 2 or 1
If the indicators do not change, Lambda sensors and associated circuit are suspect.
Reconnect the pressure regulator vacuum pipe.

MAIN RELAY

Description

The main relay is mounted on the RH side of the luggage compartment.
The relay coil is energised when the ignition is switched on, which when operated closes the relay contacts and supplies battery voltage to the ECU and the pump relay.

PUMP RELAY

Description

The pump relay is mounted in the luggage compartment adjacent to the main relay.

Battery voltage is supplied to the relay coil by the main relay. The earth circuit is via an electronic delay circuit in the ECU. The relay contacts supply battery voltage to the fuel pump. Should the ignition switch be turned 'on' but not to 'start', the delay circuit in the ECU will allow the pump to operate for one or two seconds, then the earth circuit of the relay will be broken, thus switching the pump off. When the ignition switch is turned to the start position, and the engine is cranked, the delay circuit is by-passed. The fuel pump will then operate continuously until such time as the ignition is switched off, or the engine speed signal pulses cease, indicating that the engine has stopped. This provides a safety back up to the inertia switch should the engine stop and the ignition be left in the 'on' position due to an accident situation.

Renew

Disconnect the battery earth lead.
Remove the relay from the retaining bracket.
Disconnect the multiplug connector from the relay.

MAIN RELAY AND PUMP RELAY

Test

Switch on the ignition. The pump should run for one or two seconds, then stop.

NOTE: If the pump does not run, or does not stop, check systematically as follows:

Check that the inertia switch cut-out button is pressed in.
Remove the screws, detach the inertia switch cover, and ensure that both cables are secure.

Pull the connectors from the switch and check continuity across the terminals.

Pull the button out and check for open circuit.

Remove the ohmmeter, replace the connectors, reset the button and refit the cover.

If the inertia switch is satisfactory, ground the pump relay terminal 85, switch on the ignition and check the circuit systematically as detailed below:

A Check for battery voltage at terminal 86 of the main relay.

   If yes: Proceed to Test B.

   If no: Check the battery supply from the ignition switch via the inertia switch.

B Check for battery voltage at Terminal 87 of the main relay.

   If yes: Proceed to Test C.

   If no: Check for battery voltage at the earth lead and connection from terminal 85 of the main relay. If satisfactory, renew the main relay.

C Check for battery voltage at terminal 86 of the pump relay.

   If yes: Proceed to Test D.

   If no: Open circuit between terminals 87 of the main relay and 86 of the pump relay. If satisfactory, proceed to Test D.

D Check for battery voltage at terminal 87 of the pump relay.

   If yes: Proceed to Test E.

   If no: Check for battery voltage at earth lead and connections from terminal 85 of the pump relay. If satisfactory, remove the pump relay.

E Check for battery voltage at supply lead (NS) and connections to the fuel pump.

   If yes: Faulty pump or earth connections.

   If no: Open circuit between terminal 87 of the pump relay and supply lead connection to the fuel pump.

INERTIA SWITCH

Renew

Disconnect the battery.

Remove the trip reset cable rubber knob and trim panel securing screws.

Disconnect the trip reset cable from the bracket and carefully displace the driver's side dash liner.

Remove the switch cover.

Disconnect the switch block connector.

Remove the two screws securing the switch and remove the switch.

On refitting the switch, ensure that it is correctly reset.

FUEL MAIN FILTER

Renew

Depressurise the fuel system.

Disconnect the battery.

Remove the spare wheel cover and the spare wheel.

Fit pipe clamps to the outlet and inlet hoses.

Disconnect the filter hoses (1, Fig. 36).

Fit plugs to the disconnected hoses and to the filter.

Slacken the filter clamp screw (2, Fig. 36) and remove the filter (3, Fig. 36).

On refitting, ensure the hose connections are secure.

Fig. 36

FUEL PUMP

Renew

Depressurise the fuel system.

Disconnect the battery earth lead.

Remove the spare wheel.

Remove the two screws securing the fuel pump cover to the battery tray.

Peel back the floor carpet, release the fuel pump cover from the floor clips.

Fit clamps to the inlet and outlet hoses.

Remove the nuts and washers securing the fuel pump retaining band to the mounting, retrieve the spacer washers and remove the fuel pump assembly.

Remove the electrical block connector from the fuel pump.

NOTE: Place a suitable container beneath the car to collect spilled fuel. Release the clips securing the hoses and disconnect the hoses. Separate and remove the two halves of the fuel pump retainer band. Remove the foam rubber insulation bands.

FUEL TANK

Renew

WARNING: FUEL IS EXTREMELY FLAMMABLE. GREAT CARE SHOULD BE TAKEN WHEN DRAINING THE FUEL TANK AND THE FOLLOWING PROCEDURES SHOULD BE STRICTLY ADHERED TO:

NO SMOKING ALLOWED NEAR THE AREA,

'NO SMOKING' WARNING SIGN POSTED ROUND THE AREA,

DISCONNECT THE BATTERY LEADS,

A CO_2 FIRE EXTINGUISHER IS CLOSE AT HAND,

DRY SAND IS AVAILABLE TO SOAK UP ANY SPILLAGE,

THE WORKING AREA IS WELL VENTILATED,

FUEL IS DRAINED INTO AN AUTHORISED EXPLOSION PROOF CONTAINER,

DISCARDED TANK NOT TO BE DISPOSED OF UNTIL RENDERED SAFE FROM EXPLOSION.

Remove the spare wheel.

Depressurise the fuel system and then

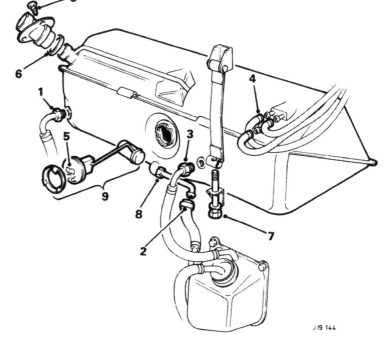

Fig. 37

disconnect the battery.

Remove the two screws securing the fuel pump cover to the battery tray.

Peel back the floor carpet, release the fuel pump cover from the floor clips.

Clamp the inlet hose to the fuel pump.

Loosen the hose clip securing the inlet hose to the fuel pump and disconnect the hose from the pump.

Remove the grommet from the vent hole in the luggage compartment floor, insert a suitable pump into the pump inlet hose, secure with a hose clip and push the pipe through the vent hole in the luggage compartment floor.

Place a suitable container beneath the car and the protruding pipe to collect the fuel.

Remove the fuel filter cap, release the clamp from the fuel pump inlet hose and drain the tank.

Remove the battery.

Clamp the fuel return hose and disconnect the union nut at the fuel tank (1, Fig. 37).

Release the hose clip securing the expansion tank supply hose to pipe and disconnect the hose (2, Fig. 37).

Release the union nut securing the expansion tank return pipe to fuel tank (3, Fig. 37).

Release the three hose clips securing the vent hoses to the fuel tank vent pipes.

Note the position of and disconnect the hoses (4, Fig. 37).

Note the position of and disconnect the electrical connectors from the fuel gauge tank unit (5, Fig. 37).

Release the hose clips securing the fuel filler assembly to the body and remove the fuel filler assembly (6, Fig. 37).

Remove the two bolts and release the fuel tank securing straps (7, Fig. 37).

Remove the LH and RH luggage compartment trim pads.

Remove the seven screws, plain washers and four fibre washers securing the LH side luggage compartment side panel to body lower panel.

Remove the seven screws, plain washers and four fibre washers securing the RH side panel to body, disconnect the earth leads from the relay bracket and lower the panel.

Remove the fuel tank.

Remove the expansion tank supply pipe (8, Fig. 37) and remove the fuel gauge tank unit (9, Fig. 37).

Refitting

Reverse the above procedure ensuring that all earth and electrical connections are secure and tight.

Check for fuel leaks.

CABRIOLET
FUEL TANK

Renew

Fuel filler cap and gasket
Renew

Special Tool No 18G 1001

WARNING: FUEL IS EXTREMELY FLAMMABLE. GREAT CARE SHOULD BE

TAKEN WHEN DRAINING THE FUEL TANK AND THE FOLLOWING PROCEDURES SHOULD BE STRICTLY ADHERED TO:

NO SMOKING ALLOWED NEAR THE AREA,

'NO SMOKING' WARNING SIGN POSTED ROUND THE AREA,

DISCONNECT THE BATTERY LEADS,

A CO_2 FIRE EXTINGUISHER IS CLOSE AT HAND,

DRY SAND IS AVAILABLE TO SOAK UP ANY SPILLAGE,

THE WORKING AREA IS WELL VENTILATED,

FUEL IS DRAINED INTO AN AUTHORISED EXPLOSION PROOF CONTAINER,

DISCARDED TANK NOT TO BE DISPOSED OF UNTIL RENDERED SAFE FROM EXPLOSION.

Remove the spare wheel.

Disconnect the battery leads.

Remove the fuel pump cover securing screws, release the cover from the floor clips and remove cover.

Remove the grommet from below the auxilliary fuel tank drain plug.

Fit a drain tube to the auxilliary fuel tank drain plug.

Place a suitable container beneath the drain tube.

Open the filler cap.

Open the drain plug and drain fuel from the fuel tank.

Remove the filler cap securing screws and the fuel type label.

Loosen the filler neck upper securing clip (1, Fig. 38).

Loosen the filler cap breather pipe securing clip (2, Fig. 38).

Carefully displace and remove the filler cap assembly complete with gasket.

Clean the filler cap sealing area.

Remove the boot side trim panels.

Remove the LH upper panel securing screws and displace the panel for access.

Remove the screws securing the service interval counter and displace the interval counter from the panel.

Note and disconnect the boot lamp switch cables.

Remove the nut and bolt securing the bulb failure unit.

Displace the unit from the panel.

Unclip the cable harness from the panel cable clips and remove the panel.

Remove the relays and the interface unit from the RH side panel mounting lugs.

Remove the screws securing the RH upper panel and displace the panel for access.

Note and and disconnect the boot lamp cables.

Remove the nut and bolt securing the bulb failure unit.

Displace the unit from the panel.

Remove the cable harness from the panel cable clips and remove the panel.

Remove the battery.

Remove the auxilliary fuel tank cover securing screws, displace and remove the cover.

Remove the auxilliary fuel tank securing bolts.

Remove the fuel pump bracket securing bolts.

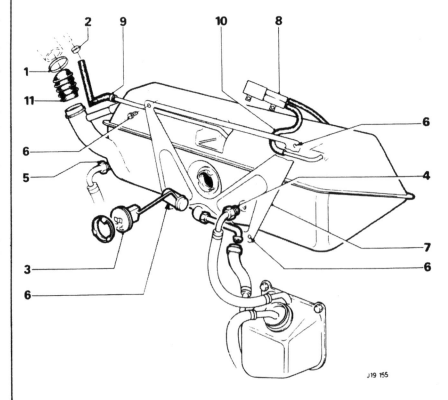

J19 155

Displace the auxilliary fuel tank and the fuel pump assembly for access.

Remove the nuts and bolts securing the battery carrier.

Remove the carrier drain hose from the boot floor, displace and remove the battery carrier.

Remove the screws securing the fuel hose clamp plate and remove the plate.

Note and disconnect the fuel tank unit electrical cables using tool No. 18G 1001, rotate the locking ring anti-clockwise to clear the lugs in the tank (3, Fig. 38).

Remove the locking ring and withdraw the tank unit.

Clamp the fuel tank to auxilliary tank hose, release the union nut at the fuel tank, disconnect and displace the hose (4, Fig. 38).

Clamp the fuel tank return hose, release the union nut and displace the hose from the tank (5, Fig. 38).

Remove the fuel tank retaining bracket securing bolts (6, Fig. 38).

Loosen the retaining bracket aligning plate securing bolts.

Displace and remove the retaining bracket assembly (7, Fig. 38).

Remove the aligning plate packing strips.

Remove the nuts securing the vapour separator to the rear bulkhead, displace the separator and disconnect the hoses (8, Fig. 38).

Displace and carefully manoeuvre the fuel tank from its location.

Remove the breather hose (9, Fig. 38), vapour separator hose (10, Fig. 38), and the filler neck hose (11, Fig. 38) from the fuel tank

Refitting is the reversal of the above operations.

Ensure the foam pads are glued to the side of the tank.

Fit new gaskets to the fuel tank unit and the filler cap assembly.

Ensure all electrical components operate satisfactory.

FUEL SYSTEM

Pressure Test

Depressurise the fuel system.

Slacken the clip on the cold start injector fuel hose and disconnect the hose.

Connect the pressure gauge to the disconnected hose.

Disconnect the leads from the negative terminal of the ignition coil.

Connect a jumper lead from terminal 85 of the pump relay to earth.

Switch on the ignition and check the pressure gauge reading. The reading should be between 2.0 — 2.2 kgf/cm² (28.5 — 30.8 lbf/in²).

Should the reading be high, check the fuel return pipe for blockage. If the reading is low, check the fuel pump for correct operation, blockage in the supply fuel line or the pump suction pipes, or a choke filter.

Switch the ignition off.

Depressurise the fuel system and remove the pressure gauge.

Reconnect the fuel hose to the cold start injector.

Switch the ignition on and check for leaks.

Remove the jumper lead from the terminal 85 of the pump relay.

Reconnect the leads to the negative terminal of the ignition coil.

This page is intentionally blank

CONTENTS

Operation	Operation No.	Page No.

Thermostat closed

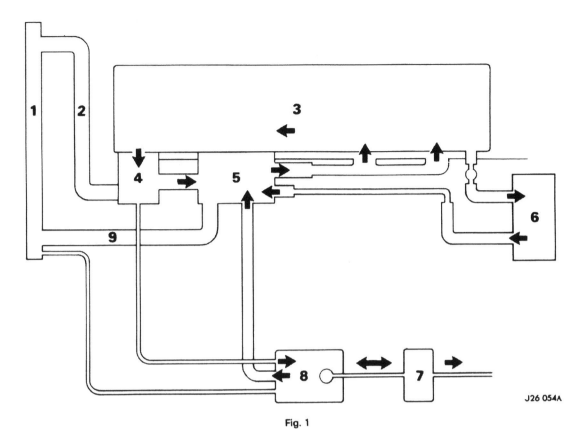

J26 054A

Fig. 1

Thermostat open

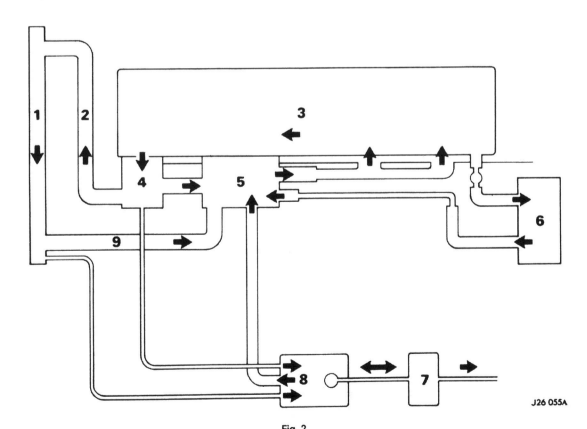

J26 055A

Fig. 2

KEY

1. Radiator
2. Top hose
3. Engine

4. Thermostat housing
5. Water pump
6. Heater

7. Atmospheric catchment tank
8. Header tank
9. Bottom hose

COOLING SYSTEM FLOW

DESCRIPTION

The cooling system consists of a crossflow radiator matrix, a water pump belt driven by the engine crankshaft, a header tank and an expansion tank located in the near side wing, and a thermostatic valve to ensure rapid warm up. The radiator is cooled by a viscous coupled fan and also an electric driven fan, this fan is thermostatically controlled and it is possible under very hot conditions for it to operate after the engine is switched off. The fan will switch off automatically when the coolant drops below 90°C.

Under cold start condition coolant is forced by the water pump through the cylinder block and cylinder head to the thermostat housing. The valve is closed and the coolant is therefore returned to the water pump suction inlet.

Cars fitted with automatic transmission have a cooling tube included in the centre section of the radiator matrix.

We use and recommend BP Type H21 or Union Carbide UT184 or Unipart Universal anti-freeze which should be used at the specified concentration whenever the cooling system is refilled. For topping-up purposes, only reputable brands of anti-freeze, formulated and approved for 'mixed metal' engines should be used.

IMPORTANT NOTE: THE CONCENTRATION OF ANTI-FREEZE MUST NOT BE ALLOWED TO FALL BELOW THE RECOMMENDED STRENGTH AS SEDIMENT MAY BE FORMED IN THE COOLING SYSTEM BY CERTAIN TYPES OF ANTI-FREEZE AT LOW CONCENTRATES.

A 40% solution by volume in the United Kingdom (55%, U.S.A./Canada and all other countries) must be used at all times, either by topping-up or replenishing the cooling system. For maximum corrosion protection, the concentration should never be allowed to fall below 25%. Always top-up with recommended strength of anti-freeze, NEVER WITH WATER ONLY.

In countries where it is unnecessary to use anti-freeze, Marston SQ 35 Corrosion Inhibitor must be used in the cooling system in the proportion of 1 part SQ 36 to 24 parts water. CHANGE COOLANT EVERY TWO YEARS. The system should be drained, flushed and refilled with fresh anti-freeze (or Corrosion Inhibitor), mixed with 1 sachet of 'Barrs Leaks'.

An alternative coolant known as CARBUROL FORLIFE is recommended where temperatures below 10°C (14°F) are not encountered. Before Carburol Forlife is used, the coolant already present in the system must be drained out and the system flushed before filling with Carburol Forlife. Once in use the system should be topped-up with Carburol Forlife only, and a label giving this information should be affixed in an appropriate and prominent position.

Fault diagnosis

Symptom	Reason	Remedy
Overheating	Insufficient water in cooling system	Top up radiator
	Fan belt slipping	Tighten fan belt to recommended tension or replace if worn
	Radiator core blocked or radiator grille obstructed	Reverse flush the radiator, remove obstruction from grille
	Thermostat not opening properly	Remove and fit new thermostat
	Ignition advance and retard incorrectly set (accompanied by loss of power and perhaps misfiring)	Check and reset ignition timing
	Incorrect fuel/air mixture	Check CO/HC level
	Exhaust system partially blocked	Check exhaust pipe for obstruction
	Oil level in sump too low	Top up to correct level
	Blown cylinder head gasket (water/steam being forced down the radiator overflow pipe under pressure)	Remove cylinder head and fit new gasket
	Engine not yet 'run-in'	Run-in slowly and carefully
	Brakes binding	Check brake calipers for sticking pistons and seized brake pad pins
Engine running 'cold'	Thermostat jammed open	Remove and renew thermostat
	Incorrect grade of thermostat fitted	Remove and replace with correct type of thermostat
	Thermostat missing	Check and fit correct thermostat
Leaks in system	Loose clips on water hoses	Check and tighten clips
	Top or bottom water hoses perished	Check and replace any faulty hoses
	Radiator leaking	Remove radiator and repair
	Thermostat gasket leaking	Inspect and renew gasket
	Pressure cap spring worn or seal ineffective	Renew pressure cap
	Cylinder wall or head cracked	Replace either cylinder block or cylinder head

COOLING SYSTEM

Pressure Test

Ensure that the engine is warm from recent operation. Carefully remove the pressure cap from the expansion tank and check that it is the correct type for the vehicle.

Select the correct adaptors for the neck of the header tank (1, Fig. 3) and the Tester (2, Fig. 3) and fit it to the header tank.

Pump up the pressure to a pressure equal to the pressure cap poundage. Watch the gauge for ten seconds and if the pressure drops check the water pump and external connections for leaks. If the pressure remains constant for ten seconds, maintain the pressure and visually check for pin point leaks. Tighten all connections and replace worn or leaky hoses, internal leaks may be cured by torqueing the cylinder head bolts.

A more severe test may be carried out by using the above procedure with the engine running. Absence of external leaks accompanied by fluctuations in pressure usually indicates a 'blown' cylinder head gasket, however, under no circumstances should the pressure be allowed to exceed the upper limit of the cap.

COOLING SYSTEM

Drain and Refill

Place a drain tray in position under the radiator drain tap (1, Fig. 4). Remove the expansion tank filler cap, open the radiator drain tap (2, Fig. 4) and drain the coolant. Remove the transmission cooler pipes bracket from the drain plug. Remove the rear drain block plug and drain the coolant. Refit the cylinder block drain plug and close the radiator drain tap, reposition the cooler pipes and refit clamp. Fill the cooling system with a solution of anti-freeze and water, move the heater valve control lever to the hot position, refit the expansion tank cap and run the engine until the thermostat is open. Switch off engine, remove expansion tank cap and refill with water.

PRESSURE CAP

Test

Ensure that the cooling system is cold before removing the cap. Determine the correct adaptors for the cap to be checked and attach the adaptors to the tester, check and examine the rubbers (1, Fig. 5) for cracks and wear. Rinse the cap in water to remove any sediment and apply the cap (2, Fig. 5) to the tester in the wet condition. Pump up the tester until the gauge pointer stops rising. This pressure should read between 93,15 - 120,75 kPa. Reject the cap if it does not reach or exceeds this pressure. Reject the cap which will not hold the rated pressure for 10 seconds without additional pumping.

NOTE: When renewing a defective pressure cap with a new unit, never fit a cap of a higher pressure than of the original fitment.

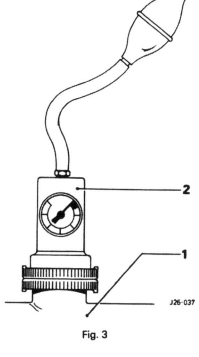

Fig. 3

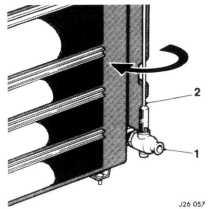

Fig. 4

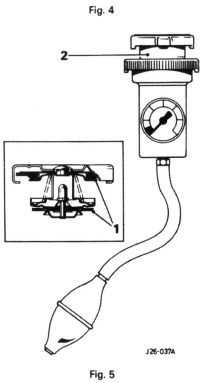

Fig. 5

THERMOSTAT

Test

The thermostat may be checked by suspending it, with the thermostat heat control unit facing down, in a small pan or dish (1, Fig. 6) containing a solution of ethylene glycol coolant containing a thermometer (2, Fig. 6). Neither the thermostat or the thermometer should rest on the bottom of the pan because of the uneven concentration of heat at this point when the pan is heated. The thermostat (3, Fig. 6) should open when a temperature of 82°C is attained. When the coolant reaches 97°C the valve should be fully open.

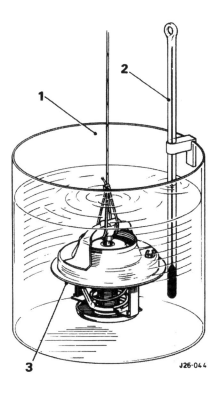

Fig. 6

THERMOSTAT

Renew

Drain the coolant.
Remove the thermostat cover securing bolts (1, Fig. 7) and remove the cover (2, Fig. 7). Remove the thermostat (3, Fig. 7) and sealing ring (4, Fig. 7) from the housing (5, Fig. 7), remove the gasket and clean the mating faces.

Fit a new sealing ring and place the new thermostat into the housing ensuring that the face marked 'to RAD' is facing out, and the position marked 'TOP' is at the top.

Smear the gasket face with 'Hylosyl 101', fit the thermostat cover, tighten the securing bolts and refill the cooling system.

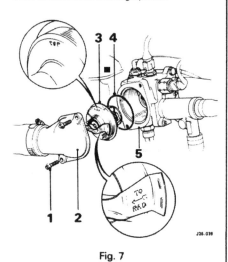

Fig. 7

THERMOSTATIC SWITCH

Renew

Drain the coolant, note the position of the switch wires (1, Fig. 8) before disconnecting, remove the switch (2, Fig. 8).

Fit and tighten the new switch and refit the connections. Fill the system with coolant.

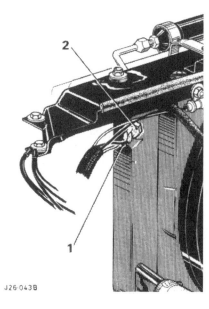

J26·043B

Fig. 8

THERMOSTAT HOUSING REAR OUTLET GASKET

Renew

Drain the coolant. Remove the water pump to thermostat housing hose. Remove the rear outlet cover securing bolts (1 Fig. 9) and remove the cover (2, Fig. 9). Clean the rear outlet gasket faces. Smear the faces with 'Hylosyl 101', fit the rear outlet cover secured with two bolts. Refit the water pump to thermostat housing hose and refill the system with coolant.

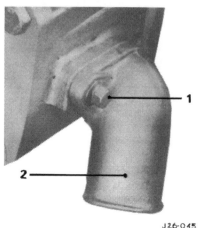

J26-045

Fig. 9

THERMOSTAT HOUSING GASKET

Renew

Drain the coolant, disconnect the temperature transmitter, water temperature sensor and the thermotime switch feed wires (1, Fig. 10). Disconnect the hoses from the thermostat housing.

Remove the thermostat housing securing bolts (2, Fig. 10) and remove the housing (3, Fig. 10). Clean the thermostat housing gasket faces.

Coat the faces with 'Hylosyl 101' and refit the thermostat housing and secure with bolts ensuring that the earth lead is attached to the top bolt. Refit the hoses to the housing and tighten the clips. Reconnect the water temperature transmitter, water temperature sensor and thermotime switch wires and refill the cooling system.

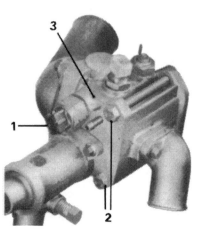

Fig. 10

THERMOSTAT HOUSING

Renew

Remove the thermostat housing cover (1, Fig. 11), thermostat (2, Fig. 11) and 'O' ring (3, Fig. 11). Remove the auxiliary air valve (4, Fig. 11) and discard the gasket. Remove the rear elbow (2, Fig. 9), the water temperature switch (6, Fig. 11), thermotime switch (7, Fig. 11), the thermostatic switch (8, Fig. 11) and discard the sealing washers.

Clean the gasket and mating faces and fit the sensors and switches with new sealing washers. Smear the rear elbow gasket face with 'Hylosyl 101' and fit the elbow. Fit the auxiliary air valve with a new gasket. Fit the thermostat and cover.

Fit the thermostat housing as instructions in 'Thermostat housing gasket renew'.

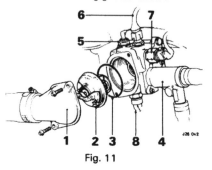

J26 042

Fig. 11

WATER PUMP HOUSING

Renew

Drain the coolant. Remove the air cleaner element, the torquatrol unit and the fan. Remove all the hoses from the housing. Remove the water pump housing securing bolts (1, Fig. 12) and remove the housing (2, Fig. 12). Clean the water pump and gasket faces. Smear the gasket face with 'Hylosyl 101', align and fit the pump to the housing and tighten the securing bolts. Fit the housing and pump assembly back to the engine and secure with the mounting bolts. Refit all the hoses and tighten clips. Refit the fan belt and retension by moving the alternator. Refit the fan, the torquatrol unit and the air cleaner element.

Refill the system with coolant.

J26-060

Fig. 12

WATER PUMP GASKET

Renew

Drain the coolant, remove the air cleaner element, remove the torquatrol unit.
Pivot the alternator towards the engine and remove the fan belt.
Remove the bolts (1, Fig. 13) securing the water pump (2, Fig. 13) to the housing and pull the pump clear of the housing. Remove the gasket and clean the faces (3, Fig. 13).

Smear the pump gasket face with 'Hylosyl 101', fit and align the pump to the housing and secure with the bolts. Refit the fan belt and retension by moving the alternator.
Fit the torquatrol unit and air cleaner element. Refill the system with coolant.

ATMOSPHERIC CATCHMENT TANK

Renew

Jack up and place the front of the vehicle on stands.
Remove the front left-hand road wheel.
Remove the screws securing the wheel arch rear panel and remove the panel.
Unclip the catchment tank (1, Fig. 16).
Unclip the overflow pipe (2, Fig. 16) from the body and displace the tank from its mounting position.
Disconnect and remove the catchment tank from the joining hose.
On refitting apply new body sealant to the wheel arch panel seal.

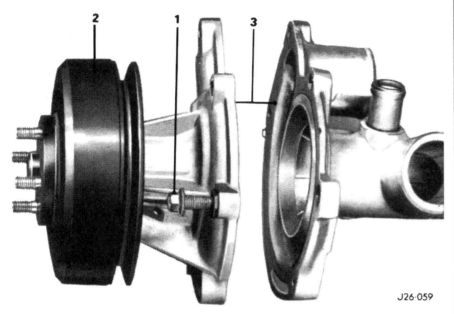

Fig. 13

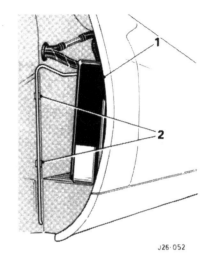

Fig. 16

EXPANSION TANK

Renew

Remove the air cleaner and box, disconnect the low coolant probe sensor. Place a suitable container below the tank lower outlet (1, Fig. 14) and remove all the hoses (2, Fig. 14) from the tank. Drain the expansion tank, remove the securing bolts (3, Fig. 14) and remove the tank (4, Fig. 14) from the vehicle. Remove the pressure cap, retaining ring, coolant probe, coolant probe grommet and anti-vibration rubber, from the tank.
Assemble all fittings to replacement tank and refit tank to inner wing bracket. Refit hoses and refill with correct anti-freeze/water solution.
Refit air cleaner element and box.

COOLANT LEVEL PROBE

Renew

Release the pressure cap (1, Fig. 15) from the expansion tank (2, Fig. 15), disconnect the probe feed wire (3, Fig 15), place a suitable water container in position and remove probe (4, Fig. 15) and sealing washer.
Lubricate probe seal and seat it on the expansion tank, fit and seat the probe to the seal, connect the feed wire, top-up the coolant as required and refit the pressure cap.

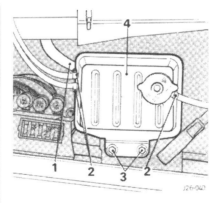

Fig. 14

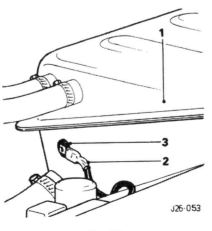

Fig. 15

HOSE EXPANSION TANK TO ATMOSPHERIC CATCHMENT TANK

Renew

Jack up and place the front of the vehicle on stands.
Remove the left-hand front wheel.
Remove the screws securing the wheel arch rear panel and remove the panel.
Unclip the catchment tank and overflow pipe from the body.
Displace the tank from its mounting position.
Slacken the hose clip and remove the catchment tank complete with the joining hose from the vehicle.
Remove the hose clip from the hose.
Displace the hose from its mounting position and remove the hose from the header tank.
On refitting apply new body sealant to the wheel arch seal.

RADIATOR

Renew

Drain the cooling system, remove the torquatrol unit and fan. Disconnect the expansion tank to radiator hose (1 Fig. 17), disconnect the thermostatic switch feed wires (2, Fig. 17) and move the fan cowl away from the top rail. Remove the receiver drier securing nuts (3, Fig. 17) and bolts, move drier (4, Fig. 17) clear of top rail. Release the earth leads, 2 left-hand side (5, Fig. 17), 1 right-hand side (6, Fig. 17).

Disconnect the amplifier leads, the distributor harness, the amplifier to coil harness and remove the top rail/amplifier assembly (7, Fig. 17).

Disconnect the top hose (8, Fig. 17) from the radiator.

Disconnect the bottom hose (9, Fig. 17) and lift the radiator (10, Fig. 17) from its lower mounting, remove the fan cowl (11, Fig. 17) and remove the radiator (11, Fig. 17).

Carefully remove the radiator sealing rubbers, the thermostatic switch and the drain tap.

Fit the drain tap, the thermostatic switch and the sealing rubbers to the new radiator and reverse the removal procedure noting to refill the cooling system.

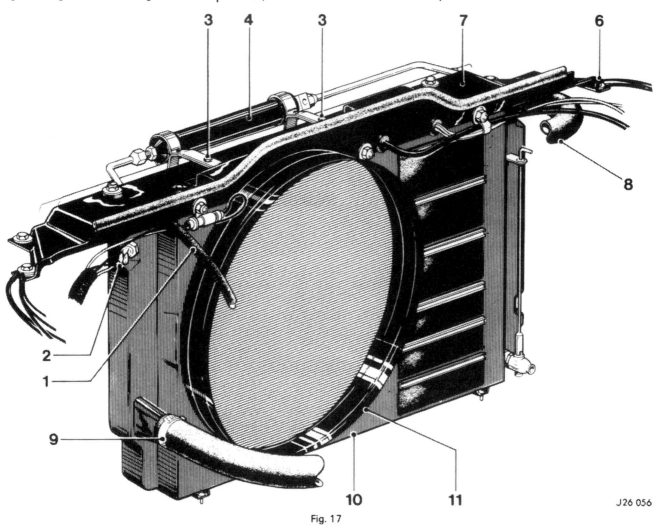

Fig. 17

J26 056

TORQUATROL UNIT

Renew

Slacken the fan blade to torquatrol unit securing nuts (1, Fig. 18) and bolts (2, Fig. 18). Pull the torquatrol unit (3, Fig. 18) clear of fan pulley face and remove the securing nuts, remove the fan and torquatrol unit clear of pulley and remove the fan (4, Fig. 18) from the torquatrol unit.

Refit the fan to the torquatrol unit and bolt the assembly to the fan pulley face. Fill the system with coolant and fit filler cap.

CYLINDER BLOCK WATER TRANSFER PIPE GASKETS

Renew

Drain the coolant and depressurise the fuel system. Remove the air box and element. Remove the induction manifold. Slacken the hose securing clip (1, Fig. 19), cut and remove the harness to water pipe securing clip. Remove the water pipe securing bolts (2, Fig. 19) and remove the pipe (3, Fig. 19). Clean the gasket faces (4, Fig. 19) and then coat them with 'Hylosyl 101', fit a new pipe, fit a new harness securing clip, refit the induction manifold and the air cleaner element. Refill the cooling system.

HOSE SET

Renew

Drain the cooling system, slacken hose clips as required and carefully remove the hoses. To ease assembly of the hoses to attachment points, a liberal application of soft soap may be applied to the inside of the hoses. Position and tighten the securing clips and refill the system with coolant.

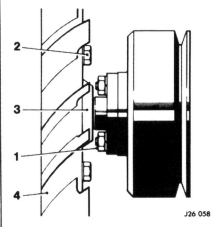

Fig. 18

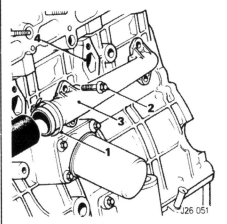

Fig. 19

CONTENTS

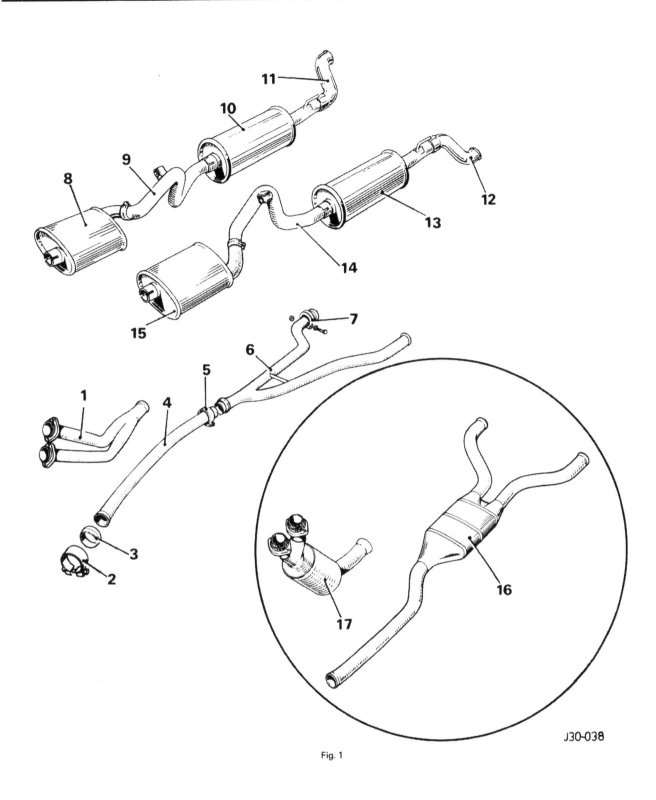

Fig. 1

J30-038

1. Down pipe	7. Clamp	13. LH Rear silencer
2. Clamp	8. RH Silencer	14. LH Intermediate pipe
3. Olive	9. RH Intermediate pipe	15. LH Silencer
4. Front pipe	10. RH Rear silencer	16. Branch pipe/catalyst assembly
5. Clamp	11. RH Tail pipe/trim	17. Down pipe/catalyst assembly
6. Branch pipe	12. LH Tail pipe/trim	

Note: Items 16 and 17 are only used on USA Federal market vehicles only.

TORQUE WRENCH SETTINGS

ITEM	SPANNER SIZE	TIGHTENING TORQUE			
		DESCRIPTION	Nm	kgf/cm	lbf/ft
Intake tube — inlet manifold	3 mm (Allen key)	6 mm bolt	9.5 to 12.2	0,98 to 1,26	7 to 9
Inlet manifold to cylinder head	10 mm	8 mm bolt	23 to 27	3,21 to 3,76	17 to 19
Inlet manifold to cylinder head	13 mm	8 mm nut	12 to 16	1,26 to 1,66	9 to 12
Exhaust manifold to cylinder head	13 mm	10 mm nut	49 to 54	6,82 to 7,52	36 to 40

AIR CLEANER

Renew

Release the outer air box clips (1, Fig. 2) and remove the air cleaner together with the air box (2, Fig. 2).

Hold the new air cleaner and the box together and feed them into the bottom air box flange, pivot air box towards engine and secure with the 'over centre' clips (1, Fig. 2).

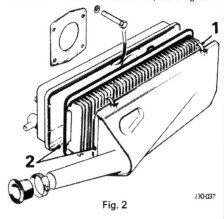

Fig. 2

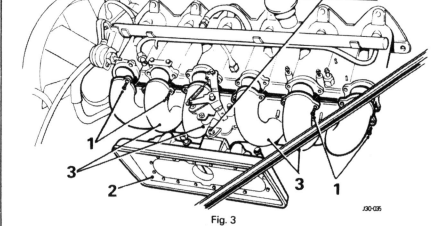

Fig. 3

RAM PIPES

Renew Set

Release the four securing screws per ram pipe (1, Fig. 3) and remove pipes No. 1 and 6.

Remove the air cleaner backplate (2, Fig. 3) for ram pipes 2, 3, 4 and 5 (3, Fig. 3).

Note for No. 4 ram pipe the throttle pedestal (4, Fig. 3) will have to be removed.

Refit replacement pipes and secure with screws.

EXHAUST MANIFOLDS

Renew

Disconnect the spark plug leads, release the two manifold to down pipe flanges (1, Fig. 4). Remove the manifold securing nuts (2, Fig. 4) and remove the manifolds (3, Fig. 4).

Fit new exhaust manifold gaskets, position the manifolds and secure with the nuts. Reposition the manifold flanges and secure with the nuts and refit the spark plug leads.

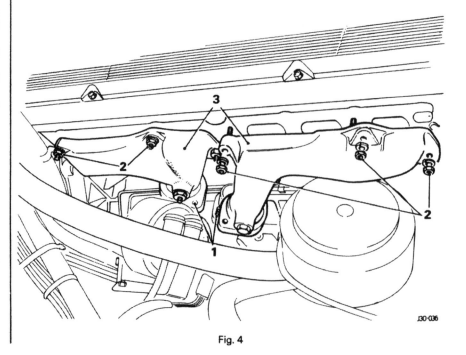

Fig. 4

EXHAUST SYSTEM AND MOUNTINGS

NOTE: For federal vehicles disregard any reference to the front pipe and refer to Fig. 6.

Renew

Open the boot and lift the carpet to expose the rear exhaust mounting securing nuts. Remove the rear exhaust mounting nuts and mountings. Remove the mounting spacer washers.

Slacken the tail pipe and silencer to rear intermediate pipe clamp nut (1, Fig. 5) and bolt. Split the rear silencer assembly (2, Fig. 5) from the rear intermediate pipe (3, Fig. 5). Remove the exhaust pipe trim (4, Fig. 5).

Remove the intermediate pipe to silencer clamp nuts (5, Fig. 5) and bolts and support the rear silencers (6, Fig. 5). Remove the front pipe to manifold securing nuts (7, Fig. 5). Carefully disconnect the front pipe (8, Fig. 5) from the manifold and support the

system. Remove the clamps (9, Fig. 5) from the rear joints and separate the silencer (6, Fig. 5) from the branch (10, Fig. 5) and front pipe (11, Fig. 5). Remove clamp (12, Fig. 5 and separate the frontpipe (11, Fig. 5) (where fitted) and the branch pipe (10, Fig. 5). Separate the silencers (6, Fig. 5) from the intermediate pipes (3, Fig. 5) and remove the intermediate pipes (3, Fig. 5).

Remove the intermediate pipe mounting bar, spacers and securing nuts, repeat for second pipe and remove the front sealing rings.

Remove the rear mounting to rear suspension securing nuts and bolts, remove the mounting and repeat for the second rear mounting.

Fit new exhaust system brackets to vehicle ensuring that the tail pipes and rear intermediate pipes are fitted to the vehicle first.

Assemble complete system loosely and align to clear of all other fittings ensure that all joints are smeared with exhaust sealant prior to assembly and finally tighten all mountings and clamps.

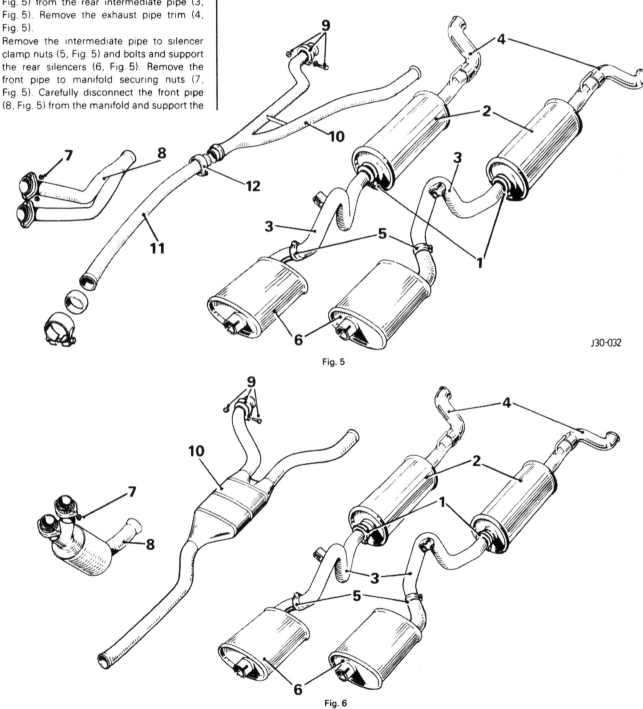

Fig. 5

J30-032

Fig. 6

J30-033

INDUCTION MANIFOLD GASKETS

Renew

Depressurise the fuel system, by disconnecting the relay and cranking the engine over, disconnect the auxiliary air valve hoses, identify and disconnect the air valve feed wires and disconnect the throttle potentiometer block connector.

Disconnect the return hose (1, Fig. 7) from the pressure regulator (2, Fig. 7) and fit blanking plugs to the hose and the regulator.

Disconnect the fuel rail feed hose (3, Fig. 7) from the rail.

Disconnect the kickdown cable (4, Fig. 7) from the throttle linkage and remove the securing bracket (5, Fig. 7).

Disconnect the brake servo hose from the inlet manifold (6, Fig. 7), the air conditioning econocruise pipes (7, Fig. 7) and the ECU vacuum pipe from the manifold (8, Fig. 7).

Disconnect the solenoid air switch vacuum pipe and distributer advance/retard pipe (9, Fig. 7), the ambient air temperature sensor (10, Fig. 7) and injector block connectors (11, Fig. 7).

Remove the breather hose (12, Fig. 7) from the cam cover and the hoses from the air cleaner back plate and manifold.

Release the securing bolt and pull the oil filler pipe (13, Fig. 7) clear of its lower housing.

Remove the dipstick tube securing bolt and tube, the manifold securing nuts and bolts, lifting eyes and filler pipe bracket.

Release the engine earth lead from the rear of the manifold.

Disconnect the solenoid air switching valve and remove.

Remove the manifold (14, Fig. 7) and at the same time feed the injector harnesses and connectors through the manifold. Remove the manifold gaskets (15 Fig. 7) and discard the oil filler tube lower seal.

Clean the filler tube seal and gasket faces and fit new gaskets and seal.

Position the injector, sensor, switch wires and distributer advance/retard pipes through the relevant apertures in the inlet manifold.

Position the manifold onto the cylinder head studs. Refit the filler tube bracket and lifting eyes and reconnect the air solenoid switch to the manifold.

Fit and tighten the manifold securing nuts and bolts ensuring that the earth lead is connected to its correct bolt hole.

Reposition the dipstick tube and secure with the bolt and 'P' Clip.

Fit the filler tube into the lower housing and tighten the tube upper securing bolt. Fit the

breather hose to the filler tube and cam cover. Connect the hoses to the manifold and air cleaner back plate and the injector and air sensor block connectors.

Connect the distributor advance/retard pipes, the air solenoid vacuum pipe, the ECU vacuum pipe, the air conditioning and econocruise vacuum pipes and the brake servo hose to the manifold.

Fit the throttle kickdown cable bracket, the throttle cable and position the clip over the nipple and reconnect the kickdown cable to the linkage.

Remove the plugs from the fuel rail feed hose and refit the hose.

Remove the plugs from the pressure regulator valve return hose and reconnect the hose. Connect the throttle potentiometer block connector and the air switching valve feed wires.

Connect the auxiliary air valve hoses and tighten the clips.

Refit the air cleaner element and box.

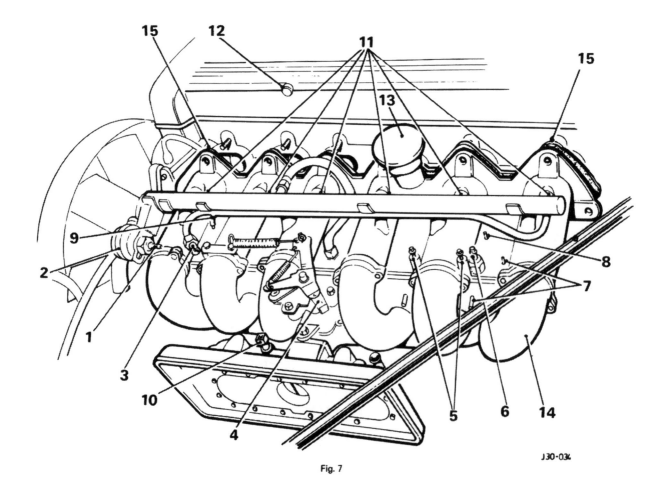

Fig. 7

J30-034

CONTENTS

TORQUE WRENCH SETTINGS

ITEM	DESCRIPTION	SPANNER SIZE	TIGHTENING TORQUE		
			Nm	kgf/cm	lbf/ft
Clutch cover to flywheel	W/F* Setscrew M8	10mm	33 to 27	2,4 to 2,8	17 to 20
Slave cylinder to bellhousing	Nut M8	13mm	23 to 27	2,4 to 2,8	17 to 20
Hydraulic pipe to slave cylinder	Hex End Fitting M12	17mm	34 to 40	3,5 to 5,6	25 to 30
Damper to bracket	W/F* Setscrew M6	8mm	9 to 10	0,9 to 1,0	6.5 to 7.5
Damper bracket to bulkhead	Setscrew $\frac{5}{16}$in UNF	$\frac{1}{2}$ in AF	15 to 17,5	1,5 to 1,8	11 to 13
Hydraulic pipe to damper	Male tube nut M12	13mm	20 to 24	2 to 2,5	14.5 to 17.5
Hydraulic pipe to master cylinder	Male tube nut M12	13mm	20 to 24	2 to 2,5	14.5 to 17.5
Master cylinder to bulkhead	Socket head Setscrew $\frac{5}{16}$ in UNC	$\frac{7}{32}$ Allen key	15 to 17,5	1,5 to 1,8	11 to 13

*Washer faced

CLUTCH

Description

The clutch is a single plate, diaphragm type, and is operated hydraulically by a slave cylinder mounted to the gearbox bell housing. This is in turn operated by a clutch master cylinder through a series of hydraulic pipes and a damper. The piston in the master cylinder is moved via a push rod by the clutch pedal, which is mounted to the bulkhead and is positioned to the left of the brake pedal.

CAUTION: The hydraulic fluid used in the clutch hydraulic system is injurious to car paintwork. Utmost precautions MUST at all times be taken to prevent spillage of fluid. Should fluid be accidentally spilled onto the paintwork, wipe fluid off immediately with a cloth moistened with denatured alcohol (methylated spirits).

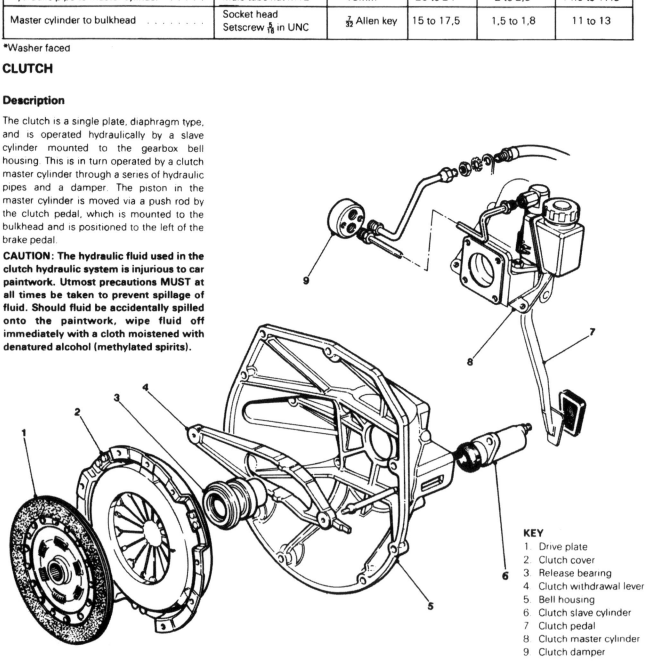

KEY
1. Drive plate
2. Clutch cover
3. Release bearing
4. Clutch withdrawal lever
5. Bell housing
6. Clutch slave cylinder
7. Clutch pedal
8. Clutch master cylinder
9. Clutch damper

CLUTCH FAULT DIAGNOSIS

SYMPTOM	POSSIBLE CAUSE	CHECK	REMEDY
Grabbing clutch. (Harsh engagement from standing start, often followed by clutch judder).	Operating mechanism faulty.	Check operating for wear and binding which usually indicates a binding withdrawal race thrust bearing.	Free off bearing. Replace as necessary.
		Check pedal for sticking.	Free off pedal and check for damaged and distorted parts including return spring. Replace as necessary.
	Clutch unit faults.	Check for oil on friction faces.	Clean off cover. Replace driven plate. Rectify oil leak.
		Check clutch plate and flywheel for wear. Check flywheel run-out. Check also for glazing on driven plate linings.	Reclaim or replace as applicable.
		Check for driven plate hub splines sticking on pinion shaft.	Free driven plate and check for wear and distortion. Check pinion shaft for wear.
		Check for broken or weak pressure springs. Check torque damper springs in clutch driven plate.	Replace as necessary.
	Engine mountings faults.	Check for damaged or deteriorated engine mountings. Check fixings for looseness.	Replace or tighten as applicable.
Slipping clutch (Indicated by vehicle road speed not responding to engine speed increases).	Faulty driving technique.	Ensure that none of the Remedy conditions prevail.	Do not increase engine speed with clutch partially engaged. Do not drive with left foot resting on clutch pedal.
	Operating mechanism faulty.	Check for binding withdrawal lever.	Free lever and check for wear and distortion.
		Check for binding of clutch pedal moving components.	Free off seized or binding components.
	Clutch unit faults.	Check for oil on friction faces.	Clean off metal faces. Replace driven plate. Rectify oil leak.
		Check for broken or weak pressure springs.	Replace cover as necessary.
		Check clutch plates and flywheel for wear and distortion.	Reclaim or replace as applicable clutch plate flywheel.
		Check clutch driven plate for fractures and distortion. Damage may be caused by accidental loading during assembly of gearbox to engine. Always support gearbox weight during refitting.	Replace driven plate and check mating components for damage.
Dragging or spinning clutch	Clutch unit faults.	Check for primary pinion bearing seized.	Rectify or replace as necessary.
		Check clutch driven plate hub for binding on primary pinion splines. Check for too thick friction linings. Ensure linings are good.	Replace as necessary.
		Check for distorted clutch pressure plate and clutch cover.	Replace as necessary.
		Check for foreign matter in clutch unit.	Clean and replace components as necessary.

CLUTCH FAULT DIAGNOSIS

SYMPTOM	POSSIBLE CAUSE	CHECK	REMEDY
	Hydraulic system defects.	Check fluid level in reservoir.	Replenish as necessary and bleed system if necessary.
		Check for air in the system.	Bleed the system.
Rattling clutch.	Operating mechanism faults.	Check for defective pedal return spring.	Replace as necessary.
	Clutch unit faults.	Check for damaged pressure plate.	Replace pressure plate.
		Check splines on clutch driven plate and primary pinion shaft for wear.	Replace as necessary clutch plate, or primary pinion.
		Check clutch driven plate for loose or broken springs and for warping.	Replace driven plate.
		Check for wear in the clutch withdrawal mechanism.	Replace as necessary.
		Check for worn primary pinion bearing.	Replace as necessary.
Squeaking clutch	Primary pinion bearing fault.	Check for seizing on primary shaft or in flywheel.	Lubricate or replace as necessary.
Vibrating clutch or clutch judder (often preceded by clutch grab)	Clutch unit faults.	Check the clutch driven plate for distortion and damage and for loose or broken torque springs.	Replace driven plate.
		Check for oil and other foreign matter on the clutch friction linings.	Replace driven plate and clean related parts.
		Check for incorrectly fitted clutch pressure plate.	Dismantle from clutch and refit, where applicable.
		Check that contact witness on friction linings is evenly distributed.	Replace driven plate as necessary.
	Defects other than in clutch unit.	Check for loose flywheel fixings.	Tighten to correct torque loading. Check flywheel run-out.
		Check for loose engine mountings	Tighten mounting nuts and bolts.
		Check for worn propeller shaft universal joints.	Replace as necessary.
		Check for bent primary pinion shaft.	Replace as necessary.
Stiff clutch operation.	Operating linkage fault.	Check for damaged moving parts in operating linkage.	Replace as necessary.
		Check for seized linkage.	Lubricate linkage and recheck operation.
Clutch knocks	Clutch unit fault.	Check for worn clutch driven plate hub splines.	Replace driven plate.
	Primary pinion bearing.	Check for wear in bearing.	Replace as necessary.
Fractured clutch plate	Incorrect fitting method.	Damage may be caused by accidental loading during fitting. Always support gearbox weight during fitting.	Replace driven plate. Check mating components for damage.
Excessive lining wear	Overloading vehicle.	Refer to Owner's handbook for permissible load details.	Fit replacement clutch assembly.
	Driving with left foot resting on clutch pedal.	Check as described under 'slipping clutch'.	Fit replacement clutch assembly.

HYDRAULIC PIPES

Renew

Remove the pipes or hoses as required and replace with new units, bleed the hydraulic system as necessary with the recommended brake fluid.

SLAVE CYLINDER

Overhaul

Remove the slave cylinder from the vehicle, dismantle the cylinder, and unscrew the bleedscrew.

The new parts in the kit will indicate which used parts should be discarded. Clean the remaining parts and the cylinder thoroughly with unused brake fluid of the recommended type and place the cleaned parts onto a clean sheet of paper.

Examine the cylinder bore and the pistons for signs of corrosion, ridges or score marks. Provided the working surfaces are in perfect condition, new seals from the kit can be fitted, but if there is any doubt as to the condition of the parts then a new cylinder must be fitted.

Fit the new seal to the piston with the flat back of the seal against the shoulder (see inset Fig. 1).

Lubricate the seal and the cylinder bore with unused brake fluid of the recommended type and reassemble the cylinder (Fig. 1). Before fitting the dust cover, smear the sealing areas with rubber grease. Squeeze the remainder of the grease from the sachet into the cover to help protect the internal parts.

Refit the slave cylinder to the vehicle and bleed the system.

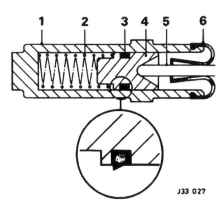

J33 027

Fig. 1

KEY TO DIAGRAM

1. Body
2. Spring
3. Seal
4. Piston
5. Operating rod
6. Rubber boot

CLUTCH SLAVE CYLINDER

Renew

Disconnect the pipe from the slave cylinder; plug or tape the pipe to prevent the ingress of any dirt.

Remove the nuts (1, Fig. 2) and spring washers securing the slave cylinder to the gearbox.

Slide the slave cylinder (1, Fig. 3) off the mounting studs; slide the rubber boot along the push rod, withdraw the cylinder from the push rod (2, Fig. 3).

To reassemble, reverse the removal operations and bleed the clutch hydraulic system.

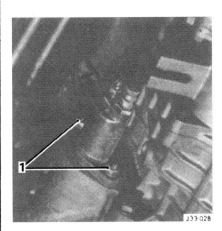

Fig. 2

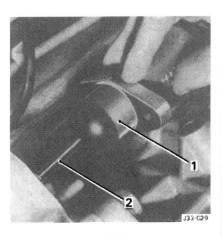

Fig. 3

CLUTCH RELEASE BEARING

Renew

Remove the gearbox from the vehicle and remove the bell housing. Remove the withdrawal mechanism from the bell housing.

Press off the bearings and refit the new bearings to the housing.

Grease all the release arm pivots with Molykote FB 180 Grease, fit the assembly to the release arm and grease the withdrawal assembly.

Refit the bell housing and gearbox, check the oil level and road test the vehicle.

CLUTCH WITHDRAWAL LEVER

Renew

Remove the gearbox and the bell housing. Remove the clutch release bearing assembly (1, Fig. 4) and remove the withdrawal lever assembly pivot nut and remove the lever (2, Fig. 4).

Remove the spire clip (3, Fig. 4) securing the pivot to the withdrawal lever and separate the pivot from the lever.

Fit a new withdrawal lever to the pivot and secure with the spire clip.

Fit the assembly to the bell housing, fit and tighten the securing nut.

Refit the release bearing assembly (after applying Molykote FB 180 Grease) to the bell housing.

Refit the bell housing to the engine and the gearbox to the bell housing.

Fig. 4

CLUTCH ASSEMBLY

Renew

Remove the gearbox from the vehicle and remove the bell housing from the rear of the engine.

Lock the flywheel and remove the clutch cover securing bolts (1, Fig. 5), remove the clutch cover (2, Fig. 5) and drive plate.

Clean the flywheel face and dowels, fit a new plate and cover the flywheel, align the drive plate with a dummy input shaft and torque up the cover securing bolts.

Remove the dummy input shaft, refit the bell housing and gearbox assembly, carry out road test.

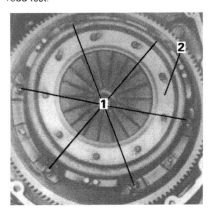

Fig. 5

CLUTCH PEDAL

Renew

Remove the pedal box, see 33.10.07.
Using a large screwdriver to disengage the return spring (1, Fig. 6) from the pedal (2, Fig. 6). Remove the spring clip (3, Fig. 6), washer (4, Fig. 6), and clevis pin (5, Fig. 6) securing the pedal arm to the master cylinder push rod. Remove the locating pin.
Tap the clutch shaft (6, Fig. 6) from the pedal box casting and recover the pedal return spring and washers.
Locate the return spring on the pedal boss.
Smear the clutch shaft with suitable grease.
Tap the clutch shaft through the casting, locating the groove to the outside edge, positioning a washer at either side of the pedal boss.
Fit the pedal arm to the master cylinder push rod and secure using a clevis pin, plain washer and new split pin.
Fit the pedal box and bleed the hydraulic system, see 33.15.01.

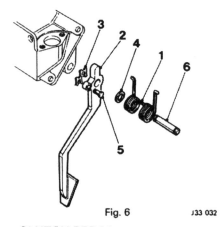

Fig. 6   J33 032

CLUTCH PEDAL

Overhaul

Remove the clutch pedal.
Using a suitable mandrel, press the bush (1, Fig. 7) from the pedal boss (2, Fig. 7).
Using a suitable mandrel, press the new bearing bushes in from each side. Press until the bush is flush with the sides of the pedal.
Lightly ream the bush to the size using the pedal shaft to check fit.
Smear bushes with a suitable grease.
Fit the clutch pedal and fit a new rubber (3, Fig. 7).

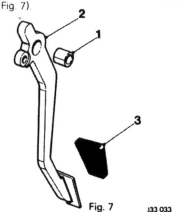

Fig. 7   J33 033

MASTER CYLINDER

Renew

Remove the pedal, see 33.30.02
Note the fitted position of the return spring, disconnect the spring from the pedal.
Remove and discard the split pin.
Withdraw the clevis pin, recover the plain washer.
Remove nuts and spring washers securing the master cylinder to pedal box.
RH drive cars only.
Remove the setscrew nut (1, Fig. 8) and washers securing the master cylinder (2, Fig. 8) to the pillar.
Withdraw the brake fluid reservoir bracket; lift off the master cylinder and shims (if fitted).
To refit the new cylinder, reverse the removal procedure and bleed the hydraulic system.

CAUTION: The hydraulic fluid used in the clutch hydraulic system is injurious to car paintwork. Utmost precautions MUST be taken at all times to prevent spillage of fluid. Should fluid be accidentally spilled on paintwork, wipe fluid off immediately with a cloth moistened with denatured alcohol (methylated spirits).

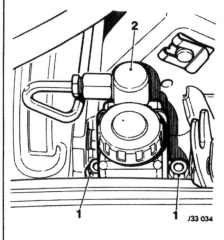

Fig. 8

MASTER CYLINDER

Overhaul

WARNING: USE ONLY CLEAN BRAKE FLUID OR DENATURED ALCOHOL (METHYLATED SPIRITS) FOR CLEANING. ALL TRACES OF CLEANING FLUID MUST BE REMOVED BEFORE REASSEMBLY. ALL COMPONENTS SHOULD BE LUBRICATED WITH CLEAN BRAKE FLUID AND ASSEMBLED USING THE FINGERS ONLY.

Dismantling

Remove the master cylinder, see 33.20.01.
Detach the rubber boot (1, Fig. 9) from the end of the barrel and move the boot along the push rod (2, Fig. 9).
Depress the push rod and remove the circlip (3, Fig. 9).
Withdraw the push rod, piston (4, Fig. 9), piston washer (5, Fig. 9), main cup (6, Fig. 9), spring retainer (7, Fig. 9), and spring (8, Fig. 9).
Remove the secondary cup (9, Fig. 9) from the piston.

Inspection

Examine the cylinder bore for scoring.
Thoroughly wash out the reservoir and ensure the by-pass hole in the cylinder bore is clear. Dry using compressed air or a lint-free cloth.
Lubricate the replacement seals with clean brake fluid.

Reassembling

If necessary, fit end plug on the new gasket.
Fit the spring retainer to the small end of the spring, if necessary, bend over the retainer ears to secure.
Insert the spring, large end leading, into the cylinder bore; follow with the main cup, lip foremost. Ensure the lip is not damaged on the circlip groove. Using the fingers only, stretch the secondary cup on to the piston with the small end towards the drilled end and groove engaging the ridge. Gently work round the cup with the fingers to ensure correct bedding.
Insert the piston washer into the bore, curved edge towards main cup.
Insert the piston in the bore, drilled end foremost.
Fit rubber boot to the push rod.
Offer the push rod to the piston and press into the bore until the circlip can be fitted behind the push rod stop ring.

CAUTION: It is important to ensure that the circlip is fitted correctly into the groove.

Locate the rubber boot in the groove.

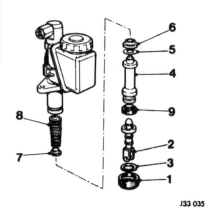

J33 035

Fig. 9

CLUTCH HYDRAULIC SYSTEM

Bleed

CAUTION: Only use Castrol-Girling brake fluid in the hydraulic system.

Remove reservoir filler cap. Top up the reservoir to the correct level with hydraulic fluid.

Attach one end of a bleed tube (1, Fig. 10) to the slave cylinder bleed nipple (2, Fig. 10).

Partially fill a clean container with hydraulic fluid. Immerse the other end of the bleed tube in the fluid.

Slacken the slave cylinder bleed nipple.

Pump the clutch pedal slowly up and down, pausing between each stroke.

Top up the reservoir with fresh hydraulic fluid after every three pedal strokes.

CAUTION: Do not use fluid bled from the system for topping up purposes as this will contain air. If the fluid has been in use for some time it should be discarded. Fresh fluid bled from the system may be used after allowing it to stand for a few hours to all the air bubbles to disperse.

Pump the clutch pedal until the pedal becomes firm, tighten the bleed nipple.

Top up the reservoir, refit the filler cap.

Apply the working pressure to the clutch pedal for two to three minutes and examine the system for leaks.

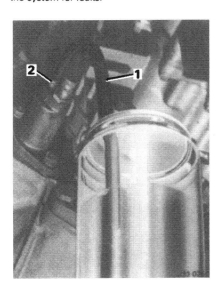

Fig. 10

CLUTCH PIPE DAMPER

Renew

Disconnect the hydraulic pipes (1, Fig. 11).

Remove the securing bolts (2, Fig. 11) and remove the damper (3, Fig. 11).

Refit the new damper and secure with the bolts (2, Fig. 11), reconnect the hydraulic pipes (1, Fig. 11) and bleed the system.

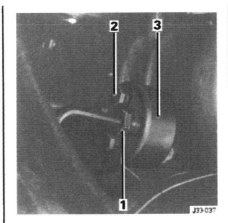

Fig. 11

This page is intentionally blank

CONTENTS

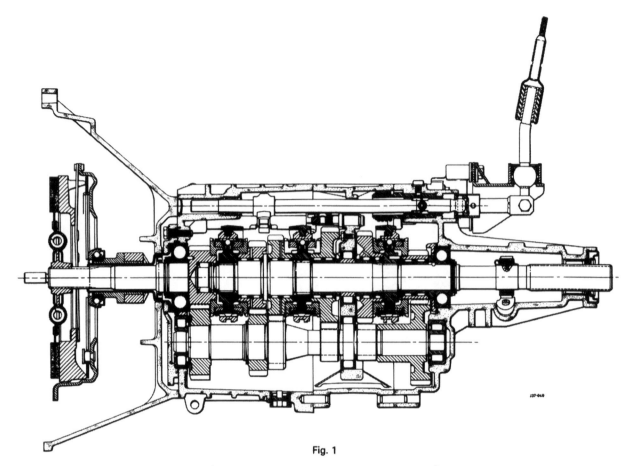

Fig. 1

5 SPEED GETRAG MANUAL GEARBOX

The 5-speed transmission fitted to this vehicle incorporates a synchromesh on all gears. Gear selection is by a centrally mounted lever, connected to the transmission selector shaft via a pivoting joint.

All the gears are engaged by a single selector shaft operating three rods which move the selector forks.

The drive pinion is supported at the rear by a duplex ball bearing situated in the front casing, and at the front a spigot engages in a needle roller bearing fitted in the flywheel.

The mainshaft is supported at the front by a caged roller bearing situated in the drive pinion counterbore in the centre by a roller bearing supported by the intermediate casing and at the rear by a duplex bearing in the transmission rear casing.

Each of the forward speed mainshaft gears incorporates an integral synchromesh mechanism with the clutch hubs splined to the mainshaft and situated between each pair of gears.

The countershaft is supported at the front by a roller bearing situated in the front casing in the centre by a roller bearing in the intermediate casing and at the rear by a roller bearing situated in the tail housing.

The reverse idler gear is supported by two caged roller bearings and is in constant mesh and is situated on a stationary shaft. Longitudinal location of the idler gear is controlled by a spacer abutting on the shaft.

RATIOS
1 3.573:1
2 2.056:1
3 1.391:1
4 1.000:1
5 0.76 :1
R 3.463:1

| TORQUE WRENCH SETTINGS | DESCRIPTION | SPANNER SIZE | TIGHTENING TORQUE | | |
ITEM			Nm	kgf/cm	lbf/ft
Bell housing to adaptor plate	W/F* Bolt M10	13mm	49 to 54	5,0 to 5,5	36 to 40
Front cover to gearbox — top three bolts	Setscrew M8	13mm	23 to 27	2,4 to 2,8	17 to 20
Front cover to gearbox — lower four bolts only	Setscrew M8	13mm	16 to 20	1,6 to 2,1	12 to 15
Gearbox to bell housing	Nut M12	19mm	70 to 80	7,2 to 8,2	52 to 59
Gear lever housing to pedestal	Bolt M8	13mm	23 to 27	2,4 to 2,8	17 to 20
Gear lever pivot bolt	Nyloc Nut M8	13mm	23 to 27	2,4 to 2,8	17 to 20
Speed transducer to gearbox	W/F* Setscrew M6	8mm	9,5 to 12	1,3 to 1,6	7 to 9
Pedestal to gearbox	W/F* Bolt M8	10mm	23 to 27	2,4 to 2,8	17 to 20
Pedestal to gearbox	Bolt M8	13mm	23 to 27	2,4 to 2,8	17 to 20

*Washer faced

BELL HOUSING

Renew

Remove the gearbox from the vehicle and remove the sound deadening pad, remove the exhaust bracket from the bell housing (1, Fig. 2) and remove the clutch slave cylinder (2, Fig. 2).

Remove the bell housing/starter motor securing bolts (3, Fig. 2) and nuts and remove the bell housing (4, Fig. 2).

Remove the release arm securing nut and remove the arm, remove the slave cylinder securing studs.

Fit the slave cylinder studs to the replacement bell housing, the clutch release arm and securing nut.

Fit the bell housing to the adaptor plate and the exhaust bracket to the bell housing.

Refit the starter motor.

Reposition the sound deadening pad and refit the gearbox to the vehicle.

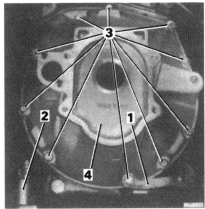

Fig. 2

OIL SEAL — REAR

Renew

Remove the propellor shaft and remove the rear oil seal (1, Fig. 3) from the rear of the gearbox, taking care not to damage the output shaft on the gearbox case.

Carefully position a new seal at the rear of the gearbox and carefully tap the seal into the gearcase, ensuring that the seal is square and flush with the housing.

Lubricate the output shaft and carefully fit the propellor shaft and secure the rear flange to the differential drive flange.

Top up the gearbox and road test the vehicle.

Fig. 3

FRONT OIL SEAL

Renew

Remove the gearbox from the vehicle and drain the oil, remove the seven bolts (1, Fig. 4) securing the thrust plate (2, Fig. 4) to the gearbox and remove the plate.

Prise the seal (3, Fig. 4) from the housing.

Clean the thrust plate and remove the spiral spring from the seal prior to installing the seal in the housing, press the seal into the plate and refit the spiral spring to the seal.

Coat the lip of the seal with grease and the transmission case cover with roller bearing grease and fit the plate to the gearbox ensuring that any shims removed are replaced.

Fit the four lower bolts first followed by the top three.

Prior to fitting, coat all the bolts with Loctite 573.

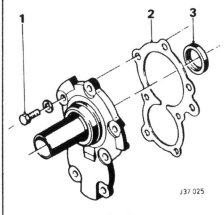

Fig. 4

GEAR SELECTOR SHAFT

Overhaul

Remove the remote control assembly and slide the selector shaft universal joint cover back (1, Fig. 5), remove the selector shaft retaining pin (2, Fig. 5) and collect the shaft (3, Fig. 5).

Remove the pin securing the universal joint and remove the joint.

Clean all components, examine for wear or damage and replace as necessary.

Lubricate all new or components to be re-used and fit the universal joint to the rear selector shaft, secure with the retaining pin, assemble to the gearbox and secure with the retaining pin. Reposition the retaining cover and refit the remote control assembly.

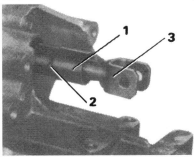

Fig. 5

GEAR LEVER DRAUGHT EXCLUDER

Renew

Remove the gear lever knob (1, Fig. 6) remove the gear lever gaiter, remove the draught excluder securing screws (2, Fig. 6) and ring (3, Fig. 6) and remove the draught excluder (4, Fig. 6).

Renew the draught excluder and fit over the gear lever, secure with the ring and screws, refit the gear lever gaiter and refit the knob.

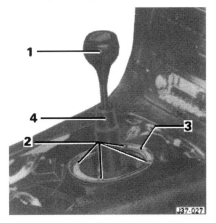

Fig. 6

GEAR LEVER

Renew

Remove the remote control assembly (1, Fig 7) lever securing snap ring (2, Fig. 7), collar (3, Fig. 7), spring (4, Fig. 7) and the lever (5, Fig. 7). Remove the spacer from the lever, remove the lower nylon cup (6, Fig. 7) and the upper nylon spacer (7, Fig. 7).

Clean the gear lever and all other component parts, check for wear and distortion and replace as necessary.

Fit the nylon cup to the mounting and grease the new gear lever ball, fit the lower spacer, spring and collar, compress the spring and secure with the snap ring

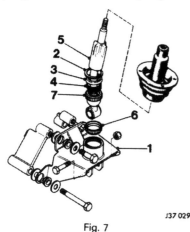

Fig. 7   J37 029

GEAR CHANGE REMOTE CONTROL ASSEMBLY

Overhaul

Slacken the torquatrol to water pump securing nuts, turn the torquatrol assembly and remove the torquatrol from the pulley.

Fit Service Tool MS 53 (1, Fig. 8) across the wing channels, and fit to the rear lifting eye and take the engine weight.

Fig. 8

Remove the front pipe clamp nut/bolt.

Disconnect exhaust from the front pipe and remove the sealing olive.

Remove intermediate heat shield.

Remove rear mounting centre securing nut (1, Fig. 9), the crash bracket to gearbox securing nuts/bolts (2, Fig. 9) and remove the bracket (3, Fig. 9).

Take the weight of the rear mounting with a jack.

Remove the rear mounting to body securing bolts (4, Fig. 9), lower the jack and remove the mounting assembly (5, Fig. 9).

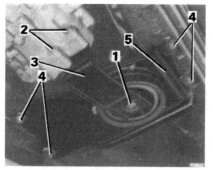

Fig. 9

Remove the spacers from the rear mounting assembly, remove the rear heat shield securing screws, carefully ease the exhaust down and remove the heat shield.

Slacken MS 53 hook nut and slacken, the gear lever to rear shaft securing nut/bolt (1, Fig 10). Also remove the control mounting securing bolts (2, Fig. 10) and remove the gear lever bolt (1, Fig. 10).

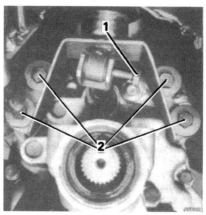

Fig. 10

Pull the gear change lever (1, Fig 11) from the selector shaft (2, Fig 11) and remove the remote control/gear lever assembly. Ensure that the draught excluder (3, Fig 11) is not moved from its position. Once moved, invert the assembly (gear lever down) to aid removal.

Remove the mounting rubbers and washers.

Remove the snap ring (4, Fig. 11) and the gear change lever assembly.

Remove the collar (5, Fig 11) and spring (6, Fig 11) from the lever, remove the gear lever lower nylon cup from the control.

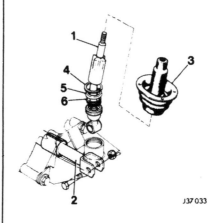

Fig. 11   J37 033

Clean all the component parts and inspect for wear or damage, and replace as necessary.

Fit the lower nylon cup.

Grease the gear lever ball.

Fit the spring and collar to the lever.

Compressing the spring with the gear lever assembly, secure the snap ring.

Lubricate the gear lever to ease fitment to the draught excluder.

Lubricate the lower spacer and position the assembly into the mounting.

Enter the lever into the gaiter and position the lever into the yoke, ensure the lever is fitted correctly.

Fit but do not tighten the gear lever securing nut and bolt.

Fit the lower LH mounting rubber and washer. Align mounting bracket and fit the remaining spacer and rubber.

Fit but do not tighten the mounting securing bolt.

Repeat the above procedure for the remaining rubbers and finally tighten the securing bolts.

Finally tighten the gear lever securing nut and bolt.

Tighten MS 53 hook nut to raise the gearbox into position.

Fit the gearbox mounting to the body.

Ensure that the spring is fully seated in the gearbox spring pan and fit but do not fully tighten the mounting bolts.

Align the mounting assembly, finally tighten the securing bolts, remove the engine retaining Service Tool MS 53 and fit the rear mounting spacers.

Fit the crash bracket and tighten the securing nuts and bolts.

Fit the rear heat shield.

Fit the intermediate heat shield.

Clean the exhaust sealing olive and smear the olive with sealant. Fit the sealing olive to the intermediate pipe.

Connect the exhaust system to the front pipe, position the clamp over the union and tighten the clamp bolt.

Reposition the fan/torquatrol assembly, fit but do not tighten torquatrol securing nuts, fully seat the assembly onto the studs and finally tighten the assembly securing nuts.

SPEEDOMETER DRIVE PINION

Renew

Release the speed transducer knurled ring (1, Fig. 12) and remove the transducer (2, Fig. 12).

Remove the drive gear locking clamp (3, Fig. 12) and remove the drive gear from the gearbox and remove the 'O' ring.

Fit and 'O' ring to the drive, lubricate, fit the assembly to the gearbox and secure with a clamp.

Refit the transducer and secure with the knurled ring.

Fig. 12

GEARBOX ASSEMBLY

Renew

Select third gear.

Remove the torquatrol unit from the front pulley. Fit Service-Tool MS 53 (1, Fig. 13) to he rear engine lifting eye and take the weight of the engine.

Fig. 13

Release the clamp (1, Fig. 14) securing the front pipe from the intermediate pipe and collect the olive.

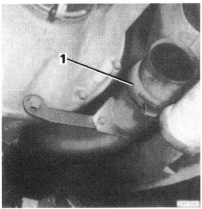

Fig. 14

Remove the intermediate heat shield.

Remove the gearbox rear mounting nut (1, Fig. 15) and the crash bracket (2, Fig. 15). Take the weight of the gearbox and remove the rear mounting complete (3, Fig 15) and lower the gearbox.

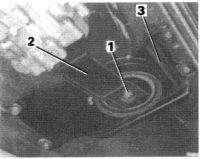

Fig. 15

Remove the rear heat shield and the propeller shaft and fit a blanking plug to the rear of the gearbox.

Disconnect the speed transducer block connector (1, Fig. 16), remove the clutch slave cylinder securing nuts and pull the cylinder clear of the bell housing, disconnect the reverse light feed wires from the switch (2, Fig. 19).

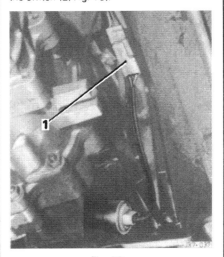

Fig. 16

Remove the selector bracket securing bolts (1, Fig. 17), and collect the mounting rubbers and washers (2, Fig. 17).

Remove the selector lever to yoke securing bolt and nut (3, Fig. 17) and disconnect the lever.

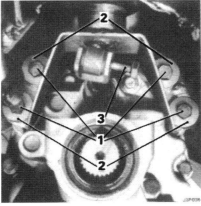

Fig. 17

Take the weight of the gearbox, remove the bell housing securing bolts (1, Fig 18 & 19) and remove the gearbox from the vehicle.

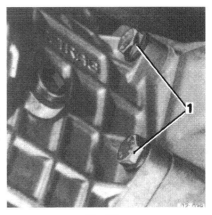

Fig. 18

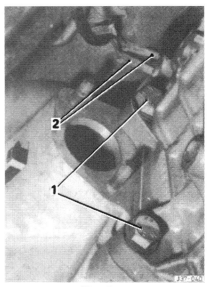

Fig. 19

Check the alignment of the clutch plate

Select third gear on the new gearbox, fit the reverse light switch, lift the gearbox into position and secure with the bell housing bolts.

Fit a new 'O' ring to the speed transducer, fit the transducer and secure with the clamp.

Fit the rear mounting assembly to the gearbox, lift the gearbox into position at the rear and secure with the mounting nuts and washers.

Grease the selector universal joint and align and fit the selector assembly to the shaft, secure with the pin and reposition the universal joint cover

Fit the securing rubbers to the selector lever mounting bracket and fit the lever to yoke securing bolt.

Fit the selector mounting washers to the rubbers, align the mounting holes and fit the bracket securing bolts.

Reconnect the reverse light switch and speed transducer feed wires.

Fit the crash bracket to the rear mounting and secure with the bolts.

Fit and tighten the slave cylinder securing nuts.

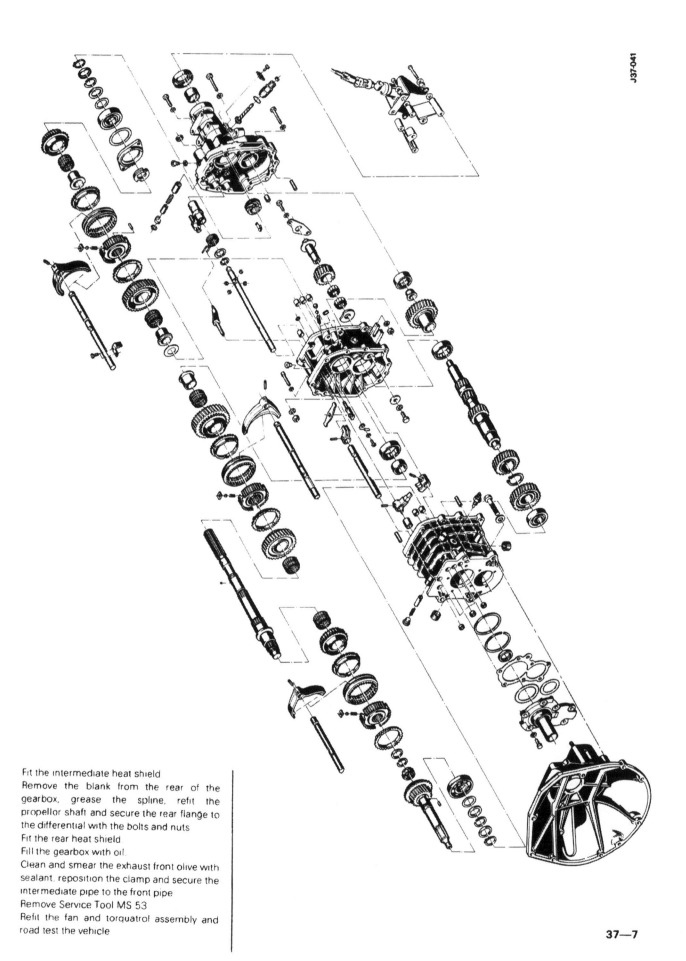

J37-041

Fit the intermediate heat shield
Remove the blank from the rear of the
gearbox, grease the spline, refit the
propellor shaft and secure the rear flange to
the differential with the bolts and nuts
Fit the rear heat shield
Fill the gearbox with oil.
Clean and smear the exhaust front olive with
sealant, reposition the clamp and secure the
intermediate pipe to the front pipe
Remove Service Tool MS 53
Refit the fan and torquatrol assembly and
road test the vehicle

37—7

CONTENTS

TORQUE WRENCH SETTINGS

Item	Spanner Size	Description	Tightening Torque		
			Nm	kgf/m	lbf/ft
Drive shaft to drive unit (Cleveloc)	$\frac{11}{16}$ in AF	$\frac{7}{16}$ in UNF nut	66,4 to 74,5	6,78 to 7,6	49 to 55
Drive shaft to hub carrier	$1\frac{1}{8}$ in AF	$\frac{3}{4}$ in UNF nut	136 to 163	13,8 to 16,6	100 to 120
Propeller shaft flange bolts	$\frac{9}{16}$ in AF	$\frac{3}{8}$ in UNF bolts and nuts	36,7 to 43,4	3,74 to 4,42	27 to 32

SERVICE TOOLS

Tool No.   Description
JD1D   Hub remover

DRIVE SHAFTS AND PROPELLER SHAFT

Description

The drive shafts replace the half shafts of a conventional rear axle, and in addition serve as upper transverse members to locate the rear wheels; their inner universal joints are attached to the final drive unit by bolts which also carry the brake discs, but the brakes are not disturbed in drive shaft removal. The outer joints are integral with the hub driving shafts, and the hubs must therefore be separated from the drive shafts before they can be removed.

The propeller shaft is a two universal joint type, at the front end of which is a reverse spline fitting coupled to the gearbox and at the rear a flange bolted to the input drive flange of the final drive unit.

When fitting a propeller shaft it is essential to ensure that the universal joints operate freely; any stiffness, even in a single joint, will initiate propeller shaft vibration.

DRIVE SHAFT

Renew

Service tool: Hub remover JD 1D.

To remove a drive shaft it is necessary to detach the hub and to swing one suspension unit aside to clear the inner joint.

Ensure that car is securely supported on stands before removing the wheel. Release clips (1, Fig. 1) and before removing nut from drive shaft in hub, slide inner shroud along shaft.

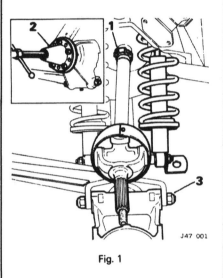

Fig. 1

Remove grease nipple from hub carrier, and using tool JD 1D (2, Fig. 1), withdraw hub from shaft. Allow the hub carrier to pivot about wishbone pin. Before detaching inner joint, release lower end of rear spring/damper unit (3, Fig. 1) and swing aside to clear joint. Collect and retain any camber setting shims fitted between inner joint and brake disc.

Refitting

Replacement drive shafts are supplied without shrouds, oil seal track or spacer; remove these items and transfer them to the new shaft. Seal shroud joints with underseal. Ensure that chamfer on oil seal track clears radius on shaft, and apply Loctite to spline before refitting hub. Tighten all nuts and bolts to the correct torque. Check and if necessary adjust hub bearing end-float and ensure that camber angles of the wheels are correct.

DRIVE SHAFT

Overhaul

Dismantling

Remove drive shaft.

Remove grease nipples (1, Fig. 2), place shaft in vice and remove two opposed circlips (2, Fig. 2).

NOTE: Tap bearings slightly inwards to assist removal of circlips.

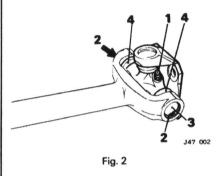

Fig. 2

Tap one bearing inwards to displace opposite bearing (3, Fig. 2).

Trap displaced bearing in vice and remove shaft and joint from bearing.

Replace shaft in vice, displace second bearing by tapping joint spider across and extract second bearing.

Remove two grease seals (4, Fig. 2).

Detach spider, with end section of shaft, from centre section of shaft.

Place end section of shaft in vice and repeat above operations.

Remove spider from end section of shaft.

Repeat above operations on opposite end of shaft.

Inspection

Wash all parts in petrol.

Check splined yoke for wear of splines.

Examine bearing races and spider journals for signs of looseness, load markings, scoring or distortion.

NOTE: Spider or bearings should not be renewed separately, as this will cause premature failure of the replacement.

It is essential that bearing races are a light drive fit in yoke trunnion.

Reassembling

Remove bearing assemblies from one replacement spider; if necessary, retain rollers in housings with petroleum jelly. Leave grease shields in position.

Fit spider to one end section of shaft.

Fit two bearings and circlips in end section trunnions. Use a soft round drift against bearing housings.

Insert spider in trunnions of centre section of shaft.

Fit two bearings and circlips in centre section trunnions.

Fit grease nipple to spider.

Repeat above operations on opposite end of drive shaft.

Grease joints with hand grease gun.

Refit drive shaft.

PROPELLER SHAFT

Renew

Slacken the front to intermediate exhaust pipe clamp $\frac{7}{16}$ in AF nut and bolt, (1, Fig. 3) spread the clamp (2, Fig. 3) and push it clear along the pipe. Pull the intermediate pipe down and remove the olive (3, Fig. 3).

Remove the four self-tappers and washers (1, Fig. 4) and pull the heat shield (2, Fig. 4) forward and down and remove from the vehicle.

Mark the propshaft flange relative to the differential flange, remove the four $\frac{9}{16}$ in AF flange securing bolts and nuts and remove the propshaft from the vehicle.

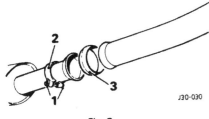

Fig. 3

Refit the propshaft in the opposite procedure to the removal noting that the propshaft flange nuts are torqued to 27 - 32 lbf/ft (3,7 - 4 kgf/m) and the propshaft alignment marks are observed.

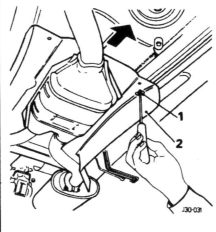

Fig. 4

PROPELLER SHAFT

Overhaul

To overhaul the propeller shaft universal joints, remove the snap-rings from the grooves (1, Fig. 5).

NOTE: If difficulty is encountered, tap the bearing cup (2, Fig. 5) inwards to relieve the pressure on the snap-ring.

Hold the flange yoke and tap the yoke with a soft-faced hammer. The bearing cup should gradually emerge and can be finally removed.

Alternatively, secure the propeller shaft in a vice. Using a suitable soft metal drift, drift down on a bearing cup to displace the opposite cup. Remove the propeller shaft from the vice, hold the displaced cup in the vice and separate from the propeller shaft by pulling and twisting.

Repeat the above operations for the opposite bearing cup, and the remaining bearing cups at each end of the shaft.

Reassembling

Using new universal joint assemblies if necessary, insert the spider into the flange, tilting it to engage in the yoke bores. Ensure that all the needle rollers are in position; fill each bearing cup one-third full of grease of the recommended type.

Fit one of the bearing cups (2, Fig. 5) in the yoke bore, and using a suitable soft metal drift, tap the bearing cup fully home.

Fit a new snap-ring (1, Fig. 5) ensuring it is correctly located in the groove.

Assemble the other spiders and bearing cups, and fit new snap-rings, to retain the bearing cups.

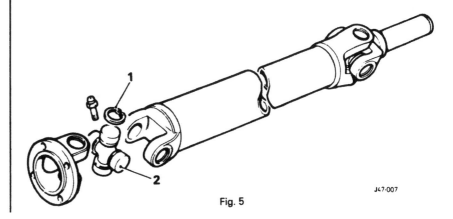

Fig. 5

This page is intentionally blank

CONTENTS

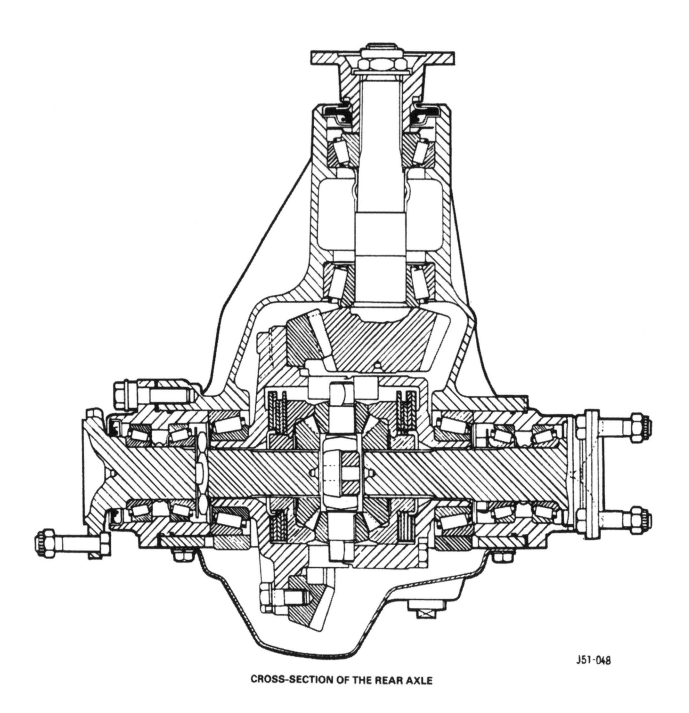

J51-048

CROSS-SECTION OF THE REAR AXLE

TORQUE WRENCH SETTINGS

Item	Spanner Size	Description	Tightening Torque		
			Nm	kgf/m	lbf/ft
Caliper mounting bracket to unit	$\frac{5}{8}$ in AF	$\frac{7}{16}$ in UNC setbolts	81,3 to 93	8,3 to 9,54	60 to 69
Differential bearing caps	$\frac{3}{4}$ in AF	$\frac{1}{2}$ in UNC setbolts	85,4 to 97	8,71 to 9,95	63 to 72
Drive pinion nut	$1\frac{1}{8}$ in AF	$\frac{3}{4}$ in UNF nut	244 to 256	24,92 to 26,34	180 to 190
Drive gear to differential flange	$\frac{5}{8}$ in AF	$\frac{7}{16}$ in UNF setbolts	102 to 118	10,78 to 12,16	77 to 88
Powr-Lok differential case	$\frac{9}{16}$ in AF	$\frac{3}{8}$ in UNC setbolts	58,3 to 67	5,95 to 6,9	43 to 50
Rear cover attachment	$\frac{1}{2}$ in AF	$\frac{5}{16}$ in UNC setbolts	20,5 to 27	2,1 to 2,76	15 to 20
Ring gear attachment	$\frac{11}{16}$ in AF	$\frac{7}{16}$ in UNF Rippbolt	136 to 151	13,8 to 15,46	100 to 111

SERVICE TOOLS

Tool No.	Description
18G 120 5	Flange Holder
18G 134 (MS 550, 550, SL 550)	Adaptor Handle
SL 550-1	Outer Pinion Cup Remover
47 (MS 47, SL 14)	Hand Press
⎰ SL14-3/2	Differential Side Bearing Remover
⎱ SL14-3/1	Differential Side Bearing Remover Button
⎰ SL 3	Clock Gauge Tool
⎱ 4 HA	Pinion Height Setting Gauge
SL 550-9	Pinion Inner Bearing Cup Replacer
SL 550-8/1	Pinion Outer Bearing Cup Replacer
⎰ SL 47-1/1	Pinion Head Bearing Remover
⎱ SL 47-1/2	Pinion Head Bearing Replacer
18G 1428	Rear Oil Seal Replacer
SL 15A	Spanner
18G 681 CBW 548	Torque Driver
⎰ SL 47-3/1	Output Shaft Outer Bearing Remover
⎱ SL 47-3/2	Output Shaft Outer Bearing Replacer
JD 14	Dummy Shaft

⎰
⎱ Items marked thus are sold as sets.

DESCRIPTION

The standard transmission unit is a Salisbury 4HU final drive, incorporating a 'Powr-Lok' differential when specified; this is identified by the letters 'PL' on a tab under a cover bolt. A Powr-Lok differential differs from a conventional bevel gear unit by the addition of plate clutches loaded by input torque to oppose rotations of the output shafts relative to the differential cage. Clutch plates are splined to the cage, and their mating discs to the output bevels; the loading between plates and disc increases with input torque due partly to the separating forces of bevels and also to the bevel pinion cross-shafts being carried on ramps instead of being positively located in the cage. Increase in output torque causes the cross-shafts to move 'up' the ramps and, by pressing plates and discs together, to 'lock' the differential; this gives the effect of a differential-less axle at maximum torque without increasing the disadvantages of this type of axle in low-torque conditions. Some low-torque stiffness, to reduce one-wheel spin on ice, is provided by forming the outer plates as Belleville washers to produce compression between plates and discs; if one wheel is held and the propeller shaft is disconnected, a torque of between 5,6 and 9,6 kgf/m (40 to 70 lbf/ft) is required to turn the other wheel.

The final drive unit is rigidly attached to a fabricated sheet steel cross-beam which is flexibly mounted to the body structure by four rubber and metal sandwich mountings. Noises coming from the vicinity of the final drive unit usually originate from incorrect meshing of drive gear and pinion, or from bearings on differential or pinion shafts developing play. Operation procedures for the correction of these noise sources are fully covered in operation 51.25.19, but a noise occurring at low speeds only, under braking, could be caused by loss of pre-load in the output shaft bearings. Bearing inspection involves the removal and renewal of an oil seal before resetting pre-load, and is covered in operation 51.20.04, while if inspection indicates that bearing renewal is advisable this is detailed in operation 51.10.22.

TO CHECK THE TOOTH CONTACT PATTERN

Sparingly paint eight or ten of the drive gear teeth with a stiff mixture of marking raddle or engineers blue. Move the painted gear teeth in mesh with the pinion until a good impression of the total contact is obtained. The result should conform with the ideal tooth contact pattern (Fig. 1).

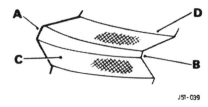

J51-039

Fig. 1 Ideal tooth contact pattern.

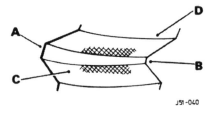

J51-040

Fig. 2 High tooth contact pattern.

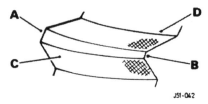

J51-042

Fig. 3 Low tooth contact pattern.

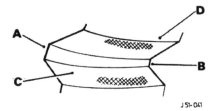

J51-041

Fig. 4 Toe contact pattern.

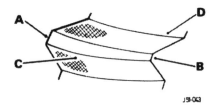

J51-043

Fig. 5 Heel contact pattern.

A The HEEL is the larger outer end of the tooth.

B The TOE is the small or inner end of the tooth.

C The DRIVE side of the drive gear tooth is convex.

D The COAST side of the drive gear tooth is concave.

FAULT DIAGNOSIS

TOOTH PATTERN	REMEDY
The ideal tooth bearing impression on the drive and coast sides of the gear teeth is evenly distributed over the working depth of the tooth profile and is located nearer to the toe (small end) than the heel (large end). This type of contact permits the tooth bearing to spread towards the heel under operating conditions when allowance must be made for deflection.	———
In High Tooth Contact it will be observed that the tooth contact is heavy on the drive gear face or addendum. To rectify this condition, move the pinion deeper into mesh, that is, reduce the pinion cone setting distance, by adding shims between the pinion inner bearing cup and the housing and fitting a new collapsible spacer.	Move the drive pinion deeper into mesh, i.e. reduce the pinion cone setting.
In Low Tooth Contact it will be observed that the tooth contact is heavy on the drive gear flank or dedendum. This is the opposite condition from that shown in High Tooth Contact and is therefore corrected by moving the pinion out of mesh, that is, increase the pinion cone setting distance by removing shims from between the pinion inner bearing cup and housing and fitting a new collapsible spacer.	Move the drive pinion out of mesh. i.e. increase the pinion cone setting
Toe Contact occurs when the bearing is concentrated at the small end of the tooth.	Move the drive gear out of mesh, that is, increase backlash, by transferring shims from the drive gear side of the differential to the opposite side.
Heel Contact is indicated by the concentration of the bearing at the large end of the tooth.	Move the drive gear closer into mesh, that is, reduce backlash, by adding shims to the drive gear side of the differential and removing an equal thickness of shims from the opposite side. **NOTE:** It is most important to remember when making this adjustment to correct a heel contact that sufficient backlash for satisfactory operation must be maintained. If there is insufficient backlash the gears will at least be noisy and have a greatly reduced life, whilst scoring of the tooth profile and breakage may result. Therefore, always maintain a minimum backlash requirement of 0,10 mm. (0.004 in).

DRIVE PINION SHAFT OIL SEAL

Renew

Service tools: Torque screwdriver 18G 681, Oil seal replacer 18G 1428.

Detach the four bolts (1, Fig. 6) securing propeller shaft to final drive flange; support propeller shaft rear end and clean flange and nose of final drive.

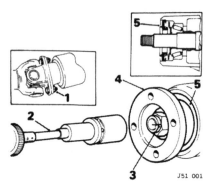

Fig. 6

Accurately measure torque required to turn flange through backlash, using torque screwdriver 18G 681 (2, Fig. 6) with a suitable adaptor and socket.

NOTE: Set screwdriver initially to 0,057 kgf/m (5 lbf/in) and increase setting progressively until torque figure is reached at which flange commences to move. Flange MUST be turned fully anti-clockwise through backlash between each check.

Mark nut and pinion shaft so that in refitting, nut may be returned to its original position on shaft (3, Fig. 6).
Unscrew nut and remove washer and place both washer and nut aside for refitting.
Draw flange (4, Fig. 6) off pinion shaft using extractor.
Prise oil seal (5, Fig. 6) out of final drive casing.

Refitting (using original bearings)

Thoroughly clean splines on pinion shaft and flange. Clean oil seal recess and coat internally with Welseal liquid sealant. Using tool No. 18G1428 tap new oil seal squarely into position with sealing lip facing to rear (1, Fig. 7).

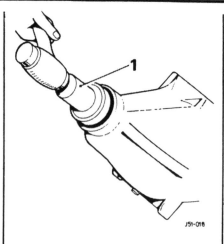

Fig. 7

Smear sealing lip with grease.
Apply grease lightly to outer two thirds of pinion shaft splines.
Lightly tap flange back on pinion shaft, using wooden mallet.
Refit washer and nut and tighten nut until it exactly reaches position previously marked.
Re-check turning torque. Torque required to turn pinion shaft through backlash should exceed by 0,7 to 1,4 kgf/m (5 to 10 lbf/in) the torque recorded earlier. If, however, torque required to turn pinion shaft exceeds 0,52 kgf/m (45 lbf/in), final drive overhaul, operation 51.25.19 MUST be carried out.

Lift propeller shaft into position, replace bolts, fit and tighten nuts to correct torque.
Check oil level in final drive unit and top up if necessary.
Remove car from ramp and road test.
If final drive is noisy, an overhaul must be carried out.

FINAL DRIVE REAR COVER GASKET

Renew

Remove the fourteen $\frac{1}{2}$ in AF bolts and setscrews (1, Fig. 8) securing the bottom tie-plate to the cross-beam and inner fulcrum brackets.
Drain the oil from the final drive.
Remove the ten $\frac{1}{2}$ in AF setscrews (1, Fig. 9) and remove the rear cover (2, Fig. 9) noting the position of the identification tabs.
Clean off any gasket or sealant from the rear cover and the hypoid housing.
Smear the rear cover flange with Wellseal jointing compound and place the gasket on the casing.
Refit the rear cover and secure with the ten setscrews, prior to fitting coat the threads of the bolts with Loctite.
Refill with new oil.

NOTE: The vehicle must be on level ground before checking the oil level.

Replace the bottom tie-plate and tighten the bolts and setscrews to the correct torque.

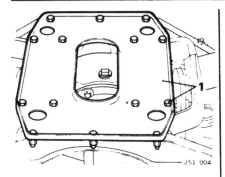

Fig. 8

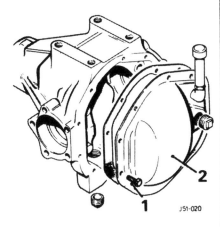

Fig. 9

OUTPUT SHAFT ASSEMBLY (One Side)

Renew

To remove an output shaft it is necessary to detach the inboard end of the drive shaft, the forward attachment of the radius rod, and to remove the brake caliper and disc (1, Fig. 10).

These operations are detailed in Section 70, the Brake System.

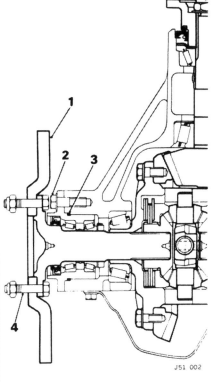

Fig. 10

Cut locking wire and remove five set bolts (2, Fig. 10) securing caliper mounting flange to final drive. Withdraw complete output shaft assembly and discard 'O' ring (3, Fig. 10).

Before fitting, ensure that four bolts (4, Fig. 10) are in position, and that new 'O' ring (3, Fig. 10) is fitted. Lightly oil splines and outside of bearing with final drive oil, insert assembly; fit bolts with spring washers, tighten to 8.4 to 9.66 kgf/m (60 to 69 lbf/ft), tightening the bolt nearest to the input flange first, and wire lock bolt heads together so that wire tension is tending to tighten bolts.

Replace brake caliper and disc as described in Brake System section; check camber angle of rear wheels, and adjust if necessary, refer to Section 64 for the correct procedure.

OUTPUT SHAFT BEARINGS

Renew

Service tools: 47 Press, Torque screwdriver 18G 681, Adaptor, Spanner SL 15A or 15, Output shaft bearing remover/replacer SL 47-3/1, SL 47-3/2.

Remove output shaft assembly incorporating bearing to be removed.

Clean assembly and clamp caliper mounting bracket between suitably protected jaws of vice.

Turn down tabs of lock washer and remove nut (1, Fig. 11) from shaft, using spanner SL 15A (Fig. 12).

Remove and discard lock washer.

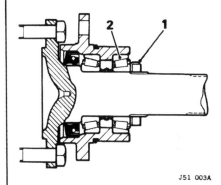

Fig. 11

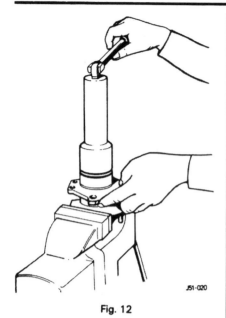

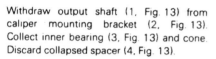

Fig. 12

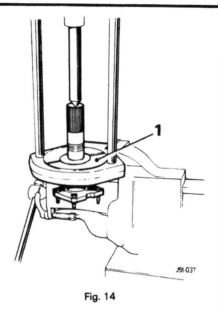

Fig. 14

Withdraw output shaft (1, Fig. 13) from caliper mounting bracket (2, Fig. 13). Collect inner bearing (3, Fig. 13) and cone. Discard collapsed spacer (4, Fig. 13).

NOTE: If outer bearing remains on shaft and pushes oil seal out of caliper mounting bracket on withdrawal, remove it from shaft using tool SL 47-3/1, 47 (1, Fig. 14).

Prise oil seal from caliper mounting bracket. Collect outer bearing and cone. Discard oil seal.

Using a suitable drift, gently tap bearing cups (5, Fig. 13) out of housing

Remove caliper mounting bracket from vice and carefully clean internally

NOTE: When bearings are to be renewed, always replace complete bearings. Never fit new cone and roller assemblies into used cups.

Before fitting, bearings should be lightly greased, but it is most important that at least 4 cc of hypoid oil is added to the cavity between the bearings during assembly, and that the oil seal is lubricated by packing the annular space between its sealing edges with grease. This prevents premature seal or bearing wear before oil flow begins from the axle centre.

Refitting

Press cups of replacement bearings into housing, using suitble press and adaptors to ensure that cups are pressed fully home in housing.

Place roller and cone assembly of outer bearing (already greased) in position.

Press replacement oil seal into position (1, Fig. 15) ensuring that spring-loaded sealing edge is adjacent to bearing. Load seal with grease between sealing edges.

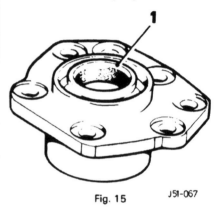

Fig. 15   J51-067

Clamp caliper mounting bracket between protected jaws of vice.

Check that four special bolts for brake disc are in position in output shaft flange and enter shaft through seal and outer bearing.

Fit new collapsible spacer and fill the space between bearings with Hypoid EP 90 oil before replacing rollers and cone of inner bearing and fitting new lock washer on shaft

Place nut on shaft, grease face next to washer and tighten finger-tight only

Using spanner SL15A and a tommy-bar at disc attachment bolts to oppose torque, tighten nut on shaft just sufficiently to almost eliminate play from bearings. Torque required to turn shaft should be 0.14 to 0.28 kgf/m (10 to 20 lbf/in)

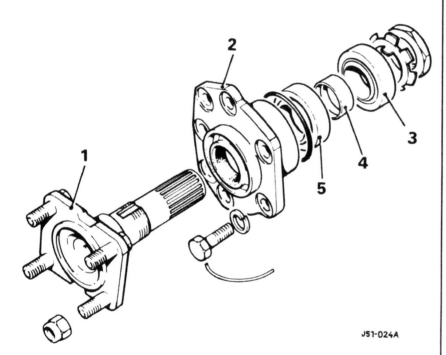

Fig. 13

Further tighten nut, very slightly (not more than a thirty-second of a turn — about 5 mm ($\frac{3}{16}$ in) at perimeter of nut) and re-check torque required to turn shaft. Continue to tighten nut in very small increments, turning shaft to seat bearings and measuring torque after each increment, until correct figure is reached.

CAUTION: If torque required to turn shaft exceeds by more than 0,28 kgf/m (20 lbf/in) torque recorded in first check, it is necessary to dismantle assembly, discard collapsed spacer and rebuild with new collapsible spacer. It is not permissible to slacken back nut after collapsing spacer as bearing cones are then no longer rigidly clamped.

Turn down tab washers in two places to lock nut and remove assembly from vice. Refit output shaft assembly to final drive unit, see operation 51.10.20.

OUTPUT SHAFT OIL SEAL

Renew

Service tools: 47 Press, torque screwdriver 18G 681, Adaptor, Spanner SL 15A or 15 Output shaft bearing remover/replacer SL 47-3/1, SL 47-3/2.

Remove output shaft assembly.
Clean assembly and clamp caliper mounting bracket between suitably protected jaws of vice.
Turn down tabs of lock washer (1, Fig. 16) and remove nut from shaft, using spanner SL15A (1, Fig. 17).

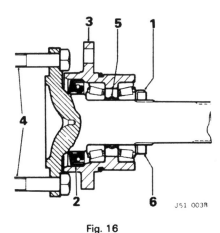

Fig. 16

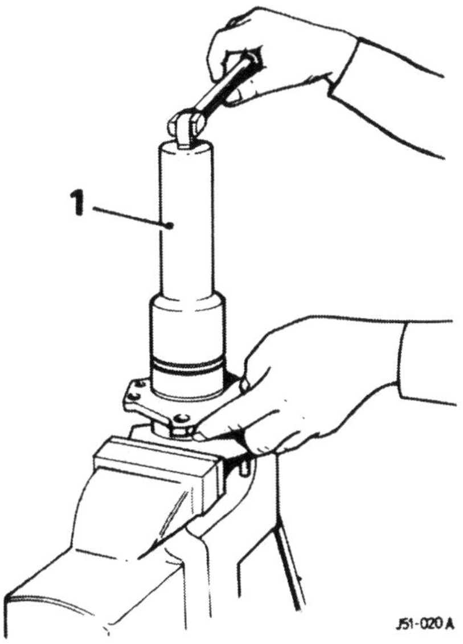

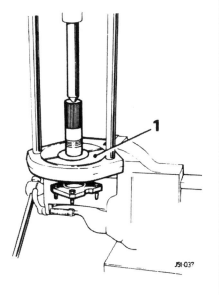

Fig. 18

NOTE: Carefully inspect taper roller bearing components before refitting. If any fault is found in either bearing, replace both complete bearings. Refer to operation 51.10.22, for full details. Never fit new cone and roller assemblies into used cups.

Fig. 17

Remove and discard lock washer.

Withdraw output shaft from caliper mounting bracket. Collect inner bearing and cone and mark for correct reassembly. Discard collapsed spacer.
Prise oil seal from caliper mounting bracket and discard. Collect outer bearing and cone. Remove caliper mounting bracket from vice and thoroughly clean internally.
If outer bearing remains on shaft and pushes oil seal out of caliper mounting bracket on withdrawal, remove it from shaft using tool SL47-3/1, 47 (1, Fig. 18).

Before fitting, bearings should be lightly greased, but it is most important that at least 4 cc of hypoid oil is added to the cavity between the bearings during assembly, and that the oil seal (2, Fig. 16) is lubricated by packing the annular space between its sealing edges with grease. This prevents premature seal or bearing wear before oil flow begins from the axle centre.

Refitting (using original bearings)

Place roller and cone assembly of outer bearing (already greased) in position.
Press replacement oil seal into position, ensuring that spring-loaded sealing edge is adjacent to bearing. Load seal with grease between sealing edges.
Clamp caliper mounting bracket (3, Fig. 16) between protected jaws of vice.
Check that four special bolts (4, Fig. 16) for brake disc are in position in output shaft flange and enter shaft through seal and fit the outer bearing using tools SL47-3/1, SL47-3/2 (1, Fig. 19).

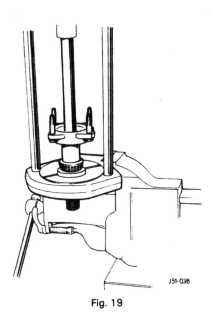

Fig. 19

Smear oil on portion of shaft in contact with seal.
Fit new collapsible spacer (5, Fig. 16) and fill the space between bearings with Hypoid EP 90 oil before replacing rollers and cone of inner bearing and fitting new lock washer on shaft.
Place nut (6, Fig. 16) on shaft, grease face next to washer and tighten finger-tight only.
Using torque screwdriver 18G 681 and adaptor check torque required to turn shaft in caliper mounting bracket against resistance of the oil seal. Record the torque.

NOTE: Set screwdriver initially to 0,05 kgf/m (4 lbf/in). Setting should then be progressively increased until torque figure is established at the point when shaft commences to turn.

Using spanner SL15A and a tommy-bar at disc attachment bolts to oppose torque, tighten nut on shaft just sufficiently to almost eliminate play from bearings. Repeat torque check. Torque required to turn shaft should be unchanged, if it has increased, slacken nut very slightly and re-check.

Further tighten nut, very slightly (not more than a thirty-second of a turn — about 5 mm ($\frac{3}{16}$ in) at perimeter of nut — and re-check torque required to turn shaft. If this torque exceeds by 0,05 to 0,10 kgf/m (4 to 8 lbf/in) the torque recorded earlier, correct bearing pre-load has been achieved, otherwise continue to tighten nut in very small increments, turning shaft to seat bearings and measuring torque after each increment, until correct figure is reached.

CAUTION: If torque required to turn shaft exceeds by more than 0,10 kgf/m (8 lbf/in) torque recorded initially, it is necessary to dismantle assembly, discard collapsed spacer and rebuild with new collapsible spacer. It is not permissible to slacken back nut after collapsing spacer as bearing cones are then no longer rigidly clamped.

Turn down tab washer in two places to lock nut and remove assembly from vice.

Refit output shaft assembly to final drive unit, refer to operation 51.10.20 for full details.

FINAL DRIVE UNIT

Service tool: Dummy shaft JD14.

Renew

The final drive unit cannot be removed from the vehicle unless it is detached as part of the rear suspension unit, removal of this item is detailed in the rear suspension section.

Drain the oil from the unit to prevent any leakage from the breather, and invert the whole assembly onto a workbench.

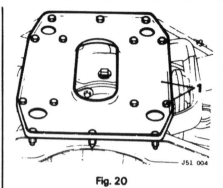

Fig. 20

Remove the fourteen $\frac{1}{2}$ in AF bolts, nuts and setscrews (1, Fig. 20) securing the bottom tie-plate to cross-beam and inner fulcrum brackets.

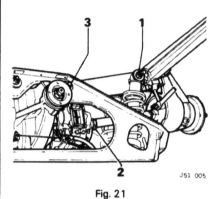

Fig. 21

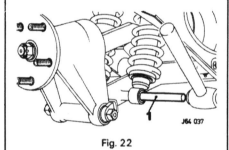

Fig. 22

Remove the $\frac{11}{16}$ in AF nuts and washers (1, Fig. 21) securing the dampers to the wishbone and drift out the retaining pins (1, Fig. 22) recover the spacers and tie-down brackets.

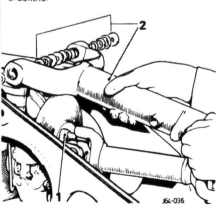

Fig. 23

Slacken the clips (2, Fig. 21) securing the inner universal joint shrouds and slide the shrouds outwards.

Remove the four $\frac{11}{16}$ in AF self locking nuts (1, Fig. 23) either side securing the drive shaft inner universal joint to the brake disc and output flange.

Remove the $\frac{3}{4}$ in AF nut (3, Fig. 21) from the inner wishbone fulcrum shaft and drift out the shaft (1, Fig. 24) collecting the spacers, seals and bearings from the wishbone pivots (2, Fig. 23).

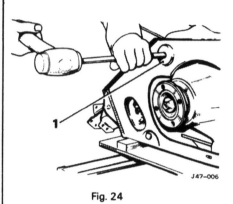

Fig. 24

Remove the drive shaft, hub and wishbone assembly from the rear suspension assembly.

Remove the camber shims from the drive shaft flange studs at the brake disc on both sides.

Remove the spacer tubes from between the lugs of the fulcrum brackets and turn the suspension assembly over on the bench.

Disconnect the brake feed pipes from the calipers, seal the ends of the pipes and the ports in the calipers. Release the brake return springs from the operating levers.

Cut the locking wire and remove the four $\frac{3}{4}$ in AF bolts (1, Fig. 25) securing the final drive to the cross-beam and lift the cross-beam off the unit (Fig. 26).

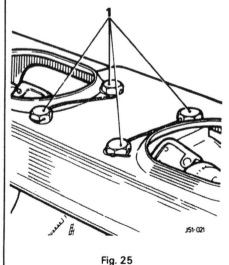

Fig. 25

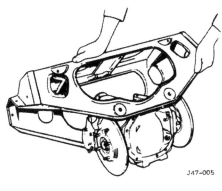

Fig. 26

Invert the unit and remove the locking wires and the $\frac{11}{16}$ in AF setscrews securing the fulcrum brackets to the final drive unit (1, Fig. 27).

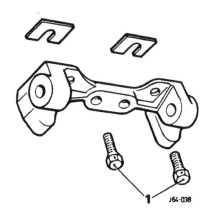

Fig. 27

Remove the brackets, noting the position and number of shims at each attachment point.

Cut the wires from the $\frac{5}{8}$ in AF caliper mounting bolts, remove the bolts and calipers (1, Fig. 28). Remove the brake discs, noting the number of shims between the discs and the flanges.

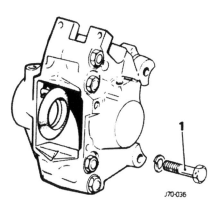

Fig. 28

Replace the shims and disc on one output shaft flange and secure with two nuts. Replace the caliper, tighten the mounting bolts and check the centering and the run out of the disc. The centering tolerance is ±0.25 mm (0.010 in), this can be rectified by transferring shims from one side of the disc to another. The disc run out should not exceed 0.15 mm (0.006 in).
Tighten the caliper bolts to a torque of 6.78 - 7.60 kgf/m 66.4 - 74.5 Nm (49 - 55 lb/ft).
Repeat the above operations on the opposite side. Remove the nuts from both discs.
Place the cross-beam over the final drive, align and replace the bolts and tighten to the correct torque and wire lock 10.4 kgf/m, 101.68 Nm (75 lb/ft). Slacken the brake feed pipes at the centre union, unseal the brake pipes and the ports in the caliper, align and fit the pipes and tighten the unions.
Replace the handbrake lever return springs and invert the assembly on the bench. Position the fulcrum brackets against the final drive unit and locate each bracket loosely with two setscrews. Replace the shims between the fulcrum brackets and the final drive unit.
Tighten the setscrews and wire lock. Refit the camber shims to the drive shaft studs on one side. Fit the drive shaft on to the studs and loosely fit the nuts, and then tighten fully. Replace the spacer tube between the lugs of the fulcrum bracket.
Clean, inspect and grease the lower wishbone bearings, thrust washer etc. Fit new seals and offer up the wishbone fulcrum bracket lugs and locate with dummy shafts.
Tool No. JD14 (1, Fig. 29).

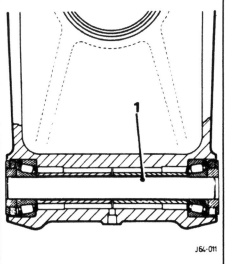

Fig. 29

Take great care not to displace any component during this operation. Drift the dummy shafts from the fulcrum bracket with the fulcrum shaft. Restrain the dummy shafts to prevent spacers or thrust washers dropping out of position.
Tighten the fulcrum shaft nuts to a torque of:

    Inner 61.0 - 67.8 Nm, 6.23 - 6.91 kgf/m (45 - 50 lb/ft).
    Outer 131 - 145 Nm, 13.4 - 14.8 kgf/m (97 - 107 lb/ft).

Reposition the drive shaft shroud and secure it with the clip. Line up the damper lugs with the wishbone bosses and replace the damper shaft, including the spacer and tie down bracket and tighten the nuts to a torque of 43.4 - 48.8 Nm, 4.43 - 4.97 kgf/m (32 - 36 lb/ft).
Replace the wishbone, drive shaft and damper shaft on the opposite side. Replace the bottom tie-plate and tighten the bolts and setscrews.
Replace the rear suspension unit.
Check the rear wheel camber. Bleed the brakes and fill the final drive with oil as necessary.

NOTE: Use Shell Super Spirax 90 or BP Gear Oil 1453 if new gears have been fitted; otherwise use a recommended refill or top up oil as specified in Section 09.

FINAL DRIVE UNIT

Overhaul

Service tools: 18G 1205, 47, SL 47-1/1, 18G 134, SL 550/1, SL 14-3/1, SL 14-3/2, SL 550-1, SL 3, 4HA, SL 550-9, SL 550-8-1, SL 47-1/1, SL 47-1/2, 18G 1428.

Dismantling

Ensure that all lubricant is drained from the unit and support the unit in a vice.

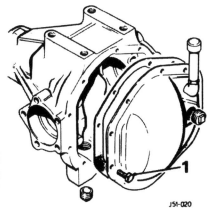

Fig. 30

Remove the ten ½ in AF rear cover securing bolts (1, Fig. 30), the cover and the gasket.
Remove the locking wire and five ⅝ in AF bolts securing the caliper mounting bracket on one side and withdraw the output shaft assembly.
Repeat for the shaft on the other side.
Remove the two ¾ in AF bolts (1, Fig. 31) securing the differential bearing cap, lift out the cap from the differential housing, repeat for the other side.

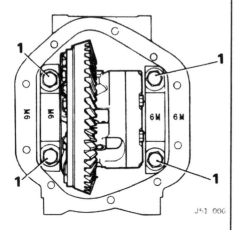

Fig. 31

Using two suitably padded levers, prise out the differential unit.
Using tool 18G 1205 (1, Fig. 32) to hold the drive flange, remove the pinion nut and washer and withdraw the flange (2, Fig. 32).

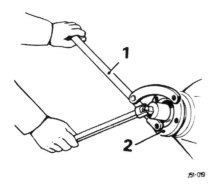

Fig. 32

Using a suitable press extract the pinion from the housing.
Using tool 18G 134 remove the oil seal, oil thrower and outer bearing cone.
Examine the inner and outer bearing cups for wear, if replacement is required extract the outer cup using tools 18G 134 and SL550/1 for inner bearing removal, carefully tap the bearing cone out with a brass punch in the cut-outs provided in the differential casing and carefully collect the shims.

Remove the pinion head bearing using tools 47 (1, Fig. 33), SL 47-1/1 (2, Fig. 33).

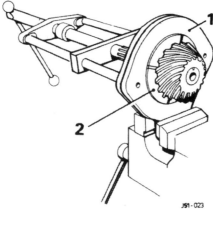

Fig. 33

Remove the differential side bearings using tool Nos. 47 (1, Fig. 34), SL 14-3/2 (2, Fig. 34) and SL 14-3/1 (3, Fig. 34), and collect the shims.

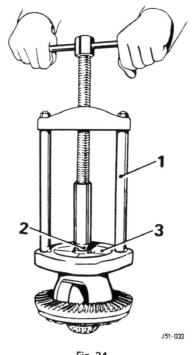

Fig. 34

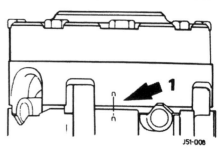

Fig. 35

In the absence of any alignment marks (1, Fig. 35), scribe a line across both halves of differential casing to facilitate reassembly.
Remove the ten 11/16 in AF crown wheel bolts (1, Fig. 36) and remove the crown wheel (2, Fig. 36).

Remove the eight 9/16 in AF bolts (1, Fig. 37), securing both halves of the differential casing (2, Fig. 37).

Remove differential side ring (3, Fig. 37).
Remove pinion side gear and pinion cross-shafts complete with gears (4, Fig. 37).
Separate cross-shafts (5, Fig. 37).
Remove remaining side gear (6, Fig. 37) and ring (7, Fig. 37).

Extract the remaining clutch discs (8, Fig. 37) and plates (9, Fig. 37).

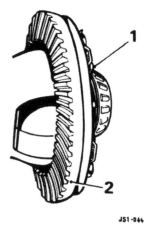

Fig. 36

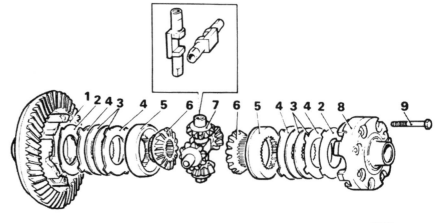

Fig. 37

Reassembling

NOTE: Before commencing assembly, check from reference numbers and letters that pinion and drive gear are a matched pair.

The same serial number must be marked on the pinion end and the outer periphery of the crown wheel (1, Fig. 38), (e.g. 7029). If these requirements are not met the unit must be exchanged.

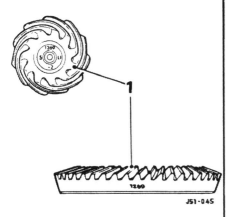

Fig. 38

Prior to reassembly coat all plates and discs with Powr-lok oil.

Refit two Belleville clutch plates (2, Fig. 39) so that convex sides are against differential casing.

Refit clutch plates (4, Fig. 39) and discs (3, Fig. 39) as shown into each half of the casing.

Fit side ring (5, Fig. 39).

Position one side gear into ring recess (6, Fig. 39).

Fit cross-shafts.

Refit pinion mating cross-shafts complete with pinion gears ensuring that ramps on the shafts coincide with the mating ramps in the differential case (7, Fig. 39).

Assemble remaining side gear (6, Fig. 39) and ring (7, Fig. 39).

Offer up right-hand half of differential case (8, Fig. 39) to flange half in accordance with identification marks and position clutch

friction plate tongues so that they align with grooves in differential case.

Assemble right-hand half to flange half of differential case using eight bolts but do not tighten at this stage (9, Fig. 39).

Tighten eight bolts to a torque of 6,05 to 6,9 kg/m (43 to 50 lb/ft) while drive shafts are in position (1, Fig. 40, 1, Fig. 41). With one drive shaft locked, the torque to turn the other (2, Fig. 41) should be between 40 lb/ft and 70 lb/ft.

e.g. hold one shaft in vice soft jaws whilst turning the other.

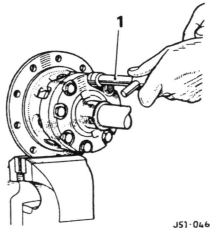

Fig. 40

NOTE: Ensure that prior to assembly the crown wheel mounting face is free from damage or burrs, particularly on the edge; should any burrs be left on the carrier they must be removed with an oil stone prior to fitment of the crown wheel.

Fit the crown wheel to the carrier diametrically using the ten bolts and tab washers, torque up the bolts to 10,78 to 12,4 kgf/m (77 to 88 lb/ft)

Thickness of shims required in the installation of the differential side bearings is determined as follows:

Fit the differential side bearings (1, Fig. 42) using tools 18G 134 (2, Fig. 42) and SL 550-1 (3, Fig. 42) without the shims onto the differential case, making sure that

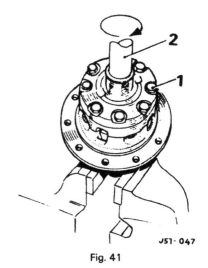

Fig. 41

the bearings and housing are perfectly clean.

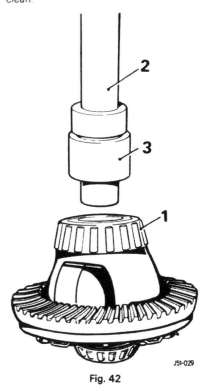

Fig. 42

Place the differential assembly with the bearings in their housing into the differential case without the pinion in position.

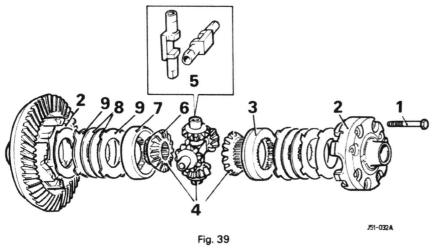

Fig. 39

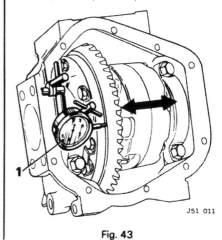

Fig. 43

Install a dial indicator gauge setting the button against the back face of the crown wheel (1, Fig. 43).

Inserting two levers between housing and the bearing cups, move the differential assembly to one side of the carrier.
Set the dial indicator to zero.

Move the assembly to the other side and record indicator reading, giving total clearance between bearings, as now assembled, and abutment faces of the gear carrier housing.

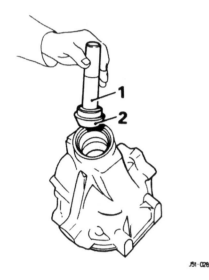

Fig. 44

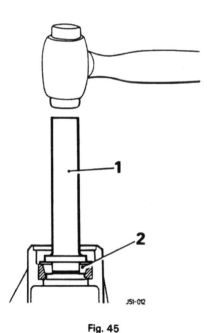

Fig. 45

Remove differential assembly from the gear carrier.
Re-install the pinion outer bearing cup using tools 18G 134 (1, Fig. 44 & 45) and SL 550-8 (2, Fig. 44 & 45).

Fit the inner bearing cup (1, Fig. 46) and shims using tools 18G 134 and SL 550-9 (2, Fig. 46).

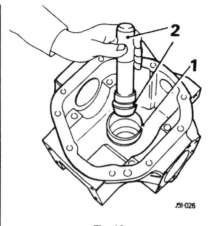

Fig. 46

Press the inner bearing cone onto the pinion using tools 47 (1, Fig. 47), SL 47-1/1 (2, Fig. 47) and SL 47-1/2 (3, Fig. 47).

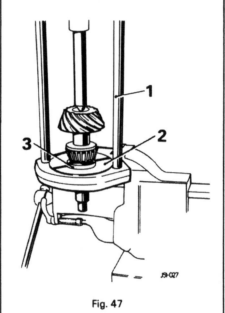

Fig. 47

NOTE: The hypoid drive pinion must be correctly adjusted before attempting further assembly, the greatest care being taken to ensure accuracy.

The correct pinion setting is marked on the ground end of the pinion. The matched assembly serial number is also marked on the periphery of the crown wheel, and care should be taken to keep similarly marked gears and pinions in their matched sets as each pair is lapped together before despatch from the factory. The letter on the left is a production code letter and has no significance relative to assembly or servicing of any axle. The letter and figure on the right refer to the tolerance on offset or pinion drop dimension, which is stamped on the cover facing of the gear carrier housing. The number at the bottom gives the cone setting distance of the pinion and may be Zero (0), Plus (+) or Minus (−) (Fig. 48).

Fig. 48

When correctly adjusted a pinion marked Zero will be at the zero cone setting distance dimension which is 66,67 mm (2.625 in) (i.e. from the centre line of the gear to the face on the small end of the pinion. A pinion marked Plus two (+2) should be adjusted to the nominal (or Zero) cone setting plus 0,0508 mm (0.002 in) and a pinion marked Minus two (−2) to the cone setting distance minus 0,0508 mm (0.002 in). Thus for a pinion marked Minus two (−2) the distance from the centre of the drive gear to the face of the pinion should be 66,619 mm i.e. 66,67 - 0,0508 mm (2.623 in i.e. 2.625 - 0.002 in) and for a pinion marked Plus three (+3) the cone setting distance should be 66,746 mm (2.628 in). Place pinion, together with inner bearing cone, into gear carrier.

  A Pinion drop 38,1 mm (1.5 in)
  B Zero cone setting 66,67 mm (2.625 in)
  C Mounting distance 108,52 mm (4.312 in)
  D Centre line to bearing housing 139,57 mm (5.495 in) to 139,83 mm (5.505 in).

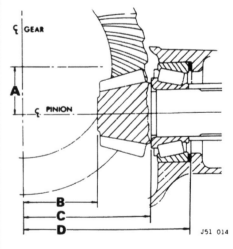

Fig. 49

Turn carrier over and support pinion with a suitable block of wood for convenience before attempting further assembly.

Fit pinion outer bearing cone, companion flange, washer and nut only, omitting the collapsible spacer, oil thrower and oil seal, and tighten nut to remove all backlash.

Check pinion setting distance by means of gauge tool SL3 (1, Fig. 50).

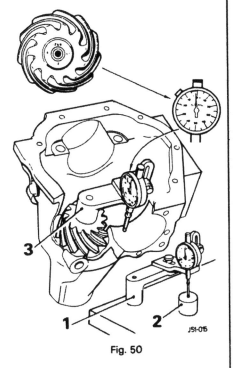

Fig. 50

Adjust bracket carrying dial indicator using 4HA setting block (2, Fig. 50) and set dial face to zero.

Check pinion setting by taking a dial indicator reading on the differentail bearing bore with the assembly firmly seated on the ground face of the pinion (3, Fig. 50). The correct reading will be the minimum obtained; that is, when the indicator spindle is at the bottom of the bore. Slight movement of the assembly will enable the correct reading to be easily ascertained. The dial indicator shows the deviation of the pinion setting from the zero cone setting and it is important to note the direction of any such deviation as well as the magnitude.

If pinion setting is incorrect it is necessary to dismantle the pinion assembly and remove the pinion inner bearing cup. Add or remove shims as required from the pack locating the bearing cup and re-install the shim pack and bearing cup. Adjusting shims are available in thicknesses of 0,076 mm, 0,127 mm and 0,254 mm (0.003 in, 0.005 in and 0.010 in). Repeat setting operations until satisfactory result is obtained.

Extract pinion shaft from gear carrier far enough to enable the outer bearing cone to be removed from the pinion.

Fit the collapsible spacer to the pinion ensuring that it seats firmly on the machined shoulder on the pinion shaft.

Insert pinion into gear carrier

Refit the outer bearing cone, oil thrower and using tool 18G 1428 (1, Fig. 51) fit the oil seal.

Lightly grease the splines of the pinion shaft and fit the flange. Fit a new washer, convex face outermost. Fit, but DO NOT tighten the flange retaining nut.

Begin tightening the flange nut, stopping at frequent intervals to check the torque required to turn the pinion, using the string and spring balance, until the required torque is obtained.

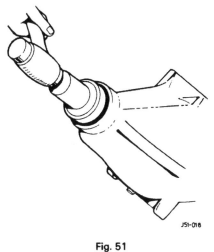

Fig. 51

The flange nut may have to be tightened to as much as 18 kgf/m (130 lbf/ft).

Torque required to turn pinion bearings and oil seal:

Old bearings — 0,20 to 0,28 kgf/m (20 to 25 lbf/in).

New bearings — 0,35 to 0,46 kgf/m (30 to 40 lbf/in).

Note the actual figure required to turn the pinion.

If the above values are exceeded a new collapsible spacer must be fitted. ON NO account must the nut be slackened off and retightened as the collapsed spacer will not then sufficiently clamp the bearing cones.

Place differential assembly complete with side bearings but less shims, in the housing. Ensure that bearings and housing are perfectly clean.

Using the shim pack previously selected, vary the shim thicknesses between each bearing cup and the carrier face to achieve a backlash of 0,15 to 0,25 mm (0.006 to 0.010 in) measured at the outer edge of the ring gear (Fig. 52).

Add an additional 0,07 mm (0.003 in) shim to each pack and carefully note from which side of the differential case the pack was removed

Remove the bearing cups and cones from the differential case using SL 14-3/2 and SL 14-3/1

Fit appropriate shim pack to the differential case and refit the bearing cone.

Ensure that the matching shim pack and cone are fitted to the same side of the differential housing that they were removed from

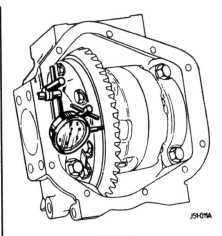

Fig. 52

Lower differential assembly into position lightly tapping the bearings home with a hide hammer.

NOTE: Ensure that gear teeth are led into mesh with those of the pinion. Careless handling at this stage may result in bruising the gear teeth. Removal of the consequent damage can only be partially successful and will result in inferior performance.

When refitting side bearing caps, ensure that position of the numerals marked on gear carrier housing face and side bearing cap coincide (1, Fig. 53).

Tighten cap bolts to a torque of 8,82 to 10,08 kg/m (63 to 72 lb/ft) (2, Fig. 53).

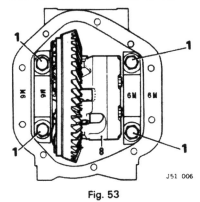

Fig. 53

Mount a dial indicator on gear carrier housing with the button against back face of gear (1, Fig. 54).

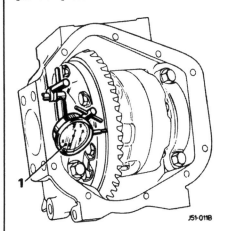

Fig. 54

Turn pinion by hand and check run out on back face of gear. Run out should not exceed 0,13 mm (0.005 in). If run out excessive, strip the assembly and rectify by cleaning the surfaces locating the drive gear. Any burrs on these surfaces must be removed.

Remount dial indicator on gear carrier housing with button tangentially against one of drive gear teeth (1, Fig. 55).

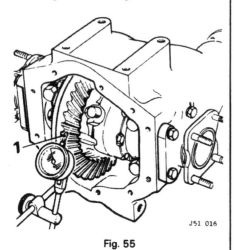

Fig. 55

Move drive gear by hand to check backlash which should be 0,15 to 0,25 mm (0.006 to 0.010 in). If backlash is not to specification, transfer the necessary shims from one side of the differential case to the other to obtain the desired setting. Check backlash in at least four positions of drive gear, ensuring that backlash is always greater than 0,15 mm (0.006 in).

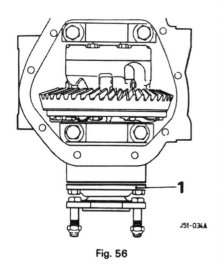

Fig. 56

Check that the torque to turn the input flange is 1,4 to 2,8 kgf/m (10 to 20 lbf/in) additional to the torque measured previously to turn the pinion (page 51-15).

Smear cover flange only with Welseal jointing compound, place gasket on final drive casing, place cover over gasket and insert two bolts to retain, coating threads with Loctite.
Replace remaining eight bolts, coating threads with Loctite and replace the tabs.
Tighten screws by diagonal selection to correct torque 2,1 to 2,8 kgf/m (15 to 20 lbf/ft).

Refit both output shaft assemblies (1, Fig. 56) and torque the bolts to 8,4 to 9,66 kgf/m (60 to 69 lbf/ft), replace the drain plug and refit the drive unit to the cross-member (1, Fig. 57).

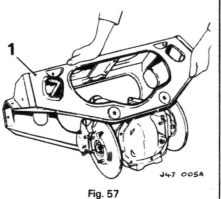

Fig. 57

Secure with bolts (1, Fig. 58) torque and lockwire (2, Fig. 58), ensuring that when lockwired, the wire is tightening the bolts.

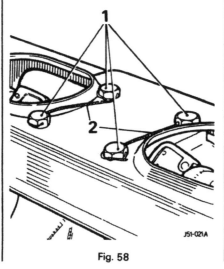

Fig. 58

After refitting the unit to the vehicle fill with new oil.

CONTENTS

TORQUE WRENCH SETTINGS

ITEM	DESCRIPTION	SPANNER SIZE	TIGHTENING TORQUE		
			Nm	kgf/cm	lbf/ft
Collet adaptor pinchbolt	$\frac{1}{4}$ in UNF	$\frac{7}{16}$ in AF	13,5 to 16,2	1,4 to 1,67	10 to 12
Collet adaptor retaining screw locknut	$\frac{1}{4}$ in UNF nut	$\frac{7}{16}$ in AF	8,1 to 9,5	0,8 to 0,98	6 to 7
Hydraulic hose to outlet port of steering pump	—	16 mm	20 to 35	2,1 to 3,6	15 to 26
Hydraulic pipe connections to steering rack	$\frac{7}{16}$ in UNF	$\frac{1}{2}$ in AF	8 to 15	0,83 to 1,55	5.9 to 11
Inner balljoint assembly to rack bar	—	—	60 to 74	6,2 to 7,54	44.3 to 54.6
Outlet port to steering pump	M16	26 mm	50 to 70	5,3 to 7,24	38 to 52
Outer balljoint locknut	—	—	52 to 68	5,5 to 6,9	38.5 to 50.4
Pinion housing to rack end housing	—	13 mm	7 to 10,5	0,75 to 1,1	5.2 to 7.7
Steering arm to stub axle carrier	M12	—	67,8 to 74,5	6,9 to 7,7	50 to 55
Steering pump adaptor to engine	M8	10 mm	23 to 27	2,4 to 2,8	17 to 20
Steering pump to adaptor	M10 bolt	13 mm	49 to 54	5 to 5,6	36 to 40
Steering rack mounting to subframe	$\frac{5}{16}$ in UNF	$\frac{1}{2}$ in AF	19 to 24,4	1,95 to 2,52	14 to 18
Steering rack damper adjuster locknut	—	—	36 to 67	3,7 to 6,85	26.5 to 49.4
Steering rack damper plate securing bolts	—	—	19 to 24,4	1,95 to 2,52	14 to 18
Universal joint pinch bolts	$\frac{5}{16}$ in UNF	$\frac{1}{2}$ in AF	19 to 24,4	1,95 to 2,52	14 to 18
Upper steering column longitudinal strut to vertical strut	—	$\frac{1}{2}$ in AF	10,8 to 13,6	1,12 to 1,4	8 to 10
Upper steering column lower mounting and longitudinal strut to body	$\frac{5}{16}$ in UNF bolt	$\frac{1}{2}$ in AF	19 to 24,4	1,95 to 2,52	14 to 18
Upper steering column to vertical strut	$\frac{5}{16}$ in UNF bolt	$\frac{1}{2}$ in AF	19 to 24,4	1,95 to 2,52	14 to 18
Upper steering column transverse strut to vertical strut .	$\frac{5}{16}$ in UNF bolt	$\frac{1}{2}$ in AF	19 to 24,4	1,95 to 2,52	14 to 18
Upper steering column vertical strut to bracket	$\frac{5}{16}$ in UNF	$\frac{1}{2}$ in AF	19 to 24,4	1,95 to 2,52	14 to 18
Upper steering column vertical strut bracket to body	$\frac{5}{16}$ in UNF	$\frac{1}{2}$ in AF	19 to 24,4	1,95 to 2,52	14 to 18
Steering rack end housing ring nut	—	—	119 to 170	12,3 to 17,6	87.8 to 125.4
Steering rack end housing locating screw	—	—	10 to 17	1,1 to 1,8	7.4 to 12.5
Steering wheel to shaft	$\frac{5}{8}$ in UNF nut	$\frac{15}{16}$ in AF	34 to 43,4	3,5 to 4,5	25 to 32

SERVICE TOOLS

Ball joint separator	JD.24
'C' nut spanner	S.355
Camber gauge	
Castor gauge	
Circlip pliers	
Compression sleeve	606603
Dial test indicator	
Drive dog remover/replacer	18G 1445
Expansion sleeve	606602
Pipe connection plugs	
Pressure gauge	JD.10
Rack centralising tool	12297 or 18G 1466
Rack checking fixture	
Seal saver	18G 1259
Suspension setting links	JD.25B
Tap	JD.10—2
Wheel alignment equipment	

DESCRIPTION

A power assisted steering rack and pinion assembly is mounted to the front crossmember with rubber damping bushes. The power assisted steering rack differs from the un-assisted type of rack, a single piston is fitted to the rack bar, operating in an enclosed cylinder. The pinion housing is replaced by a control valve/pinion assembly with a combined pinion shaft and spool assembly. This directs hydraulic pressure to the appropriate side of the rack piston, thus providing power assistance when the steering wheel is turned.

Oil flow through the control valve is continuous; when the wheels are in the straight ahead position, low oil pressure is applied to each side of the piston. As the steering wheel is turned, a small torsion bar, within the control valve/pinion assembly allows a few degrees of rotation before actually turning the pinion. This rotation is used to open and close ports in the control valve, to ensure that as the torsion bar is twisted, the hydraulic pressure directed to one side of the piston is also increased, from 2,8 kgf/cm² (40 lbf/in²) to a maximum of 84,4 kgf/cm² (1200 lbf/in²), returning to minimum when the load on the torsion bar from the steering wheel is zero. The increase in pressure being proportional to the twist in the torsion bar.

Hydraulic pressure is provided by a vane-type, non-submerged pump, driven directly from the engine auxiliary shaft.

To prevent the hydraulic pressure from exceeding 84,4 kgf/cm² (1200 lbf/in²) a flow control valve is fitted to the outlet port of the pump.

DATA

Castor angle	3½° ± ¼° positive
Camber angle — Front	½° ± ¼° negative (both front wheels to be within ¼° of each other)
Front wheel alignment	0 to 3,18 mm (0 to ⅛ in) toe in
Number of turns, lock to lock	2.75
Total stroke of rack	146,46 mm (5.77 in)

Rack 'pull through load', with a feed pressure of 2,11 kgf/cm² (30 lbf/in²) and a pump flow of 9,45 litres/min (2.08 gal/min) is to be 18,1 kg (40 lb) max, 13,6 kg (30 lb) min.

Maximum axial lift of inner ball joint	0,0127 to 0,0762 mm (0.0005 to 0.0030 in)
Rack pinion inclination	22°
Tie rod articulation, maximum force	4 Nm (36 lb in)
Rack pinion — inlet ports	0.500 in — 20 UNF
— outlet ports	0.625 in — 18 UNF
Quantity of grease from dry:	
Specification	Lithium based grease
Rack and pinion tooth area	42,5 to 57 grammes (1.5 to 2 oz)
Bellows — each	57 grammes (2 oz)

Power Steering Pump

External diameter of shaft	19,04 to 19,05 mm (0.7496 to 0.7500 in)
Internal diameter of drive dog	18,986 to 19,014 mm (0.7475 to 0.7486 in)
Relief valve operating pressure	77,34 to 84,37 kgf/cm² (1100 to 1200 lbf/in²)
Power steering fluid	S.A.E. Automatic Transmission Fluid Type A or Dexron R

DIMENSIONAL DATA — STEERING ARM (Fig. 1)

A. 82,3 to 82,8 mm (3.24 to 3.26 in)
B. 79,73 to 80,24 mm (3.139 to 3.159 in)
C. 53,34 to 53,85 mm (2.10 to 2.12 in)
D. 17,78 to 18,29 mm (0.70 to 0.72 in)
E. 58,93 to 59,44 mm (2.32 to 2.34 in)
F. 135,38 to 135,89 mm (5.33 to 5.35 in)

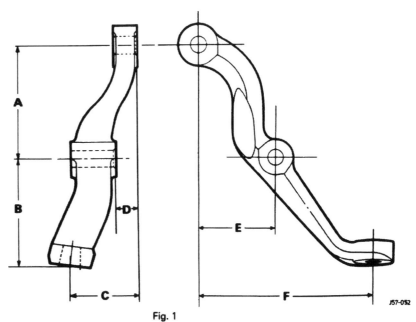

Fig. 1

SYMPTON AND DIAGNOSIS CHART

SYMPTOM	CAUSE	CURE
External oil leaks from steering rack unit.	Damaged or worn seals. Loose unions. Damaged union sealing washers.	Replace seals. Tighten unions. Replace sealing washers.
Oil leak at pump shaft.	Damaged shaft seal.	Replace shaft seal.
Oil leak at high pressure outlet union.	Loose or damaged union. Damaged pipe end.	Tighten union. Replace pipe.
Oil leak at low pressure inlet connection.	Loose or damaged hose connection.	Remove and refit or renew hose and clip.
Oil overflowing reservoir cap.	Reservoir overfull. Sticking flow control valve (closed).	Reduce level in reservoir. Remove valve, renew and refit.
Noise from hydraulic system.	Air in system.	Bleed system.
Noise from pump.	Worn drive coupling. Internal wear and damage.	Renew coupling. Overhaul pump.
Noise from rack (rattling).	Worn rack and pinion gears. Worn inner ball joints. Universal joint loose.	Adjust rack damper. Replace inner ball joints. Tighten clamping bolts.
Steering veering to left or right.	Unbalanced tyre pressures. Incorrect tyres fitted. Incorrect geometry. Steering unit out of trim.	Inflate to correct pressure. Fit tyres of correct specification. Reset geometry to correct specification. Replace valve and pinion assembly.
Heavy steering when driving.	Low tyre pressures. Tightness in steering column. Tightness in steering joints.	Inflate to correct specification. Grease or replace. Grease or renew joints.
Heavy steering when parking.	Low tyre pressures. Tightness in steering column. Tightness in steering joints. Restricted hose. Sticking flow control valve (open). Internal leaks in steering unit.	Inflate to correct specification. Grease or replace. Grease or renew joints. Replace hose. Remove and renew valve. Replace seals
Steering effort too light.	Valve torsion bar dowel pins worn. Valve torsion bar broken.	Replace valve assembly. Replace valve assembly.
Momentary increase in effort when the steering wheel is turned quickly in either direction.	Low fluid level. High internal leakage.	Top up fluid reservoir Overhaul steering pump.
Jerky steering, especially when parking.	Low fluid level. Insufficient pump pressure. Sticking flow control valve.	Top up fluid reservoir. Check flow valve/overhaul pump. Check and rectify.

SYSTEM TESTING

Service Tools: Tap JD10-2, Pressure gauge JD 10.

Before commencing the test procedure, ensure that the fluid in the reservoir is free from froth, and that the level is correct.
Fit the pressure gauge (1, Fig. 2) in the pressure line from the pump.

NOTE: The gauge must read to 100 kgf/cm² (1500 lbf/in²).

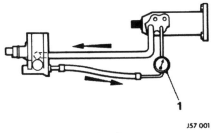

Fig. 2

Start the engine and allow it to idle. Turn the steering wheel to full lock and continue to increase steering effort, until the pressure recorded on the gauge ceases to increase.
Check that the recorded pressure is between 77.55 and 84.4 kgf/cm² (1100 and 1200 lbf/in²).

NOTE: If the pressure is below 77.5 kgf/cm² (1100 lbf/cm²) at idle, but rises to the correct value when the engine speed is increased, then a defective pump control valve, or excessive internal leakage in the rack and pinion unit is indicated.

Carry out the following test to establish the location of the fault.

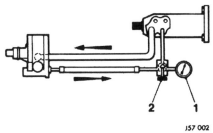

Fig. 3

Fit the tap JD 10-2 (2, Fig. 3) between the pump and the pressure gauge (1, Fig. 3), arranged as shown so that the pressure gauge will at all times be connected to the pump but the rack unit can be isolated.
With the tap in the 'Open' position, start the engine and allow it to idle. Turn the steering wheel to full lock. Check the reading on the gauge which should be in excess of 77.5 kgf/cm² (1100 lbf/in²).
If the pressure does not reach this figure, CLOSE THE TAP IMMEDIATELY, noting the reading on the gauge as the tap reaches the 'OFF' position.

CAUTION: Do NOT keep the tap closed for any longer than 5 seconds when the engine is running.

If the reading on the gauge increases when the tap is closed to at least 77.5 kgf/cm² (1100 lbf/in²), then leaks are confined to the steering unit, which requires overhauling.

If the reading on the gauge exceeds 84.4 kgf/cm² (1200 lbf/in²), remove the pump discharge port and withdraw the spring and control valve assembly.
Inspect the small hemispherical gauge filter located at the inner end of the control valve. If blocked, clean using compressed air. Refit the control valve, spring and discharge port.

POWER STEERING SYSTEM

Bleed

The presence of any air in the system will cause the steering to be lumpy and erratic: Air can be removed by the following method:
Fill the fluid reservoir to the full mark, start the engine and turn the steering wheel from lock to lock; avoid the steering lock stops.
Keep a constant check on the fluid level in the reservoir and top up as necessary.
Position the wheels in the straight ahead position and allow the engine to continue to idle for a few minutes.
Switch off the engine, allow the system to settle and recheck the fluid level; add fluid as necessary.
Use only fluid of the correct specification.

CASTOR ANGLE/CAMBER ANGLE

Service Tools: JD 25B Suspension links. Camber and castor angle checking gauges.

CAUTION: Before checking, examine all rubber/steel bushes for deterioration or distortion. Check upper and lower wishbone ball joints for excessive play. Check shock absorbers for leaks and mountings for security.

The two operations require the vehicle to be set up in a mid-laden condition. This can be done as follows:
Ensure that the car is standing on level ground and inflate the tyres to the correct pressure; check that the standing heights are equal on both sides of the car, and the front wheels are in the straight-ahead position.
Make up two front suspension tubes to the dimensions shown (Fig. 4).
Compress the front suspension and insert the setting tubes under the upper wishbones, adjacent to the rebound stop rubbers and over the brackets welded to the bottom of the 'turrets'. This locks the front suspension in the mid-laden condition.
Lock the rear suspension in the mid-laden condition using the suspension links, service tool JD 25B.

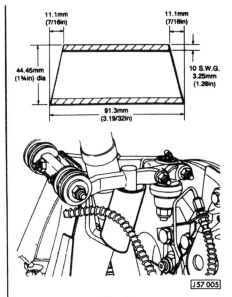

Fig. 4

For each side, compress the suspension, pass the hooked end of service tool JD 25B through the lower hole in the rear mounting and fit the looped end over the rear pivot nut (Fig. 5).

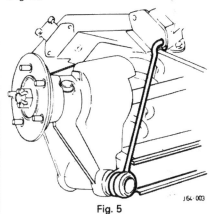

Fig. 5

Castor angle — check and adjust

Using the castor angle checking gauge, check the castor angle. Refer to the Data for correct setting.
To adjust, slacken the two bolts, on each side, securing the upper wishbone members to the upper ball joints.
Transpose shims, which can now be lifted out, from front to rear or vice versa, to reduce or increase the castor angle respectively (Fig. 6).

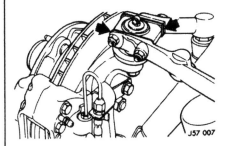

Fig. 6

Transposing one shim 1,6 mm (0.0625 in) thick will alter the castor angle by approximately $\frac{1}{4}$°.

After adjusting the castor angle to the correct figure, tighten the bolts to the correct torque. Check the front wheel alignment and adjust if necessary.

Camber angle — check and adjust

Using the camber angle checking gauge, check the camber angle. Refer to data for the correct settings.

Rotate the road wheels through 180° and re-check.

To adjust, slacken the nuts and bolts securing the upper wishbone inner pivots to the cross member turrets.

Add or remove shims between the pivot shafts and cross member turrets to reduce or increase the camber angle (Fig. 7).

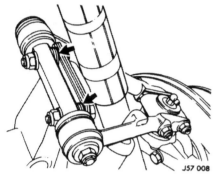

Fig. 7

Shims are available in 0,8 mm ($\frac{1}{32}$ in), 1,6 mm ($\frac{1}{16}$ in) and 3,2 mm ($\frac{1}{8}$ in) thickness. A change of 1,6 mm ($\frac{1}{16}$ in) in shim thickness will alter the camber angle by approximately $\frac{1}{4}$°.

NOTE: It is necessary to partly withdraw the bolts to change the shims, so only one bolt of a pair should be shimmed at a time. It is important that an equal thickness of shims should be changed on front and rear bolts, otherwise the castor angle will be affected.

Tighten all the bolts and nuts to the correct torque, and re-check the camber angle.
Check the front wheel alignment and adjust if necessary.

FRONT WHEEL ALIGNMENT

Check and adjust

Service Tool: Centralising tool 18G 1466 (was Jaguar Parts Issue 12297)

Check

Inflate the tyres to the correct pressures.
Set the front wheels in the straight-ahead position.

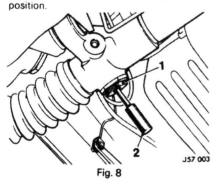

Fig. 8

Remove the grease nipple from the rack adjuster pad (1, Fig. 8).
Insert the centralizing tool (2, Fig. 8) and adjust the position of the rack until reduced tip of the tool enters the locating hole in the rack.
Check the alignment by using light beam equipment or an approved track setting gauge.

NOTE: As a front wheel alignment check is called for in the Option Service of the Maintenance Summary, very little variation from specified figures for the wheel alignment is to be expected; if, however, a discrepancy of as much as 3 mm ($\frac{1}{8}$ in) from specified limits is recorded, accidental damage to a steering lever may have occurred and the following check must be carried out, on both levers.

Remove the steering levers.
Accurately check the dimensions of each lever against those quoted in the Data and Fig. 1.
Reject for scrap and replace any lever with dimensions outside the limits quoted.

WARNING: IT IS ABSOLUTELY FORBIDDEN TO ATTEMPT TO STRAIGHTEN A DAMAGED STEERING ARM. IT MUST BE REPLACED.

If both steering levers are within limits, a discrepancy in alignment figures may be due to the distortion of the upper or lower wishbones, or the end of the stub axle carriers (vertical links.) Dimensioned drawings of these parts for checking purposes, are given in Group 60.

Adjust

Slacken the locknuts at the outer end of each tie-rod.
Release the clips securing the outer ends of the gaiters to the tie-rods.
Turn the tie-rods by an equal amount until the alignment of the wheels is correct.
Tighten the locknuts to the figure quoted in the torque wrench setting table while holding the track rod end spanner flats.
Re-check the alignment.
Ensure that the gaiters are not twisted and re-tighten the clips.
Remove the centralising tool (2, Fig. 8) and refit the grease nipple.

STEERING RACK PINION CLEARANCE

Check and adjust

Service Tools: Ball joint separator JD 24, Dial test indicator.

The correct clearance should allow smooth travel of the rack, without binding.
The maximum clearance between the rack and the pinion should not exceed 0.25 mm (0.010 in).
The clearance is measured from beneath the car.
Detach the outer ball joint on the tie rod nearest to the pinion, from the steering arm, using Service Tool JD 24 (Fig. 9).

NOTE: It may be necessary to substitute a 2 in long $\frac{1}{2}$ in UNF socket headed (grub) screw for the existing bolts of JD 24.

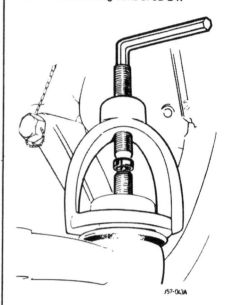

Fig. 9

Remove the grease nipple (1, Fig. 10) from the rack damper threaded plug and insert the stem of a dial test indicator through the grease nipple hole to contact the back of the rack shaft. Secure the gauge to the rack. Zero the gauge.

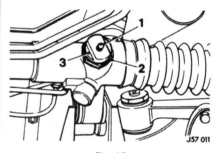

Fig. 10

Firmly grip the tie rod inner ball joint assembly, nearest to the pinion, and move it away from the pinion, towards the rack damper assembly; note the reading on the gauge.
If the clearance is excessive and adjustment is required, release the locknut (2, Fig. 10), screw in the plug (3, Fig. 10) until firm resistance is felt. Back off $\frac{1}{8}$th of a turn and tighten the locknut; re-check the pinion to rack clearance.
Move the rack through its full travel. If binding occurs at any point, slightly increase the clearance and re-check.
When the correct clearance is achieved, fully tighten the locknut and refit the grease nipple.
Connect the tie rod outer ball joint to the steering arm.
Check the front wheel alignment; adjust as necessary.

CONTROL VALVE AND PINION

Renew

Service Tool: Pinion Housing Seal Saver 18G 1259.

It is possible to remove the control valve and pinion assembly without removing the rack. Extreme care must be taken to prevent contamination from entering the rack housing whilst the pinion is removed.
Place the car on a ramp or over a pit.
Mark the relationship of the steering column universal joint to the upper and lower columns, remove the steering column universal joint.
Mark the relationship of the lower column to the rack pinion and remove the lower column.
Thoroughly clean the pinion housing.
Remove the wavy washer and the dust shield from the pinion.
Disconnect the rack feed hose from the pinion housing and drain the fluid into a suitable container.

Upper Seal

The upper seal can be renewed without removing the control valve and pinion assembly from the rack.

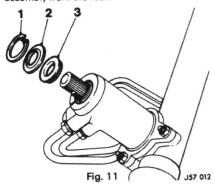

Fig. 11

Remove the circlip (1, Fig. 11) seal retainer (2, Fig. 11) and seal (3, Fig. 11) from the pinion housing.
Position the seal saver, Service Tool 18G 1259, over the serrations on the pinion. Fit a new seal, grooved face downwards. Ensure that the flange fits snugly in the recess. Fit the seal retainer with the rubber side and lip outermost. Fit the circlip, ensure that it is seated fully in the groove. Remove the Service Tool.

Control Valve and Pinion

Disconnect the hydraulic pipe unions from the pinion valve housing and remove the pipes (1, Fig. 12).
Remove the three self locking nuts (2, Fig. 12) securing the pinion valve housing to the steering rack.
Undo the rack damper pad locknut and slacken the rack damper threaded plug (3, Fig. 12) one turn.
Mark the position of the pinion shaft in relation to the valve housing (4, Fig. 12).
Remove the control valve and pinion assembly.
Do NOT move the front wheels or turn the steering wheel until the pinion assembly is refitted.

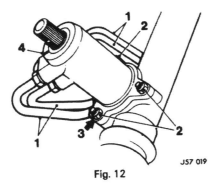

Fig. 12

Remove and discard the 'U' section seal and clean the sealing faces.
Fit a new 'U' section seal to the pinion rack housing, ensure that the grooves in the seal face upwards and that the seal flange fits snugly in the groove.
Grease the pinion teeth with the correct specification grease.
Fit the pinion valve housing to the rack, ensure that the pinion shaft is in the correct position in the housing as previously marked.
Fit and tighten the three self locking nuts securing the pinion valve housing to the rack.
Reset the rack damper adjuster plug and tighten the locknut.

Refitting

Refit the lower steering column to the rack pinion and the universal joint to the upper and lower columns. Note the previously marked positions. Fit and tighten the pinch bolts.
Refit the hydraulic pipes and the rack feed hose.
Fill the system to the correct level with the fluid of the correct specification and bleed the system.
Check for leaks.

TIE ROD BALL JOINTS

Renew

Service Tool: Ball Joint Separator JD 24.

The inner ball joint is only supplied as a complete assembly. To renew the inner ball joint, it is necessary to remove the outer ball joint and locknut first.
To assist in the initial setting of the front wheel alignment prior to dismantling, measure the distance between the centres of the inner and outer ball joints.
The front wheel alignment must be checked after renewing either ball joint, as it is difficult to ensure that the length between the ball joint centres is not altered.

Outer Ball Joint

Remove the self locking nut securing the outer ball joint to the steering arm. Use Service Tool JD 24 (Fig. 13) to detach the ball joint.

NOTE: It may be necessary to substitute a 2 in long $\frac{1}{2}$ in UNF socket headed (grub) screw, for the existing bolt of JD 24.

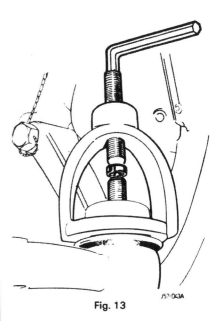

Fig. 13

Release the outer ball joint locknut. Do not run the locknut along the thread. Unscrew the ball joint from the tie rod.
Screw on the replacement ball joint up to the locknut. This gives an approximate setting prior to checking the front wheel alignment.
Refit the ball joint to the steering arm and secure using a new self locking nut.
Check and adjust the front wheel alignment.

Inner Ball Joint

Release the outer ball joint from the steering arm and remove the ball joint and locknut from the tie rod.
Release the clips (1, Fig. 14) securing the rubber gaiter to the steering rack housing and tie rod. Withdraw the rubber gaiter (2, Fig. 14).
Check for splits or perishing.

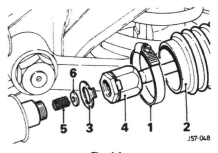

Fig. 14

Knock back the tab washer (3, Fig. 14) and unscrew the inner ball joint and tie rod assembly (4, Fig. 14) from the rack bar.
Collect the spring and packing washer (5 & 6, Fig. 14).
Fit the new inner ball joint assembly to the rack bar, with a new tab washer. Bend over the tab washer.
Coat the inner ball joint with 60 g (2 oz) of the recommended grease and refit the gaiter. Secure with the clips.
Refit the outer ball joint and secure to the steering arm.
Check and adjust the front wheel alignment.

STEERING ARM

Renew

Service Tool: Ball Joint Separator JD 24.

Jack up the front of the car and support on stands.
Remove the self locking nut (1, Fig. 16) securing the tie rod outer ball joint to the steering arm. Detach the ball joint using Service Tool JD 24 (Fig. 15).

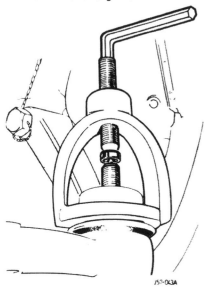

Fig. 15

NOTE: It may be necessary to substitute a 2 in long $\frac{1}{2}$ in UNF socket headed (grub) screw for the existing bolt of JD 24.

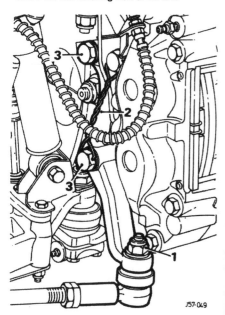

J57-049

Cut and remove the locking wire (2, Fig. 16) securing the steering arm mounting bolts, remove the bolts (3, Fig. 16). Make a record of the number and position of any shims fitted between the steering arm and the brake caliper.
Check the steering arm against the dimensions given in Fig. 16. No attempt should be made to straighten a damaged steering arm. IT MUST BE REPLACED.
Fit the steering arm and secure with the bolts. Tighten to correct torque and wirelock.

NOTE: Ensure that any shims between the steering arm and brake caliper are refitted in their original positions.

Check and adust the front wheel alignment.

UPPER STEERING COLUMN

Renew

Removing the Steering Wheel Assembly

Disconnect the battery earth cable.
Remove the three screws securing the steering column lower switch cover and remove the cover.
Slacken the locknut (1, Fig. 17) and release the grub screw (2, Fig. 17), in the collet adaptor, two complete turns.

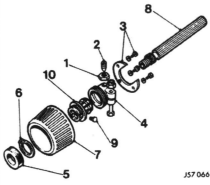

Fig. 17

Centralise the front wheels and mark the relationship of the steering wheel to the steering column upper switch cover.
Remove the clamp bolt and nyloc nut securing the collet adaptor to the steering column.
Withdraw the steering wheel complete with the lock ring collar and adjuster assembly.

Dismantling the Steering Column Adjusting Clamp

Unscrew the two self locking screws securing the horn pad to the steering wheel.
Unscrew the nylon nut from the top of the steering wheel shaft and remove it, withdrawing the horn contact tube with it.
Remove the self locking nut and plain washer securing the steering wheel to the splined shaft.
Carefully remove the steering wheel from the splined shaft. Collect both halves of the split cone.
Remove the three screws securing the 'U' plate (3, Fig. 17) to the collet adaptor (4, Fig. 17).
Unscrew the collet adaptor from the lock ring collar.
Remove the impact rubber (5, Fig. 17) from the steering wheel shaft.
Remove the circlip (6, Fig. 17) securing the lock ring collar (7, Fig. 17) to the splined steering wheel shaft (8, Fig. 17) and remove the lock ring collar.
Slacken the grub screw (9, Fig. 17) securing the split collet (10, Fig. 17) to the steering wheel shaft and remove the collet.

Check the split collet for wear in the splines. There should not be any radial movement between the collet and the splined shaft.

Removing the Upper Steering Column

To obtain access to the column upper mounting bolts, it is necessary to remove the Instrument Panel (Module) as follows:
Remove the driver's side dash lower casing.
Remove the centre securing strip and screw from the instrument module surround (1, Fig. 18).
Remove the screws securing the side pieces of the surround and remove the surround (2, Fig. 19).

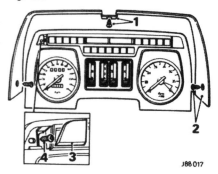

J88.017

Fig. 18

Prise off the covers from the securing screw apertures (3, Fig. 18) and remove the screws (4, Fig. 18) securing the instrument panel module to the fascia.
Ease the module forwards and disconnect the harness block connectors.
Manoeuvre the instrument panel clear of the fascia.
Remove the screws securing the ignition switch surround and remove the surround.
Disconnect the block connectors for the ignition switch and the switchgear harnesses.
Remove the pinch bolts securing the universal joint to the upper and lower steering columns.
Disconnect the horn electrical feed from the column.
Slacken the two setscrews securing the lower end of the column to the mounting struts.
Release the upper mounting bolts and recover distance pieces and washers.
Supporting the column by hand, remove the lower mounting setscrews and withdraw the column.

Dismantling the Upper Column

Remove the upper switch cover.
Slacken the screw securing the switchgear to the column. Make a note of the position of the switches and slide the assembly from the column.
Using a suitable punch, remove the two shear bolts securing the ignition switch and lock assembly to the column (Fig. 19).
Remove the bolts securing the horn feed contact assembly to the column mounting bracket.
Remove the feed and contact units from the column.

NOTE: No further overhaul of the column is possible.

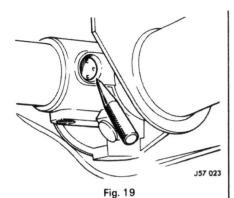

Fig. 19

Assembling the Upper Column

This operation is the reverse of the dismantling procedure. Ensure that the switch gear is fitted in the correct position. Refit the ignition switch and lock assembly using new shear head bolts. Tighten until the heads shear off.

Fitting the Upper Column

Position the column in the car. Fit but do not fully tighten the lower mounting bolts.

Fit the upper mounting bolts, spacers and washers; do not fully tighten the bolts.

Adjust the column so that the upper section is in the centre of the cutout in the fascia and the groove on the lower portion of the inner splined shaft aligns with the bolt locating hole of the universal joint.

Tighten the upper and lower mounting bolts to the correct torque.

Refit the pinch bolts securing the universal joint to the upper and lower steering columns.

Reconnect the block connectors for the ignition switch and the switchgear harnesses.

Refit the instrument panel (module) and secure to the fascia. Refit the driver's side dash lower casing.

Assembling the Steering Column Adjusting Clamp

Slightly smear the splined steering wheel shaft with grease.

Fit the split collet (1, Fig. 20) to the shaft, ensure that the screw locating hole (2, Fig. 20) aligns with the groove (3, Fig. 20) in the shaft.

Fit and tighten the grub screw and back off half a turn. The split collet should slide freely along the shaft splines.

Position the lock ring collar (4, Fig. 20) over the split collet and secure with the circlip (5, Fig. 20).

Refit the collet adaptor (6, Fig. 20) to the locking right collar. Refit the 'U' plate (7, Fig. 20) to the collet adaptor, and secure with the three screws.

Refit the impact rubber (8, Fig. 20) to the steering wheel shaft, refit the steering wheel. Ensure that the split cones (9, Fig. 20) are refitted. Fit the plain washer and fit and tighten the self locking nut.

Refitting the Steering Wheel Assembly

Centralise the horn slip ring needle (10,

Fig. 20) in the upper column and slide the steering wheel shaft into the column.

Make sure that the front wheels and the steering wheel are set in the straight ahead position. Also ensure that the tongue on the collet adaptor aligns with the groove in the nylon bush.

Tighten the grub screw (11, Fig. 20) in the collet adaptor finger tight only. Tighten the locknut (12, Fig. 20).

Fit and tigthen the clamp bolt and nut (13, Fig. 20).

Refit the horn contact rod (14, Fig. 20) to the upper column. Tighten the nylong nut onto the top of the steering wheel shaft.

Secure the horn pad to the steering wheel with the two self tapping screws.

Refit the lower switch cover to the column. Reconnect the battery earth cable.

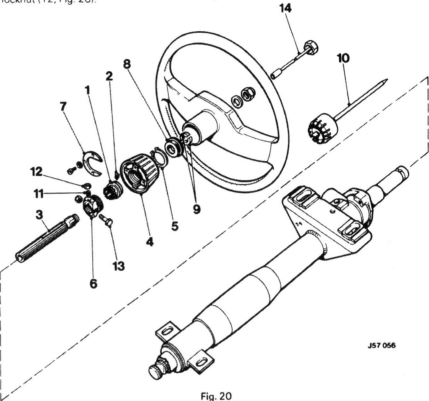

Fig. 20

LOWER STEERING COLUMN SEAL

Renew

Removing the Universal Joint

Remove the pinch bolts (1, Fig. 21) securing the steering column universal joint to the upper and lower steering columns.

Set the wheels and the steering wheel in the straight ahead position.

Remove the universal joint, (2, Fig. 21) first from the upper column, then from the lower column.

Removing the Lower Steering Column

Cut and remove the 'Oetika' clip (3, Fig. 21) securing the nylon draft (4, Fig. 21) excluder to the lower column and remove the draft excluder.

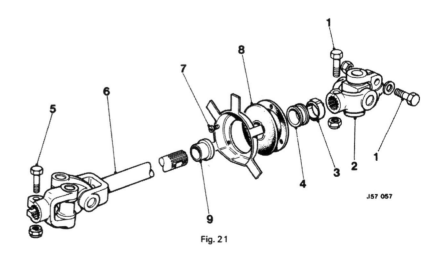

Fig. 21

Remove the pinch bolt (5, Fig. 21) securing the lower universal joint (part of the lower column) to the steering rack pinion shaft.
Remove the lower column assembly (6, Fig. 21).

Renewing the Lower Column Seal and Bush

Remove the three self tapping screws (7, Fig. 21) securing the gaiter retainer (8, Fig. 21) to the bulkhead.
Remove the retainer, gaiter and nylon bush (9, Fig. 21).
Check the bush for wear and the gaiter for splits or perishing, and renew as necessary.
Fit the gaiter to the bulkhead, secure in position with the retainer and self tapping screws.

To Refit the Lower Column and Universal Joint

Reverse the removal procedure. Fit the nylon draft excluder with the flanged end first and secure with a new 'Oetika' clip.
When fitting the universal joint, ensure that the road wheels and the steering wheel are set in the straight ahead position.
Refit and tighten the pinch bolts.

POWER STEERING RACK

Overhaul

Service Tools: Ball joint separator JD 24 +2 in long $\frac{1}{2}$ in UNF socket headed (grub) screw. Rack checking fixture JD 36A. Plugs for pipe connections. End housing 'C' nut remover S355. Pinion ring expansion sleeve 606602. Pinion ring compression sleeve 606603 (JD33). Pinion housing seal saver 18G 1259. Rack centralising tool, Jaguar Part No. 12297.

Steering Rack Remove

Slacken the power steering fluid reservoir filler cap. Raise the car and support; detach both the hoses from the pinion housing. Collect the escaping fluid in a suitable container. Blank off all ports and hoses.

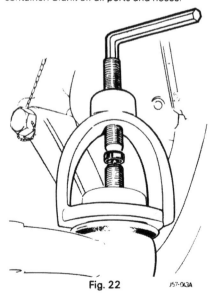

Fig. 22   J57-043A

Separate the ball joints from the steering arms, using Service Tool JD 24 (Fig. 22).

NOTE: It may be necessary to substitute a 2 in long $\frac{1}{2}$ in UNF socket headed (grub) screw, for the existing bolt of JD 24.

Remove the pinch bolt (1, Fig. 23) securing the lower steering column universal joint, to the rack pinion.

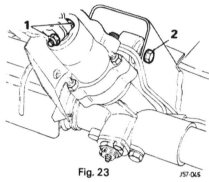

Fig. 23   J57-045

Remove the bolt, washer and self locking nut, securing the steering rack top mounting (pinion side of rack assembly) (2, Fig. 23), to the crossmember.
Remove both the rack bottom mounting bolts, washers and nuts, securing the steering rack to the crossmember.

CAUTION: Make a careful record of the number and position of the packing washers for refitting.

Release the steering rack from the crossmember and retrieve the packing washers.

Steering Rack Dismantle

Thoroughly clean the exterior of the steering rack.
Remove the blanking plugs from the pinion housing ports and purge any remaining fluid by turning the pinion gently from lock to lock. Centre the pinion gear and note the location of the pinchbolt groove.
Remove the rack mounting rubbers and sleeves.
Release the nuts securing the feed pipes to the pinion valve housing and the rack body; remove the pipes from the rack assembly.
Remove the sealing washer from the port in the pinion end rack housing.

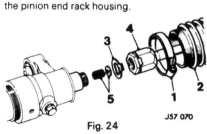

Fig. 24   J57 070

Make a note of their position and release the two large clips (1, Fig. 24) securing the tie rod gaiters to the pinion and end housings. Pull back the gaiters (2, Fig. 24) to allow access to the inner ball joint assemblies.

NOTE: Do not disturb the outer ball joints, unless replacement is necessary.

If the outer ball joints are to be renewed, measure accurately and record the total length of each tie rod, before releasing the locknuts. This will assist when re-tracking the car.

Knock back the tab washers (3, Fig. 24) securing the inner ball joint assembly locknuts to the rack.

CAUTION: Do not disturb the tab washers between the locknuts and the ball pin housings.

Hold one inner ball joint assembly (4, Fig. 24) with a suitable spanner and release the opposite one.
Protect the rack teeth and back of the rack; clamp the rack to enable the other inner ball joint to be released.
Unscrew the tie rod assemblies from the rack. Collect the springs and packing pieces (5, Fig. 24).
Release the locknut securing the rack damper; remove the nut, threaded plug, spring and rack damper pad.

NOTE: If the rack damper adjustment is satisfactory and the rack damper assembly does not require overhauling then remove the two bolts and lift off the plate; remove the 'O' ring, spring and rack damper pad (Fig. 25).

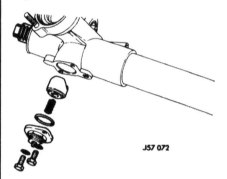

Fig. 25

Pinion Valve and Housing

Remove the three self locking nuts securing the pinion and valve assembly to the pinion end rack housing. Note the relationship of the ports to the rack and remove the complete pinion and valve assembly (1, Fig. 26).

Fig. 26

Remove the sealing ring (2, Fig. 26) the pinion seal (3, Fig. 26) and the backing washer (4, Fig. 26).
Using a suitable mallet gently tap the pinion valve from the pinion valve housing.

Remove the circlip washer and ball bearing race, from the valve assembly, if a replacement is necessary.

NOTE: The pinion valve cannot be dismantled further. This item must be replaced as a complete assembly.

Port Inserts Renew

Tap a suitable thread in the bore of the insert (1, Fig. 27).
Insert a setscrew (2, Fig. 27) with attached nut (3, Fig. 27) and distance piece (4, Fig. 27).

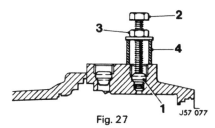

Fig. 27

Tighten the nut and withdraw the insert.
Ensure that all swarf and metal particles are completely removed.
Fit a new insert into each port and tap home squarely using a soft mandrel.

End Housing

Release the small hexagon socket grub screw (1, Fig. 28) in the end housing.
Using Service Tool S355, unscrew the ring nut from the end housing (2, Fig. 28). Remove the end housing (3, Fig. 28) from the rack tube.
Remove the air transfer pipe and sealing rings (4, Fig. 28), from both the pinion and end housings.

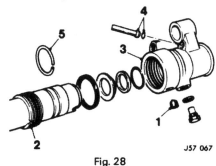

Fig. 28

Rack and Inner Sleeve

Remove the hexagon socket grub screw (1, Fig. 29) from the pinion end rack housing and collect the sealing washer (2, Fig. 29). Remove the rack complete with the inner sleeve (3, Fig. 29) from the bore of the rack tube.

NOTE: Removal of the inner sleeve over the rack teeth will destroy the seal (6, Fig. 29).

Bend up the retaining tabs on the seal cap (4, Fig. 29) and remove the cap from the inner sleeve.
Remove the seal 'O' ring (5, Fig. 29), seal (6, Fig. 29) and split bearing (7, Fig. 29).
Remove the rubber 'O' ring (8, Fig. 29) and nylon washer (9, Fig. 29) from the bottom of the rack tube.

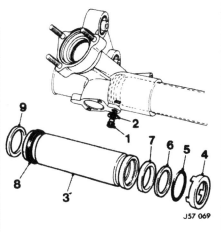

Fig. 29

The piston cannot be removed from the rack but the piston ring (1, Fig. 30) and the backing ring (2, Fig. 30) can be renewed.

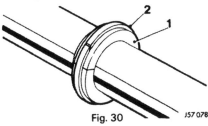

Fig. 30

The rack tube cannot be removed from the plain (pinion end) rack housing, but the ring nut (2, Fig. 28) and circlip (5, Fig. 28) can be renewed. Exercise caution, when removing and replacing the circlip over the ground sealing outer diameter of the rack tube.

Renewing Seals

Discard all the old seals, and the inner sleeve seal retaining cap.
Thoroughly clean and inspect each item for surface damage and wear.
For efficient sealing it is essential that all seal surfaces, lead chambers etc., are smooth, with no scratches or score marks.

Re-assembling — Inner Sleeve and Rack Bar

Fit a new backing ring and piston ring to the piston, and ensure that it moves freely in its groove.
Place a new seal retaining cap over the rack teeth, with the three tabs facing away from the piston.
Fit a new split bearing in the recess in the inner sleeve, ensure that it is seated correctly.
To protect the new inner sleeve seal from being damaged by the rack teeth; cover the rack teeth with a piece of suitable plastic adhesive tape, placed lengthways over the teeth.
Carefully slide the seal, with the recessed face towards the piston, over the tape and onto the rack bar.
Remove the tape.
Fit a new 'O' ring in its recess in the inner sleeve, ensure that it seats correctly.
Ensure that the ends of the split bearing are on the opposite side of the rack bar to the teeth and push the inner sleeve along the

rack bar. Carefully push the seal up against the retaining cap and in turn, against the piston.
Ensure that the inner sleeve is square to the piston; continue pushing until the seal is fully home.
Maintain the pressure against the piston and neatly bend the three tangs into the groove on the outside of the inner sleeve, securing the retaining cap.
Apply a smear of silicone to the bore of the new square section sealing ring. Fit the nylon backing washer (9, Fig. 29) and sealing ring (8, Fig. 29) into the bore of the rack tube; slide them all the way down until they contact the pinion end main housing.
Assemble the rack bar, with the inner sleeve still against the piston, into the rack tube bore. Guide the piston ring into the rack tube bore. Continue sliding the rack and inner sleeve assembly into the rack tube bore, until the inner sleeve enters the sealing ring and seats firmly against the pinion end rack housing.
Look into the hexagon socket screw hole and ensure that the retaining shoulder has passed the hole. Fit the sealing washer and socket grub screw (1 & 2, Fig. 29). After tightening, it should fit flush to slightly proud, stake in position.

End Housing

Remove the seal (1, Fig. 31) and 'O' ring (2, Fig. 31) using a suitable sharp instrument.
With a suitable soft metal drift, carefully remove the steel retaining washer (3, Fig. 31).
Fit a new 'O' ring in the recess, pushing a new seal with the groove uppermost, on to the top of the 'O' ring; replace the steel retaining washer with the spigot towards the seal and press home.
Fit a new square section sealing ring (4, Fig. 31) into the end housing; smear the sealing ring bore, with a silicone lubricant, to aid assembly.
Fit new air transfer pipe sealing rings (5, Fig. 31) to the pinion and end housings. Fit the air transfer pipe to the pinion rack housing.
Fit the end housing over the rack bar, taking care to align it, to avoid damaging the end housing seal.
Slide the end housing onto the rack tube with a slightly twisting action. Engage the air transfer pipe in to its port. Align the end housing mounting lug (6, Fig. 31) with the lower cut out of the rack tube. Tighten the 'C' nut using Service Tool S355, to the correct torque.
Refit the hexagon socket grub screw. Tighten and stake in position.

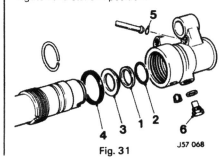

Fig. 31

Pinion Valve and Housing

To remove the seals from the pinion, use a sharp knife and cut diagonally, taking care not to damage the groove ends.

Using Service Tool 606602, to expand the seals, fit one in the groove nearest to the ball bearing race. Repeat the procedure for the other three seals.

The rings can then be compressed to their original size by fitting a sleeve over them. Use 606603. If this tool is not available, then recovery will take place naturally if left for about ¾ hour.

Fit the washer and 'U' section seal, into the pinion main housing, ensure that the grooves in the seal face upwards and that the seal flange fits snugly in the groove.

Valve Housing

Using suitable circlip pliers remove the circlip (1, Fig. 32). Remove the seal retainer (2, Fig. 32) and the seal (3, Fig. 32).

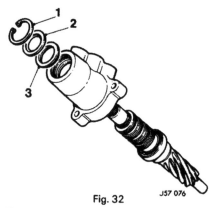

Fig. 32   *J57 076*

Fit a new seal, grooved face downwards. Ensure that the flange sits snugly in the recess. Fit the seal retainer, with the rubber side and lip outermost. Fit the circlip, ensure that it is seated fully in the groove.

Smear the seals with a little clean power steering fluid. Fit the taper seal saver 18G 1259 over the serrations on the pinion valve, and enter the pinion valve into the pinion valve housing. Press the ball bearing race fully home.

Refit the rack damper assembly. Ensure that the threaded plug is slack. Remove the grease nipple from the plug, and centralise the rack using Service Tool 18G 1466 or Jaguar part number 12297 (Fig. 33).

Refit the pinion valve assembly to the pinion rack housing, ensure that the coupling groove in the pinion is in the correct position. Fit and tighten the three self locking nuts.

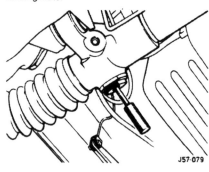

J57-079

Fig. 33

Adjust the rack damper pad assembly to obtain the correct end float. Tighten the locknut and refit the grease nipple.

Fit a new sealing washer to the port in the pinion end rack housing. Fit and tighten the feed pipes to the pinion valve housing and rack body. Do not overtighten the pipe nuts as irreversible damage could be caused to the pipes.

Tie Rods

Refit the new tab washers to the rack, dished face outermost. Screw on the tie rods. Holding one ball joint and tighten the opposite ball joint (one joint should react against the other). Do not restrain the rack assembly. Secure the tab washers in four places against the spanner flats.

Regrease the ball joint areas and replace any lost from the gaiters. Each gaiter should contain 57 gms (2 oz) of grease.

Fit the gaiters and secure with the clips, ensuring that the clips are in their correct position.

Refitting the Steering Rack

Ensure that the steering wheel is set to the straight ahead position and refit the rack.

Fit the lower coupling to the pinion. Ensure that the single rack mounting lug is shimmed so that it is central between the cross-beam brackets. This is achieved by fitting shims between the faces of the steel/rubber washers and the bracket. Check that a gap of 2.5 to 3.0 mm (0.10 to 0.12 in) exists between the face of the rubber thrust washers and the single lug of the rack.

Insert the mounting bolts, fit but do not fully tighten the nuts.

Slacken the clips securing the rubber gaiters to the rack housing, pull the gaiters (1, Fig. 34) clear of the inner ball joint assemblies.

Locate the two attachment brackets of Service Tool JD 36A on the heads of the lower wishbone fulcrum shaft bolts (2, Fig. 34).

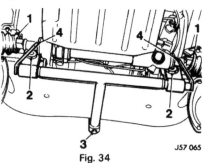

J57 065

Fig. 34

Release the locking screw (3, Fig. 34) on the forward arm of the tool and position the slide so that the slot engages with the front welded flange of the cross beam. Tighten the lock screw.

Rotate the alignment legs (4, Fig. 34) of the tool until one or both rest on the rack shaft. Adjust the position of the rack if necessary, until both legs are in contact with the rack shaft.

Tighten the nuts of the mounting bolts to secure the rack in this position. Remove Service Tool JD 36A.

Refit the rubber gaiters and secure with the clips.

Refit the ball joints to the steering arms and secure with the nyloc nuts.

Remove the blanking plugs and connect both fluid hoses to the pinion housing.

Refit the pinch bolt and nut to the lower universal coupling.

Refill the system with the recommended fluid and carry out the bleed procedure.

Check the front wheel alignment.

NOTE:

(A) It is important that the distance between the rubber faces of the thrust washers and the adjacent rack lug should in no case be less than 2.5 mm (0.1 in). This is to allow adequate 'rack' compliance in either direction.

(B) If a replacement rack unit is to be fitted it may be necessary to detach the lower column from the upper column at the universal joint, to obtain correct centralisation.

STEERING PUMP

Overhaul

Service Tools: Drive dog remover/replacer 18G 1445.

Steering pump removal

Open the bonnet.

Remove the cap from the hydraulic fluid reservoir. Disconnect the inlet hose (1, Fig. 35) from the pump, and drain the fluid into a suitable container.

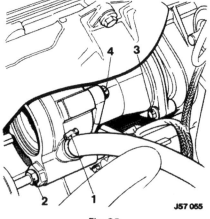

J57 055

Fig. 35

Blank off the hose and pump inlet to prevent the ingress of dust and dirt.

CAUTION: Under no circumstances must the fluid be reused.

Replace the fluid reservoir cap.

Disconnect the rack feed hose (2, Fig. 35) from the pump outlet, remove and discard the 'O' ring seal.

Remove the three bolts (3, Fig. 35) securing the pump mounting adaptor to the cylinder block and remove the pump assembly.

Blank off the hose and port to prevent the ingress of dust and dirt.

Remove the drive coupling.

Remove the three bolts (4, Fig. 35) securing the steering pump to the pump adaptor and remove the pump.

Steering Pump Dismantle

NOTE: Absolute cleanliness and extreme care are essential when overhauling the pump. This operation should not be entrusted to inexperienced mechanics.

If any doubt exists as to the necessity of the replacement of partly worn items, they should be replaced, as pump overhaul is not specified in the routine maintenance schedule.

Thoroughly clean the exterior of the pump. Remove the plugs, drain and discard the fluid.

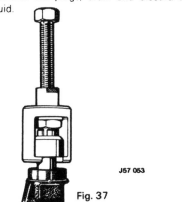

Fig. 37

Using Service Tool 18G 1445 remove the drive dog from the pump shaft (see Fig. 37).

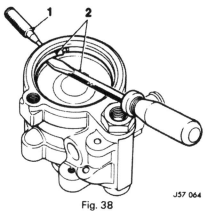

Fig. 38

Insert a suitable punch (1, Fig. 38) in the hole in the rear of the pump body and dislodge the spring ring.

Extract the ring with a screwdriver (as shown 2, Fig. 38).

If the endplate is not ejected by the spring pressure, a light tap on the casing should free it.

Extract the endplate 'O' ring seal (13, Fig. 36) from the pump body and discard.

NOTE: Examine the exposed portion of the driveshaft. If it is corroded, thoroughly clean it with crocus cloth. This will prevent damage being caused to the shaft bushing when tapping the shaft through.

If the shaft bushing is damaged, then replacement of the entire pump housing will be necessary.

Lightly tap the shaft (1, Fig. 36) throught the pump body, carrying the pump rotor assembly with it.

Extract the other 'O' ring seal (12, Fig. 36) from the pump body and discard.

Carefully separate the pump rotor components.

Remove the circlip (8, Fig. 36) and withdraw

the rotor (6, Fig. 36) and thrust plate (5, Fig. 36) from the shaft.

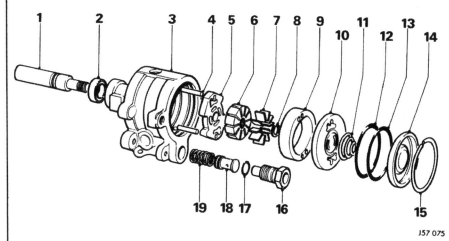

Fig. 36

KEY TO FIG. 36

1. Drive shaft	10. Pressure plate
2. Drive shaft seal	11. Pressure plate spring
3. Pump body	12. Pump ring 'O' ring seal
4. Dowel pins	13. End plate 'O' ring seal
5. Thrust plate	14. End plate
6. Pump rotor	15. End plate spring ring
7. Pump rotor vanes	16. Outlet connector
8. Circlip	17. 'O' ring seal
9. Pump ring	18. Control valve
	19. Control valve spring

Inspection

Clean all the components; carefully inspect for any signs of wear and damage.

Light scoring of the thrust and pressure plates can be removed by lapping.

If the pump ring or vanes show signs of chattering or grooving, then they must be renewed.

Scuff marks and light universal wear are acceptable.

Check the control valve for free movement; remove any burrs and renew the valve if it is at all faulty.

Check the shaft in the bush; it should run freely and no excessive sideways movement should be evident.

Measure the external diameter of the shaft and the internal diameter of drive dog. There MUST be an interference fit of 0.025 to 0.066 mm (0.001 to 0.0026 in) between the drive dog and the shaft.

Extract the drive shaft oil seal (2, Fig. 36) using a suitable drift.

Steering Pump Assemble

Fit a new shaft seal to the pump housing, lightly smear with petroleum jelly and insert the shaft, splined end first.

Fit the dowel pins (1, Fig. 39). Fit the thrust plate, ensure that the port face is facing outwards (2, Fig. 39) over the dowel pins.

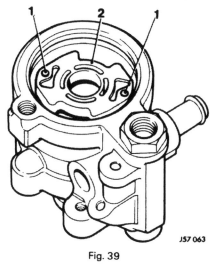

Fig. 39

continued

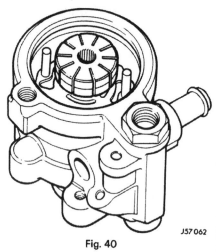

Fig. 40

J57 062

Fit the rotor (Fig. 40) counterboard·face first, to the shaft splines and secure with a new circlip. Do not overstretch the circlip.

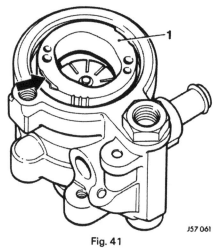

Fig. 41

J57 061

Slide the pump ring (1, Fig. 41) over the dowel pins; ensure that the rotation arrow is visible (arrowed Fig. 41).

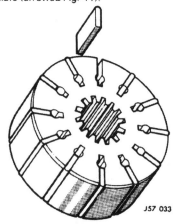

Fig. 42

J57 033

Fit the vanes in the slots in the rotor with their radiused edges outermost (Fig. 42).

Lightly smear the new pressure plate 'O' ring seal with petroleum jelly and insert in the groove of the pump housing. Fit the pressure plate (1, Fig. 43) with the spring recessed face outermost and press firmly in to the 'O' ring seal.

Lightly smear the other new 'O' ring seal with petroleum jelly and insert into the outer groove of the pump housing.

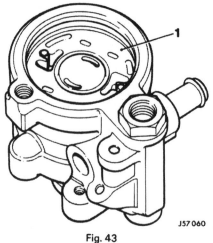

Fig. 43

J57 060

Refit the spring and place the end plate in position. Place the spring ring (1, Fig. 44) with the gap away from the extractor hole in the pump body.

Place the assembly under a press (2, Fig. 44) and carefully depress the end plate until the spring ring can be sprung into the groove.

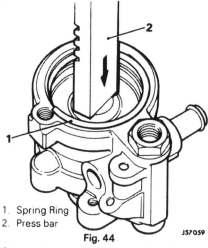

1. Spring Ring
2. Press bar

Fig. 44

J57059

Reassemble the control valve (Fig. 45) and refit to the outlet port.

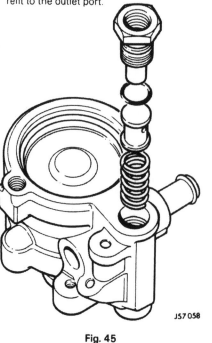

J57 058

Fig. 45

Using Service Tool 18G 1445 refit the drive dog to the shaft (see Fig. 46).

J57 054

Fig. 46

Steering Pump Refitting

Refit the steering pump mounting adaptor to the pump and secure with the three bolts. Clean the drive coupling faces, check for wear or damage and renew as necessary.

Locate the coupling in the drive dog and refit the pump assembly to the engine. Ensure that the drive coupling engages with the distributor drive shaft.

Secure the pump mounting adaptor to the cylinder block with the three bolts.

Fit a new 'O' ring seal to the pump outlet port, remove the blanking plug from the rack feed hose and secure the hose union to the pump. Tighten to the correct torque.

Remove the blanking plug from the pump inlet hose, refit the hose to the pump inlet and secure with the hose clip.

Refill the fluid reservoir with fluid of the correct specification.

Bleed the system and carry out a system check.

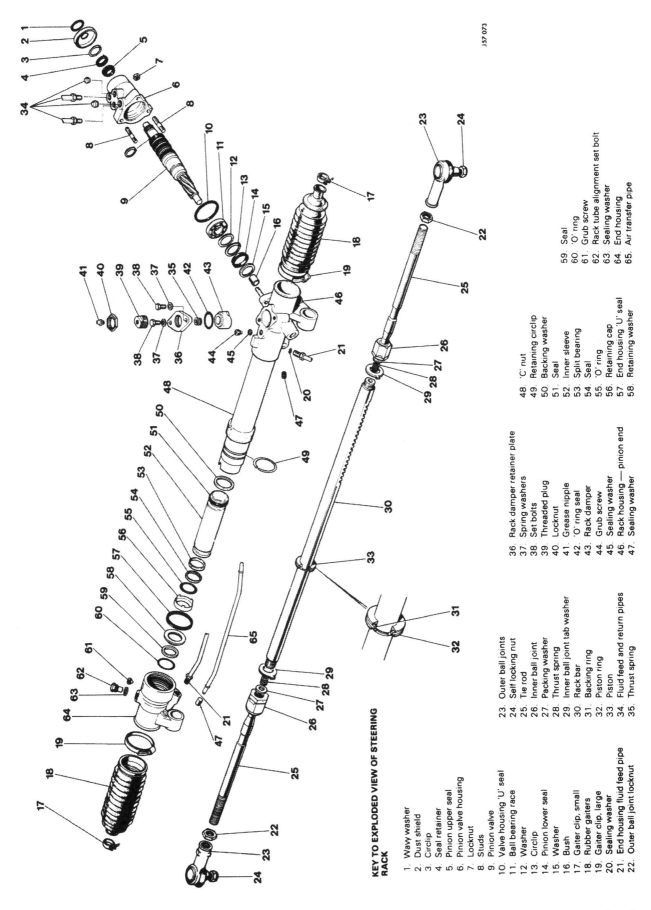

JS7 073

KEY TO EXPLODED VIEW OF STEERING RACK

1. Wavy washer
2. Dust shield
3. Circlip
4. Seal retainer
5. Pinion upper seal
6. Pinion valve housing
7. Locknut
8. Studs
9. Pinion valve
10. Valve housing 'U' seal
11. Ball bearing race
12. Washer
13. Circlip
14. Pinion lower seal
15. Washer
16. Bush
17. Gaiter clip, small
18. Rubber gaiters
19. Gaiter clip, large
20. Sealing washer
21. End housing fluid feed pipe
22. Outer ball joint locknut

23. Outer ball joints
24. Self locking nut
25. Tie rod
26. Inner ball joint
27. Packing washer
28. Thrust spring
29. Inner ball joint tab washer
30. Rack bar
31. Backing ring
32. Piston ring
33. Piston
34. Fluid feed and return pipes
35. Thrust spring

36. Rack damper retainer plate
37. Spring washers
38. Set bolts
39. Threaded plug
40. Locknut
41. Grease nipple
42. 'O' ring seal
43. Rack damper
44. Grub screw
45. Sealing washer
46. Rack housing — pinion end
47. Sealing washer

48. 'C' nut
49. Retaining circlip
50. Backing washer
51. Seal
52. Inner sleeve
53. Split bearing
54. Seal
55. 'O' ring
56. Retaining cap
57. End housing 'U' seal
58. Retaining washer

59. Seal
60. 'O' ring
61. Grub screw
62. Rack tube alignment set bolt
63. Sealing washer
64. End housing
65. Air transfer pipe

Fig. 47

187

CONTENTS

TORQUE WRENCH SETTINGS

ITEM	DESCRIPTION	SPANNER SIZE	TIGHTENING TORQUE		
			Nm	kgf/cm	lbf/ft
Anti-roll bar bracket to body	3/8 in UNF nut	9/16 in AF	37 to 43	3,74 to 4,42	27 to 32
Anti-roll bar to link	3/8 in UNF nut	9/16 in AF	19 to 24	1,94 to 2,48	14 to 18
Anti-roll bar link to lower wishbone	3/8 in UNF nut	9/16 in AF	19 to 24	1,94 to 2,48	14 to 18
Brake caliper to stub axle carrier	M12 bolt	19 mm	68 to 81	6,91 to 8,29	50 to 60
Brake disc to hub	7/16 in UNF bolt	5/8 in AF	41 to 54	4,2 to 5,54	30 to 40
Bump stop to spring pan	5/16 in UNF nut	1/2 in AF	11 to 14	1,11 to 1,38	8 to 10
Clamp and shield to stub axle carrier	1/4 in UNF nut	7/16 in AF	6 to 8	0,69 to 0,83	5 to 6
Clamp, crossbeam front mounting	1/2 in UNF nut	3/4 in AF	34 to 41	3,46 to 4,14	25 to 30
Damper mounting bracket to wishbone	3/8 in UNF nut	9/16 in AF	37 to 43	3,74 to 4,42	27 to 32
Damper mounting, lower	7/16 in UNF nut	11/16 in AF	61 to 68	6,23 to 6,91	45 to 50
Damper mounting, upper	3/8 in UNF nut	9/16 in AF	37 to 43	3,74 to 4,42	27 to 32
Front mounting bolt	3/4 in UNF nut	1 1/8 in AF	129 to 156	13,14 to 15,91	95 to 115
Fulcrum shaft, lower	9/16 in UNF nut	7/8 in AF	43 to 68	4,43 to 6,91	32 to 50
Fulcrum shaft, upper	1/2 in UNF nut	3/4 in AF	61 to 75	6,23 to 7,60	45 to 55
Lower ball joint to lower wishbone	9/16 in UNF nut	7/8 in AF	75	7,60	55
Lower ball joint to stub axle carrier	5/16 in UNF bolt	1/2 in AF	20 to 27	2,08 to 2,76	15 to 20
Rear mounting to body	3/8 in UNF bolt	9/16 in AF	30 to 35	3,05 to 3,59	22 to 26
Rear mounting to crossbeam	3/8 in UNF nut	9/16 in AF	19 to 24	1,94 to 2,48	14 to 18
Rebound stops to upper wishbone	5/16 in UNF bolt	5/8 in AF	11 to 14	1,11 to 1,38	8 to 10
Spring pan to lower wishbone	3/8 in UNF bolt	9/16 in AF	37 to 43	3,74 to 4,42	27 to 32
Steering arm to stub axle carrier	M12 bolt	19 mm	108 to 122	6,91 to 7,60	50 to 55
Stub axle to carrier	5/8 in UNF nut	15/16 in AF	37 to 43	11,1 to 12,4	80 to 90
Upper ball joint to stub axle carrier	1/2 in UNF nut	3/4 in AF	47 to 68	4,84 to 6,91	35 to 50
Upper ball joint to wishbone	3/8 in UNF bolt	9/16 in AF	35 to 43	3,60 to 4,42	26 to 32
Upper fulcrum shaft to spring turret	7/16 in UNF nut	11/16 in AF	66 to 75	6,78 to 7,60	49 to 55
Wheel nuts	1/2 in UNF stud	7/8 in AF	89 to 99	9,1 to 10,4	65 to 75

DESCRIPTION

The front suspension is fitted to the car as a complete unit. It comprises a fabricated sheet steel crossmember, mounted to the Body/Chassis structure at four points.

The two longitudinal members are attached to brackets at the front end of the chassis side member with rubber/steel mountings.

The crossmember has a turret welded at each end, which houses a coil spring. The spring is retained at its lower end by a seat pan, bolted to a lower wishbone.

The lower wishbone is a one piece forging, attached at the inner end to the crossmember, by rubber/steel bonded bushes, and at the outer end by a lower ball joint assembly, attached to a stub axle carrier.

Upper wishbone levers, steel forged, are mounted at their inner ends to a fulcrum shaft on rubber/steel bonded bushes. The fulcrum shaft is bolted to the spring turret. The outer ends of the wishbone are attached to the stub axle carrier by an upper ball joint.

The wheel hub is supported on two taper roller bearings, the inner races of which fit on the stub axle shaft, located in a tapered hole bored in the stub axle carrier.

An anti roll bar, fitted between the two lower wishbones, is attached to the chassis side members by rubber insulated brackets.

DIMENSIONAL DATA

The following dimensional drawings are provided to assist in assessing accidental damage. A component suspected of being damaged should be removed from the car, cleaned off and the dimensions checked and compared with those given in the appropriate illustration.

Components found to be dimensionally inaccurate, or damaged in any way MUST be scrapped and NO ATTEMPT made to straighten and re-use.

Dimension — Lower Wishbone

A. 225,30 to 225,81 mm (8.87 to 8.89 in)
B. 244,35 to 244,60 mm (9.62 to 9.63 in)
C. 177,67 to 177,93 mm (6.995 to 7.005 in)
D. 5,84 to 6,35 mm (0.23 to 0.25 in)
E. 60,20 to 60,45 mm (2.37 to 2.38 in)
F. 34,67 to 35,18 mm (1.365 to 1.385 in)
G. 353,82 to 354,33 mm (13.93 to 13.95 in)
H. 6° ± 0° 30'
J. 23,11 to 23,37 mm (0.91 to 0.92 in)
K. 26,67 to 27,18 mm (1,05 to 1.07 in)
L. 21,08 to 21,59 mm (0.83 to 0.85 in)
M. 33,20 to 33,45 mm (1.307 to 1.317 in)

Fig. 1

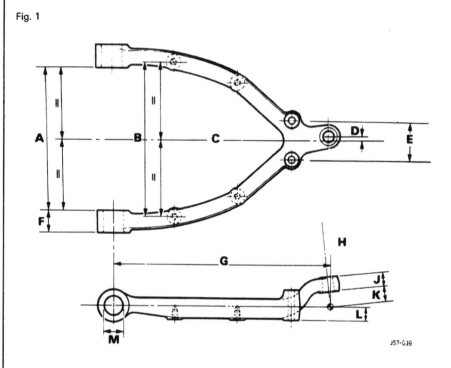

J57-039

Dimension — Stub Axle Carrier

A. 81,28 to 81,78 mm (3.26 to 3.22 in)
B. 42,68 to 43,18 mm (1.68 to 1.70 in)
C. 0,8 to 1,0 mm (0.03 to 0.04 in)
D. 19,02 to 19,07 mm (0.749 to 0.751 in)
   70 ± 0° 30' Total taper.
E. 25,4 mm (1.0 in)
F. 100,09 to 100,59 mm (3.94 to 3.96 in)
G. 5° ± 0° 30'
H. 88° ± 0° 30'
J. 148,91 to 149,41 mm (5.86 to 5.88 in)
K. 31,76 to 32,26 mm (1.25 to 1.27 in)
L. 88,65 to 89,15 mm (3.49 to 3.51 in)
M. 112,3 to 115,3 mm (4.42 to 4.54 in)
N. 70,85 to 71,35 mm (2.79 to 2.81 in)
P. 58,93 to 59,43 mm (2.32 to 2.34 in)

Fig. 2

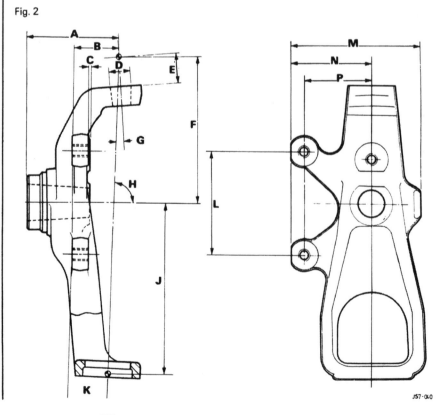

J57-040

Dimension — Upper Wishbone Arm — Front

A. 60,20 to 60,71 mm (2.37 to 2.39 in)
B. 161,44 to 161,95 mm (6.356 to 6.376 in)
C. 44,2 to 44,7 mm (1.74 to 1.76 in)
D. 31,50 to 31,75 mm (1.24 to 1.25 in)

Fig. 3

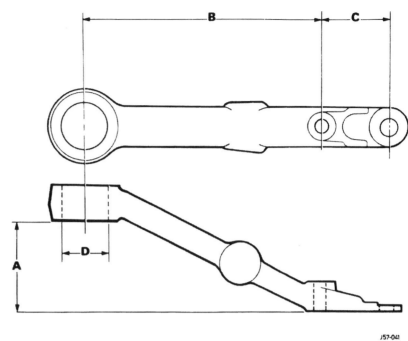

J57-041

Dimension — Upper Wishbone Arm — Rear

A. 44,2 to 44,7 mm (1.74 to 1.76 in)
B. 161,44 to 161,95 mm (6.356 to 6.376 in)
C. 44,2 to 44,7 mm (1.74 to 1.76 in)
D. 31,50 to 31,75 mm (1.24 to 1.25 in)

Fig. 4

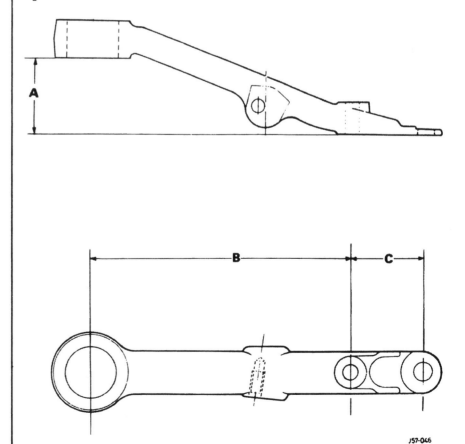

J57-046

ANTI-ROLL BAR LINK BUSHES

Renew

Jack up the front of the car and place on stands.
Remove the upper self locking nut (1, Fig. 5), special washer and rubber bush, securing each end of the anti-roll bar to the anti-roll bar links. Remove the lower self locking nut (2, Fig. 5) special washer and rubber bush, securing each end of the anti-roll bar link to the anti-roll bar support brackets.

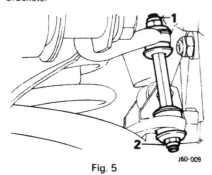

Fig. 5

Remove each link in turn, recover two spacer tubes, rubber bushes and special washers from each link.
Inspect the rubber bushes, renew if any wear or damage is evident.
Reverse the above procedure, but do not fully tighten the fixings until the car is resting on its wheels.

ANTI-ROLL BAR

Renew

Service Tools: JD24 Steering Joint Separator, and ½in UNF socket headed (grub) screw 2in long.

Jack up the front of the car and place on stands. Leave the jack in position under the front suspension crossmember.
Remove both front wheels.
Remove the plastic drive fasteners securing the wheel arch front dust shields, and

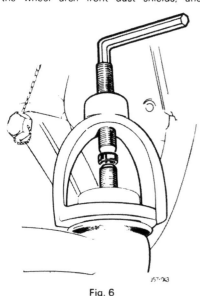

Fig. 6

screws and plastic drive fasteners securing the spoiler undertray to the body and spoiler.
Remove the nut securing the LH steering tie rod to the steering arm, and using service tool JD24 detach the tie rod from the steering arm (Fig. 6).

NOTE: When using JD24 Steering Joint Separator, substitute a 2in long ½in UNF socket headed (grub) screw for the existing bolt.

Remove the self locking nuts, special washers and rubber bushes securing each end of the anti-roll bar, to the anti-roll bar links (1, Fig. 7). Ensure that the jack is supporting the front suspension crossmember, and remove the front suspension crossmember front mounting bolts.

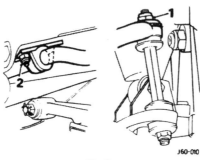

Fig. 7

Lower the jack; collect the washers, spacers and bush sleeves.
Remove the nuts and bolts securing the anti-roll bar brackets to the front crossmember (2, Fig. 7) and detach the keeper plates.
Remove the rubbers from the anti-roll bar.
Lift the anti-roll bar off the links and manoeuvre out through the LH wheel arch.

NOTE: The fitting of the rubber bushes will be greatly assisted if a proprietary rubber lubricant, or a liquid soap and water solution is used.

Manoeuvre the anti-roll bar into position across the car.
Lubricate the rubber bushes and position them on the anti-roll bar adjacent to the keeper plate locations, with the splits towards the rear of the car.
Fit the keeper plates and brackets, loosely secure to the front suspension cross frame.
Fit the anti-roll bar to the anti-roll bar links and fit the rubbers, special washers and self locking nuts.
Refit the steering tie rod to the LH steering arm and tighten the nut.
Raise the jack to locate the front suspension crossmember front mountings with the body. Fit bolts, washers, spacers and bush sleeves.
Tighten the nuts.
Refit the spoiler undertray using new plastic drive fasteners. Refit the two setscrews.
Refit the wheel arch front dust shields using new plastic drive fasteners.
Refit the front road wheels.
Lower the car and fully tighten the anti-roll bar links and mounting nuts.

CAUTION: All anti-roll bar fixings must only be tightened with the full weight of the car resting on the wheels. Premature failure of the rubber bushes could occur if this precaution is not taken.

FRONT DAMPERS

Renew

NOTE: In the event of a damper being unserviceable a replacement unit must be fitted.

Remove the locknut, nut, outer washer, rubber buffer and inner washer from the damper top mountings (1, Fig. 8), accessible from the engine compartment.

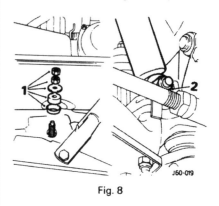

Fig. 8

To remove the LH side front damper, it is necessary to remove the air cleaner cover and element, to gain access to the top mounting.
Jack up the front of the car and place on stands.
Remove the self locking nut and bolt from the bottom mounting (2, Fig. 8). Telescope the damper and withdraw from the car.
Before fitting new dampers it is advisable to 'bleed' any air which may have accumulated in the pressure chamber due to the damper having been stored in a horizontal position. To do this, hold the damper vertically and make several strokes (not exceeding more than halfway) until there is no lost motion. Then extend the damper to its full length once or twice. Keep upright until fitted.
To fit the new dampers, reverse the above procedure, ensuring that the lower washers and rubber buffer are in position on the damper stem, before inserting through the hole in the wheel arch.
Tighten all fixings to the correct torque.

FRONT HUB BEARINGS

Renew

Jack up the front of the car, place on stands and remove the road wheel. Remove the bolts and washers securing the hub assembly to the brake disc. Access is gained through the aperture in the disc shield (1, Fig. 9). Remove the hub grease cap, extract the split pin; remove the nut retaining cap, nut and washer from the stub axle. (2, Fig. 9). Withdraw the hub by hand.

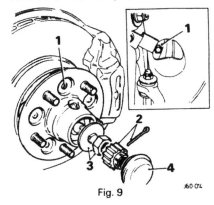

Fig. 9   J60-074

The front hub wheel studs can be replaced using a power press and suitable mandrel, once the hub assembly has been removed from the stub axle.
Remove the grease seal (1, Fig. 10). Withdraw the inner and outer bearing races (2, Fig. 10). Taking the appropriate safety precautions, use a suitable punch to drift out the inner and outer bearing races (3, Fig. 10). Cut-outs are provided in the hub assembly abutment shoulders for this purpose.

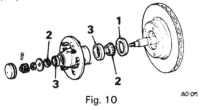

Fig. 10   J60-015

Tap replacement cups into position, ensuring that they seat squarely in the abutments.
Lubricate and fit the inner bearing. Fit a new grease seal, ensuring that it is seated squarely in position.
Pack the hub with the specified grease and fit to the stub axle.
Fit the outer bearing, washer and nut.
Adjust the bearing endfloat to 0,03 — 0,08 mm (0.001 in — 0.003 in), this is measured with a dial indicator gauge mounted with the plunger against the end of the hub.
If a gauge is not available, tighten the hub nut until there is no endfloat, i.e. when rotation of the hub is slightly restricted. (A torque of 0,691 kg/fm, (5 lb/ft), must not be exceeded or damage may be caused to the bearings and bearing tracks). Slacken the nut one flat and fit the nut retaining cap.
Fit a new split pin and bend over. Refit the grease cap, ensure that the vent hole is clear.
Refit the road wheel and lower the car.

FRONT STUB AXLE CARRIER

Renew

Service Tool: JD6G Road Spring Compressor.

Jack-up the front of the car, place on stands and remove the road wheel. Remove the brake pad retaining pin spring clips, remove the upper pin, anti-rattle springs and lower pin.
Withdraw the brake pads. Note their position for refitting.
Break the lock wire and remove the the two bolts and spring washers securing the steering arm and brake caliper to the stub axle carrier (1, Fig. 11).

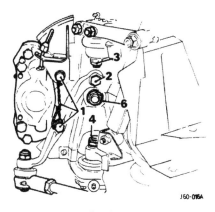

Fig. 11   J60-016A

Move aside caliper and secure with wire or strong cord to prevent damaging the brake hose.
Remove the remaining bolt and spring washer (2, Fig. 11) securing the steering arm to the stub axle carrier.

NOTE: Record the number of shims fitted between the steering arm and the brake caliper.

Remove the front hub grease cap, extract the split pin, remove the nut retaining cap, and washer from the stub axle.
Withdraw the front hub and disc assembly.
Using JD6G Spring Compressor, relieve the stub axle carrier of spring tension (Fig. 12).

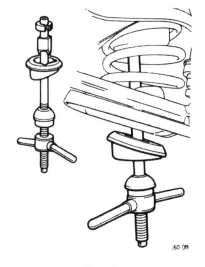

Fig. 12   J60-011

Undo and remove the nyloc nut and washer securing the upper ball joint to the stub axle carrier (3, Fig. 11), release the taper.
Undo and remove the nyloc nut and washer securing the lower ball joint to the lower wishbone (4, Fig. 11). Release the taper and remove the stub axle carrier assembly from the vehicle.
Remove the two nyloc nuts securing the clamps at the bottom of the disc shields (5, Fig. 11). Remove the attachment plate and the front and rear disc shields.
The stub axle can now be removed. Undo and remove the nyloc nut (6, Fig. 11) and washer, support the stub axle carrier, and using a suitable punch, drift out the stub axle.
If the carrier is to be renewed it is necessary to remove the lower ball joint. Bend back the tab washers and remove the four bolts securing the ball pin cap to the stub axle carrier. Discard the tab washers.

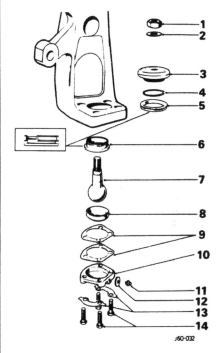

Fig. 13   J60-032

Key to Fig. 13.

1. Self-locking nut
2. Washer
3. Gaiter
4. Flexible gaiter retaining ring
5. Moulded gaiter retainer
6. Ball socket — upper
7. Ball pin
8. Ball socket — lower
9. Shim
10. Ball pin cap
11. Grease nipple
12. Washer
13. Tab washer
14. Bolts

Detach the lower ball pin cap and lower socket, remove the shims, retain for refitting, and lift out the ball pin.

Remove the gaiter flexible plastic retaining ring, and remove the gaiter.

Release the moulded plastic gaiter retainer from the upper socket. Tap the upper socket out of the stub axle carrier.

Fit the stub axle to the carrier. Secure with a new nyloc nut tightened to the correct torque.

Fit the lower ball joint upper socket to the stub axle carrier.

Fit the moulded plastic gaiter retainer, lip first, to the ball joint upper socket. Lip of the retainer must engage with the slot in the upper socket.

Fit the gaiter over the gaiter retainer and secure with the flexible plastic ring.

Fit the ball pin through the gaiter and locate in the ball joint upper socket.

Replace the shims, previously removed, fit the lower socket and ball pin cap.

Fit the four bolts and new tab washers, tighten to the correct torque. Check the adjustment of the ball pin, it should be slightly stiff in the socket when the bolts are tightened to their correct torque.

Correct adjustment of the ball pin is achieved by limiting the vertical lift of the ball pin, without lubrication, to 0,025 — 0,203 mm (0.001 — 0.008 in). When grease is added the torque required to move the ball pin must not exceed 1 Nm (9 lbf/in). This is acheived by adding or subtracting shims which are available in 0,5 mm (0.002 in) and 0,10 mm (0.004 in) thicknesses.

When this adjustment is obtained bend over the locking tabs.

Refit the disc shields and attachment plate to the stub axle assembly and refit the stub axle carrier to the vehicle.

Refit the front hub and disc assembly, ensure that the hub bearing end float is properly adjusted.

Refit the brake caliper and steering arm to the stub axle carrier, ensuring that if any shims were removed, they are refitted in their previously noted positions.

Refit the brake pads, pins, and spring clips.

Remove the service tool JD6G, refit the road wheel and lower the car.

Depress the brake pedal several times to centralise the brake pads.

Carry out a front suspension geometry check.

FRONT SUSPENSION RIDE HEIGHT

Check and Adjust

Service tools: JD6G Spring Compressor. Slip Plates.

Ensure that the car is in its kerb weight condition i.e. full tank of fuel; engine oil and water at correct levels.

Check that the tyres are at the correct pressure.

Position the front wheels on slip plates.

Press downwards on the front bumper to depress the suspension and slowly release.

Measure the distance between the lower face of the crossbeam and the ground on both sides of the car.

Obtain values for dimension 'A' (Fig. 14), left and right hand.

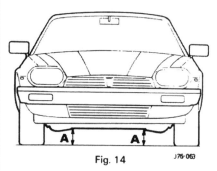

Fig. 14   J76-063

The correct height is 152 mm (6 in) minimum, plus the thickness of the slip plates. If necessary, add or remove packing plates from either end of the spring to obtain this dimension.

The packing rings are 3,18 mm (0.125 in) thick, and vary the ride height by 7.93 mm (0.3125 in).

To add or remove packing rings the front road springs must be removed.

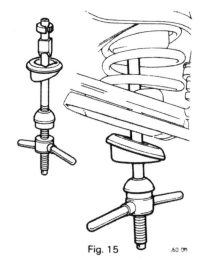

Fig. 15   .60 0n

Jack up the front of the car and place on stands. Remove the appropriate road wheel.

Fit Spring Compressor tool JD6G, as shown in Fig. 15. Compress the spring sufficiently to relieve the load on the spring pan fixings, and remove the fixings.

Slacken the spring compressor tool and remove with the spring pan, spring and packers.

NOTE: Record the quantity and position of the packers. Add or remove packers to obtain the correct ride height.

WARNING: THE MAXIMUM NUMBER OF PACKERS THAT CAN BE FITTED TO EACH SPRING MUST NOT EXCEED FIVE, COMPRISING, THREE BELOW THE SPRING IN THE SPRING PAN AND TWO ABOVE IN THE CROSSMEMBER SPRING TURRETS.

Assemble the spring, packers and spring pan; lift up into position in the spring turret. Retain in this position using the Spring Compressor tool JD6G.

Align the spring pan fixing holes with the tapped holes in the lower wishbone using the pilot studs.

Wind up the handle of the spring compressor tool sufficiently to compress the spring, locating the spring pan on the pilot studs, until the setscrews, nuts, bolts and washers can be fitted. Remove the pilot studs and fit the setscrews and washers.

Remove the spring compressor.

Refit the road wheel, and lower the car.

Recheck the suspension ride height as previously detailed.

LOWER BALL JOINT

Overhaul

Service Tools: JD6G Road Spring Compressor, JD24 Steering Joint Taper Separator, and $\frac{1}{2}$in socket headed (grub) screw 2in long.

Jack up the front of the car, place on stands and remove the road wheel.

Using service tool JD6G (as shown in Fig. 15), compress the road spring sufficiently to relieve the stub axle carrier of spring pressure.

Remove the self locking nut and washer securing the tie-rom ball joint to the steering arm, and separate using service tool JD24.

NOTE: When using JD24 Steering Joint Separator, substitute a 2in long $\frac{1}{2}$in UNF socket headed (grub) screw for the existing bolt.

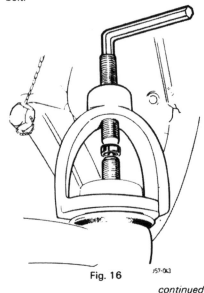

Fig. 16   J57-043

continued

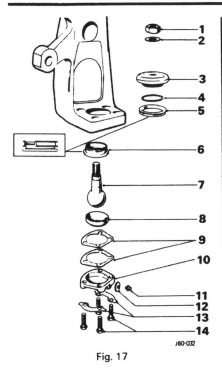

Fig. 17

Key to Fig. 17.

1. Self-locking nut
2. Washer
3. Gaiter
4. Flexible gaiter retaining ring
5. Moulded gaiter retainer
6. Ball socket — upper
7. Ball pin
8. Ball socket — lower
9. Shim
10. Ball pin cap
11. Grease nipple
12. Washer
13. Tab washer
14. Bolts

Undo and remove the upper ball joint inner nut and bolt; slacken but do not remove the outer nut and bolt.

Undo and remove the nyloc nut and washer securing the lower ball joint to the lower wishbone.

Support the hub/disc assembly and release the lower ball joint taper. Swing the assembly outwards.

Knock back the tab washers securing the ball pin cap to the stub axle carrier. Remove the four bolts and tab washers, discard the tab washers. Detach the lower ball pin cap and lower socket, remove the shims, retain for refitting, and lift out the ball pin.

Remove the gaiter flexible plastic retaining ring, and remove the gaiter.

Release the moulded plastic gaiter retainer from the upper socket.

Tap the socket out of the stub axle carrier.

NOTE: Excessive wear in the ball pin and sockets cannot be adjusted by shims, worn parts must be renewed.

Inspect all the components for wear or damage.
The lower socket has to be broken up to remove it from the ball pin cap.
Renew parts as necessary. The lower socket is a press fit in the ball pin cap.
Fit the lower ball joint upper socket to the

stub axle carrier.
Fit the moulded plastic gaiter retainer, lip first, to the ball joint upper socket. Lip of the retainer must engage with the slot in the upper socket.
Fit the gaiter over the gaiter retainer and secure with the flexible plastic ring.
Fit the ball pin through the gaiter and locate in the ball joint upper socket.
Replace the shims, previously removed, fit the lower socket and ball pin cap.
Fit the four bolts and new tab washers, tighten to the correct torque.
Check the adjustment of the ball pin, it should be slightly stiff in the socket when the bolts are tightened to their correct torque.
To obtain the correct adjustment of the ball pin, after assembling the ball joint, remove shims one by one until the ball pin is tight in its socket when the bolts are tightened to their correct torque. Remove the ball pin and add shims to the value of 0,10 — 0,15 mm (0.004 — 0.006 in).

NOTE: The correct adjustment of the ball pin is achieved by limiting the vertical lift of the ball pin, without lubrication, to 0,025 — 0,203 mm (0.001 — 0.008 in). When grease is added the torque required to move the ball pin must not exceed 1 Nm (9 lbf/in). Achieved by adding or subtracting shims, which are available in 0,05 mm (0.002 in) and 0,10 mm (0.004 in) thicknesses.

When the correct adjustment has been obtained, bend over the locking tabs; reverse the dismantling procedure.

CAUTION: The bolts securing the upper ball joint to the upper wishbone MUST be fitted from the front of the car.

Tighten all nuts and bolts to the correct torque.
Carry out a front suspension geometry check.

UPPER BALL JOINT

Renew

Service Tool: JD6G Spring Compressor.

NOTE: The upper wishbone ball joint cannot be dismantled and if worn, the complete assembly must be replaced.

Jack up the front of the car, place on stands, and remove the front road wheel. Fit service tool JD6G Spring Compressor (see Fig. 18), and relieve stub axle carrier of spring tension.
Using strong cord or wire, tie the stub axle carrier to the cross member spring turret, to prevent straining the front brake caliper hose.
Remove the two nuts bolts and plain washers securing the ball joint to the upper wishbone arms (1, Fig. 19).

CAUTION: Make a careful note of the number and position of packing and shims as these control the castor angle.

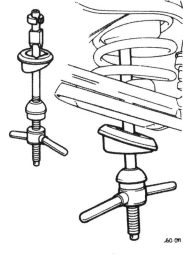

Fig. 18

Remove the self locking nut and plain washer securing the ball joint to the stub axle carrier (2, Fig. 19), release taper and remove the ball joint.

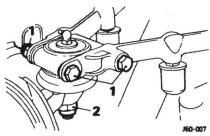

Fig. 19

To fit a new ball joint, reverse the above procedure, ensuring that the packing and shims are repositioned as previously noted.
Fit the ball joint to the stub axle carrier before securing it to the wishbone arms.

CAUTION: The bolts securing the upper ball joints to the upper wishbone, must be fitted from the front of the car.

Remove the support wire or cord. Remove service tool JD6G.
Refit the road wheel and lower the car.
Carry out a front suspension geometry check.

REBOUND STOPS

Renew
NOTE: Rebound stops must only be replaced as a pair; uneven loads will be placed on the upper wishbone arms if this is not done.

Jack up the front of the car, place on stands and remove the road wheel.
Unscrew the rebound stops from the upper wishbone arms.
To fit new rebound stops, reverse the above procedure; tighten to the correct torque.

BUMP STOP

Renew

Jack up the front of the car, place on stands and remove the road wheel.

Remove the two plain nuts and spring washers (1, Fig. 20) securing the bump stop to the spring pan seat.

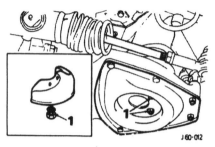

Fig. 20

Manoeuvre the bump stop clear, through the coils of the spring. It may be necessary to carefully prise apart the coils, to enable the bump stop to be removed.

To fit a new bump stop, reverse the above procedure.

Tighten nuts to the correct torque.

SUSPENSION UNIT — MOUNTING BUSHES

Renew

NOTE: A worn or damaged bush infers that undue strain has been placed upon the apparently satisfactory bush on the opposite side of the car. Bushes must therefore be renewed as a pair.

Jack up the front of the car, using a trolley jack under the front suspension crossmember (1, Fig. 21). Position axle stands under the jacking spigots (2, Fig. 21), and lower the car onto them.

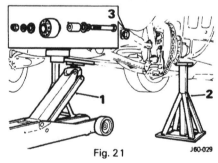

Fig. 21

Adjust the jack to release the load from the bushes, but to still remain in contact with the crossmember.

Remove the wheel arch front dust shields.

Remove the self locking nut securing one of the mounting bolts and drift the bolt clear of the bush (3, Fig. 21).

NOTE: Record the position of the plain and special washers, and if fitted the securing bracket.

Slacken the clamping nut and bolt, securing the relevant mounting bush eye.

Lower the jack SLIGHTLY to improve access to the bush, and tap the bush clear of the eye.

Repeat this procedure for the other side.

To fit new bushes, reverse the above procedure, tightening the nuts to their correct torque.

SUSPENSION UNIT — REAR MOUNTINGS

Renew

NOTE: A worn or damaged mounting infers that undue strain has been placed upon the apparently satisfactory mounting on the opposite side of the car. Mountings must therefore be renewed as a pair.

Jack up the front of the car, using a trolley jack under the front suspension crossmember (1, Fig. 22). Position axle stands under the jacking spigots (2, Fig. 22), and lower the car onto them.

Remove the nuts securing the engine front mountings.

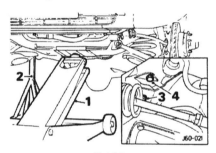

Fig. 22

Ensure that the trolley jack is supporting the crossmember, remove the self locking nut and washer (3, Fig. 22) securing each mounting to the crossmember. Carefully lower the rear of the crossmember unit, sufficiently to enable the two special setscrews and washers (4, Fig. 22) securing each mounting to the body, to be removed.

Carefully lever one side of the suspension crossmember down, to facilitate the removal of the mounting. Repeat for the other side.

To fit new mountings reverse the above procedure. Tighten all fixings to their correct torque.

NOTE: Mountings are offset and only fit one way.

LOWER WISHBONE

Overhaul

Service tool: JD6G Spring Compressor.

Jack up the front of the car, place on stands and remove the road wheel.

Remove the pinch bolt (1, Fig. 23) securing the lower steering column to the rack pinion.

Remove the bolt, washer and self locking nut, securing the rack top mounting (pinion side of rack assembly) (2, Fig. 23) to the crossmember.

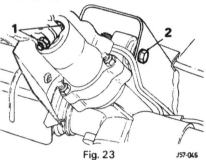

Fig. 23

Remove both the rack bottom mounting bolts, washers and nuts, securing the steering rack to the crossmember.

CAUTION: Make a careful record of the number and position of the packing washers for reassembly.

Release the steering rack from the crossmember, and retrieve the packing washers.

NOTE: Removal of the RH side lower wishbone requires the removal of the exhaust downpipe from the exhaust manifold.

Fit the Spring Compressor JD6G as shown, (Fig. 24), and compress the spring sufficiently to relieve the load on the spring seat pan fixings. Remove the bolts and nuts securing the spring seat pan to the lower wishbone.

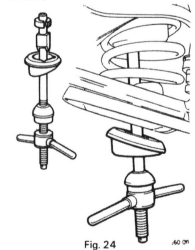

Fig. 24

Slacken the spring compressor tool and remove the spring seat pan, spring and packers.

NOTE: Record the quantity and position of the packer for reassembly.

continued

FRONT SUSPENSION

Remove the split pin, castellated nut and washer from the end of the lower wishbone fulcrum shaft; drift out the shaft.

Remove the inner bolt, nut and washer, securing the upper ball joint to the upper wishbone, and slacken the outer nut and bolt.

NOTE: Record the position of the castor shims and packing pieces BEFORE removing them.

Pull the top of the hub and stub axle carrier outwards and swivel to gain access to the self locking nut, securing the lower ball joint to the stub axle carrier.

Remove the self locking nut and plain washer; release the taper.

Remove the two bolts and nuts securing the front damper and anti-roll bar brackets to the lower wishbone.

Telescope the shock absorber and move clear.

Remove the lower wishbone and spacer washers.

Drift or press out the bush from the wishbone eye. Press a new bush into the eye, ensure that the bush projects from each side by an equal amount. Fitting the bush will be made easier by using a recommended rubber lubricant or a liquid soap solution.

The refitting procedure of the lower wishbone is the reverse of the above procedure.

CAUTION: The lower fulcrum shaft securing nut MUST NOT be fully tightened until the full weight of the car is resting on the suspension. Failure to carry out this procedure will result in undue torsional loading of the rubber bushes with possible premature failure.

UPPER WISHBONE

Overhaul

Service tool: JD6G Spring Compressor.

Jack up the front of the car, place on stands and remove the road wheel. Fit service tool JD6G, Spring Compressor, and tighten to relieve the stub axle carrier of spring tension (Fig. 25).

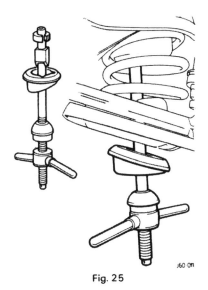

Fig. 25

Remove the two bolts, nuts and plain washers (1, Fig. 26), securing the ball joint to the upper wishbone levers. Note the relative positions of the packing pieces and shims, as these control the castor angle. Alternatively, remove the self locking nut (2, Fig. 26) securing the ball joint to the stub axle carrier, and release the taper. Tie up the stub axle carrier to the suspension crossmember, so that the flexible brake hose is not damaged.

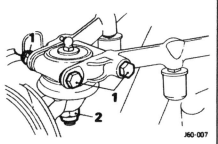

Fig. 26

Remove the two bolts which secure the upper wishbone fulcrum shaft to the suspension crossmember turret. Note the relative positions of the shims as these control the camber angle.

The upper wishbone assembly can now be removed.

Remove the self locking nuts and plain washers and withdraw the wishbone levers from the fulcrum shaft.

Release the locknuts and unscrew the rebound rubbers from the levers.

Drift out or press out the bush from the wishbone eye.

Press a new bush into the eye, ensuring that the bush projects from each side by an equal amount. Fitting of the bush will be made easier if a recommended rubber lubricant or a liquid soap solution is used.

The reassembly and refitting of the wishbone assembly, is a reversal of the above procedure.

CAUTION: The nuts securing the wishbone levers to the fulcrum shaft must not be fully tightened until the upper wishbone assembly has been fitted to the car and the full weight of the car is resting on the suspension. Failure to carry out this procedure will result in undue torsional loading of the rubber bushes with possible premature failure.

KEY TO EXPLODED VIEW OF THE FRONT SUSPENSION COMPONENTS

1. Crossmember
2. Rear mounting plate
3. Rear mounting fixings
4. Rear mounting
5. Bolt
6. Anti-roll bar
7. Rubber bush
8. Keeper plate
9. Self locking nut
10. Anti-roll bar
11. Cup wash-outer
12. Anti-roll bar to link — outer bush
13. Anti-roll bar to link — inner bush
14. Anti-roll bar link
15. Cup washers — inner
16. Lower bushes
17. Cup washer
18. Self locking nut
19. Bolt
20. Plain washer
21. Steel bush
22. Clamping bolt
23. Crossmember front mounting rubber bush
24. Plain washer
25. Self locking nut
26. Self locking nut
27. Plain washer
28. Gaiter retaining ring
29. Top ball joint gaiter
30. Gaiter retainer
31. Top ball joint
32. Castor shims
33. Self locking nut
34. Self locking nut
35. Plain washer

36. Upper wishbone mounting rubbers
37. Upper wishbone — front
38. Grease nipple
39. Plain washer
40. Fulcrum shaft — upper
41. Plain washer
42. Self locking nut
43. Camber shims
44. Bolt
45. Upper wishbone — rear
46. Top ball joint — inner bolt
47. Top ball joint — outer bolt
48. Rebound rubber
49. Locking washer
50. Grease cap
51. Nut retainer
52. Hub nut
53. Split pin
54. Plain washer
55. Outer bearing cone
56. Outer bearing track
57. Front hub
58. Inner bearing track
59. Inner bearing cone
60. Grease seal
61. Stub axle
62. Water deflector
63. Stub axle carrier
64. Plain washer
65. Self locking nut
66. Steering arm
67. Road spring
68. Spring packer
69. Locknut
70. Nut
71. Plain washer

72. Rubber bush
73. Cup washers
74. Rubber bush
75. Plain washer
76. Front damper
77. Damper mounting bracket
78. Anti-roll bar link mounting bracket
79. Nut and bolt
80. Self locking nut
81. Plain washer
82. Lower ball joint gaiter
83. Gaiter retaining ring
84. Gaiter retainer
85. Upper socket
86. Ball pin
87. Lower socket
88. Shims
89. Ball pin cap
90. Tab washers
91. Bolts
92. Grease nipple
93. Plain washer
94. Bolts
95. Spring washer
96. Nut
97. Spring pan
98. Lower wishbone
99. Lower fulcrum shaft
100. Bush
101. Washer-serrated
102. Washer-plain
103. Castellated nut
104. Split pin
105. Bump stop rubber
106. Shims

198

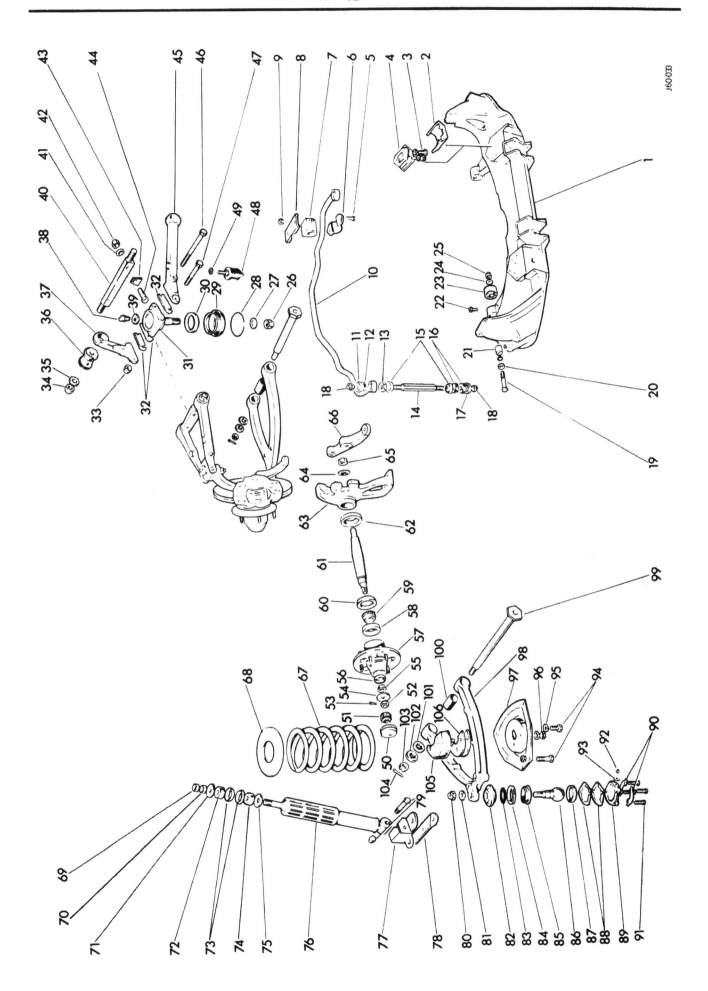

J60-033

CONTENTS

TORQUE WRENCH SETTINGS

ITEM	DESCRIPTION	SPANNER SIZE	TIGHTENING TORQUE		
			Nm	kgf/cm	lbf/ft
Bumpstop rubbers to body	$\frac{5}{16}$ in UNF nut	$\frac{1}{2}$ in AF	10,8 to 13,6	1,12 to 1,4	8 to 10
Drive shaft to rear hub	$\frac{3}{4}$ in UNF nut	$1\frac{1}{8}$ in AF	136 to 163	13,8 to 16,6	100 to 120
Inner fulcrum mounting attachment	$\frac{7}{16}$ in UNC bolt	$\frac{11}{16}$ in AF	81,3 to 88	8,4 to 9,1	60 to 65
Inner fulcrum shaft nuts	$\frac{1}{2}$ in UNF nuts	$\frac{3}{4}$ in AF	61 to 68	6,23 to 6,9	45 to 50
Outer fulcrum shaft nuts	$\frac{5}{8}$ in UNF nuts	$\frac{15}{16}$ in AF	131 to 145	13,5 to 15	97 to 107
Radius arm and safety strap to body	$\frac{7}{16}$ in UNF bolt	$\frac{5}{8}$ in AF	61 to 68	6,23 to 6,9	45 to 50
Radius arm to lower wishbone	$\frac{1}{2}$ in UNF bolt	$\frac{3}{4}$ in AF	88 to 95	9,1 to 9,8	65 to 70
Rear damper attachment	$\frac{7}{16}$ in UNF nut	$\frac{11}{16}$ in AF	43 to 49	4,5 to 5,04	32 to 36
Safety strap to floor panel	$\frac{3}{8}$ in UNF nut	$\frac{9}{16}$ in AF	37 to 43	3,8 to 4,5	27 to 32
Tie plate to crossbeam & inner fulcrum	$\frac{5}{16}$ in UNF nuts & bolts	$\frac{1}{2}$ in AF	19 to 24	1,95 to 2,52	14 to 18
'Vee' mountings to body	$\frac{3}{8}$ in UNF nut	$\frac{9}{16}$ in AF	37 to 43	3,8 to 4,5	27 to 32
'Vee' mountings to crossbeam	$\frac{5}{16}$ in UNF nut	$\frac{1}{2}$ in AF	19 to 24	1,95 to 2,52	14 to 18
Wheel studs to rear hub	$\frac{1}{2}$ in UNF	—	75 to 81,3	7,66 to 8,4	55 to 60

REAR SUSPENSION
Description

The complete independent rear suspension system is mounted to a pressed steel crossbeam, which is attached to the body/chassis structure via four 'Vee' rubber mountings.

The system is geometrically similar to that of a double wishbone set-up, but a drive shaft takes the place of a conventional upper wishbone, and the lower wishbone is replaced by a lower link.

One end of the link pivots at the crossmember, and the other end at a hub carrier. The lower link is strengthened torsionally to resist dive and braking loads, which are partially transmitted to the body structure via radius rods.

The aluminium alloy hub carrier houses a rear hub, which is splined to the drive shaft and runs in taper roller bearings.

Two gas filled shock absorber and spring assemblies are mounted between each lower link, and the crossbeam to regulate bump and rebound forces.

Plates are attached between the lower rear pick up points of the dampers and the hub carriers to provide lash-down points when transporting the vehicle.

WARNING: ON CARS FITTED WITH A 'POWR-LOK' LIMITED SLIP DIFFERENTIAL UNIT; UNDER NO CIRCUMSTANCES MUST THE ENGINE BE RUN WITH THE CAR IN GEAR AND ONLY ONE WHEEL OFF THE GROUND. IF IT IS FOUND NECESSARY TO TURN THE TRANSMISSION WITH THE CAR IN GEAR, THEN BOTH REAR WHEELS MUST BE RAISED CLEAR OF THE GROUND.

DIMENSIONAL DATA

The dimensional drawings below (Fig. 1) are provided to assist in assessing accidental damage. A component suspected of being damaged should be removed from the car and cleaned off, the dimensions should then be checked and compared with those given in the appropriate illustration.

Dimension

A. 15,75 to 16,26 mm (0.62 to 0.64 in)
B. 519,43 to 519,94 mm (20.45 to 20.47 in)
C. 150,62 to 151,13 mm (5.93 to 5.95 in)
D. 270,05 to 270,31 mm (10.632 to 10.642 in)
E. 155,45 to 155,70 mm (6.12 to 6.13 in)

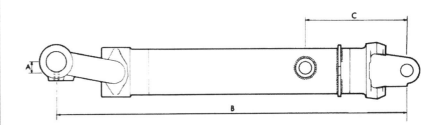

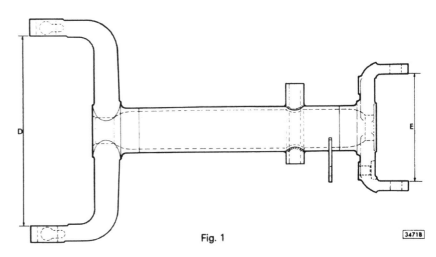

Fig. 1

3471B

REAR SUSPENSION RIDE HEIGHT

Check

Ensure that the car is in its kerb weight condition i.e. full tank of fuel, engine oil and water at correct levels.

Check that the tyres are at the correct pressure.

Roll the car forward three lengths on a perfectly level surface, and measure the distance between the lower surface of the rear crossbeam and the ground on both sides of the car.

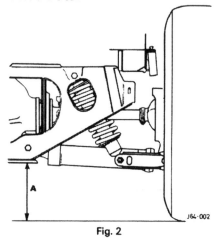

Fig. 2

Dimension A (Fig. 2) must be 190,6 ± 6,4 mm, 7.55 ± 0.25 in.

If this dimension is not correct, check all bushes and bearing points of the rear suspension. If the cause is not found, then the rear road springs must be changed; all four must be replaced as a complete set.

REAR SUSPENSION CAMBER ANGLE

Check and Adjust

Service Tools: Setting Links JD25B, Camber Gauge.

Check that the tyre pressures are correct.
Position the car on a level surface.
Fit the Setting Links JD25B as shown in Fig. 3. Engage the hook end of JD25B in the lower hole of the rear mounting and depress the body until the other end can be slid over the outer fulcrum nut. Repeat on the other side of the car.

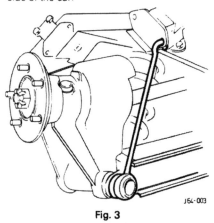

Fig. 3

Position the camber gauge (Fig. 4) against each rear wheel in turn, and make a note of the camber angle reading.

The correct reading should be $\frac{3}{4}° \pm \frac{1}{4}°$ negative. If these limits are not met, then the camber angle must be adjusted.

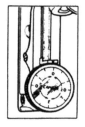 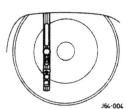

Fig. 4

To adjust the camber angle

Remove the setting links, jack up the rear of the car and place on stands. Remove the appropriate road wheel.

Remove the lower wishbone outer fulcrum grease nipple (to prevent damaging the nipple).

Slacken the clip (1, Fig. 5) securing the universal joint cover in position, and slide the cover clear of the joint.

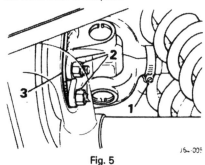

Fig. 5

Remove the four self-locking nuts (2, Fig. 5) securing the drive shaft flange to the brake discs (3, Fig. 5).

Remove the drive shaft flange from the disc to enable the number of shims to be altered.

NOTE: Each shim is 0,5 mm (0.020 in) thick and will alter the camber angle by $\frac{1}{4}°$.

Add or remove the necessary number of shims to obtain the correct camber angle.
Refit the drive shaft flange to the brake disc and secure with the four self-locking nuts.
Position and secure the drive shaft flange cover.
Refit the outer fulcrum grease nipple.
Refit the road wheel and lower the car.
Carry out the camber angle checking procedure as previously described.

REAR HUB BEARING END-FLOAT

Check and Adjust

Service Tools: Backlash Gauge JD13A, Hub Remover JD1D.

The hub bearing end-float is controlled by a spacer located next to the universal joint on the drive shaft.

Spacers are available in thickness from 2,77 to 3,84 mm (0.109 to 0.151 in) in 0,076 mm (0.003 in) increments.

End-float should be between 0,025 to 0,076 mm (0.001 to 0.003 in) and MUST be rectified if it exceeds 0,102 mm (0.004 in) by changing the spacer for a thicker one.

Checking

Jack up the rear of the car and place on stands. Remove the road wheel and gently tap the hub inwards.

Clamp the Backlash gauge JD13A to the hub carrier as shown (1, Fig. 6) so that the stylus of the dial gauge (2, Fig. 6) contacts the hub flange.

Make a note of the reading of the dial gauge.

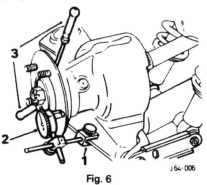

Fig. 6

Using two screwdrivers (3, Fig. 6), or similar levers, move the hub outwards to its fullest extent. Take care not to damage the water thrower.

Make a note of the new readings on the dial gauge.

NOTE: The difference between the dial gauge readings represents the end-float of the hub bearings. If this exceeds 0,102 mm (0.004 in) then carry out the adjusting procedure.

Adjusting

Remove the split pin, nut and washer from the end of the drive shaft.

Remove the fulcrum shaft grease nipple from the hub carrier.

Fit the Hub Puller JD1D to the hub (Fig. 7) and secure using the road wheel nuts.

Withdraw the hub and carrier from the drive shaft and remove the hub puller.

Remove the spacer from the drive shaft and measure the thickness using a micrometer.

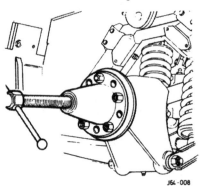

Fig. 7

Determine the thickness of the spacer required to remove end-float to the specified amount 0,025 to 0,076 mm (0.001 to 0.003 in) i.e. if the end-float measurement was 0,203 mm (0.007 in), the replacement spacer would need to be 0,127 mm (0.005 in) thicker than the one removed. This will give an end-float of 0,051 mm (0.002 in).

As the spacers are only supplied in 0,075 mm (0.003 in) increments, then a spacer 0,152 mm (0.006 in) must be used, thus reducing the end-float to 0,025 mm (0.001 in).

Clean any dried Loctite from the drive shaft splines.

Place the selected spacer on the drive shaft.

Apply Loctite grade AAV to the outer two-thirds of the drive shaft splines, using a small brush.

Introduce the half shaft into the hub, and engage the splines.

Drift the hub onto the shaft, fit the washer and tighten the nut to the correct torque. Fit a new split pin.

Re-check the end-float.

Refit the fulcrum shaft grease nipple to the hub carrier.

Remove the service tool JD13A, and refit the road wheel.

REAR HUB AND CARRIER ASSEMBLY

Overhaul

Service Tools: Hub Remover JD1D, Dummy Shaft JD14, Hand Press 47 or Hydraulic/Workshop Press and Adaptor Plate MS370, Press Tools Adaptors JD16C, Press Tool JD20A and Adaptors JD20A-1, Dial Gauge JD13A, Master Spacer JD15.

Jack up the rear of the car and place on stands. Remove the road wheel.

Remove the fulcrum shaft grease nipple (1, Fig. 8) from the hub carrier; this is to avoid damaging the grease nipple when the hub is pulled off.

Remove the split pin, nut and washer (2, Fig. 8) from the end of the drive shaft.

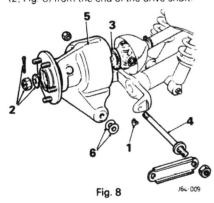

Fig. 8

Fit the Hub Puller JD1D to the hub (Fig. 9) and secure using the wheel nuts.

Using the service tool, withdraw the hub and carrier assembly (5, Fig. 8) from the drive shaft.

Remove the hub puller.

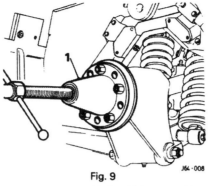

Fig. 9

Remove the spacer (3, Fig. 8) from the drive shaft and retain.

Remove one of the self-locking nuts securing the hub carrier assembly fulcrum shaft to the wishbone. Drift out the fulcrum shaft (4, Fig. 8) using a suitable drift.

Remove the hub and carrier assembly from the car; make a note of any shims and spacers that may be disturbed.

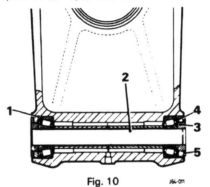

Fig. 10

Key to Fig. 10

1. Bearing spacer ring
2. Dummy shaft (JD14)
3. Seal tracks
4. Felt seal
5. Seal retaining ring

Remove the oil seal retainers from the fulcrum shaft housing.

Remove seals, bearings, spacers, distance tube and shims (Fig. 10) from the fulcrum shaft housing.

Transfer the hub carrier to a suitable press; press out the hub assembly from the carrier (Fig. 11) and place to one side.

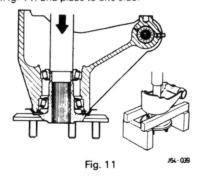

Fig. 11

Secure the hub carrier in a soft jawed vice. Using a suitable brass drift and observing the proper safety precautions, drift the bearing tracks out of the fulcrum shaft housing.

Drift out the hub inner bearing track, together with the seal and bearing from the hub carrier.

Drift out the hub outer bearing track from the hub carrier.

Fit Adaptors JD16C-1 to the hub outer bearing, position the assembly in the Hand Press 47 (Fig. 12) or Hydraulic Press Adaptor MS370, and remove the bearing.

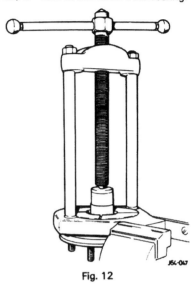

Fig. 12

Examine all the bearing cones, rollers and tracks for wear or damage, and renew as necessary. Inspect the oil seal track on the hub for damage; a minute score can considerably shorten oil seal life.

Press the outer bearing cone into position on the hub shaft and liberally grease the bearing with Retinax 'A'.

Press the hub bearing outer races into the hub carrier using service tool JD20-A and Adaptors JD20A-1 (Fig. 13).

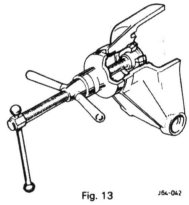

Fig. 13

Fit a new outer oil seal to the hub carrier; lower the carrier onto the hub shaft and outer bearing assembly.

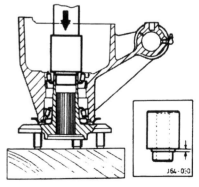

Fig. 14

Place the inner bearing into position, position the Master Spacer, service tool JD15 (as shown in Fig. 14), and press the inner bearing into position until the Master Spacer contacts the hub. This will ensure a certain amount of end-float.

Remove the hub and carrier from the press and secure in a vice, in order to measure the end-float.

Hub Bearing End-Float

With the inner end of the hub uppermost and the Master Spacer JD15 in position, fit the Dual Gauge JD13A to the hub as shown in Fig. 15. Tap the hub carrier downwards and set the dial gauge to zero. Using two screwdrivers or similar levers between the hub and the hub carrier, lift the hub carrier to its fullest extent. Note the reading on the dial gauge.

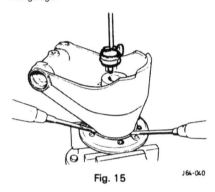

Fig. 15   J64-040

CAUTION: Take care not to damage the water thrower with the levers.

Having determined the measured end-float, a spacer washer must be fitted in place of the Master Spacer tool JD15 to give an end-float of 0,025 to 0,076 mm (0.001 to 0.003 in).

The Master Spacer tool has a diameter of length 'A' equivalent to a spacer of 3,81 mm (0.15 in).

Spacers are supplied in thicknesses of 2,75 to 3,84 mm (0.109 to 0.151 in) in increments of 0,076 mm (0.003 in) and are lettered A to R (less letters I, N and O). See table below.

SPACER LETTER	THICKNESS mm	inches
A	2,77	0.109
B	2,85	0.112
C	2,92	0.115
D	3,00	0.118
E	3,07	0.121
F	3,15	0.124
G	3,23	0.127
H	3,30	0.130
J	3,38	0.133
K	3,45	0.136
L	3,53	0.139
M	3,61	0.142
P	3,68	0.145
Q	3,76	0.148
R	3,84	0.151

Example: Assume the end-float measured was 0,66 mm (0.026 in). Subtract the required nominal end-float of 0,050 mm (0.002 in) from the measured end-float i.e. 0,66 mm (0.026 in) minus 0,050 mm

(0.002 in) equals 0,61 mm (0.024 in). Since the Master Spacer JD15 represents a thickness of 3,81 mm (0.150 in) then the thickness of the spacer to be fitted will be 3,81 mm (0.150 in) minus 0,61 mm (0.024 in) which equals 3,20 mm (0.126 in). The nearest spacer is 3,23 mm (0.127 in) — letter G. This spacer must be fitted to obtain the correct end-float.

Make a note and place to one side.

Remove the Master Spacer and fit a new inner bearing oil seal to the hub carrier.

Fulcrum Shaft Bearings

Fit the fulcrum shaft bearing outer tracks to the hub carrier.

The fulcrum shaft bearing pre-load adjustment is effected by shims fitted between the fulcrum shaft spacer tube and bearings. The correct bearing adjustment is 0,025 mm (0.001 in) nominal pre-load.

A simple jig should be made out of a piece of plate steel approximately 18 cm x 10 cm x 9,5 mm (7 x 4 x 0.375 in).

Drill and tap a hole suitable to receive the outer fulcrum shaft.

Secure the steel plate in a vice and screw in the fulcrum shaft. Slide an oil seal track onto the shaft.

Place the hub carrier assembly into position on the fulcrum shaft.

Omit the felt grease seals, but add an excess number of shims between the spacer tube and one of the bearings.

Place an inner wishbone fork, outer thrust washer onto the fulcrum shaft, so that it abuts the other oil seal track.

Fill the remaining spacer on the shaft with washers and secure with a nut, tightened to the correct torque.

Press the hub carrier assembly towards the steel plate using a slight twisting motion to settle the taper rollers.

Maintain a steady pressure against the hub carrier, and use feeler gauges to measure the amount of clearance between the large diameter washer and the hub carrier (Fig. 16). Make a note of this reading 'A'.

Pull the hub carrier towards the large diameter washer, twisting it to settle the taper rollers. Maintain a steady pressure on the carrier, and use feeler gauges to again measure the clearance between the washer and the hub carrier. Make a note of this reading 'B'.

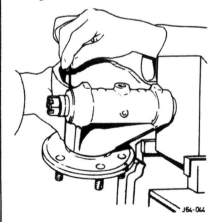

Fig. 16   J64-044

By subtracting clearance 'A' from clearance 'B' an end-float measurement is obtained for the fulcrum shaft bearing assemblies.

Dismantle the hub carrier assembly from the jig plate, and remove sufficient shims to obtain a pre-load reading of 0,05 mm (0.002 in).

Example: Assume the end-float found by the above method was 0,25 mm (0.010 in). Shims to the value of 0,25 mm (0.010 in) + 0,05 mm (0.002 in) = 0,030 mm (0.012 in) must be removed to give the correct pre-load.

Pack the fulcrum bearing housing with the specified grease and reassemble the components. Ensure that a spacer ring is fitted between each taper roller bearing track and the grease seal retainer. (See Fig. 17).

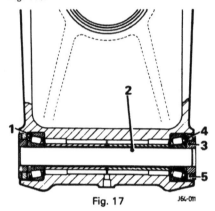

Fig. 17   J64-011

Key to Fig. 17

1. Bearing spacer ring
2. Dummy shaft (JD14)
3. Seal tracks
4. Felt seal
5. Seal retaining ring

Insert Dummy Shaft JD14, fit the grease seal tracks and new felt grease seals.

Locate the fulcrum shaft boss of the hub carrier, between the jaws of the lower wishbone. Press the hub carrier against one of the jaws of the wishbone and using feeler gauges, measure the distance between the other jaw and the oil seal track. (See Fig. 18).

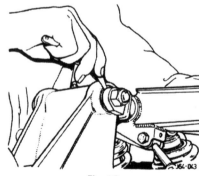

Fig. 18

Select shims to this value and divide into two; fit these either side of the hub carrier to centralise the carrier between the jaws of the wishbone.

These also prevent the jaws of the wishbone from closing inwards.

continued

Shims are available in 0,75 mm (0.003 in) and 0,179 mm (0.007 in) thicknesses and 22,2 mm (0.875 in) diameter.

Fit a seal retaining ring and selected shims to each side of the fulcrum bearing assembly. Position the hub carrier assembly to the lower wishbone and chase dummy shaft, JD14, through the wishbone, with the fulcrum shaft.

Fit and tighten the nuts to the correct torque.

Thoroughly clean and degrease the splines of the drive shaft and hub. Fit the selected spacer washer to the drive shaft.

Using a small brush, sparingly apply Loctite grade AAV to the outer two-thirds of the drive shaft splines.

Assemble the hub carrier to the drive shaft, fit the washer and nut. Tighten to the correct torque and fit a new split pin.

Refit the fulcrum shaft grease nipple.

Refit the road wheel and lower the car.

RADIUS ARM BUSHES

Renew

Service Tool: Press Mandrel JD21.

Jack up the rear of the car; support the body on stands forward of the radius arm anchorage points.

Remove the appropriate road wheel.

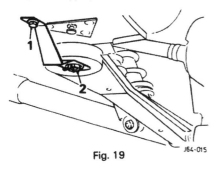

Fig. 19

Remove the special bolt and spring washer, securing the safety strap to the body (1, Fig. 19).

Remove the locking wire and bolt securing the radius arm to the body (2, Fig. 19); remove the safety strap.

Remove the self-locking nut and plain washer (1, Fig. 20) securing the front damper assembly to the damper lower mounting pin.

Drift the damper mounting pin (2, Fig. 20) far enough through the wishbone to clear the front damper and spacer (3, Fig. 20). Recover the spacer.

Knock back the tabs of the lockwasher (4, Fig. 20) and remove the bolt securing the

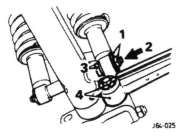

Fig. 20

radius arm to the wishbone. Remove the radius arm.

Using service tool JD21 and a suitable press, press the front and rear bushes from the radius arm (Fig. 21).

Press a new bush into the rear bush housing, centralising the bush in the radius arm.

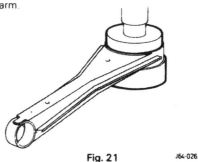

Fig. 21

Press a new bush into the front bush housing; ensure that the holes in the bush rubber are in line with the centre line of the radius arm and the ring of the bush is flush with the housing.

When pressing in the bushes, ensure that the small hole in the bush is uppermost.

To refit the radius arm, reverse the above procedure.

Renew the locking wire and tab washer.

Tighten all the bolts to their correct torques.

ROAD SPRING AND DAMPER

Renew

Service Tools: Hand Press 47, Adaptors JD11B.

NOTE: The roadspring and hydraulic damper assemblies can be removed from the car with the rear suspension assembly in position.

Jack up the rear of the car; place the body on stands and remove the road wheel.

Position a trolley jack to support the lower wishbone.

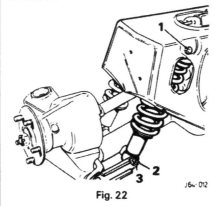

Fig. 22

Remove the bolt and self-locking nut (1, Fig. 22), securing the top of each damper assembly to the crossbeam.

Remove the nuts and washers (2, Fig. 22) securing the damper assemblies to the lower wishbone.

Drift out the damper mounting pin (3, Fig. 22).

Recover the spanner from the forward end of the damper mounting pin.

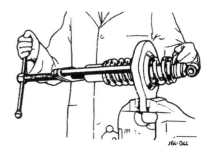

Fig. 23

Withdraw the spring and damper assemblies from the car.

Using Hand Press 47 and Adaptors JD11B, as illustrated in Fig. 23, compress the road spring sufficiently to allow the collets and spring seat to be removed.

Release the spring and separate the damper from the spring.

NOTE: In some cases, a spring packer may be fitted. Note position for refitting.

To refit, compress the road spring, using service tools JD11B and 47, sufficiently to allow the damper to be passed through the road spring. Fit the packing ring, spring seat and split collets.

Ensure that the spring and collets are fully seated before releasing the pressure on the road spring.

Fit the road spring and damper assemblies to the car by reversing the removal procedure.

WISHBONE/INNER FULCRUM MOUNTING BRACKET

Overhaul

Service Tool: Dummy Shafts JD14.

Dismantling

Jack up the rear of the car; support the body on stands and remove the appropriate road wheel.

Remove the self-locking nut from one end of the outer fulcrum shaft (1, Fig. 24) and drift out the shaft.

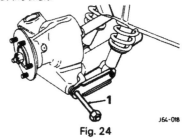

Fig. 24

Fit the Dummy Shaft JD14 to the hub carrier assembly.

Retain the shims and oil seal retaining washers on each side of the hub carrier with adhesive tape.

NOTE: Make a record of the number and position of any shims fitted between the wishbone and the hub carrier assembly. (This is to ensure that they are reassembled in their correct positions.)

continued

Raise the hub carrier and drive shaft clear of the wishbone; secure to the crossbeam using wire or strong cord.

Remove the six bolts and nuts (1, Fig. 25), securing the tie-plate to the crossbeam.

Remove the bolt securing the radius arm safety strap to the body. Remove the locking wire and unscrew the radius arm retaining bolt from the body. Remove the safety strap and detach the forward end of the radius arm from the mounting on the body.

Remove the eight setscrews (2, Fig. 25) securing the tie-plate to the inner fulcrum brackets.

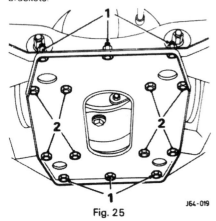

Fig. 25   J64-019

Remove the nuts and washers (1, Fig. 26) securing the spring and damper assemblies to the wishbone.

Drift out the damper mounting pin (2, Fig. 26) and recover the spacer (3, Fig. 26) from the front damper.

Remove the rear nut from the inner fulcrum shaft and drive the shaft clear of the inner fulcrum mounting bracket and wishbone.

Withdraw the wishbone and radius arm assembly from the car. Collect the four outer thrust washers (1, Fig. 27) and inner thrust washers (2, Fig. 27); grease seals (3, Fig. 27) and retainers (4, Fig. 27).

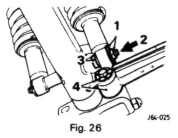

Fig. 26   J64-025

Bend back the tab washer (4, Fig. 26) and remove the bolt securing the radius arm to the wishbone.

Remove the two bearing tubes (5, Fig. 27).

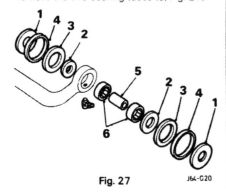

Fig. 27   J64-020

Use a suitable drift to carefully tap the needle roller cages (6, Fig. 27) from the wishbone.

If the inner fulcrum mounting bracket is to be removed, tap the spacing tube (1, Fig. 28) from between the lugs of the mounting bracket.

Remove the locking wire from the setscrews (2, Fig. 28) securing the mounting bracket to the final drive unit.

Remove the setscrews and withdraw the fulcrum bracket.

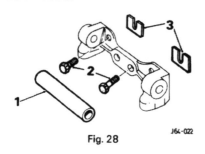

Fig. 28   J64-022

NOTE: Make a record of the number and position of the shims (3, Fig. 28) at each attachment point.

Examine all the components for wear and damage; renew as necessary.

Assembly

Press the needle roller cages into position in the wishbone arms. The engraved edge of each bearing needle cage facing outwards.

Position the inner fulcrum mounting bracket against the final drive unit and loosely secure with the two setscrews.

Pass the inner fulcrum shaft through the crossbeam and mounting bracket.

Position the shims previously removed between the final drive unit and mounting bracket.

Using feeler gauges, as shown in Fig. 29, determine whether the shimming is sufficient to remove the clearance. Adjust the shim pack as necessary. Shims are available in thicknesses of 0,127 mm (0.005 in) and 0,178 mm (0.007 in).

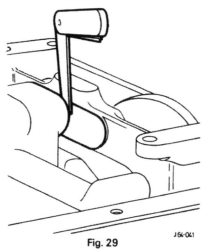

Fig. 29   J64-041

Remove the fulcrum shaft; position the selected shims and tighten the setscrews to the correct torque.

Wirelock the setscrews to tension in a clockwise direction.

Tap the spacing tube into position between the mounting bracket lugs.

Fit new grease seals to the wishbone inner fulcrum arms; relocate the seal retainers and thrust washers with grease.

Offer up the wishbone to the inner fulcrum mounting bracket complete with bearing tubes, needle roller bearings, inner and outer thrust washers, grease seals and grease seal retainers. Ensure that the radius arm mounting bracket is towards the front of the car.

Align the holes and spacers, press a Dummy Shaft JD14 (1, Fig. 30) through each side of the crossbeam and wishbone.

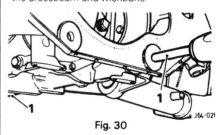

Fig. 30   J64-021

The dummy shafts locate the wishbone, spacers, crossbeam and inner fulcrum mounting bracket, and facilitate the refitting of the inner fulcrum shaft.

Smear the inner fulcrum shaft with grease and gently tap the shaft through the crossbeam, wishbone and inner fulcrum mounting bracket. As the fulcrum shaft is tapped into position, the short dummy shafts will be displaced from the opposite side. It will be found advantageous to keep a slight amount of pressure exerted on the dummy shafts as they emerge from the crossbeam. This will reduce the possibility of the dummy shafts being knocked out of position and allowing a spacer or thrust washer to become displaced.

If a washer or spacer does become displaced, it will be necessary to remove the inner fulcrum shaft, dummy shafts and wishbone, and repeat the operation.

When the fulcrum shaft is in position, fit and tighten the self-locking nuts to the correct torque.

Locate the radius arm to the wishbone, fit the bolt and new tabwasher. Tighten to the correct torque and bend over the tabwasher.

Raise the wishbone, refit the damper mounting pin, spacer and tie-down bracket; tighten the nuts to the correct torque.

Raise the radius arm; clean and lightly grease the spigot. Fit the safety strap. Fit the radius arm securing bolt; tighten to the correct torque and wirelock.

Fit and tighten the bolt securing the safety strap to the body.

Release the wire or strong cord suspending the hub assembly to the crossbeam.

Remove the adhesive tape attaching the shims and washers to the hub carrier.

Fit new grease seals. Refit the grease seal retainers and shims.

Align the hub carrier with the wishbone outer fulcrum and chase out the Dummy Shaft JD14 with the outer fulcrum shaft. Fit and tighten the self-locking nuts to the correct torque.

continued

Refit the tie-plate and insert the eight setscrews (2, Fig. 31).
Refit the six bolts and nuts (1, Fig. 31). Tighten all tie-plate fixings to the correct torque.

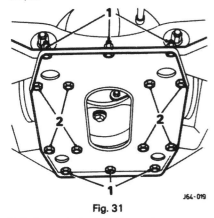

Fig. 31

Refit the road wheel and lower the car.
Carry out a rear suspension camber-angle check and adjust as required.

REAR SUSPENSION MOUNTINGS

Inspect

Drive the vehicle onto a ramp. Raise the ramp and position a ramp jack under the jacking point, in front of the rear radius arm body mounting.
Lower the ramp sufficiently to allow either the rear wheel to clear the ramp, or until the distance between the lower edge of the rear quarter valance and the ramp is 34 cm (13.5 in) DO NOT exceed this distance.
Visually inspect the condition of the rubber and the rubber/metal bonding. If the rubber shows signs of cracking, or there is unbonding of the rubber to a depth greater than 3.175 mm (0.125 in), then the mounting must be replaced.
If a visual inspection is not conclusive, insert a lever between the two 'V's' of the mounting and apply pressure.
Check the rubber for cracking and the rubber/metal bonding.
Repeat this procedure for the other side.

BUMP STOP

Renew

Jack up the car support the body on stands. and remove the appropriate road wheel.
Remove the two self-locking nuts and washers securing the bump stop to the body, and remove the bump stop. (1, Fig. 32).

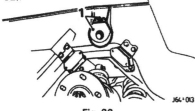

Fig. 32

To fit a new bump stop, reverse the above procedure.
Tighten all fixings to their correct torque.

REAR SUSPENSION UNIT

Remove and Refit

Jack up the rear of the car and support on stands forward of the radius arm anchorage points.
Remove the clamps securing the left and right hand intermediate exhaust pipes to the tail pipes and main silencers.
Disengage each intermediate pipe mounting pin from the rubber mounting and manoeuvre the pipes clear of the rear suspension unit.
Position suitable packing between each exhaust tail pipe and the rear bumper to support the silencer and tail pipe assemblies in their fitted positions.
Remove the special bolt and spring washer (1, Fig. 33) securing each radius arm safety strap to the body.
Remove the locking wire and bolt (2, Fig. 33) securing each radius arm to the body; remove both the safety straps.

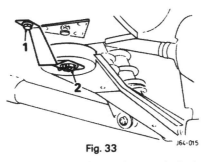

Fig. 33

Disconnect the brake pipe hose at the body mounting bracket. Plug the ends to prevent the ingress of dirt.
Remove the bolts and self-locking nuts securing the propeller shaft flange to the differential unit pinion flange.
Ensure that the handbrake is in the fully 'OFF' position. Lift the carpet adjacent to the handbrake mounting for access to the cable adjusting nuts. slacken the cable lock and adjusting nuts.
Release the springs (1, Fig. 34) from the caliper operating arms. Move the right-hand arm (2, Fig. 34) (when viewed from the front of the car) towards the centre line of the car, to enable the end of the handbrake operating cable (3, Fig. 34) to be detached from the arm.

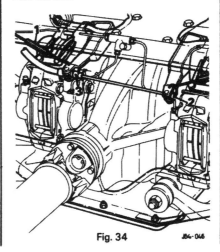

Fig. 34

Position the handbrake cable clear of the suspension unit.
Position a jack underneath the tie-plate and raise sufficiently to take the weight.
Remove the eight bolts and self-locking nuts, and the four nuts, securing the mounting brackets to the crossbeam. Carefully lower the jack and suspension unit.
To refit reverse the above procedure.
Bleed the hydraulic brake system and re-adjust the handbrake cable.

Key to the Exploded View of the Rear Suspension (Fig. 35)

1 Crossmember
2 Rubber mounting
3 Inner fulcrum mounting
4 Shims
5 Tie-plate
6 Inner fulcrum shaft
7 Distance piece
8 RH wishbone
9 Bearing tube
10. Needle roller bearing
11. Thrust washer
12. Grease seal
13. Seal retaining ring
14. Thrust washer
15. Grease nipple
16. Spacer tube
17. Shims
18. Taper roller bearing
19. Grease seal track
20. Bearing spacer
21. Grease seal retainer
22. Grease seal
23. Grease seal retaining washer
24. Shims
25. Outer fulcrum shaft
26. Lashdown bracket
27. Self locking nut
28. Hub carrier
29. Grease nipple
30. Grease retaining cap
31. Hub outer bearing
32. Outer grease seal
33. Grease seal track
34. Rear hub
35. Hub inner bearing
36. Spacer
37. Inner grease seal
38. Seating ring
39. Drive shaft
40. Camber shims
41. Inner joint cover
42. Outer joint cover
43. Coil spring
44. Spring packing ring
45. Shock absorber (Damper)
46. Dust shield
47. Rubber bush
48. Spring seat
49. Split collars
50. Damper mounting pin
51. Radius arm
52. Rubber bush
53. Safety strap
54. Rubber bush

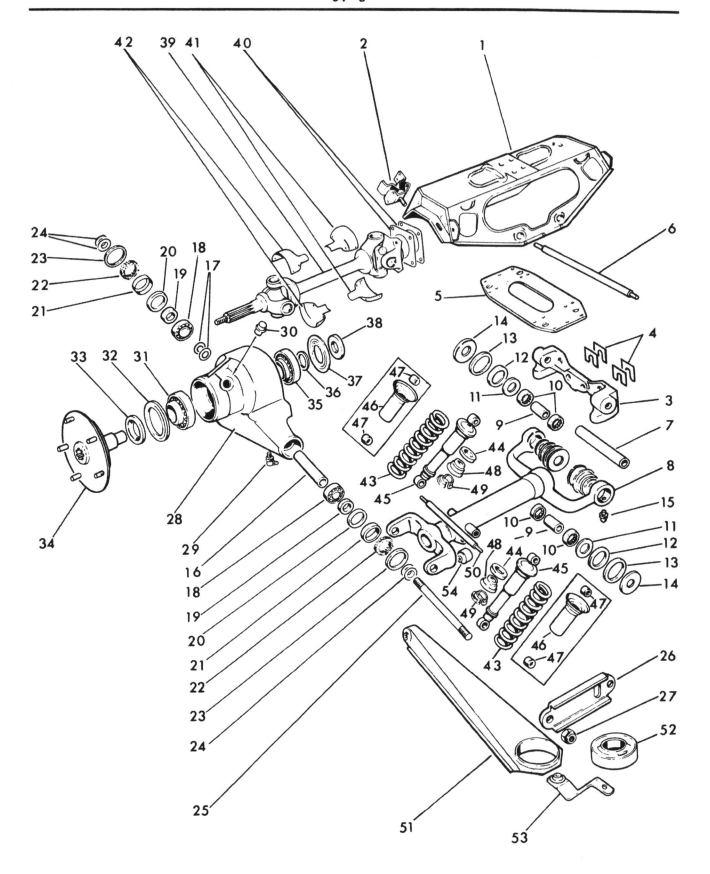

Fig. 35

J64-045

CONTENTS

TORQUE WRENCH SETTINGS

ITEM	DESCRIPTION	SPANNER SIZE	Nm	TIGHTENING TORQUE	
				kgf/cm	lbf/ft
Brake fluid reservoir to bracket	$\frac{1}{4}$ in UNF nut	$\frac{7}{16}$ in AF	27 to 34	0,28 to 0,34	2 to 2.5
Brake pedal pivot pin	$\frac{3}{8}$ in UNF nut	$\frac{9}{16}$ in AF	19 to 24	1,95 to 2,49	14 to 18
Hydraulic connections for $\frac{3}{16}$ in pipe	UNF		8 to 9,5	0,83 to 0,97	6 to 7
	M 12		16 to 19	1,66 to 1,95	12 to 14
	M 10 male		8 to 11	0,83 to 1,12	6 to 8
	M 10 female		11 to 13	1,12 to 1,4	8 to 10
Handbrake to body	$\frac{5}{16}$ in UNF bolt	$\frac{1}{2}$ in AF	19 to 24	1,95 to 2,52	14 to 18
Handbrake lever pivot to body	$\frac{5}{16}$ in UNF bolt	$\frac{5}{8}$ in AF	19 to 24	1,95 to 2,52	14 to 18
Hose bracket to rear X-beam	$\frac{5}{16}$ in UNF nut	$\frac{1}{2}$ in AF	19 to 24	1,95 to 2,52	14 to 18
Hoses to bracket & body	M 10 nut	15 mm	13 to 16	1,4 to 1,66	10 to 12
Master cylinder to servo	M 10 nut	15 mm	22 to 27	2,24 to 2,8	16 to 20
Pedal to body	$\frac{5}{16}$ in UNF nut $\frac{5}{16}$ in UNF bolt	$\frac{1}{2}$ in AF	15 to 17,5	1,55 to 1,81	11 to 13
Servo to pedal box	M 8 nut	13 mm	11 to 13	1,12 to 1,4	8 to 10
Vacuum pipe clips to wing stays	No 10 UNF nut	$\frac{5}{16}$ in AF	5,5 to 6	0,56 to 0,62	4 to 4.5
Vacuum tank bracket to body	$\frac{5}{16}$ in UNF nut	$\frac{1}{2}$ in AF	11 to 13	1,12 to 1,4	8 to 10

DATA

Front brakes—make and type . . . . Girling, ventilated disc
Rear brakes—make and type . . . . Girling, inboard disc
Handbrake—type . . . . Mechanical, operating on rear disc
Disc diameter—front . . . . 283,8 mm (11.175 in)
　　　　　　—rear . . . . 263,8 mm (10.385 in)
Disc thickness—front . . . . 24,0 mm (0.945 in)
　　　　　　—rear . . . . Normal 12,7 mm (0.5 in)
　　　　　　　　　　　　　　　　　Min. permissible 11,43 mm (0.45 in)
Master cylinder bore diameter . . . . 23,8 mm (0.937 in)
Hydraulic fluid specification . . . . Castrol Girling Code 1735 (SAE J1703)
Main brake friction pad specification . . . . Ferodo 3401
Handbrake friction pad specification . . . . Mintex M68/1
Servo unit make . . . . Girling

CLEANING SOLVENTS

WARNING: NEVER USE METHYLATED SPIRIT (DENATURED ALCOHOL) FOR CLEANING PURPOSES. USE ONLY CASTROL/GIRLING BRAKE CLEANING FLUID.
THROUGHOUT THE FOLLOWING OPERATIONS ABSOLUTE CLEANLINESS MUST BE OBSERVED TO PREVENT GRIT OR OTHER FOREIGN MATTER CONTAMINATING THE BRAKE SYSTEM. IF THE SYSTEM IS TO BE FLUSHED OR CLEANED THROUGH, ONLY GIRLING BRAKE CLEANER MUST BE USED. BRAKE SYSTEM COMPONENTS MUST BE WASHED AND ALL TRACES OF CLEANER REMOVED BEFORE REASSEMBLY.

ALL BRAKE SYSTEM RUBBER COMPONENTS MUST BE DIPPED IN CLEAN BRAKE FLUID AND ASSEMBLED USING THE FINGERS ONLY.

BRAKE FLUID

WARNING: DURING OPERATIONS WHICH NECESSITATE THE HANDLING OF BRAKE FLUID, EXTREME CARE MUST BE OBSERVED; BRAKE FLUID MUST NOT BE ALLOWED TO CONTACT THE CAR PAINTWORK. IN INSTANCES WHERE THIS HAS OCCURRED THE CONTAMINATED AREA MUST IMMEDIATELY BE CLEANED, USING A CLEAN CLOTH AND WHITE SPIRIT. THIS SHOULD BE FOLLOWED BY WASHING THE AREA WITH CLEAN WATER. METHYLATED SPIRIT (DENATURED ALCOHOL) MUST NOT BE USED TO CLEAN THE CONTAMINATED AREA.

DESCRIPTION

The brake system comprises the following main components:
Pedal box
Servo unit
Tandem master cylinder
Remote fluid reservoir
Four disc brake assemblies
Two handbrake calipers

The above components provide the car with a dual braking system in which the front and rear caliper assemblies are totally independent of each other. In the event of a brake line fracture, or partial loss of fluid, one pair of brake calipers will remain operative.

Operation (Fig. 1)

When the brake pedal is depressed, the servo unit, which is directly coupled to the master cylinder, transfers increased pedal pressure to the master cylinder primary piston 'A', causing this piston to move forward past the by-pass port 'B', and establish a rear brake line pressure in chamber 'C'. Pressure from the primary piston return spring 'D', combined with the rear brake line brake line pressure in chamber 'C', forces the secondary piston 'E' forward past the by-pass port 'F' to establish front brake line pressure in chamber 'G'.

Brake pressure entering the caliper 'H' forces the pistons 'J' (four on each front disc, two on each rear) out to act on the friction pads 'K', clamping the brake disc 'L'. When the brake pedal is released, brake line pressure collapses, allowing the piston seals 'N' to retract the pistons into the calipers sufficiently for the friction pads to be in a relaxed position away from the disc. This provides automatic adjustment for brake pad lining wear.
If the brake servo unit becomes inoperative, front and rear braking systems will still operate, but at a greatly reduced brake line pressure.
A divided brake fluid reservoir ensures that in the event of fluid loss to front or rear brake systems, one pair of calipers will remain operative.
The fluid reservoir contains a float/switch assembly 'P' which, when actuated, illuminates a warning light on the fascia, should the level of fluid in the reservoir fall to an unsatisfactory level.

Metrication

The examples shown in Fig. 2, Fig. 3 and Fig. 4 are intended as an aid to identification of brake components in metric form.

All metric pipe nuts, hose ends, unions and bleed screws are coloured black. The hexagon area of pipe nuts are indented with the letter 'M'.

Metric and UNF pipe nuts are different in shape and the female nut is always used with a trumpet flared pipe, the male nut always having a convex flared pipe.

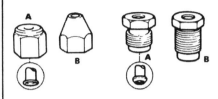

J70-002

A = Metric   B = UNF
Fig. 2

Hose ends differ slightly between metric and UNF.

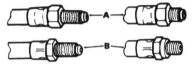

J70-004

A = Metric   B = UNF
Fig. 3

Copper gaskets are not used with metric hose and a gap exists between the hose end and cylinder.

J70-003

A = Metric   B = UNF
Fig. 4

Metrication does not apply to the following brake components:
1. Rear calipers.
2. Handbrake calipers.
3. Feed pipes from rear three-way connector to rear calipers.
4. Three-way connector.

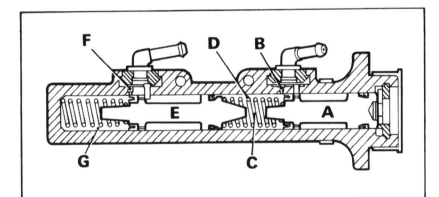

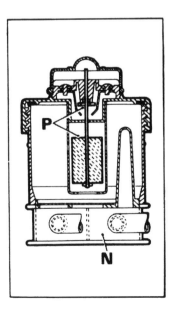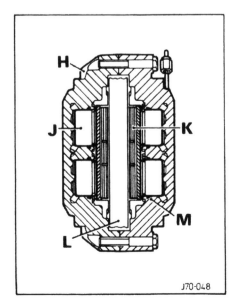

J70-048

Fig. 1

SYMPTOM AND DIAGNOSIS CHART FOR HYDRAULIC BRAKE SYSTEM

SYMPTOM	DIAGNOSIS	ACTION
Fade	Incorrect pads. Overloaded vehicle. Excessive braking. Old hydraulic fluid.	Replace the pads, decrease vehicle load or renew hydraulic fluid as necessary.
Spongy pedal	Air in system. Badly lined pads. Weak master cylinder mounting.	Check for air in the system, and bleed if necessary. Check the master cylinder mounting, pads and discs and replace as necessary.
Long pedal	Discs running out pushing pads back. Distorted damping shims. Misplaced dust covers.	Check that the disc run out does not exceed 0,101 mm (0.004 in). Rotate the disc on the hub. Check the disc/hub mounting faces.
Brakes binding	Handbrake incorrectly adjusted. Seals swollen. Seized pistons. Servo faulty.	Check and adjust handbrake linkage. Check for seized pistons. Repair or replace as necessary. Carry out servo test procedure. Replace servo if faulty.
Hard pedal—poor braking	Incorrect pads. Glazed pads. Pads wet, greasy or not bedded correctly. Servo unit inoperative. Seized caliper pistons. Worn shock absorbers causing wheel bounce.	Replace the pads or if glazed, lightly rub down with rough sandpaper. Carry out servo test procedure. Replace servo if faulty. Check caliper for damage and repair as necessary. Fit new shock absorbers.
Brake pulling	Seized pistons. Variation in pads. Unsuitable tyres or pressures. Worn shock absorbers. Loose brakes. Greasy pads. Faulty discs, suspension or steering.	Check tyre pressures, seized pistons, greasy pads or loose brakes; then check suspension, steering and repair or replace as necessary. Fit new shock absorbers.
Fall in fluid level	Worn disc pads. External leak. Leak in servo unit.	Check the pads for wear and for hydraulic fluid leakage. Carry out servo test procedure. Replace servo if faulty.
Disc brake squeal—pad rattle	Worn retaining pins. Worn discs. Worn pads. Broken anti-chatter spring.	Renew the retaining pins, or discs. Fit new pads, or anti-chatter spring.
Uneven or excessive pad wear	Disc corroded. Disc badly scored. Incorrect friction pads.	Check the disc for corrosion, or scoring and replace if necessary. Fit new pads with correct friction material.
Brake warning light illuminated	Fluid level low. Short in electrical warning circuit.	Top up reservoir. Check for leaks in system and pads for wear. Check electrical circuit.

GENERAL FITTING INSTRUCTIONS

Hoses

Thoroughly clean the unions of the hose to be removed.
Ensure the pipe sealing plugs are at hand.
Fully release the unions (1, Fig. 5) securing each end of the hose to the fluid pipes.

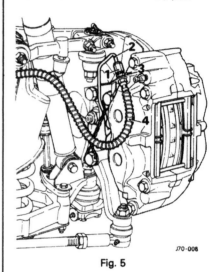

J70-006

Fig. 5

Withdraw the pipe unions (2, Fig. 5) from the hose ends. Plug the ends of the pipes to prevent the loss of fluid and the ingress of dirt.
Remove the locknut (3, Fig. 5) securing the hose to the mounting bracket and remove the hose.
Thoroughly clean the hose and examine it for any signs of deterioration or damage. Renew the hose if there is any doubt about its condition.
Using compressed air, thoroughly clean the bore of the hose.
To refit the hose, reverse the above procedure. Before fully tightening the locknut (3, Fig. 5) ensure that the hose is neither kinked nor twisted.
Carry out the brake bleeding procedure.

Pipes

Clean the unions of the pipe to be removed. Ensure that sealing plugs are at hand. Fully release the pipe unions.
Withdraw the pipe from the car. Plug the open end of the pipe remaining on the car, to prevent the loss of fluid or the ingress of dirt.
Thoroughly clean and examine the pipe for any signs of damage or deterioration. Renew the pipe if there is any doubt about its condition.
Using compressed air, thoroughly clean the bore of the pipe.
To refit, reverse the above procedure and carry out the brake bleed procedure.

BRAKE SERVO TEST PROCEDURE

NOTE: The following tests on the vacuum system should only be carried out if it is known that the hydraulic brake system is in a satisfactory condition.

Jack up the front of the car and check that one wheel will rotate freely. Start the engine, allow the vacuum to build up, and apply the brake pedal several times. Check that the wheel is free to rotate immediately the pedal is released. If the brakes bind, then the servo is faulty.

With the engine running, apply the brake pedal several times, and check the operation of the pedal. If the response is sluggish, check the condition of the vacuum hoses and the servo unit air filter.

Allow the vacuum to build up, and then switch off the engine and operate the brake pedal. The pedal should be vacuum assisted for approximately three applications; any less indicates a leaking vacuum system or an inoperative non-return valve.

With the engine switched off, operate the brake pedal several times to evacuate any vacuum left in the system.

Whilst maintaining light pressure on the footbrake pedal, start the engine. If the servo unit is operating correctly, the pedal will fall under the existing foot pressure. If the pedal remains stationary, then there is a leak in the vacuum system.

BLEEDING THE BRAKE SYSTEM

Bleeding the brake system (expelling the air) is not a routine maintenance operation and should only be necessary when part of the system has been disconnected, or air has contaminated the fluid. The presence of air in the system will cause the brake pedal to feel 'spongy' when applied.

During the bleeding operation, **it is important** that the level of the fluid in the fluid reservoir is kept topped up, to avoid drawing more air into the system.

Attach a bleed tube to the left hand rear bleed screw, immerse the open end of the tube in a small jar, partially filled with clean, fresh brake fluid.

Position the gear selector lever in neutral; start the engine and allow it to idle.

Slacken the left-hand rear bleed screw.

An assistant is required to slowly operate the brake pedal through its full stroke. Continue to do this, until the fluid pumped into the jar is free from air bubbles. When this is achieved, keep the pedal depressed and close the bleed screw. Release the pedal.

Repeat the operation for the right-hand rear brake caliper and the two front brake calipers.

Check the tightness of all the bleed screws and fit protective caps.

Top up the reservoir as necessary.

CAUTION: DO NOT 'top up' the fluid reservoir with fluid which has been bled through the system, as it will have become aerated. Always use fresh, clean fluid from a new tin.

Apply a normal 'working' load to the brake pedal, for several minutes. If the pedal moves or feels spongy, then further bleeding of the system is required. Also check the entire system for any signs of fluid leakage.

DRAINING AND FLUSHING THE BRAKE SYSTEM

Service Tool: Brake piston retractor, Girling Part No. 64932392

Draining

Place the car on stands and remove all the road wheels.

Attach a bleed tube to the rear left-hand caliper bleed screw with the open end in a suitable container. Slacken the bleed screw and slowly operate the brake pedal through its full stroke, until the rear brake section of the fluid reservoir is empty and fluid ceases to come out of the bleed tube.

Remove the rear left-hand caliper friction pads.

WARNING: DO NOT OPERATE THE BRAKE PEDAL WHILE THE FRICTION PADS ARE REMOVED.

Using Service Tool 64932392, as shown in Fig. 6 A and B, lever the pistons into their bores, expelling any remaining fluid.

Replace the friction pads.

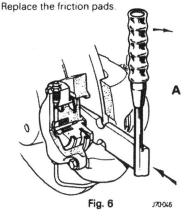

Fig. 6   J70-046

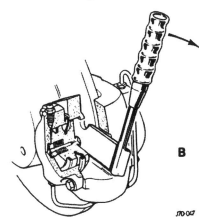

J70-007

NOTE: It is not necessary to replace the retaining pins and spring clips at this time, as the friction pads will have to be removed again.

Close the bleed screw.

Discard the expelled fluid.

Repeat the draining procedure for the rear right-hand and front brake calipers.

FLUSHING

Fill the fluid reservoir with Castrol/Girling Flushing Fluid.

Attach a bleed tube to the rear left-hand caliper bleed screw with the open end in a suitable container.

Slacken the bleed screw and slowly operate the brake pedal, through its full stroke; continue until clear flushing fluid is expelled from the tube.

NOTE: The fluid reservoir must be kept 'topped-up' with fresh flushing fluid.

Close the bleed screw and operate the brake pedal two or three times.

Repeat the flushing procedure on the remaining brake calipers.

Carry out the draining procedure to expell all the flushing fluid from the rear brake calipers. Secure the friction pads with the retaining pins and spring clips. Do not forget to fit the anti-rattle springs to the front calipers.

Close all the bleed screws.

Discard the expelled flushing fluid.

Fill the brake fluid reservoir with new brake fluid of the correct specification.

Carry out the brake bleeding procedure.

Ensure that the brake fluid expelled through the bleed tube is completely free of flushing fluid.

Refit the road wheels and remove the car from the stands.

SERVO ASSEMBLY

Overhaul

The servo assembly is a sealed unit and cannot be overhauled. Should the operation of the servo unit deteriorate to an extent where braking efficiency is affected, or a fault develops in the unit, then a replacement servo assembly must be fitted. Servo unit replacement is covered in the pedal box overhaul procedure.

DISC SHIELDS — FRONT

Renew

Jack up the front of the car, place on stands and remove the road wheel.

Slacken the upper bolt securing the steering arm to the stub axle carrier.

Remove the locking wire securing the caliper mounting bolts and remove the upper mounting bolt.

Remove the self locking nuts which retain the disc shield securing brackets (1, Fig. 7) to the lower portion of the stub axle carrier.

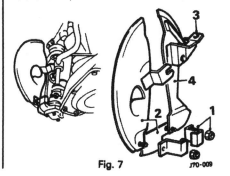

Fig. 7   J70-009

Withdraw the front and lower disc shields (2, Fig. 7).

Disconnect the brake caliper feed pipe from the flexible hose. Remove the locknut securing the flexible hose union to the bracket (3, Fig. 7) on the rear disc shield. Remove the rear disc shield (4, Fig. 7).

Plug the ends of the brake pipes to prevent the loss of fluid and the ingress of dirt.

To refit the disc shields, reverse the above procedure.

Ensure that the flexible hose is not twisted or kinked when secured to the rear disc shield.

Tighten all fixings to their correct torque figures.

Wirelock the caliper mounting bolts.

Refit the road wheel.

Bleed the brakes.

DISC — FRONT

Renew

Service Tool: Brake piston retractor, Girling Part No. 64932392

Jack up the front of the car, place on stands and remove the road wheel.

Remove the spring clips (1, Fig. 8), securing the brake pad retaining pins (2, Fig. 8) and withdraw the retaining pins.

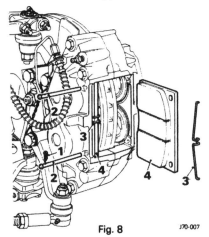

Fig. 8   J70-007

Recover the anti-rattle springs (3, Fig. 8), and withdraw the brake pads (4, Fig. 8).

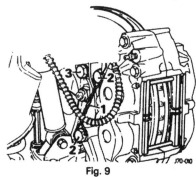

Fig. 9

Break the lock wire (1, Fig. 9) and remove the two bolts and spring washers (2, Fig. 9), securing the brake caliper to the stub axle carrier.

Slacken the bolt and washer (3, Fig. 9) securing the steering arm to the stub axle carrier.

NOTE: Make a careful record of the shim(s) fitted between the steering arm and the brake caliper.

Gently ease the caliper aside and secure with wire or strong cord, to prevent damaging the brake hose.

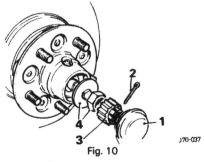

Fig. 10   J70-037

Remove the front hub grease cap (1, Fig. 10), extract the split pin (2, Fig. 10), remove the hub retaining cap (3, Fig. 10), nut and washer (4, Fig. 10) from the stub axle. Withdraw the front hub and disc assembly.

Remove the five bolts and spring washers securing the brake disc to the front hub assembly, and withdraw the brake disc.

Inspect the disc for cracks and heavy scoring; light scratches are not detrimental. If any doubt exists, a new disc must be fitted.

Fit a new disc to the hub assembly, secure with five bolts and spring washers, tightened to the correct torque.

Refit the hub and disc assembly to the stub axle; adjust the hub bearing endfloat to 0,03 - 0,08 mm (0.001 - 0.003 in). This is measured using a dial test indicator gauge, mounted with the plunger against the hub. Tighten the nut until the correct endfloat is obtained, refit retainer, split pin and hub cap.

If a gaugef is not available, the following procedure can be used:

Tighten the nut until there is no endfloat i.e. when the rotation of the hub is slightly restricted.

CAUTION: A torque of 0,691 Kgf/m (5 lbf/ft) must not be exceeded or damage may be caused to the bearings and bearing tracks.

Slacken the nut one flat and fit the nut retaining cap. Fit a new split pin and bend over. Refit the grease cap, ensure that the vent hole is clear.

Refit the caliper to the stub axle carrier; ensure that the correct number of shims are fitted between the steering arm and the brake caliper. Tighten and wire lock the two caliper securing bolts. Tighten the steering arm securing bolt.

When a new disc has been fitted, it is recommended that new brake pads are also fitted. If the thickness of the pad material is less than 4 mm (0.16 in), then NEW pads MUST be fitted.

NOTE: It is advisable to reduce the level of brake fluid in the reservoir before fitting the new brake pads.

Lever the caliper pistons into the cylinder bores using Girling tool 64932392.

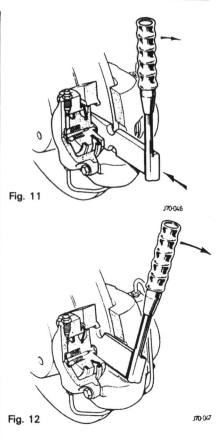

Fig. 11   J70-046

Fig. 12   J70-047

Fit the new brake pads to the caliper, fit the new anti-rattle springs. Secure in position with the retaining pins. Fit the spring clips. Refit the road wheel and lower the car.

Run the engine and apply the brake pedal several times until the pedal feels solid.

DISC — REAR

Renew

Service Tool: Brake piston retractor, Girling Part No. 64932392

Handbrake Mechanism

Jack up the rear of the car and place on stands.

Remove the nuts and bolts securing the tie-plate to the rear suspension unit; remove the tie-plate.

Ensure that the handbrake lever is in the fully 'OFF' position. Lift the carpet adjacent to the handbrake lever mounting for access to the cable adjusting nuts (1, Fig. 13); Slacken the cable lock and adjusting nuts.

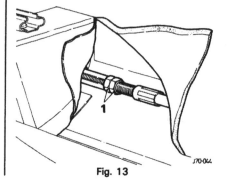

Fig. 13   J70-044

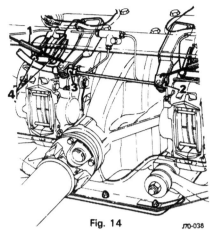

Fig. 14   J70-038

Release the springs (1, Fig. 14) from the handbrake caliper arms. Move the right-hand arm on RHD vehicles, or the left-hand arm on LHD vehicles (when viewed from the front of the car), to the centre line of the car, enabling the end of the inner (operating) cable (2, Fig. 14) to be detached from the operating arm. Slide the rubber cover from the outer cable (3, Fig. 14) and release the outer cable from the other handbrake operating arm.

Position the handbrake cable clear of the caliper.

Bend back the locking tabs (4, Fig. 14), securing the handbrake caliper mounting bolts. Remove the two mounting bolts, tab washer and retraction plate.

Slide the handbrake caliper around the brake disc, and withdraw through the gap exposed by the removal of the tie-plate.

Rear Caliper

Remove the brake pads from the caliper.

Slacken the caliper feed pipe union at the three way connector, and disconnect the feed pipe from the caliper.

Plug the ports to prevent the loss of fluid and ingress of dirt.

Break the locking wire and remove the caliper mounting bolts.

CAUTION: DO NOT under any circumstances remove the bolts securing the halves of the caliper together.

Slide the caliper around the brake disc and withdraw through the gap exposed by the removal of the tie-plate.

Rear Disc

Remove the road wheel, adjacent to the brake disc to be renewed.

Remove the hub carrier fulcrum shaft grease nipple (1, Fig. 16) to prevent damaging the nipple.

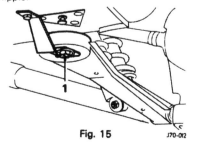

Fig. 15   J70-012

Remove the locking wire securing the radius arm bolt (1, Fig. 15) and remove the bolt.

Lever the radius arm from the spigot anchor on the body.

Position a jack, or support blocks, under the hub carrier assembly.

Remove the nuts and washers securing the rear dampers to the lower wishbone.

Drift out the damper mounting pin (2, Fig. 16); recover the front damper spacer collar.

Slacken the clip (3, Fig. 16) securing the drive shaft inner universal joint cover and slide the cover along the drive shaft clear of the joint.

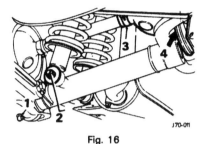

Fig. 16

Remove the nuts (4, Fig. 16) securing the driveshaft flange to the brake disc.

Carefully separate the drive shaft flange from the brake disc; collect the camber angle shims located on the disc mounting bolts.

NOTE: Record the number and value of the camber shims.

Remove the brake disc from the mounting bolts.

NOTE: Do NOT disturb the shims mounted between the final drive flange and the brake disc.

Inspect the disc for cracks and heavy scoring; light scratching is not detrimental. If any doubt exists, a new disc MUST be fitted.

Fit the new disc on to the mounting bolts. Replace the camber angle shims.

Fit the drive shaft flange over the shims on to the mounting bolts.

Fit and tighten the nuts to secure the drive shaft flange to the brake disc.

Check the disc for run-out. Clamp a dial test indicator to the suspension unit cross-member, position the indicator button against the brake disc face and zero the dial. Run-out must not exceed 0,10 mm (0.004 in).

Offer the brake caliper to its mountings and secure with the mounting bolts.

Check that the brake disc is central in the caliper by measuring the gap between the caliper abutments and the disc faces.

The gap on opposite sides of the disc may differ by up to 0,25 mm (0.010 in), but the gap between the upper abutment face and the disc, and the lower abutment face and the disc, on the same side should be equal.

To adjust, remove the caliper and disc, add or remove shims from between the axle unit output shaft flange and the brake disc. Make a note of the thickness of the shims added or removed.

On completion of the centralisation operation, add or withdraw camber shims to the same value as those used in the above adjustment operation.

e.g. If a 2,15 mm (0.06 in) shim was ADDED between the axle unit output flange and the brake disc, to centralise the disc, then REMOVE the same value of the shims from the camber angle shim pack, fitted between the brake disc and the drive shaft flange.

If shims were removed to centralise the disc, then ADD the same value of shims to the camber angle shim pack. This ensures that the camber angle, prior to the renewal of the disc, is retained.

When the disc is central in the caliper, tighten the caliper mounting bolts to the correct torque and wire lock.

Reposition the radius arm to the body spigot, refit the bolt, tighten to the correct torque and wire lock.

NOTE: Prior to fitting the radius arm, wire brush the spigot and smear with grease.

Slide the inner universal joint cover into place and secure with the clip.

Examine the brake pads for wear and damage. If the lining thickness is less than 4 mm (0.16 in), then new brake pads must be fitted.

Refit the brake pads to the caliper. Locate the retaining pins and fit the spring pins.

Examine the handbrake pads for wear. If the lining thickness is less than 4 mm (0.16 in) then new pads must be fitted.

If new handbrake pads are fitted, adjust the handbrake caliper.

Hold one pad carrier and unwind the other one until the distance between the pad faces is 19 mm (0.75 in).

Position the handbrake caliper, retraction plate and new tab washer. Secure with the mounting bolts and bend up the tab washer. Operate the actuating lever until the adjuster ratchet ceases to click. The handbrake pads are now adjusted to the correct clearance. Refit the springs to the operating arms.

Fit the brake feed pipe to the caliper and the three-way connector.

Refit the handbrake cable and adjust so that with the handbrake fully 'OFF' there is a slight amount of slack within the cable.

NOTE: If the cable is adjusted so that all the slack is removed, binding of the handbrake caliper may result.

Tighten the locknut and replace the sill carpet.

Refit the tie-plate to the suspension unit.

Bleed the brakes.

Refit the road wheel and lower the car.

Check and, if necessary, adjust the camber angle.

BRAKE CALIPER — FRONT

Overhaul

Jack up the front of the car, place on stands and remove the road wheel.

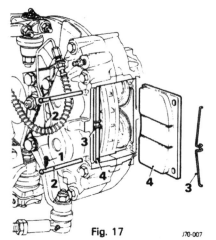

Fig. 17   J70-007

Remove the spring clips (1, Fig. 17) securing the brake pad retaining pins (2, Fig. 17) and withdraw the retaining pins. Recover the anti-rattle springs (3, Fig. 17) and withdraw the brake pads (4, Fig. 17).

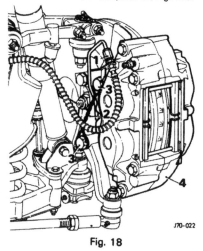

Fig. 18   J70-022

Slacken the caliper feed pipe union (1, Fig. 18) at the support bracket and disconnect the feed pipe from the caliper, plug the pipe to prevent the loss of fluid and ingress of dirt.
Break the lock wire (2, Fig. 18) and remove the two bolts and spring washers (3, Fig. 18) securing the brake caliper and the stub axle carrier.

NOTE: Make a careful record of the shim(s) fitted between the steering arm and the brake caliper.

Withdraw the caliper (4, Fig. 18) from the car.
Thoroughly clean the caliper using ONLY Castrol/Girling Brake Cleaning Fluid.

CAUTION: DO NOT under any circumstances remove, or attempt to remove, the bolts securing the halves of the caliper together.

Remove the spring clips (1, Fig. 19) retaining the piston dust covers (2, Fig. 19) and remove the dust covers from the pistons (3, Fig. 19).

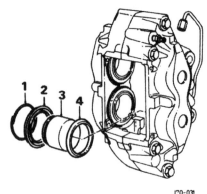

Fig. 19   J70-031

To expel the pistons, carefully feed compressed air into the caliper fluid inlet port; remove the pistons from the caliper.

WARNING: EXTREME CARE MUST BE TAKEN NOT TO DAMAGE THE CYLINDER BORES WHEN EXTRACTING THE SEALS.

Carefully prise each seal (4, Fig. 19) from the recess in each piston cylinder bore.
Using only Castrol/Girling Brake Cleaning Fluid, thoroughly clean all the components.
Examine the pistons and cylinder bores for signs of abrasion, 'scuffin', scratches or corrosion. If any doubt exists as to the condition of a component then a new one must be fitted.
Coat the new seals in Castro/Girling Brake Lubricant, or new, clean brake fluid. Fit the new seals into the recesses in the cylinder bores (1, Fig. 20) using ONLY finger pressure.

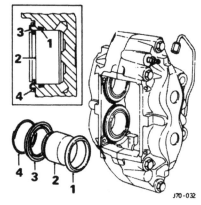

Fig. 20   J70-032

Lubricate the pistons (2, Fig. 20) with brake lubricant, or new, clean brake fluid, and enter them into the cylinder bores.
Fit new dust covers (3, Fig. 20) over the pistons and locate in the outer groove of the piston. Push the pistons fully home and locate the dust cover in the outer grooves of the cylinder bores.
Secure the dust covers with the spring clips (4, Fig 20).
Refit the caliper to the stub axle carrier; ensure that the correct value of shim(s) is fitted between the brake caliper and the steering arm. Tighten the bolts to the correct torque and wire lock.
Reconnect the caliper feed pipe to the caliper and tighten the pipe union to the support bracket.

NOTE: Examine the brake pads for wear and damage. If the lining thicknes is less than 4 mm (0.16 in), then new brake pads must be fitted.

Refit the brake pads. Bleed the brakes.
Refit the road wheel and lower the car.

BRAKE CALIPER — REAR

Overhaul

Service Tool: Piston Clamp 18G672

Jack up the rear of the car and remove the handbrake and rear brake calipers as described on page 70—6 in the rear disc renewal procedure.
Thoroughly clean the caliper using ONLY Castrol/Girling Brake Cleaning Fluid.

CAUTION: DO NOT under any circumstances remove or attempt to remove the bolts securing the caliper halves together.

Fit the piston clamp, Tool No. 18G672, to one half of the caliper, to retain one piston in position, whilst carefully feeding compressed air into the caliper inlet port, to expel the other piston (3, Fig. 21).

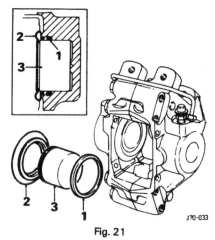

Fig. 21   J70-033

Remove the dust seal (2, Fig. 21) from the piston and cylinder bore.

WARNING: EXTREME CARE MUST BE TAKEN NOT TO DAMAGE THE CYLINDER BORE WHEN EXTRACTING THE SEAL.

Carefully prise the seal (1, Fig. 21) from the recess in the cylinder bore.
Using only Castrol/Girling Brake Cleaning Fluid, thoroughly clean the cylinder bore and piston.
Examine the components for signs of abrasion, 'scuffing', scratches or corrosion. If any doubt exists as to the condition of a component, then a new one MUST be fitted.
Coat the new seal (1, Fig. 21) in Castrol/Girling Brake Lubricant, or in new clean brake fluid, and enter it onto the cylinder bore.
Fit a new dust cover (2, Fig. 21) over the piston and locate it in the outer groove of the piston. Push the piston (3, Fig. 21) fully home and locate the dust cover in the outer groove of the cylinder bore.

Release the piston clamp tool and fit it to the other half of the caliper.
Repeat the applicable operations on the other piston.
Refit the caliper assembly to the car.

HANDBRAKE LEVER ASSEMBLY

Renew

Disconnect the battery.
Remove the driver's seat assembly. This operation is fully covered in the Body Section of this Manual Book 8, Section 76, page

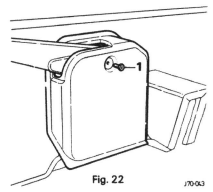

Fig. 22

Remove the setbolt (1, Fig. 22) securing the handbrake lever mechanism cover. Raise the handbrake lever and slide off the cover.
Disconnect the electrical lead (1, Fig. 24) from the brake warning light switch.

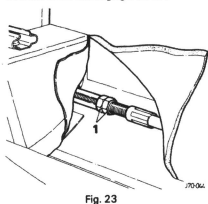

Fig. 23

Lift the carpet adjacent to the handbrake lever mounting for access to the cable adjusting nuts (1, Fig. 23); slacken the cable, lock and adjusting nuts.

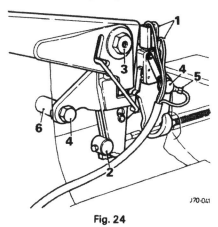

Fig. 24

Remove the adjustable nipple (2, Fig. 24) from the end of the handbrake cable.
Remove the pivot bolt (3, Fig. 24) and the bolts (4, Fig. 24) securing the handbrake assembly to the inner sill.
Detach the earth lead (5, Fig. 24). Recover the distance pieces (6, Fig. 24) fitted between the handbrake assembly and the inner sill.
Disengage the handbrake assembly and remove from the car.
Remove the warning light switch.
Fit the warning light switch to the new handbrake lever assembly; reverse the above procedure to fit the handbrake lever assembly to the car.
Reconnect the battery. Adjust the handbrake warning light switch so that when the handbrake is in the 'OFF' position the warning light just goes out. Tighten the securing bolts.
Check that the warning light comes 'on' when the handbrake is applied and goes 'off' when the handbrake is released. Readjust as necessary.
Adjust the handbrake cable; a slight amount of slack should be evident with the handbrake in the 'OFF' position.
Refit the handbrake cover. Refit the driver's seat assembly.

HANDBRAKE CABLE

Renew

Remove the driver's seat. This operation is fully covered in the Body Section of this Service Manual, Book 8, Section 76, page

Remove the setbolt securing the handbrake lever mechanism cover. Raise the handbrake lever and slide off the cover.

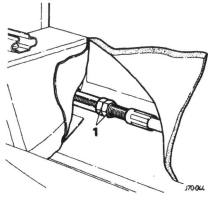

Fig. 25

Lift the carpet adjacent to the handbrake lever mounting for access to the cable adjusting nuts (1, Fig. 25). Slacken the cable lock and adjusting nuts.
Remove the adjustable nipple (1, Fig. 26) from the end of the handbrake cable.
Disconnect the nipple and inner cable from the operating arm of the handbrake caliper (2, Fig. 26).
Disconnect the outer cable from the other handbrake operating arm (3, Fig. 26).
Cut the clips securing the protective sleeve (4, Fig. 26) to the outer cable; slide the sleeve off the cable.

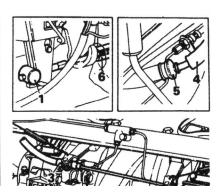

Fig. 26

Remove the protective cover (5, Fig. 26) from the point where the outer cable goes through the body.
From inside the car, pull the cable assembly free from the body guide tube (6, Fig. 26); remove the cable from the car.
Remove the adjusting nut, locknut and guide tube from the cable.
To refit the cable assembly, reverse the above procedure.
Adjust the cable so that with the handbrake fully 'OFF' there is a slight amount of slack within the cable.

NOTE: If the cable is adjusted so that all the slack is removed, the handbrake calipers may bind on the discs.

Tighten the locknut and replace the sill carpet.

MASTER CYLINDER

Overhaul

Peel back the rubber cover (1, Fig. 27) from the top of the brake fluid reservoir and disconnect the wires from the fluid level indicator switch.
Remove the reservoir cap and filter.
Using a suitable syringe, remove the brake fluid from the reservoir.

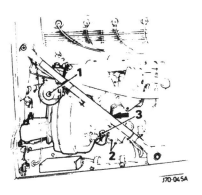

Fig. 27

To remove the fluid reservoir:

Slacken the clips securing the master cylinder feed pipe hoses to the fluid reservoir and disconnect the hoses from the reservoir. Use a piece of rag, or a suitable container, to collect any remaining fluid.

Remove the nuts, bolts and plain washers securing the fluid reservoir to the mounting bracket; remove the reservoir.

Plug all the hose and pipe ends to prevent any further loss of fluid and the ingress of dirt.

To remove the master cylinder

Slacken the clips securing the feed pipe hoses (2, Fig. 27) to the master cylinder, and disconnect the hoses from the master cylinder. Use a piece of rag, or a suitable container, to collect any remaining fluid.

Plug all the hose and pipe ends to prevent any further loss of fluid or ingress of dirt.

CAUTION: Before removing the master cylinder, it is imperative that the brake pedal is operated at least ten times to ensure that there is no vacuum left to operate the servo.

Operation of the servo, when the master cylinder is not fitted, can cause the servo mechanism to travel beyond its normal limit, making the servo unit irrepairable.

Remove the nuts and washers (3, Fig. 27) securing the master cylinder to the servo unit; remove the master cylinder.

NOTE: The overhaul of the master cylinder should be carried out with the work area, tools and hands in a clean condition.

To dismantle the master cylinder:

Carefully prise out the master cylinder inlet pipe adaptors (1, Fig. 28) from the sealing grommets.

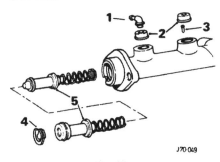

Fig. 28

Using a suitable screwdriver, lever out the sealing grommets (2, Fig. 28) from the master cylinder.

Press in the primary piston to relieve the pressure on the secondary piston stop pin (3, Fig. 28); remove the stop pin from the front grommet sealing housing. Maintain the pressure on the primary piston; remove the circlip (4, Fig. 28).

Tap the flange (open) end of the master cylinder on a wooden block to remove the primary and second pistons and spring assemblies (5, Fig. 28). It may be necessary to feed compressed air into the master cylinder front delivery port.

NOTE: Once the piston assemblies have been withdrawn, it is important that the appropriate piston and spring are kept together. In the event of the springs being mixed, then the secondary piston spring is slightly thicker and longer than the primary spring.

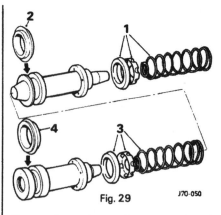

Fig. 29

Remove the spring, spring seat and seal from the front end of the secondary piston (1, Fig. 29).

Carefully prise the seal (2, Fig. 29) from the rear of the secondary piston.

Remove the spring, spring seat and seal from the front end of the primary piston (3, Fig. 29).

Carefully prise the seal (4, Fig. 29) from the rear of the primary piston.

Discard all the old seals and associated items, which will be replaced by those contained within the service kit.

Clean all the components with Castrol/Girling Cleaning Fluid and dry with a lint free cloth.

Examine the piston and bore of the master cylinder for visible signs of scoring, ridges and corrosion. If any doubt exists as to the condition of a component, then it must be renewed.

CAUTION: To help prevent damage, it is essential that generous amounts of clean brake fluid are used at all stages of the seal fitting.

Carefully fit the rear seal in its groove in the secondary piston; ensure that the lip of the seal faces towards the primary piston.

Fit the front seal, spring seat and spring to the front end of the secondary piston; ensure that the lip of the seal faces away from the primary piston.

Carefully fit the rear seal in its groove in the primary piston; ensure that the lip of the seal faces forwards i.e. away from the circlip groove.

Fit the seal, spring seat and spring to the front end of the primary piston, with the lip of the seal facing forwards i.e. away from the circlip grooves.

CAUTION: Adherence to the following instructions is vitally important. Failure to comply will result in damaged piston seals.

Carefully secure the master cylinder in a vice and generously lubricate the piston seals with new, clean brake fluid. Offer the secondary piston assembly to the master cylinder until the front seal rests centrally in the mouth of the cylinder. Ensure that the seal is not trapped; slowly rotate and rock the piston assembly whilst GENTLY introducing the piston into the cylinder bore. Once the front seal has entered the bore of the cylinder, SLOWLY push home the piston in one continuous movement.

Repeat this method for the primary piston assembly.

Press the primary piston into the bore and fit the circlip.

Fully push home the primary piston and fit the secondary piston stop pin.

Fit the sealing grommets to the master cylinder. Lubricate and press the hose adaptors into the sealing grommets.

Refit the master cylinder to the servo.

Reconnect the fluid feed pipe hoses to the master cylinder.

Fill the reservoir with new fluid of the correct specification and carry out the brake system bleed procedure as detailed on page 70—5.

RESERVAC TANK

Renew

The reservac tank is situated behind the front right-hand wheel arch front dust shield. The dust shield must be removed to gain access to the tank.

The dust shield is secured to the wheel arch with plastic drive fasteners.

CONTENTS

DESCRIPTION

Cast alloy wheels fitted with 215/70 VR15 tyres are used as standard equipment on all models. The tyre type used have tread wear indicators moulded into the tread pattern to provide indication when the tread depth remaining is 1,5 mm (0.6 in). In addition, all tyres on cars exported to the USA are moulded with maximum pressure and load ratings to conform with USA Federal Regulations.

TYRES

General

Tyres of the correct type, dimensions and at the correct cold inflation pressures are an integral part of the vehicle's design. Regular maintenance of tyres, therefore, not only contributes to safety but retains road-holding, steering and braking qualities.
Tyres of the same type and size have widely varying characteristics depending on manufacturer. It is recommended, therefore, that only the tyres specified are fitted.
The radial ply tyres specified are designed to meet the high speed performance of which this car is capable.

Pressure

The tyre inflation pressures specified provide optimum ride and handling characteristics for all normal conditions.

DATA

Wheels

Type — All countries excluding USA/Canada	6 x 15 perforated alloy
— USA/Canada only	6.5 x 15 starfish alloy
Fixing	Five studs and nuts

Tyres (normal)

Make	Pirelli or Dunlop
Type — Pirelli	215/70 VR15 P5 Cinturato
— Dunlop	215/70 VR15 SP Sport Super D7

Pressure	Front	Rear
Speeds above 160 km/hr	2.20 bars	2.20 bars
(100 mph) under all	2.25 kg/cm²	2.25 kg/cm²
load conditions	32 lb/in²	32 lb/in²

For maximum comfort in countries where speeds are not in excess of 160 km/hr (100 mph) the above inflation pressures may be reduced by 0,42 kg/cm² (6 lb/in², 0.41 bars).

Valves

Make	Bridgeport
Type	Screw in
Torque	3.39 to 4.52 Nm (30 to 40 lbs/in)

Tyres (snow)

Make	Dunlop
Type	Weathermaster 185 SR15 SP M & S (Mud and Slush)
Fitting	Complete set with inner tubes Weathermaster only

Pressure	Front	Rear
Speeds up to 137 km/hr	1.79 bars	1.79 bars
(85 mph) under all	1.83 kg/cm²	1.83 kg/cm²
load conditions	26 lb/in²	26 lb/in²
Speeds up to 161 km/h	2.35 bars	2.35 bars
(100 mph) under all	2.39 kg/cm²	2.39 kg/cm²
load conditions	34 lb/in²	34 lb/in²

NOTES:

1. All tyre inflation pressures are applicable to cold tyres only.
2. Snow chains may be fitted over snow tyres of rear wheels only.
3. Studs may be fitted to all snow tyres, but a maximum speed restriction of 121 km/hr (75 mph) is imposed.

Tyre pressures should be checked and, if necessary, adjusted weekly with the tyres cold i.e. not immediately following a run as pressure increases with temperature due to road friction. 'Bleeding' a warm tyre to the recommended pressure will result in under-inflation which can be both dangerous and depreciate tyre life. Pressure loss with time is normal, but investigation should be made if a pressure loss in excess of 0,14 kg/cm² (2 lb/in²) is encountered during a period of one week.

It is an offence in the UK to use a vehicle on public roads with tyres improperly inflated. Furthermore, incorrect inflation accelerates wear and causes excessive heating which can result in tyre failure due to blow out.

CAUTION: When inflating a tyre, it is important to ensure that a pressure of 2,8 kg/cm² (40 lb/in², 3.1 bars) is not exceeded otherwise serious tyre damage may result.

Wear

All tyres fitted as standard have a tread wear indicator (Fig. 1) moulded into their tread pattern to provide indication when the tread depth remaining is 1,5 mm (0.6 in). Each indicator appears on the tread surface as bars which connect the tread pattern across the full width of the tyre. It is illegal in the UK and certain other countries to continue using tyres after the tread has worn to less than 1 mm (0.039 in) over three quarters of the tread width around the entire circumference.

The properties of many tyres alter progressively with wear. In particular, 'wet grip' and aquaplaning resistance properties are gradually, but substantially, reduced. Extra care and speed restriction should therefore be exercised on wet roads as the effective tread depth diminishes.

Should either front or rear tyres only show excessive wear, new tyres must be fitted to replace worn ones. Under no circumstances interchange tyres from front to rear or vice versa as individual tyre wear produces unique characteristics which adversely affect performance.

Damage

Excessive localised distortion, sometimes caused by severe contact with kerbs or stones, can cause the tyre casing to fracture and may lead to premature tyre failure. Tyres should, therefore, be periodically examined and any tyre having distortion, cracks and/or cuts should be replaced. In addition, all tread imbedded objects, such as stones and glass, should be withdrawn and all contamination, i.e. oil and grease, removed using a suitable solvent.

CAUTION: Paraffin (kerosene) must not be used as a cleansing agent on tyres.

Heat

Tyres should not be subjected to excessive heat such as that inherent of paint drying/baking ovens. It is recommended, therefore, that all wheels be removed or at least the tyres be relieved of vehicle body weight.

Repairs

All minor tyre and tube repairs must be vulcanised in accordance with the vulcanising equipment manufacturer's operating instructions.

Valves

When a new tubeless tyre is fitted, the valve should be renewed.

USA Gradings

The following information relates to the tyre grading system developed by the national Highway Traffic Safety Administration which grades tyres by tread wear, traction and temperature performance.

Treadwear

The treadwear grade is a comparative rating based on the wear rate of the tyre when tested under controlled conditions on a specified government test course. For example, a tyre graded 150 would wear one and a half times less on the government course than a tyre graded 100. The relative performance of the tyres depends upon the actual conditions of their use, however, and may depart significantly from normal due to variations in driving habits, service practices and differences in road characteristics and climate.

Traction — A, B, C

The traction grades, from highest to lowest, are A, B and C, and they represent the tyres ability to stop in wet conditions measured on specified government test surfaces of asphalt and concrete.

WARNING: THE TRACTION GRADE IS BASED ON BRAKING (STRAIGHT AHEAD) TRACTION TESTS AND DOES NOT INCLUDE CORNERING (TURNING) TRACTION.

Temperature — A, B, C

The temperature grades are A (the highest), B and C, and they represent the tyre's resistance to the generation of heat and its ability to dissipate heat when tested under controlled conditions on a specified indoor laboratory test wheel. Sustained high temperature can cause the material of the tyre to degenerate and reduce tyre life, and excessive temperature can lead to sudden tyre failure. Grade C corresponds to a level of performance which all passenger car tyres must meet under Federal Motor Vehicle Safety Standard No. 109. Grades B and A represent higher levels of performance on the laboratory test wheel than the minimum required by law.

WARNING: THE TEMPERATURE GRADE IS ESTABLISHED FOR A TYRE THAT IS PROPERLY INFLATED AND NOT OVERLOADED. EXCESSIVE SPEED, UNDERINFLATION, OR EXCESSIVE LOADING, WHETHER SEPARATELY OR IN COMBINATION, CAN CAUSE HEAT BUILD UP AND POSSIBLE TYRE FAILURE.

WHEELS

Misalignment and road camber effects

It is important that correct wheel alignment be maintained. Misalignment causes tyre tread to be scrubbed off laterally because the natural direction of the wheel differs from that of the car.

A sharp 'fin' protrusion on the edge of each pattern rib is a sure sign of misalignment and it is possible to determine from the position of the 'fins' whether the wheels are toeing in or toeing out.

'Fins' on the inside edges of the pattern ribs, particularly on the nearside tyre, indicate toe-in. 'Fins' on the outside edges, particularly on the offside tyre, indicate toe-out.

With minor misalignment, the evidence is less noticeable and sharp pattern edges

J74 001

Fig. 1

may be caused by road camber even when the wheel alignment is correct. In such cases it is better to make sure by checking with an alignment gauge. Road camber affects the direction of the car by imposing a side thrust and, if left to follow its natural course, the car will drift towards its nearside. This is instinctively corrected by steering towards the road centre and, as a result, the car runs crabwise as illustrated in an exaggerated form in Fig. 2. The diagram shows why nearside tyres are very sensitive to too much toe-in and offside tyres to toe-

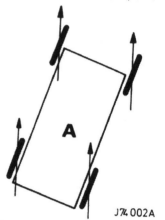

J74 002A

(A) Wheels parallel in motion: tyre wear equal

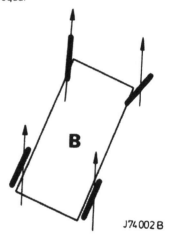

J74 002 B

(B) Wheels toed-out in motion: RH front tyre wears faster

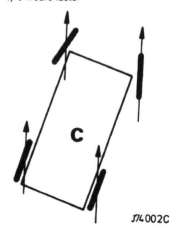

J74 002 C

(C) Wheels toed-in in motion: LH front tyre wears faster.

Fig. 2 Exaggerated diagram of the way in which road camber affects a car's progress.

out. It also shows why sharp 'fins' appear on one tyre but not on the other, and why the direction of misalignment can be determined by noting the position of the 'fins'. Severe misalignment produces clear evidence on both tyres.

The front wheels on a moving car should be parallel. Tyre wear can be affected noticeably by quite small variations from this condition. It will be noted from the diagram that even with parallel wheels, the car is still out of line with its direction of movement, but there is less tendency for the wear to be concentrated on one tyre.

The near front tyre sometimes persists in wearing faster and more unevenly than the other tyres, even when the mechanical condition of the car and tyre maintenance are satisfactory. The more severe the average road camber, the more marked this tendency will be.

Alignment Precautions

Wheels and tyres vary laterally within their manufacturing tolerances, or a result of service, and alignment figures obtained without moving the car are unreliable. The following precautions should, therefore, be observed:

1. The car should have come to rest from a forward movement. This ensures, as far as possible, that the wheels are in natural running positions.
2. It is preferable for alignment to be checked with the car laden.
3. With a conventional base bar tyre alignment gauge, measurements should be taken in front of and behind the wheel centres at the same position on the tyre and rim flanges. This is achieved by marking the tyres where the first reading is taken and moving the car forwards approximately half a road wheel revolution before taking the second reading at the same points. With an optical gauge, two or three readings should be taken with the car moved forwards to different positions — 180° road wheel turn for two readings and 120° for three readings. An average figure should then be calculated.

TYRE AND WHEEL BALANCE

Static Balance

In the interests of smooth riding, precise steering and the avoidance of high speed 'tramp' or 'wheel hop', all tyres are balance checked to predetermined limits. To ensure the best degree of tyre balance, the covers are marked with white spots on one bead and these indicate the highest part of the cover.

Some tyres are slightly outside standard balance limits and are corrected before issue by attaching special patches to the inside of the covers at the crown. These patches contain no fabric, they do not affect the local stiffness of the tyre and should not

be mistaken for repair patches. They are embossed 'Balance Adjustment Rubber'.

The original degree of balance is not necessarily maintained and it may be affected by uneven tread wear, by cover and tube repair, by tyre removal or refitting, or by wheel damage or eccentricity. The car may also become sensitive to unbalance due to normal wear of moving parts. If roughness or high speed steering troubles develop and mechanical investigation fails to disclose a possible cause, wheel and tyre balance should be suspected.

Recommended Tyre Balancing Equipment can be found in the BL STEP programme manual.

WARNING: IF BALANCING EQUIPMENT IS USED WHICH DYNAMICALLY BALANCES THE ROAD WHEELS ON THE CAR, ALWAYS JACK BOTH REAR WHEELS OFF THE GROUND WHEN REAR WHEEL BALANCING OTHERWISE DAMAGE MAY BE CAUSED TO THE DIFFERENTIAL. THIS IS DOUBLY IMPORTANT IN THE CASE OF CARS FITTED WITH A 'POWR-LOK' DIFFERENTIAL AS, IN ADDITION TO POSSIBLE DAMAGE TO THE DIFFERENTIAL, THE CAR MAY DRIVE ITSELF OFF THE JACK OR STAND.

Dynamic Balance

Static unbalance can be measured when the tyre and wheel assembly is stationary. There is another form known as dynamic unbalance which can be detected only when the assembly is revolving.

There may be no heavy spot, that is, there be no natural tendency for the assembly to rotate about its centre due to gravity, but the weight may be unevenly distributed each side of the centre tyre line. Laterally, the eccentric wheels give the same effect. During rotation, the offset weight distribution sets up a rotating couple which tends to steer the wheel to the right and left alternately.

Dynamic unbalance of the tyre and wheel assemblies can be measured on suitable tyre balancing equipment, and corrections made when cars show sensitivity to this form of unbalance. Where it is clear that a damaged wheel is the primary cause of severe unbalance, it is advisable for the wheel to be replaced.

CONTENTS

SYMBOL	MEASUREMENT TAKEN FROM	cm	in
A	Front suspension mounting point to datum line	7,70	3.05
B	Inner face of front suspension mounting point to centre line of car	39,50	15.56
C	Rear suspension front lower mounting point to datum line	11,50	4.54
D	Rear suspension rear lower mounting point to datum line	11,00	4.34
E	Front suspension, front mounting point to rear suspension front lower mounting point	278,23	109.54
F	Rear suspension, front lower mounting point to rear suspension rear lower mounting point	33,05	13.01
G	Distance between inner faces of front suspension mounting points	79,04	31.12
H	Radius arm mounting to rear suspension front mounting point	26,65	10.50
J	Wheelbase	259,10	102.00
K	Track (front)	148,30	58.40
L	Track (rear)	149,60	58.90
M	Distance between inner faces at rear of front chassis members	34,10	13.43
N	Horizontal datum line	—	—
O	Centre line of car	—	—
P	Overall width of car	179,30	70.60
Q	Front bumper mounting to datum line	38,23	15.05
R	Distance between front bumper mountings	76,46	30.10
S	Front suspension mounting to front cross member mounting	40,59	15.98
T	Radius arm mounting to rear suspension front mounting	28,57	11.25
U	Radius arm mounting to datum line	56,03	22.06
V	Distance between radius arm mountings	112,06	44.12
W	Rear suspension front mounting to datum line	49,60	19.53
X	Distance betwen rear suspension front mounting	99,20	39.06
Y	Rear bumper mounting to datum line	46,38	18.26
Z	Distance between rear bumper mountings	92,76	36.52

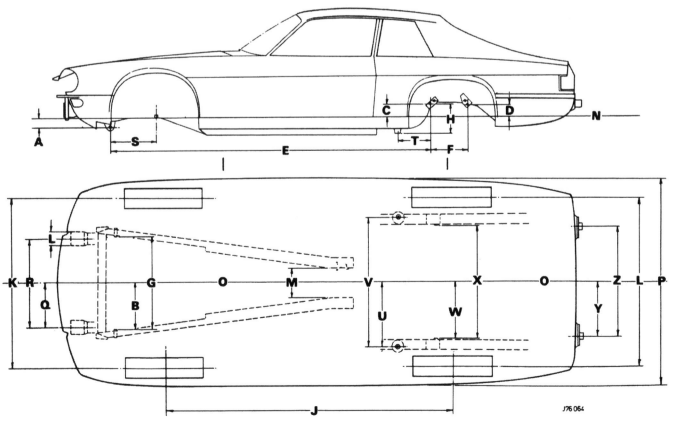

Fig. 1

ELECTRIC DOOR MIRROR

Renew

Remove the screws securing the door mirror and displace the mirror to gain access to the cable harness block connector.

Remove the block connector securing strap, and disconnect the block connector.

Remove the mirror and gasket

On refitting, ensure the connections are clean and tight.

Secure the block connector with a new strap.

FUEL FILLER FLAP

Renew

Open the fuel filler flap.

Remove the bolts securing the flap and hinge mechanism to the body.

Remove the flap assembly.

On refitting adjust the flap to the correct position before tightening the securing bolts.

FRONT SPOILER AND LOWER GRILLE

Renew

Drive the vehicle on a ramp and raise the ramp.

Remove the two screws retaining the oil cooler grille and detach the grille (1, Fig. 2).

Prise out and discard the plastic drive fasteners securing the spoiler undertray to the body and spoiler (2, Fig. 2).

Remove the screws and detach the spoiler undertray (3, Fig. 2).

Remove the screws securing the spoiler (4, Fig. 2).

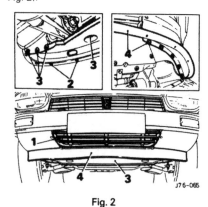

Fig. 2

Refitting

Offer the spoiler up and align the holes.

Replace and tighten the screws securing the spoiler.

Offer up the spoiler undertray and secure new drive fasteners.

Replace and tighten the screws securing the spoiler undertray.

Fit the oil cooler grille and secure with the two screws

'A' POST TRIM PAD

Renew

Unclip the section of the crash roll adjacent to the 'A' post trim pad.

Remove the screws securing the trim pad to the 'A' post (1, Fig. 3).

Lift the trim pad clear.

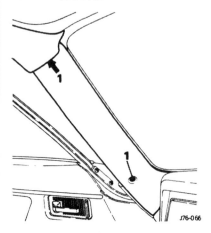

Fig. 3

'B' POST TRIM PAD

Renew

Move the seat forward and tilt the headrest.

Unclip the cantrail trim pad to clear the 'B' post trim pad.

Remove the seat belt anchorage cover and remove the anchorage bolt (1, Fig. 4).

Displace the belt from the 'B' post, the spacers and wavy washers.

Remove the trim pad (2, Fig. 4).

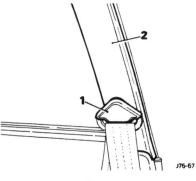

Fig. 4

REAR QUARTER TRIM CASING (Lower) FRONT SEAT BELT

Renew

Remove the rear seat cushion and squab (1, Fig. 5).

Remove the upper 'B' post trim pad.

Remove the front seat belt lower and upper anchorage bolt, spacing washers and spring (2, Fig. 5).

Displace the 'Furplex' trim from the door flange at the trim position (3, Fig. 5).

Carefully displace the trim casing flap from the door flange and remove the casing front securing screws.

Remove the remaining casing securing screws and displace the casing from the mounting position for access to the rear speaker cables.

Disconnect the rear speaker cables.

Reposition the seat belt through the casing and remove the casing assembly.

Remove the bolt securing the seat belt rod and remove the reel.

Remove the screws securing the arm rest and remove the arm rest.

Remove the companion box lower securing rivets.

Remove the trim clips and displace the upper securing tabs.

Remove the companion box assembly.

Replace the tabs securing the rear seat belt blank or surround (if fitted) and the front seat belt guides.

Remove the guides.

Refitting

On refitting, ensure the speaker cables are reconnected correctly and the trim is securely adhered to the body flange.

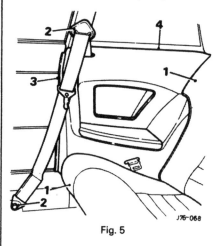

Fig. 5

REAR QUARTER TRIM PAD (Upper)

Renew

Remove the rear seat cushion and squab.

Remove the blanking screws from the rear parcel shelf, unclip and remove the shelf, (1, Fig. 6). Unclip the cantrail crash roll from the trim pad location (2, Fig. 6).

Remove the trim pad securing screws and unclip the trim pad (3, Fig. 6).

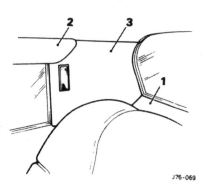

Fig. 6

Note and disconnect the interior lamp cable harness.

Remove the trim pad.

Remove the interior lamp from the trim pad. On refitting, ensure the lamp is connected correctly and the connections are clean.

COMPANION BOX LINER — REAR QUARTER LOWER

Renew

Remove the rear seat cushion and squab.

Remove the 'B' post trim pad.

Remove the rear quarter lower trim casing assembly and remove the rear speaker.

Remove the speaker grille by releasing the companion box securing tabs.

Remove the companion box liner.

FACING TRIM PAD — COMPANION BOX — REAR QUARTER (Lower)

Renew

Remove the rear seat cushion and squab.

Remove the upper 'B' post trim pad.

Remove the rear quarter lower trim casing assembly.

Remove the companion box assembly and arm rest.

Remove the trim pad lower securing rivets and clips.

Displace the upper securing tabs and remove the trim pad.

BONNET AND GRILLE

Renew

Remove the screws securing the stay to the bonnet and wing valance (1, Fig. 7).

Remove the stay, retaining the back plates and seating blocks.

Remove the screws securing the radiator grille to the bonnet and remove the grille (2, Fig. 7). Retain the washers and distance pieces.

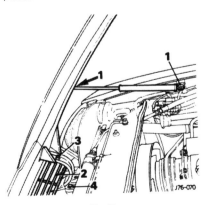

Fig. 7

Remove the four bolts securing the bonnet to the hinges (3, Fig. 7) and lift the bonnet from the car.

When fitting the bonnet to the hinges, do not fully tighten the securing bolts. Close the bonnet and if necessary, adjust the position to centralise the bonnet between the wing valances.

Open the bonnet and fully tighten the securing bolts.

BONNET HINGES

Renew

With the bonnet assembly removed.

Mark the location of the hinges to the crossmember.

Remove the four bolts securing the hinges to the crossmember (4, Fig. 7) and remove the hinges.

Position the new hinges to the marked location, fit and tighten the securing bolts.

When refitting the bonnet to the hinges, do not fully tighten the securing bolts.

Close the bonnet, adjust the position to centralise the bonnet between the wing valances.

Open the bonnet and tighten the securing bolts.

BONNET LOCK — LH

Renew

Slacken the bolt securing the control cable to the bonnet release handle nipple and release the cable (1, Fig. 8).

Slacken the bolt securing the control cable to lock pivot and release the cable (2, Fig. 8).

Remove the four screws securing the lock assembly to the bulkhead (3, Fig. 8).

Remove the lock assembly complete with the lock to release handle cable (4, Fig. 8).

Remove the two bolts securing the lock to the backplate and remove the backplate (5, Fig. 8).

Remove the lock to release the handle control cable.

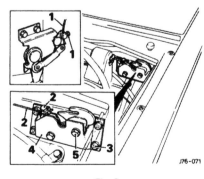

Fig. 8

BONNET LOCK — RH

Renew

Remove the two screws securing the start relay cover to the RH wing valance and remove the cover.

Remove the four screws securing the lock assembly to the bulkhead.

Slacken the bolt securing the control cable to the LH bonnet lock pivot and release the cable.

Remove the lock assembly complete with the control cable then remove the control cable.

Remove the two bolts securing the lock to the backplate and remove the backplate.

When fitting new lock, fit but do not tighten the two screws securing the lock to the backplate.

Raise or lower the lock as required then pinch up the screws to secure the lock.

Close the bonnet, check the adjustment by operating the bonnet release handle. Raise the bonnet, readjust as necessary.

Fully tighten the securing screws.

BONNET LOCK CONTROL CABLE

Adjust

Slacken the bolt securing the control cable at the LH bonnet lock (1, Fig. 9).

Adjust the control cable as required by tightening or slackening in the nipple.

Tighten the control cable securing bolt.

Close the bonnet and check the control cable adjustment by operating the release handle.

Readjust if necessary.

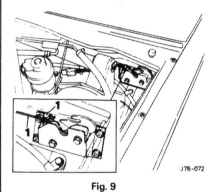

Fig. 9

BONNET LOCK TRIGGER RELEASE HANDLE

Renew

Remove the driver's footwell side trim pad.

Move the bonnet release handle to the correct open position.

Slacken the control cable damping screw and release the control cable (1, Fig. 10).

Move the bonnet release handle to the bonnet closed position (2, Fig. 10).

Remove the three bolts and washers securing the release handle mounting bracket to the body (3, Fig. 10).

Remove the bonnet release handle assembly.

On refitting, check the bonnet alignment and adjust the control cable.

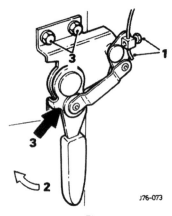

Fig. 10

BOOT LID

Renew

Disconnect the battery earth lead.

Peel back the boot carpet to gain access to the electrical cable harness (1, Fig. 11).

Disconnect the cable harness block connector (2, Fig. 11) and remove the cable clips securing the harness to the RH boot lid hinge (3, Fig. 11).

Remove the four bolts and washers securing the boot lid to the hinges (4, Fig. 11), then remove the boot lid.

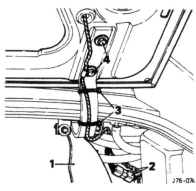

Fig. 11

On fitting new boot lid, fit but do not tighten the four bolts securing the boot lid to the hinges.

Close the boot lid to check the alignment, adjust by moving the boot lid as required on the hinges elongated bolt holes.

Fully tighten the hinge securing bolts.

Clip the cable harness to RH boot lid hinge and ensure the cable harness connector is clean and securely connected.

Check the electrical components for correct operation.

BOOT LID HINGES

Renew

With the boot lid removed.

Remove the rear seat cushion and the rear seat squab.

Carefully prise the rear parcel shelf from the rear parcel shelf panel.

Remove the bolt and washer securing the boot hinges to the rear parcel shelf plate (1, Fig. 12).

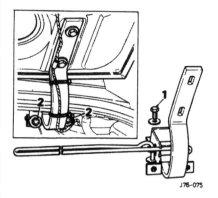

Fig. 12

Remove the two bolts and washers securing the hinges to the mounting brackets (2, Fig. 12).

Remove the hinge assemblies.

Refitting is the reversal of the above operations.

Ensure the cable harness connections are clean and secure, and the electrical components operate correctly.

BOOT LID SEAL

Renew

Remove the screws securing the finisher to the boot aperture valance and lift the panel from the boot (1, Fig. 13).

Remove the tape joining the ends of the seal and ease the remainder of the seal from the flange (2, Fig. 13).

On fitting new seal, take care not to stretch the seal and ensure it is correctly bedded down.

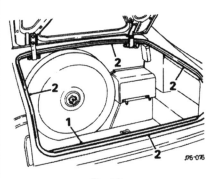

Fig. 13

BOOT LID LOCK

Renew

Remove the clip securing the boot lid handle operating lever to the boot lid lock operating rod and disconnect the rod from the lever (1, Fig. 14).

Remove the three bolts and washers securing the lock assembly to the boot lid (2, Fig. 14).

Remove the lock assembly (3, Fig. 14).

Release the spring clip securing the operating rod to the lock operating lever and remove the operating rod.

On fitting the new lock, smear the operating mechanism with a suitable grease.

Fit the operating rod to the lock operating lever using a new spring clip.

Fit the lock assembly to its location but do not overtighten the three securing bolts and washers.

Use a new spring clip to secure the operating rod to the boot lid handle operating lever.

Close the boot lid, check the adjustment of the lock and adjust as necessary before fully tightening the three securing bolts.

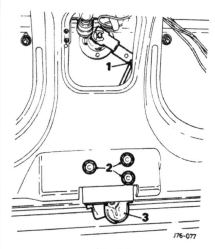

Fig. 14

BOOT LID LOCK STRIKER

Renew

Remove the screws securing the finisher panel to the boot aperture valance and lift the panel from the boot (1, Fig. 15).

Mark the striker legs along the top face of the clamp plate for a reference when refitting.

Slacken the bolts securing the striker clamp plate (2, Fig. 15) and slide the striker free from the clamp (3, Fig. 15).

On fitting new striker, align scribe marks to top face of the clamp and tighten the clamp bolts.

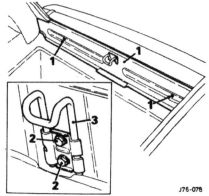

Fig. 15

Adjust the position of the striker, should more than a push effort be required to close the boot lid.

Refit the finisher panel to the boot aperture valance.

BOOT LID HANDLE/ LOCK ASSEMBLY

Renew

Disconnect the battery earth lead.

Disconnect the number plate and reverse lamp cable harness connectors.

Remove the bolts securing the lamp assembly and ease the assembly clear of the boot lid.

Remove the spring clip securing the boot lid handle operating lever to the boot lid lock operating rod and disconnect the rod from the lever (1, Fig. 16).

Remove the two screws and spring washers securing the handle/lock assembly to the boot lid (2, Fig. 16).

Remove the plate from the rear of the assembly (3, Fig. 16) and manoeuvre the handle/lock assembly through the boot lid aperture (4, Fig. 16).

On fitting new assembly, ensure electrical connections are clean and secure.

Check the lamps for correct operation.

J76-079

Fig. 16

BUMPER — REAR CENTRE SECTION — BLADE

Renew

Displace the boot side carpeting for access to quarter bumper securing bolts.

Remove the quarter bumper securing bolts and nuts securing the blade to the body flange (1, Fig. 17).

Remove the blade, quarter bumper and rubber buffer assembly.

NOTE: It is advisable to carry out this operation with two men so as to avoid damage to paintwork.

Remove nuts and bolts securing the quarter bumpers to the blade (2, Fig. 17).

Remove the quarter bumpers and rubber joint finishers (3, Fig. 17).

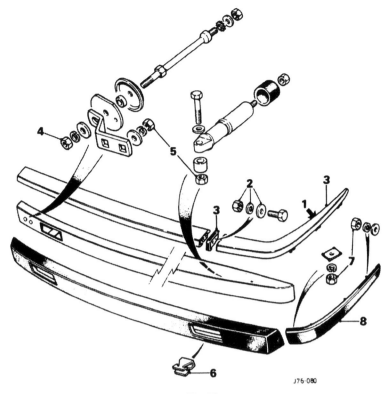

J76-080

Fig. 17

ENERGY ABSORBING BEAM — REAR

Renew

Remove the nuts securing the fog guard lamps to the rear beam and displace the lamps from the mounting position (where fitted).

Slacken the beam bumper blade to body securing nuts (4, Fig. 17).

Remove the beam to the bracket securing nuts (5, Fig. 17).

Carefully remove the beam assembly ensuring the rear fog guard lamps (if fitted) are not damaged.

Remove the clips securing the beam cover (6, Fig. 17) and remove the cover.

QUARTER BUMPER — REAR — RUBBER BUFFER

Renew

With the rear bumper centre blade assembly removed.

Remove the nuts securing the rubber buffer (7, Fig. 17) and remove the buffer (8, Fig. 17).

QUARTER BUMPER — FRONT

Renew

Remove the quarter bumper trim finisher (1, Fig. 18).

Remove the quarter bumper blade securing nuts (2, Fig. 18).

Remove the quarter bumper blade to the centre blade securing nuts and bolts (3, Fig. 18).

Remove the quarter bumper blade and quarter bumper finisher (4, Fig. 18).

ENERGY ABSORBING BEAM — FRONT

Renew

Disconnect the battery.

Disconnect the front flasher lamps.

Remove the nuts securing the beam mounting brackets to body struts (5, Fig. 18).

Remove the beam, cover and lamp assembly (6, Fig. 18).

Remove the nuts securing the brackets to the beam and remove the brackets (7, Fig. 18).

Remove the flasher lamp securing screws and remove the lamps with the sealing rubbers.

Remove the bolts and clips securing the beam cover to the beam and remove the cover (8, Fig. 18).

Remove the flasher lamp securing cage nuts.

On refitting, ensure that the beam assembly is aligned before all the mounting bracket securing nuts are fully tightened.

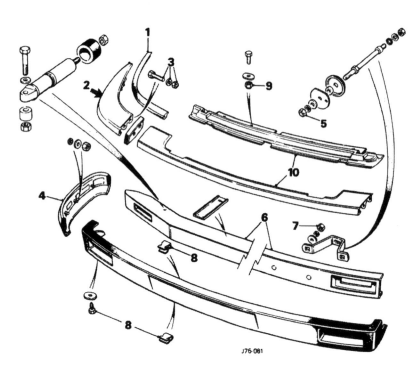

Fig. 18

FRONT BUMPER — CENTRE SECTION — BLADE

Renew

Remove the upper radiator grille.

Remove the nuts and bolts securing the bumper blade to the quarter bumpers (3, Fig. 18).

Remove the nuts and bolts securing the finisher to the body (9, Fig. 18).

Remove the blade and the finisher assembly (10, Fig. 18).

Drill out the pop rivets securing the finisher to the blade and remove the finisher.

On refitting, ensure the beam is aligned before the mounting nuts are fully tightened. Also align the fog lamps before the securing nuts are fully tightened.

CONSOLE ASSEMBLY — FRONT

Renew

Remove the air conditioning switch knobs (1, Fig. 19) and the control panel securing bezel. Displace the panel (2, Fig. 19).

Remove the console side casing securing screws (3, Fig. 19).

Remove the footwell vents and the console side casing.

Unscrew and remove the gear lever knob (4, Fig. 19).

Remove the screws securing the console finisher panel (5, Fig. 19) and displace the panel over the gear lever (6, Fig. 19).

Note and disconnect the cable connectors from the window lift switches and the cigar lighter.

Remove the console finisher panel.

Remove the screws securing the stop light failure sensor and displace the sensor (7, Fig. 19).

Remove the screws securing the front of the console (8, Fig. 19).

Move the seats fully forward and remove the screws securing the rear of the console. Reposition the seat backrest.

Position the air conditioning control panel through the aperture.

Displace and remove the console assembly.

Remove the centre glove box lid, check strap securing screws (1, Fig. 20).

Remove the centre glove box lid hinge screws and remove the glove box lid (2, Fig. 20).

Remove the glove box lid catch and liner.

Remove the rear outlet duct and the ashtray assembly.

Refitting new console assembly is the reversal of the above procedure.

Ensure the electrical connections are clean and tight.

Check stop lamps, window lift motors and the cigar lighter for correct operation.

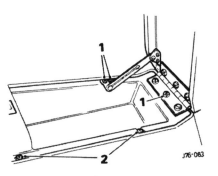

Fig. 20

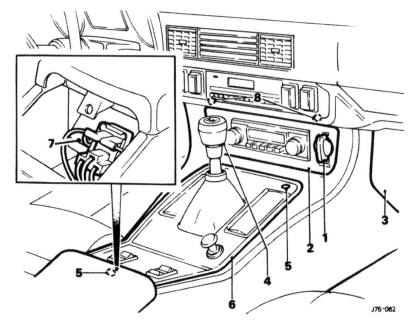

Fig. 19

DOOR

Renew

NOTE: The door MUST be adequately supported in a half open position during this operation.

Remove and swing the inertia switch aside (driver's door only).
Remove the three screws securing the footwell side trim pad and remove trim pad.
Remove the screws securing the door lock solenoid relay and ease the relay to one side (1, Fig. 21).
Note and disconnect the door cable harness connectors and the earth lead from the solenoid relay mounting bracket (2, Fig. 21).
Disconnect the radio speaker connectors.

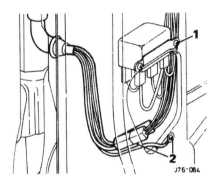

Fig. 21

Remove the six nuts and washers securing the hinge plates to the body and remove the hinge plates (1, Fig. 22).
With assistance, remove the door assembly.

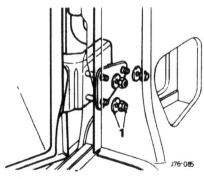

Fig. 22

Refitting

Locate the door assembly and hinges to the body, secure using one nut and washer on each hinge, do not tighten fully.
Remove the support, close the door, check the alignment and align as required.
Fit the remaining nuts and washers.
Tighten fully.
Ensure all electrical connections are secure and the button on the inertia switch is depressed.

DOOR HINGE (DOOR TRIM PAD)

Renew

NOTE: Ensure the door is adequately supported in the half open position.

Remove the door mirror switch bezel.
To avoid damage to the arm rest and the stitching, apply masking tape to the arm rest.
Carefully displace the chrome finisher on to the masking tape (1, Fig. 23).
Remove the arm rest lower securing screw (2, Fig. 23).
Remove the screws securing the door pillar switch plate and remove the plate (3, Fig. 23).
Carefully unclip and raise the door trim pad to release it from the door (4, Fig. 23).
Note and disconnect the puddle light and the radio speaker cables (5, Fig. 23).
Carefully remove the trim pad assembly (6, Fig. 23).

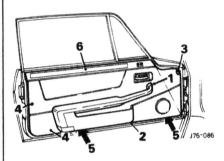

Fig. 23

Remove and swing the inertia switch to one side (driver's door only).
Remove the three screws securing the underscuttle casing and remove the casing.
Remove the two screws securing the footwell side trim pad and remove the trim pad.
Remove the two screws securing the door lock relay and ease the relay aside (1, Fig. 21).
Remove the three nuts and washers securing the hinge plate to the body. Remove the hinge plate.
Remove the nuts and washers securing the hinge plate to the door (four on the top hinge and three on lower hinge) (1, Fig. 24).
Remove the hinge plate and hinge (2, Fig. 24).

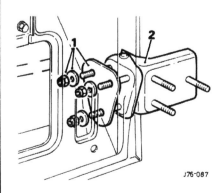

Fig. 24

Refitting

Locate the new hinge and hinge plate to the body, secure with one nut and washer, do not fully tighten.

Remove the support.
Close the door, check the alignment, and readjust the door as required.
Fit the remaining nuts and washers, then fully tighten.

On refitting the door trim ensure the electrical components operate correctly.

DOOR GLASS

Renew

With door trim pad removed and the glass in the raised position, remove the two screws and washers securing the window channel to the door (1, Fig. 25).
Ease the rubber clear of the channel (2, Fig. 25) and remove the window channel (3, Fig. 25).
Remove the two bolts, plain and shockproof washers securing the window stop to the base of the door (4, Fig. 25).
Remove the window stop (5, Fig. 25).
Lower the door glass.

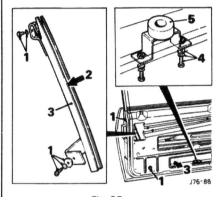

Fig. 25

NOTE: When lowering the window glass, care should be taken to guide the glass by hand.

Disconnect the window lift remote control mechanism from the window glass by tilting the glass forward (1, Fig. 26).
With the glass still in the tilted position, carefully lift the glass clear of the door (2, Fig. 26).
Refitting is the reversal of the above procedure.

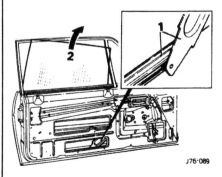

Fig. 26

QUARTERLIGHT — FRONT

Renew

Remove the two screws securing the quarter light from the door panel (1, Fig. 27).
Lower the door glass, ease the door glass rubber to one side (2, Fig. 27) and remove the two screws securing the quarterlight to the door glass frame (3, Fig. 27).
Remove the quarterlight assembly.
Remove the quarterlight glass and rubber from the frame (4, Fig. 27).

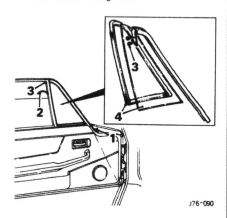

Fig. 27

QUARTERLIGHT — REAR

Renew

Remove the rear quarter upper and lower trim pads.
Using a suitable tool, prise the rear quarter glass and rubber from the body.
Remove the seal from the glass (1, Fig. 28).

Refitting

Apply sealant into the locating grooves of the new seal and fit the glass into the seal.
Fit the glass and seal to the quarterlight location.

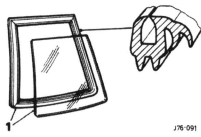

Fig. 28

DOOR GLASS WEATHER STRIP

Renew

Remove the front quarterlight and the door trim pad.
With the glass in the raised position (1, Fig. 29), remove the two bolts and washers securing the window stop to the base of the door (2, Fig. 29).
Remove the window stop and lower the door glass.

Remove the three screws securing the quarterlight channel and remove the channel (3, Fig. 29).
Release the clip securing the door finisher at the rear edge of the door (4, Fig. 29).
Remove the eight screws securing the door finisher to the door and remove the door finisher (5, Fig. 29).
Remove the weather strip from the door finisher (6, Fig. 29).

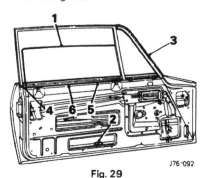

Fig. 29

DOOR ARM REST

Renew

With the door trim pad removed, remove the two screws securing the arm rest retaining bracket and remove the bracket (1, Fig. 30).
Remove the split rivets securing the arm rest (2, Fig. 30), unclip and remove the arm rest
Unclip and remove the chrome finishers.

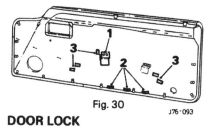

Fig. 30

DOOR LOCK

Adjust

WARNING: IF ANY OF THE FOLLOWING SYMPTOMS BECOME EVIDENT IMMEDIATE REMEDIAL ACTION MUST BE TAKEN AS OUTLINED BELOW:

A Door fails to fully close.
B Door fails to open with the operation of the inside door handle.
C Door opens upon the initial movement of the inside door handle.

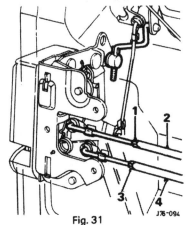

Fig. 31

D Door fails to lock upon operation.

NOTE: When symptom A, B or C are evident proceed as follows:

Remove the door trim pad.
Slacken the link rod adjusting sleeve lock nut (1, Fig. 31). Adjust sleeve (2, Fig. 31) as required until the door opens when the handle is in the halfway position when operated and the door closes fully.
Re-tighten the lock nut.

NOTE: If symptom D is evident:

Slacken the link rod adjusting sleeve lock nut (3, Fig. 31). Adjust the sleeve (4, Fig. 31) as required. close door and check for correct operation of lock lever.

DOOR LOCK

Renew

Remove the door trim pad.
Disconnect the inner handle to lock connecting rods from the anti-rattle and connecting clips (1, Fig. 32).
Ease the plastic sheeting to one side.
Disconnect the outer handle to lock connecting rods (3, Fig. 32).
Remove the four screws securing the outer lock unit to the door (3, Fig. 32).
Remove the outer lock unit (4, Fig. 32) and the inner lock unit (Fig. 32).
Refitting is the reversal of the above procedure.

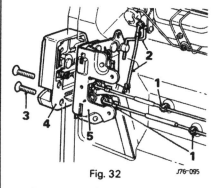

Fig. 32

DOOR LOCK STRIKER PLATE

Renew

Remove the rear seat cushion and rear seat squab.
Remove the 'B' post upper trim pad and the rear quarter lower trim casing.
Remove the screws securing the striker plate (1, Fig. 33) and remove the striker plate halves (2, Fig. 33).
On refitting, adjust the striker plate until the door closes with the minimum push effort.

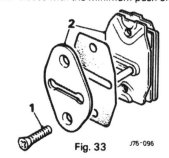

Fig. 33

DOOR — VENEER PANEL

Renew

Disconnect the battery.
Remove the door mirror switch mounting panel and the door trim pad assembly.
Remove the nuts securing the veneer panel to the trim pad and remove the veneer panel.

DOOR OUTSIDE HANDLE

Renew

Remove the door trim pad.
Remove the door glass.
Remove the three clips securing the control arms to the outside handle.
Disconnect the control arms from the outside handle (1, Fig. 34).
Remove the two nuts and washers securing the door handle (2, Fig. 34).
Remove the securing bracket (3, Fig. 34).
Remove the outside handle assembly (4, Fig. 34).

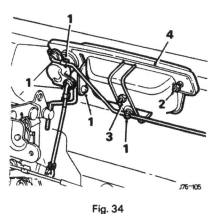

Fig. 34

DOOR INSIDE HANDLE

Renew

Remove the door trim pad.
Remove the three screws securing the handle to the door (1, Fig. 35).
Disconnect the control arms from the anti-rattle clips and lock clips (2, Fig. 35).
Ease the handle clear of the door panel and release the lock control arm from the plastic clip (3, Fig. 35).
Remove the interior handle assembly (4, Fig. 35).

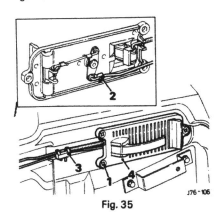

Fig. 35

FASCIA VENEER PANEL — PASSENGER SIDE
FASCIA VENEER PANEL — DRIVER'S SIDE
FASCIA VENEER PANEL — CENTRE

Renew

For the centre and passenger side veneer panels, open the glove box for access. By using a suitable long thin bladed instrument, carefully release the veneer panel retaining clips and remove the appropriate panel.

DRIVER'S UNDERSCUTTLE CASING

Renew

Remove the screws securing the casing to the fascia and the scuttle (1, Fig. 36).
Carefully lower the casing sufficiently to give access to the rheostat (2, Fig. 36).
Disconnect the electrical cable from the rheostat and lift the casing from the car (3, Fig. 36).

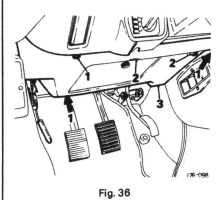

Fig. 36

PASSENGERS' UNDERSCUTTLE CASING

Renew

Remove the screws securing the casing to the fascia panel and the underscuttle (1, Fig. 37).
Carefully manoeuvre casing from its location (2, Fig. 37).

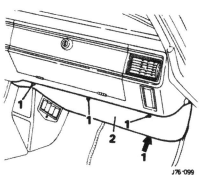

Fig. 37

GLOVE BOX LID — VENEER PANEL

Renew

Remove the screws securing the vanity mirror moulding to the rear of the glove box lid (1, Fig. 38).
Remove the assembly (2, Fig. 38).
Remove the screws securing the veneer panel to the glove box lid and remove the panel.
On refitting, ensure that the veneer panel is correctly aligned before fully tightening the securing screws.

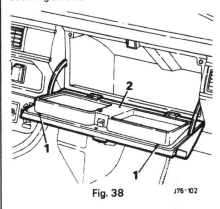

Fig. 38

GLOVE BOX LINER

Renew

Remove the screws securing the liner to the fascia (1, Fig. 39).
Carefully prising the liner around the latch withdraw the liner from the fascia (2, Fig. 39).

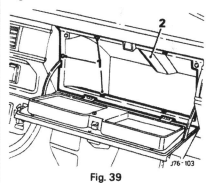

Fig. 39

FASCIA SWITCH PANEL — VENEER PANEL

Disconnect the battery.
Remove the electric clock, or trip computer (1, Fig. 40) and disconnect the leads (2, Fig. 40).
By using a long thin bladed instrument, carefully unclip and displace the switch panel (3, Fig. 40).
Note and disconnect the switch block connectors and the switch illuminator bulb holders.
Remove the panel assembly.
Unclip and remove the switches from the panel.
Remove the spire nuts securing the bulb holders to the panel and remove the bulb holders.

On refitting, use new spire nuts to secure the bulb holders.

Ensure the electrical connectors are secure and the correct operation of components.

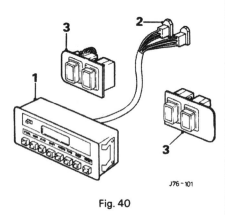

Fig. 40

FASCIA

Renew

Disconnect the battery earth lead.

Slacken the steering wheel adjusting collar and pull the steering wheel out to its fullest extent (1, Fig. 40).

Remove the screws securing the lower and upper shroud to the steering column and remove the shrouds from the column (2, Fig. 40).

Slacken the steering wheel grub screw locknut and remove the grub screw.

Remove the steering wheel pinch bolt and nut.

Note the position of and remove the steering wheel (3, Fig. 40).

Remove the screws securing the fascia side trim and remove the side trims.

Remove the screws securing the driver's dash liner, note and disconnect the cables from the rheostat, then remove the dash liner (4, Fig. 40).

Slacken the screws securing the light switch surround, remove the lower screws and displace the shroud.

Depress the light switch knob retaining peg and remove the switch knob (5, Fig. 40).

Remove the lower and slacken upper screws securing the ignition switch surround.

Remove the screws securing the instrument panel surround and remove the centre finisher (6, Fig. 40).

Displace the switch surround.

Remove the instrument panel side finishers.

Remove the switch surround upper securing screws and nuts.

Ease the surrounds from the mounting positions.

Displace the fibre optic strands and remove the surrounds.

Remove the instrument panel blanking plates and remove the screws securing the instrument panel (7, Fig. 40).

Displace the panel and disconnect the cable harness block connectors then remove the instrument panel.

Displace the interior lamps from the fascia, disconnect the cables and remove the interior lamps (8, Fig. 40).

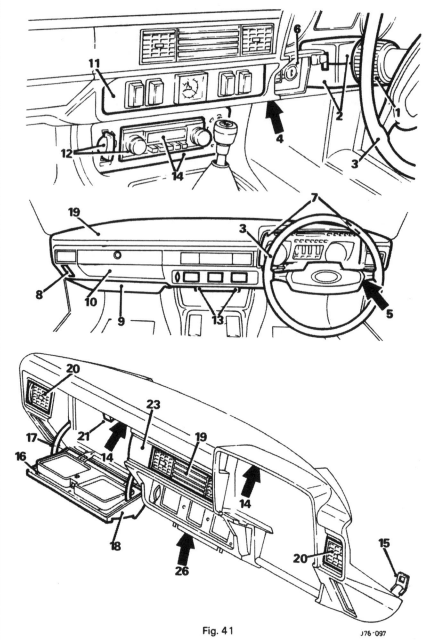

Fig. 41

J76-097

Remove the screws securing the passenger side dash liner and remove the dash liner (9, Fig. 40).

Remove the screws securing the glove liner, displace and remove the liner (10, Fig. 40).

By using a long thin bladed instrument, carefully unclip and displace the switch panel (11, Fig. 40).

Note and disconnect the switch block connectors and the switch illuminator bulb holders.

Remove the panel assembly.

Remove the air conditioning and radio control knobs.

Remove the air conditioning and radio control panel securing bezels (12, Fig. 40).

Displace the panel and remove the fascia securing screws (13, Fig. 40).

Slacken the fascia to fascia rail securing nuts (14, Fig. 40).

Remove the fascia side securing nuts and bolts (15, Fig. 40).

Remove the nut securing the light switch to the fascia and displace the switch from the fascia.

Displace the fascia from its location and reposition over the column stalks.

Disconnect the air conditioning in car temperature sensor tube and remove the fascia assembly.

Remove the screws securing the glove compartment lid liner and remove the liner (16, Fig. 40).

Remove the nuts securing the glove box lid stay and remove the stay (17, Fig. 40).

Remove the screws securing the lid to the fascia and remove the lid (18, Fig. 40).

Unclip and remove the fascia centre veneer panel and remove the centre vent (19, Fig. 40).

Unclip and remove the side veneer panel and remove the side vents (20, Fig. 40).

Remove the air conditioning sensor elbow halves.

Remove the light and ignition switch brackets.

Remove the glove box lid catch (21, Fig. 40).

Remove the interior light backing mouldings.

HEADLINING

Renew

WARNING: THIS OPERATION SHOULD NOT BE ATTEMPTED BY PERSONS KNOWN TO BE ALLERGIC TO GLASS FIBRE (FIBREGLASS). SHOULD SKIN AREAS DEVELOP A RASH OR IF ITCHING OCCURS, WASH AFFECTED AREA WITH WATER AND SEEK MEDICAL ADVICE IMMEDIATELY. ALWAYS WEAR GLOVES, FACE MASK AND GOGGLES WHEN HANDLING HEADLINING.

Remove the sun visors (1, Fig. 42) and sun visor retaining brackets (2, Fig. 42).
Remove the interior mirror (3, Fig. 42).
Ease the interior lamp clear of its location, disconnect the electrical leads and remove the lamp (4, Fig. 42).
Remove the cantrail crash roll from both sides.
Remove the rear seat squab and the rear shelf.
Remove the rear quarter upper trim pads from both sides.
Bend back the six tabs securing the headlining (5, Fig. 42) and remove the headlining assembly via the passenger door.

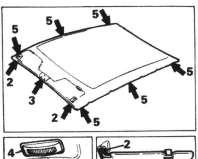

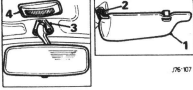

Fig. 42

Refitting

Introduce the replacement headlining through the passenger door, raise the headlining into position and secure with the clips.
Ease the top forward edge of the rear window surround over the rear side of the headlining.
Refitting the trim etc. is the reversal of the dismantling procedure.

REAR PARCEL SHELF AND REAR SEAT CUSHION AND SQUAB

Renew

Move the front seats and tilt the front seat backrests forward.
Remove the screws securing the rear seat cushion (1, Fig. 43).
Displace and remove the rear seat cushion assembly (2, Fig. 43).

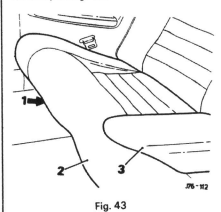

Fig. 43

Remove the screw and nut securing the centre trim panel and remove the panel (3, Fig. 43).
Displace the sound deadening flaps.
Remove the screws securing the seat squab (1, Fig. 44).
Lift the seat squab upwards and carefully remove.

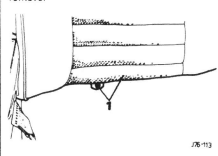

Fig. 44

Remove the blanking plug from the child seat anchorage.
Remove the screws securing the parcel shelf assembly.
Unclip and remove the parcel shelf (1, Fig. 45).

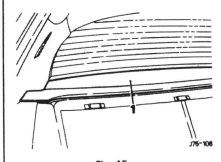

Fig. 45

REAR SEAT ARM REST

Renew

With the rear seat cushion and squab removed:
Remove the upper 'B' post trim pad.
Remove the lower rear quarter trim pad.
Remove the lower rear quarter trim casing.
Remove the screws securing the arm rest to the trim casing and remove the arm rest.

FRONT ASHTRAY

Renew

Open the ashtray cover and withdraw the ash container.
Withdraw the screws securing the ash container holder to the console.
Withdraw the holder and the securing bracket from the console.

Refitting

Secure the bracket with one screw to the holder unit.
Fit the holder and the bracket to the console.
Align the unsecured portion of the bracket with the hole in the holder and fully tighten both the screws.
Fit the ash container to the holder.,

REAR ASHTRAY

Renew

Remove the screws securing the console stowage lid check strap.
Remove the screws securing the stowage lid hinge and lift the lid from the console.
Remove the screws securing the stowage liner to the console, prise the liner lock catch from the console and remove the liner.
Open the ashtray, push down and remove the ash container from the holder.
Straighten the upper clips securing the holder to the console.
Lift the holder and the bezel surround from the console.

REAR SEAT BELT

Renew

With the rear seat cushion and squab removed (1, Fig. 46).
Remove the bolt and washer securing the rear seat belt buckle assembly (2, Fig. 46) then remove the assembly (3, Fig. 46).
Remove the rear quarter lower trim pad for access (4, Fig. 46).
Thread the seat belt through the trim pad (5, Fig. 46).
Remove the bolt and washer securing the seat belt reel mechanism then remove the reel.

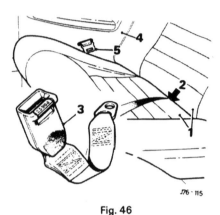

Fig. 46

FRONT SEAT

Renew

With the seat pushed to the full rear position (1, Fig. 47), remove the two Allen head screws securing the seat runner assembly to the floor mountings (2, Fig. 47).
Push the seat to the full forward position and remove the two Allen headed screws securing the rear of the runner assembly to the floor mountings (3, Fig. 47).
Lift the seat assembly from the car and recover the distance pieces from the front mountings.

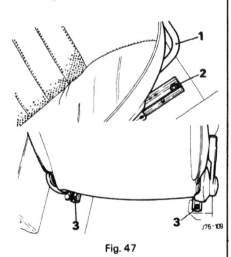

Fig. 47

FRONT SEAT RUNNER AND ADJUSTER ASSEMBLY

Renew

With the front seat removed, operate the runner lever and push the runners to the rear, exposing screws securing the runners to the front of the seat (1, Fig. 48).
Remove the securing screws (2, Fig. 48).
Align the runners to the full forward position and remove the screws securing the runners to the rear of the seat (3, Fig. 48).
Lift the runner and adjuster assembly from the seat.
Retain the washers.

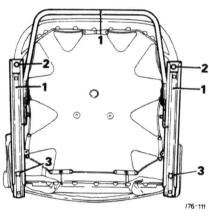

Fig. 48

FRONT SEAT CUSHION

Renew

With the front seat and adjuster assembly removed, remove the screw securing the rake angle adjustment lever, lift the lever and the plastic disc from the seat (1, Fig. 49).
Carefully prise the trim plate from the seat (2, Fig. 49).
Remove the screw securing the trim plate to the side of the seat and prise the trim plate from the seat (3, Fig. 49).
Remove the bolts securing the seat cushion and separate the cushion from the back rest.

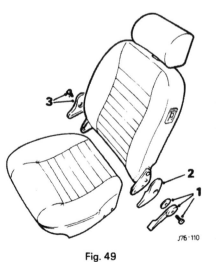

Fig. 49

FRONT SEAT BELT BUCKLE — LH AND RH

Renew

Remove the centre console rear securing screws and raise the rear console for access.
Disconnect the seat belt buckle cable harness block connector.
Remove the seat belt buckle anchorage bolt cover and remove the bolt.
Tilt the seat back for access and remove the buckle assembly with the spacer.

REAR STOWAGE COMPARTMENT

Renew

Tilt and push the driver and passenger seats fully forward.
Unlock and open both stowage container lids.
Remove the container slip mats.
Remove the five screws securing the stowage container.
Displace the passenger side rear courtesy lamp, note and disconnect the electrical cables.
Remove the lamp and return the cables to inside the aperture.
Extend and displace the passenger side seat belt.
Tilt the passenger side of the container upwards past the courtesy light aperture, tilt towards the front of the vehicle and manoeuvre the stowage container through the passenger door.
Invert the stowage container for access.
Remove the nuts securing the rail and remove the rail.
Refitting is the reversal of the above operations. Ensure care is taken manoeuvring the container into position.

HOOD CANOPY GUIDES

Renew

Release the hood locking catch.
Release the headlining trim to 'D' post 'velcro' fastening (1, Fig. 50).
Fold the hood to the rearward position.
Remove the canopy guide securing screws and remove the guides (2, Fig. 50).
On fitting new guides fit but do not tighten the canopy guide securing screws.
Seat the hood and secure the top catches on the roll bar mountings.
Lock the hood catch.
Position guides to the hood guide pin and tighten the guide securing screws.

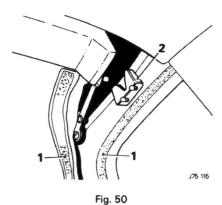

Fig. 50

HOOD CANOPY

Renew

Release the hood locking catch.
Release the headlining trim from the 'D' post velcro fastening (1, Fig. 50).
Release the headlining from the double velcro fastening and the rear of the parcel shelf.
Fold the hood to rearward position (Fig. 51).

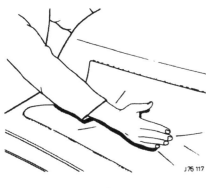

Fig. 51

Release the outer cover velcro fastening (1, Fig. 52) and with a $\frac{5}{32}$ allen key or a suitable tool remove the hood to body securing screws (2, Fig. 52).

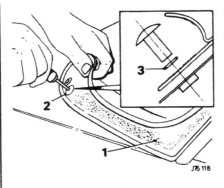

Fig. 52

Push the rear of the hood upwards and release the headlining fasteners (1, Fig. 53).

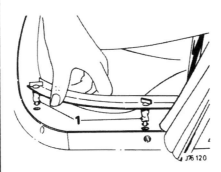

Fig. 53

Manoeuvre the hood and carefully remove the hinge plate securing screws (1, Fig. 54), displace and remove the hood assembly.

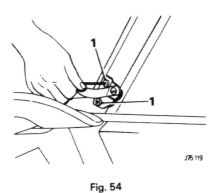

Fig. 54

Refitting

Place the new hood on to the car body.
Fit rubber spacers to the headlining fasteners, fit and tighten the headlining fastener (Fig. 53).
Fit new 'O' rings to the hood securing screws (3, Fig. 52).
Fit and secure the hood with the securing screw using appropriate tool (2, Fig. 52).
Manoeuvre the hood to align the hinge plate to the securing position.
Secure and tighten the securing screws (Fig. 54).

Raise the hood and lock in position, reposition the headlining to the 'D' post and to the rear of the parcel shelf, secure with the velcro fastening.
Position the outer cover, secure with the velcro and stud fasteners.

HOOD CANOPY LATCHES

Renew

Release the hood locking catch.
Release the headlining trim from the 'D' post velcro fastening.
Fold the hood to the rearward position.
Remove the screws securing the canopy latches and remove the latches from the roll bar.
When refitting new catches fit but do not fully tighten the securing screws.
Raise the hood and position in the guides, lock the hood locking catch, adjust the hood latches and tighten the securing screws.
Reposition the headlining to the 'D' post velcro fastening.

HOOD CANOPY SEAL

Renew

Release the hood locking catch.
Release the headlining trim from the 'D' post velcro fastening (1, Fig. 55).
Fold the hood to the rearward position.
Release the outer hood velcro fastening.
Remove the three outer hood securing screws from each side of the hood canopy and displace the hood for access.
Remove the screws securing the hood latches and remove the latches (2, Fig. 55).
Displace and remove the seal (3, Fig. 55).
Ensure the sealing surfaces are clean.
Fit the new seal ensuring great care is taken to avoid splitting the seal when pushing the seal into position at the corners of the roll bar and side panels.
Fit and secure the retaining clamps.
Fit new 'O' rings to the hood securing screws, secure the hood and refit the velcro fastening.
Secure and lock the hood to the roll bar and reposition the headlining to the 'D' post velcro fastening.

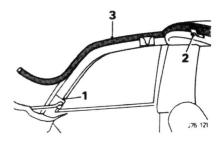

Fig. 55

TARGA TOP SEAL

Renew

Unlock and remove.
Remove screws securing the targa top latches and remove the latches.
Displace and remove the seal.
On refitting new seal ensure the seal seating surfaces are clean.
Place the seal into aperture with the red marker to the front central position. To avoid splitting the seal carefully push the seal fully into position.
Refit the latches but do not fully tighten the securing screws.
Fit and engage the latches to the targa tops, adjust the latches into the correct position then tighten the latch securing screws.

FRONT HEADLINING

Renew

Unlock and remove the targa tops.
Remove the screws securing the two 'A' post trims and remove the trims.
Remove the sun visors and the sun visor guides.
Remove the screws securing the interior rear view mirror and remove the mirror.
Displace the interior lamp, note and disconnect the electrical cables and remove the lamp.
Remove the two front targa top catches.
Displace the forward edge of the targa top seal.
Remove the front screws securing the cant rail trims and displace the front of the trims.
Remove the screws securing the headlining trim and remove the trim.
Refitting is a reversal of the above procedure.

ROLL BAR TRIM

Renew

Unlock and fold back the hood canopy.
Unlock and remove the targa tops.
Remove the screws securing the targa top catches and the hood canopy catches, remove the catches.
Remove the screws securing the 'B' post trim and the seat belt anchor bolt, remove the 'B' post trim.
Displace the canopy hood and the targa top seal from the roll bar.
Displace and remove the roll bar trim.
Refitting is the reversal of the above procedure.

CANT RAIL TRIM

Renew

Unlock and remove the targa tops.
Remove the 'A' post trim.
Remove the 'B' post trim.
Displace the targa top seal.
Remove the roll bar trim.
Remove the screws securing the cant rail trim and remove the trim.

CONTENTS

WARNING: EXTREME CARE SHOULD BE EXERCISED IN HANDLING THE REFRIGERANT. LIQUID REFRIGERANT AT ATMOSPHERIC PRESSURE BOILS AT −29°C (−20°F). SERIOUS DAMAGE OR BLINDNESS MAY OCCUR IF REFRIGERANT IS ALLOWED TO CONTACT THE EYES.
Goggles and gloves must be worn while working with Refrigerant.

FIRST AID: If refrigerant should contact the eyes or skin, splash the eyes or affected area with cold water for several minutes. Do not rub. As soon as possible thereafter, obtain treatment from a doctor or eye specialist.

SPECIAL TOOLS AND EQUIPMENT FOR SERVICING AIR CONDITIONING SYSTEM ON JAGUAR SERIES III

1 Pektron test unit
1 Charging station
1 Leak detector
1 Temperature test box
1 Compressor service tool kit
1 Setting jig for temperature differential control, 18G1363.
1 Voltmeter
1 Ohmmeter

TORQUE LEVELS FOR THE AIR CONDITIONING HOSE CONNECTIONS

Item	Nm	Kgf/m	lbf/ft
1. Compressor/Condenser (Compressor End)	40,67 to 47,45	4,15 to 4,84	30 to 35
2. Condenser/Compressor (Condenser End)	28,47 to 36,30	2,90 to 3,73	21 to 27
3. Condenser/Receiver Drier (Condenser End)	20,34 to 27,12	2,10 to 2,76	15 to 20
4. Receiver Drier/Condenser (Receiver Drier End)	40,67 to 47,45	4,15 to 4,84	30 to 35
5. Receiver Drier/Evaporator (Receiver Drier End)	40,67 to 47,45	4,15 to 4,84	30 to 35
6. Evaporator/Receiver Drier (Evaporator End)	14,91 to 17,62	1,52 to 1,80	11 to 13
7. Expansion Valve/Evaporator (Expansion Valve End)	20,34 to 27,12	2,10 to 2,76	15 to 20
8. Evaporator/Compressor (Evaporator End)	28,47 to 36,60	2,90 to 3,73	21 to 27
9. Compressor/Evaporator (Compressor End)	40,67 to 47,45	4,15 to 4,84	30 to 35

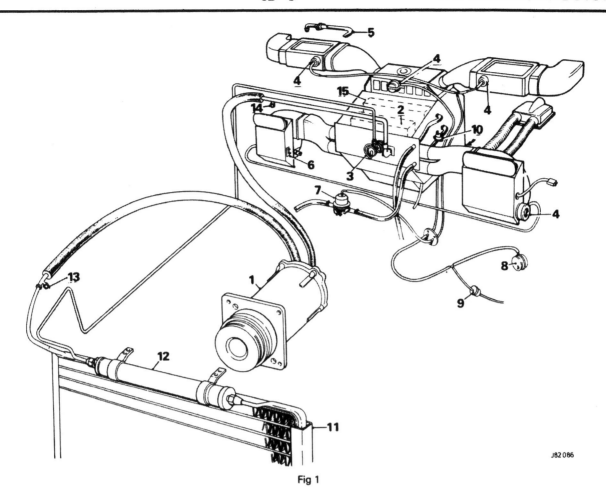

Fig 1

J82 086

KEY TO COMPONENTS (Fig. 1)

1. Compressor
2. Evaporator
3. Expansion valve
4. Vacuum valve
5. In-car sensor
6. Ambient temperature sensor
7. Water control valve
8. Vacuum reservoir
9. Non-return valve
10. Water valve temperature switch
11. Condenser
12. Receiver-drier
13. High pressure schrader valve
14. Low pressure schrader valve
15. Heater matrix

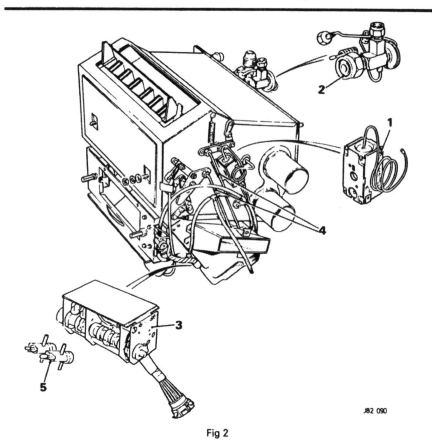

1. Ranco thermostat
2. Expansion valve
3. Servo control unit
4. Control rod
5. Vacuum valve

J82 090

Fig 2

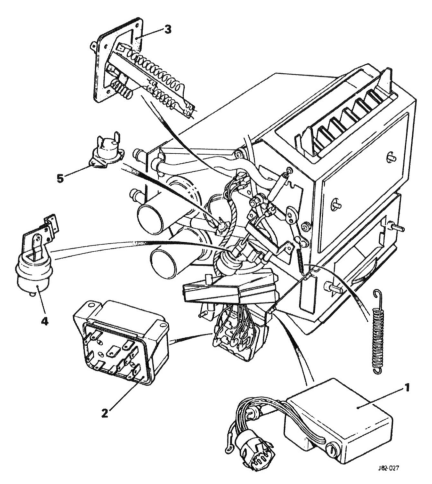

KEY TO COMPONENTS (Fig. 3)

1. Amplifier
2. Relays
3. Fan speed resistance
4. Vacuum actuator motor
5. Water thermostat

J82-027

Fig 3

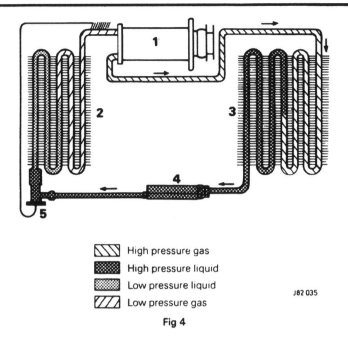

NNNN High pressure gas
■■■■ High pressure liquid
▓▓▓▓ Low pressure liquid
//// Low pressure gas

J82 035

Fig 4

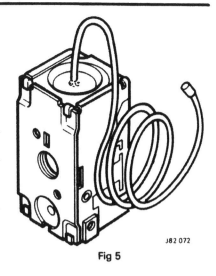

J82 072

Fig 5

The switch breaks the electric feed to the compressor magnetic clutch winding and the refrigeration cycle ceases. When the evaporator matrix temperature rises above 2°C (33.8°F) the thermostat switch closes and the refrigeration cycle re-starts.

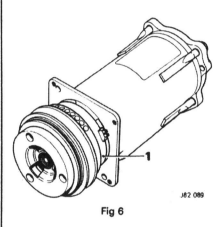

J82 089

Fig 6

REFRIGERATION CYCLE

Description   82.00.00

A belt-driven compressor (1, Fig. 4) draws in superheated refrigerant vapour at low pressure and compresses it.

The pressure forces the refrigerant round the refrigeration system.

The pressurized refrigerant is forced into a condenser (3, Fig. 4) located in front of the engine cooling radiator. The condenser is a matrix of tubes surrounded by fins. The refrigerant vapour travelling inside the tubes gives up its heat to the air-flow through the condenser. With the heat removed the vapour condenses to a cool liquid. The dimensions of the condenser determine that further heat transference occurs and the liquid becomes sub-cooled. Complete condensation has occurred.

The sub-cooled refrigerant still under pressure is forced into a receiver/drier (4, Fig. 4). The receiver/drier has several functions. It is a reservoir for the liquid; a filter to remove any particles which would contaminate the liquid; and it contains a quantity of molecular sieve desiccant to soak up any moisture in the liquid. Moisture would impair the efficiency of the refrigerant and cause damage at a later stage. The clean 'dry' liquid now passes into an expansion valve (5, Fig. 4) located at the inlet to the air conditioning unit. The liquid refrigerant is metered by the expansion valve so that the correct quantity is allowed to an evaporator matrix (2, Fig. 4) located in the air-conditioning unit. The metering orifice of the expansion valve is protected by a gauze filter located in the inlet union. The size of the metering orifice is controlled by the temperature sensed by a capillary at the evaporator outlet. If the temperature of the outlet pipe falls, the expansion valve closes to cut down the flow of refrigerant to the evaporator. As the temperature of the outlet

rises, a further quantity of refrigerant, metered by the expansion valve, enters the evaporator. The evaporator is a low pressure area so the refrigerant suddenly expands and the temperature drops. When the temperature falls below 0.6°C (33°F) it boils (i.e. vaporizes) and as any liquid requires a large amount of heat to change to vapour, the temperature of the evaporator matrix falls. Heat is taken from the air passing through the matrix on its way into the car.

Heat transfer continues until the vapour becomes low pressure super-heated vapour. The cycle recommences as the compressor draws in the super-heated low pressure vapour.

NOTE: Moisture from the cooled air passing over the fins of the evaporator condenses. The water is drained from the bottom of the evaporator by rubber tubes, and may form a pool of water under the vehicle when standing. This is normal and does not indicate a malfunction.

COMPONENT DESCRIPTION

RANCO THERMOSTAT

Ice formation on the evaporator fins due to moisture in the air is possible. Icing would damage the evaporator, so a thermostatic device to prevent this is fitted.

The Ranco thermostat (Fig. 5) is a temperature-operated switch which is normally closed in all functions and modes. It opens only when the temperature sensor capillary probe inserted in the evaporator matrix falls below 2°C (33.8°F).

It is important that the end of the capillary tube is inserted 10 cm and is in contact with the evaporator finning.

The magnetic clutch coil is mounted on the end of the compressor (1, Fig. 6) and the electrical connections are made to the coil terminals. The clutch permits the compressor to be engaged or disengaged as required for the air conditioning operation. When current passes through the clutch coil, the armature clutch plate assembly, keyed to the compressor shaft, is drawn rearwards against the belt driven pulley that is free wheeling upon the same shaft. This locks pulley and armature plate together to drive the compressor. When current ceases to flow, springs in the armature plate draw the clutch face from the pulley. The compressor comes to rest and the pulley continues to free wheel.

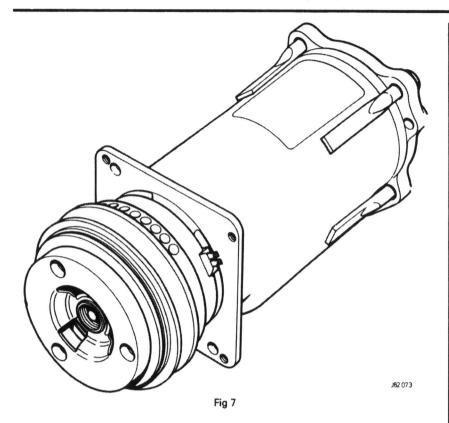

Fig 7

J82 073

COMPRESSOR

The compressor (Fig. 7) is a six-cylinder, reciprocating piston type of special design in which three sets of double acting pistons are actuated by a swash plate on the compressor shaft so that the pistons move back and forth in the cylinders as the shaft is rotated. There are in effect three independent cylinders at each end of the compressor and reed valves are provided for each cylinder at both ends of the compressor. Internal 'cross-over' passages for suction and discharge are provided within the compressor so that the high and low service fittings on the rear end of the compressor control refrigerant flow to and from all the cylinders. A gear type oil pump located in the rear head provides for compressor lubrication.

CONDENSER

The condenser (Fig. 8) consists of a refrigerant coil mounted in a series of thin cooling fins to provide a maximum of heat transfer in a minimum amount of space. It is usually mounted directly in front of the car radiator so that it receives the full flow of RAM AIR. Ram air is the air flow induced by the forward motion of the car and the suction of the cooling fan.

The condenser receives heat laden high pressure refrigerant vapour from the compressor.

The refrigerant enters the inlet at the top of the condenser as a high pressure very hot vapour and as this hot vapour passes down through the condenser coils, heat will follow its natural tendency and move from the hot refrigerant vapour into the cooler ram air as it flows across the condenser coils and fins.

When the refrigerant vapour reaches the temperature and pressure that will induce a change of state a large quantity of heat will be transferred to the outside air and the refrigerant will change from a high pressure HOT VAPOUR to a high pressure WARM LIQUID.

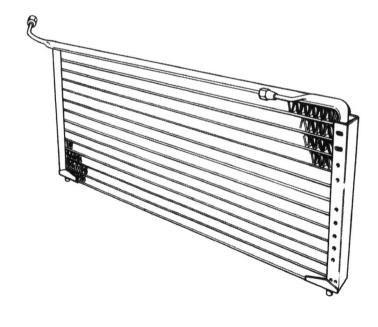

Fig 8　　　　　　J82 074

SIGHT GLASS

The sight glass located on the output side of the receiver-drier through which the refrigerant flows is used to indicate the condition of the refrigerant charge. A clear sight glass (Fig. 9) normally indicates the system has a correct charge of refrigerant. It may also indicate the system has a complete lack of refrigerant; this will be accompanied by a lack of any cooling action by the evaporator. Also the system may be overcharged; this must be verified with test gauge readings.

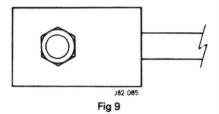

Fig 9

Foam or a constant stream of bubbles (Fig. 10) indicates the system does not contain sufficient refrigerant. Occasional bubbles when the system is first started is normal.

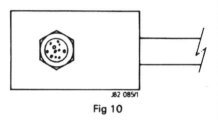

Fig 10

Foam or a heavy stream of bubbles (Fig. 11) indicates the refrigerant is very low.

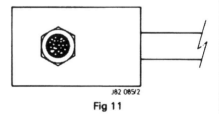

Fig 11

Oil streaks on the sight glass (Fig. 12) indicates a complete lack of refrigerant.

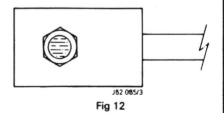

Fig 12

A cloudy sight glass (Fig. 13) indicates that the desiccant contained in the receiver-drier has broken down and is being circulated through the system.

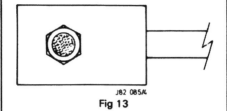

Fig 13

RECEIVER-DRIER

The receiver-drier (Fig. 14) is a storage tank which receives the high pressure warm refrigerant liquid from the condenser through an inlet line and delivers the refrigerant to the thermostatic expansion valve through the outlet line. The receiver-drier has two separate functions:

Acts as a storage tank for liquid refrigerant since the amount of refrigerant required by the evaporator varies widely under the different operating conditions.

Contains a filter and desiccant to remove and retain foreign particles and moisture from the refrigerant which would be harmful to the system if allowed to circulate with the refrigerant.

EVAPORATOR

The evaporator (Fig. 15) consists of a refrigerant coil mounted in a series of thin fins to provide a maximum amount of heat transfer in a minimum amount of space. It is usually mounted in a housing under the cowl where warm air from the passenger compartment is blown across the coils and fins.

The evaporator receives refrigerant from the thermostatic expansion valve as a low pressure cold atomized liquid. As this cold liquid refrigerant passes through the evaporator coils, heat will follow its natural tendency and move from the warm air into the cooler refrigerant.

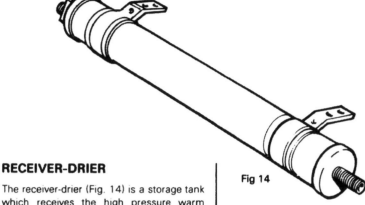

Fig 14

When the liquid refrigerant reaches a temperature and pressure that will induce a change of state, a large quantity of heat will move from the air into the refrigerant and the refrigerant will change from a low pressure COLD ATOMIZED LIQUID to a low pressure COLD VAPOUR.

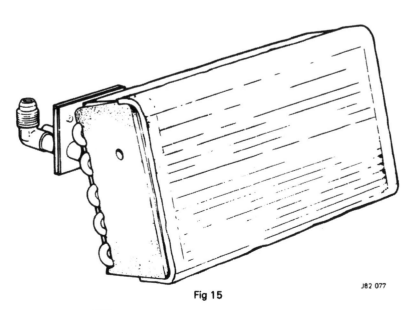

Fig 15

EXPANSION VALVE

The expansion valve (Fig. 16) is the dividing point between the high and low pressure sides of the system, and automatically meters the high pressure, high temperature liquid refrigerant through a small orifice, controlled by a metering valve, into the low presure, cold temperature side of the evaporator matrix. The refrigerant must be controlled to obtain the maximum cooling while assuring complete evaporation of the liquid refrigerant within the evaporator. To do this, the valve senses the outlet pipe temperature, the inlet pipe pressure, and increases or decreases the flow of refrigerant liquid to maintain the outlet temperature constant.

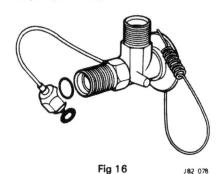

Fig 16   J82 078

The thermostatic expansion valve continually meters the exact amount of refrigerant required to supply some liquid refrigerant throughout the evaporator coil while ensuring that all of the refrigerant will be vapourized at the evaporator outlet. The refrigerant vapour then returns to the low (suction) side of the compressor.

AMPLIFIER

Automatic control is achieved by comparing car interior temperature and the temperature selected. This comparison provides an error signal to the air conditioning control unit, demanding an increase or decrease in car interior temperature. When the selected temperature is reached, the control unit will maintain it.

The error signal is detected across a Wheatstone Bridge circuit; two arms of which are fixed resistors, one arm contains the in-car thermistor and the fourth arm the temperature selection potentiometer. An error signal will be detected if car interior temperature is above or below that set on the temperature selection potentiometer. This signal is fed into the amplifier, (Fig. 17) amplified, and via relays, switches the servo motor to run clockwise or anti-clockwise. The position of the servo motor cam shaft directly determines the heating or cooling effect of the air conditioning system. Full heating and full cooling, are at opposite extremes of camshaft travel.

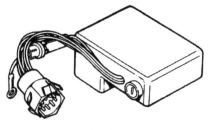

Fig 17   J82 079

The ambient thermistor in the Wheatstone Bridge circuit modifies the effect of the in-car thermistor. The result is a slightly colder interior temperature on hot days, and vice versa. A potentiometer driven by the servo motor is connected into the bridge circuit, modifying dynamic response. This provides control system damping, preventing excessive fluctuations in discharge air temperature.

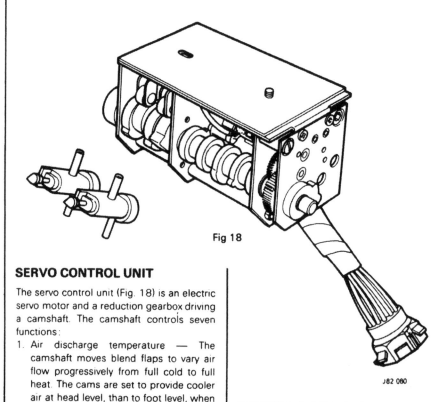

Fig 18

SERVO CONTROL UNIT

The servo control unit (Fig. 18) is an electric servo motor and a reduction gearbox driving a camshaft. The camshaft controls seven functions:

1. Air discharge temperature — The camshaft moves blend flaps to vary air flow progressively from full cold to full heat. The cams are set to provide cooler air at head level, than to foot level, when the unit is in the low-medium heating mode. This prevents stuffiness at head level.
2. Fan speeds — The camshaft alters fan speed progressively to increase air flow at full cold or full heat positions. Four fan speeds are available on cooling, three on heating. On low heating or cooling the camshaft selects a low fan speed, preventing noise and excessive air movement.
3. Mode — The camshaft controls a vacuum switch so that the distribution of air in the car is automatically controlled by a vacuum operated flap. Cold air is distributed from the face level vents, and hot air is distributed mainly from foot level vents with a bleed of air from screen vents.

4. Fresh/Recirculated Air — To improve performance the camshaft selects recirculated air on maximum cooling. Fresh air is selected for all other requirements.
5. Water Valve — On maximum cold, the camshaft controls a second vacuum switch to switch off the water valve controlling flow through the heater block.
6. Water Thermostat — A thermostat is fitted to prevent the system operating until engine water is hot enough to produce warm air. When on cooling mode the camshaft overrides this switch, and allows the system to operate immediately.
7. Evaporator thermostat — A thermostat is fitted to prevent icing of the evaporator. Under conditions where icing would be impossible and maximum cooling performance is required, the thermostat is overriden by the camshaft.

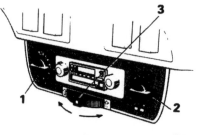

J82 080

MANUAL CONTROLS

Temperature Selector

The left hand control (1, Fig. 19) is the potentiometer to select the temperature from 18°C (65°F) to 29°C (85°F) that is to be maintained automatically in the car.

Fig 19   J82 093

MODE SWITCH

The right hand switch (2, Fig. 19) has five positions. When the switch is in the 'Off' position, the system is off and the fresh air intakes are closed. The 'Auto' position operates the system automatically. The high and low positions operate the fan high or low speed independently from that selected by the automatic control. The defrost position directs 90% of the air flow to the screen, closing the lower heater flap and opening the bleed flap to the screen outlets. At the same time, an additional resistor is switched into the Wheatstone Bridge circuit to ensure that the servo motor camshaft runs to full heat position.

Air Distribution Temperature Control

The thumbwheel (3, Fig. 19) can be used to alter the temperature of the air being distributed through the face level vents. It is most effective when the main controls are set at Auto and 75, and the system has been allowed to stabilise. To increase the temperature of the air being delivered through the vents, move the thumbwheel to the right; this will open the upper heater flap, allowing the increased air temperature to the face level vents. To decrease the air temperature, move the thumbwheel to the left; the upper flap will close and the air temperature to the vents will be lower.

METHOD OF TEMPERATURE VARIATION

Full Cooling

All air passes through the evaporator matrix in which the air is cooled·and dehumidified. After leaving the evaporator, four blend flaps control the degree of heat added by the heater matrix. On maximum cooling, the cooler flaps are fully open and the heater flaps are fully closed. Cold air only flows into the car (Fig. 20). A larger area of the cooling matrix is exposed to the upper flap than is exposed to the lower outlets, and most of the cooling output is directed out through the centre face level grille.

A vacuum switch on the camshaft is closed so that the water valve closes to prevent hot water flowing to the heater matrix.

The water temperature thermostat is overridden in the cooling mode. (Water temperature thermostat prevents the fans operating before water reaches 40°C.)

The evaporator thermostat (Ranco) is overridden by the camshaft cam in this mode as normally full cooling is only required when ambient conditions would prevent the evaporator from icing up, i.e. hot days.

Full Heating

When full heating is selected the camshaft moves to the full heat position. The camshaft mechanically operates the four flaps in the air conditioning unit so that the upper and lower cooling flaps are fully closed preventing cold air reaching the car interior direct. The heater flaps are fully

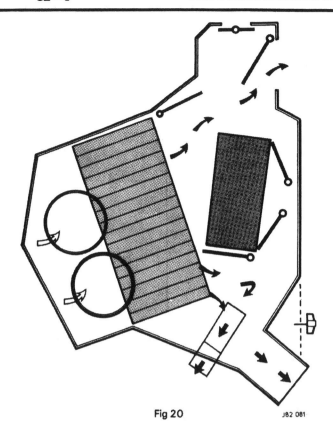

Fig 20   J82 081

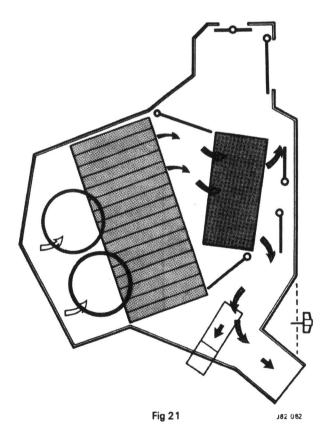

Fig 21   J82 082

open and the cool, de-humidified air blown through the evaporator now flows through the heater matrix and via the open heater flaps, to the screen rail end outlets and the front and rear footwells (Fig. 21).

The face level outlet is closed by the camshaft closing the vacuum switch so that

no hot air is delivered from the centre face level output. 90% of all air passing through the unit now passes out of the front and rear footwell outlets.

NOTE: Screen outlets are only open in defrost mode, although a slight air bleed is permitted for defrost purposes.

Air Blend

The system automatically maintains any temperature selected, irrespective of external ambient conditions by blending hot and cold air to maintain the temperature selected. Both heating and cooling flaps are progressively positioned so that the correct blend is obtained.

The illustration (Fig. 22) shows possible positions the flaps could adopt to give correct in-car temperature. It can be seen that both heating and cooling flaps are in operation.

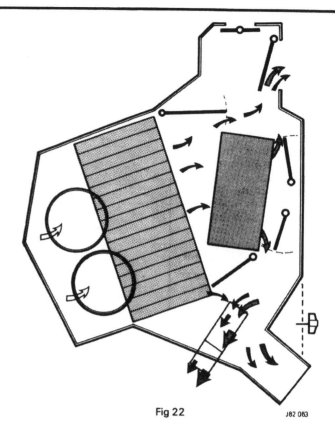

Fig 22   J82 063

Defrost

When defrost is selected the camshaft will travel to full heating. The vacuum switch on the right hand control will be closed to vacuum allowing the defrost flaps to open to pass air on to the windscreen. The left hand actuator will relax and allow the lower heating flap to close to direct 90% of all air through the unit to the windscreen (Fig. 23).

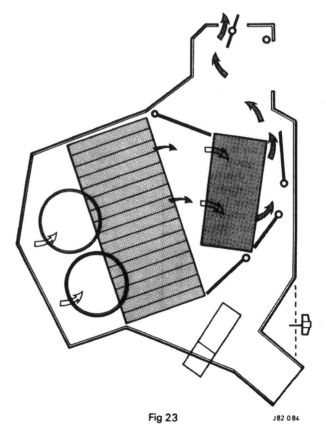

Fig 23   J82 084

GENERAL SECTION

This section contains safety precautions, general information, good practice and standards that must be followed when working upon the air conditioning system. A fault-finding and rectification section is included.

Safety precautions

The air conditioning equipment is manufactured for use only with Refrigerant 12 (dichlorodifluoromethane) and extreme care must be taken NEVER to use a methyl-chloride refrigerant.

The chemical reaction between methyl-chloride and the aluminium parts of the compressor will result in the formation of products which burn spontaneously on exposure to air, or decompose with violence in the presence of moisture. The suitable refrigerant is supplied under the following trade names.

  Freon 12; Arcton 12; Isceon 12 or any refrigerant to specification 12.

Goggles and gloves must be worn while working with the refrigerant.

WARNING: EXTREME CARE SHOULD BE EXERCISED IN HANDLING THE REFRIGERANT. LIQUID REFRIGERANT AT ATMOSPHERIC PRESSURE BOILS AT −29°C (−20°F). SERIOUS DAMAGE OR BLINDNESS MAY OCCUR IF REFRIGERANT IS ALLOWED TO CONTACT THE EYES.

FIRST AID: If refrigerant should contact the eyes or skin, splash the eyes or affected area with cold water for several minutes. Do not rub. As soon as possible thereafter, obtain treatment from a doctor or eye specialist.

Good practice

1. The protective sealing plugs must remain in position on all replacement components and hoses until immediately before assembly.

2. Any part arriving for assembly without sealing plugs in position must be returned to the supplier as defective.

3. It is essential that a second backing spanner is always used when tightening all joints. This minimises distortion and strain on components or connecting pipes.

4. Components must not be lifted by connecting pipes, hoses or capillary tubes.

5. Care must be taken not to damage fins on condenser or evaporator matrices. Any damage must be rectified by the use of fin combs.

6. Before assembly of tube and hose joints, use a small amount of clean new refrigerant oil on the sealing seat.

7. Refrigerant oil for any purpose must be kept very clean and capped at all times. This will prevent the oil absorbing moisture.

8. Before assembly the condition of joints and flares must be examined. Dirt and even minor damage can cause leaks at the high pressure encountered in the system.

9. Dirty end fittings can only be cleaned using a cloth wetted with alcohol.

10. After removing sealing plugs and immediately before assembly, visually check the bore of pipes and components. Where ANY dirt or moisture is discovered, the part must be rejected.

11. All components must be allowed to reach room temperature before sealing plugs are removed. This prevents condensation should the component be cold initially.

12. Before finally tightening the hose connections ensure that the hose lies in the correct position, is not kinked or twisted, and will not be trapped by subsequent operations, e.g. closing bonnet, refitting bonnet.

13. Check that the hose is correctly fitted in clips or strapped to the sub-frame members.

14. The Frigidaire compressor must be stored horizontally and sump down. It must not be rotated before fitting and charging. Do not remove the shipping plate until immediately before assembly. Always use new 'O' ring seals beneath union housing plate, and in those pipe joints which incorporate them.

15. Components or hoses removed must be sealed immediately after removal.

16. After a system has been opened the receiver/drier must be renewed.

Before commencing checks, run the engine until normal running temperature is reached. This ensures that sufficient vacuum is available for tests. For cooling tests the engine must be running for the compressor clutch to operate.

SPECIAL TOOLS AND EQUIPMENT FOR SERVICING AIR CONDITIONING SYSTEM ON JAGUAR SERIES III

1 Pektron test unit
1 Charging station
1 Leak detector
1 Temperature test box
1 Compressor service tool kit
1 Setting jig for temperature differential control, 18G1363.
1 Voltmeter
1 Ohmmeter

The Pektron Climatic Control tester (Fig. 24) is recommended for testing the air conditioning electrical system.

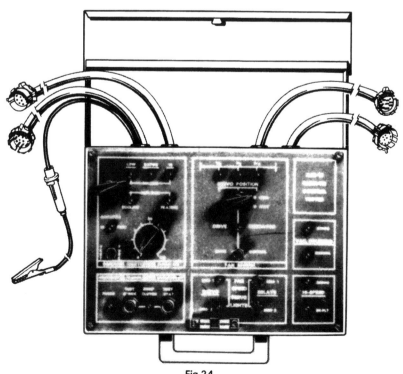

Fig 24

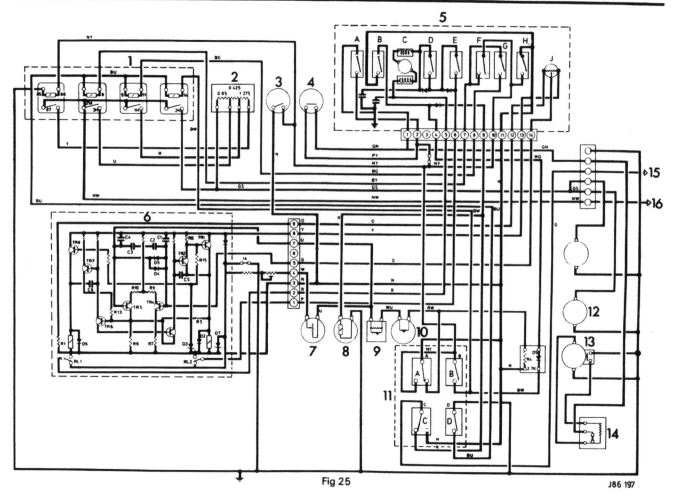

Fig 25

J86 197

KEY TO DIAGRAM

1. Blower motor relay
2. Blower motor resistor
3. Water temperature transmitter
4. Thermostat
5. Servo
6. Amplifier
7. Temperature selector
8. Vacuum valve
9. Ambient sensor
10. In-car sensor
11. Mode control switch
12. Blower motors
13. Compressor clutch
14. Thermal fuse
15. To fuse
16. To fuse

KEY TO SERVO UNIT (5, Fig. 25)

A Thermo override
B Recirc & Highspeed
C Servo motor
D Cool limit switch
E Heat limit switch
F Med 1 switch
G Med 2 switch
H Temperature bypass switch
J Feedback potentiometer

Mode switch functions (11, Fig. 25)

Micro switch		Off	Lo	Auto	Hi	Def
A	Defrost	NC	NC	NC	NC	NO
B	High Speed	NO	NC	NO	NC	NC
C	On/Off	NO	NC	NC	NC	NC
D	Low Speed	NO	NO	NC	NC	NC

NC = Normally Closed
NO = Normally Open

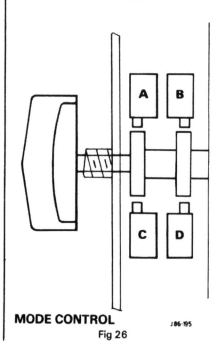

MODE CONTROL

Fig 26

J86-195

KEY TO MODE CONTROL

A Defrost micro switch
B High speed micro switch
C Low speed micro switch
D On/off micro switch

AIR CONDITIONING SERVO UNIT

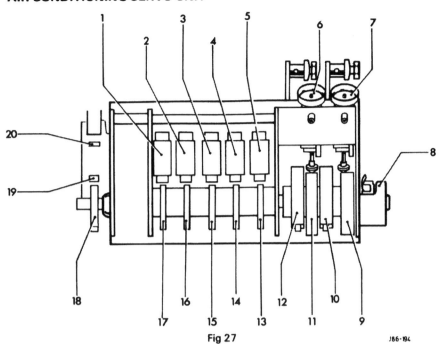

Fig 27

J86-194

KEY TO SERVO UNIT

1. Full cooling micro switch
2. Full heating micro switch
3. Med 1 micro switch
4. Med 2 micro switch
5. Temperature override micro switch
6. Water valve vacuum switch
7. Centre flap vacuum switch
8. Feedback potentiometer
9. Cam S9
10. Cam S8
11. Cam S7
12. Cam S6
13. Cam S5
14. Cam S4
15. Cam S3
16. Cam S2
17. Cam S1
18. Cam S10
19. Ranco override micro switch
20. Recirculate Hi Speed switch

Test Procedure

Allow coolant temperature to stabilize to ambient by not running the engine for at least two hours, and open all of the vehicle windows for this period.

Set the car mode control to off, and the temperature selector to approximately the ambient temperature in the vehicle. Ensure that the ignition is switched off, then disconnect the plug at the servo control unit. Insert the tester 15 way socket into the servo input and join the harness socket to the tester 15 way plug.

Disconnect the plug and socket at the amplifier. Insert the tester 12 way socket into the amplifier input, and join the harness to the tester 12 way plug. Connect the tester earth lead to a good earth on the vehicle. Carry out the following operations and note the effects.

ACTION	EFFECT	WHEN INCORRECT
A Switch on Car Ignition	'VAC. SOLENOID MANUAL' will illuminate. No other lights 'ON'	Check Ignition Supply, Fuse, etc. Ensure that A/C Mode Control is 'OFF'. Check Wiring to Vacuum Solenoid
B Start the Car Engine and switch car mode Control to LOW. Then switch SERVO POSITION to HEAT and press the DRIVE CONTROL until the DRIVE INDICATOR goes out (Note. It only comes ON when DRIVE CONTROL is Pressed).	1. The Fan Speed $\frac{1}{4}$, $\frac{1}{2}$ and $\frac{3}{4}$ lights should be ON for Servo Position.	If the DRIVE INDICATOR did not light, switch SERVO POSITION to COOL and press the DRIVE CONTROL. When this also does not light the DRIVE INDICATOR check the SERVO motor and servo components and wiring
	2. The vehicle cooling fans should be OFF	Ensure that the engine is not yet warmed up, then disconnect the water temperature switch in the car, and if fans continue to run suspect wiring and the Micro Switch Water Temp. override in the Servo Control Unit (5, Fig. 27).
	3. The VAC. SOLENOID MANUAL lights should go out	Check Switch C in mode control unit (Fig. 26) and wiring

ACTION	EFFECT	WHEN INCORRECT
C. Run the car engine fast to warm up the cooling water to working temperature.	1. The fans will start to run at low rate and the SERVO MED 1 light will operate (FAN SPEED SERVO LIGHT). The majority of air will be directed through the floor vents	Check water temperature switch. Check low speed relay Check blower resistors R1 & 2 Check fans. Check wiring
	2. The compressor clutch will operate as indicated by RANCO COMP. CLUTCH light.	(A) Check 10A fuse, which if faulty will be indicated by the RANCO FUSED light being ON. (B) Check Ranco thermostat by shorting out at the component terminals and monitoring the COMP clutch light
D. Switch the SERVO POSITION to COOL and press DRIVE CONTROL until '$\frac{3}{4}$' light goes out then release.	MED 1 FAN SPEED SERVO light goes out.	Check MED 1 micro switch (3, Fig. 27) and wiring to SERVO Control Unit.
E. Press DRIVE CONTROL until '$\frac{1}{2}$' light goes out then release.	The air emission is evenly distributed between face level and floor vents.	Check adjustment of blend flaps, or vacuum system.
F. Press DRIVE CONTROL until '$\frac{1}{4}$' light goes out then release.	MED 1 FAN SPEED SERVO light is 'ON'.	Check MED 1 micro switch (3, Fig. 27) and wiring to SERVO Control Unit.
G. Press DRIVE CONTROL until DRIVE INDICATOR goes out, then release.	HIGH SPEED SERVO light ON.	Check HI-SPEED/RECIRC micro switch and wiring to SERVO Control Unit.
	COMP. CLUTCH light stays on when TEST O'RIDE switch is pressed	Check ranco override micro switch (19, Fig. 27) and wiring to SERVO Control Unit.
	MED 2 FAN SPEED SERVO light is 'ON'	Check MED 2 Micro Switch (4, Fig. 27) and wiring to Servo Control Unit.
	VAC.-SOLENOID SERVO light 'ON'	Check Diode D3 in Servo control Unit harness
H. Switch car mode control to AUTO. Drive servo to '$\frac{1}{1}$' position by selecting DRIVE CONTROL unit '$\frac{1}{4}$' and '$\frac{1}{2}$' lights are 'ON'.	This has given the cooling compressor the protection of its freezing sensing thermostat.	— — — — — — — — —
J. Press MED 1 switch (FAN SPEED SERVO lights).	Car Fan Speed increases.	Check Main Relay Check Resistor R3. Check Wiring.
K. Keeping MED 1 pressed, operate MED 2 (FAN SPEED SERVO light).	Car Fan Speed increases further.	Check Main Relay Check Resistor R2. Check Wiring
L. Release MED 1 and MED 2 switches. Select HI on car mode control.	Car fan speed increases to maximum.	Check High Speed Micro Switch at mode control Check Main Relay Check Wiring
	Ensure D5-FLT does not light.	Check Diode D5 in mode control harness Check Wiring
M. Select AUTO on car mode control.	To reduce fan speed to low rate.	— — — — — — — — —

ACTION	EFFECT	WHEN INCORRECT
N. Press TEST on AMPLIFIER fuse.	AMPLIFIER Fuse light should be 'ON'. Feed to amplifier good.	Check feed to amplifier.
P. Switch to AMPLIFIER on sensing Switching System (Tester) Rotate control 0-100 fully clockwise then fully anti-clockwise alternately.	Towards the '100' point HEATING light should come ON, then towards the 'O' point COOLING light should come ON.	Check Wiring. Replace Amplifier.
O. Ensure that the car temperature setting control is at approximately the ambient temperature. Switch to SENSOR on Sensing Switching System (Tester). Monitor LOW, DATUM, and HI lights and adjust rotary control 0-100 until only DATUM is illuminated.	If it is not possible to 'balance' the DATUM light, then the ambient temperature may be incorrectly set on the vehicle temperature selector, or out of its range, or there is a fault in the sensors, wiring or Temperature Control. If OK proceed to Item R.	Check sensors and wiring. Check Micro Switch overriding sensing circuit in the mode control, and wiring. Check temperature SELECTOR and wiring.
R. Increase or decrease the car TEMPERATURE SELECTOR by 5°F from its set point, whichever is convenient.	If increased the HI light will come ON in addition to the DATUM.	Check TEMPERATURE SELECTOR and wiring.
S. Adjust the rotary 0-100 control to cancel the HI or LOW light obtained in Item R. Return the TEMPERATURE SELECTOR to its original point.	If decreased to its original point the LOW light will come ON in addition to the DATUM. If increased to its original point the HI light will come ON in addition to the DATUM.	Check TEMPERATURE SELECTOR and wiring.
T. Adjust the rotary 0-100 control to cancel the HI or LOW light obtained in Item S. Switch the mode control from AUTO to DEF.	The HI light operates in addition to the DATUM.	Check Micro Switch override sensing circuit at mode control and resistance unit in mode control harness.
U. Select the OFF on the Mode control. Switch off the vehicle engine and ignition circuit.	All tester indicators are OFF.	Check vehicle ignition switch. Check relays and wiring.
V. Remove tester connectors and return the vehicle wiring and plugs and sockets to standard.	The complete system can now be tested following any corrective action taken as a result of the checks.	Identify the problem area, and after carrying out the preliminary procedure of tester connection, repeat only the relevant parts of the schedule.

Familiarity with the tester should be easily acquired, and then the Operator will find the flexibility of control offered by having access to test each sub-assembly will lead to quick identification of faults, and a system knowledge which allows him to extend this scope of the scheduled checks.

'IN-CAR' FAULT FINDING CHART

Equipment required

1. Voltmeter capable of covering 0 to 13 volts d.c.
2. Continuity tester.
3. Ohmmeter capable of covering 0 to 20K ohms.
4. Vacuum gauges (not essential) to check vacuum level.

The battery should be disconnected whenever an electrical unit is being removed or refitted.

TEST 1   **R.H. OFF**   **L.H. 75°**

1. A

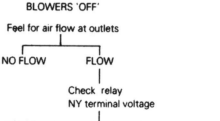

1. B

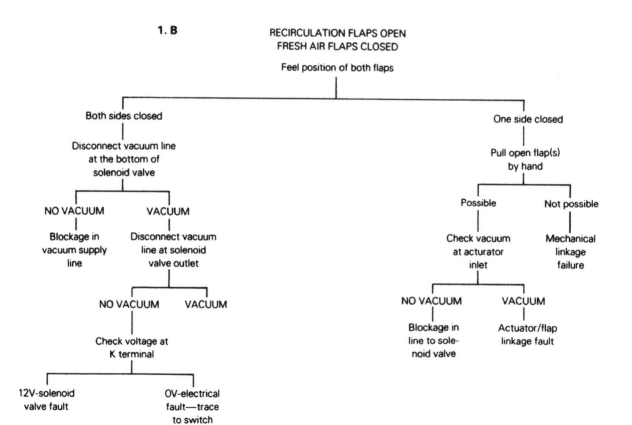

TEST 2   **R.H. 'DEFROST'**   **L.H. 75°**

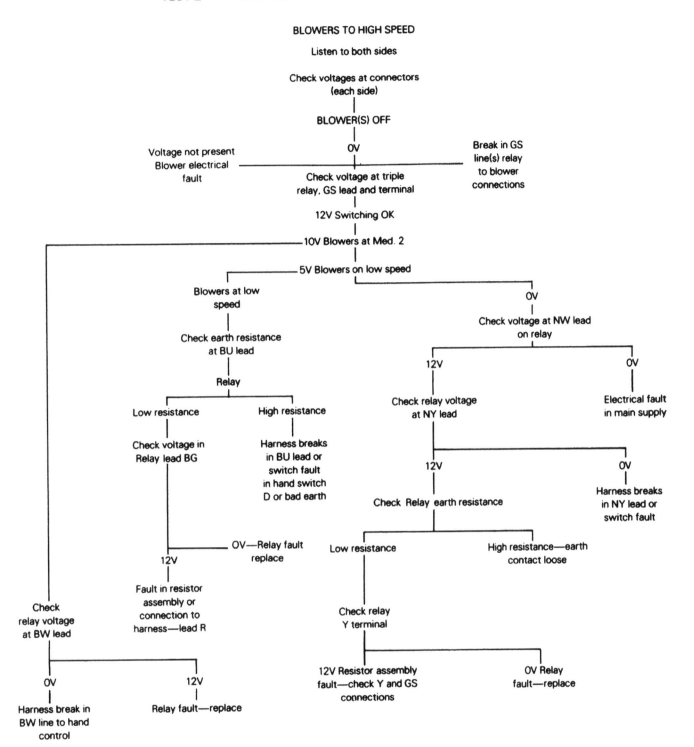

2B   HOT AIR TO SCREEN

Feel outlets both sides
(Close side face-level vents)

| COLD AIR FOOT-LEVEL | COLD AIR FOOT-LEVEL | NO AIR TO SCREEN |
| OR COLD AIR SCREEN | COLD AIR SCREEN | (AIR TO FOOT-LEVEL ONE/BOTH SIDES) |

Mechanical linkage/
flap fault
see Test 6

Servo has not traversed—
see Test 5
Or water valve failure—
see Test 6

Check vacuum at
'T' piece under crash
padding (green line)

NO VACUUM   VACUUM

Actuator/mechanical   Vacuum switch C fault
flap fault   or hand control

R.H. AUTO—HI   L.H. HEATING MODE (HIGHER THAN AMBIENT)

TEST 3

3A   BLOWER SPEED HIGH

Listen to blowers

BLOWERS OFF

Water temperature switch faulty or harness break in
N and NY lines

3B   SCREEN FLAPS CLOSE
AIR TO FOOT-LEVEL

Feel for air flow
at screen outlets

HIGH FLOW AT SCREEN
OR COLD AIR AT FOOT-LEVEL

Check vacuum at 'T' piece (green)
below crash padding

NO VACUUM   VACUUM

Faulty vacuum switch C   Actuator fault or flap
or line blockage   mechanical failure

TEST 4   **R.H. AUTO—LOW**   **L.H. 85°**

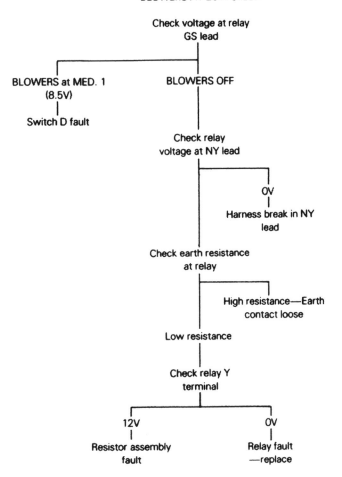

BLOWERS AT LOW SPEED

Check voltage at relay
GS lead

BLOWERS at MED. 1
(8.5V)

Switch D fault

BLOWERS OFF

Check relay
voltage at NY lead

0V

Harness break in NY
lead

Check earth resistance
at relay

High resistance—Earth
contact loose

Low resistance

Check relay Y
terminal

12V

Resistor assembly
fault

0V

Relay fault
—replace

TEST 5   AMPLIFIER/SERVO RESPONSE

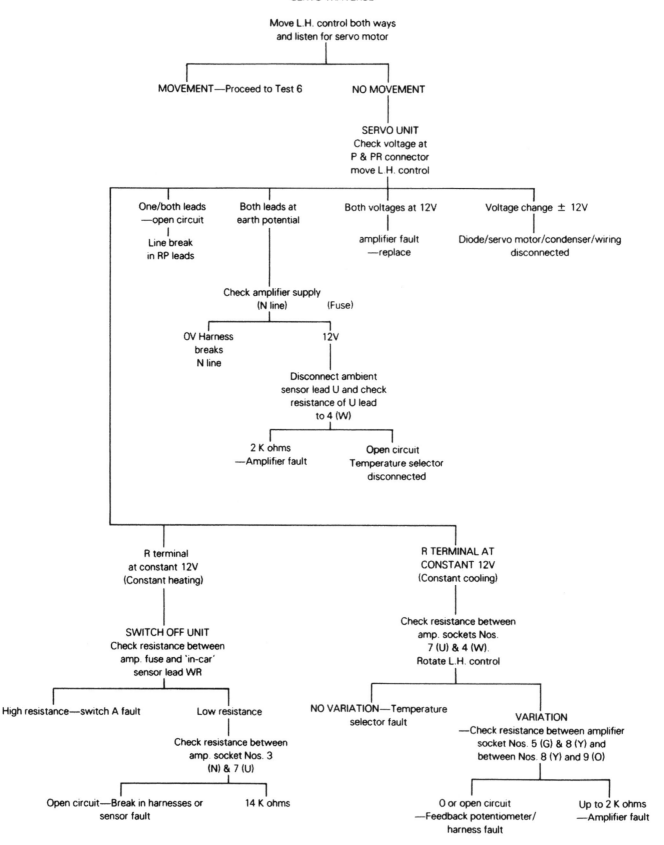

SERVO TRAVERSE

Move L.H. control both ways
and listen for servo motor

MOVEMENT—Proceed to Test 6   NO MOVEMENT

SERVO UNIT
Check voltage at
P & PR connector
move L.H. control

One/both leads   Both leads at   Both voltages at 12V   Voltage change ± 12V
—open circuit   earth potential
                                              amplifier fault   Diode/servo motor/condenser/wiring
Line break   —replace   disconnected
in RP leads

Check amplifier supply
(N line)   (Fuse)

0V Harness   12V
breaks
N line   Disconnect ambient
                    sensor lead U and check
                    resistance of U lead
                    to 4 (W)

                    2 K ohms   Open circuit
                    —Amplifier fault   Temperature selector
                                           disconnected

R terminal   R TERMINAL AT
at constant 12V   CONSTANT 12V
(Constant heating)   (Constant cooling)

                                          Check resistance between
                                          amp. sockets Nos.
SWITCH OFF UNIT   7 (U) & 4 (W).
Check resistance between   Rotate L.H. control
amp. fuse and 'in-car'
sensor lead WR

High resistance—switch A fault   Low resistance   NO VARIATION—Temperature   VARIATION
                                                       selector fault   —Check resistance between amplifier
                                   Check resistance between   socket Nos. 5 (G) & 8 (Y) and
                                   amp. socket Nos. 3   between Nos. 8 (Y) and 9 (O)
                                   (N) & 7 (U)

Open circuit—Break in harnesses or   14 K ohms   0 or open circuit   Up to 2 K ohms
sensor fault   —Feedback potentiometer/   —Amplifier fault
                                                       harness fault

6.1   AUTOMATIC FUNCTIONS

LIMITED COOLING/HEATING

SERVO WILL NOT TRAVERSE

Could be damaged—check Test 5

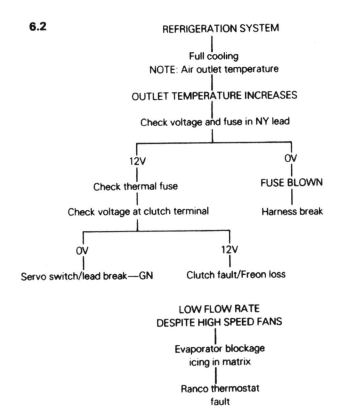

6.2

REFRIGERATION SYSTEM

Full cooling
NOTE: Air outlet temperature

OUTLET TEMPERATURE INCREASES

Check voltage and fuse in NY lead

12V — Check thermal fuse — Check voltage at clutch terminal

0V — Servo switch/lead break—GN

12V — Clutch fault/Freon loss

0V — FUSE BLOWN — Harness break

LOW FLOW RATE
DESPITE HIGH SPEED FANS

Evaporator blockage
icing in matrix

Ranco thermostat
fault

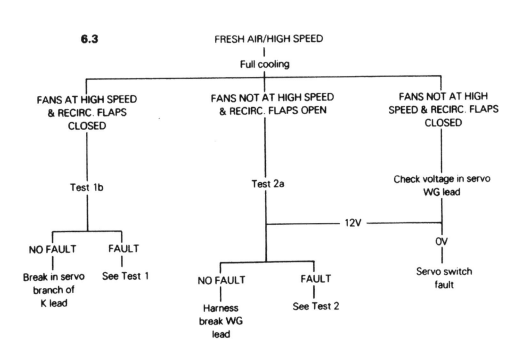

6.3   FRESH AIR/HIGH SPEED

Full cooling

FANS AT HIGH SPEED & RECIRC. FLAPS CLOSED — Test 1b — NO FAULT (Break in servo branch of K lead) / FAULT (See Test 1)

FANS NOT AT HIGH SPEED & RECIRC. FLAPS OPEN — Test 2a — NO FAULT (Harness break WG lead) / FAULT (See Test 2)

FANS NOT AT HIGH SPEED & RECIRC. FLAPS CLOSED — Check voltage in servo WG lead — 12V / 0V (Servo switch fault)

6.4

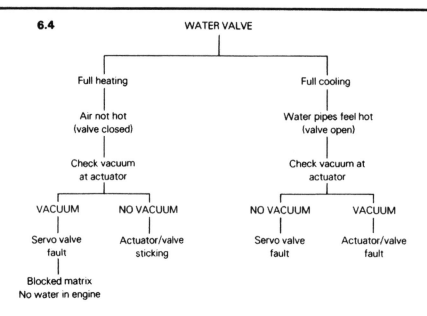

WATER VALVE

Full heating

Air not hot
(valve closed)

Check vacuum
at actuator

VACUUM — Servo valve fault — Blocked matrix / No water in engine

NO VACUUM — Actuator/valve sticking

Full cooling

Water pipes feel hot
(valve open)

Check vacuum at actuator

NO VACUUM — Servo valve fault

VACUUM — Actuator/valve fault

6.5

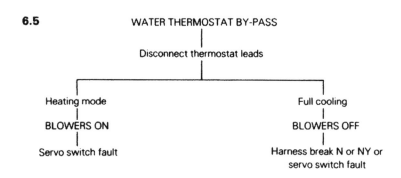

WATER THERMOSTAT BY-PASS

Disconnect thermostat leads

Heating mode
BLOWERS ON
Servo switch fault

Full cooling
BLOWERS OFF
Harness break N or NY or servo switch fault

6.6

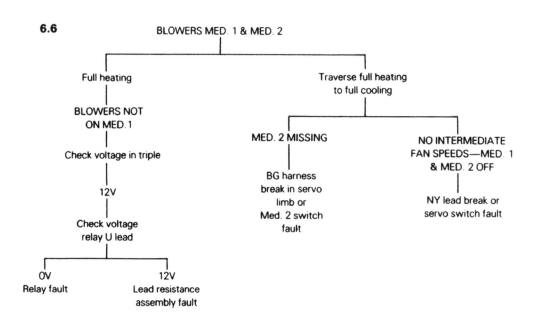

BLOWERS MED. 1 & MED. 2

Full heating
BLOWERS NOT ON MED. 1
Check voltage in triple
12V
Check voltage relay U lead
0V — Relay fault
12V — Lead resistance assembly fault

Traverse full heating to full cooling

MED. 2 MISSING — BG harness break in servo limb or Med. 2 switch fault

NO INTERMEDIATE FAN SPEEDS—MED. 1 & MED. 2 OFF — NY lead break or servo switch fault

6.7

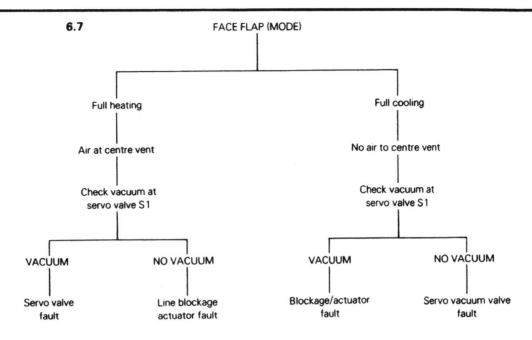

FACE FLAP (MODE)

Full heating — Air at centre vent — Check vacuum at servo valve S 1

- VACUUM — Servo valve fault
- NO VACUUM — Line blockage actuator fault

Full cooling — No air to centre vent — Check vacuum at servo valve S 1

- VACUUM — Blockage/actuator fault
- NO VACUUM — Servo vacuum valve fault

6.8

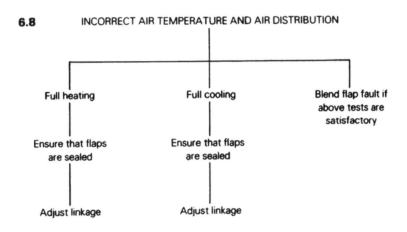

INCORRECT AIR TEMPERATURE AND AIR DISTRIBUTION

Full heating — Ensure that flaps are sealed — Adjust linkage

Full cooling — Ensure that flaps are sealed — Adjust linkage

Blend flap fault if above tests are satisfactory

CHARGING AND TESTING EQUIPMENT

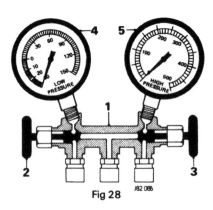

Fig 28

J82 088

The charging and testing equipment consists of a charging manifold (1, Fig. 28) fitted with two stop valves (2 & 3, Fig. 28). One compound gauge (4, Fig. 28) reading both vacuum and pressure, and it is connected to the suction side of the system. The other gauge is a high pressure gauge (5, Fig. 28) and is connected to the delivery side of the system.

WARNING: FOR SAFETY REASONS, THE ACCURACY OF BOTH GAUGES MUST BE CHECKED AT FREQUENT INTERVALS.

Gauge Manifold

The manifold is designed to control refrigerant flow. As shown in the following illustration, when the manifold test set is connected into the system, pressure is registered on both gauges at all times. During all tests, both the low and high side hand valves are in the closed position (turned inward until the valve is seated). Refrigerant will flow around the valve stem to the respective gauges and register the system low side pressure on the low side gauge, and the system high side pressure on the high side gauge. The hand valves isolate the low and high side from the central portion of the manifold.

Low Side Gauge

This gauge (4, Fig. 28) has a dial reading from 0 to 150 psi (pressure scale) in a clockwise direction, and from 0 to 30 inches of Mercury (vacuum scale) in a counter-clockwise direction. This low side gauge is called a Compound Gauge and has a dual purpose, to register both Pressure and Vacuum. This gauge is used to measure evaporator outlet pressure.

High Side Gauge

This gauge (5, Fig. 28) has a dial reading from 0 to 500 psi in a clockwise direction. The high side gauge is a Pressure gauge only.

A test hose connected to the fitting directly under the low side gauge is used to connect the low side of the test manifold into the low side of the system, and a similar connection is found on the high side.

Two hose connectors must be fitted with depressors to operate the schrader valves on the high and low pressure sides of the system.

CAUTION: Do not open the high side hand valve while the air conditioning system is in operation. Under no circumstances should this be done. If the high side hand valve should be opened while the system is operating, high pressure refrigerant will be forced through the high side gauge and to the refrigerant can if it is attached. This high pressure can rupture the can or possibly burst the fitting at the safety can valve, resulting in much damage (including physical injury).

With the engine switched off, remove the protective caps from the schrader valves. Ensure both high and low side hand valves are in the closed position. Connect the high pressure gauge hose to the high pressure schrader valve (1, Fig. 29) and connect the low pressure or compound valve hose to the low pressure schrader valve (2, Fig. 29).

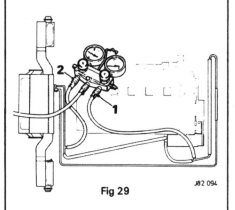

Fig 29   J82 094

PURGING TEST HOSES

Using System Refrigerant

Be sure high and low side hoses are properly connected to service valves (all hose connections tight).

Now, purge the high side test hose by opening the hand valve on the high side gauge for 3 - 5 seconds. This allows the system's refrigerant to force air through the test hoses and out of the centre service hose. Immediately close the high side gauge hand valve.

Purge the low side test hose in the same manner, using the hand valve of the low side gauge. Close hand valve after 3 - 5 seconds.

Stabilizing the System

The manifold gauge set is attached to the system, and the test hoses purged of air. You must now operate the system for a few minutes to stabilize all pressures and temperatures throughout the system in order to obtain accurate test gauge readings. Stabilize the system as follows:

Place all test hoses, gauge set and other equipment away from all engine moving parts. Also keep hoses from touching hot manifolds. Start the engine and adjust engine speed to fast idle.

Turn air conditioner controls to maximum cooling. Set blower fan on high speed.

Open doors and/or windows (to quickly eliminate interior heat).

Operate system under these conditions for 5 to 10 minutes and the system will be stabilized and ready for test readings.

Test Conditions

1. Use a large fan to substitute for normal ram air flow through the condenser.
2. Car adjusted to normal fast idle speed.
3. All conditions equivalent to 30 mph

TORQUE LEVELS FOR THE AIR CONDITIONING HOSE CONNECTIONS

Item	Nm	Kgf/m	lbf/ft
1. Compressor/Condenser (Compressor End)	40,67 to 47,45	4,15 to 4,84	30 to 35
2. Condenser/Compressor (Condenser End)	28,47 to 36,30	2,90 to 3,73	21 to 27
3. Condenser/Receiver Drier (Condenser End)	20,34 to 27,12	2,10 to 2,76	15 to 20
4. Receiver Drier/Condenser (Receiver Drier End)	40,67 to 47,45	4,15 to 4,84	30 to 35
5. Receiver Drier/Evaporator (Receiver Drier End)	40,67 to 47,45	4,15 to 4,84	30 to 35
6. Evaporator/Receiver Drier (Evaporator End)	14,91 to 17,62	1,52 to 1,80	11 to 13
7. Expansion Valve/Evaporator (Expansion Valve End)	20,34 to 27,12	2,10 to 2,76	15 to 20
8. Evaporator/Compressor (Evaporator End)	28,47 to 36,60	2,90 to 3,73	21 to 27
9. Compressor/Evaporator (Compressor End)	40,67 to 47,45	4,15 to 4,84	30 to 35

PRESSURE — TEMPERATURE RELATIONSHIP

NOTE: Pressures shown are under exact conditions (see Test Conditions below) and are not necessarily true for every car checked.

Ambient Temperature is given as the temperature of the air surrounding the condenser and is taken 2 inches in front of the condenser.

Ambient Temperature °F	High Pressure Gauge Reading
60	95-115
65	105-125
70	115-135
75	130-150
80	150-170
85	165-185
90	175-195
95	185-205
100	210-230
105	230-250
110	250-270
115	265-285
120	280-310

Low Pressure Gauge Reading	Evaporator Temperature °F
10	2
12	6
14	10
16	14
18	18
20	20
22	22
24	24
26	27
28	29
30	32
35	36
40	42
45	48
50	53
55	58
60	62
65	66
70	70

Normal operating ranges shown by dotted line boxes.

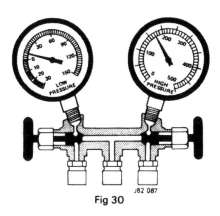

Fig 30

Complaint

Little or no cooling.

Condition

1. Low side gauge reading too low. Should be 15 - 30 psi.
2. High side gauge reading too low. Should be 185 - 205 psi at ambient temperature of 95°F.
3. Stream of bubbles evident in sight glass.
4. Discharge air from evaporator only slightly cool.

Diagnosis

System low on refrigerant. May be caused by small leak.

Correction

1. Leak test system.
2. Discharge refrigerant from system if necessary to replace units or lines.

3. Repair leaks.
4. Check compressor oil level. System may have lost oil due to leakage.
5. Evacuate system using vacuum pump.
6. Charge system with NEW Refrigerant.
7. Operate system and check performance.

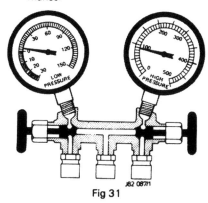

Fig 31

Complaint

Cooling is not adequate.

Condition

1. Low side gauge reading is very low. Should be 15 - 30 psi.
2. High side gauge reading very low. Should be 185 - 205 psi at ambient temperature of 95°F.
3. No liquid and no bubbles evident in sight glass.
4. Discharge air from evaporator is warm.

Diagnosis

System excessively low of refrigerant. Serious leak indicated.

Correction

1. Leak test system.

NOTE: Add partial refrigerant charge before leak testing to ensure a leak test indication. Leak test compressor seal area very carefully.
2. Discharge refrigerant from system.
3. Repair leaks.
4. Check compressor oil level. System may have lost oil due to leakage.
5. Evacuate system using vacuum pump.
6. Charge system with NEW Refrigerant.
7. Operate system and check performance.

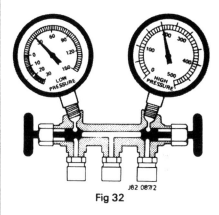

Fig 32

Complaint

Cooling is not adequate.

Condition

1. Low side gauge reading is constant and does not drop. Pressure should drop until compressor cycles (thermostat control).
2. High side gauge reading slightly high (or slightly lower especially if large fan used to substitute ram air). High side guage reading should be 185 - 205 psi at ambient temperature of 95°F.
3. Sight glass free of bubbles or only shows occasional bubble.
4. Discharge air from evaporator only slightly cool.

Diagnosis

Non condensables present in system. Air or moisture present instead of full refrigerant charge.

Correction

1. Leak test system.
Leak test compressor seal area very carefully.
2. Discharge refrigerant from system.
3. Repair leaks as located.
4. Replace receiver-drier. Drier probably saturated with moisture.
5. Check compressor oil level.
6. Evacuate system using vacuum pump.
7. Charge system with NEW Refrigerant 12.
8. Operate system and check performance.

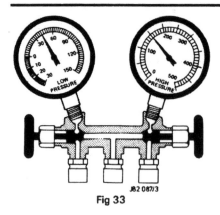

Fig 33

Complaint

Cooling is not adequate.

Condition

1. Low side gauge reading too high. Should be 15 - 30 psi.
2. High side gauge reading too low. Should be 185 - 205 psi at temperature of 95°F.
3. Sight glass free of bubbles (system is fully charged).
4. Discharge air from evaporator not sufficiently cool.

Diagnosis

Internal leak in compressor.

Correction

1. Discharge the system and replace compressor and receiver-drier.
2. Evacuate the system using a vacuum pump.
3. Charge the system with new refrigerant.
4. Operate system and check performance.

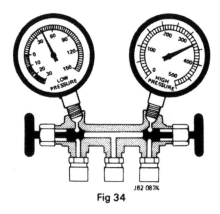

Fig 34

Complaint

Little or no coolant. Engine overheating may also be noted.

Condition

1. Low side gauge reading excessively high. Should be 15 - 30 psi.
2. High side gauge reading excessively high. Should be 185 - 205 psi at 95°F.
3. Bubbles may appear occasionally in sight glass. Liquid line very hot.
4. Discharge air from evaporator is warm.

Diagnosis

Improper condenser operation with lack of cooling caused by too high a high side pressure. System may have either normal or overcharge of refrigerant.

Correction

1. Check for loose or worn driver belts causing excessive compressor head pressures.
2. Inspect condenser for clogged air passages, bug screen, or other obstructions preventing air flow through condenser.
3. Inspect condenser mounting for proper radiator clearance.
4. Inspect clutch type fan for proper operation.
5. Inspect radiator pressure cap for correct type and proper operation.

After making the above checks:
Operate system and check performance.

If condition not corrected:

1. Inspect system for overcharge of refrigerant and correct as follows:
   (a) Discharge refrigerant until stream of bubbles appears in sight glass and both high and low gauge readings drop below normal.
   (b) Add new Refrigerant 12 until bubbles disappear and pressures are normal, then add $\frac{1}{4}$ - $\frac{1}{2}$ lb of additional refrigerant.
2. Operate system and check performance.

If gauge readings still too high:

1. Discharge system.
2. Remove and inspect condenser for oil clogging. Clean and flush condenser to ensure free passage of refrigerant or replace condenser.
3. Replace receiver-drier.
4. Evacuate system using vacuum pump.
5. Charge system with NEW Refrigerant 12.
6. Operate system and check performance.

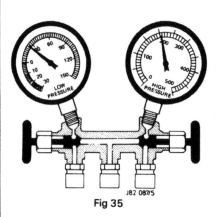

Fig 35

Complaint

Cooling is not adequate during hot part of day.

NOTE: Cooling may be satisfactory during early morning or late evening hours but is not adequate during hot part of the day.

Condition

1. Low side gauge reading (15 - 30 psi) but may drop into vacuum during testing.
2. High side gauge reading normal (205 psi at 95°F) but will drop when low side gauge reading drops into vacuum.
3. Sight glass may show tiny bubbles.
4. Discharge air from evaporator is sharp and cold but becomes warm when low side gauge reading drops into a vacuum.

Diagnosis

Excessive moisture in system. Desiccant agent saturated with moisture which is released during high ambient temperatures. Moisture collects and freezes in expansion valve and stops refrigerant flow.

Correction

1. Discharge refrigerant from system.
2. Replace receiver-drier.
3. Evacuate system with vacuum pump.
4. Charge system with NEW Refrigerant 12.
5. Operate system and check performance.

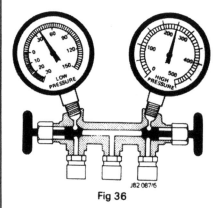

Fig 36

Complaint

Little or no cooling.

Condition

1. Low side gauge reading too high. Should be 15 - 30 psi.
2. High side gauge reading too high. Should be 185 - 205 psi at ambient temperature of 95°F.
3. Occasional bubbles in sight glass.
4. Discharge air from evaporator is not cool.

Diagnosis

Air in system. Refrigerant contaminated by non-condensables (air and/or moisture).

Correction

1. Discharge refrigerant from system.
2. Replace receiver-drier which may be saturated with moisture.
3. Evacuate system using vacuum pump.
4. Charge system with NEW Refrigerant.
5. Operate system and check performance.

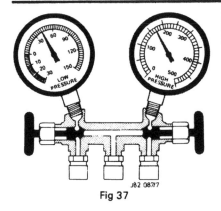

Fig 37

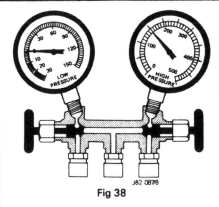

Fig 38

Correction

1. After cleaning expansion valve screen, or replacing expansion valve if necessary, and properly mounting temperature sensing bulb on evaporator outlet pipe, proceed as follows:
   (a) Evacuate system using vacuum pump.
   (b) Charge system with NEW Refrigerant 12.
   (c) Operate system and check performance.

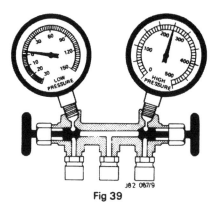

Fig 39

Complaint

Little or no cooling.

Condition

1. Low side gauge reading too high. Should be 15 - 30 psi.
2. High side gauge reading normal or slightly low. Should be 185 - 205 psi at an ambient temperature of 95°F.
3. Discharge air from evaporator warm.
4. Suction hose and evaporator show heavy sweating.

Diagnosis

Expansion valve allowing excessive flow of refrigerant through evaporator causing flooding of evaporator coils.

Testing

Check for expansion valve stuck open of incorrect mounting or temperature sensing bulb as follows:
(a) Set air conditioner for maximum cooling and operate the system.
(b) Spray liquid Refrigerant 12 on head of valve or capillary bulb, not low side gauge reading. Low side gauge should drop into a vacuum.
(c) If low side vacuum reading obtained, warm expansion valve diaphragm chamber with hand, then repeat test (step 'b').

Correction

1. If expansion valve test indicates valve operation is satisfactory, proceed as follows:
   (a) Clean contact surface of evaporator outlet pipe and temperature sensing bulb, clamp bulb securely in contact with pipe.
   (b) Operate system and check performance.
2. If expansion valve test indicates valve is defective, proceed as follows:
   (a) Discharge system.
   (b) Replace expansion valve.
   (c) Evacuate system using vacuum pump.
   (d) Charge system with NEW Refrigerant.
   (e) Operate system and check performance.

Complaint

Cooling is not adequate

Condition

1. Low side gauge reading too low (0 psi or a vacuum). Should be 15 - 30 psi.
2. High side gauge reading too low. Should be 185 - 205 psi at ambient temperature of 95°F.
3. Discharge air from evaporator only slightly cool.
4. Expansion valve inlet may show heavy sweating or frost.

Diagnosis

Expansion valve restricting refrigerant flow due to clogged screen, stuck valve, or temperature sensing bulb having lost its charge.

Testing

1. If expansion valve inlet is cool to touch, proceed as follows:
   (a) Set air conditioner for maximum cooling and operate the system.
   (b) Spray liquid Refrigerant 12 on head of valve of capillary bulb, note low side gauge reading. Low side gauge should drop into a vacuum.
   (c) If low side vacuum reading obtained, warm expansion valve diaphragm chamber with hand, then repeat test (step 'b')
   (d) If expansion valve test indicates valve operation is satisfactory, clean contact surface of evaporator outlet pipe and temperature sensing bulb, clamp bulb securely in contact with pipe. Proceed with correction procedure (below).
2. If expansion valve inlet shows 'sweating' or frost proceed as follows:
   (a) Discharge system
   (b) Disconnect inlet at expansion valve, remove and inspect filter.
   (c) Clean and replace filter, reconnect inlet line.
   (d) Proceed with correction, procedure (below)
3. If expansion valve test (step '1' preceding) indicates valve is defective, proceed as follows:
   (a) Discharge system.
   (b) Replace expansion valve, then proceed with correction procedure.

Complaint

Cooling is not adequate.

Condition

1. Low side gauge reading too low. Should be 15 - 30 psi.
2. High side gauge reading will build excessively high. Should be 185 - 205 psi at an ambient temperature of 95°F.

NOTE: An overcharged system, or a Condenser or receiver-drier that is too small, will cause high side gauge reading to be normal or excessively high.

3. Discharge air from evaporator only slightly cool.
4. Liquid line cool to the touch, line or receiver-drier may show heavy sweating or frost.

Diagnosis

Restriction in receiver-drier or liquid line with compressor removing refrigerant from evaporator faster than it can enter resulting in a 'starved' evaporator.

Correction

1. Discharge system.
2. Remove and replace receiver-drier, liquid lines, or other defective parts.
3. Evacuate system using vacuum pump.
4. Charge system with NEW Refrigerant 12.
5. Operate system and check performance.

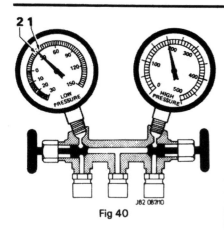

Fig 40

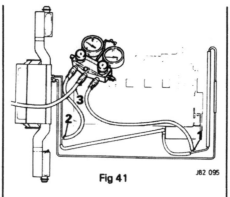

Fig 41

Complaint

Compressor cycles (cuts in and out) too rapidly. Compressor cycles on 34 psi (1, Fig. 40). Compressor cycles off 28 psi (2, Fig. 40).

Condition

1. Low side pressure cycle too high with insufficient range between OFF and ON. Cycle should be:
   Cycle 'Off' — 12 - 15 psi.
   Cycle 'On' — 36 - 39 psi.
   Cycle Range — 24 - 28 psi.
2. High side gauge reading Normal (200 psi). Should be 185 - 205 psi at ambient temperature of 95°F.

Diagnosis

Ranco thermostat faulty.

Correction

1. Stop car engine, turn air conditioning off and disconnect the battery.
2. Remove and discard old thermostatic switch, install new switch of same type.
3. When installing new thermostatic switch, make certain that capillary tube installed in same position and to same depth in evaporator core as old switch tube.

CAUTION: Do not kink or bend capillary tube too sharply — tube is gas filled.

Operate system and check performance of new thermostatic switch.

AIR CONDITIONING SYSTEM

Depressurise

Observe all safety precautions and do not smoke while carrying out the following procedure.

With the engine switched off, remove the protective caps from the schrader valves.

Connect the manifold gauge set with the red hose to the high pressure side (1, Fig. 41) and the blue hose to the low pressure side (2, Fig. 41).

Place the free end of the centre hose (3, Fig. 41) into a suitable container. Slowly open the high or low side manifold hand valve and adjust the valve for a smooth refrigerant flow. Watch for any signs of escaping oil and adjust the hand valve so that no oil escapes. If oil is lost during the discharge, the compressor oil level will have to be checked and topped up.

As the discharge rate slows down, open the other manifold hand valve. Refrigerant will now flow from the high and low pressure sides of the system. Constantly adjust the hand valves to ensure that oil does not flow. When a zero reading is shown on both the high and the low pressure gauges the system is discharged. Close both hand valves.

Evacuate

Once a system has been opened for repairs, or is found low of refrigerant, it must be fully evacuated with a vacuum pump to remove all traces of moisture before a new refrigerant charge is added.

Moisture may collect and freeze in the expansion valve which will block the refrigerant flow and stop the cooling action. Moisture will also react with Refrigerant 12 and cause corrosion of the small passages and orifices in the system.

The desiccant in the receiver-drier can absorb only a limited amount of moisture before it becomes saturated, therefore it is important to prevent moisture entering the system, and to remove any moisture which may have entered the system through a leak or an open connection.

Unwanted air and moisture are removed from the system by a vacuum pump. A vacuum pump is the only piece of equipment designed to lower the pressure sufficiently so that the moisture boiling temperature is reduced to a point where the water will vaporise and then can be evacuated from the system.

The compressor cannot be used as a vacuum pump because the refrigeration oil circulates with the refrigerant. The compressor depends to a large extent on the refrigerant distributing oil for lubrication and damage to the compressor may result due to lack of refrigerant which carries the oil for lubrication.

After the system has been fully discharged, and with the test gauge set still connected, attach the centre hose of the manifold gauge set, to the inlet fitting of the vacuum pump.

Open the low and high side manifold hand valves to their maximum positions.

Open the discharge valve on the vacuum pump or remove the dust cap on the discharge outlet whichever is appropriate.

Turn the vacuum pipe on, note the low side gauge to make certain that a vacuum is being created in the system by the vacuum pump.

From the time the lowest vacuum is attained, continue to operate the vacuum pump for a few minutes to be sure complete evacuation has been performed.

Close both gauge hand valves, turn off the vacuum pump, note the low side gauge reading. The gauge needle should remain stationary at the point where the pump was turned off. Should the gauge needle return towards zero a leak exists in the air conditioning system.

If a leak exists charge the system with R12 refrigerant, locate the leak with a leak detector, discharge the refrigerant from the system. Repair the leak and repeat the evacuation procedure.

If the gauge needle remains stationary and vacuum is maintained for 3 to 5 minutes, close both the high and low manifold hand valves, and disconnect the hose from the vacuum pump.

The air conditioning system is now ready for charging.

Flushing

If contamination of the expansion valve and associated pipeworks occurs it is essential that the whole of the air conditioning is fully flushed out using Freon 12, or other suitable charging gas.

Discharge the system.

Disconnect the inlet (low pressure) and the outlet (high pressure) pipes from the compressor.

Fit a suitable blanking plate over the end of the high pressure pipe and retain the plate with a suitable G clamp, also remove the schrader valve from the charging connection on the high pressure pipe.

Place the low pressure pipe into a suitable metal container and cover.

Disconnect the high pressure pipe from the expansion valve and carefully remove the conical filter, then reconnect the pipe on the expansion valve.

Carefully remove the thermal bulb coil attached to the evaporator outlet pipe and allow the thermal bulb to remain in the ambient air.

This will prevent the expansion valve closing when refrigerant is flushed through it.

Connect a suitable hose, from the liquid connection of a recommended refrigerant canister, to the charging connection on the high pressure pipe.

Open the canister (the pressure in the canister should be approximately 4.22 kgf/cm² (60 lbf/in²)) and allow the refrigerant to flush through the air conditioning system for approximately 30 seconds, or until a steady liquid flow is observed from the low pressure pipe.

Turn the refrigerant canister off and remove the connections.

IMPORTANT: On re-assembling the system. Fit a new receiver-drier.

Check the compressor oil level.

Thoroughly clean and refit the expansion valve filter.

Refit the thermal bulb on to the evaporator outlet pipe.

Refit the schrader valve into the high pressure pipe.

Refit all the pipe connections and recharge the system.

Charge

Charging the air conditioning system is the process of adding a specific quantity of refrigerant to the circuit. Before attempting the charging operation the system **must** have been evacuated and, if necessary, flushed through immediately beforehand. No delay between evacuation and charging procedures is permissible. The equipment should be fitted with a means of accurately weighing the refrigerant during the charging process. Great care must be taken to charge correctly, as undercharging will result in very inefficient operation, and overcharging will result in very high pressures and possible damage to components.

Evacuate the system with hoses (1 & 2, Fig. 42) connected as shown.

Connect the centre hose of the charging manifold (3, Fig. 42) to a supply of refrigerant. The supply available must be at least 3,3 kg (7.2 lb) weight.

Open the refrigerant supply valve.

Purge the centre hose by momentarily cracking the connection at the manifold block: retighten the connector.

Record the weight of refrigerant supply source. Open both valves on the charging manifold and allow the refrigerant source pressure to fill the vacuum in the system.

Between 0,23 kg and 0,45 kg ($\frac{1}{2}$ lb to 1 lb) weight will enter the system.

Record the quantity.

NOTE: The quantity drawn in will vary with ambient temperature.

Close the high pressure side valve on the manifold block.

Ensure that all is clear and start the vehicle engine. Run the engine at 1500 rev/min.

Set the air conditioning system blower speed control to **'Fast'**.

NOTE: This engages the compression clutch to start system circulation, and runs the blower motors at fast speed to heat the evaporator coil. Vapour will be turned to liquid in the condenser and stored in the receiver-drier.

Control the flow of refrigerant with the suction side valve on the charging manifold, and allow a total weight of 1,13 kg $\pm$ 0,028 kg ($2\frac{1}{2}$ lb $\pm$ 2 oz) refrigerant to enter system.

Close the suction side valve.

NOTE: Alternatively, observe the sight glass on receiver-drier until the sight glass clears, and no bubbles or foam are visible.

Re-open the suction valve for 2 to 5 minutes (2 minutes if the ambient temperature is low, 5 minutes if high).

This will allow an additional 0,11 kg ($\frac{1}{4}$ lb) of refrigerant to enter the system.

Run the system for 5 minutes, observing the sight glass.

If foaming is very slight, switch off the engine.

NOTE: It is normal for there to be slight foaming if the ambient air temperature is 21°C (70°F) or below.

Close the refrigerant supply valve, disconnect the hose.

Quickly disconnect the hoses from the schrader valves.

Fit protective sealing caps.

Switch on the engine and check the function of the air conditioning system.

Switch off the engine; flush the engine compartment and interior of the vehicle with shop compressed air line.

Conduct a leak test on the installation.

SUPERHEAT SWITCH AND THERMAL FUSE

Description

The superheat switch and a thermal fuse are included in the clutch circuit to provide a compressor protection system. This guards against low refrigerant charge and blockages causing extreme superheated inlet gas conditions and resulting compressor damage.

The superheat switch is located in the rear of the compressor in contact with the suction side gas, whose pressure drops and temperature rises with low refrigerant charge (ie Freon leak). This condition closes the superheat switch contacts.

The thermal fuse is a sealed unit containing a heater and meltable fuse. The superheat switch brings in the heater which melts the fuse and disconnects the compressor clutch and heater. The compressor stops and damage from insufficient lubrication will be avoided.

CAUTION: After a thermal fuse melt, establish and rectify the cause before replacing the thermal fuse unit complete.

Thermal fuse melt:
Temperature: 157 to 182°C (315 to 360°F)
Time: 2 minutes — 14V system voltage
     5.5 minutes — 11,5V system voltage
Heater resistance, cold: 8 to 10 ohms:

Air Conditioning Superheat Switch

Testing

If the refrigerant level is satisfactory and there is not a blockage in the air conditioning system but the thermal fuse persists in melting.
Carry out the following checks.

Test Procedure 'A' — for use with a cold engine and at ambient temperatures below 30°C (86°F).

Connect a test lamp in series with the superheat switch (Fig. 44).

NOTE: With the test lamp connected in the circuit it will prevent the thermal fuse from operating as a safety device therefore care should be taken when carrying out the test.

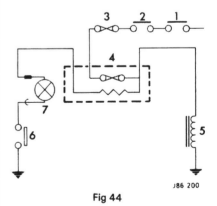

Fig 44

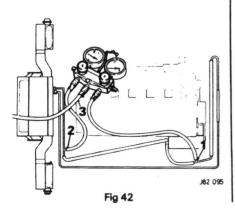

Fig 42

KEY TO DIAGRAM (Fig. 43)

1. To compressor clutch
2. Superheat switch
3. Thermal fuse
4. + Feed cable

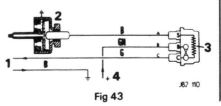

Fig 43

KEY TO DIAGRAM (Fig. 44)

1. Air conditioning switch
2. Ambient switch
3. Compressor clutch fuse
4. Thermal fuse
5. Compressor clutch coil
6. Superheat switch
7. Test lamp

With the ignition and air conditioning switched on.

Ensure a serviceable thermal fuse is fitted.

Evacuate the air conditioning system and then close the taps.

The test lamp should not light. If the test lamp does light then follow test procedure 'B'.

With the lamp not illuminated start and run the engine at about 2000 rpm. After a few minutes the lamp should light. As soon as the test lamp lights, open the taps to allow refrigerant to charge into the system. As the air conditioning system becomes charged the lamp should go out.

If the above lamp functions do not occur, replace the superheat switch.

After checking remove the test lamp from the circuit and reconnect the superheat switch lead onto the terminal.

Operate and check the system.

Test Procedure B — for use with a hot engine or at ambient temperatures above 30°C (86°F).

Connect the test lamp in series with the superheat switch.

Switch the ignition and air conditioning on.

Ensure a serviceable thermal fuse is fitted.

Evacuate the air conditioning system and then close the taps.

The test lamp should light. (If the lamp does not light carry on checks as in Procedure 'A').

With the test lamp illuminated open the taps and allow refrigerant to charge the system. As the system becomes charged the test lamp should go out.

If the lamp functions do not occur, then replace the superheat switch.

After checking remove the test lamp and reconnect the superheat switch lead onto the terminal.

Operate and check the system.

SUPERHEAT SWITCH

Renew

Discharge system.

Disconnect harness connector from superheat switch.

Remove suction (low pressure) and out-put (high pressure) hoses.

Remove superheat switch retaining circlip and remove switch by pulling out of the compressor housing.

Remove the superheat switch 'O' ring located in the compressor housing.

Lightly lubricate the new 'O' ring seal and fit into compressor housing.

Locate the replacement superheat switch into the compressor housing and gently push switch into housing until seated.

Fit new circlip and secure.

Connect the suction (low pressure) and out put (high pressure) hoses to the compressor. Evacuate and recharge system and check system for leaks using suitable leak detection equipment.

COMPRESSOR

Remove and Refit

WARNING: BEFORE COMMENCING WORK, REFER TO THE GENERAL SECTION. DO NOT OPERATE THE COMPRESSOR UNTIL THE SYSTEM IS CORRECTLY CHARGED.

NOTE: Ensure that clean, dry male and female caps are to hand.

Disconnect the battery earth lead.

Depressurize the system.

On NAS vehicles, remove the air pump.

Note the position of the hoses.

Remove the clamping plate securing the high and low pressure hoses (1, Fig. 45).

Displace the hoses (2, Fig. 45). Fit blanking caps to the hoses and the compressor.

Remove the superheat switch cable connector (3, Fig. 45).

Slacken the compressor front and rear pivot bolts (4, Fig. 45).

Slacken the adjusting link locking and adjusting nuts.

Remove the bolts securing the adjusting link and remove the link.

Displace the drive belt. Disconnect the clutch cable connector (5, Fig. 45).

Remove the nuts securing the cruise control actuator and displace the actuator unit.

Remove the compressor pivot bolts and displace the compressor.

Manoeuvre the compressor from the engine compartment, keeping it horizontal and the sump down.

If a new compressor is being fitted, remove the mounting brackets from the old compressor and fit to new unit.

On refitting, ensure that new 'O' sealing rings are fitted.

Ensure the compressor drive belt is adjusted to the correct tension.

Correct tension as follows:

A load of 2.9 kg (6.4 lb) must give a total belt deflection of 4.32 mm (0.17 in) when applied at mid-point of the belt.

Recharge the air conditioning system.

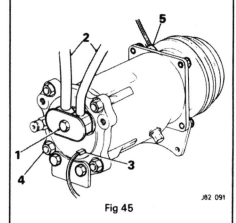

Fig 45

J82 091

CAUTION: After recharging, cycle the clutch in and out 10 times by selecting OFF-LOW, AUTO-OFF on the mode selector switch with the engine running. This ensures that the pulley face and the clutch plate are correctly bedded-in before a high demand is made upon them.

Check the system for correct operation.

Check the cruise control for correct operation.

COMPRESSOR OIL — CHECKING PROCEDURE

The following procedure should be adopted when checking the amount of oil in a compressor prior to its being fitted to a car:

1. Remove drain plug from compressor sump and drain oil into container having capacity of at least 285 cc (10 fl oz).

2. Remove pressure plate across inlet and outlet ports at rear of compressor; more oil may flow from sump plug hole.

3. With pressure plate still removed, set compressor on its rear end so that inlet and outlet ports are over container; slowly rotate drive plate through several revolutions both clockwise and anti-clockwise. Oil may flow from ports.

4. Measure quantity of oil drained out; make this up to 199 cc (7 fl oz) and re-fill compressor with this amount of 525 viscosity refrigerant oil.

If the compressor is not to be fitted immediately, it is important that the pressure plate be refitted over the ports and secured there, to prevent ingress of foreign matter.

Should it be suspected that the compressor oil level is low, on a car in service, the checking procedure detailed should be followed after the car engine has been run for at least 10-15 minutes with the air conditioning system switched on; this will cause the refrigerant oil to be returned to the compressor sump.

Should a new receiver-drier bottle, condenser or evaporator be fitted, without the car engine being run as above, immediately before dismantling, the following quantities of 525 viscosity refrigerant oil must be added to the system:

(a) For a new receiver-drier bottle — add 28 cc (1 fl oz).

(b) For a new condenser — add 85 cc (3 fl oz).

(c) For a new evaporator — add 85 cc (3 fl oz).

Additional oil is not needed after renewal of hose assemblies.

Oil may be added to the system either directly into the compressor or into the compressor charging port.

Compressor Servicing Procedure

To enable the servicing of the air conditioning compressor the following components are now available. The following servicing procedures should be adopted in the event of a malfunction of the compressor which involves any of the parts listed, as opposed to the replacement of the compressor unit.

Part Description

Pulley Bearing
Superheat Switch
Pressure Relief Valve
'O' Ring Suction Discharge Port
$\frac{1}{4}$ Pint 525 Viscosity Oil
Clutch Driver Assembly
Shaft Nut
Woodruff Key
Coil and Housing Clutch
Pulley Bearing Assembly
Retainer Ring Kit
Body of Compressor less Clutch, Pulley and Coil Housing Assembly
Shaft Kit for Seal
Bearing Retaining Ring

The specialist tool kit required to service the compressor unit in conjunction with the following procedures are available from KENT MOORE.

Tool Kit	10500
Hub Holding Tool	10418
Thin Walled Socket	10416

Tool Kit Contents

Pulley Extractor Kit
Pulley Bearing Remover and Installer Kit
Seal Assembly Remover and Installer
Hub Drive Plate Remover Kit
Hub and Drive Plate Assembly Installer
'O' Ring Remover
'O' Ring Installer
Snap Ring Installer
Ceramic Seal Remover and Installer, and Shaft
Seal Protector
Hub Holding Tool
Thin Walled Socket

When servicing the compressor, remove only the necessary components that preliminary diagnosis indicates are in need of service.

Seven service operations may be performed on the GM 6 cylinder compressor.

(i) Replacement of compressor assembly.
(ii) Replacement of clutch drive and pulley assembly.
(iii) Replacement of pulley bearing.
(iv) Replacement of clutch coil and housing assembly.
(v) Replacement of shaft seal.
(vi) Replacement of superheat switch.
(vii) Replacement of compressor cylinder and shaft assembly (less clutch drive, coil housing and pulley).

General Instructions During Servicing Operations

(i) Discharge system prior to removal of compressor unit.

(ii) During removal, maintain the compressor positioned so that the sump is downward. Do not rotate compressor shaft.

(iii) If the compressor is being replaced due to a component failure within the main body of the compressor, the clutch coil housing and clutch plate drive and hub assembly must be removed from the original compressor unit and fitted to the replacement unit. This also applies when fitting a replacement compressor body.

(iv) If the original compressor is being reinstalled following servicing, replace with the right quantity of 525 viscosity oil.

(v) Discard 'O' rings from suction and discharge ports of compressor and replace and with new 'O' rings.

(vi) Install compressor and adjust drive belt tension to service manual specifications.

(vii) Lubricate 'O' rings with refrigerant oil and attach suction and discharge hose connections and retaining plate to compressor torque to 2,764 - 3,455 kgfm (20 - 25 ft lbs).

Replacement of Clutch Drive Plate and Hub Pulley, Clutch Coil and Housing Assemblies.

Discharge the system.
Remove the compressor from the engine.
Using suitable mounting jig or vice, secure compressor.
Holding the hub of the clutch drive plate with the hub holding tool. Using the thin walled $\frac{9}{16}$ in socket remove the shaft nut. Refer to Fig. 46.

Fig 46

Screw the threaded hub puller to the hub. Hold the body of the hub puller with a suitable spanner, tighten centre screw of hub pulley (Fig. 47), until drive plate, hub and woodruff key can be removed (Fig. 48).

Fig 47

Fig 48

Using suitable circlip pliers remove the bearing to head retainer ring (Fig. 49).

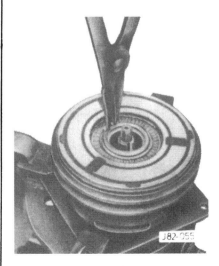

Fig 49

Fig 50

Remove the absorbent felt sleeve retainer ring to enable the location of the pulley extraction tool.

Using the pulley extraction tool locate the puller pilot on hub of front head and remove the pulley assembly (Fig. 50).

NOTE: The next operation details removal of pulley bearing. DO NOT remove the pulley bearing unless it is to be replaced. Removal may cause the bearing to be damaged.

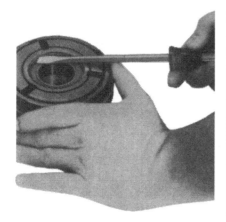

Fig 51

Fig 52

Remove bearing to pulley retaining ring with small screwdriver (Fig. 51). Drive out the bearing using bearing remover and handle (Fig. 52).

Fig 53

Mark position of the coil and housing assembly in relationship to the shell of the compressor. Remove the coil housing retainer ring using suitable circlip pliers (Fig. 53) and lift off the coil and housing assembly (Fig. 54).

Examine coil for loose or distorted terminals and cracked insulation. Check that the current consumption is 3.2 Amps at 12 volts. The resistance should be 3.75 Ohms at room temperature.

Fig 54

Reassemble coil and housing assembly by reversing the dismantling procedure. Be sure coil and housing assembly markings line up.

NOTE: If the pulley assembly is going to be reused, clean the friction surface with suitable solvent cleaner.

Fig 55

Drive the new bearing into the pulley assembly with the bearing installer and handle. The bearing installer will ride cn the outer race of the bearing (Fig. 55).

Fig 56

Lock the bearing in position with the bearing to pulley retainer ring.

Press or tap the pulley assembly into the hub of front head using installer tool and handle (Fig. 56).

Check the pulley for binding or roughness, and that the pulley rotates freely.

Using suitable circlip pliers lock pulley assembly in position with bearing to head retainer ring (flat side of retainer ring should face towards pulley).

Install square drive woodruff key in the key way of the clutch drive hub.

Wipe frictional surface of clutch plate and pulley clean. Using a suitable solvent.

Place clutch plate and hub assembly on shaft, aligning shaft key way with key in hub (refer to Fig. 48 dismantling procedure).

NOTE: The woodruff key is made with a slight curvature to help hold it in the plate hub during assembly.

IMPORTANT: To avoid damage to the compressor, undue force should not be applied to the hub or shaft. This could misplace axial plate on shaft, resulting in damage to the compressor.

Fig 57

Place spacer on hub. Thread clutch plate and hub assembly installer tool onto end of the shaft (Fig. 57).

Hold the head of the bolt and turn tool body several revolutions to press hub partially on shaft. Remove clutch plate and hub assembly installer and spacer.

Check alignment of woodruff key with key way in shaft. If alignment is correct, replace installer tool and continue to press hub into shaft until there is approximately 2,38 mm ($\frac{3}{32}$ in) air gap between the frictional surfaces of pulley and clutch plate. Remove installer tool and spacer.

Install a new shaft lock nut with the small diameter boss of the nut against the hub using a thin wall $\frac{9}{16}$ in socket. Hold clutch with holding tool and tighten nut to 2,07 kg/fm (15 ft/lbs), using a 3.455 kg/fm (25 ft/lbs) torque wrench. The air gap between the frictional surfaces of pulley and clutch plate should now be approximately 0,56 mm to 1,45 mm (0.022 in to 0.057 in).

Shaft Seal Leak Detection

A compressor shaft seal should not be changed because of an oil line on the underside of the bonnet. The seal is designed to seep some oil for lubrication purposes. Only change a shaft seal when a leak is detected by the following procedures:

Ensure there is refrigerant is in the system. Turn off the engine.

Blow off compressor clutch area with compressed air. Blow out clutch vent holes to completely remove any freon and oil deposits.

Allow car to stand for 5 minutes, without operating compressor.

Rotate the compressor clutch drive plate by hand until one of the vent holes is at the lower side of drive plate. Using leak detector, sense through vent hole at lower side of drive plate only.

Some compression shaft seal leaks may be the result of misplacement of the axial plate on the compressor shaft. The mispositioning of the axial plate may be caused by improper procedures used during pulley and driven plate removal, undue force collisions, or dropping the compressor.

Replacement of Shaft Seal

Remove clutch driven plate and hub assembly as previously described.
Remove compressor absorbent felt retaining ring and felt sleeve.

Fig 58

Thoroughly clean the area inside the compressor neck surrounding the shaft, the exposed portion of the seat and the shaft itself of any dirt or foreign material. This is absolutely necessary to prevent any such material from getting into the compressor.

Remove the seal seat retaining circlip (Fig. 58) using suitable circlip pliers.

Fig 59

Remove the ceramic seal seat using the seal seat remover and installer tool (Fig. 59). Position tool into seal seat recess, grasp flange of shaft seal seat and pull straight out.

Fig 60

Using the seal remover and installer tool grip the seal by inserting the tool into the seal recess. Turning clockwise. Withdraw the tool and seal (Fig. 60).

Fig. 61

Remove the seal seat 'O' ring (Fig. 61) using the 'O' ring remover tool.

Recheck the inside of the compressor neck and the shaft. Be sure these areas are perfectly clean and free of burrs before installing new parts.

Coat shaft and 'O' ring with clean compressor 525 viscosity oil.

Fig 62

Place 'O' ring on 'O' ring installer (Fig. 62) and insert tool and 'O' ring into seal recess. Release 'O' ring by sliding down tool hook, and remove tool.
(Fig. 63) illustrates the tool being removed following 'O' ring installations.

Fig 63

Place the seal protective sleeve over the compressor shaft and fit new shaft seal. Gently twisting the tool clockwise to engage the seal housing flats onto the compressor shaft. Withdraw the tool by pressing downwards and twisting the tool anti-clockwise.
Coat the seal face of the new ceramic seal seat with clean 525 viscosity oil. Mount the seal seat on to the remover and installer tool and carefully guide the seal into the compressor neck gently twisting it into the 'O' ring seal.
Disengage and remove tool, and compressor shaft protective sleeve.
Install new circlip with the flat side against seal seat, and press home.
Install the new absorbent sleeve by rolling the material into the cylinder, overlapping the ends and slipping it into the compressor neck with the overlap at the top of the compressor. Using a small screwdriver or similar tool carefully spread the sleeve so that in its final position, the ends butt together at the top vertical centre line.

Install the new absorbent sleeve retainer so that its flange face will be against the front end of the sleeve, press and tap with a mallet setting the retaining ring and absorbent sleeve until the outer edge of the sleeve retainer is recessed approximately 0,8 mm ($\frac{1}{32}$ in) from the face of the compressor neck.
Lightly lubricate absorbent felt sleeve with 525 viscosity oil.
Refit clutch drive plate and hub assembly.
Check compressor oil level.
Refit compressor to vehicle, and connect the suction (low pressure) and discharge (high pressure) hoses using new 'O' ring seals. Prior to fitment of compressor drive belt, rotate the compressor drive plate clockwise several revolutions to prime lubrication pump.
Evacuate and recharge system.

NOTE: During charge procedure check compressor seals for leaks using suitable leak detection equipment.

Leak Test

A high proportion of all air conditioning work will consist of locating and repairing leaks. Many leaks will be located at points of connections and are caused by vibration. They may only require the retightening of a connection or clamp. Occasionally a hose will rub on a structural part of the vehicle and create a leak, or a hose will deteriorate and require a replacement. Any time the system requires more than $\frac{1}{2}$ lb of refrigerant after a period of operation, a leak is indicated which must be located and rectified.
The 'Robinair Robbitek 30001 Leak Detector' is designed for speedy detection of leaks. The leak detector is small and portable, and is battery operated. This instrument will indicate leaks electronically by sounding an alarm signal. Provision is made to plug in an earphone, which is useful in a noisy workshop; and it has the recommended sensitivity of 0,45 kg (1 lb) in 32 years.

FLAP LINKAGE

Adjust Air

Service Tools: 18G 1363, Setting Jig (Fig. 64)

Remove the console right hand panel and underscuttle trim panels to gain access to the air conditioning unit flap linkages. Note: On LH drive cars it is necessary to remove the glovebox compartment.
Remove the footwell outlet vent from the air conditioning unit.
Switch on the ignition, position the right hand control knob to 'DEF'. When the servo has reached its full heat position, switch off the ignition and disconnect the battery.
Disconnect the linkage rods (1, Fig. 66) from the servo lever connections.
Set the link bolt adjuster (2, Fig. 66) in its mid position.
Gently pull the wire link (3, Fig. 66) to detach it from the grommet in link (4, Fig. 66).

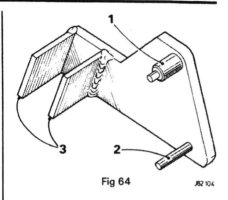

Fig 64

Move the thumbwheel (Fig. 65), located in the radio console panel, fully to the right.
Using the jig setting tool 18G 1363, locate peg (1, Fig. 64) into the hole (7, Fig. 66) on the linkage protection bracket, and peg (2, Fig. 64) in the hole in linkage (4, Fig. 66), from which link (3, Fig. 66) was removed. The parallel end guides on the setting jig tool (3, Fig. 64), should locate over the linkage assembly (8, Fig. 66) so that the linkage is in a straight line. If linkage (8, Fig. 66) is not in line adjust the distribution temperature control cable (9, Fig. 66), until the linkage is straight. Tighten the cable clamp (6, Fig. 66).
With the jig setting tool in position adjust the linkage (10, Fig. 66) until post (11, Fig. 66) is at the top of the slot.
Remove the jig setting tool.
Position the setbolt adjuster (12, Fig. 66) at its furthest point away from fulcrum (13, Fig. 66). Refit the link rod (3, Fig. 66) to linkage (4, Fig. 66).
Reconnect the servo linkage rods (1, Fig. 66) to the servo motor levers, ensure that the servo lever cam followers locate against the servo cams.
Reconnect the battery and switch on the ignition. Motor the system to the full cooling position.
Switch off the ignition.
Check that the linkage (14, Fig. 66) abuts against the snail cam (15, Fig. 66).
The lower heat flap should now be fully sealed; check by manually pushing the snail cam, no movement should be evident.

Fig 65

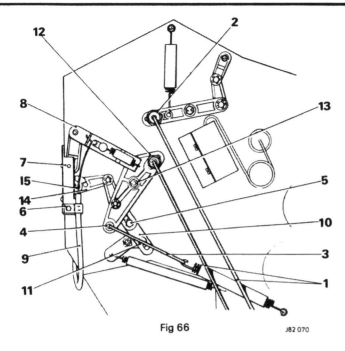

Fig 66   J82 070

If there is movement, switch on the ignition and motor the system to its full heat position, switch off the ignition. Slacken the set bolt adjuster (12, Fig. 66) and move it slightly towards the fulcrum point (13, Fig. 66). Reset the servo lever linkage rods (1, Fig. 66) so that the servo lever cam followers locate against the servo cam.

Switch on the ignition and motor the system to its full cooling position (AUTO 65). Check that linkage (14, Fig. 66) abuts against the snail cam (15, Fig. 66). Check manually by pressing the snail cam, if movement is evident, repeat the procedure in the previous paragraph.

If no movement is evident, then the lower heat flap is sealing correctly and the flap linkages and distribution temperature control are correctly set.

Ensure that all linkages and adjustments are secure.

Refit the footwell outlet vent, the underscuttle trim panels and console right hand side panel.

CONDENSER UNIT AND RECEIVER-DRIER

Renew

Before commencing this operation, ensure that suitable clean, dry sealing plugs and caps are to hand.

Disconnect the battery earth lead.

Depressurise the air conditioning system.

Remove the nuts and washers securing the fan cowl to the radiator (1, Fig. 67) top rail and pull the fan cowl clear of the mounting studs.

Disconnect the pipes from the receiver-drier (2, Fig. 67) and the pipe from the compressor to the condenser (3, Fig. 67). Fit blanking plugs to all the disconnected pipes to avoid contamination.

Remove the nuts and washers securing receiver-drier (4, Fig. 67).

Remove the receiver-drier (5, Fig. 67).

Note the connections and disconnect the cable harness to the ignition amplifier.

Remove the nuts and washers securing the condenser mounting bracket to the radiator top rail (6, Fig. 67).

Remove the four bolts and washers securing the radiator top rail to the wing valances (7, Fig. 67).

Ease the top rail clear of the condenser and lift the condenser clear of the car.

On refitting, reverse the above operations and fit a new receiver-drier.

NOTE: If the system is opened, even for a short time, the receiver-drier must be renewed. Do not remove the protective sealing caps from the new unit until it has been fitted and is ready for the pipes to be connected.

AMBIENT TEMPERATURE SENSOR

Renew

Disconnect the battery earth lead.

Remove the right hand underscuttle casing.

Remove the component panel securing screws and displace the component panel.

Note the position of the electrical connections and disconnect the cables from the sensor.

Remove the two screws securing the sensor and remove the sensor.

IN CAR TEMPERATURE SENSOR

Renew

Disconnect the battery earth lead.

Remove the screws securing the passengers underscuttle casing and remove the casing.

Remove the screws securing the glove box liner and the glove box latch. Remove the latch and carefully withdraw the liner.

Carefully manoeuvre the elbow hose from the sensor outlet.

Disconnect and remove the sensor assembly (1, Fig. 68) from the air pick-up tube (2, Fig. 68).

Remove the sensor assembly from the elbow hose (3, Fig. 68).

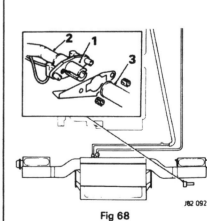

J82 092

Fig 68

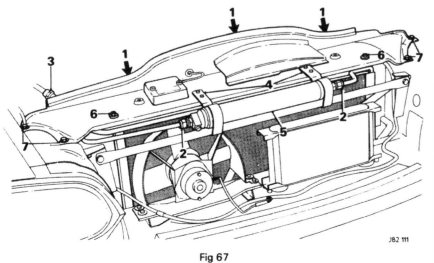

J82 111

Fig 67

TEMPERATURE SELECTOR

Renew

MODE SELECTOR

Renew

Disconnect the battery lead.
Carefully prise off the switch knobs from the temperature and fan controls.
Unscrew the fasteners from behind the control knobs and detach the control panel.
Disconnect the optical fibre elements, remove the control panel and the radio escutcheon assemblies.
Remove the right and left hand side pads.
Unscrew the gear selector control knob.
Remove the screws securing the switch panel to the centre console, ease the switch panel from the centre console, note the position of the wiring connectors, disconnect the connectors and remove the switch panel.
Remove the screws securing the front and rear ends of the console.
Displace the stop lamp bulb failure sensor and remove the console assembly.
Remove the bolts and washers securing the stays to the transmission tunnel (1, Fig. 69).
Ease the stays aside to give access to the switch panel.
Remove the nuts and washers securing the switch panel to the left hand of the unit (2, Fig. 69).
Remove the screws securing the switch cover to the switch panel (3, Fig. 69). Note the position and disconnect the vacuum pipes.
Remove the nut and washer securing the switch cover to unit at lower right hand side.
Release the harness and remove the switch cover.
Remove the nut and washer securing the switch panel at the upper right hand side of the panel (5, Fig. 69).
Ease the panel clear of the mounting studs, remove the two screws securing the temperature selector (4, Fig. 69), note the

position of the cable connections, disconnect the cables at the connections (6, Fig. 69) and remove the selector.
Note the position of and disconnect the cables from the mode selector micro-switches (1, Fig. 70).
Note the position of the micro-switches, remove the two screws and nuts securing the switches (2, Fig. 70).
Remove the switches and retain the distance pieces.
Remove the screws securing the vacuum switch mounting bracket (3, Fig. 70) and remove the vacuum switch assembly.
Remove the circlip securing the cam assembly and remove the cam assembly.

NOTE: Care must be taken to ensure that correct replacement parts are used, and that the items are replaced in the correct position.

When refitting the cams, ensure that the vacuum switch operating rod is pressed back to allow the camshaft into position.

THERMOSTAT

Renew

Disconnect the battery earth lead.
Remove the right hand underscuttle casing and the right hand side casing.
On left hand drive cars remove the glove box liner.
Remove the nut securing the thermostat to the bracket (1, Fig. 71).
Note the position of and disconnect the cables from the lucar connectors (2, Fig. 71).
Carefully remove the thermostat by withdrawing the capillary tube from the air conditioning unit (3, Fig. 71).

NOTE: Ensure the replacement thermostat capillary tube is formed to the exact dimensions of the unserviceable unit, ensuring that the capillary tube makes contact with the evaporator matrix.

BLOWER MOTOR RESISTANCE UNIT

Renew

Disconnect the battery earth lead.
On right hand drive cars, remove the left hand underscuttle casing and the glove box liner.
On left hand drive cars, remove the left hand underscuttle casing and the left hand side casing.
Note the position of and disconnect the cables from the resistance unit lucar connectors (1, Fig. 72).
Remove the screw securing the vacuum hose clip, move the hose to one side (2, Fig. 72).
Remove the screws securing the resistance unit and withdraw the unit from the air conditioning unit case (3, Fig. 72).
On refitting, ensure the cable connectors are secure and connected correctly.

WATER VALVE TEMPERATURE SWITCH

Renew

Disconnect the battery earth lead.
Remove the left hand underscuttle casing and the console side casing.
On right hand drive cars remove the glove box liner.
Disconnect the cables at the lucar connectors on the switch (1, Fig. 73).
Withdraw the securing screws and remove the switch (2, Fig. 73).
On refitting, ensure the connectors are clean and tight.

Figs 69 — 73

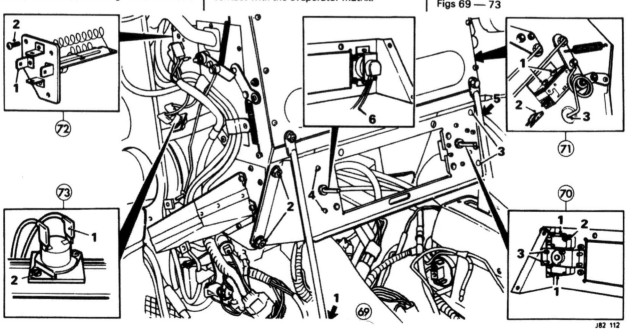

J82 112

BLOWER MOTOR RELAY

Renew

Disconnect the battery earth lead.
Remove the left hand console side casing.
Remove the screws securing the footwell air outlet duct and remove the duct.
Note the position of the cable connectors.
Disconnect the block connector, the lucars and the main feed cable from the relay (1, Fig. 74).
Remove the nuts securing the relay and remove the relay (2, Fig. 74).
On fitting replacement relay, ensure the cables are secure and connected correctly.

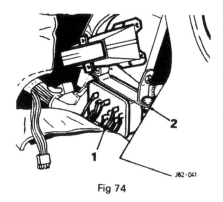

Fig 74

WATER VALVE

Renew

Remove the engine coolant filter and header tank caps, then open the radiator drain tap. Allow the coolant to drain from the system. Slacken the securing clip on the water valve to cylinder head hose and disconnect the water valve from the hose (1, Fig. 75).
Disconnect the vacuum hose from the water valve (2, Fig. 75).
Reposition the valve for access to the water valve to heater hose clip (3, Fig. 75).
Slacken the clip and remove the water valve.

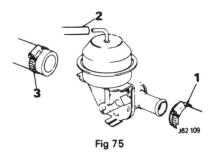

Fig 75

On refitting, ensure the cooling system is refilled with coolant to the correct specification.

THERMAL FUSE

Renew

Disconnect the cable block connector from the thermal fuse (1, Fig. 76) assembly located to the front of the right hand wing valance.
Remove the nut and screw securing the thermal fuse (2, Fig. 76).
Remove the thermal fuse.

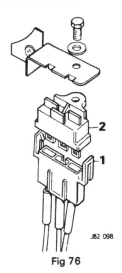

Fig 76

EXPANSION VALVE

Renew

Depressurize the air conditioning system.
Partially drain the engine coolant.
Disconnect the hose unions (1, Fig. 77) and seal with clean blanking caps.
Release the clip securing the water valve to the cylinder head hose, disconnect the water valve from the hose, and move the water valve clear of the expansion valve.
Remove the padding from the capillary tube (2, Fig. 77).
Disconnect the capillary tube at the union (3, Fig. 77).
Release the valve by unscrewing the union nut (4, Fig. 77).

NOTE: To avoid straining the joint or the pipe, ensure the valve is held firmly as the union is unscrewed.

Slacken the two screws securing the capillary tube clear of the clamp.
Remove the valve assembly carefully, manoeuvering the capillary tube clear of the clamp.
On fitting replacement unit, ensure new 'O' rings are fitted, the cooling system is refilled, and the air conditioning system is recharged.

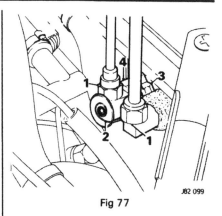

Fig 77

BLOWER ASSEMBLY

Remove and refit RH Unit

The blower fans are heavy duty motors with impellors attached. Speed is varied by controlled switching of resistances in series with the motors. The right hand unit has the ambient temperature sensor mounted in the inlet duct. Air flow control flaps are operated by a vacuum actuator situated in the side of the inlet duct.

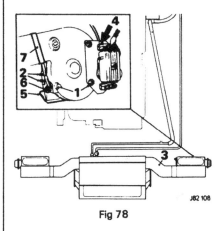

Fig 78

Disconnect the battery earth lead.
Remove the right hand underscuttle casing and right hand console casing.
On left hand drive cars, remove the glove box liner.
Remove the nuts securing the component panel to the blower (1, Fig. 78) assembly, secure the component panel clear for access to the blower assembly mounting bolts.
Disconnect the ambient temperature sensor leads at the lucar connectors (2, Fig. 78).
Disconnect the pliable trunking from the stub pipes on the side of the air conditioning unit (3, Fig. 78).
Disconnect the blower motor harness block connector.
Disconnect the vacuum pipe from the flap operating servo on blower assembly (4, Fig. 78).
Open the recirculation flap and fit a wedge to retain it in the open position (5, Fig. 78).
Remove the bolts securing blower assembly to mounting brackets (6, Fig. 78).
Ease the blower assembly from its location.
Remove the tape securing the ducting to the assembly (7, Fig. 78).

AIR CONDITIONING UNIT

Remove and refit

Removing

Disconnect the battery earth lead.

Withdraw the steering wheel and the adjuster assembly from the upper steering column.

Remove the left and right hand underscuttle casing.

Remove the instrument panel module and carefully remove switch panel.

Withdraw the air conditioning knobs from the air conditioning selector switches, remove the radio, remove the screws securing the facia and the console to the air conditioning unit.

Remove the glove box liner.

Slacken the nuts securing the top rear portion of the facia to the bulkhead, remove the bolts securing the sides of the facia to the bulkhead, remove the nut securing the main light switch, displace the switch and carefully remove the facia from the car.

Disconnect the air conditioning hoses at the bulkhead connectors to the expansion valve on the engine compartment (1, Fig. 79).

Disconnect the coolant hoses at the heater bulkhead connectors in the engine compartment.

Remove the nuts securing the air conditioning unit to the bulkhead (2, Fig. 79).

Unclip the main harness from the securing clips on the screen rail (3, Fig. 79).

Remove the bolts securing the demist duct support rail to the body mounting points and remove the support rail (4, Fig. 79).

Disconnect the pliable ducting between the air conditioning unit and the blower motors from the stub pipes (5, Fig. 79).

Remove the rear compartment ducts (6, Fig. 79).

Remove the nuts and bolts securing the unit support stays (7, Fig. 79); recover the stays.

Remove the automatic gearbox selector quadrant cover.

Remove the bolts securing the upper steering column to the mounting bracket; remove the spacers and the packing washers (8, Fig. 79).

Remove the bolts securing the earth leads and the support stays to the steering column mounting bracket. Retain the washers (9, Fig. 79).

Remove the bolt securing the mounting bracket to the screen rail (10, Fig. 79) and retain the bracket.

NOTE: To facilitate refitting, it is advised that the position of all the electrical multi-pin connectors are noted and marked. The position and the routes of all the vacuum pipes noted and marked.

Disconnect the blower motor flap vacuum pipes at the 'T' piece (11, Fig. 79), and the demister duct vacuum pipe at the servo (12, Fig. 79).

Disconnect the main panel harness electrical connectors and remove the harness from the securing clips.

Remove the nuts securing the air conditioning switch panel to the air conditioning unit (13, Fig. 79) and remove the screws securing the mode switch cover, retain the switch cover (14, Fig. 79)

Disconnect the mode switch vacuum pipes and the mode switch electrical connectors.

Disconnect the earth cable and the motor harness multi-pin at the air conditioning main harness.

Disconnect the remaining block connectors including the multi-pin connector of the windscreen wiper motor harness at the bulkhead.

Disconnect the ambient and in car sensors.

Ease the drain tubes clear of the grommets in the transmission tunnel, ease the main panel harness clear of the unit and ease the demist duct vane securing studs from the screen rail.

Retain the demist duct assembly.

Remove the screw securing the air conditioning unit to the top rail (15, Fig. 79).

Manoeuvre the unit from its location taking great care to prevent damage to the unit or to the surrounding components.

With the unit on a workbench, remove the face level vent, the brackets and the demist duct assembly from the unit.

Refit by reversing the above procedure noting that the receiver-drier must be replaced.

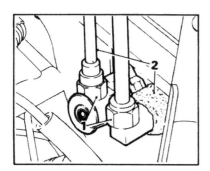
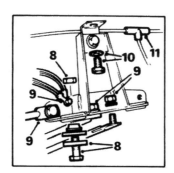
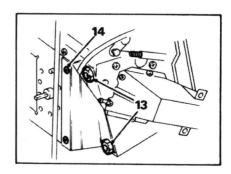

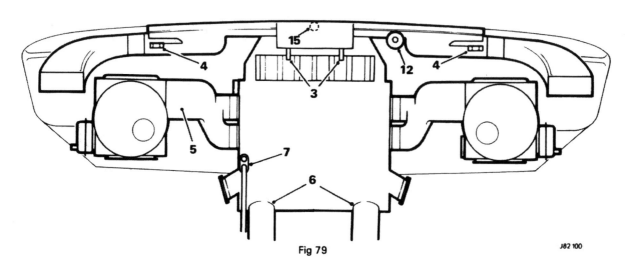

Fig 79

J82 100

BLOWER ASSEMBLY

Remove and refit LH Unit

Removing

Disconnect the battery earth lead.

Remove the left hand underscuttle casing and the left hand console casing.

On right hand drive cars, remove the glove box liner.

Remove the nuts securing the compartment panel (1, Fig. 80) to the blower motor assembly, ease the panel clear and secure for access to blower assembly (2, Fig. 80).

Disconnect the pliable ducting from the stub pipes at the side of the air conditioning unit (3, Fig. 80).

Disconnect the blower motor harness at the block connector.

Disconnect the vacuum pipe from the flap operating servo on the blower assembly (4, Fig. 80).

Open the recirculation flap in the base of the blower assembly and hold open with a suitable wedge (5, Fig. 80).

Remove the bolts securing the blower assembly to the mounting brackets, and ease the blower assembly from its location (6, Fig. 80).

Remove the tape securing the ducting to the assembly and remove the ducting (7, Fig. 80).

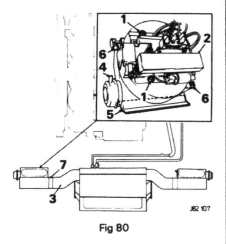

Fig 80

VENTILATORS

Remove and refit — Centre
Right hand
Left hand

Removing

For the centre and passengers side veneer panels, open the glove box for access. By using a suitable long thin-bladed instrument, carefully release the veneer panel, retaining clips, and remove the appropriate panel.

Withdraw the appropriate ventilator.

HEATER MATRIX

Renew

With the air conditioning unit removed and located on a workbench, the heater matrix can be removed.

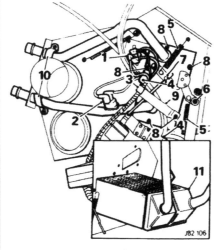

Fig 81

IMPORTANT: It is recommended that the positions of all the operating rods are marked with a scriber, or a similar methods.

Note and disconnect all the cables from the blower motor resistance unit (1, Fig. 81) and the water valve temperature switch (2, Fig. 81).

Remove the screws securing the cable harness clip and the bracket (3, Fig. 81).

Remove the screws securing the vacuum pipe clips (4, Fig. 81).

Disengage the return springs from the operating levers (5, Fig. 81), remove the screw securing the lower flap operating lever to flap hinge (6, Fig. 81) and remove the lever.

Slacken the screw securing the operating rod from the vacuum servo to the flap operating lever on the matrix cover (7, Fig. 81) and release the lever from the rod.

Remove the screws securing the matrix cover plate to the unit (8, Fig. 81).

Remove the screws securing the heater matrix pipes retaining bracket to unit and remove the bracket (9, Fig. 81).

Remove the pipe clips (10, Fig. 81).

With a straight pull, ease the matrix clear of the unit (11, Fig. 81).

Remove the sleeve from the top pipe, the cover plate and the water valve temperature switch from the lower pipe.

EVAPORATOR

Renew

With the air conditioning unit removed from the car and placed on a workbench.

Remove the screws securing the heater matrix pipe retaining bracket to the unit and remove the bracket.

Remove the screws securing the back plate to the unit, ease the rubber pad from the back plate (1, Fig. 82), remove the screws securing the expansion valve mounting plate to the back plate (2, Fig. 82), and ease the back plate (3, Fig. 82) over the expansion valve.

NOTE: Take care to prevent damage to the capillary tube.

Remove the thermostat (4, Fig. 82) by disconnecting the cables and removing the fixing nut. Carefully ease the thermostat capillary tube from the air conditioning unit. Ease the evaporator clear of the air conditioning unit (5, Fig. 82).

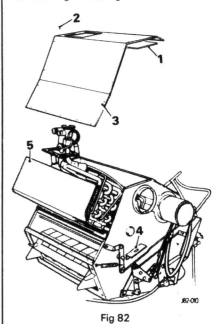

Fig 82

SERVO AND CONTROL UNIT

Remove and refit

Disconnect the battery.

Remove the RH console side casing.

Remove the RH footwell vent by withdrawing the four securing screws (1, Fig. 83).

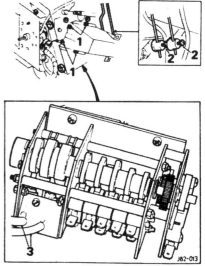

Fig 83

Disconnect the two flap operating rods from the cam followers, marking the rods to facilitate correct refitting (2, Fig. 83).

Mark the vacuum tubes for identification before disconnecting them from the vacuum switches (3, Fig. 83).

Disconnect the cable harness at the multi-pin plug and socket.

Remove the servo unit chrome dome nut and ease the servo clear of the unit.

Refit by reversing the above procedure.

SERVO AND CONTROL UNIT ASSEMBLY

Overhaul

CAUTION: No attempt must be made to dismantle the servo motor from the gearbox. 12 volts must never be applied direct to the motor connections. The motor will over-run the limit switches and could strip the gear assembly. Do not attempt to dismantle the camshaft assembly.

The servo and control unit must not be serviced under warranty.

Dismantling

To remove the Ranco thermostat and recirculation over-ride micro-switches, withdraw the two securing screws and take the switch from the end plate (1, Fig. 84).

The other micro-switches can now be removed by easing the friction washers from the ends of the micro-switch locating rods (2, Fig. 84). Push the rods through the micro-switch pack (3, Fig. 84) and ease the micro-switches from the assembly (4, Fig. 84).

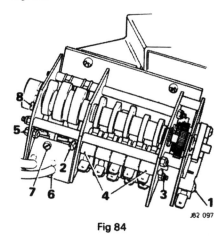

Fig 84

The vacuum switches can be removed by removing the two nuts and screws retaining the vacuum switch mounting bracket (5, Fig. 84). Pull the bracket from the assembly (6, Fig. 84). Remove the nut and screw clamping mounting plates together. Remove the plates to free the switches (7, Fig. 84).

The feedback potentiometer is removed by withdrawing the two securing screws (8, Fig. 84). Note the position of cables and unsolder.

Reassembling

Re-solder the cables to the potentiometer and secure with the two fixing screws.

Reposition the vacuum switches to the clamping plates. Fit the bracket to the assembly.

Secure with the nuts and screws.

Ease the micro-switch pack into the assembly.

Push the locating rods through the micro-switch pack and ease the friction washers onto the end of the rods.

Refit the Ranco thermostat and recirculation over-ride micro-switch with the two securing screws.

VACUUM SOLENOID

Renew

Remove the left hand console side pad (1, Fig. 85).

Remove the screws securing the footwell outlet vent to air conditioning unit and remove the vent.

Remove the nut securing the earth leads to the mounting bolt (2, Fig. 85).

Remove the nut and bolt securing the vacuum solenoid (3, Fig. 85).

Disconnect the vacuum pipes and electrical cables from the solenoid (4, Fig. 85).

Remove the vaccum solenoid.

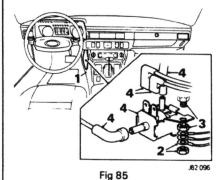

Fig 85

AMPLIFIER UNIT

Renew

Disconnect the battery.

Remove the left-hand console side panel.

Remove the screws securing the footwell vent to the air-conditioning unit and remove the duct (1, Fig. 86).

Remove the nut securing the blower motor relay to the mounting bracket on the air conditioning unit (2, Fig. 86).

Displace the vacuum solenoid from its location and swing aside (3, Fig. 86).

Disconnect the amplifier cable harness multi-pin plug and socket (4, Fig. 86).

Displace the amplifier from the spring clip under the unit and move the harness aside.

Remove the nylon strap securing the harnesses and remove the amplifier (5, Fig. 86).

Fig 86

BLOWER ASSEMBLY

LH
Overhaul RH

NOTE: The blower assembly must not be dismantled under warranty.

Dismantling

Remove the three self tapping screws from the air intake casing (1, Fig. 87).

Part the air inlet casing (2, Fig. 87) from the motor assembly (3, Fig. 87) and disconnect electrical connections at the lucar connectors (4, Fig. 87).

NOTE: It is recommended at this stage that the positions of the various components are marked either with paint or a scriber. This will facilitate reassembly.

One cable Lucar has a raised projection which matches the aperture in the motor casing. This ensures that the connections are replaced correctly and that the rotation of the motor is not altered.

Remove the bolts securing the motor mounting bracket to the fan housing (5, Fig. 87).

Remove the motor and fan assembly from the fan housing.

Remove the mounting bracket from the motor.

Using the appropriate Allen key, remove the impeller fan from the spindle.

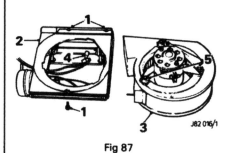

Fig 87

Reassembling

Refit the fan to the motor and secure it to the spindle.

Refit the mounting bracket to the motor.

Locate the fan and mounting assembly into the fan-housing.

Fit and tighten the bolts securing the assembly to the fan housing.

Place the flap box assembly to the fan housing and reconnect the electrical connections.

Fit and tighten the screws securing the flap box to the housing.

Raise the recirculation flap, fit and tighten the remaining screw.

Refit the blower motor assembly.

Reconnect the battery.

CHARGING VALVE CORE

Renew

A possible reason for very slow charging is a bent or damage schrader valve depressor. Do not attempt to straighten. The valve core must be replaced.

If excessive leakage is detected from the schrader valve cores at the rear of the compressor, use a soap solution to ensure that the valve core itself is at fault. If the valve core is leaking replace it by following this procedure.

Ensure replacement clean dry valve core is to hand before commencing operation.

Depressurize the system.

Remove the valve core using a schrader removing tool.

NOTE: Do not overtighten when refitting, then charge the system.

AIR CONDITIONING

Test operation

NOTE: During the following tests windows should be closed and footwell fresh air vents shut 'off'.

Warm the engine up and check operation of thermostatic cut-out and low speed override. RH control to 'auto'.

With the engine cold, turn the LH knob to 'full heat'. Start the engine and run at 1000 rpm. If, after any previous running the camshaft has turned to the cold position, the servo will operate for a few seconds and then shut down. As the water temperature reaches 40°C the system will start up, the centre outlet will close if not already closed, and the fans will slip up to speed 2. This can be checked by turning the RH knob to low, when a drop in speed should be noticed.

Sequence of operation check. RH control to 'auto'

With the engine warm, turn the LH knob to 65°. Operate the cigar lighter or other heat source and hold the heated unit about 1 in below the sensor inlet hole, which is situated below the centre parcel shelf. The unit should then go through the following sequence in approximately 20 seconds.

Blower speeds will drop to low.

Temperatures will decrease, the upper temperature dropping more quickly than the footwell temperature.

After approximately ten seconds the centre outlet flap will open.

Approximately one second after this the fan speeds will shift up to a medium 1.

A further one second later the fan speed will shift up to medium 2.

Another one second later the fan speeds will shift to maximum, at the same time the fresh air vents will close and the recirculating flaps will open. The rush of air into the air boxes will be felt along the bottom edge of the lower trim panels. Turn the RH knob to 'LOW' which should cause the fan speeds to drop. Return RH knob to auto setting.

On some cars in which the servo action is

fairly fast the separation of the fan speeds may not be discernible.

Aspiration and intermediate position check

Remove the heat source from the sensor. Within ten seconds, depending upon ambient conditions, the unit should shift off recirculation and the blowers will drop to one of the intermediate speeds. This test can be carried out on the road since thermistor aspiration will be better and hence the test will be performed more quickly. In certain high ambient conditions the system will be reluctant to come off recirculation, in which case the intermediate modes can be checked by inching the servo through these positions. This is done by turning the LH knob slightly clockwise until the servo motor is heard to operate, and then returning it to a lower position to stop the servo motor at the desired position.

Defrost and fan vibration check

Turn the RH knob to defrost. The centre outlet flap should close and the screen outlets open. Air to the footwells should be cut off leaving air to the upper ducts only. The fans should shift to maximum speed and hot air should issue from the upper ducts. Fan vibration is best assessed under these conditions. Tests in accordance with the defrost schedule can be carried out at this point if desired.

Outlet vent valve check

Check that air can be cut off from the outer face-level vents by rotating the wheels beneath the outlets.

Settled mid-range and High speed over-ride check

Set the RH knob to 'Auto'. Set the LH knob to 75° and wait for the unit to settle. The fans should now be on low speed. Turn the RH knob to 'HIGH'. Maximum fan speeds should now be engaged.

DEFROST AND DEMIST TESTS

Purpose

To ascertain that the heating/air conditioning system is functioning correctly in the 'Defrost' mode, and that adequate airflow is maintained in the heat mode to ensure that the windscreen remains mist-free.

Method

Set the LH control to '85°C'.
Set the RH control to 'Defrost'.
Close the end of dash outlets.
Start the engine and run it for seven minutes at 1500 rpm.
During the running period measure the airflow from each screen outlet using checking ducts and velometer. Ensure that the centre dash outlet is closed and that it seals satisfactorily. The velocity from the screen outlets should be 1550 ft/min.
Also during the running period turn the RH

control to 'HIGH' and open the end of the dash outlets. Using the screen outlet and end of dash checking ducts measure the resulting air velocity. This should be:

Minimum velocity (ft/min)
Screen   End of dash
500   850

At the end of seven minutes running at 1500 rpm check that the water temperature gauge indicates 'Normal'. Using mercury in glass thermometers check that the following minimum screen outlet temperatures are achieved.

Plenum inlet		Screen outlet (minimum)	
°C	°F	°C	°F
10	50	54	129.2
12	53.6	55	131
14	57.2	55.5	131.9
16	60.8	56.5	133.7
18	64.4	57	134.6
20	68	58	136.4
22	71.6	58.5	137.3
24	75.2	59.5	139.1

Conclusions

If the above minimum requirements are met, then it can be assumed that:
(a) The thermostats are opening correctly.
(b) The water valve is opening fully.
(c) The flaps and linkages are correctly adjusted for the heating mode.
(d) The fans give adequate airflow at maximum speed.
If the above criteria are not met, the causes may be related to:

Thermostats

The water temperature guage will not achieve 'Normal' position within seven minutes and the air outlet temperature remains low. The thermostat(s) must be removed and checked for sticking open.

Water valve

The temperature gauge reads 'Normal' but the air outlet temperature remains low. Check that the vacuum-operated water valve is subjectd to at least 21,6 cmHg (8.5 inHg) of vacuum. If the valve is under adequate vacuum, change the valve. However, if the vacuum is low, check that the vacuum is being supplied to the whole system, that the water valve vacuum actuator is operational and that the water valve vacuum switch is operational. (See that the supply from the switch to valve is not pinched or trapped).

Flaps and linkages

Inadequate flap sealing will result in low air velocity at the screen outlets. Check that the centre facia flap closes fully on 'Defrost' and that only a small air bleed to the footwells occurs. These leaks can be detected by hand and may be rectified by adjusting the linkage. Excessive air-flow from the screen outlets in heat mode may be caused by the demist control flap sticking open.

Blowers

If following flap inspection the air flow is still low, investigations should be carried out into the blower assemblies. Check that full voltage is being received on maximum speed and that the units are correctly wired for rotation. If all is correct the only remaining procedure is to change the fan assembly.

NOTE: The engine must be running for this check.

Check that the compressor drive belt is correctly adjusted and is not slipping at higher engine speeds, at idle speed, or on sudden acceleration of the engine, with the compressor clutch speed.

Observe the sight glass on the receiver-drier and check for frothing or bubbles with engine running at 1000 rpm.

Slowly increase engine speed and repeat check at 1800 rpm.

NOTE: It is normal for there to be slight foaming if ambient air temperature is below 21°C (70°F).

Check for frosting on the connector union housing; the region around the suction part is normally cold, and slight frosting is permissible.

Check by feel along pipe lines for sudden temperature changes that would indicated blockage at that point.

Place a thermometer in the air outlet louvres.

Run the vehicle on the road and note the drop in temperature with air conditioning system switch on or off.

Ensure that the condenser matrix is free of mud, road dirt, leaves or insects that would prevent free air-flow. If necessary, clear the matrix.

If the foregoing checks are not met satisfactorily, refer to rectification and fault-finding procedures.

System check

The following check must be carried out to ensure that the system is basically functional. These checks may also be used to ensure satisfactory operation after any rectification has been done. If the system proves unsatisfactory in any way, refer to fault finding.

Check that blower fans are giving an air flow expected in relation to control switch position.

Check that air delivered is equal at both outlets.

Check that compressor clutch is operating correctly, engaging and releasing immediately control switch is set to an 'on' position.

NOTE: the engine must be running and the thermostat control set fully cool.

Check that the radiator cooling fan starts operating when the compressor clutch engages.

CONTENTS

WINDSCREEN WIPERS

Description

The windscreen wiper motor and gearbox assembly is a two speed self parking unit driving two wiper arm wheel boxes via a flexible drive. The motor is a two pole permanent magnet type, the field assembly comprising two ceramic magnets housed in a cylindrical yoke. A worm gear formed on the extended armature shaft drives a moulded gear within the gearbox. The rotary motion of the motor being converted to a linear movement by a connecting rod actuated by a crankpin carried on the gear. The gearbox incorporates the self park mechanism which automatically parks the wiper blades at the end of the wiping cycle.

Two speed operation is obtained by switching the positive feed to the third brush when the higher speed is selected.

The following description of the Windscreen Wiper Operation should be studied with the accompanying circuit diagram.

OFF The Load Relay is energised by selection of 'IGN ON' and applies 12 volts through fuse 9 to terminal 1 on the wiper switch. 12 volts is internally fed via the wiper switch contacts to terminal 7 and from there passes to terminal 1 on the Wiper Motor. This applies 12 volts to one side of the motor winding and, via the closed 'PARK' switch to the closed contacts of the Wiper Timer at terminal 31 b1. Terminal 31 b2 of the timer passes the 12 volts via terminal 6 of the switch and internal connection to terminal 5. From terminal 5 the 12 volt supply is applied to terminal 5 on the motor and thereby the opposite brush to terminal 1, completely stalling the motor.

SLOW The 12 volts on terminal 1 of the switch is applied to terminal 5 and to terminal 5 on the wiper motor. An earth connection to terminal 2 on the switch is

internally applied to terminal 7 and thereby to terminal 1 on the motor. The motor then runs in a forward direction at a slow speed.

FAST The earth connection detailed in 'SLOW' is maintained but the 12 volt supply is moved via terminal 4 on the switch and terminal 3 on the motor to the high speed brush. The motor then runs in a forward direction at a high speed.

OFF When 'OFF' is selected the 'RUN' position of the Park/Run contact in the motor applies an earth at connection 4 of the motor via connection 2 and the timer contacts to position 6 on the switch. Then via the internal connection to 5 on the switch applying the earth to terminal 5 on the motor and the brush previously supplied with 12 volts. The motor therefore stops. Meanwhile 12 volts is applied to the previously earthed motor contact 1 via switch contacts 1 and 7, the motor therefore immediately stops the sweep and runs in a reverse direction to the limit of its travel. At that point the internal Park/Run switch moves to the 'PARK' position, removes the earth at 4 and applies 12 volts via the timer contact to the opposite brush. The motor instantly stalls.

Single Sweep Operation

A single sweep of the wiper is obtained by pulling the lever towards the steering wheel and releasing.

When this position is selected and released, 12 volts are applied both to the coil of the timer unit, operating its contact, and to the terminal 1 on the motor. An earth is applied to the opposite brush of the motor, terminal 5 via the timer contact 31 b2 and an internal connection to terminal 31. The motor therefore starts and runs in a reverse direction. As the supply to the timer is

applied then instantly removed the contacts operate then relax to the 'AT REST' position. The motor earth on the brush of terminal 5 is then achieved through the Park/Run contacts at the 'RUN' position until the motor reaches the end of the sweep. The Park/Run switch then returns to the 'PARK' position, applying 12 volts to brush 5 as previously described. The motor therefore stalls

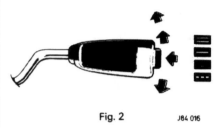

Fig. 2   J84 016

Intermittent Operation

When position 'D' is selected 12 volts are applied to the operating coil of the timer and simultaneously to terminal 1 of the wiper motor. Terminal 5 of the wiper motor is connected via terminal 5 and 6 of the switch to the timer contact 31 b2(1). When 12 volts are applied to the timer its operation consists of an instantaneous 'flip flop' then a delay of approximately 5 seconds and another 'flip flop'. This action applies instantaneous earth to terminal 5 of the motor, starting it running in a reverse direction. The motor then obtains its own earth via the Park/Run contact at 'RUN' to sustain it for one sweep. As the motor is running in a reverse direction, the 'PARK' switch closes at the end of the sweep and applies 12 volts via the new closed contacts of the Timer to the opposite brush to stall the motor in the 'PARK' position.

WINDSCREEN WIPER CIRCUIT

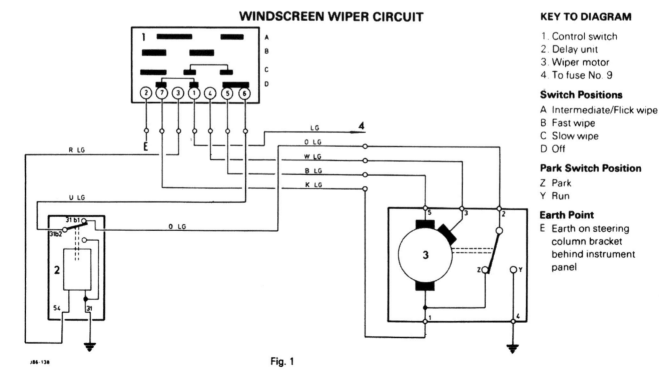

Fig. 1

KEY TO DIAGRAM

1. Control switch
2. Delay unit
3. Wiper motor
4. To fuse No. 9

Switch Positions

A Intermediate/Flick wipe
B Fast wipe
C Slow wipe
D Off

Park Switch Position

Z Park
Y Run

Earth Point

E Earth on steering column bracket behind instrument panel

After approximately 5 seconds the Timer again 'flip flops' removing the 12 volts from the motor brush at 5 and replaces it with an earth. The cycle already described then repeats itself until the Wiper Switch is restored to the 'OFF' position cancelling the operation of the timer contacts. The motor then stalls as previously described.

Fault Finding

Check the fuse and all connections ensuring the earth connections are clean and tight.

With the ignition switched on battery voltage should be obtained at terminals 1, 2 and 5 of the wiper motor. Battery voltage should also be obtained at terminals 31 b1 and 31 b2 on the delay unit. With the wiper switch in the slow run position and the ignition switched on battery voltage should be obtained at terminal 5 of the wiper motor. The wiper motor earth circuit is via terminals 7 and 2 of the wiper switch. In the fast run position battery voltage should be obtained at terminal 3 of the wiper motor. The earth circuit is the same as the slow run position.

In the intermittent wipe position battery voltage should be obtained at terminal 1 of the wiper motor and terminal 54 on the delay unit. The earth is switched intermittently via the delay unit and the Park/Run switch in the wiper motor.

To operate the windscreen washer, press the knob on the end of the control.

The washer reservoir should be filled with soft water where possible. If soft water is not available and continued use of hard water is necessary occasional attention should be paid to washer jet outlet holes. It is permissible to clear deposits from outlets with thin wire when necessary. The washer bottle should also be cleaned out, and the filter flushed occasionally. Windscreen washer additives should be confined to proprietary brands or a mild detergent.

CAUTION: Denatured alcohol or methylated spirit must **NOT** be used.

Washer jets can be adjusted with a screwdriver as illustrated (Fig. 3), the jet should be adjusted to strike the top of the windscreen.

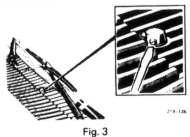

Fig. 3

WINDSCREEN WIPER ARMS AND BLADES

Renew

Raise the plastic cover from the spindle nut (1, Fig. 4).
Remove the arm retaining nut from the spindle (2, Fig. 4).
Remove the arm assembly from the spindle (3, Fig. 4).

NOTE: The position of arm in relationship to spline should be noted at this point.

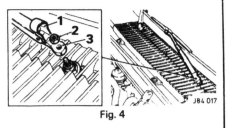

Fig. 4

Depress the blade retaining clip (1, Fig. 5) and remove the blade (2, Fig. 5).

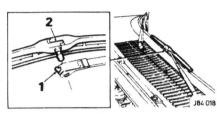

Fig. 5

On refitting ensure the wiper arms and blades are fitted and positioned as shown at (A, Fig. 6) RH Drive or (A, Fig. 7) LH Drive.
Prior to testing the operation of windscreen wipers ensure the windscreen is wet.

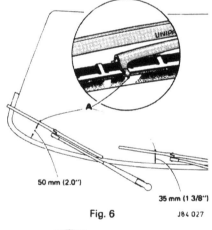

50 mm (2.0")
35 mm (1 3/8")
Fig. 6   J84 027

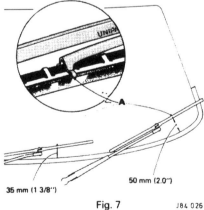

35 mm (1 3/8")
50 mm (2.0")
Fig. 7   J84 026

WIPER MOTOR

Renew

Disconnect the battery earth lead.
Remove the wiper arms (1, Fig. 8).
Remove the air inlet grille retaining nuts, bolts and washers (2, Fig. 8).

Manoeuvre and raise the grille (3, Fig. 8).
Disconnect the washer tube at the jet assembly and remove the multi-plug (4, Fig. 8) connector at the bulkhead.
Remove the four 7 mm bolts securing the motor mounting bracket to the grille (5, Fig. 8).
Remove the wheel box spindle nuts and remove the complete assembly from the grille (6, Fig. 8).
Remove the three 8 mm nuts securing the brackets to the motor and remove the brackets (7, Fig. 8).
Remove the rack cover plate retaining bolts (8, Fig. 8) and lift off the cover (9, Fig. 8) and remove the rack assembly.

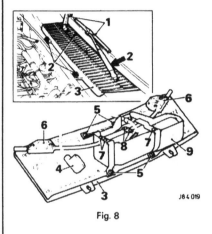

Fig. 8

Remove the screws securing the gearbox cover (1, Fig. 9) and remove the cover (2, Fig. 9).
Remove the circlip washer retaining the gear assembly (3, Fig. 9).
Remove the gear and crankpin assembly noting the position of the belled washer (4, Fig. 9).

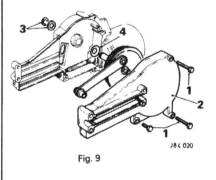

Fig. 9

Refitting is the reversal of the above operations.

WINDSCREEN WIPER MOTOR RACK DRIVE AND WIPER WHEEL BOXES

With the wiper motor removed.
Remove the wheel box back plate nuts (1, Fig. 10).

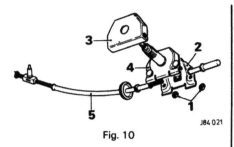

Fig. 10

J84 021

Remove the wheel box back plate (2, Fig. 10) and the shroud (3, Fig. 10). Pull the wheelbox (4, Fig. 10) clear of the rack drive. Remove the bundy tube from rack (5, Fig. 10).

WINDSCREEN WIPER/WASHER CONTROL SWITCH

Renew

Disconnect the battery earth lead.

Slacken the steering wheel adjustment ring (1, Fig. 11) and extend to its maximum travel.

Remove the screws securing the steering column lower shroud (2, Fig. 11) and remove the shroud.

Remove the screws securing the horn pad (3, Fig. 11) and remove the pad (4, Fig. 11).

Adjust the wheels to the straight ahead position, remove the ignition key to lock the steering.

Remove the horn contact rod from the upper column.

Remove the nut securing the steering wheel to the upper column and gently tap the steering wheel withdrawing the wheel from the column.

Remove the screws securing the upper shroud to the bracket on the steering column (5, Fig. 11).

Slacken the pinch screw securing the switch assembly to steering column (6, Fig. 11).

Ease the switch assembly and upper shroud off the steering column.

Remove the shroud from the switch assembly (7, Fig. 11).

Remove the spire and screws (8, Fig. 11) securing the switch mounting.

Disconnect the earth cable at the snap connector (9, Fig. 11).

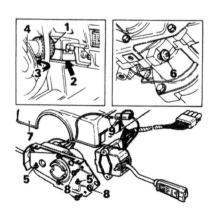

Fig. 11

J84 022

WINDSCREEN WIPER MOTOR HARNESS — BULKHEAD CONNECTOR

Renew

Disconnect the battery earth lead.

Remove the wiper arms and the fresh air intake (1 & 2, Fig. 12).

Disconnect the harness at the multi-pin connection (3, Fig. 12). Disconnect the washer tube.

Remove the bulkhead connector retaining screws (4, Fig. 12) and ease the connector from its location (5, Fig. 12).

Disconnect the panel harness connector (6, Fig. 12) and secure the harness.

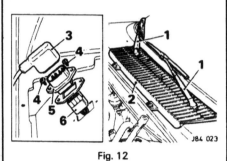

Fig. 12

J84 023

WIPER MOTOR DELAY UNIT

Renew

Disconnect the battery.

Remove the passenger's side dash casing.

Remove the delay unit from the retaining socket behind the auxilliary fuse box (1, Fig. 13).

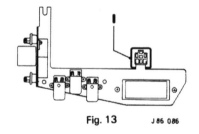

Fig. 13   J 86 086

WASHER PUMP

Renew

Disconnect the battery earth lead.

Displace the washer reservoir from the reservoir bracket (1, Fig. 14).

Disconnect the harness connector from the washer motor (2, Fig. 14). Ensure the reservoir is empty of water.

Loosen the clip retaining the washer tube to motor and remove the tube (3, Fig. 14).

Withdraw the motor from the reservoir (4, Fig. 14).

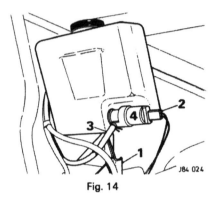

Fig. 14

J84 024

WASHER JETS

Renew

Disconnect the battery earth lead.

Remove the wiper arms (1, Fig. 15).

Remove the air inlet grille retaining nuts, bolts and washers (2, Fig. 12). Raise the grille and the wiper motor assembly (3, Fig. 15).

Disconnect the washer pump to jet tube at the jet (4, Fig. 15).

Remove the jet assembly retaining nut and shake proof washer (5, Fig. 15).

Remove the jet assembly (6, Fig. 15)

Fig. 15

J84 025

HEADLAMP WASH WIPE

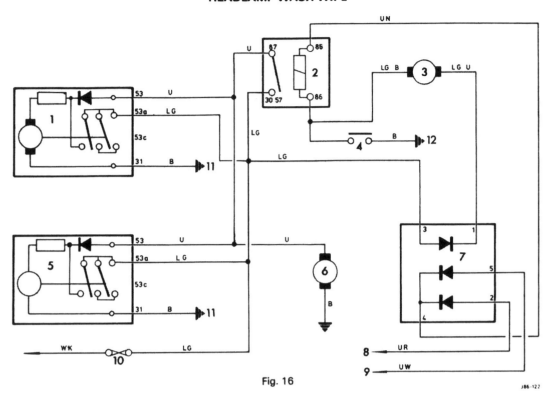

Fig. 16

KEY TO DIAGRAM

1. RH headlamp wiper motor
2. Wash wipe relay
3. Windscreen washer pump
4. Windscreen washer switch
5. LH headlamp wiper motor
6. Headlamp washer pump
7. Diode pack
8. To headlamp dip beam
9. To headlamp main beam
10. Fuse No. 11
11. Earth point on radiator crossmember
12. Earth point on steering column bracket

HEADLAMP WASH WIPE

Description

The headlamp wash wipe circuits will only be activated with headlamps switched to main or dip beam. When the headlamps are switched on power is applied to the wash wipe relay via a diode in the wash wipe diode unit. Power is also supplied to the windscreen washer motor via a diode in the wash wipe diode unit. When the windscreen washer switch is operated the circuit is completed to earth energising the wash wipe relay and the headlamp washer motor. Power is now supplied to the headlamp wiper motors via the relay contacts. When the windscreen washer switch is released the relay is de-energised thus switching off the headlamp washer motor. The headlamp wiper motors will continue to operate via power being applied to terminal 53a on the wipers until the wiper internal switch contacts open.

Fault finding

Check that all connections are clean and tight. Check the fuse.
Ensure the earth connections are clean and tight.
With the ignition switched on. Battery voltage should be obtained at terminals 1 and 3 of the diode module. If battery voltage is obtained at terminal 3 but a zero reading at terminal 1 a faulty diode is indicated in the diode module.
With the headlamps switched to main beam. Battery voltage should be obtained at terminals 5 and 4 of the diode module. With the headlamps switched to dip beam. Battery voltage should be obtained at terminals 2 and 4 of the diode module. A zero reading at terminal 4 in either main or dip beam position indicates a faulty diode in the module.
With the ignition switched on. Battery voltage should be obtained at terminal 30/51 of the wipe/wash relay. With headlamps switched on. Battery voltage should be obtained at terminals 85 and 86 of the relay. When the windscreen washer is operated the terminal 86 of the relay should drop to zero, and battery voltage should then be obtained at terminal 87 of the relay. If the voltage remains at 12 volts at terminal 86 of the relay. Check the wash/wipe switch, and wiring. If the terminal 86 voltage drops to zero voltage but the terminal 87 voltage remains at zero, replace the relay.
With the ignition switched on. Battery voltage should be obtained at terminal 53a of the headlamp wiper motor. With headlamps switched on and the washer button pressed, battery voltage should be obtained at terminal 53 of the wiper motor. Terminal 31 of the motor should be earthed. If the voltage reading and the earth are satisfactory remove the wiper motor and bench check.

HEADLAMP WIPER MOTOR

Renew

Disconnect the battery earth lead.
Raise the plastic covers from the wiper motor spindle nuts and remove the wiper arm retaining nuts.
Remove the wiper arm assembly from the spindles.
Remove the screws securing the top finisher to the top of the lamp housing, then ease the top of the finisher away from the lamp unit and lift the bottom locating spigots from the housing.
Depress the nylon securing tabs retaining the lamp unit.
Withdraw the unit from its housing and disconnect the cable harness at the block connectors behind the unit.
Remove the two nuts and washers securing the wiper motor assembly bracket (1, Fig. 17).
Manoeuvre the motor assembly clear of its location.
Disconnect the motor assembly clear of its location.
Disconnect the cable harness block connector and remove the motor.
Remove the two nuts and washers securing the bracket to the motor (2, Fig. 17) and remove the bracket (3, Fig. 17).

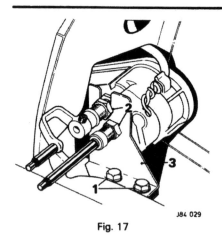

Fig. 17

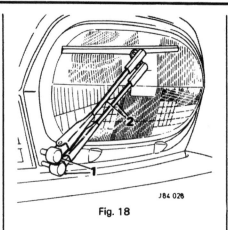

Fig. 18

HEADLAMP WASH/WIPE RELAY

Renew

Disconnect the battery earth lead.
Remove the relay from the retaining tab on the radiator cross member.
Disconnect the cable harness block connector and retain the relay.

HEADLAMP WASH/WIPE DIODE PACK

Renew

Disconnect the battery earth lead.
Remove the diode pack from the retaining tab on the radiator top cross member.
Disconnect the cable harness block connector and retain the diode.

HEADLAMP WIPER ARM ASSEMBLY

Renew

Raise the plastic covers from the headlamp wiper motor spindle nuts and remove the nuts from the spindles (1, Fig. 18).
Remove the wiper arm assembly (2, Fig. 18) from the spindles.
Remove the washer tube from the headlamp washer jet located on the arm assembly and recover the arm.

HEADLAMP WIPER BLADE

Renew

Lift the wiper arm assembly from the headlamp.
Prise the blade from the arm.

HEADLAMP WASHER PUMP

Renew

On cars fitted with the headlamp wash/wipe facility the washer reservoir is of larger capacity than the normal windscreen washer only reservoir. The reservoir contains both the headlamp and windscreen washer pumps.
Disconnect the battery earth lead.
Remove the reservoir retaining bar securing nuts and remove the retaining bar.
Displace the reservoir and ensure the reservoir is empty of water.
Disconnect the cable harness connectors from the washer pump motors (2, Fig. 14).
Loosen the clip retaining the washer tube to the washer motor and remove the tube.
Withdraw the motor from the retaining clip and the reservoir.

CONTENTS

INTRODUCTION

Fault diagnosis is the method of locating faults while the electrical equipment is still fitted to the vehicle. In the interests of efficiency and economy, the diagnosis must be carried out in the shortest possible time. It is the aim of this manual to present a series of tests that may be carried out in order to achieve this objective. The electrical systems of Jaguar Cars are sophisticated and, of necessity, complex. Besides the main wiring diagram, this manual divides the circuitry into more readily understood subsystems. With each system diagram is a description of the circuit and a test procedure to ensure the fast diagnosis of electrical faults.

Some of the tests described require the use of specialised equipment, and it is in the interest of efficiency and economy that this is used.

GENERAL PRACTICE

1. Always disconnect the battery earth lead before disconnecting any components.

2. Always disconnect the battery earth lead before connecting an ammeter into a circuit.

3. When connecting electrical components, always ensure that a good contact is made by the connectors. Ensure that earth connections are made to a clean metal surface, and are tightly fastened using the correct screws and washers.

WARNING

1. When using an arc or spot welder on the vehicle, disconnect the battery and remove any electrical equipment in the close proximity of the welding.

2. Do not disconnect the battery with the engine running.

3. Do not use a high speed charger as a starting aid.

4. When using a high speed charger to charge the battery, the battery must be disconnected from the rest of the vehicle's electrical system.

5. When installing, ensure that the battery is 12 volts and is connected with the right polarity, i.e. negative earth.

6. Ensure that the battery is kept in an upright position when removing or refitting.

7. All car batteries generate hydrogen gas which is highly inflammable. If ignited by a spark or flame, the gas may explode, causing spraying of acid, fragmentation of the battery, and possible personal injuries. Wear safety glasses when working near batteries. In case of contact with acid, flush immediately with water.

ALTERNATOR

ALTERNATOR CIRCUIT

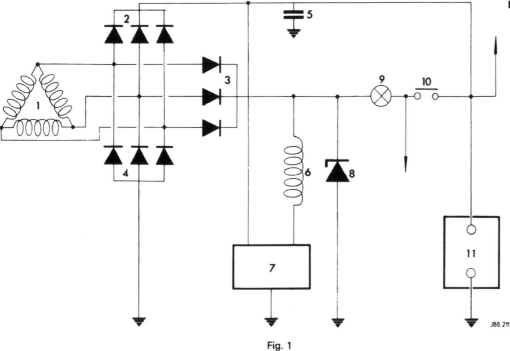

KEY TO CHARGING CIRCUIT

1. Stator
2. Output diodes
3. Auxiliary diodes
4. Negative diodes
5. Suppressor
6. Rotor
7. Regulator
8. Surge protection diode
9. Warning lamp
10. Ignition switch
11. Battery

Fig. 1

Description

The A133 alternator is a three-phase machine with a delta wound stator, twelve pole rotor, full wave rectification, and a 15TR voltage regulator.

The alternator is machine sensed with an externally fitted radio suppression capacitor. When the ignition switch is turned on, the rotor winding (6, Fig. 1) is connected to the battery via the warning lamp (9, Fig. 1).

The small current flowing through the rotor (6, Fig. 1) to earth via the voltage regulator (7, Fig. 1) produces a magnetic field which is sufficient to begin the build-up of the alternator output voltage through the output diodes (2, Fig 1) and the auxiliary diodes (3, Fig. 1). As the voltage builds-up, the same voltage will be applied to both sides of the warning lamp and the lamp will be extinguished. The action of the regulator is similar to that of the vibrating contact type of voltage control unit, but the switching of the field circuit is achieved by transistors instead of vibrating contacts. When the battery voltage reaches approximately 14 volts, the transistors located in the control box switch off and on very quickly in order to maintain the 14 volts.

A surge protection device is connected between the 'IND' terminal and the frame which is an avalanche diode (8, Fig. 1). This device protects the alternator by absorbing the high transient voltages caused by faulty connections, or the removal of the battery leads while the engine is running.

Testing

NOTE: Check that all connections are clean and tight. Check the fan belt. A load of 1,5 kg (3.3 lb) must give a total belt deflection of 4,4 mm (0.17 in) when applied at the mid-point of the belt. Check the battery hydrometer readings.

Test 1

Remove the connectors from the alternator. Switch the ignition on.

Connect the voltmeter between a good earth and each of the disconnected leads in turn (Fig. 2). The voltmeter should indicate battery voltage.

If the voltmeter indicates a zero reading when connected to the main output lead, check the wiring to the starter solenoid and battery.

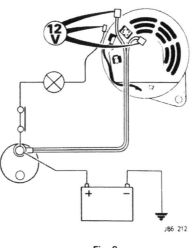

Fig. 2

If the voltmeter indicates a zero reading when connected to the 'IND' lead, check for earth or open-circuit between the warning light and the alternator connector. Check the warning light bulb and all connections to the warning light.

Test 2

Refit the alternator connectors.
Switch the ignition on.
Connect the voltmeter between a good earth and the 'IND' terminal (Fig. 3).
The voltmeter should indicate approximately 2 volts.

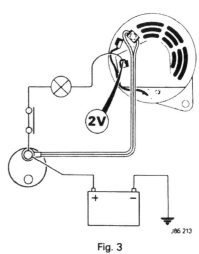

Fig. 3

If the voltmeter indicates a zero reading, the surge protection diode is suspect and should be checked.
If the voltmeter indicates battery voltage the brushes, rotor, or regulator are suspect. Proceed to the next test.

NOTE: If the warning light operates with the ignition off but goes out when the ignition is switched on, check the voltage at the 'IND' terminal with the ignition switched 'off'. If battery voltage is indicated, the diode pack is faulty.

Test 3

Connect the voltmeter between a good earth and the metal link on the regulator (Fig. 4). Switch the ignition on. The voltmeter should indicate approximately 0.5 volt. If 12 volts is indicated, the regulator is faulty.

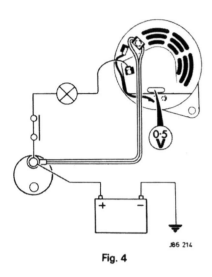

Fig. 4

Test 4

Start and run the engine at a constant 2500 rev/min.

Connect the voltmeter to a good earth and the 'IND' terminal; note the voltage.
Connect the voltmeter to the main output terminal; the voltmeter readings should be the same (Fig. 5). If there is a difference of more than 0.5 volt, the diode pack is suspect.

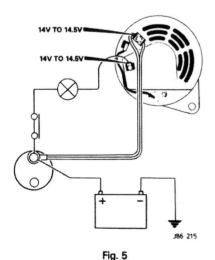

Fig. 5

Test 5

Connect the voltmeter between the battery insulated terminal and the alternator main output terminal (Fig. 6).
Start and run engine at approximately 2500 rev/min. The voltmeter should not exceed 0.5 volt.
If the voltmeter reading is higher than 0.5 volt, check the wiring from the alternator to the battery for loose or dirty connections.

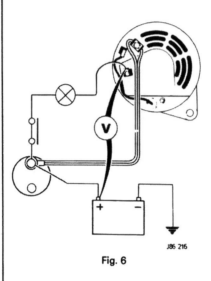

Fig. 6

NOTE: The warning light glowing while the engine is running at normal charging speeds usually indicates a faulty diode pack or dirty or loose connections in the wiring from alternator to battery.

Test 6

Disconnect the battery earth lead.
Disconnect the alternator.
Connect an ammeter between the main output terminal of the alternator and the disconnected output lead.
Connect a jumper lead between the 'IND' lead and 'IND' terminal (Fig. 7).
Re-connect the battery.
Switch on all load (except wipers) for one minute.

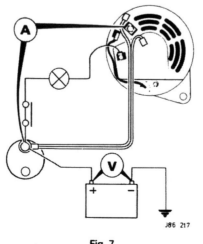

Fig. 7

Start and run the engine at normal charging speed. The ammeter should indicate the maximum output for the alternator.
If the output is low, short the metal link on the regulator to earth with a jumper lead and repeat the test.
If maximum output is now indicated on the ammeter, the regulator is suspect.
Should the output still be low, the stator windings are suspect.
Disconnect the battery earth lead.
Connect the ammeter in series with the alternator main output cable and the starter solenoid.
Re-connect the battery.
Connect the voltmeter across the battery terminals.
Start and run engine at normal charging speed until the ammeter reads less than 10A.
The voltmeter should read 13.6 to 14.4 volts.
An incorrect reading indicates that the regulator is faulty.

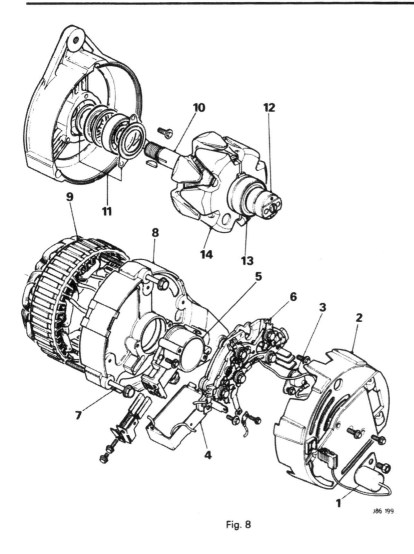

KEY TO ALTERNATOR

1. Capacitor
2. Cover
3. Surge protective diode
4. Regulator
5. Brush box assembly
6. Rectifier pack
7. Through bolts
8. Slip-ring end bracket
9. Stator
10. Rotor shaft
11. Bearing kit
12. Slip-ring
13. Slip-ring end bearing
14. Rotor

J86 199

Fig. 8

Specification	
Voltage	12 volts
Maximum rev/min	15,000 rev/min
Maximum output	75 amps
Regulated voltage	13.6 — 14.4 volts
Rotor resistance	2.46 ohms
Stator winding resistance per phase	0.144 ohms
Maximum brush length	20 mm (0.79 in)
Minimum brush length	10 mm (0.39 in)
Warning lamp bulb	2.2 watts

ALTERNATOR OVERHAUL

Dismantle

Disconnect the capacitor Lucar connector.
Remove the capacitor securing screw and remove the capacitor (1, Fig 8)
Remove the two screws securing the cover and remove the cover (2, Fig 8)
Remove the surge protection diode (3, Fig 8)
Note the arrangement of the regulator leads, disconnect the leads and remove the regulator (4, Fig 8)

Remove the two screws securing the brush box assembly and remove the brush box (5, Fig 8)
Apply a hot iron to the stator lead terminal tags on the rectifier pack and prise out the stator leads when the solder melts.
Remove the remaining two screws securing the rectifier pack assembly (6, Fig 8) and lift the pack from the slip-ring end bracket (8, Fig 8)

Remove the three through bolts (7, Fig 8) and lift the slip-ring end bracket (8, Fig 8) from the stator (9, Fig 8) using a mallet if necessary
Note the position of the stator leads relative to the alternator fixing lugs, and then lift the stator (9, Fig 8) from the drive end bracket.
Remove the shaft nut, washer, pulley, cooling fan, woodruff key and spacers from the rotor shaft (10, Fig 8)
Press the rotor shaft from the drive end bearing (11, Fig 8).
To replace the slip-ring end bearing (13, Fig 8) unsolder the outer and inner slip-rings (12, Fig 8) then prise the slip-rings gently off the rotor shaft.
Using a suitable extractor withdraw the bearing from the rotor shaft.

NOTE: Care should be taken not to damage the insulation on the rotor leads when removing or refitting the slip-rings. Use a resin covered solder ensuring a build-up of solder does not occur on the upper face of the inner slip-ring.
Check all the components using normal procedures. Referring to the resistance values and brush lengths as detailed.
Re-assembly is the reversal of the dismantling procedure ensuring the brushes move freely in the brush box, also ensure the slip-rings are clean and smooth.

ROTOR

Testing

Connect an ohmmeter to the slip-rings (Fig. 9) and the ohmmeter should give a

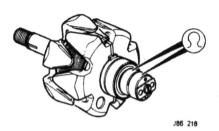

Fig. 9

reading of 2.46 ohms. With the ohmmeter connected to one slip-ring and the rotor body (Fig. 10), the ohmmeter should register infinity.

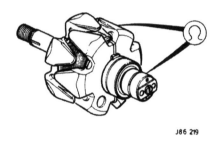

Fig. 10

STATOR

Testing

Visually inspect the stator windings for signs of damage due to overheating.

Check the insulation with a 110 volt test lamp. Connect the test leads to the laminated yoke and to each of the stator leads in turn (Fig. 11). If the lamp lights at any one of the stator leads, the stator is defective.

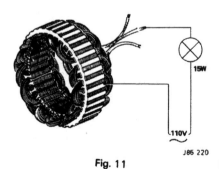

Fig. 11

DIODES

Testing

With diode pack disconnected and isolated from the alternator, connect a battery lead in series with a test lamp to a diode plate.

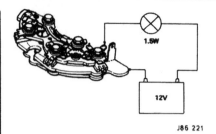

Fig. 12

Connect the other test lead to a pin (Fig. 12). Reverse the test lamp connections to the diode plate and diode pin. The lamp should illuminate when in one direction only. Should the lamp illuminate when connected in both ways or not illuminate at all, the diode pack is defective. The remainder of the diodes in the diode pack assembly can be checked the same way.

ALTERNATOR — DRIVE BELT

Renew and adjust

Slacken the pivot bolts securing the air conditioning compressor (1, Fig. 13).
Slacken the adjusting link securing bolt (2, Fig. 13) and the trunnion block bolt (3, Fig. 13).
Slacken the adjusting link lock nut and adjust the compressor towards the engine (4, Fig. 13) until the compressor drive belt can be removed.
Slacken the alternator pivot nut and bolt (5, Fig. 13).
Slacken the bolt securing the adjusting link (6, Fig. 13) and the trunnion block bolt (7, Fig. 13).
Slacken the adjusting link lock nut; adjust the alternator towards the engine by means of the adjusting nut (8, Fig. 13).
Remove the trunnion block nut; push the alternator towards the engine until the drive belt can be removed from the pullies.

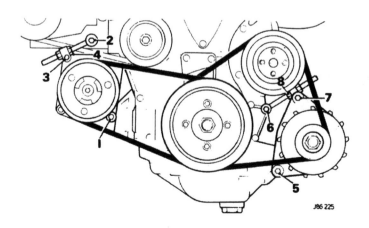

Fig. 13

On fitting new belt, ensure the drive belts are adjusted to the correct tension. A load of 1,5 kg (3.3 lb) must give a total belt deflection of 4,4 mm (0.17 in) when applied at mid point of the belts.

ALTERNATOR

Renew

Disconnect the battery earth lead.
Remove the air cleaner for ease of access.
Slacken the alternator pivot bolt (1, Fig. 14).
Slacken the bolt securing the adjusting link (2, Fig. 14), and the trunnion block bolt (3, Fig. 14).
Slacken the adjusting link lock nut (4, Fig. 14) and adjust the alternator towards the engine by means of the adjusting nut (5, Fig. 14).

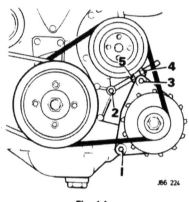

Fig. 14

Remove the bolt securing the trunnion block; ease the belt off the pulley.
Remove the pivot nut and bolt and manoeuvre the alternator from the engine compartment.
On refitting, ensure that the drive belt is adjusted to the correct tension. A load of 1,5 kg (3.3 lb) must give a total belt deflection of 4,4 mm (0.17 in) when applied at mid point of the belt.

BATTERY

The battery is of advanced design and has improved performance characteristics.
High and low electrolyte level marks are moulded into the case.
Topping-up should be carried out when the electrolyte level falls below the top of the separators or the low level mark on the case.
The vent cover should be left in position at all times, except during the topping-up procedure.
When battery charging is carried out, the vent cover should be left in position to allow gas to escape or flooding of electrolyte may result.

Testing

The electrolyte consists of a mixture of sulphuric acid and water in given proportions.
The electrolyte becomes weaker as the cell discharges, and this weakening effect is directly proportional to the amount of electricity given up by the cell. Therefore the specific gravity of the electrolyte gives a direct indication of the condition of the battery.
When the tube of a hydrometer is inserted into the electrolyte and the rubber bulb is pressed and released, a small quantity of electrolyte is drawn into the hydrometer. The specific gravity of the electrolyte determines the depth of the float in the liquid. With the float in a high position, the specific gravity is high. If the specific gravity is low, the float sinks to a lower position. The specific gravity readings are taken when the liquid level crosses the scale on the float, and this gives an accurate indication of the state of charge of the battery.
The volume of electrolyte and hence its specific gravity varies with temperature. Therefore readings of electrolyte taken at temperatures other than 15°C (60°F) (Fig. 15) should be corrected to correspond with the equivalent reading at 15°C (60°F).

Electrolyte Temperature Correction

For every 10°C below 15°C subtract 0.007 from the hydrometer reading and for every 10°C above 15°C add 0.007 to the hydrometer reading.

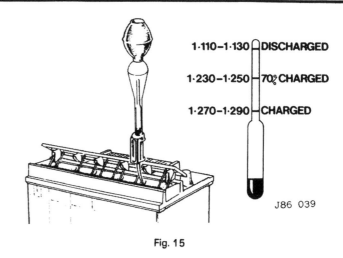

1.110–1.130 DISCHARGED
1.230–1.250 70% CHARGED
1.270–1.290 CHARGED

J86 039

Fig. 15

Example:
Specific gravity reading = 1.250
The temperature = 5°C
The equivalent specific gravity at 15°C
    = 1.250 −0.007
    = 1.243

For every 10°F below 60°F subtract 0.004 from the hydrometer reading and for every 10°F above 60°F add 0.004 to the hydrometer reading.

Example:
Specific gravity reading = 1.250
The temperature = 50°F
The equivalent specific gravity at 60°F
    = 1.250 −0.004
    = 1.246

Heavy discharge test

This test can be carried out as a further check of the battery condition. A heavy discharge tester should be applied to the battery terminals as shown in (Fig. 16).
The tester should be set to discharge the battery at three times the ampere hour rate (20 hour rate) for 15 seconds. On a battery with a capacity of 68 ampere hour rate, the tester should be set to 204 amps. Observe the voltmeter during the 15 seconds the battery is being discharged. If the voltage drops below 9.6 volts, the battery is suspect, but if the voltmeter reading is above 9.6 volts, the battery can be considered satisfactory.

Starting with jump leads and booster battery

Both booster and discharged battery should be treated carefully when using jumper cables. Follow exactly the procedure outlined below, being careful not to cause sparks:

1. Apply hand brake and place transmission in neutral. Turn off lights, heater and other electrical loads.

2. Attach one end of one jumper cable to the positive terminal of the **booster battery** and the other end of same cable to positive terminal of **discharged battery.** DO NOT PERMIT vehicles to touch each other as this could establish an earth connection and counteract the benefits of this procedure.

3. Attach one end of the remaining negative cable to the negative terminal of the **booster battery** and the other end to earth at least 305 mm (12 inches) from the battery of the vehicle being started. (DO NOT CONNECT DIRECTLY TO THE NEGATIVE POST OF THE DEAD BATTERY).

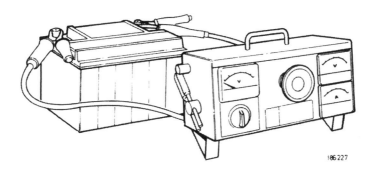

186 227

Fig. 16

Renew

Remove the battery cover by releasing the two fasteners and easing the cover from beneath the retaining fold on the battery clamp (1, Fig. 17).
Ease back the battery terminal cover from the terminals; slacken the clamp bolts and disconnect the battery.
Release the battery filler cover securing strap and remove the cover (2, Fig. 17).
Release the battery clamp securing nuts (3, Fig. 17), and remove the clamp, then remove the battery from the luggage compartment.
On refitting, smear the battery terminals with petroleum jelly before re-connecting the battery leads.
Ensure the battery is kept level to prevent spillage of electrolyte.

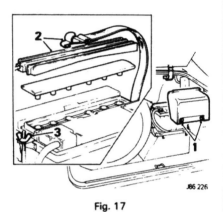

Fig. 17

ELECTRICALLY OPERATED WINDOWS

Description

Power is supplied to the relay contacts from the main battery supply via a thermal circuit breaker. The window lift relay is activated when the ignition is switched on. When the control switch is operated to wind the windows down the contacts AD and CF are connected. This allows current to flow to the motor via contacts AD and the circuit to earth is completed via contacts CF. When the control switch is operated to wind the windows up, the contacts BD and CE are connected. This allows the current to flow through the motor in the opposite direction via contacts CE and the circuit to earth is completed via contacts BD.

Fault finding

Check the fuse and all connections, ensuring the earth connections are clean and tight.
Check the thermal circuit breaker by joining the circuit breaker leads together. Switch on the ignition and operate the window lift switches. Should the windows operate satisfactorily replace the circuit breaker.
With the ignition switched off battery voltage should be obtained at terminal C1 of the window lift relay. With the ignition switched on battery voltage should also be obtained at terminals W1 and C2 of the relay. If battery voltage is obtained at terminals W1 and C1 but a zero reading at terminal C2 replace the relay.

With the ignition switched on battery voltage should be obtained at the brown and blue lead terminal of the LH window lift switch. Operate the switch. Battery voltage should now be obtained at the red and green lead terminal when the switch is operated in one direction, or the green and red lead terminal when the switch is operated in the opposite direction. Should a zero reading be at either test point replace the switch.
The same checks apply for the RH window lift switch. Noting the switch cable colours are red and blue for one direction. Green and blue for the reverse direction.
If the checks prove satisfactory, check the window lift motor wiring for continuity. Should the wiring prove satisfactory remove the window lift motor for bench checks.

WINDOW LIFT MOTOR

Renew

Disconnect the battery earth lead.
Remove the screw securing the arm rest to the door trim pad, and release the arm rest finisher.
Slacken the two screws securing the interior light switch plate and pull the door pad clear of the striker plate.
Remove the screw securing the bottom of the door pad, unclip the door trim pad, disconnect the speaker and remove the door pad assembly.
Remove the inner door handle securing screws (1, Fig. 19) and disengage the remote control levers from the handle (2, Fig. 19).

WINDOW LIFT CIRCUIT

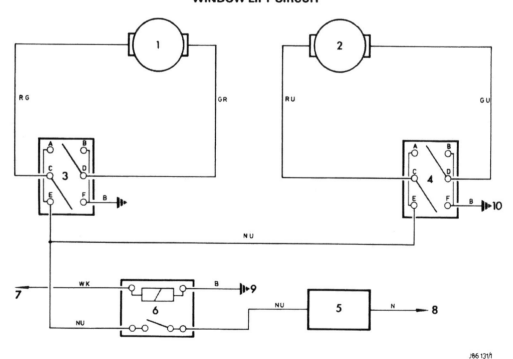

KEY TO DIAGRAM

1. LH window lift motor
2. RH window lift motor
3. LH switch
4. RH switch
5. Thermal circuit breaker
6. Window lift relay
7. To ignition switch
8. Terminal post
9. Earth on passenger side of air conditioning unit
10. Earth behind fascia panel

Fig. 18

Completely lower the window, ease the waterproof sheet clear of the motor connections and disconnect the motor (3, Fig. 19).

Remove the bolts securing the motor (4, Fig. 19), ease the window lift assembly towards the rear of door, and disengage regulator from the locating channel.

Raise the glass and wedge at its highest point, withdraw the seven securing bolts (5, Fig. 19) and remove the mounting plate.

Remove the motor and quadrant assembly from the door.

Before disengaging the regulator from the motor, ensure that the lifting arm and quadrant of the regulator are clamped in a vice.

NOTE: This prevents the spring disengaging suddenly and causing possible damage.

Remove the bolts securing the regulator to the motor.

Reverse the above operations when fitting new unit.

Fig. 19

WINDOW LIFT SWITCHES

Renew

Disconnect the battery earth lead.
Withdraw the three screws securing the centre panel in the console (1, Fig. 20).
Remove the gear lever knob.
Lift the panel for access to the rear of the switches; note the position of the cables and the switches (2, Fig. 20).
Disconnect the cables from the appropriate switch, depress the switch securing tags and push the switch through the panel (3, Fig. 20).
On refitting, ensure the cable connections are clean and tight.

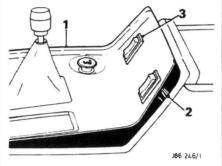

Fig. 20

WINDOW LIFT RELAY AND CIRCUIT BREAKERS

Renew

Disconnect the battery earth lead.
Remove the passenger's side dash liner.
Remove the nuts securing the component panel to the fan motor assembly, displace the component panel and remove the relay located behind the panel (1, Fig. 21).

The circuit breakers are located on the component panel (2, Fig. 21) secured by two screws.

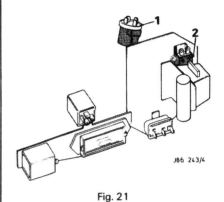

Fig. 21

ELECTRICALLY OPERATED DOOR LOCKS

Description

The electric door lock circuit is activated from either of the two doors if the key is turned in either door lock, the two door solenoids will activate into the lock position, or the unlock position.

The two interior door lock control levers will also operate the two door solenoids into the locked or into the unlocked positions. The 12 volt supply is taken from the terminal post through a thermal circuit breaker to the

DOOR LOCK CIRCUIT

KEY TO DIAGRAM

1. Lock relay
2. Trigger unit
3. Unlock relay
4. Trigger unit
5. Thermal circuit breaker
6. LH unlock solenoid
7. LH lock solenoid
8. RH unlock solenoid
9. RH lock solenoid
10. Fuse No. 8
11. Terminal post
    Earth points behind the fascia panel

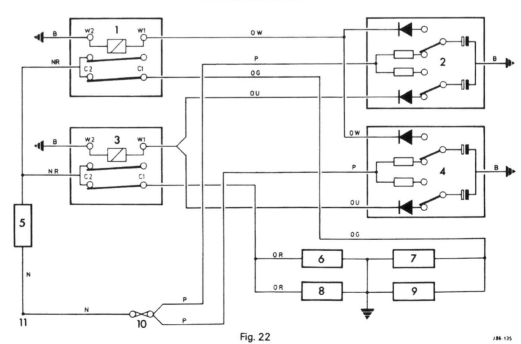

Fig. 22

297

lock and unlock relays. 12 volts is also taken from fuse No 8 to the resistors in the RH and LH trigger units. The trigger units contains 2 diodes, 2 resistors, 2 electrolytic capacitors and 2 micro switches. The capacitors are charged to the battery potential via the resistors.

If one of the trigger units is turned to the unlocked position (by the door key or by the interior levers) the micro switches are moved to the unlock position, allowing the capacitor to discharge through a diode a surge of current to the unlock relay coil. This action closes the relay contacts thus completing the circuit to the door solenoids. After the initial discharge from the capacitor the unlock relay will de-activate reverting to its normal contacts open position. The second trigger unit will be moved to the unlocked position mechanically by the action of the solenoid to which it is attached. While the trigger units are kept in the unlocked position the door lock capacitors are kept in a charged state.

The lock and unlock circuits are protected from each other by the diodes.

Fault finding

Should the door locks not operate electrically check that all the connections are secure ensuring that the earth connections are clean and tight. Check the fuse.

Check the thermal circuit breaker by joining the thermal breaker leads together. If the door locks now operate replace the thermal circuit breaker.

To check the lock or unlock relay, battery voltage should be obtained at the C1 terminal of the appropriate relay.

Disconnect the lead from the W1 terminal and connect a voltmeter between the disconnected lead and a good earth. Operate the locks and battery voltage will be obtained from the disconnected lead which will gradually drop. If a zero reading is obtained at the disconnected lead check that battery voltage is obtained on the purple lead of the trigger unit, if satisfactory replace the door lock solenoid assembly.

DOOR LOCK SOLENOID

Renew

Ensure that the window is fully closed. Disconnect the battery earth lead. Remove the arm rest and door trim pad. Remove the solenoid securing screws (1, Fig. 23) disconnect the cable harness block connectors (2, Fig. 23), unhook the solenoid operating piston (3, Fig. 23) from the door lock push rod and remove the solenoid.

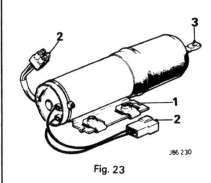

Fig. 23

DOOR LOCK SOLENOID RELAY

Renew

Disconnect the battery earth lead.
Remove the appropriate side dash liner and footwell side casing.
Remove the screws securing the relay mounting bracket to the body side panel (1, Fig. 24).
Note the position of the cable connectors and disconnect the cables (2, Fig. 24).
Remove the relay from the mounting bracket.

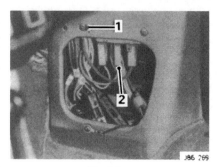

Fig. 24

HORNS

Description

Twin horns are fitted, mounted on the front lower cross-member behind and beneath the front bumper. With the ignition switched on, current is supplied to the horn relay coil via fuse No. 11 and the earth circuit is completed via the horn push.

With the horn relay energised, current will flow to operate both horns simultaneously via the relay contacts.

Fault finding

With the ignition switched on, battery voltage should be obtained at the terminals 85, 86 and 87 of the relay. If battery voltage is obtained at terminal 85 but not at terminal 86 the relay is faulty.

When the horn push is operated the terminal 86 voltage should drop to zero. If terminal 86 remains at battery voltage check the horn push and the wiring to the horn push. With the horn push pressed and the relay terminal 86 voltage at zero, battery voltage should be obtained at the terminal 30/51 of the relay. A zero reading at terminal 30/51 indicates a faulty relay.

Should the above tests prove satisfactory check the wiring to the horns and both horns.

HORN CIRCUIT

KEY TO DIAGRAM

1. Horn push
2. Horn relay
3. To terminal post
4. Horn
5. Horn
6. Fuse No. 11
7. Earth point on steering column bracket
8. Earth point on radiator top rail

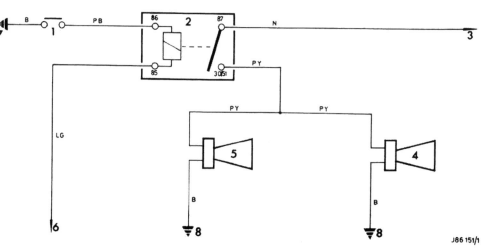

Fig. 25

HORN PUSH

Renew

Disconnect the battery earth lead.
Slacken the steering column adjustment ring and extend the column to its full extent (1, Fig. 26).
Withdraw the two screws from behind the steering wheel (2, Fig. 26).
Remove the horn push from the steering wheel (3, Fig. 26).
Ease the trim pad from the horn push (4, Fig. 27) and recover the motif (5, Fig. 27).

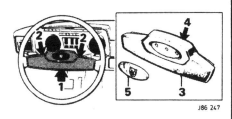

Fig. 26

HORN RELAY

Renew

The horn relay is located at the front LH side of the engine compartment.
Remove by disconnecting the cable harness block connector and withdraw the relay from its locating tab.

HORNS

Renew

Disconnect the battery earth lead.
Remove the nut securing the horns to the mounting beneath the front bumper apron (1, Fig. 27).
Lower the horns and retain the distance pieces and washers.
Disconnect the supply leads and recover the horns (2, Fig. 27).

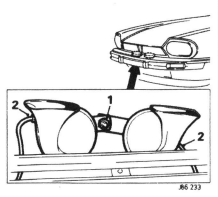

Fig. 27

IGNITION SYSTEM

IGNITION SYSTEM CIRCUIT

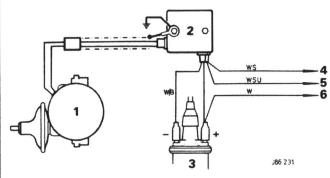

Fig. 28

SYSTEM DESCRIPTION

The constant energy electronic ignition system employs output current limiting and variable dwell for optimum performance. A long dwell is provided at high speeds for adequate energy storage in the coil, and a dwell is provided at low speeds for minimum power dissipation. The output current limiting function of the amplifier maintains the coil storage energy for spark, and consequently the system open circuit output voltage constant over a wide engine speed range. It eliminates the need for a ballast resistor whilst ensuring that the correct coil current flows at all times, even when cranking. No current flows through the HT coil when the ignition is switched on but the engine is stationary.

The distributor incorporates a standard automatic advance system, anti-flash shield, rotor arm and cover but the contact breaker is replaced by a reluctor and pick-up assembly.

The reluctor is a gear-like component and it is mounted on the distributor shaft in place of the cam. The pick-up consists of a winding around a pole piece attached to a permanent magnet.

The distributor is prewired with two leads terminating in a moulded two-pin inhibited connector.

When a reluctor tooth passes across the pick-up limb, the magnetic field strength around the pick-up winding is intensified creating a voltage in the winding. The rise and fall of this voltage is sensed by the amplifier and is used to trigger the transistorised output stage of the amplifier which switches on and off the current flowing in the primary winding of the ignition coil.

The amplifier assembly consists of a solid state electronic amplifier module, a zenor diode to protect the amplifier in the event of a current surge, a suppression capacitor, and a moulding containing two resistors.

The amplifier module is a sealed unit containing 'BERYLIA'.

KEY TO DIAGRAM

1. Distributor
2. Amplifier
3. HT coil
4. To ECU
5. To tachometer
6. To ignition switch

WARNING: THIS SUBSTANCE IS EXTREMELY DANGEROUS IF HANDLED. DO NOT ATTEMPT TO OPEN THE AMPLIFIER MODULE.

The engine speed signal for the electronic control unit and the tachometer is derived from the HT coil negative terminal. The voltage wave form at this point can reach as much as 400 volts when a spark is generated. It is desirable to suppress this voltage before feeding it into the wiring harness to the electronic control unit and the tachometer. The two resistors located in the moulding serve this purpose (3 and 4, Fig 30).

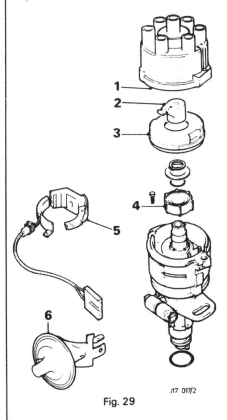

Fig. 29

KEY TO DISTRIBUTOR (45 DM6)

1	HT cover	4	Reluctor
2	Rotor	5	Pick-up
3	Flash shield	6	Vacuum

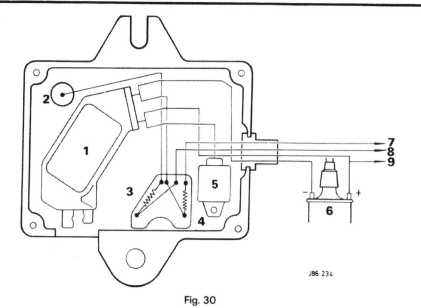

Fig. 30

J86 234

Fig. 33

KEY TO AMPLIFIER ASSEMBLY

1. Amplifier module
2. Surge protection diode
3. Resistor to ECU
4. Resistor to tachometer

5. Suppression capacitor
6. HT coil
7. To tachometer
8. To ECU
9. To ignition switch

Test 4

With the ignition switched on, the voltage at the HT coil negative terminal should be battery voltage or not more than one volt less than battery voltage (Fig. 34).

TESTING

Test 1

Battery Hydrometer Readings

Take specific gravity readings of the electrolyte in each cell. A reading of below 1.230 — Recharge battery or substitute with a charged battery.

Test 2

Remove the HT lead from the centre of the distributor cap and hold the lead approximately 6 mm (0.25 in) from the engine (Fig. 31). Crank the engine. If a good spark is obtained, check the HT leads, spark plugs and distributor cap.

To check the rotor, remove the distributor cap and hold the HT lead 3 mm (0.13 in) from the rotor arm (Fig. 32). Crank the engine. If a spark is obtained, the rotor is defective and should be renewed.

Fig. 34

If a zero or low reading is obtained, disconnect the lead to the amplifier from the negative terminal of the HT coil (Fig. 35). If the reading is still too low, or zero, a faulty HT coil primary winding is indicated. Should the reading be battery voltage, check the wiring to/from the ECU, the tachometer if found to be satisfactory the amplifier is suspect.

Fig. 31

Fig. 32

Test 3

With the ignition switched on, the voltage at the HT coil positive terminal (Fig. 33) should be battery voltage. If the voltage is more than one volt less than the battery voltage, check the wiring to/from the ignition switch, the ignition switch and connections.

Fig. 35

Test 5

Disconnect the distributor pick-up leads from the amplifier and measure the resistance of the distributor pick-up coil (Fig. 36). The resistance should be 2.2 to 4.8 K ohms. An incorrect reading indicates a faulty distributor pick-up coil.

Fig. 36

Test 6

Connect a voltmeter between the positive terminal of the battery, and the negative terminal of the HT coil (Fig. 37). Switch the ignition on and the voltmeter should indicate a zero reading. A 12 volt reading indicates an earth on the tachometer or ECU leads or a faulty amplifier. Crank the engine and the voltmeter should rise to between 2 and 3 volts. If the voltmeter remains at zero the amplifier is suspect.

Fig. 37

During normal service the air gap between the reluctor and the pick-up module does not alter and will only require re-setting if it has been tampered with. If it is necessary to adjust the gap, then it should be set such that the minimum clearance between the pick-up and the reluctor teeth is not less than 0,20 mm (0.008 in). The gap should not be set wider than 0,35 mm (0.014 in). The air gap is measured between a reluctor tooth and the pick-up module (1, Fig 38) and should be checked with a plastic feeler gauge.

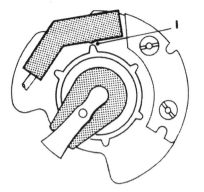

Fig. 38

The use of a metal feeler gauge may result in a misleading gauge reading due to the pick-up module contacts being magnetic. However, their use will not affect the electrical operation of the pick-up module.

AMPLIFIER

Renew

Disconnect the battery earth lead.
Remove the two bolts securing the amplifier to the radiator top rail.
Disconnect the distributor pick-up multi-plug from the amplifier.
Disconnect the tachometer and ECU cable connectors. Disconnect the amplifier Lucar connectors from the HT coil.
Remove the clip holding the cable harness to the radiator top rail, carefully pull the cables through the grommet in the top rail and remove the amplifier (Fig. 39).

Fig. 39

IGNITION SWITCH

Renew

Disconnect the battery earth lead.
Remove the driver's side dash casing.
Remove the screw securing the shroud to the fascia (1, Fig 40) and the instrument module surround.
Slacken the screws securing the shroud to the mounting bracket.

Ease the shroud clear of its location to gain acccess to the grub screw securing the switch.
Slacken the grub screw (2, Fig. 40) ease the switch and harness clear of location.
Disconnect the block connector and remove the switch unit (3, Fig. 40).

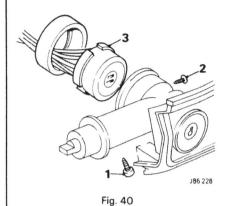

Fig. 40

IGNITION COIL

Renew

Disconnect the battery earth lead.
Disconnect the HT lead from the centre of the ignition coil.
Note the position of and disconnect the LT leads from the positive and negative terminals of the ignition coil.
Remove the bolts securing the coil to the wing valance and remove the coil.

DISTRIBUTOR

Renew

Disconnect the battery earth lead.
Remove the distributor cap and note the position of the rotor arm.
Disconnect the vacuum pipe and the amplifier cable harness multi-plug.
Remove the mounting plate securing screw and carefully withdraw the distributor.
On refitting ensure that the rotor arm is in the position noted when removed.
To check the ignition timing, run the engine until the normal operating temperature is reached. Disconnect the vacuum pipe. Run the engine at 2000 rpm with the aid of a stroboscope, adjust the timing to 18° BTDC NAS or 21° BTDC European.

Overhaul

With the distributor removed from the vehicle
Remove the rotor arm (1 Fig. 41) and the flash shield (2, Fig. 41).
With the use of circlip pliers, remove the circlip (3, Fig. 41).
Lift off the washer, 'O' ring, reluctor and coupling ring (4, Fig. 41).

Remove the two screws securing the vacuum unit and with a downward movement detach the vacuum unit (5, Fig. 41) from the peg on the underside of the pick-up plate.

Remove the two screws holding the base plate in position and lift off the plate complete with pick-up and leads.

On re-assembly, lightly smear the bearing surfaces of base and pick-up plate with grease.

After re-assembly, apply one or two drops of clean engine oil to the felt pad in the top of the reluctor carrier.

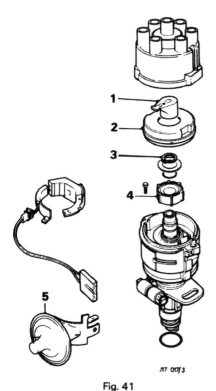

Fig. 41

HEADLAMPS
North American specification

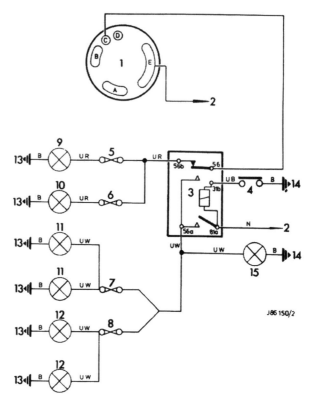

J86 1SQ/2

Fig. 43

KEY TO DIAGRAM

1. Master light switch
2. To battery terminal post
3. Headlamp dip relay
4. Dip switch
5. RH dip beam fuse
6. LH dip beam fuse
7. RH main beam fuse
8. LH main beam fuse

9. RH dip beam
10. LH dip beam
11. RH main beam
12. LH main beam
13. Earth point on radiator top rail
14. Earth point on steering column bracket
15. Main beam warning light

IGNITION CONTROLLED RELAY

Renew

Disconnect the battery earth lead.
Remove the driver's side dash liner.
Locate and remove the ignition controlled relay from the component panel (1, Fig. 42).

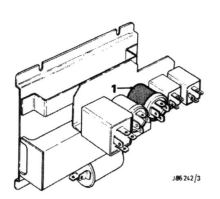

J86 242/3

Fig. 42

HEAD AND FOG LAMPS

With the master light switch in the headlamp position the contacts A, B, C and E are connected together to supply power to the headlamp relay. The headlamp flash switch activates the headlamp relay which in turn selects the main or dip headlamp filaments.

When the rear fog guard lamps are selected contacts A, B, D and E are connected together supplying power to the rear fog guard lamps.

The headlamp inhibit relay is also energised supplying current to the headlamp dip filaments only.

The headlamp flash switch activates the headlamp relay, supplying current to the headlamp main filaments.

Fault finding

Check the fuses and all connections. The earth connections should be clean and tight. With the master light switch in the off position battery voltage should be obtained at terminal 81a of the headlamp relay. Short terminal 31b to earth and battery voltage should then be obtained at terminal 56a the headlamp main beam terminal. Should a zero reading be obtained replace the relay.

With the master light switch in the headlamp on position battery voltage should be obtained at terminal 56. By shorting terminal 31b on and off to earth, the relay should switch battery voltage alternately to terminals 56a and 56b. The main and dip beam headlamp terminals.

With master light switch in the fog-lamp position battery voltage should be obtained at terminals 85, 87 and 30/51 of the headlamp inhibit relay. If battery voltage is present at terminals 85 and 30/51 but not at terminal 87 replace the relay.

HEAD AND FOG LAMPS
European specification

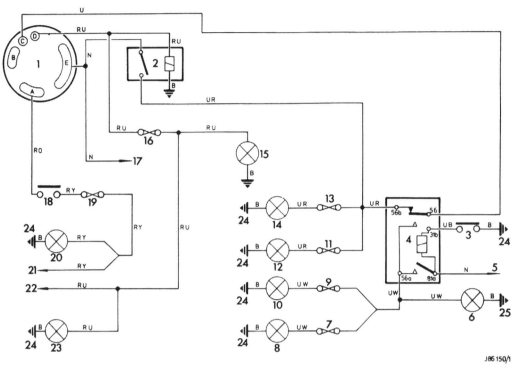

Fig. 44

KEY TO DIAGRAM

1. Master light switch
2. Headlamp inhibit relay
3. Headlamp flash switch
4. Headlamp relay
5. Terminal post
6. Main beam warning lamp
7. Fuse No. 2
8. LH main beam
9. Fuse No. 4

10. RH main beam
11. Fuse No. 3
12. LH dip beam
13. Fuse No. 5
14. RH dip beam
15. Fog lamp warning lamp
16. Fuse No. 6
17. To terminal post
18. Fog lamp switch

19. Fuse No. 1
20. LH front fog lamp (where fitted)
21. RH front fog lamp (where fitted)
22. LH rear fog lamp
23. RH rear fog lamp
24. Earth point radiator
    top rail
25. Earth point on steering
    column bracket

HEADLAMP
European specification

Renew

Disconnect the battery earth lead.
Remove the screws securing the top finisher to the top of the lamp housing (1, Fig. 45), then ease the top of the finisher away from the lamp unit and lift the bottom locating spigots from the housing (2, Fig. 45).
Depress the nylon securing tabs retaining the lamp unit (1, Fig. 46).
Withdraw the unit from its housing and disconnect the cable harness at the block connectors behind the unit (2, Fig. 46).

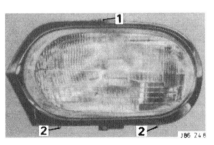

Fig. 45

Fig. 46

HEADLAMP
North American specification

Renew

Remove the three screws securing the finisher and remove the finisher.
Remove the six screws securing the assembly and disconnect the cable harness at the block connectors at the rear of the sealed beam units.
Remove the assembly.

For European cars, to renew bulb, release the clips securing the defective bulb (Fig. 47) and withdraw the bulb.
On refitting the new bulb, ensure the bulb is not touched by hand or contaminated with oil or grease.

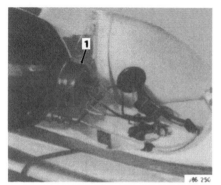

Fig. 47

The light unit on NAS cars is removed by slackening the three screws securing the light unit retaining ring (1, Fig. 48).

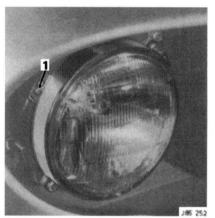

Fig. 48

Remove the ring by turning until it releases from the locating slots.
Withdraw the light unit and disconnect the cable harness block connector from the rear of the unit (1, Fig. 49).

Fig. 49

HEADLAMP ALIGNMENT

Headlamp beam setting should only be carried out with approved beam setting apparatus.

Adjustment

The adjustment screws are set diagonally opposite each other.
The upper screw is for vertical alignment and the lower screw is for horizontal alignment.
With the headlamp rim finisher removed, turn the top screw anti-clockwise to lower the beam, clockwise to raise the beam.
Turn the lower screw anti-clockwise to move the beam to the left, clockwise to move the beam to the right.

HEADLAMP RELAY

Renew

Disconnect the battery earth lead.
Note the position of the cables at the Lucar connectors on the relay, then disconnect the cables.
Remove the screws securing the relay, and retain the relay.
On refitting, ensure that all connectors are clean and tight.

MASTER LIGHTING SWITCH

Renew

Disconnect the battery earth lead.
Remove the driver's side dash liner and remove the screw securing the switch shroud to the fascia (1, Fig. 50).
Slacken the screws securing the shroud to the lower mounting bracket, and ease the shroud clear to give access to the spring loaded pin retaining the switch knob (2, Fig. 50).
Depress the pin and withdraw the knob.
Remove the shroud, remove the nut securing the switch (3, Fig. 50).
Remove the switch and disconnect the cable harness block connector.

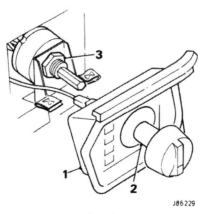

Fig. 50

HEADLAMP INHIBIT RELAY

Renew

Disconnect the battery earth lead.
Remove the driver's side dash liner.
Locate the relay and remove the relay from the component panel (1, Fig. 51).

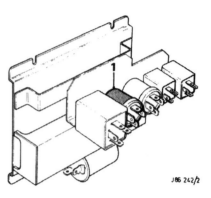

Fig. 51

FRONT AND REAR PARKING LAMPS

With the master light switch in the parking lamp on position current flows to the lamp via the switch contacts A and B, the bulb failure units and fuses.
The current flowing through the bulb failure units will cause the bulb failure warning lamp to glow for 15 to 30 seconds. If the warning lamp fails to go out, then there is a bulb failure or a circuit fault in the front parking lamp, rear lamps, or number plate lamps.
The contact C on the master light switch supplies current to the panel, cigar lighter and selector illumination lamps via fuse (5, Fig. 52).

Fault finding

Check the fuses and all connections, ensuring the earth connections are clean and tight.
With the master light switch in the parking lamp on position, battery voltage should be obtained at the B and the L terminals of the bulb failure unit.
If battery voltage is obtained at the B terminal but a zero reading at the L terminal replace the bulb failure unit.

FRONT AND REAR PARKING LAMP CIRCUIT

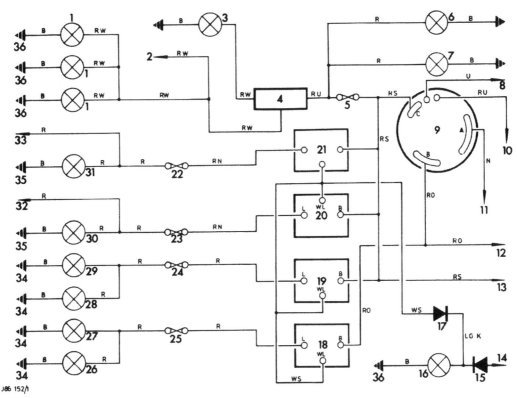

Fig. 52

KEY TO DIAGRAM

1. Switch and clock lamps
2. To panel lamps
3. Fibre optic lamp
4. Panel light rheostat
5. Fuse
6. Cigar lighter lamp
7. Gearbox selector lamp
8. To headlamp relay
9. Master light switch
10. To headlamp inhibit relay
11. To terminal post
12. To caravan socket

13. To caravan socket
14. To stoplamp bulb failure unit
15. Warning lamp blocking diode
16. Warning lamp
17. Warning lamp blocking diode
18. RH rear bulb failure unit
19. LH rear bulb failure unit
20. LH front parking lamp bulb failure unit
21. RH front parking lamp bulb failure unit
22. Fuse
23. Fuse
24. Rear lamp in-line fuse

25. Rear lamp in-line fuse
26. RH number plate lamp
27. RH rear lamp
28. LH number plate lamp
29. LH rear lamp
30. LH front parking lamp
31. To side marker lamp (if fitted)
32. RH front parking lamp
33. To side marker lamp (if fitted)
34. Earth point in luggage compartment between the battery and wheel arch
35. Earth point radiator top rail

PARKING LAMP BULB

Renew

European specification.
Disconnect the battery earth lead.
Remove the headlamp assembly, withdraw the bulb holder from the mounting in the reflector (1, Fig. 53) and remove the bulb.

Fig. 53

PARK LAMP WARNING SENSOR

Renew

Disconnect the battery earth lead.
Remove the passenger side dash liner.
Locate and remove the sensor (1, Fig. 53A).

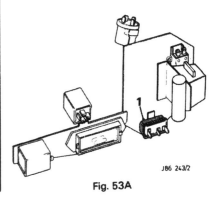

Fig. 53A

FLASHER INDICATOR CIRCUIT

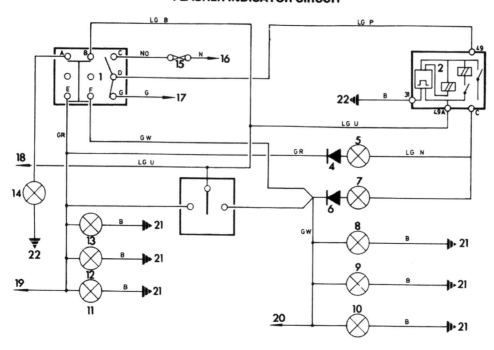

Fig. 54

J86 121/1

KEY TO DIAGRAM

1. Hazard switch
2. Flasher unit
3. Flasher switch
4. Blocking diode
5. LH warning lamp
6. Blocking diode
7. RH warning lamp
8. RH repeater lamp

9. RH front indicator lamp
10. RH rear indicator lamp
11. LH rear indicator lamp
12. LH front indicator lamp
13. LH repeater lamp
14. Hazard warning lamp
15. Fuse No. 2
16. To terminal post

17. To fuse No. 5
18. To acoustic unit
    Australia only
19. Caravan socket
20. Caravan socket
21. Earth points on radiator top rail
22. Earth point on steering
    column bracket

FLASHER LAMPS

Description

With the ignition switched on and the LH flasher lamps selected, current flows at the appropriate flash rate from fuse No. 5 to the flasher lamps via the hazard switch, flasher unit and the flasher switch. The warning light is supplied with flashing signal via the C terminal on the flasher unit. The circuit to earth for the warning light is diode 6 and the RH flasher lamps.

When the hazard lamps are selected terminals C and D in the hazard switch are connected. The terminals D and G are disconnected. Terminals A, B, E and F are connected together. This allows current to flow to all the flasher lamps via fuse No. 2, the hazard switch and the flasher unit.

Fault finding

Check the fuse and all connections ensuring the earth connections are clean and tight.

With the ignition switched on battery voltage should be obtained at terminal 49 on the flasher unit. If a zero reading is obtained check the wiring to/from the hazard lamp switch and the hazard lamp switch itself.

With the ignition switched on and the flasher lamps switched on to either right or left hand. Bridge terminals 49 and 49A together with an ammeter. If the lamps now illuminate and the ammeter registers approximately 3.5 amps replace the flasher unit.

Should the lamps still fail to illuminate check the flasher switch and wiring.

HAZARD FLASHER/TURN SIGNAL UNIT

Renew

Disconnect the battery earth lead.
Remove the driver's side dash liner.
Locate the flasher unit (1, Fig. 55) and remove from the multi-pin cable harness connector.

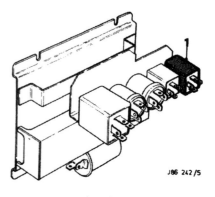

J86 242/5

Fig. 55

SIDE MARKER LAMP ASSEMBLY LENS AND BULB

Renew

Disconnect the battery earth lead.
Withdraw the lens retaining screw; remove the lens and the bulb.
Remove the securing nuts and washers from the captive retaining screws inside the wheel arch.
Disconnect the cable from the snap connectors and remove the assembly from the wing (Fig. 56).

Fig. 56

TAIL STOP AND FLASHER LAMP ASSEMBLY LENS AND BULBS

Renew

Disconnect the battery earth lead.
Remove the securing screws from the lens, and remove the lens (Fig. 57).
Remove the appropriate bulb.
Remove the three nuts and washers from the rear of the assembly.
Withdraw the assembly to gain access to the cable connectors.
Disconnect the cable harness and remove the assembly.

Fig. 57

FRONT FLASHER LAMP ASSEMBLY BULB

Renew

Disconnect the battery earth lead.
Remove the two screws securing the lamp assembly (Fig. 58).
Carefully ease the lamp from the energy absorbing beam.
Disconnect the cable harness block connector and remove the lamp assembly.
The bulb can be remove by turning and withdrawing the bulb holder from the rear of the lamp assembly.
Remove the bulb from the holder.

Fig. 58

FRONT FLASHER REPEATER LAMP ASSEMBLY — LENS AND BULB (where fitted)

Renew

Disconnect the battery earth lead.
Remove the lens and the bulb (Fig. 59).

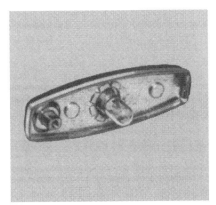

Fig. 59

Remove two screws, two nuts and bolts securing the front wheel splash guard, then remove the splash guard.
Remove the two nuts and washers from the captive retaining screws (Fig. 60).
Disconnect the cables from the snap connectors and remove the lamp from the wing

Fig. 60

NUMBER PLATE AND REVERSE LAMP ASSEMBLY BULBS AND LENS

Renew

Disconnect the battery earth lead.
Disconnect the cable harness connectors (1, Fig. 61).
Displace the grommets from the boot lid and feed the cable harness through them.
Remove the bolts securing the lamp assembly to the boot lid (2, Fig. 61) and ease the assembly clear of the boot.
To replace a bulb, remove the screws retaining the lens and remove the lens for access.
On refitting the lens, ensure the seal is in good condition.

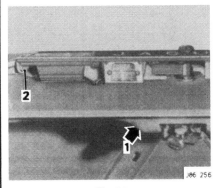

Fig. 61

INTERIOR LAMP CIRCUIT

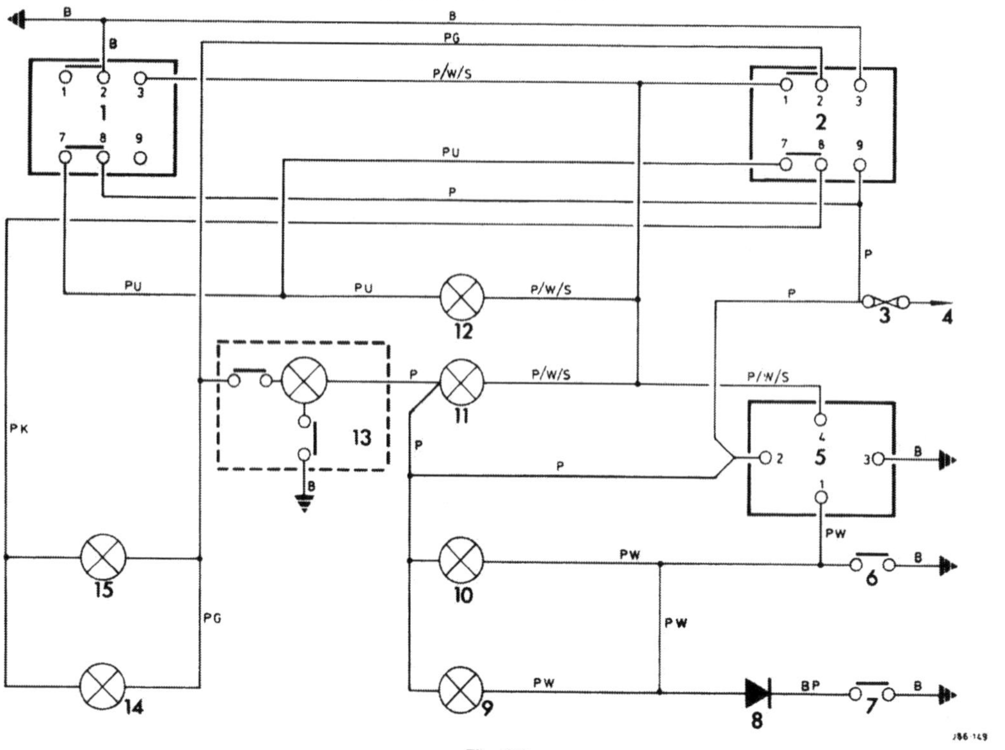

KEY TO DIAGRAM

1. Map light switch
2. Interior light switch
3. Fuse No. 13
4. To terminal post
5. Delay unit
6. LH door switch
7. RH door switch
8. Blocking diode
9. Puddle lamp
10. Puddle lamp
11. Front passenger lamp
12. Driver's lamp
13. Roof lamp
14. Rear passenger lamp
15. Rear passenger lamp
   Earth points behind
   fascia passenger's side

Fig. 62

INTERIOR LAMPS

Description

The map lights each side of the fascia lower panel are controlled by the door switches, or when the doors are closed, the passenger side is operated by depressing the map light switch (1, Fig. 61).

The rear passenger lamps are controlled by the door switches, or when the doors are closed, by depressing the interior light switch (2, Fig. 61).

The rear passenger lamps and the map lights will remain on for approximately 10 seconds after the doors are closed. The roof lamp has 3 positions. When pressed rearwards, the light will be on all the time, irrespective of the door positions. The light will remain off when the lamp is in the centre position. In the forward position, the light will come on when either door is opened.

The courtesy lamp delay unit controls the operation of the vehicle interior lamps so that they remain on for approximately 10 seconds after the doors are closed. The puddle lamps are not affected by the delay unit and will switch on and off as the doors are opened and closed.

The delay unit is polarity conscious, but a reverse polarity connection will not result in damage to the unit

With terminal 2 connected to a positive supply via the fuse No. 13 and terminal 3 connected to earth. When terminal 1 is earthed via a door switch a transistor charges a capacitor in a timing circuit which joins terminals 3 and 4 together via an internal relay. When terminal 1, earth circuit, is broken (a door closed) the capacitor commences to discharge turning off the relay at the end of the prescribed period which in turn switches off the interior lamps.

Power is supplied to the rear passenger lamps via the map and interior lamp switches. Power supplied to the driver's lamp via the map lamp switch. The roof lamp, front passenger lamp and the puddle lamps are supplied with power from the fuse. The circuit to earth for the rear passenger lamps is through the interior lamp switch and the delay unit. The circuit to earth for the driver and passenger lamps is through the delay unit. With the interior lamps switched on by the panel switch the delay unit is by-passed and therefore the delay unit will not operate.

Fault finding

Check the fuse and all connections ensuring the earth connections are clean and tight.

Battery voltage should be obtained at terminal 2 of the delay unit. If the voltage is satisfactory bridge terminal 1 of the delay unit to earth. If the interior lamps now illuminate, check the door switches, and the wiring to the door switches. If the lamps still operate unsatisfactorily bridge terminals 3 and 4 together on the delay unit. Should the lamps now illuminate replace the delay unit.

ROOF LAMP ASSEMBLY AND BULB

Renew

Disconnect the battery earth lead.

Prise the lamp assembly from the mounting in the headlining and clear of the aperture.

Remove the shroud from the rear of the lamp, and remove the bulb.

Disconnect the electrical connectors from the lamp terminals and remove the lamp assembly (Fig. 63).

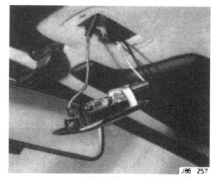

Fig. 63

REAR PASSENGER AND MAP LAMPS AND BULBS

Renew

Disconnect the battery earth lead.

Prise the appropriate lamp from its location and clear of the aperture.

Remove the bulb.

Disconnect the electrical connections and remove the lamp (Fig. 63).

LUGGAGE COMPARTMENT LAMPS AND BULBS

Renew

Disconnect the battery earth lead.
Prise the lamp from its retaining bracket.
Remove the bulb.
Disconnect the electrical connections and remove the lamp.

OPTICELL UNIT AND BULB

Renew

Disconnect the battery earth lead.
Remove the three screws securing the console centre panel and lift the panel clear of the console.
To allow full movement of the panel, note the position of the cables to the switches and the cigar lighter, and disconnect the cables.
The bulb holder can now be prised from the rear of the opticell unit and the bulb removed (1, Fig. 64).
Remove the nuts securing the opticell mounting bracket and lift the assembly clear (2, Fig. 64).
Disconnect the fibre elements and the opticell electrical cables (3, Fig. 64).
Remove the two screws retaining the opticell unit and remove the unit from the bracket.

INSTRUMENT ILLUMINATION BULB

Renew

Disconnect the battery earth lead.
Remove the instrument panel module.
Pull the appropriate bulb holder from the module and remove the bulb.

CLOCK ILLUMINATION BULB

Renew

Disconnect the battery earth lead.
Prise the clock from the fascia.
Pull the bulb holder from the clock and remove the bulb.

PANEL SWITCH ILLUMINATION BULB

Renew

Disconnect the battery earth lead.
Carefully prise the switch mounting panel from the fascia.
Pull the bulb holder from the diffuser and remove the bulb.

STARTER SYSTEM

Starter circuit test

Checking for excessive voltage drop in the starter circuit

If tests have proved that the battery and the battery connections are satisfactory, a moving coil voltmeter (0 to 20 volt range) should be used to determine whether there is excessive voltage drop in the circuit.

NOTE: During the voltmeter checks, the starter should crank the engine, without starting it.

The low-tension circuit of the ignition coil should be disconnected between the coil and distributor.
To prevent fuelling while the engine is cranked the pump relay should be disconnected.

Test 1

Check all connections ensuring the earth connections are clean and tight.

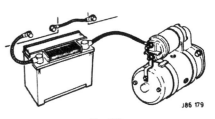

Fig. 66

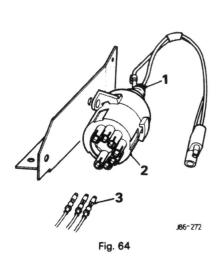

Fig. 64

STARTER SYSTEM CIRCUIT

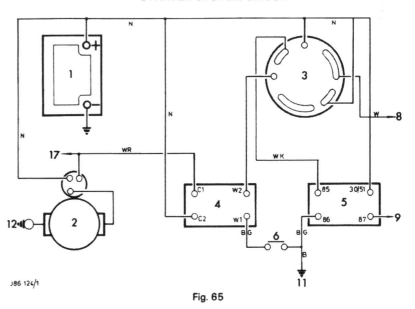

Fig. 65

KEY TO DIAGRAM

1. Battery
2. Starter motor
3. Ignition/starter switch
4. Starter relay
5. Ignition controlled protection relay
6. Gearbox safety switch
7. To the ECU
8. To the HT coil
9. To fuse No. 12
10. Earth point in luggage compartment
11. Earth point on steering column bracket
12. Earth point through the unit

Test 2

Checking the battery terminal voltage under load conditions

Connect the voltmeter across the battery terminals (Fig. 67) and operate the starter switch. The reading should be about 10.0 volts. Proceed to Test 3.

A low voltage reading would indicate excessive current flow in the circuit. The starter should be removed for bench testing.

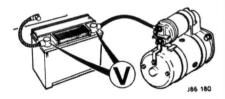

Fig. 67

NOTE: If the solenoid operates intermittently during the test or the engine is cranked at a low or irregular speed, there is insufficient voltage at the solenoid operating winding terminal or the solenoid is faulty.

To check the switching circuit for high resistance, connect the voltmeter between the solenoid operating winding terminal and earth (commutator end bracket) (Fig. 68). When the switch contacts are closed the reading on the voltmeter should be slightly less than the reading in Test 2. A satisfactory reading will indicate that there is a negligible voltage drop in the circuit and that the fault is in the solenoid.

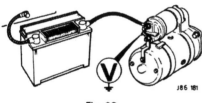

Fig. 68

If the reading is appreciably lower than in Test 2, check the starter relay, wiring and connection in the starter motor switching circuit.

Test 3

Checking the starter terminal voltage under load conditions

Having ascertained the battery voltage under load, the voltage across the starter is checked with the voltmeter connected between the starter input terminal and earth

(commutator end bracket) (Fig. 69). When the operating switch is closed, the reading should be not more than 0.5 volt below that obtained in Test 2.

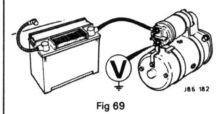

Fig 69

If the reading is within this limit, the starter circuit is satisfactory. If there is a low reading across the starter, but the voltage at the battery is satisfactory, it indicates a high resistance in the cable or at the solenoid contacts. Proceed to Test 4.

Test 4

Checking the voltage drop on the insulated line

The voltage drop on the insulated line is then checked with the voltmeter connected between the starter input terminal and the battery (insulated) terminal (Fig. 70).

Fig. 70

When the operating switch is open, the voltmeter should register battery voltage. When the operating switch is closed, the voltmeter reading should be practically zero. A high voltmeter reading indicates a high resistance in the starter circuit. All insulated connections at the battery, solenoid and starter should be checked. Proceed to Test 5.

Test 5

Checking the voltage drop across the solenoid contacts

To check the voltage drop across the solenoid contacts, connect the voltmeter across the two main solenoid terminals (Fig. 71). Crank the engine.

A zero or fractional reading on the voltmeter indicates that the high resistance deduced in Test 4 must be due either to high resistance starter cables or soldered connections.

A high reading (similar to that in Test 4) indicates a faulty solenoid or connections.

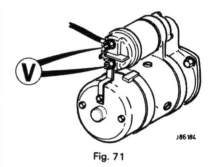

Fig. 71

Test 6

Checking the voltage drop on the earth line

Finally, check the voltage drop on the earth line. Connect the voltmeter between the battery earth terminal and the starter earth (commutator end bracket) (as shown in Fig. 72). When the operating switch is closed, the voltmeter reading should be practically zero.

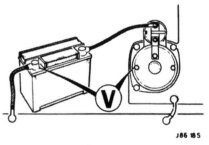

Fig. 72

STARTER RELAY

Renew

Disconnect the battery earth lead.
Remove the relay cover.
Remove the screws securing the relay and mounting plate to the wing valance.
Remove the nuts and washers securing the earth terminal and the relay to the mounting plate.
Note the position of the electrical connectors and disconnect from the Lucars.
Recover the relay.

STARTER MOTOR

Renew

Disconnect the battery earth lead.
Remove the air cleaner element and the air cleaner housing.
Remove the LH tie bar securing bolt and displace the tie bar over the manifold.
On RH drive cars, displace the windscreen washer for access.

Remove the oil dipstick bracket assembly securing bolt and remove oil dipstick assembly.
Remove the oil filter bracket securing bolt, disconnect the hose and remove the oil filter tube.
Disconnect the cables at the starter motor solenoid.
Remove the starter motor upper securing bolt, then the lower bolt, and ease the starter motor from the engine compartment. On refitting, ensure the electrical connections are clean and tight, fit new air cleaner housing gasket.

STARTER MOTOR

Overhaul

Remove the nuts and washers securing the starter to solenoid connecting link (1, Fig. 75).
Remove the two bolts and washers securing the solenoid to the starter fixing bracket (2, Fig. 75).
Lift the terminal end of the solenoid clear of the connecting link and withdraw the solenoid body (3, Fig. 75). Remove the plunger by applying an upward lift at the front end of the plunger (4, Fig. 75).

Test

Check the continuity of the solenoid windings by connecting a 12 volt battery operated test lamp between the solenoid main terminal STA and an earth point on the solenoid body (as shown in Fig. 73).

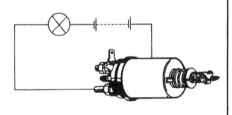

Fig. 73

If the lamp lights, it indicates that both windings are satisfactory.
Check that the contacts open and close satisfactorily by connecting a 12 volt battery and a high wattage test lamp between the main solenoid terminals (as shown in 1, Fig. 74). The lamp should not light. Close

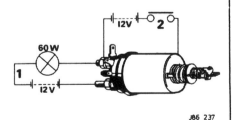

Fig. 74

the switch (2, Fig. 74) and energise the solenoid windings. The solenoid should be heard to operate, and satisfactory closing of the contacts will be indicated by the lamp lighting with full brilliance.

Starter Motor Dismantle

Remove the mutator end cap seal (5, Fig. 75).
Remove the spire retaining ring using a suitable tool to remove some of the claws on the retaining ring (6, Fig. 75).

NOTE: Discard the spire ring, ensure a new ring is fitted on reassembling the starter motor.

Remove the two through bolts and washers (7, Fig. 75).
Withdraw the commutator end cover, taking care when disengaging the two field coil brushes (8, Fig. 75) from the brush box moulding.
Withdraw the armature and drive assembly from the field coil assembly.
Withdraw the drive engagement lever pin (9, Fig. 75) from the fixing bracket.
The armature assembly comprising the roller clutch drive and the lever assembly can now be separated from the fixing bracket.
The roller clutch drive and the lever assembly is removed from the armature shaft as a complete unit.
Use a tubular tool to drive the thrust collar squarely off the jump ring. Remove the jump ring from its groove in the armature shaft. Slide the collar and the roller clutch drive with the lever assembly off the shaft.

Brushes

Check that the brushes move freely in their respective guides in the brush box mouldings. A sticking brush can be cleaned with a petrol moistened cloth.
Brushes which are worn to approximately 9.5 mm (0.375 in) must be renewed. To renew the brushes, cut the worn brush flexible lead from the field coil leaving approximately 6 mm (0.25 in) of flexible lead each side of the field coil end.

Solder the new brushes to the ends of the old leads to ensure a good connection. Also ensure the soldered connection is insulated from the starter motor body.
Replace the remaining two brushes complete with terminal link. Ensure the brushes are positioned exactly as originally fitted.

Armature

Check the armature for signs of the core folding the pole shoes. This indicates worn bearings or a distorted shaft. A damaged armature must be replaced and no attempt should be made to machine the armature core or to straighten a distorted shaft. Check for signs of thrown solder or lifted commutator segments. This indicates overspeeding of the motor and the operation of the roller clutch should be checked.
The condition of the armature should be checked as follows:
Test the armature insulation by means of a 110 volt ac 15 watt test lamp. Connect the lamp between one commutator segment and the armature shaft (as shown in Fig. 76). The test lamp should not light.
If it does the insulation has broken down, and the armature must be replaced.

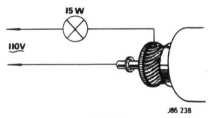

Fig. 76

If the commutator needs servicing, the copper may be skimmed to a minimum thickness of 3.55 mm (0.140 in) before a replacement armature is necessary. The surface should then be polished with fine emery cloth, and finally cleaned with a petrol-moistened cloth. The insulation between the commutator segments MUST NOT BE UNDERCUT.

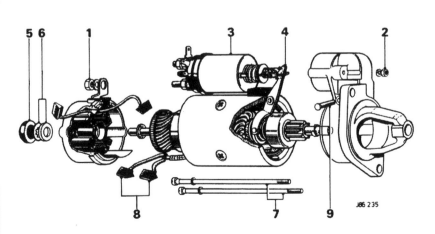

Fig. 75

Field winding, Continuity

Check the winding for continuity by means of a 12 volt test lamp and battery. Connect the test lamp between each of the brushes in turn, and a clean part of the yoke. If the test lamp does not light, an open-circuit in the field winding is indicated and a replacement must be fitted (Fig. 77).

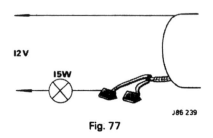

Fig. 77

Insulation

To make a positive check of the insulation between the field winding and yoke, it will be necessary to disconnect the riveted earth connection at the yoke.

To avoid disturbing this connection unnecessarily, first inspect the inside of the yoke for obvious signs of insulation breakdown, and if so, rectify or replace the field winding assembly as necessary.

The field winding insulation can be checked, after disconnecting the end of the winding at the yoke by connecting a 110 volt ac 15 watt test lamp between the disconnected end of the winding and a clean part of the yoke (Fig. 78). If the test lamp lights, it indicates an earth at some point on the yoke or pole shoes and a replacement field winding is necessary. Check that the earth connection, brush flexibles and brushes are not contacting the yoke before suspecting the field windings.

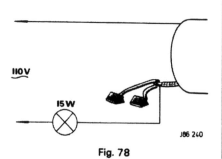

Fig. 78

Roller clutch and drive operating mechanism

The roller clutch drive assembly, if in good condition will provide instantaneous take-up of the drive in one direction and rotate smoothly and easily in the other. The assembly should move freely round and along the armature shaft splines without roughness or tendency to bind. The armature shaft splines and moving parts of the engagement lever should be liberally smeared with Shell SB2628 (home and cold climates), Retinax 'A' (hot climates).

The roller clutch mechanism is a sealed unit. During production it is pre-packed with sufficient grease to last the life of the starter motor. In the unlikely event of the clutch becoming faulty, it is not possible to rectify the fault and the whole drive assembly should be renewed.

Bearings

Both end brackets are fitted with porous bronze bearing bushes. New bushes should be allowed to stand for 24 hours at room temperature completely immersed in clean light engine oil. Alternatively, the bush may be immersed in the above lubricant at 100°C for two hours and allowed to cool before removal.

New bushes **must not be** reamed after fitting as the porosity of the bush will be impaired.

The bushes must be renewed when there is excessive side-play of the armature shaft. Fouling of the pole shoes by the armature, or inefficient operation of the starter motor is likely to occur when the inner diameter of the bushes exceeds the following dimensions:

Commutator end cover bush 11,20 mm (0.441 in).
Drive end fixing bracket bush 12,09 mm (0.476 in).

Reassembly

Reassembling the starter motor is the reversal of the dismantling procedure.

Ensure that the internal thrust washer is fitted at the commutator end of the armature shaft.

When the starter motor is assembled, drive on the NEW SPIRE RETAINING RING to the armature shaft into a position which provides a maximum of 0,25 mm (0.010 in) clearance between the retaining ring and the bearing brush shoulder.

Finally, fit the end cap seal to the commutator end.

COMBINED HEADLIGHT/ DIRECTION/FLASHER/DIP SWITCH

Renew

Disconnect the battery earth lead.
Slacken the steering column adjustment ring and extend the column to its fullest extent (1, Fig. 83).
Remove the screws securing the steering column lower shroud and remove the shroud (2, Fig. 83).
Remove the steering wheel.
Remove the screw securing the upper shroud to the bracket on the steering column.
Slacken the pinch screw (3, Fig. 83) securing the switch assembly to the steering column, and ease the switch assembly complete with the shroud.
Remove the shroud from the switch assembly and disconnect the cable harness at the block connectors.
Remove the spire nuts and screws securing the wiper switch to the mounting plate (4, Fig. 83).
Disconnect the earth cable at the snap connector (5, Fig. 83) and remove the wiper washer switch from the assembly.

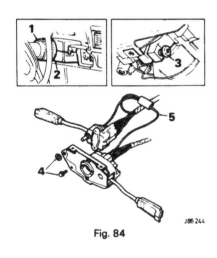

Fig. 84

PANEL SWITCHES

Renew

NOTE: This operation applies to the following: Interior light switch, map light switch, back light switch, hazard warning switch.

Disconnect the battery earth lead.
Prise the switch panel from the fascia (1, Fig. 85).
Disconnect the cable harness block connector from the appropriate switch.
Depress the retaining clips on the top and bottom of the switch (2, Fig. 85).
Push the switch through the panel.

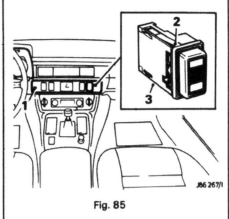

Fig. 85

PANEL LIGHT RHEOSTAT

Renew

Disconnect the battery earth lead.
Remove the driver's side dash liner.
Disconnect the cables at the Lucar connectors on the rheostat (1, Fig. 86).

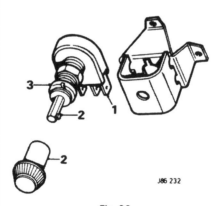

Fig. 86

Remove the control knob by depressing the spring loaded pin on the shaft and pulling the knob off the shaft (2, Fig. 86).
Slacken the nut securing the rheostat to the mounting bracket and remove the rheostat (3, Fig. 86).

REVERSE LIGHT SWITCH

Renew

Drive the vehicle onto a ramp and raise the ramp.
Disconnect the battery earth lead.
Disconnect the lucar cable connectors (1, Fig. 87).
Unscrew and remove the switch (2, Fig. 87).

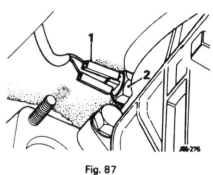

Fig. 87

DOOR PILLAR SWITCH AND LUGGAGE COMPARTMENT LIGHT SWITCH

Renew

Disconnect the battery earth lead.
Remove the screws securing the switch.
Withdraw the switch and disconnect the cable at the Lucar connector.

WARNING LIGHT FAILURE UNIT AND STOP LAMP BULB FAILURE UNIT CIRCUIT

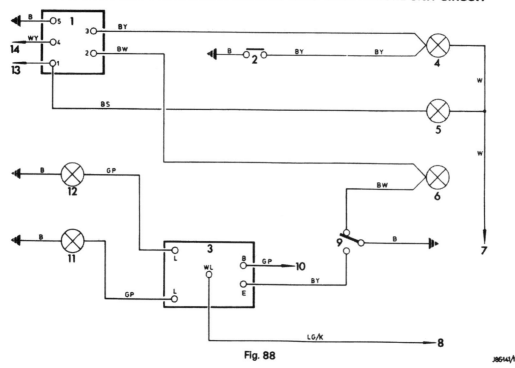

Fig. 88

J86141/1

KEY TO DIAGRAM

1. Warning light failure unit
2. Reservoir brake switch
3. Brake lamp warning sensor
4. Brake warning lamp
5. Oxygen warning lamp
6. Handbrake warning lamp
7. To ignition switch
8. To bulb failure warning lamp via blocking diode
9. Handbrake switch
10. To footbrake switch
11. LH stop lamp
12. RH stop lamp
13. Stop lamp bulb failure unit
14. Starter relay
   Earth point for failure unit
   on the steering column bracket

WARNING LIGHT FAILURE UNIT

Description

When the starter motor is activated a voltage supply to the warning light failure unit switches a transistor circuit to earth which completes the circuit of the oxygen warning lamp, brake warning lamp, and the handbrake warning lamp, causing the lamps to glow. This indicates that the warning lamps are operating satisfactorily. A failure of a warning lamp should be investigated immediately.

Fault finding

To test the warning light failure unit, switch the ignition on and short terminal 1 of the brake light failure unit to earth. The oxygen sensor warning lamp should glow. Should the warning lamp fail to glow check the warning lamp bulb, and the warning lamp supply voltage.

Repeat the test at terminal 2 for the handbrake warning lamp, and terminal 3 for the brake warning lamp. If the above checks prove satisfactory replace the failure unit.

STOP LAMP BULB FAILURE UNIT

Description

The stop lamps are supplied with current via a bulb failure unit. Should the bulb failure warning lamp glow with the master light switch off, the ignition switched on, the handbrake released and the footbrake depressed, a circuit fault or a faulty bulb is indicated.

Fault finding

To test the stop lamp failure unit switch on the parking lamps. The bulb failure warning lamp should glow for 30 seconds this proves the warning lamp is satisfactory. Switch off the parking lamps and switch off the parking lamps and switch the ignition on.

Remove a stop lamp bulb, release the handbrake, and depress the foot brake. The warning lamp should glow. Should the warning lamp fail to glow replace the bulb failure unit.

STOP LIGHT BULB FAILURE SENSOR

Renew

Disconnect the battery earth lead.
Remove the screws securing the centre panel and raise the panel for access to the bulb failure unit.
Note the position of the electrical connections to the sensor unit and disconnect.
Remove the screws securing the sensor and remove the sensor.

WARNING LAMP BULB FAILURE UNIT

Renew

Disconnect the battery earth lead.
Remove the driver's side dash liner.
Remove the bulb failure unit from the component panel (1, Fig. 89).

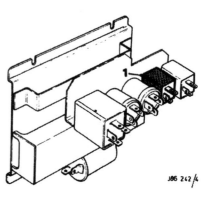

J86 242/4

Fig. 89

WARNING LAMP BULB(S)

Renew

Slacken the steering column adjustment ring and extend to its fullest extent.
Prise out the covers from each end of the warning light strip.
Remove the screws securing the warning lamp cover and remove the cover.
Remove the defective bulb.

STOP LIGHT SWITCH

Renew

Disconnect the battery earth lead.
Disconnect the electrical connectors from the Lucar connectors on the switch.
Remove the switch securing bolt and remove the switch.
On refitting, reconnect battery and switch on the ignition.
Adjust the switch position until the stop lights operate when the brake pedal is depressed, and off when the pedal is fully released.

HANDBRAKE WARNING SWITCH

Renew

Disconnect the battery earth lead.
Remove the screw securing the handbrake mechanism cover and slide the cover clear of the mechanism.
Disconnect cable harness from the Lucar connectors on the switch.
Remove the bolts securing switch to the handbrake assembly.
Remove the bolt and spacer.
Remove the switch.
On refitting, with the battery connected and the ignition switched on, adjust the switch with the handbrake until the warning light just goes out.
Tighten the securing bolts.
Check the light comes on with handbrake applied and goes off when the handbrake is released.

SEAT BELT WARNING SYSTEM CIRCUIT

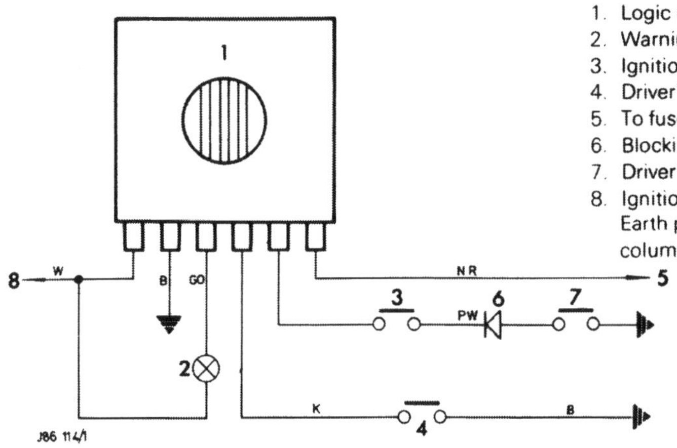

KEY TO DIAGRAM

1. Logic unit
2. Warning lamp
3. Ignition switch
4. Driver's seat belt switch
5. To fuse No. 2
6. Blocking diode
7. Driver's door switch
8. Ignition switch
   Earth point on the steering column bracket

Fig. 90

SEAT BELT WARNING SYSTEM

Description

The seat belt logic unit will operate an audible signal if the driver's door is open and the ignition key is inserted into the ignition switch. The signal will cease to operate if the door is closed or the key is withdrawn from the ignition.
The unit also operates two timing circuits; one of the circuits will cause the warning light to illuminate for 10 seconds when the ignition is switched on, whether the seat belt is in use or not.
The other timed circuit operates an audible signal when the ignition is switched on which will cease when the seat belt is fastened or after 10 seconds have elapsed.

SEAT BELT SWITCHES

Renew

Disconnect the battery earth lead.
Push the seat forward as far as possible.
Remove the bolt securing the seat belt unit, raise the unit and ease the connector leads clear of the carpet.
Disconnect the cable block connector and remove the belt switch unit.

SEAT BELT LOGIC UNIT

Renew

Disconnect the battery earth lead.
Remove the passenger side dash liner.
Remove the nut and screw securing the unit, disconnect the cable harness block connector and remove the unit (1, Fig. 91).

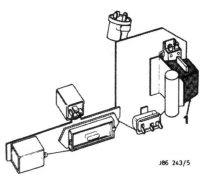

J86 243/5

Fig. 91

COOLING FAN CIRCUIT

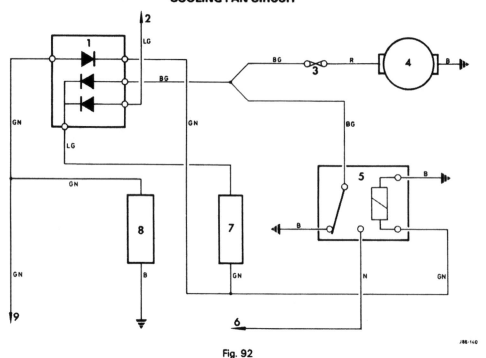

Fig. 92

J86-140

KEY TO DIAGRAM

1. Diode pack
2. Fuse No. 11
3. In-line fuse
4. Cooling fan
5. Fan relay
6. To fuse No. 1
   (Headlamp fusebox)
7. Radiator fan thermostat
8. Compressor clutch
9. To air conditioning
   Earth points on the
   radiator top rail

COOLING FAN

Description

With the coolant temperature cool the thermostat contacts are open which prevents the cooling fan relay from being energised, this in turn prevents the cooling fan from being activated.

With the engine running and the coolant warm the thermostat contacts close. This allows the relay coil circuit to be completed via a diode in the diode pack and the thermostat contacts. The cooling fan circuit is now completed via the relay contacts.

When the ignition is switched off but the coolant temperature is still hot the cooling fan will still be activated via the relay contacts. The relay will remain energised via a diode in the diode pack, the thermostat contacts, and the relay contacts.

When the coolant temperature cools sufficiently the thermostat contacts will open and the relay will become de-energised. The relay contacts will then switch an earth to the positive brush of the cooling fan motor.

FAN MOTOR

Renew

Disconnect the battery earth lead.

Remove the nuts securing the receiver/drier to the top rail. Also remove the nut and bolt securing the RH body stay.

Displace the receiver/drier and the body stay for access.

Remove the cable harness clip from the fan motor frame.

Disconnect the fan motor cable harness block connector.

Remove the bolts securing the fan assembly and remove the assembly.

Remove the fan blades from the fan motor shaft.

Remove the fan motor from the frame assembly.

DIODE PACK AND COOLING FAN RELAY

Renew

Disconnect the battery earth lead.

Disconnect the cable harness block connector from the appropriate unit, and withdraw the unit from its locating tab on the LH wing valance in the engine compartment.

LOW COOLANT CONTROL UNIT

Description

With a positive supply to the white wire on the control unit and the black wire earthed, the sensor will partially earth through the

LOW COOLANT CONTROL UNIT

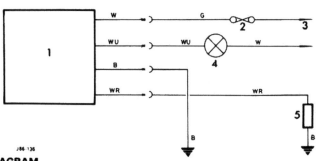

Fig. 93

J86-136

KEY TO DIAGRAM

1. Warning light control unit
2. Fuse No. 4
3. To ignition switch
4. Warning lamp
5. Coolant temperature sensor
   Earth point on the side of the
   air con unit

coolant and the warning lamp will glow for a few seconds then go out. If the coolant falls below the sensor level the partial earth circuit will be broken and the warning lamp will flash on and off.

Fault finding

Should the warning lamp fail to glow when the ignition is switched on, check all connections and the fuse, ensuring the earth connections are clean and tight.

With the ignition switched on battery voltage should be obtained from the white lead on the control unit. With the white and blue lead from the warning lamp shorted to earth, switch the ignition on and the low coolant warning lamp should glow. Should the lamp fail to glow check the warning lamp bulb and wiring.

With the white and red lead from the sensor unit disconnected, the ignition switch on the warning lamp should flash on and off. Should the lamp fail to flash replace the control unit.

LOW COOLANT WARNING CONTROL UNIT

Renew

Disconnect the battery earth lead.
Remove the passenger side dash liner.
Remove the unit from its locating clip and disconnect the cable harness block connector (1, Fig. 94).

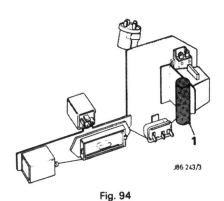

Fig. 94

ELECTRICALLY OPERATED DOOR MIRRORS

Description

The RH and LH mirror controls are located on the driver's door. The horizontal and the vertical motors are operated from one control switch. With the control switch in the up position the current flow to motor 3 is via the outer ring A and contact D. The earth circuit is via contact F and the inner ring B.

With the switch in the down position the circuit is completed in the opposite direction via the outer ring A and contact F. The circuit to earth is completed via the contact D and the inner ring B. Similar current flow takes place when the control is operated in a horizontal position using the contact C, E and the motor 2.

FUEL CUT-OFF INERTIA SWITCH

Reset

The inertia switch is fitted in the electrical supply to the fuel pump. Should the car be subjected to heavy impact forces, the switch will operate, isolating the fuel pumps and ensuring fuel is not pumped to a potentially dangerous area. The switch is located on the side of the fascia on the driver's side 'A' post. Press the button mounted on top of the switch to reset after operation.

INERTIA CUTOUT SWITCH

Renew

Disconnect the battery earth lead.
Remove the cover from the inertia switch.
Disconnect the cables from the switch.
Withdraw the screws securing the switch and remove the switch.

ELECTRIC DOOR MIRROR CIRCUIT

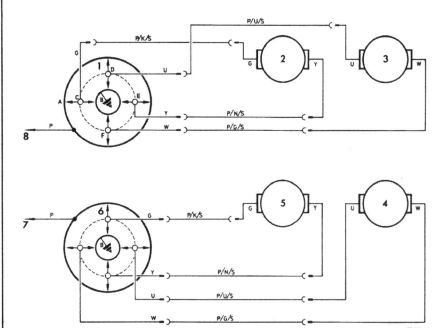

Fig. 95

KEY TO DIAGRAM

1. RH mirror switch
2. RH mirror vertical motor
3. RH mirror horizontal motor
4. LH mirror horizontal motor
5. LH mirror vertical motor
6. LH mirror switch
7. To fuse
   Earth points behind the fascia

OVERCHARGE WARNING LIGHT CIRCUIT

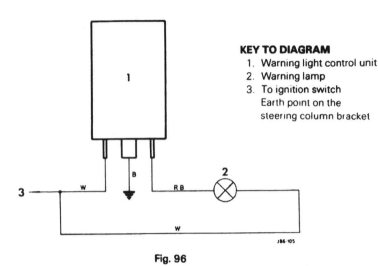

KEY TO DIAGRAM
1. Warning light control unit
2. Warning lamp
3. To ignition switch
   Earth point on the
   steering column bracket

Fig. 96

HEATED REAR SCREEN CIRCUIT

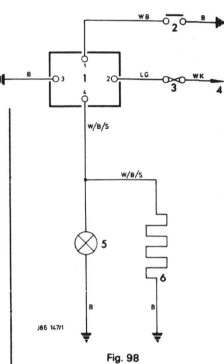

Fig. 98

KEY TO DIAGRAM
1. Delay unit
2. Control switch
3. Fuse
4. To ignition switch protection relay
5. Warning lamp
6. Heated rear screen
   Earth point for heated rear screen
   adjacent to the battery

OVERCHARGE WARNING LIGHT CONTROL UNIT

Description

With the ignition switched on the earth circuit of the overcharge voltage lamp is completed via the internal circuit of the overcharge voltage control unit.

The control unit circuit will illuminate the warning light to indicate a malfunction in the alternator. The alternator control box will be suspect by not controlling the output of the alternator and cause the alternator to overcharge the battery.

OVERCHARGE CONTROL UNIT

Renew

Disconnect the battery earth lead.
Remove the driver's side dash liner.
Disconnect the cable harness from the unit.
Remove the nut and screw securing the unit.
Remove the warning light control unit (1, Fig. 97).

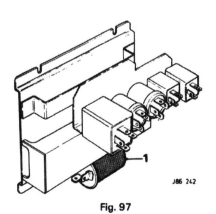

Fig. 97

HEATED REAR SCREEN

Description

The timer unit switches off the heated rear screen circuit approximately ten minutes after the heated rear screen circuit has been switched on. The timer resets whenever the circuit is switched off or when the ignition is switched off.

A small relay, and the electronic components are mounted on a circuit board inside a yellow thermoplastic cover.

The unit is polarity conscious but reverse polarity connection will not result in damage.

The terminal 2 is connected to a positive supply via fuse 12, terminal 3 is connected to earth. When terminal 1 is switched to earth by the control switch the relay is energised connecting terminals 2 and 4 together, thus supplying power to the heated rear screen, heated door mirrors and the warning light. When the timing cycle is completed the relay is de-energised and the relay contacts open.

Fault finding

Check the fuse and all connections ensuring the earth connections are clean and tight.
With the ignition switched on and a voltmeter connected between a good earth and terminal 2 of the relay, the voltmeter should indicate battery voltage. With the ignition switched on, the heated rear screen switched on and the voltmeter connected between terminals 1 and 2 of the relay, the voltmeter should indicate battery voltage. If a satisfactory reading was obtained in the first check but a zero reading is obtained in the second check, check the wiring to/from the control switch and the control switch itself. With the ignition and the heated rear screen switched on battery voltage should be obtained at terminal 4 of the relay. If terminal 4 gives a zero reading and the previous tests are satisfactory replace the relay.

HEATED REAR SCREEN DELAY UNIT

Renew

Disconnect the battery earth lead.
Remove the driver's side dash liner.
Disconnect the multi plug connector and remove the delay unit (1, Fig. 99).

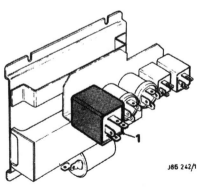

Fig. 99

CIGAR LIGHTER

Renew

Disconnect the battery earth lead.

Remove the screws securing the centre panel to the console.

Raise the panel for access to the cigar lighter assembly and disconnect the electrical connectors from the assembly (1, Fig. 100).

Remove the lamp holder by depressing the sides and unclipping from the cigar lighter (2, Fig. 100).

Unscrew and remove the cigar lighter lower sleeve (3, Fig. 100).

Remove the cigar lighter (4, Fig. 100).

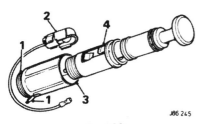

J86 245

Fig. 100

AERIAL

Circuit

The aerial maintains its extended position for a period of 10 seconds after the radio has been switched off. The aerial will then retract. If the radio is switched on before the 10 seconds has elapsed, the aerial will remain in the extended position.

With terminal 1 of the relay connected to a positive supply via fuse 5 and terminal 2 connected to earth. The positive supply from the radio to terminal 3 of the relay turns on a transistor which charges a capacitor in a timing circuit. It also activates a relay which connects terminals 1 and 5. This allows current to flow to the aerial motor and thus drive the aerial to the extended position. On switching the aerial off the aerial will remain extended for a period of 10 seconds then the relay will become de-activated the contact between terminals 1 and 5 will become disconnected, but the contact will be made between terminals 1 and 4. This will allow current to flow to the aerial motor to retract the aerial.

Fault finding

Should the aerial fail to extend when the radio is switched on.

Check the fuse and all connections ensuring the earth connections are clean and tight. Battery voltage should be obtained at terminals 1 and 3 of the delay unit. If battery voltage is obtained bridge the leads from terminals 1 and 5 together. Should the aerial now extend replace the delay unit.

RADIO, CLOCK AND LUGGAGE COMPARTMENT LAMPS CIRCUIT

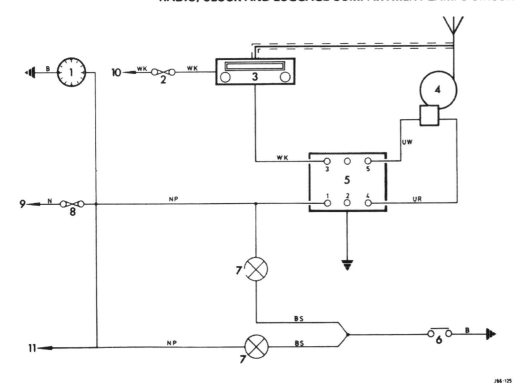

KEY TO DIAGRAM

1. Clock
2. Line fuse
3. Radio
4. Aerial motor
5. Delay relay
6. Boot lamps switch
7. Boot lamps
8. Fuse No. 3
9. Terminal post
10. To ignition switch
    Earth point adjacent
    to the battery

Fig. 101

COMPONENT LOCATION — CAR INTERIOR

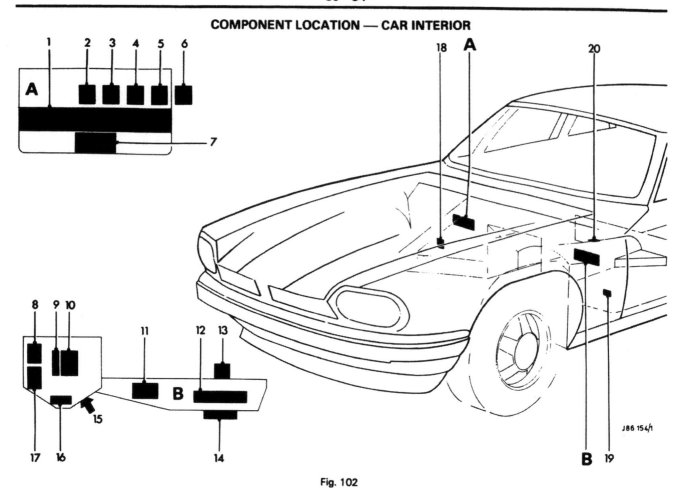

Fig. 102

COMPONENT LOCATION KEY

1. Main fuse box	
2. Heated rear screen delay unit	
3. Headlamp inhibit relay	Component
4. Ignition protection relay	Panel-A
5. Warning light bulb check unit	Driver's Side
6. Direction indicator flasher unit	
7. Voltage over-charge warning light unit	
8. Door lock thermal circuit breaker	
9. Low coolant warning light control unit	
10. Seat belt logic unit	
11. Interior light delay unit	
12. Auxiliary fuse box	Component
13. Windscreen wiper delay unit	Panel-B
14. Speed control unit (if fitted)	Passenger Side
15. Window lift relay	
16. Front parking lamp bulb failure unit	
17. Window lift thermal circuit breaker	
18. Door lock relay	
19. Door unlock relay	
20. Stop lamp bulb failure unit	

COMPONENT LOCATION — LUGGAGE COMPARTMENT

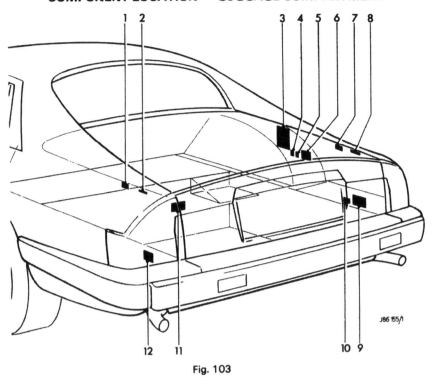

Fig. 103

COMPONENT LOCATION KEY

1. LH rear bulb failure unit
2. LH rear lamp in-line fuse
3. Electronic control unit (ECU)
4. Main relay
5. Pump relay
6. Interface unit
7. RH rear bulb failure unit
8. RH rear lamp in-line fuse
9. Aerial motor
10. Aerial motor delay unit
11. Fuel pump
12. Caravan socket

COMPONENT LOCATION — ENGINE COMPARTMENT

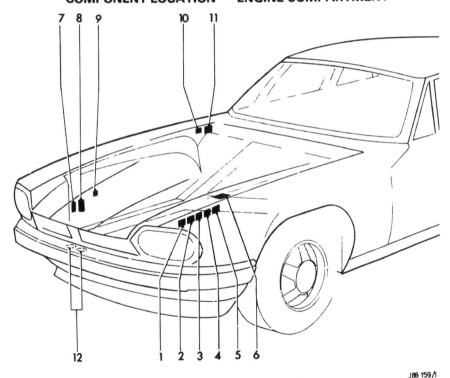

Fig. 104

COMPONENT LOCATION KEY

1. Headlamp relay
2. Cooling fan diode pack
3. Horn relay
4. Cooling fan relay
5. Overdrive inhibit relay
6. Headlamp fuse box
7. Ballast resistors
8. HT coil
9. Super heat switch thermal fuse
10. Feedback relay
11. Starter relay
12. Horns

MAIN FUSE BOX — LH Steering

Fuse No.	Protected Circuit	Fuse Capacity	Unipart Number
1.	Front fog lights	20A	GFS240
2.	Hazard warning, seat belt logic	15A	GFS415
3.	Clock, aerial, caravan, boot lamp	35A	GFS435
4.	Panel instruments, reverse light	10A	GFS410
5.	Direction indicators, stop lamps, auto kick down switch	15A	GFS415
6.	Fog rear guard	10A	GFS410
7.	Panel/cigar lighter/selector illumination	10A	GFS410
8.	Door locks, electric mirrors	3A	GFS43
9.	Wipers	35A	GFS435
10.	Air conditioning motors	50A	GFS450
11.	Air conditioning controls, horn, washers, radiator cooling fan	35A	GFS435
12.	Heated rear screen, heated mirrors	35A	GFS435

MAIN FUSE BOX — RH Steering

Fuse No.	Protected Circuit	Fuse Capacity	Unipart Number
1.	Cigar lighter	20A	GFS420
2.	Hazard warning, seat belt logic	15A	GFS415
3.	Clock, aerial, caravan, boot lamp	35A	GFS435
4.	Panel instruments, reverse light	10A	GFS410
5.	Direction indicators, stop lamps, auto kick down switch	15A	GFS415
6.	Fog rear guard	10A	GFS410
7.	Panel/cigar lighter/selector illumination	10A	GFS410
8.	Door locks, electric mirrors	3A	GFS43
9.	Wipers	35A	GFS435
10.	Air conditioning motors	10A	GFS410
11.	Air conditioning controls, horn, washers, radiator cooling fan	35A	GFS435
12.	Heated rear screen, heated mirrors	35A	GFS435

HEADLAMP FUSE BOX

Fuse No.	Protected Circuit	Fuse Capacity	Unipart Number
1.	Radiator auxiliary cooling fan motor relay	25A	GFS425
2.	LH main beam	25A	GFS425
3.	LH dip beam	10A	GFS410
4.	RH main beam	25A	GFS425
5.	RH dip beam	10A	GFS410

AUXILIARY FUSE BOX — RH Steering

Fuse No.	Protected Circuit	Fuse Capacity	Unipart Number
13.	Interior and map lights	10A	GFS410
14.	LH side lights	3A	GFS43
15.	RH side lights	3A	GFS43
16.	Front fog lights	20A	GFS420

AUXILIARY FUSE BOX — LH Steering

Fuse No.	Protected Circuit	Fuse Capacity	Unipart Number
13.	Interior and map lights	10A	GFS410
14.	LH side lights	3A	GFS43
15.	RH side lights	3A	GFS43
16.	Cigar lighter	20A	GFS420

BULB CHART

	Watts	Lucas Part No.	Unipart No.	Notes
Headlamps — not France or USA — main/dip	60/55	472	GLB 472	Halogen H4 base bulb
— France only — main/dip	60/55	476	GLB 476	Yellow Halogen H4 base bulb
—USA only — outer	37.5/60			Quartz Halogen
—USA only — inner	50			Quartz Halogen
Headlamp pilot bulb — not USA	4	233	GLB 233	
Front flasher lamp	21	382	GLB 382	Not USA
Front flasher and side lamp	21/5	380		USA only
Stop lamp	21	382	GLB 382	
Tail lamp	5	207	GLB 207	
Rear flasher	21	382	GLB 382	
Reversing lamp	21	273	GLB 273	Festoon bulb
Number plate lamp	6	254	GLB 254	Festoon bulb
Sidemarker	4	233		USA only
Flasher side repeaters — when fitted	4	233	GLB 233	
Rear fog guard lamps	21	382	GLB 382	Not USA
Interior/map lamps	6	254	GLB 254	Festoon bulb
Roof lamp	10			Festoon bulb
Boot lamp	5	239	GLB 239	Festoon bulb
Fibre optic light source	6	254	GLB 254	
Instrument illumination	2.2	987	GLB 987	
Warning lights	1.2	286	GLB 286	
Automatic selector illumination	2.2	987	GLB 987	
Cigar lighter illuminator	2.2	987	GLB 987	
Door puddle lamp	5		GLB 501	

COMPONENT LOCATION — WIRING DIAGRAMS

COMPONENT	No.	Grid ref.	COMPONENT	No.	Grid ref.
Alternator	1	A1	Interior light switch	59	C1
Aerial motor	185	C1	Line fuse	67	
Aerial motor relay	186	C1	Low coolant control unit	303	B1
Air conditioning blower	33	C3	Low coolant sensor	309	B1
Air conditioning blower relay	189	C3	Low coolant warning light	320	B1
Air conditioning blower resistor	188	C3	Main beam warning light	10	A2
Air conditioning compressor clutch	190	*	Main light switch	6	A2
Air conditioning control switch	192	C3	Map light	102	C1
Battery	3	A1	Map light switch	101	C1
Battery condition indicator	146	B1	Number plate lamp	15	A2
Blocking diode — brake warning	256	A3	Oil pressure gauge	48	B1
Blocking diode — direction indicators	289	A3	Oil pressure switch	42	B1
Boot light	66	C1	Oil pressure transmitter	147	B1
Boot light switch	65	C1	Oil pressure warning lamp	43	B1
Brake failure warning light	323	A3	Overdrive inhibit relay	360	B3
Brake fluid level switch	182	B1	Overvoltage CU	302	B1
Brake fluid level warning light	159	B1	Overvoltage W/L	321	B1
Buzzer alarm	168	C1	Panel lamps	14	B1
Cigar lighter	57	C2	Panel lamps rheostat	13	B1
Cigar lighter illumination	208	B1	Park lamp failure sensor	304	A2
Clock	56	C1	Park lamp failure warning light	32	A2
Computer	357	C1	Pulse generator	359	C2
Computer interface unit	358	C1	Puddle lamps	20	C1
Cold start injector	300	**	Radiator cooling fan motor	179	C3
Direction indicator switch	26	A3	Radiator cooling fan relay	177	C3
Direction indicator warning light	27	A3	Radiator cooling fan thermostat	178	C3
Distributor	40	A1	Radiator diode pack	174	C3
Door lock solenoid	257	B3	Radio	60	C1
Door lock solenoid relay	258	B3	Rear fog guard lamp	288	A2
Door lock switch	260	B3	Rear fog guard warning lamp	287	A2
Door thermal cut-out	259	B3	Rear window delay unit	340	
Fibre optics illumination lamp	255	B1	Rear window demist switch	115	B2
Flasher unit (part of 154)	25	A3	Rear window demist unit	116	B2
Flasher lamp RH front	28	A3	Rear window demist warning lamp	150	B2
Flasher lamp LH front	29	A3	Reverse lamps	50	A3
Flasher lamp RH rear	30	A3	Reverse lamps switch	49	A3
Flasher lamp LH rear	31	A3	Revolution counter	95	B1
Fog lamp RH (if fitted)	54	A2	Roof light	280	C1
Fog lamp LH (if fitted)	55	A2	Seat belt switch — driver	198	C1
Fuel injection control unit (ECU)	292	**	Seat belt warning control unit	290	C1
Fuel injection main relay	312	**	Seat belt warning lamp	202	C1
Fuel level warning light	319	B1	Side lamp RH or (headlamp pilot lamp)	11	A2
Fuel gauge	34	B1	Side lamp LH or (headlamp pilot lamp)	12	A2
Fuel gauge tank unit	35	B1	Service interval counter switch	277	B2
Gearbox relay	362		Service interval counter warning light***	278	B2
Gearbox solenoid	361		Speed control actuator	347	C2
Handbrake switch	165	A3	Speed control inhibit switch	344	C2
Handbrake warning lamp	166	B1	Speed control master switch	346	C2
Hazard warning flasher unit (includes 25)	154	A3	Speed control set switch	345	C2
Hazard warning lamp	152	A3	Speed control unit	342	C2
Hazard warning switch	153	A3	Speedometer	360	B1
Headlamp dip switch	7	A2	Starter motor	5	A1
Headlamp dip beam	209	A2	Starter solenoid	4	A1
Headlamp inhibit relay	164	A2	Starter solenoid	194	A1
Headlamp inner RH 'NAS'	113	A2	Stop lamps	16	A3
Headlamp inner LH 'NAS'	114	A2	Stop lamps failure sensor	301	A3
Headlamp outer RH	8	A2	Stop lamp switch	18	A3
Headlamp outer LH	9	A2	Tail lamp RH	17	A2
Headlamp pilot lamp (see sidelamp)	11	A2	Tail lamp LH	22	A2
Headlamp relay	231	A2	Thermal circuit breaker door lock	259	B3
Horn	23	B2	Water temperature gauge	46	B1
Horn push	24	B2	Water temperature transmitter for gauge	47	B1
Horn relay	61	B2	Window lift motor	220	B3
Ignition amplifier	183	A1	Window lift safety relay	221	B3
Ignition coil	39	A1	Window lift switch RH front	216	B3
Ignition protection relay	204	A1	Window lift switch LH front	218	B3
Ignition switch	38	A1	Windscreen washer pump	77	B2
Ignition warning lamp	44	A1	Windscreen washer switch	78	B2
Inertia switch	250	B1	Windscreen wiper motor	37	B2
Interior light	280	C1	Windscreen wiper switch	36	B2
Interior light rear passenger	111	C1			

(Speed control actuator 347, Speed control inhibit switch 344, Speed control master switch 346, Speed control set switch 345, Speed control unit 342 — Certain markets only)

* See Air Conditioning Wiring Diagram   ** See Fuel Injection Wiring Diagram

*** The service interval counter operates the catalyst warning light

J86 241/1

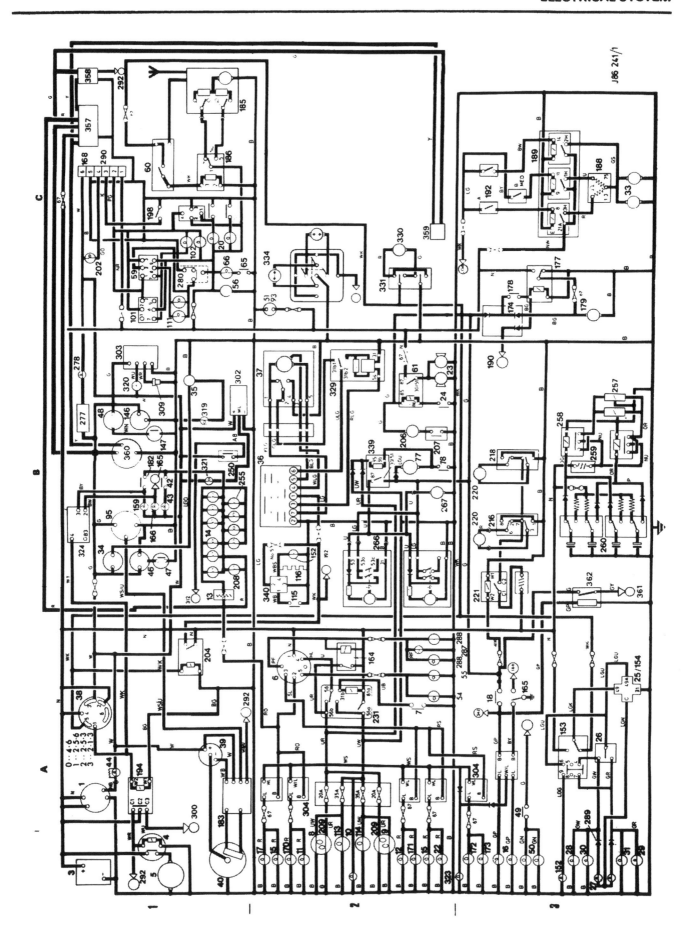

CABLE COLOUR CODE

N. Brown	P. Purple	W. White
U. Blue	G. Green	Y. Yellow
R. Red	L. Light	B. Black
K. Pink	S. Slate	O. Orange

When a cable has two colour code letters, the first denotes the Main colour and the second the Tracer Colour.

SYMBOLS USED

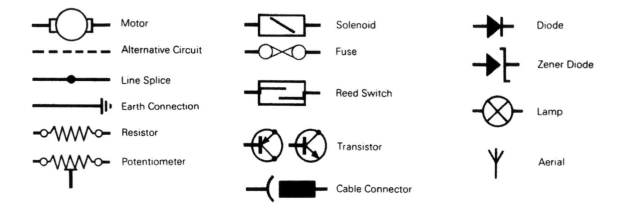

- Motor
- Alternative Circuit
- Line Splice
- Earth Connection
- Resistor
- Potentiometer
- Solenoid
- Fuse
- Reed Switch
- Transistor
- Cable Connector
- Diode
- Zener Diode
- Lamp
- Aerial

CONTENTS

ELECTRONIC SPEEDOMETER

Description

The pulse signal required to operate both the speedometer (and where fitted, the service interval counter) is controlled by a speed transducer situated in the transmission unit in place of the angle drive. The service interval counter, NAS markets only, is situated in the boot compartment and is located by removing the rear detachable boot trim panel.

It is important to note that should the harness controlling the pulse input to the speedometer become disconnected at the speedometer, the service interval counter will also CEASE TO OPERATE. The control for resetting the speedometer is now situated in the speedometer fascia and is operated by depressing the control button.

INSTRUMENTS

Description

The fuel gauge, temperature gauge, oil gauge and battery gauge, are all air cored instruments. An air cored instrument can be considered as a magnet in a magnetic field. A bar magnet pivoted in its centre is mounted in the centre of three coils. To cause the magnet and therefore the pointer to move, the current flow in one of the three coils is changed, which in turn alters the magnetic strength in that coil.

The transmitter or sensors are all variable resistors, and are connected across or in parallel with one of the coils. As the resistance varies in the transmitter the current flow through the coil will also vary. This will alter the magnetic field strength causing the pointer to move.

TACHOMETER

The engine speed signal received by the tachometer is derived from the ignition coil negative terminal. The voltage wave-form at this point can reach as much as 400 volts when a spark is generated, and it is desirable to suppress this voltage before allowing it into the wiring harness. If this is not done interference may occur with other signals. A resistor connected into the lead from the negative terminal of HT coil serves this purpose, and it is located inside the ignition amplifier. The battery voltage required to operate the instruments is derived from fuse No. 4.

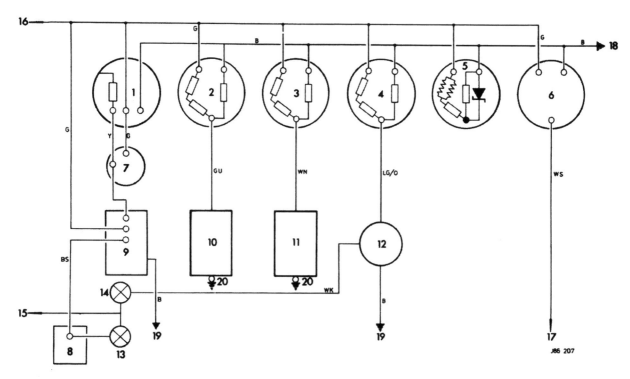

Fig. 1

KEY TO DIAGRAM

1. Speedometer
2. Temperature gauge
3. Oil pressure gauge
4. Fuel gauge
5. Battery condition gauge
6. Tachometer
7. Speed transducer
8. Bulb failure unit
9. Service interval counter (NAS)
10. Coolant temperature transmitter
11. Oil pressure transmitter
12. Fuel tank unit
13. Oxygen sensor warning lamp (NAS)
14. Low fuel warning lamp
15. To ignition switch
16. To fuse No. 4
17. To ignition amplifier
18. Earth point on steering column bracket behind instrument panel
19. Earth point between battery and wheel arch in luggage compartment
20. Earth through body of unit

Testing

Instrument tester 'Fast Check' SMD4045 (Fig. 2) has been developed to enable the operator to diagnose faults on the vehicle instrumentation with the minimal disruption to the wiring or the fascia panel. The tester will enable a functional check of the fuel, oil pressure and temperature gauges on the vehicle. The tachometer and the speedometer can also be checked. A signal from the output terminal of the tester will produce a preset reading on the gauge to which it is connected thus starting the test procedure at the transmitter of the respective gauge which will determine whether the fault lies in the gauge, the transmitter or the interconnecting harness.

To test the tachometer check the fuse and all the connections. Insert the power take off plug into the cigar lighter socket.

J88 018

Fig. 2

Disconnect the white and slate lead to the tachometer from the ignition amplifier. Connect the black lead from the tester to a good earth, the other lead to the tachometer harness lead. Select tacho on the tester and switch on the vehicle ignition. If the tachometer is operating correctly and the related circuit is satisfactory, the tachometer should register 4000 rpm. Should the tachometer not operate correctly, move the tester lead from the amplifier end of the harness to the tachometer terminal. This will establish whether the fault is in the tachometer or the harness.

The same procedure can be used when checking the electronic speedometer. Connect the tester to the disconnected lead from the speed transducer. Select speedo on the tester, switch the ignition on and the speedometer should register 60 mph. If the fault still exists, then the fault is in the related harness or the speedometer. This can be checked using the same method as with the tachometer. When checking the other gauges the system is just the same, the tester simulates the top and bottom gauge readings by switching the tester to high or low.

INSTRUMENT PANEL (MODULE)

Remove and refit

Removing

Disconnect the battery earth cable.
Remove the drivers side dash casing.
Remove the centre securing strip and screw from instrument module surround (1, Fig. 3).
Remove the screws securing the side pieces of the surround and remove the surround (2, Fig. 3).
Prise off the covers from the securing screw apertures (3, Fig. 3) and remove the screws securing the instrument panel to the fascia (4, Fig. 3).
Ease the module forward and disconnect the cable harness at the block connectors.
Slacken the steering wheel adjustment ring and extend the steering column to its maximum travel.
Manoeuvre the instrument panel clear of the fascia.

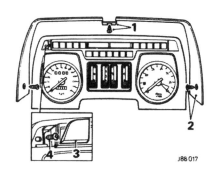

J88 017

Fig. 3

PRINTED CIRCUIT AND INSTRUMENTS

Renew

After removing the instrument panel module.
Remove the four instrument illumination bulb holders from the rear of the instrument panel (1, Fig. 4).
Note the position of all the fixing screws, connectors etc.
Remove the warning lamp retaining bar/earth (2, Fig. 4).
Remove the three nuts connecting the printed circuit to the tachometer (3, Fig. 4).
Remove the three nuts connecting the printed circuit to the speedometer (4, Fig. 4).
Remove the seven screws retaining the lens assembly and with the lens assembly removed it is now possible to withdraw the tachometer and/or the speedometer from the instrument panel.
Remove the screws connecting the printed circuit to the centre gauges (5, Fig. 4) it is now possible to remove a gauge by prising off the retaining clips and withdrawing the appropriate gauge.
Remove the seventeen warning lamp holders (6, Fig. 4).
Ease the printed circuit clear of the locating spigots.

NOTE: On refitting care should be taken to ensure the printed circuits are not torn or deformed, and all the connecting tags are correctly positioned under the terminals of components.

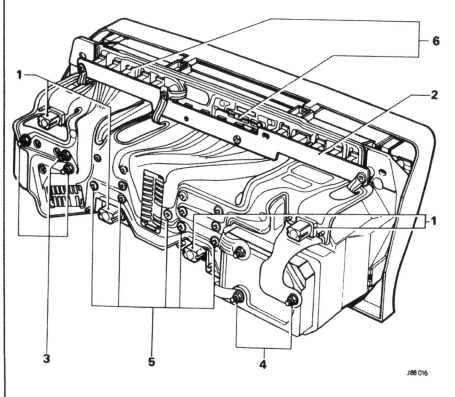

J88 016

Fig. 4

OIL AND TEMPERATURE GAUGE TRANSMITTERS

Renew

Disconnect the battery earth lead.
Depressurise the cooling system for the renewal of the temperature transmitter.
Disconnect the lead from the appropriate transmitter and withdraw the transmitter.

Fig. 5 Temperature gauge transmitter.
Fig. 6 Oil pressure gauge transmitter.
Fig. 7 Oil pressure warning light switch.

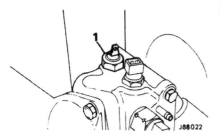

Fig. 5

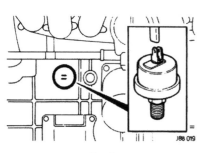

Fig. 6

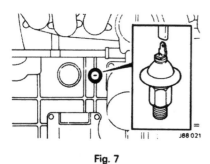

Fig. 7

FUEL GAUGE TANK UNIT

Renew

Service tool 18G 1001

Disconnect the battery earth lead.
Remove the spare wheel for access. Drain the fuel tank.
Turn the locking ring anti-clockwise using service tool 18G1001.
Remove the locking ring.
Turn the tank unit 180° and manoeuvre the unit from the tank.
Ensure new seal and locking ring is fitted when replacing the tank unit.

SPEED TRANSDUCER

Renew

Disconnect the battery earth lead.
Place an oil drip tray beneath the gear box.
Operate the left-hand seat release lever and push the seat to its full rear position.
Remove the left-hand side console pad.
Peel down the carpet from the centre console to expose the plate for access to the speed transducer (Fig. 8).
Remove the four screws retaining the plate and remove the plate.
Remove the bolt securing the speed transducer and withdraw the transducer from the gearbox housing.
Disconnect the cable harness block connector and remove the transducer (Fig. 8).

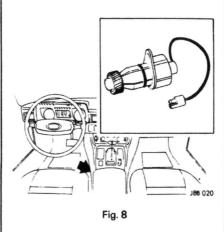

Fig. 8

On refitting smear sealant on the access plate to ensure a water tight seal.
Check the gearbox oil level.

TRIP COMPUTER (where fitted)

Description

The trip computer records fuel usage, time and distance. By storing the three sets of information and relating one to another it computes fuel consumption, both average consumption for the journey or a current consumption figure updated every three seconds. The information may be displayed either in miles and gallons or litres and kilometres.

Computer Controls

There are nine controls on the computer face, the use of each is described followed by examples of their use.

mls/km	— Use this switch to display metric or imperial / US units.
RESET	— Press for 5 seconds to switch all function displays to zero.
DISP	— Press to switch display off (function updating continues).
TIME	— Press to display time of day — press again to display elapsed time since reset — after 6 seconds, display will revert to time of day.
AV. SPD	— Press to display average speed since reset.
DIST	— Press to display distance travelled since reset.
AVE	— Press to display average fuel consumption since reset.
INST	— Press to display the fuel consumption at that time.
FUEL	— Press to display fuel consumed since reset.

To show which function is on display the relevant button will be illuminated. When the vehicle lights are switched on the computer illumination is dimmed but the legend plate is illuminated.

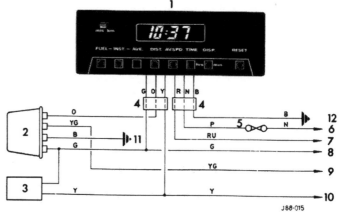

Fig. 9

KEY TO DIAGRAM

1. Trip Computer
2. Interface Unit
3. Speed Transducer
4. Connector Blocks
5. Inline Fuse (2 amps)
6. Terminal Post
7. To Sidelamps
8. To Fuse No. 4 (10 amps)
9. To ECU
10. To Speedometer
11. Earth point between battery and wheel arch in the luggage compartment
12. Earth point on steering column bracket behind the instrument panel

The signals required to operate the trip computer are picked up from the ECU via the interface unit (2, Fig. 9) and the pulse generator (3, Fig. 9). A 12 volt supply is via the fuse (5, Fig. 9). This supply voltage enables the clock to function and for the computer to retain information it has received when the ignition is switched off.
A second 12 volt supply is via fuse (8, Fig. 9) this supply enables the computer to display information when the ignition is switched on. The third 12 volt supply is via the red and blue lead (4, Fig. 9). This supply voltage enables the display and the buttons to dim when the sidelamps are switched on. The legend strip is also illuminated.

Fault Diagnosis

Check the fuses and all connections. Ensure the earth connections are clean and tight. With the ignition switched off 12 volts should be obtained on the purple lead to the trip computer.
The voltmeter should give the following readings with the ignition switched on:
12 Volts at the green lead to the trip computer, the green lead to the pulse generator, the green lead and the yellow/green lead to the interface unit.
With the engine running a voltage should be obtained at the orange lead to the computer. A zero reading indicates a faulty interface unit or lack of continuity in the wiring between the computer and the interface unit. Re-check at the interface unit located in the luggage compartment.
With the rear of the vehicle jacked up, and on stands. Start the engine and put the vehicle into drive. A voltage should then be obtained at the yellow lead to the computer. A zero reading indicates a faulty pulse generator or lack of continuity in the wiring between the pulse generator and the computer.

TRIP COMPUTER

Renew

Disconnect the battery earth lead.
Prise the computer from the switch panel.
Disconnect the cable block connectors and remove the trip computer.

INTERFACE UNIT

Renew

Disconnect the battery earth lead.
Remove the right-hand luggage compartment trim panel.
Lift the interface unit from its bracket (1, Fig. 10) and disconnect the cable harness block connector.
Remove the interface unit.

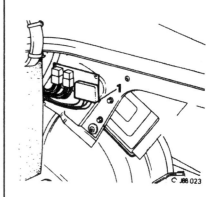

Fig. 10

FAULT	ACTION
Computer inoperative Screen blank All voltages correct	Replace Computer
Time of day displayed Average speed and distance displayed Fuel characteristics zero Speedometer operating All battery voltages correct Zero voltage on orange lead with the engine running	Replace interface unit
Time of day displayed All other functions zero Speedometer not operating All battery voltages correct Zero voltage on yellow lead with the vehicle on stands and the engine running and the vehicle in drive	Replace the pulse generator (Speed transducer)
Trip computer does not dim with the sidelamps switched on Battery voltage at the red/blue cable connection	Trip computer faulty
More than one light emitting diode illuminated at the same time All battery voltages correct	Trip computer faulty

CLOCK (where fitted)

Renew

Disconnect the battery earth lead.
Prise the clock from the switch panel.
Remove the illumination bulb holder and disconnect the supply leads.
Remove the clock.

SERVICE TOOLS — Section 99
ENGINE

MS53A

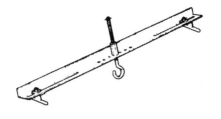

Engine support bracket

18G134-8

Rear main oil seal installer

MS204

Valve seat cutter

18G106A

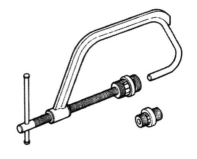

Valve spring compressor

MS150-8

Expandable pilot

18G1432

Valve guide remover/replacer

18G1434

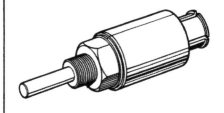

Jackshaft inner bearing
remover/replacer
(2-part tool)

18G1435

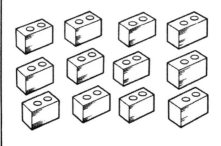

Dummy camshaft caps

18G1436

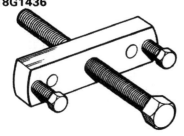

Timing chain tensioner restrainer
and crankshaft pulley remover

18G1437

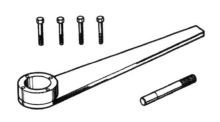

Front pulley lock

18G55A/38U3

Piston ring clamp

MS76

Basic handle set

18G1433

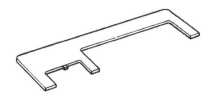

Camshaft timing tool

18G1468

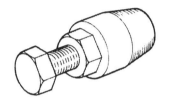

Auxiliary drives oil seal remover

18G1469

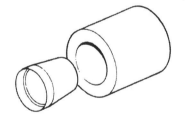

Auxiliary drives oil seal replacer

PROPSHAFT AND FINAL DRIVE

SL14-3

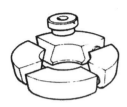

Adaptor remover differential bearing cone

SL15A

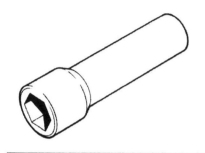

Remover/replacer drive shaft bearing cone

JD1D

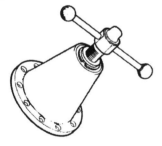

Hub remover

SL47-1

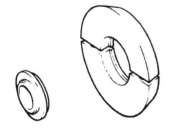

Pinion bearing cone remover/adaptor

SL47-3

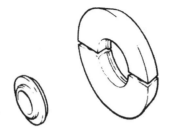

Axle shaft bearing remover/ replacer adaptor

SL3

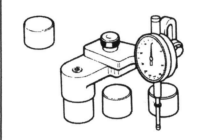

Pinion setting gauge

SL550-1

Replacer diffential bearing cone

SL550-8

Adaptor/replacer drive pinion outer bearing cup

SL550-9

Adaptor/replacer drive pinion inner bearing cup

18G1205

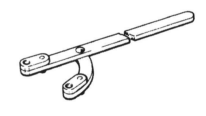

Propellor shaft flange wrench

18G1428

Pinion oil seal installer

STEERING AND FRONT SUSPENSION

JD10

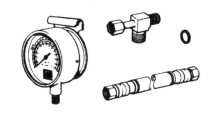

Power steering test set

JD10-2

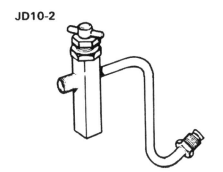

Adaptor hydraulic pressure test

JD36A

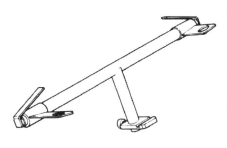

Steering rack checking fixture

S355

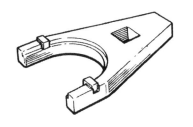

Pow-A-Rak nut wrench

18G1259

Assembly pilot steering rack pinion

606602

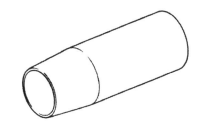

Ring expander for steering box valve and worm seals

606603 (JD33)

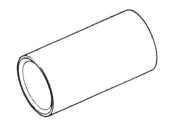

Ring compressor for steering box valve and worn seals

JD24

Steering joint taper separator

JD6G

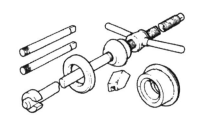

Front coil spring compressor

18G1259

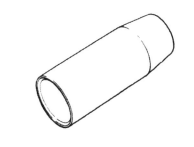

Assembly pilot steering rack pinion

18G1445

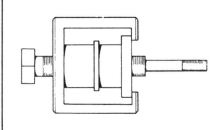

Power steering pump drivedog — remover/replacer

18G1446

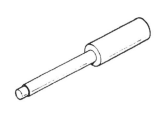

Steering rack alignment pin

REAR SUSPENSION

JD11B

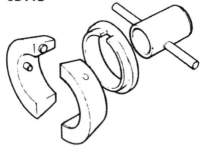

Adaptor dismantler dampers spring unit

JD13A

Rear hub end float gauge

JD14

Rear wishbone pivot dummy shaft

JD25B

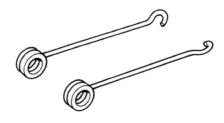

Rear camber setting links

JD16C

Remover/replacer rear hub outer bearing

JD15

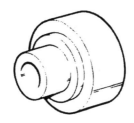

Replacer — Rear hub master spacer and bearing

JD1D

Hub remover

JD20A

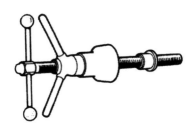

Bearing remover main tool

JD20A-1

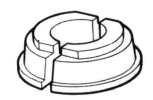

Rear hub inner and outer cup remover/replacer adaptor

JD21

Torque arm bush remover/replacer

SERVICE TOOLS

BRAKES

18G672

Replacer disc brake piston seal

AIR-CON

18G1363

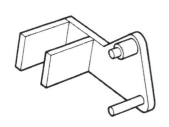

Air-con linkage setting tool

ELECTRICAL

18G1001 (600964) (P9074A)

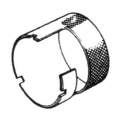

Spanner for fuel tank unit

GENERAL

47

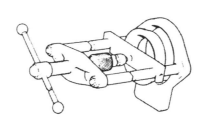

Multi-purpose hand press

370

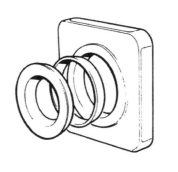

Adaptor plates

MS68

Torque screw wrench

18G134 (550)

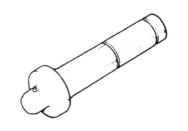

Driver handle

18G681 (CBW548) & (CBW548-1)

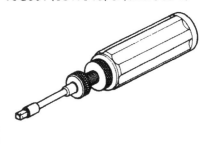

Torque screwdriver

All service tools listed are available from:
V. L. Churchill & Co. Limited
P.O. Box 3
Daventry
Northamptonshire NN11 4NF

This page is intentionally blank

JAGUAR
XJ-S

XJ-S 4.0

SERVICE MANUAL SUPPLEMENT

Published by
Jaguar Cars Ltd

Supplement AKM 9063BB

SUPPLEMENT • XJ-S 4.0

INTRODUCTION

This supplement is to be used in conjunction with XJS Repair Operation Manual Part No. AKM 9063.

It covers later vehicles, mainly dealing with those modifications brought about in line with the introduction of the automatic transmission, 4 Litre engine and related systems, ABS brake systems, body modifications, and more.

Please note that, due to the ever increasing sophistication of the later versions of these vehicles, it is necessary to cover Electrical Circuits in-depth in a separate publication. There are a limited selection of circuits printed at the rear of this Supplement, but full coverage will only be available by reference to the 1992 Model Year-on Service Manual (Part Number JJM 10 04 06 / 02) and the relevant Model Year's 'Electrical Guide'.

SPECIFICATION

Users are advised that the specification details set out in this book apply to a range of vehicles and not to any one. For the specification of a particular vehicle purchasers should consult their Dealer.

Jaguar Cars Ltd. reserve the right to vary their specifications with or without notice and at such times and in such a manner as they think fit. Major as well as minor changes may be involved in accordance with the Manufacturer's policy of constant product improvement. Whilst every effort is made to ensure the accuracy of the particulars contained in this supplement, neither Jaguar Cars Ltd., nor the Dealer, by whom this book is supplied, shall in any circumstances be held liable for any inaccuracy or the consequences thereof.

COPYRIGHT

XJS 4.0 Service Manual Supplement Contents

GENERAL SPECIFICATIONS

Engine - 4.0 Litre

Compression Pressure	160 to 170 lbf/in²
Compression differential between cylinders	10 lbf/in²
Bore	91 mm
Stroke	102 mm
Capacity	3980 cm³
Power (catalyst version)	237 bhp @ 4700 rpm
Torque (catalyst version)	282 lbf ft @ 4000 rpm
Maximum engine speed	5500 rpm
Compression ratio	9.5:1
Valve clearances (inlet and exhaust)	0.30 to 0.36 mm
Idling speed (auto in neutral) 650 to 750 rpm	
Idling speed (manual)	750 to 850 rpm
Fuel pressure	3 bar
Exhaust gas reading at idle	0.5% CO (catalyst)
	Maximum 2% CO (non catalyst)

Transmission - System

Final drive ratio (automatic)	3.58:1
Final drive ratio (automatic) (later vehicles)	3.54:1
Final drive ratio (manual)	3.54:1

Automatic Transmission

Ratios		
	1st	2.48:1
	2nd	1.48:1
	3rd	1.00:1
	4th	0.73:1
	Reverse	2.09:1
Fluid Capacity		10.2 litres
Fluid Type		Dexron IID
Weight (inc. fluid & converter)		82.1 kg

Chapter 1

4.0 LITRE ENGINE

Description

The engine differs only very slightly from the 3.6 litre unit fitted to earlier models.

Internally the bore and stroke dimensions have changed to facilitate the increase in capacity to 3980 cm^3. This has resulted in an increase in power to 237 bhp and torque to 282 lbf.ft. in the later, enhanced AJ16 version, slightly less for the AJ6 (shown in Fig. 1).

The fault-finding charts, procedures, materials, and torque specifications given in the existing Section 12 of this Manual should be followed when overhauling the unit.

Fig. 1

Chapter 2

EMISSION CONTROL SYSTEM 4.0 LITRE

EVAPORATIVE EMISSION CONTROL, DESCRIPTION

In order to accommodate up to 10% fuel expansion, the maximum fuel level is limited by the inclusion of an expansion tank located within the fuel tank.

The fuel tank is vented through a system of vapour pipes and a liquid/vapour separator to a charcoal canister located under the left hand front wheel arch.

The flow of vapour to the canister is controlled by the vacuum purge control (Rochester) valve. The vapour absorbed in the canister is purged by the engine via an electrically operated purge control valve which is activated by the ECU (Engine Control Unit).

The rate of purge flow is dependent upon engine speed, load and temperature.

Canister purging does not occur until the engine coolant exceeds 43°C; this function is controlled by a thermal vacuum valve.

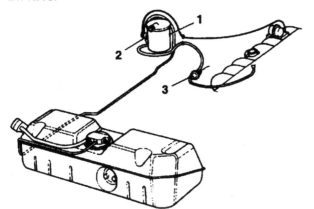

Fig. 1 Fuel and Evaporative Emission Control System

Key to Fig. 1

1. *Charcoal canister*
2. *Vacuum purge control (Rochester) valve*
3. *Electrical purge vacuum valve*

CHARCOAL CANISTER, RENEW

Jack up the front of the vehicle and support on two stands.
Remove the left hand front road wheel.
Remove the under arch access cover.

Note the positions of the hoses that connect to the Rochester valve (1 Fig. 2) and those to the canister stub pipes.

Remove the hoses by slackening the hose clips.
Undo the canister clamp securing nut (2 Fig. 2) and remove.
Prise open the clamp and remove the canister.
Fit the new canister within the clamp.
Fit and tighten the clamp securing nut.
Reconnect the hoses to the Rochester valve and canister stub pipes.
Tighten the hose clips.
Refit the access cover.

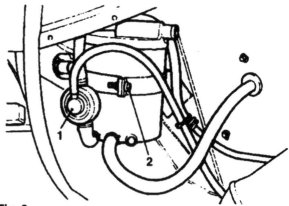

Fig. 2

Fit the road wheel.
Support the vehicle front end and remove the stands and lower the vehicle to the ground.

Note: On vehicles that are not fitted with ELC (Evaporative Loss Control) equipment, the Rochester valve is located in the luggage compartment.

VACUUM PURGE CONTROL (ROCHESTER) VALVE, RENEW

Slacken and slide back the two hose clips above the valve (1 Fig. 2) and similarly the one below, and remove the valve.
Fit a new valve and reconnect the vacuum hoses.
Reposition and tighten the hose clips.

PURGE CONTROL VALVE (ELECTRICAL), RENEW

Open the bonnet.
The purge control valve is fitted in the line of the vacuum hose between the charcoal canister and the engine (3 Fig. 3).
Unplug the electrical connector from the purge valve.
Disconnect the vacuum hose from each side of the valve and remove the valve.
Fit a new valve and reconnect the two hoses.
Plug in the electrical connector.

CATALYTIC CONVERTERS, DESCRIPTION

Catalytic converters are fitted into the exhaust system to reduce carbon monoxide and hydrocarbon emissions. Four ceramic monolith converters are mounted in the exhaust system, one pair in the down pipe close to the engine and two in an under body location.

In order for the catalytic converters to function correctly, it is necessary to control very closely the oxygen concentration in the exhaust gas entering the catalyst. This is achieved by the engine management system which continuously monitors the oxygen content of the exhaust gas by means of an oxygen sensor which is fitted to the rear of the down pipe catalytic converter. This sensor adjusts the fuelling level to obtain the required oxygen content.
Unleaded fuel must be used on catalyst equipped vehicles and labels are displayed to indicate this. The filler cap is designed to accommodate unleaded fuel pump nozzles only.

Note: The procedures for renewing down pipe and under floor catalytic converters and details of the various catalytic converter and exhaust systems are detailed in Section 30.

CRANKCASE VENTILATION SYSTEM, DESCRIPTION

To ensure that piston blow by gases do not escape to atmosphere from the engine crankcase, a depression is maintained in the crankcase under all operating conditions. This is achieved by scavenging from the crankcase and the camshaft case via the oil sump filler tube.

The crankcase and camshaft case emissions that are collected are then fed into the engine intake manifold through a part throttle control orifice and the engine air intake elbow.

To prevent possible icing up of the control orifice during cold weather the orifice is heated in a water heated housing and a breather heater element, located in the hose connected from the oil filler and air intake elbow.

The heater element is controlled by the power wash jet heater sensor, via a relay.

PART THROTTLE ORIFICE CONTROL HOUSING, RENEW

CAUTION:

This operation must only be carried out on a cold or cool engine.

Open the bonnet.
Carefully remove and refit the pressure cap from the remote header tank to release coolant system pressure.
Slacken the housing to breather pipe hose clip (1 Fig. 4) and disconnect breather hose from the housing.
Remove the bolts securing the housing (2 Fig. 4) and displace the housing above engine level for access.
Remove and discard the housing gasket.
Slacken the hose clips and disconnect the hoses from the housing.
Position the hoses above engine level to prevent spillage.

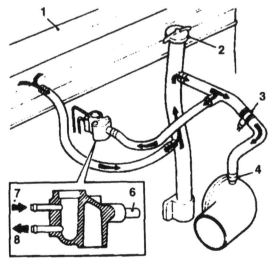

Fig. 3 Fuel and Evaporative Emission Control System

Key to Fig. 3

1. Camshaft cover
2. Oil filler
3. Heater element
4. Hose connection
5. Air intake elbow
6. Inlet
7. From thermostat
8. To coolant hose

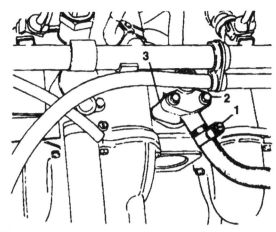

Fig. 4

Remove the housing (3 Fig. 4) and clean the housing and gasket faces.
Fitting a new housing is the reverse of the removal procedure, ensuring that the coolant is topped up to the correct level.

ENGINE BREATHER HEATER ELEMENT, RENEW

Open the bonnet and disconnect the heater element cable harness block connector.
Slacken the breather hose to element securing clip and disconnect the hose from the element.
Slacken the intake elbow to heater element securing clip. Disconnect the hose.
Remove the heater element (Fig. 5).
Fitting a new element is the reverse of the removal procedure.

BREATHER HEATER ELEMENT TO INTAKE ELBOW HOSE, RENEW

Open the bonnet.
Slacken the clip securing the hose to the breather element and disconnect the hose from the element.
Slacken the clip securing the hose to the intake elbow and remove the hose from the elbow.
Remove the clips from the hose.
Fitting a new hose is the reverse of the removal procedure.

OIL FILLER TO HEATER ELEMENT HOSE, RENEW

Open the bonnet.
Slacken the clip securing the hose to the oil filler tube and disconnect the tube from the tube.
Slacken the clip securing the hose to the part throttle control orifice housing and disconnect the hose from the housing.
Slacken the clip securing the hose to the heater element. Disconnect and remove the hose from the element.
Remove the clips from the hose.
Fitting a new hose is the reverse of the removal procedure.

Fig. 5

Chapter 3

ENGINE MANAGEMENT SYSTEM - 4.0 LITRE

INTRODUCTION

The engine management system maintains optimum engine performance over its entire operating range by metering the fuel into each cylinder's inlet tract and adjusting the ignition timing angle at the sparking plugs. Each of these functions is performed by an electronic control unit (ECU) which, from data received from sensors located on and around the engine (Fig. 1), evaluates optimum ignition timing and fuel metering parameters relative to engine load and speed.

Additional functions of the system include:

Fuel pump control to prevent fuel flooding and/or spillage when the engine is stationary with the ignition switch in position 'II'.

Cold start control to ensure sufficient fuel exists in the inlet manifold to create a combustible air / fuel mixture in the combustion chamber.

Idle speed control to compensate for varying engine temperature and load conditions, e.g. the engaging or disengaging of the transmission (4 litre automatic only) and / or the air conditioning clutch.

Fuel cut-off during engine overrun conditions to improve fuel economy by minimising the quantity of unburnt fuel discharged into the exhaust system.

Engine overspeed control to impose a maximum engine speed limit of 5800 rev/min (automatic transmission) or 6100 rev/min (manual transmission).

Air pollution control to reduce exhaust contamination to levels which comply with different country exhaust regulations.
Fuel monitoring to provide precise fuelling information to the trip computer.

Fault monitoring to provide data to the trip computer for display.

The system also incorporates a 'limp home' feature which permits continued engine operation on certain sensor failure, the sensor failure being indicated on the trip computer.

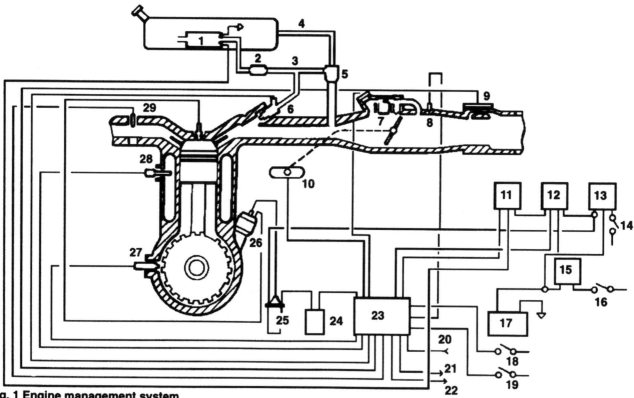

Fig. 1 Engine management system

Key to Fig. 1

1. Fuel pump.
2. Fuel filter.
3. Fuel rail.
4. Fuel return.
5. Fuel pressure regulator.
6. Injector.
7. Idle speed control valve.
8. Air thermistor.
9. Airflow meter

10. Throttle potentiometer.
11. Fuel pump relay.
12. Main relay
13. Ignition 'on' relay.
14. Inertia switch.
15. Starter motor relay
16. Ignition switch.
17. Battery.
18. Gearbox sensor (automatic transmission).
19. Air conditioning clutch sensor.

20. Idle override.
21. Trip computer (fuel failure).
22. Trip computer (fuel used).
23. Electronic control unit (ECU).
24. Ignition amplifier.
25. Ignition coil.
26. HT distributor.
27. Crankshaft sensor.
28. Coolant temperature sensor.
29. Lambda (oxygen) sensor (where fitted).

Fuel Metering

An electric pump located inside the fuel tank force feeds the engine mounted fuel rail via an in-line filter. The six fuel injectors are connected to the fuel rail and are electro-mechanically operated to inject fuel into each inlet port.

A fuel pressure regulator is fitted to the fuel rail to control the fuel line pressure so that a constant pressure is maintained across the injectors. This is accomplished by applying the intake manifold pressure to a diaphragm within the pressure regulator. Excess fuel is returned to the fuel tank via a pipe connected to the fuel pressure regulator.

The fuel pump is energised only when the ECU senses an engine cranking signal or an engine running signal. This prevents the engine being flooded with fuel in the event of an injector sticking open.

Injector operation is by means of an electrical pulse which actuates a solenoid valve within the injector body. The duration of the pulse and hence the quantity of fuel injected, is determined by the ECU on the basis of intake air (engine load) and engine speed information derived from airflow meter and crankshaft sensors. This information is used to access mapped data stored in 256 memory locations containing injector pulse durations pertaining to 16 engine loads at 16 different speeds.

Correction factors are imposed on the basic injector pulses to compensate for varying conditions. The resultant pulses are then normally applied to the injectors twice per engine cycle, i.e. once per crankshaft revolution, with only half the required amount of fuel being injected at each injector 'on' (open) time.

Injector pulse adjustments are necessary to provide:

Cranking enrichment during starting.

Temperature enrichment during starting and warm-up.

After-start enrichment during warm-up.

Demand corrections during idle, full power and acceleration.

Voltage corrections for variations in the electrical system voltage.

Contamination corrections for emission control.

Cranking enrichment

Cranking enrichment is provided, appropriate to coolant temperature only, when the starter motor is activated. This is achieved by increasing the injector operating frequency from one pulse to three pulses per crankshaft revolution and is implemented by the ECU in response to an input from the crankshaft sensor. Cranking enrichment is terminated at an engine speed of 200 rev/min above a cooling temperature of 75º C (167º F) and at 500 rev/min below a cooling temperature of 75º C (167º F).

Temperature enrichment

Temperature enrichment is provided during starting and warm-up. This is achieved by increasing the injector 'on'-time above that of basic requirements and is implemented by the ECU in response to an input from a coolant temperature sensor.

After-start enrichment

After-start enrichment is provided, dependent upon coolant temperature, to supply added fuel during warm-up. This is achieved by the ECU which increases the injector 'on' time upon release of the starter above that of basic requirements which then decreases over a short time period at a fixed rate.

Demand corrections

Corrections are provided for idle, acceleration, deceleration and full power demands. This is achieved by decreasing/increasing the injector 'on' time above that of basic requirements and is implemented by the ECU in response to an input from a throttle potentiometer.

Voltage corrections

The ECU constantly monitors the electrical system voltage, i.e. the state of the battery and electrical load, because the time taken for an injector to open is affected by voltage and results in a corresponding change in the quantity of fuel delivered. The ECU compensates for any voltage change detected by adjusting the injector 'on' time accordingly.

Contamination corrections

On certain vehicles exhaust pollution is reduced to a minimum by the use of an exhaust catalyst, used in conjunction with a sensor monitoring the oxygen content of the exhaust which in turn maintains the intake air / fuel ratio constant at approximately 14.7:1. This is achieved by the ECU in response to an input from the Lambda (oxygen) sensor.

Ignition timing

Ignition pulses are applied, via a separate ignition amplifier, to the ignition coil which generates high energy pulses for the sparking plugs via the distributor. Unlike conventional ignition systems, however, the distributor has no centrifugal or vacuum advance mechanisms and no LT circuit exists. The distributor is only required, therefore, to distribute the high energy pulse to the spark plugs.

Ignition timing is controlled by the ECU from information supplied by the airflow meter and crankshaft sensor. This information is used to access mapped data stored in 256 memory locations containing ignition timing angles pertaining to 16 engine loads at 16 different speeds.

Depending on the monitored engine speed and supply voltage, the dwell-period (dwell angle) is determined and the ignition energy is adapted to that required at any instant in time. The unnecessary consumption of energy in the ignition coil is thus prevented. In addition, a peak coil current cut-off facility limits current flowing through the ignition coil below a specific engine speed and therefore prevents the battery being discharged and the ignition coil overheating.

Description

The ECU contains a microprocessor and discrete electronic component circuits that provide interfacing to peripheral input/output devices, engine locations of which are shown in Fig. 1.

The microprocessor performs all system control functions and has a memory pre-loaded with system parameters that are accessed under software control. System response times are, therefore, reduced to a minimum and the system is rendered operational immediately the ignition switch is set to position 'II'.

System peripherals are as follows:

Fuel injectors: solenoid actuated devices which spray quantities of fuel into the inlet manifold.

Ignition amplifier: an electronic device which, through the ignition coil, generates high energy electrical pulses for distribution to the spark plugs.

Airflow meter: a hot wire sensing device which monitors inlet manifold air flow for optimum fuelling and ignition control.

Idle speed control valve: a stepper motor driven device which controls the volume of air entering the engine to maintain a correct idle speed.

Crankshaft sensor: a device which generates engine speed and crankshaft position information for precise ignition timing and fuelling control.

Throttle potentiometer: a device which interprets the throttle position to identify idle, acceleration and full power demands.

Gearbox sensor: two linear gearbox selector actuated switches that induce idle speed corrections which compensates for engine load changes that occur when 'N' or 'P' is engaged or disengaged. On vehicles fitted with manual transmission units, the switch input to the ECU is connected to logic earth.

Fuel pump relay: a device which implements controlled fuel pump switching.

Coolant temperature sensor: a thermal device which monitors the coolant temperature to induce cold starting and warm-up enrichment.

Lambda sensor: a device which monitors the oxygen content of the exhaust to induce contamination corrections.

Air conditioning sensor: an air conditioning compressor clutch actuated switch that induces idle speed corrections which compensate for engine load changes that occur when the clutch is engaged or disengaged.

Idle fuel potentiometer: a device which provides fine adjustment of engine idle fuel. This device is set during vehicle manufacture and only requires adjustment when the airflow meter assembly, of which it forms an integral part, is renewed.

Air temperature sensor: a thermal device which monitors the air inlet temperature to retard the ignition under certain conditions.

Other system devices and communications are as follows:

Ignition 'on' sensing: an input taken from contact 87 of the ignition 'on' relay which, in addition to applying power to the ignition coil, energises the main and fuel pump relays through the ECU. A timer is also initiated which, through the fuel pump relay, permits fuel pump operation for approximately 0.5 seconds to ensure the fuel rail is pressurised prior to cranking.

Cranking sensing: an input which induces cranking enrichment when the starter solenoid is activated.

Battery voltage sensing: an input taken from contact 87 of the main relay which, in addition to applying power to the airflow meter, fuel injectors and the fuel pump relay, induces fuelling corrections that compensate for battery voltage fluctuations.

Idle override: an input derived from the road speed sensor on the final drive unit which, in addition to displaying the vehicle road speed on the instrument pack, inhibits engine idle control at road speeds greater than 4,8 km/h (3 mph).

Trip computer (fuel failure): an output applied to the instrumentation system to provide indication of an engine management system fault on the trip computer.

Trip computer (fuel used): a dual purpose output applied to the instrumentation system to:

- Identify individual engine management system faults when the ignition switch is initially set to position 'II' and,
- provide precise fuelling conditions to the trip computer when the engine is running.

OPERATION

A circuit diagram of the engine management system detailing interconnections between the ECU and the various input / output peripheral devices is shown in Fig. 1.

Ignition On

Power is applied to the various devices through contacts L14-87/30 of the main relay which is energised via contacts B17-87/30 of the ignition 'on' relay, when the vehicle ignition switch is in position 'II', i.e. when 0 volts is applied to B17-85 of the ignition 'on' relay.

Application of power to the ECU at L13-1 and L12-10 generates a system reset which causes the microprocessor to perform an initialization routine (a software program stored in the memory), condition input circuits for sensor scanning and initiates fuel pump operation to pressurise the fuel rail prior to cranking. The fuel pump is activated via contacts L15-87/30 of the fuel pump relay which is energised by an output from the ECU at L12-7.

On completion of the initialization routine, sensor scanning commences and analogue information from the sensors is converted into digital data for storage. The data thus stored is then evaluated to determine prevailing engine conditions and memory locations are accessed to obtain appropriate ignition timing and fuel metering parameters.

In addition, the following functions are performed:

The stepper motor of the idle speed control valve is driven, via L12-15/16 and L12-18/19, to set the valve at a position suitable for engine starting. Rapid valve positioning is achieved for all conditions by extending or retracting its stem from a mid-travel position restored each time the ignition switch is set to 'off'.

Fuel pump operation is terminated, i.e. the output at L12-7 is removed after approximately 0.5 seconds to eliminate possible fuel flooding should an injector be stuck in the open position. This feature is also used to eliminate possible fuel spillage creating a hazardous condition at an accident scene if fuel line fracture occurs due to impact damage.

A number relating to a fault which may have occurred during the last period that the engine was run is recalled from memory and displayed on the trip computer.

A breather pipe heater is actuated, via a breather pipe heater relay, under control of a screenwash jet heater sensor.

Engine Cranking

Fuel pump operation is reinstated and any fault number displayed on the trip computer is erased when the engine is cranked, provided that the ECU detects a crankshaft sensor input at L13-13 and 14. The Lambda sensor heater is also activated to heat the sensor on vehicles fitted with exhaust emission control.

As the engine rotates under the influence of the starter motor, fuel is introduced into the air flow of the inlet manifold by the fuel injectors which are actuated by the ECU under control of the crankshaft sensor input. Injector triggering occurs, through L12-12. 13 and 25, three times per crankshaft revolution with appropriate correction factors imposed. The ignition amplifier is also actuated at the correct times, through L12-1, to fire the spark plugs.

Engine Running

When the engine starts, cranking enrichment is terminated at 600 rev/min, i.e. injector pulses occur once per crankshaft revolution and both warm-up and after-start enrichment is imposed.

The engine idle speed is controlled, through L12-15/16 and L12-18/19, by the ECU which drives the idle speed control valve stepper motor from the coolant temperature, throttle position and crankshaft speed information. Compensation is also provided for fluctuating engine load by monitoring air conditioning clutch and gearbox (automatic only) sensor inputs at L13-21 and 3 respectively.

The engine idle speeds are set as follows:

Cold in neutral - can be up to 900 rev/min at - 20° C.
Hot in neutral - 700 rev/min.
Cold in drive - 780 rev/min.
Hot in drive - 580 rev/min.

Fuel and ignition requirements are calculated at each sensor scanning to maintain optimum engine running efficiency over its entire range, including full power and acceleration demands. Should a sensor failure be detected, an output is applied to the instrument pack to provide indication of the failure on the fascia. The nature of the fault is also stored in the ECU memory and is output as a fault number the next time the ignition switch is set to position 'II'.

Exhaust emission regulation is implemented by the ECU under control of the Lambda sensor (if fitted) which monitors the oxygen content of the exhaust system and provides a corresponding signal to the ECU at L13-17. This signal imposes corrections on the basic injector 'on' times to reduce exhaust emissions to a minimum. Power to the Lambda sensor heater is supplied from contact L15-87/30 of the fuel pump relay and therefore, exhaust emission control is only operational during periods that fuel pump operation is permitted.

Fault Identification

In the event of an engine management fault occurring the 'check engine' warning light will be illuminated and the words 'Check Engine' will be permanently displayed on the trip computer until the ignition is switched OFF.

When a fault has been signalled the likely area responsible for the malfunction can be indicated when the vehicle is stationary. Switch off the ignition, wait at least 5 seconds then turn the ignition switch to position 'II' (do not start the engine), the relevant failure code will be displayed.

Check Engine Fault Codes

All fault codes are, when triggered, stored in the ECU memory. In addition, the 'check engine' warning light is illuminated and will remain illuminated (when the ignition is switched on) until the fault is rectified and the ECU memory cleared by JDS or by grounding the relevant ECU pin (see notes below for exceptions). The fault codes are displayed in order of priority, one at a time. The next code is only displayed when the preceding one is rectified or cleared.

Note
1. Catalyst Specification:
   Only the following fault codes will cause the check engine warning light to illuminate - 11, 12, 14, 16, 17, 18, 19, 68, 23, 26 and 44.
2. Non-Catalyst Specification:
   As catalyst Specification except that codes 23, 26 and 44 are not active.

Definitions of fault numbers displayed on the trip computer are as follows:

Fault Code	Description	Function
11	Idle pot or associated wiring.	Looks for idle trim pot out of normal operating range.
12	Airflow meter or associated wiring.	Looks for airflow meter signal out of normal operating range.
14	Coolant thermistor or associated wiring.	Looks for coolant thermistor static during engine warm up.
16	Air thermistor or associated wiring.	Looks for air thermistor resistance out of range.
17	Throttle pot or associated wiring.	Looks for throttle pot resistance out of range.
18	Throttle pot / airflow meter calibration.	Looks for low throttle pot signal at high air flow.
19	Throttle pot / airflow meter calibration.	Looks for high throttle pot signal at low air flow.
22	Fuel pump drive.	Looks for ECU output to fuel pump relay.
23	Fuel supply.	Looks for poor feedback control in rich direction.
24	Ignition drive.	Looks for ECU output to ignition amplifier module.
26	Air leak.	Looks for poor feedback control in lean direction.
29	ECU self check.	Checks microprocessor function.
33	Injector drive fault.	Checks for ECU output to injectors.
34	Injector drive and associated wiring.	Looks for injector 'dribble'.
44	Lambda sensor or associated wiring.	Looks for feedback out of control rich or weak.
46	Idle speed control coil 1 drive.	Looks for ECU output to idle speed control stepper motor.
47	Idle speed control coil 2 drive.	Looks for ECU output to idle speed control stepper motor.
48	Idle speed control motor or valve.	Looks for stepper motor being wildly out of position with engine temperature hot and less than 30°C (86°F).
68	Road speed sensor or associated wiring.	Looks for road speed indicating less than 5 km/h (3 mph) at high engine air flow.
69	Drive / neutral switch or wiring.	Looks for cranking in 'D' or high air flow in 'N'.

INJECTORS, DESCRIPTION

Fig. 2

Each fuel injector (Fig. 2) contains a needle valve rigidly attached to the plunger of a solenoid which, when energised by an electrical pulse, moves the needle away from its seat against a helical pressure spring. This action permits a measured quantity of pressurised fuel to enter the inlet manifold, via an integral filter and nozzle, each time an injector pulse is applied.

Valve lift is approximately 0,15 mm (0.006 in.) for the fully open position which is reached in a response time of approximately 1 millisecond. The amount of fuel delivered is dependent only on the period of time the injector is open, the precise open time being determined by the ECU and is typically 2.75 milliseconds at engine idle.

CRANKSHAFT SENSOR, DESCRIPTION

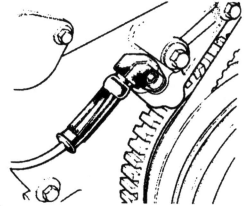

Fig. 3

The crankshaft sensor (Fig. 3) provides both engine speed and crankshaft position information to the ECU. It comprises a toothed gear and a variable reluctance coil.

CRANKSHAFT SENSOR TOOTHED GEAR, DESCRIPTION

The toothed gear is attached to the crankshaft and has 57 symmetrically spaced rectangular segments around its periphery. These segments induce a signal in the reluctance coil when the engine is running, the frequency of the signal being proportional to engine speed. Three omitted segments, which would give a total segment complement of 60, interrupt the induced signal to provide crankshaft position references.

COOLANT TEMPERATURE SENSOR, DESCRIPTION

Fig. 4

The coolant temperature sensor (Fig. 4) is located on top of the thermostat housing and comprises a temperature-sensitive resistor having a negative temperature coefficient, i.e. its electrical resistance decreases conversely with increasing temperature.

AIR TEMPERATURE SENSOR, DESCRIPTION

The air temperature sensor measures the air inlet temperature such that the ignition timing can be retarded under certain ambient and engine load conditions.

AIRFLOW METER, DESCRIPTION

Fig. 5

The airflow meter (Fig. 5) provides a measurement to the ECU of the air flow in the duct to the inlet manifold.

The unit contains two sourcing elements; one at inlet temperature and the other which is heated by an electrical current.

Both elements are situated in a by-pass, the heated element being subjected to cooling as air is drawn into the engine. The mass of air is thus determined by measuring the electrical current required to maintain the temperature differential of the elements constant. The by-pass method of air flow measurement has a number of advantages which include:

> Protection for the delicate sensing elements from shock loads.

> Reduction in the scale of measurement error due to severe flow pulsations.

> Minimal contamination of the sensing elements by dirt and backfire flows that exist in the main duct.

A potentiometer is incorporated in the airflow meter to provide idle fuel adjustment. This potentiometer operates completely independent of the airflow meter elements but is incorporated in the airflow meter for ease of access to its adjustment screw which is sealed with a plastic plug.

THROTTLE POTENTIOMETER, DESCRIPTION

The twin track throttle potentiometer (Fig. 6) is mechanically coupled to the throttle butterfly spindle and provides reference voltages for both the ECU and electronic automatic transmission controller (where fitted) dependant on throttle position.

Fig. 6

The automatic transmission controller (where fitted) detects a similar signal and will adapt gearshift points to different driving styles as indicated by the throttle input.

LAMBDA SENSOR, DESCRIPTION

The Lambda sensor (1 Fig. 7) is installed in the exhaust system downpipe on vehicles fitted with a catalytic converter.

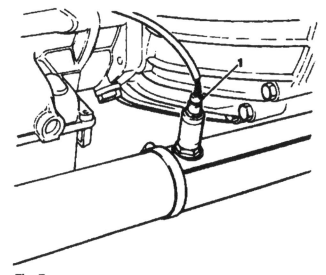

Fig. 7

The purpose of the sensor is to provide an input proportional to the oxygen content of the exhaust gases monitored to the ECU.

GEARBOX SENSOR, DESCRIPTION

The gearbox sensor is installed on vehicles fitted with automatic transmission units only. The sensor comprises two switches which provide indication to the ECU when the linear gearbox selector is in position 'N' or 'P'.

AIR CONDITIONING SENSOR, DESCRIPTION

The air conditioning sensor provides indication to the ECU when the air conditioning clutch is engaged and disengaged.

INJECTORS, RENEW

Depressurise the fuel system prior to disconnecting any fuel pipe connections or removing components as follows:
Open the luggage compartment lid and disconnect the battery earth lead.
Remove the luggage compartment right-hand trim panel.

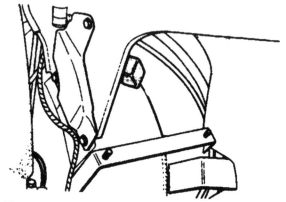

Fig. 8

Displace the relays from the securing tags at the back of the hockey stick panel (Fig. 8).
Identify the fuel pump relay; silver relay on a yellow-black base connector.
Remove the relay from the base connector.
Reconnect the battery earth lead.
Open the driver's door and crank the engine for 10 seconds to depressurise the fuel system.
Close the driver's door.
Disconnect the battery earth lead, refit the relay and reposition the base to the securing tag.
Reposition the trim panel.
Open the bonnet.

Fig. 9

Note the position and disconnect the injector connectors.

Remove the fuel rail securing bolts (Fig. 9) and release the fuel feed hose union nut. Disconnect the fuel feed hose.
Fit protective plugs to the fuel rail and hose. Release the fuel rail to pressure regulator union nut.
Disconnect the fuel rail from the induction manifold and pressure regulator. Place the fuel rail on a work bench.

Remove the injector retaining clips and withdraw each injector from the fuel rail.
Clean the fuel rail apertures and remove the injector protective caps. Insert the new injectors into the fuel rail apertures and fit the injector retaining clips.
Fit the fuel rail assembly to the induction manifold ensuring that the injectors are fully seated.
Connect the fuel rail to the pressure regulator and tighten the union nut. Remove the protective caps from the fuel rail and hose.
Connect the fuel hose to the rail and tighten the union nut.
Fit the injector connectors to the injectors as noted.
Reconnect the battery earth lead.
Start the engine; allow it to run at idling speed and check for signs of fuel leakage. Stop the engine.
Close the bonnet and luggage compartment lid.

SPARK PLUGS, CLEAN AND ADJUST

Open the bonnet.
Ensure that all plug leads are clearly identified by cylinder number (No. 1 at front of engine) before detaching them.
Disconnect the plug leads from the spark plugs.
Ensure the engine is clean around the spark plugs then release and remove the spark plugs using a suitable spark plug spanner (Fig. 10).

Fig. 10

With the spark plugs removed ensure that the plug apertures are clean.
Clean the plug electrodes using suitable equipment to remove all carbon deposits.
Set the electrode gap to 0,9 mm (0.035 in.).
Refit the spark plugs to the engine and tighten to the correct torque.
Connect the plug leads to the spark plugs.
Close the bonnet.

Special Tools
Torque wrench

Torque Figures
Spark plugs 22-30 Nm

SPARK PLUGS, RENEW

Open the bonnet.
Ensure that all plug leads are clearly identified by cylinder number (No. 1 at front of engine) before detaching them.
Disconnect the plug leads from the spark plugs.
Ensure the engine is clean around the spark plugs then release and remove the spark plugs using a suitable spark plug spanner.
With the spark plugs removed ensure that the plug aper-

tures are clean.
Set the electrode gap on the new spark plugs to 0,9 mm (0.035 in).
Fit the new spark plugs to the engine and tighten to the correct torque.
Connect the plug leads to the spark plugs.
Close the bonnet.

Special Tools
Torque wrench

Torque Figures
Spark plugs 22-30 Nm

DISTRIBUTOR CAP, RENEW

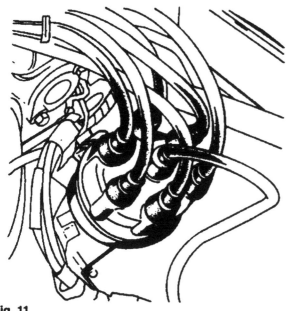

Fig. 11

Open the bonnet.
Release the distributor cap retaining clips and displace the cap (Fig. 11).
Note the positions of the plug leads and remove the leads.
Remove the HT lead from the cap and remove the cap.
Remove the carbon brush from the cap and discard the cap.
Fit the carbon brush to the new cap.
Fit the HT and plug leads to the cap as noted.
Fit the cap to the distributor and engage the retaining clips.
Close the bonnet.

DISTRIBUTOR LEADS, RENEW

Open the bonnet.
Disconnect the HT lead from the ignition coil.
Release the distributor cap retaining clips and displace the cap (Fig. 11).
Detach the lead retaining bracket from the camshaft cover and disconnect the leads from the spark plugs.
Remove the lead and cap assembly to a work bench.
Note the positions of the leads and disconnect the leads from the cap.
Remove the lead retaining brackets from the leads and discard the leads.
Fit the lead retaining brackets to the new leads.
Secure the lead retaining bracket to the camshaft cover and connect the leads to the spark plugs.
Fit the cap to the distributor and engage the retaining clips.
Connect the HT lead to the ignition coil.
Close the bonnet.

DISTRIBUTOR CARBON BRUSH, RENEW

Open the bonnet.
Release the distributor cap retaining clips and displace the cap (Fig. 11).
Remove the carbon brush from the distributor cap and discard the brush.
Fit the new carbon brush to the distributor cap.
Fit the cap to the distributor and engage the retaining clips.
Close the bonnet.

DISTRIBUTOR ROTOR ARM, RENEW

Open the bonnet.
Release the distributor cap retaining clips and displace the cap (Fig. 11).
Remove the rotor arm from the distributor and discard the arm.
Fit the new rotor arm to the distributor, fit the cap to the distributor and engage the retaining clips.
Close the bonnet.

DISTRIBUTOR, RENEW

Open the bonnet.
Jack up the front of the vehicle using a trolley jack under the front suspension cross-member (Fig. 12) for access to the engine TDC indicator.

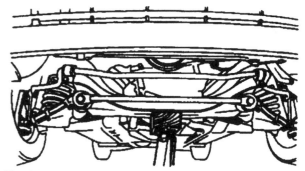

Fig. 12

Position axle stands under the jacking spigots and lower the vehicle onto them.
Leave the jack in position as a precautionary measure.
Release the distributor cap retaining clips and displace the cap (Fig. 11).
Turn the engine to TDC on number one cylinder.
Remove the distributor clamp bolt.
Note the position of the rotor arm and remove the distributor.
Note the position of the plug leads in the distributor cap and disconnect the leads from the cap.
Discard the old cap, remove the cap from the new distributor and fit the leads to the new cap as noted.
Ensure that No. 1 piston is set at TDC.
Fit the distributor and position it so that the rotor arm centre line (with the distributor cap in position) is 12.5° after the centre line of No. 1 cylinder distributor cap electrode.
Fit and tighten the distributor clamp bolt (2 Fig. 13).

Note: There is a remote possibility that a misfire could be induced at engine speeds in excess of 5000 rev/min.

Rotor setting procedure.
Rotate the engine clockwise until No. 1 cylinder piston is at TDC on the compression stroke.
The rotor arm should be approximately in the position shown in Fig. 13.
Loosen the clamp bolt (2 Fig. 13) and if necessary reset the position of the fixing plate so that the fixing plate

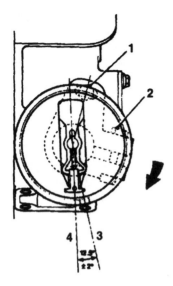

Fig. 13

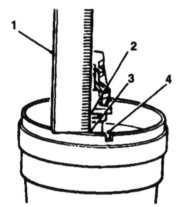

Fig. 14

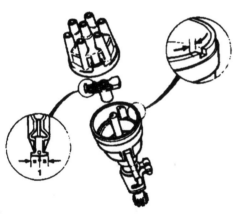

Fig. 15

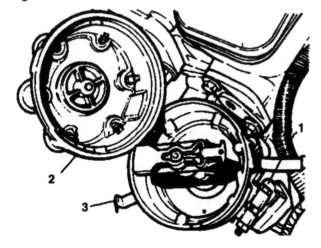

Fig. 16

setscrew (1 Fig. 13) is in the centre of the adjusting slot.
Tighten the setscrew sufficiently to enable the distributor body to be rotated by hand pressure.
The distributor cap location slot should be approximately at the 5 o'clock position.
The centre line of the rotor (4 Fig. 13) should be 12.5° after the centre line of the distributor cap location slot (3 Fig. 13).
Check the position as follows:

1. Measure in a clockwise direction from the distributor cap location slot 5,5 mm (4 Fig. 14) and scribe a mark on the shoulder of the distributor body (3 Fig. 14).
2. Mark the centre of the rotor arm (1 Fig. 15) with a pencil.
3. Using a straight edge (1 Fig. 14), align the mark on the rotor (2 Fig. 14) with that scribed on the distributor body (3 Fig. 14), rotating the distributor body as necessary.
4. Tighten the clamp bolt (1 Fig. 16) to the correct torque and the fixing plate setscrew ensuring that the setting is not altered. Re-check the setting after tightening.

Fit the distributor cap (2 Fig. 16) to the distributor and engage the retaining clips (3 Fig. 16).
Take the weight of the vehicle with the trolley jack, remove the axle stands and lower the vehicle.
Close the bonnet.

Special Tools
Torque wrench

Torque Figures
Distributor clamp bolt 9,5-12,5 Nm

IGNITION COIL, RENEW

The ignition coil is a Bosch type TC 1 and is secured to the right-hand inner wing valance.
Open the luggage compartment lid and disconnect the battery earth lead.
Open the bonnet.
Displace the rubber cover from the top of the coil.
Displace the LT terminal rubber protective covers and disconnect the LT terminal connections from the coil (Fig. 17).

Remove the coil retaining bracket inner and outer securing bolts.
Displace the earth lead and suppression capacitor.
Disconnect the HT lead and remove the coil assembly.
Remove the bracket from the coil and discard the coil.

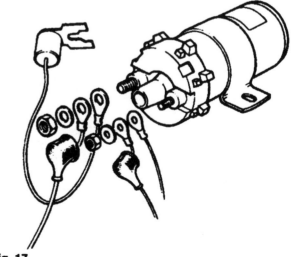

Fig. 17

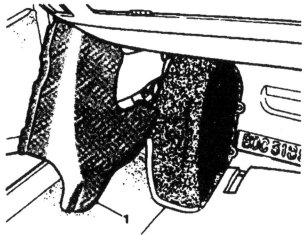

Fig. 18

Fit the bracket to the new coil, position the coil and fit the bracket inner and outer securing bolts ensuring that the suppression capacitor bracket and the earth lead are below the inner bolt.

Tighten the bracket securing bolts and connect the LT leads.

Fit the LT terminal rubber protective covers and connect the HT lead.

Securely fit the rubber cover to the top of the coil.

Reconnect the battery earth lead.

Close the bonnet and luggage compartment lid.

ELECTRONIC CONTROL UNIT, RENEW

Open the luggage compartment lid and disconnect the battery earth lead.

Open the passenger door and remove the footwell carpet.

Displace the draught welting from the 'A' post.

Remove the 'A' post lower trim pad (1 Fig. 18).

Remove the nuts/bolts securing the ECU cover and displace the cover.

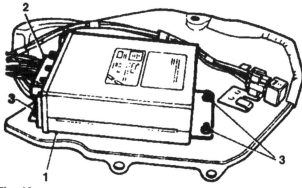

Fig. 19

Disconnect the ECU multi-plug connectors (2 Fig. 19).

Remove the ECU / cover assembly (1 Fig. 19).

Remove the four nuts (3 Fig. 19) securing the ECU to the cover.

Remove the ECU.

Fit the new ECU to the cover and secure with the four nuts.

Position the cover in the vehicle.

Connect the multi-plug connectors.

Fit the ECU/cover into position.

Fit and tighten the nuts/bolts.

Refit the trim pad and draught welting.

Reposition the carpet.

Reconnect the battery earth lead.

Close the luggage compartment lid.

IGNITION AMPLIFIER, RENEW

Open the luggage compartment lid and disconnect the battery earth lead.

Open the bonnet.

Remove the coil retaining bracket inner and outer securing bolts and displace the coil, earth lead and suppression capacitor.

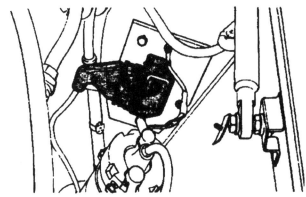

Fig. 20

Release and remove the two screws securing the amplifier to its mounting plate (Fig. 20).

Disconnect the amplifier multi-plug connector and remove the amplifier.

Connect the amplifier multi-plug connector and position the amplifier on the mounting plate.

Secure the amplifier to the mounting plate with the two screws.

Position the coil and fit the bracket inner and outer securing bolts ensuring that the suppression capacitor bracket and the earth lead are below the inner bolt.

Reconnect the battery earth lead.

Close the bonnet and luggage compartment lid.

COOLANT TEMPERATURE SENSOR, RENEW

Note: This procedure must only be performed on a cold or cool engine.

Open the luggage compartment lid and disconnect the battery earth lead.

Open the bonnet.

Disconnect the coolant temperature connector (brown) from the coolant temperature sensor (Fig. 21).

Carefully remove the coolant pressure cap from the remote header tank to release any cooling system residual pressure. Replace the cap tightly.

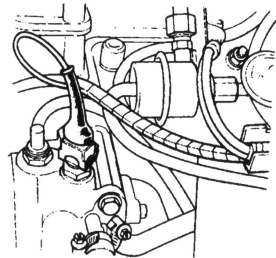

Fig. 21

Note: The replacement component is prepared at this point and the transfer made as quickly as possible.

Ensure sealing washer is located on the replacement temperature sensor and coat threads with a suitable sealing compound.
Remove the existing temperature sensor from the thermostat housing.
Screw the replacement temperature sensor into the thermostat housing.
Refit the electrical connector to the temperature sensor.
Reconnect the battery earth lead.
Check the coolant level at the remote header tank. if necessary, top up with the correct concentration of coolant solution.
Close the bonnet and luggage compartment lid.

COOLANT TEMPERATURE SENSOR, TEST

Open the luggage compartment lid and disconnect the battery earth lead.
Open the bonnet.
Disconnect the coolant temperature connector (brown) (Fig. 21).
Connect suitable ohmmeter between terminals, note resistance reading.
The reading is subject to change according to temperature and should closely approximate to the relevant resistance value given in the table.
Disconnect ohmmeter.
Check resistance between each terminal in turn and body of sensor. A very high resistance reading (open circuit) must be obtained.
Refit electrical connector to temperature sensor.
Re connect battery earth lead.
Close the bonnet and luggage compartment lid.

COOLANT TEMPERATURE	RESISTANCE (kilohms)
- 10	9.2
0	5.9
+ 10	5.9
+ 20	2.5
+ 30	1.7
+ 40	1.18
+ 50	0.84
+ 60	0.60
+ 70	0.435
+ 80	0.325
+ 90	0.250
+100	0.190

CRANKSHAFT SENSOR, RENEW

Open the luggage compartment lid and disconnect the battery earth lead.
Open the bonnet.
Remove the clips securing the sensor cable.

Disconnect the multi-pin cable connector (Fig. 22).
Remove the bolts securing the sensor housing and remove the housing complete with the sensor.
With a suitable Allen key remove the bolt securing the sensor and remove the sensor from its housing.
Fitting a new crankshaft sensor is the reversal of the removal procedure ensuring the connector is secure, clean and tight.

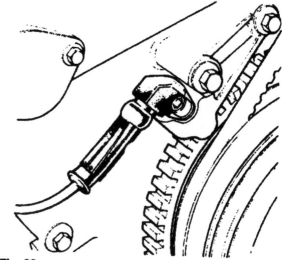

Fig. 22

Reconnect the battery earth lead.
Close the bonnet and luggage compartment lid.

CRANKSHAFT SENSOR TOOTHED GEAR, RENEW

Open the luggage compartment lid and disconnect the battery earth lead.
Open the bonnet.

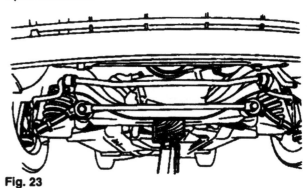

Fig. 23

Jack up the front of the vehicle using a trolley jack under the front suspension cross-member (Fig. 23).
Position axle stands under the jacking spigots and lower the vehicle onto them.
Leave the jack in position as a precautionary measure.
Slacken the air conditioning compressor adjusting nut, link arm pivot and trunnion bolts.
Slacken the compressor pivot bolts and pivot the compressor towards the engine.
Displace the air conditioning compressor drive belt from the crankshaft damper.
Slacken the alternator adjusting nut, link arm securing nut and link arm trunnion bolt.

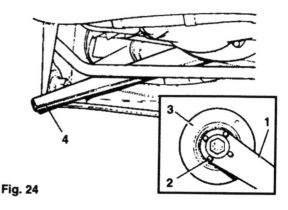

Fig. 24

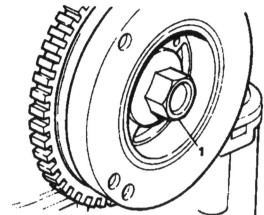

Fig. 25

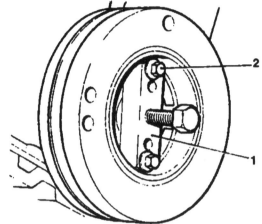

Fig. 26

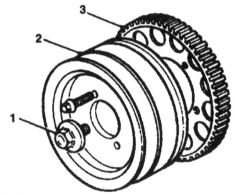

Fig. 27

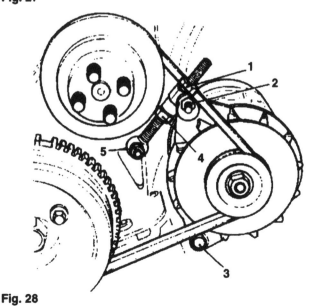

Fig. 28

Pivot the alternator towards the engine.

Displace the alternator drive belt from the crankshaft damper.

Fit Service Tool 18G 1437 (1 Fig. 24) and adaptor 18G 1437-1 to the crankshaft pulley / damper (3 Fig. 24) and tighten the two securing bolts (2 Fig. 24).

Wedge the tool (4 Fig. 24) and remove the pulley / damper retaining bolt (1 Fig. 25).

Remove Service Tool 18G 1437 from the crankshaft pulley / damper.

Fit Service Tool 18G 1436 (1 Fig. 26) to the pulley / damper and tighten the two securing bolts (2 Fig. 26).

Tighten the centre bolt, withdraw the pulley / damper from the crankshaft and remove the tool.

Remove the bolts securing the toothed gear (3 Fig. 27) to the pulley / damper (2 Fig. 27) and remove the gear from the pulley / damper.

Remove the roll pin from the pulley / damper.

Discard the pulley / damper and toothed gear, if necessary.

Fit the roll pin to the pulley / damper.

Fit the toothed gear to the pulley / damper.

Fit and tighten the securing bolts.

Fit the pulley / damper assembly to the front of the crankshaft on the engine.

Fit the pulley / damper retaining bolt (1 Fig. 27).

Fit Service Tool 18G 1437 and adaptor 18G 1437-1 to the crankshaft pulley / damper and tighten the two securing bolts.

Wedge the tool and tighten the pulley / damper retaining bolt to the correct torque.

Remove Service Tool 18G 1437 from the crankshaft pulley / damper.

Reposition the alternator drive belt over the pulley.

Tighten the lower adjusting nut (4 Fig. 28) to obtain the correct belt tension. A load of 1,5 kg must give a total deflection of 4,4 mm when applied at the mid point of the belt.

Tighten the upper adjusting nut (1 Fig. 28), trunnion block securing bolt (2 Fig. 28), adjusting rod to water pump securing nut (5 Fig. 28) and the pivot bolt locknut (3 Fig. 28).

Reposition the compressor drive belt over the pulley.

Tighten the upper adjusting nut (6 Fig. 29) to obtain the correct belt tension. A load of 2,9 kg must give a total deflection of 4,4 mm when applied at the mid point of the belt.

Tighten the lower adjusting nut (5 Fig. 29), trunnion nut and bolt (4 Fig. 29), sliding link nuts and bolts (7 Fig. 29), pivot bolts (1 Fig. 29) and mounting bracket nut and bolt (2 Fig. 29).

Take the weight of the vehicle with the trolley jack, remove

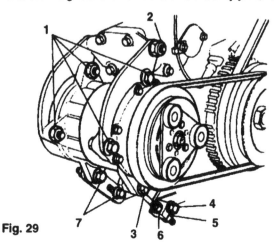

Fig. 29

Fig. 30

the axle stands and lower the vehicle.
Reconnect the battery earth lead.
Close the bonnet and luggage compartment lid.

Special Tools
Crankshaft pulley remover 18G 1436
Crankshaft pulley/damper locking tool 18G 1437
Crankshaft pulley/damper locking tool adaptor 18G 1437-1

Torque Figures
Crankshaft pulley/damper to crankshaft 203,5 Nm
Toothed ring gear to pulley/damper 16-19 Nm

AIRFLOW METER, RENEW

Open the luggage compartment lid and disconnect the battery earth lead.
Open the bonnet
Release the two spring clips securing the air cleaner to the airflow meter (Fig. 30).
Release the airflow meter to intake elbow hose clip.
Remove the air intake elbow hose securing nut.
Disconnect the airflow meter connector (black).
Remove the airflow meter.
Position the new airflow meter between the air cleaner and the intake elbow hose.
Connect the airflow meter connector (black).
Fit the airflow meter to the air cleaner and engage the securing clips.
Fit the airflow meter to the intake elbow hose.
Position the intake elbow on the locating stud, fit and tighten the nut.
Tighten the intake elbow hose clip.
Reconnect the battery earth lead.
Close the bonnet and luggage compartment lid.

THROTTLE POTENTIOMETER, RENEW

Open the luggage compartment lid and disconnect the battery earth lead.
Open the bonnet.
Slacken breather and extra air valve hose clips at intake elbow and displace hoses from elbow.
Slacken throttle housing to elbow hose clip at throttle housing and displace hose.
Slacken airflow meter to elbow hose clip and displace hose.
Remove intake elbow mounting securing nut and remove elbow for access.
Disconnect the throttle potentiometer multi-plug.

Remove the potentiometer securing bolts and remove the potentiometer (Fig. 31).

Fig. 31

Fit and fully seat the new throttle potentiometer to the throttle housing.
Fit but do not fully tighten potentiometer securing bolts.

Note: The throttle potentiometer must be adjusted and set using JDS equipment.

Following adjustment finally tighten the potentiometer securing bolts.
Refit the air intake elbow assembly and tighten hose clips.
Reconnect the battery earth lead.
Close the bonnet and luggage compartment lid.

LAMBDA SENSOR, RENEW

Open the luggage compartment lid and disconnect the battery earth lead.
Disconnect the Lambda sensor harness multi-plugs, engine compartment right-hand side.
Release the harness from the three retaining clips.
Raise the vehicle on a ramp.

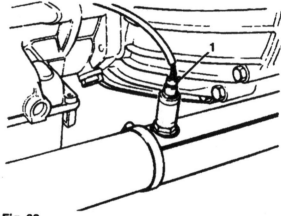

Fig. 32

Unscrew and remove the sensor (1 Fig. 32) from the exhaust downpipe catalyst assembly.
Clean the downpipe aperture.
Smear the threads of the new sensor with anti-seize grease, fit the sensor and tighten.
Lower the vehicle.
Connect the Lambda sensor multi-plugs and reposition the harness to the retaining clips.
Reconnect the battery earth lead.
Close the bonnet and luggage compartment lid.

FUEL PUMP RELAY, RENEW

Open the luggage compartment lid and disconnect the battery earth lead.
Remove the luggage compartment right-hand trim panel.

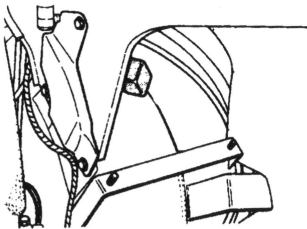

Fig. 33

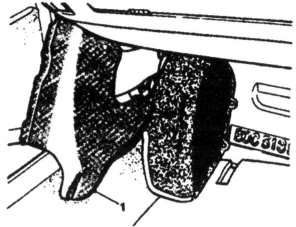

Fig. 34

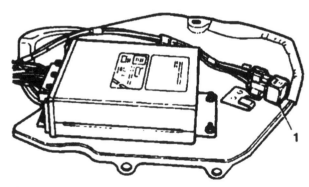

Fig. 35

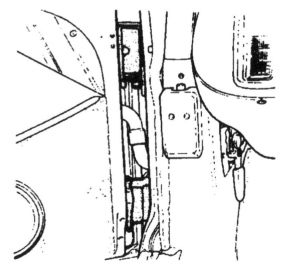

Fig. 36

Displace the relays from the securing tags at the back of the hockey stick panel (Fig. 33).
Identify the fuel pump relay; silver relay on a yellow-black base connector.
Remove the relay from the base connector.
Fit the new relay and reposition the base to the securing tag.
Reposition the trim panel.
Reconnect the battery earth lead.
Close the luggage compartment lid.

FUEL PUMP RELAY, TEST

Switch on ignition. Pump should run for one to two seconds, then stop.

Note: If pump does not run, or does not stop, check systematically as follows:

Check inertia switch cut out button is pressed in.
Detach inertia switch cover and ensure both cables are secure.
Pull electrical connectors from switch and check continuity across terminals.
Pull button out and check open circuit.
Remove ohmmeter, replace electrical connectors, re set button and refit cover.
If inertia switch satisfactory, earth (ground) pump relay terminal 85, switch on ignition and check circuit systematically as detailed below.

a.   Check for battery voltage at terminal 86 of main relay. If yes: Proceed to test b. If no: Check battery supply from ignition switch via inertia switch.

b.   Check for battery voltage at terminal 87 of main relay. If yes: Proceed to test c. If no: Check for battery voltage at earth (ground) lead and connection from terminal 85 of main relay. If satisfactory renew main relay.

c.   Check for battery voltage at terminal 86 of pump relay. If yes: Proceed to test d. If no: Open circuit between terminals 87 of main relay and 86 of pump relay. If satisfactory proceed to test d.

d.   Check for battery voltage at terminal 87 of pump relay. If yes: Proceed to test e. If no: Check for battery voltage to earth (ground) lead and connections from terminal 85 of pump relay. If satisfactory, renew pump relay.

e.   Check for battery voltage at supply lead (NS) and connections to fuel pump. If yes: Faulty pump or earth (ground) connections. If no: Open circuit between terminal 87 of pump relay and supply lead connection to fuel pump.

MAIN PI RELAY, RENEW

Open the luggage compartment lid and disconnect the battery earth lead.
Open the passenger door and remove the footwell carpet.
Displace the draught welting from the 'A' post.

Remove the 'A' post lower trim pad (1 Fig. 34).
Remove the nuts / bolts securing the ECU cover and displace the cover.

Displace the relay (1 Fig. 35) from the securing tag,
Remove the relay from the base connector.

Fit the new relay and reposition the base to the securing tag.
Fit the ECU/cover into position.
Fit and tighten the nuts / bolts.
Refit the trim pad and draught welting.
Reposition the carpet.
Reconnect the battery earth lead.
Close the luggage compartment lid.

INERTIA SWITCH, RENEW

The inertia switch (Fig. 3) is located on the passenger's door pillar adjacent to the footwell. It is a gravity actuated safety device which inhibits fuel pump operation should the vehicle be subjected to heavy impact force.

Open the luggage compartment lid and disconnect the battery earth lead.
Open the passenger's door.
Remove the switch cover.
Displace the draught welting from the 'A' post.
Remove the two screws securing the switch.
Disconnect the cable harness multi-plug connector.
Remove the inertia switch.
Connect the multi-plug connector to the new switch.
Position the switch and secure with the two screws.
Refit the draught welting to the 'A' post.
Fit the switch cover.
Reconnect the battery earth lead.
Close the luggage compartment lid.

IDLE SPEED CONTROL ACTUATOR, RENEW

The idle speed control actuator (Fig. 37) is attached to the inlet manifold and contains a stepper motor driven valve which permits intake air to bypass the throttle butterfly valve. The volume of air passed by the actuator is controlled by the ECU so that correct idle speeds are maintained irrespective of engine temperature and load.
Open the luggage compartment lid and disconnect the battery earth lead.
Open the bonnet.
Disconnect the idle speed actuator multi-plug.
Unscrew and remove the two bolts (1 Fig. 37) securing the actuator to the actuator housing.
Remove the actuator and remove then discard the gasket.
Fit a new gasket and actuator to the actuator housing.
Fit and tighten the two securing bolts.
Connect the idle speed actuator multi-plug.
Reconnect the battery earth lead.
Close the bonnet and luggage compartment lid.

Note: After fitting a new actuator, the base idle speed MUST be set.

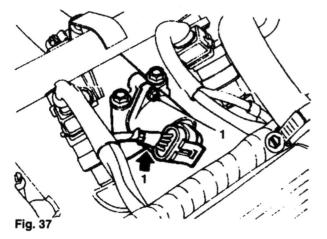

Fig. 37

BASE IDLE SPEED, SETTING

Note: This should be set using JDS Equipment, however should this not be available then the manual method described below may be used.

When using the manual method, ensure the manual instructions are adhered to as the two procedures are not the same.
Open the bonnet.
Run the engine until it reaches normal operating temperature (more than 85° C).
Switch the ignition off.
Switch the ignition on, wait 5 seconds and then disconnect the connector to the idle speed controller.
Switch the ignition off.
Wait 15 seconds then reconnect the idle speed controller.
Repeat the above operations twice, on the last occasion do not reconnect the idle speed controller.
Start the engine.
Check the base idle speed, this should be between approximately 550 rev/min and 600 rev/min. If the idle speed is not within these limits then adjust the air bypass screw to bring the idle speed within limits.
Switch the ignition off and reconnect the idle speed controller.
Start the engine and check to ensure that the idle speed is between 650 rev/min and 800 rev/min.

AIR FLOW SENSOR, DIAGNOSTIC CHECK

Disconnect the airflow meter plug (Fig. 38).

Connect pin 36 to the negative side of a suitable power supply and pin 9 to the positive side.
Set the power supply to 14V.
Connect pin 6 to the negative side and select a suitable voltage range, say 0 - 10V.

TEST 1: (Airflow meter stationary, i.e. no air flowing through it and preferably with the inlet and outlet ports covered).

Voltmeter Reading	Diagnosis
0.4V + - 0.2V	OK
OV	Output open circuit.
Voltage greater than 0.7V.	Output drifted, module inoperative, hot or cold sensor has become detached from the posts.

TEST 2: (Carried out by blowing one hard blast through the airflow meter [in the direction of air flow] thus causing the volt meter to rise and fall).

Voltmeter Reading	Diagnosis
Raise to 1.5V, then fall back to test 1 voltage.	OK
OV	Output open circuit.
Same voltage as in Test 1.	Module inoperative, hot or cold wire detached.
Rise to greater than 5V.	Module inoperative, calibration drifted.

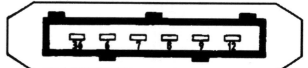

Fig. 38

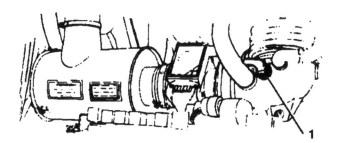

Fig. 39

Disconnect the voltmeter and power supply.
Reconnect the airflow meter plug.

AIR TEMPERATURE SENSOR, RENEW

Open the luggage compartment lid and disconnect the battery earth lead.
Open the bonnet.
Disconnect the air temperature connector (brown) from the air temperature sensor (1 Fig. 39).
Remove the air temperature sensor and copper sealing washer from the air intake elbow.
Ensure sealing washer is located on the replacement temperature sensor.
Screw the replacement temperature sensor into the air intake elbow.
Refit the electrical connector to the temperature sensor.
Reconnect the battery earth lead.
Close the bonnet and luggage compartment lid.

INTAKE ELBOW, RENEW

Open the luggage compartment lid and disconnect the battery earth lead.
Open the bonnet.
Release the two spring clips (1 Fig. 40) securing the air cleaner to the airflow meter.
Release the airflow meter to intake elbow hose clip (2 Fig. 40).
Remove the air intake elbow hose securing nut (6 Fig. 40).

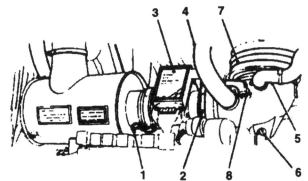

Fig. 40

Disconnect the airflow meter connector (black) and remove the airflow meter (3 Fig. 40).
Release the intake elbow to idle speed control valve hose clip and disconnect the hose from the elbow (4 Fig. 40).
Release the intake elbow to filler tube hose clip and disconnect the hose from the elbow (5 Fig. 40).
Disconnect the air temperature connector (brown) from the air temperature sensor (8 Fig. 40).
Remove the air temperature sensor and copper sealing washer from the air intake elbow.
Release the elbow to throttle housing hose clip and remove the elbow from the hose (7 Fig. 40).
Release the elbow to airflow meter hose clip and remove

the hose from the elbow.
Fit the airflow meter hose to the new elbow and tighten the hose clip.
Fit the elbow to the throttle housing hose and tighten the hose clip.
Position the airflow meter between the air cleaner and the intake elbow hose.
Connect the airflow meter connector (black).
Fit the airflow meter to the air cleaner and engage the securing clips.
Fit the airflow meter to the intake elbow hose.
Position the intake elbow on the locating stud, fit and tighten the nut.
Tighten the intake elbow hose clip.
Ensure sealing washer is located on the air temperature sensor.
Screw the air temperature sensor into the air intake elbow.
Refit the electrical connector to the air temperature sensor.
Fit the filler tube hose to the elbow and tighten the hose clip.
Fit the idle speed control hose to the elbow and tighten the hose clip.
Reconnect the battery earth lead.
Close the bonnet and luggage compartment lid.

INTAKE ELBOW TO THROTTLE HOSE, RENEW

Open the luggage compartment lid and disconnect the battery earth lead.
Open the bonnet.
Release the two spring clips (1 Fig. 40) securing the air cleaner to the airflow meter. Release the airflow meter to intake elbow hose clip (2 Fig. 40).
Remove the air intake elbow hose securing nut (6 Fig. 40).
Disconnect the airflow meter connector (black) and remove the airflow meter (3 Fig. 40).
Disconnect the air temperature connector (brown) from the air temperature sensor (8 Fig. 40).
Release the intake elbow to idle speed control valve hose clip and disconnect the hose from the elbow (4 Fig. 40).
Release the intake elbow to filler tube hose clip and disconnect the hose from the elbow (5 Fig. 40).
Release the elbow to throttle housing hose clip and remove the elbow from the hose.
Release the throttle housing hose clip and remove the hose from the housing (7 Fig. 40), remove the clips and discard the hose.
Fit the hose clips to the new throttle housing hose, fit the hose to the throttle housing and tighten the hose clip.
Fit the elbow to the throttle housing hose and tighten the hose clip.
Position the airflow meter between the air cleaner and the intake elbow hose.
Connect the airflow meter connector (black).
Fit the airflow meter to the air cleaner and engage the securing clips.
Fit the airflow meter to the intake elbow hose.
Position the intake elbow on the locating stud, fit and tighten the nut.
Tighten the intake elbow hose clip.
Refit the electrical connector to the air temperature sensor.
Fit the filler tube hose to the elbow and tighten the hose clip.
Fit the idle speed control hose to the elbow and tighten the hose clip.
Reconnect the battery earth lead.
Close the bonnet and luggage compartment lid.

Chapter 4

FUEL SYSTEM - 4.0 LITRE

SAFETY PRECAUTIONS

WARNINGS:

OPERATIONS INVOLVING THE FUEL SYSTEM RESULT IN FUEL AND FUEL VAPOUR BEING PRESENT IN THE ATMOSPHERE.

FUEL VAPOUR IS EXTREMELY FLAMMABLE, THEREFORE GREAT CARE MUST BE TAKEN WHEN WORK IS CARRIED OUT ON THE FUEL SYSTEM.

THE FOLLOWING PRECAUTIONS MUST BE STRICTLY ADHERED TO:

1. SMOKING MUST NOT BE ALLOWED NEAR THE AREA.
2. 'NO SMOKING' WARNING SIGNS MUST BE POSTED ROUND THE AREA.
3. A CO_2 FIRE EXTINGUISHER MUST BE CLOSE AT HAND.
4. DRY SAND MUST BE AVAILABLE TO SOAK UP ANY SPILLAGE.
5. EMPTY THE FUEL FROM THE TANK INTO AN AUTHORISED EXPLOSION PROOF CONTAINER USING SUITABLE FIREPROOF EQUIPMENT.
6. FUEL MUST NOT BE EMPTIED INTO A PIT.
7. THE WORKING AREA MUST BE WELL VENTILATED.
8. THE BATTERY MUST BE DISCONNECTED BEFORE CARRYING OUT WORK ON THE FUEL SYSTEM.
9. THE DISCARDED TANK MUST NOT BE DISPOSED OF UNTIL IT IS RENDERED SAFE FROM EXPLOSION.
10. MAINTENANCE PERSONNEL SHOULD HAVE UNDERTAKEN SPECIALIST TRAINING BEFORE BEING ALLOWED TO REPAIR COMPONENTS ASSOCIATED WITH THE FUEL SYSTEM.
11. DEPRESSURISE THE FUEL SYSTEM PRIOR TO DISCONNECTING ANY FUEL PIPE CONNECTIONS OR REMOVING COMPONENTS.

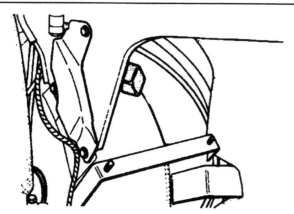

Fig. 1

DEPRESSURISING THE FUEL SYSTEM

Depressurise the fuel system prior to disconnecting any fuel pipe connections or removing components as follows;

1. Open the luggage compartment and remove the right-hand trim liner.
2. Locate the relays mounted at the rear of the right-hand 'hockey stick' panel.
3. Displace relays from their mounting and identify the fuel pump relay (silver relay on a black / yellow base) (Fig. 1).
4. Disconnect the relay from the base.

5. Crank the engine to depressurise the system.

6. Reposition the relay and trim liner, close the luggage compartment.

FUEL SYSTEM, DESCRIPTION

The fuel system used is the recirculatory type. The fuel tank is mounted across the car behind the rear passenger seats.
Fuel is drawn from the fuel tank into the in-tank module by the fuel pump, with the fuel passing through a venturi on the fuel pump pressure side.

A 70 micron filter is incorporated at the inlet port of the module to prevent ingress of particle contaminants. A further 400 micron rock filter is included at the inlet port of the fuel pump to prevent the passage of smaller participle contaminants into the fuel pump and the rest of the fuel system.

Fuel is pumped from the base of the fuel tank through a flexible hose to an in-line filter. From the filter, fuel flows through under-floor pipework to the front of the vehicle where a flexible hose connects the under-floor feed pipe to the inlet port of the fuel rail.

The fuel rail is mounted on the engine and has a fuel pressure regulator attached at the return port. The fuel pressure regulator is used to control fuel line pressure and maintains a constant delivery of fuel to the fuel injectors across the fuel rail. This is accomplished by applying inlet manifold pressure to a diaphragm within the pressure regulator.

Pressure is kept constant within the range 241.30 - 310.25 kN/m^2 (35 - 45 lbf/in^2).

Unused fuel from the engine is returned to the in-tank module through return hoses and under-floor piping. A 70 micron filter is used in the return inlet of the module to capture debris that may be present in the fuel rail or return line at initial start-up and so prevent it from entering the system.

Both the feed and return ports of the module use non-return valves for safety.

Electronic Fuel Injection

A digital type electronic control unit (ECU) with an integral manifold pressure sensor, governs the amount of fuel injected. The manifold pressure and speed signal derived from ignition pulses, provide the main control for the fuel injected. Additional sensors are used to monitor engine temperature, inlet air temperature and throttle position, thereby ensuring optimum fuelling is computed for all engine operating conditions.

Fuel Control System

The metering of fuel is controlled by regulating the time that the injectors are open during each engine cycle. The frequency of the injectors is dependent on engine speed and conditions. The basic pulse length is mapped against speed and intake manifold pressure which is sensed by a transducer located in the electronic control unit and linked to the intake manifold by a pipe. Information on engine speed is derived from the ignition trigger pulses in the ignition system.

The injectors are fired six times per engine cycle, this operation being triggered from the output of the ECU.

The injectors are energized for a time proportional to the figure given on the base map plus a constant of proportionality which is varied according to secondary control parameters, i.e. engine coolant temperature, inlet air temperature, throttle movement and position, and battery voltage. The fuel pump is energised only when the ECU senses an engine cranking signal or an engine running signal. This prevents the engine being flooded with fuel in the event of an injector sticking open.

Cranking Fuelling

Cranking fuelling provides six injections per engine cycle instead of the normal two. This reduces after a set number of injections.

Coolant Temperature Enrichment

Temperature enrichment is provided during starting and warm-up. This is achieved by increasing the injector 'on' time above that of basic requirements and is implemented by the ECU in response to an input from the coolant temperature sensor. The enrichment is reduced with increasing engine speed and load.

After-start Enrichment

After-start enrichment is provided to supply added fuel during warm-up. The enrichment is coolant temperature dependant (the colder the temperature, the more fuel is supplied). This is achieved by the ECU which increases the injector 'on' time above that of basic requirements and then decreases the amount of additional fuel supplied at a fixed rate over a number of engine revolutions.

Temperature Sensors

The temperature of the air taken into the engine through the inlet manifold and the temperature of the coolant in the cylinder block are constantly monitored. The information is fed directly to the ECU.

The air temperature sensor has a small effect on the injector pulse width, and should be regarded as a trimming rather than a control device. It ensures that the fuel supplied is directly related to the weight of air drawn in by the engine. Therefore, as the weight (density) of the air charge increases with a falling temperature, so the amount of fuel supplied is also increased to maintain optimum fuel/air ratio.

The coolant temperature sensor has a greater control although it functions mainly while the engine is initially warming-up.

Full Load Fuelling

To obtain maximum engine power it is necessary to inhibit the 'closed loop' system and simultaneously increase the fuelling level. This is determined by throttle position and engine speed.

Flooding Protection System

When the ignition is switched on, but the engine is not cranking, the fuel pump will run for two seconds to raise the pressure in the fuel rail; it is then automatically switched off by the ECU. Only after cranking has started is the fuel pump switched on again. Switching control is built into the ECU circuitry. This system prevents flooding if any injectors become faulty (remain open) when the ignition is left on.

Engine Load Sensing

The driver controls engine power output by varying the throttle opening and therefore the flow of air into the engine. The airflow determines the pressure that exists within the plenum chamber, this pressure therefore is a measure of the demand upon the engine. The pressure is also used to provide the principle control of fuel quantity, being converted by the manifold pressure sensor in the ECU into an electrical signal. This signal varies the width of the injector operating pulse as appropriate. The pressure sensor is fitted with a separate diaphragm system that compensates for ambient barometric variations.

Fuel Pump / In-tank Module

The fuel pump is fitted inside the in-tank module (1 Fig. 2), which is fixed to the centre base of the fuel tank, via a rubber cradle (2 Fig. 2) and steel bracket (6 Fig. 2). The rubber cradle behaves as an acoustic baffle to deaden pump

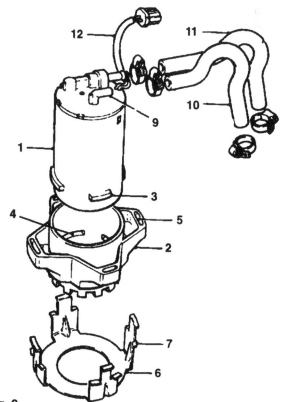

Fig. 2

361

vibration and noise. There are three large tabs (3 Fig. 2) and one small tab on the module casing, which locate into slots in the centre recess of the rubber mounting (4 Fig. 2), this ensures that the pump can only be mounted in the correct orientation. The periphery of the rubber mounting incorporates four protrusions each having a locating slot (5 Fig. 2), three slots are large and one is small, which locate onto tabs on the steel support bracket (7 Fig. 2).

Note: It is important to ensure correct orientation and fitting of fuel pump module with rubber mounting and steel support bracket.

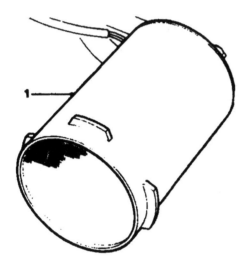

Fig. 3

The fuel pump draws fuel into the module from the fuel tank through a filter in the base of the module (1 Fig. 3).
Fuel is pumped out of the module through a port at the top of the module (9 Fig. 2). There are two rubber moulded hoses connected to the top of the module, the short hose (11 Fig. 2) connects from the module's outlet port to the under-floor fuel feed pipe at the base of the fuel tank. The long hose (10 Fig. 2) connects the fuel return pipe to the module's return port.

Connections between the hoses and solid pipes are underneath the car. The only direct connection between pump module and the tank is at the evaporative loss flange via a two wire electrical connection (12 Fig. 2). The pump delivers approximately 120 Litres/min. @ 3 Bar.
The complete fuel pump module is replaced when renewing the fuel pump.

Fuel Filter

The main in-line fuel filter is mounted under-floor on the left-hand side of the vehicle forward of the axle carrier. The unit is made from stainless steel.

```
CAUTION:

Directions for fuel flow are shown on the filter,
it is important that the filter is connected the
correct way round.
```

Evaporative Loss Flange

The flange is mounted to the tank by a seal and locking ring, it has three external outlet ports and one electrical connection. The large port (1 Fig. 4) allows the tank to vent

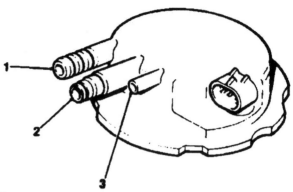

Fig. 4

vapour during refuelling and is directly connected to the fuel filler neck. The intermediate port (2 Fig. 4) has a 'Legris' quick-fit connector moulding and allows for over pressurisation; it is connected to atmosphere by a nylon pipe the outlet of which is underneath the vehicle.
The small outlet port (3 Fig. 4) is used for vehicles incorporating evaporative loss control and uses a nylon pipe to connect the tank to the charcoal canister located at the front of the vehicle. For cars not fitted with evaporative loss control, the pipe is vented directly to atmosphere via a hose passing underneath the car. In both evaporative loss control vehicles and non-evaporative loss control vehicles, a Rochester valve is included in the line, controlled by manifold pressure.

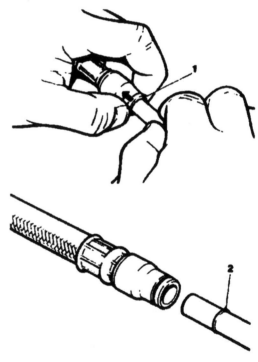

Fig. 5

Connectors and Hoses

Legris quick-fit connectors are used throughout the fuel system (Fig. 5). At connections between hoses and the under-floor pipes, the pipes have a ridge or paint line to indicate how far the connector should be pushed onto the pipe end.

```
CAUTION:

Pipes and component stubs must be fully inserted
into connector ends to ensure proper fit;
failure to do so could cause fuel leakage.
```

There are two sizes of connector, 8mm and 10mm, this is to ensure that the feed pipe is not connected as the return pipe and vice versa. The 10mm connector is used for the feed line. To release connectors, push and hold the locking ring (1 Fig. 5) towards the connector fitting while moving the connector and pipe end apart.

Always ensure pipe connectors are pushed fully home when refitting by having the pipe shoulder (2 Fig. 5) abut the locking ring.

In-tank fuel hoses

Service tool JD 175 is available for the tightening of jubilee clips which attach feed and return hoses to the in-tank module.

Throttle Potentiometer

To ensure the vehicle road performance is satisfactory, with good throttle response, acceleration enrichment is necessary. Signals are provided by the throttle potentiometer (Fig. 6), which is mounted on the throttle spindle and indicates the throttle position to the ECU. When the throttle is opened the fuelling is enriched and when closed the fuelling is weakened. When the throttle is opened very quickly, all the injectors are simultaneously energised for one pulse.

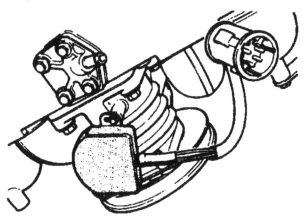

Fig. 6

This ensures that there is enough fuel available at the inlet ports for the air admitted by the sudden opening of the throttle. The duration of this extra pulse is controlled by the engine temperature signal, and is longer with a cold engine.

Lengthening the normal injection pulses is done in proportion to the rate at which the throttle is opened and it takes a short time to decay when the throttle movement stops. Enrichment in this way is also varied according to the engine temperature. The fuel cut off function is controlled by the throttle potentiometer, and the conditions under which it occurs are programmed into the ECU memory.

Injectors

Each fuel injector (Fig. 7) consists of a solenoid operated needle valve with the movable plunger rigidly attached to the nozzle needle. In the closed position, a helical compression spring holds the needle against the valve seat.
The injectors have a solenoid winding mounted in the rear section of the valve body, with a guide for the nozzle needle in the front section.

The injectors are operated in two stages, initially they are operated via a pull in circuit, then when the injectors are open a change to a hold on circuit is made via current limit-

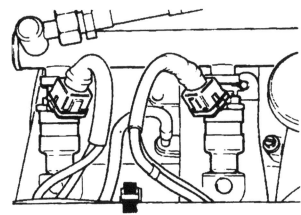

Fig. 7

ing resistors for the remainder of the injection period as determined by the ECU. In this way the heating effect on the output transistors of the ECU is reduced. It also ensures a rapid response from the injectors.

The metering of fuel is controlled by regulating the time that the injectors are open during each engine cycle. The frequency of the injectors is dependent on engine speed and conditions. The basic pulse length is mapped against speed and intake manifold pressure which is sensed by a transducer located in the electronic control unit and linked to the intake manifold by a pipe. Information on engine speed is derived from the ignition trigger pulses in the ignition system.

The injectors are energized for a time proportional to the figure given on the base map plus a constant of proportionality which is varied according to secondary control parameters, i.e. engine coolant temperature, inlet air temperature, throttle movement and position and battery voltage.

Fuel Pressure Regulator

The fuel pressure regulator (Fig. 8) is attached to the fuel rail and consists of a metal housing containing a spring loaded diaphragm. When the pressure setting of the regulator is exceeded, the diaphragm moves, exposing an opening to an overflow duct which allows excess fuel to return to the fuel tank, causing a drop in fuel pressure. The reduced fuel pressure allows the diaphragm to move back to its original position, thereby closing the fuel return outlet. This sequence of events is repeated as long as the pump is running. In this way, fuel pressure is held constant as

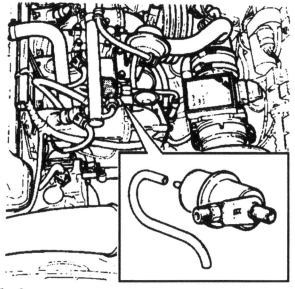

Fig. 8

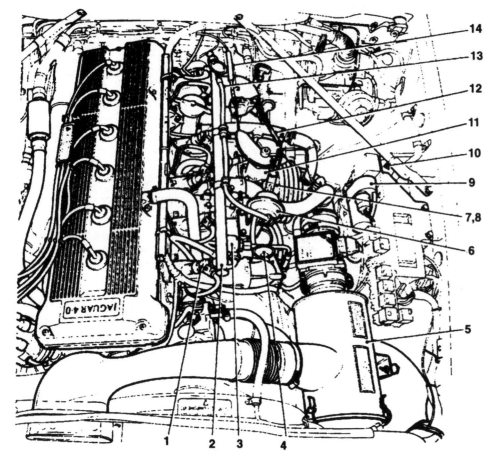

Fig. 9 Engine Compartment Component Location

Key to Fig. 9

1. Injector
2. Fuel rail
3. Pressure regulator
4. Fuel return pipe
5. Air cleaner assembly
6. Throttle actuator
7. Throttle potentiometer
8. Throttle housing assembly
9. Vacuum pump
10. Vacuum dump valve
11. Throttle pedestal assembly
12. Throttle linkage
13. Fuel feed pipe
14. Throttle cable

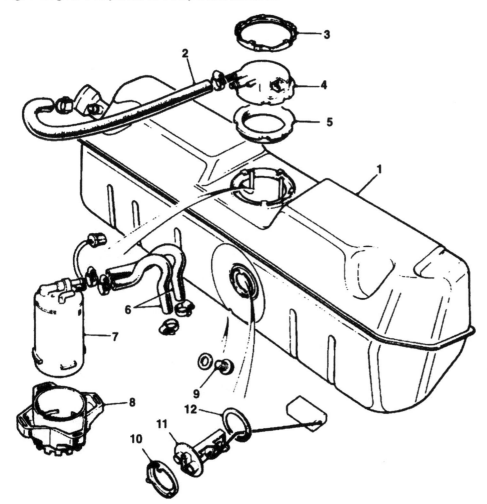

Fig. 10 Fuel Tank Assembly

Key to Fig. 10

1. Tank unit
2. Tank breather hose
3. Flange locking ring
4. Evaporative loss flange
5. Flange seal
6. Internal feed and return hoses
7. Pump module
8. Mounting rubber
9. Drain plug
10. Gauge unit locking ring
11. Fuel gauge sender unit
12. Gauge unit seal

fuel demand varies. The pressure setting is adjusted to the correct value during production when the outer spring housing is compressed until the correct spring load is obtained. This is not adjustable in service.

The spring housing of the regulator is sealed and connected to the engine inlet manifold by a small bore pipe. By allowing it to sense inlet manifold depression, the pressure drop across the injector nozzle remains constant because the fuel pressure will alter as manifold depression alters. This arrangement ensures that the amount of fuel injected is only dependent on the duration of injector open time.

System fuel pressure is 3 bar (43.5 lb/in^2) above manifold pressure.

Fuel Rail

The fuel rail is mounted and secured to the inlet manifold casting. The six injectors are directly fitted to the rail via 'O' ring seals and secured by retaining clips.

Fuel is fed into the rear of the rail and fuel flow pressure across all injectors is controlled by the pressure regulator valve mounted at the front rail.

FUEL TANK, RENEW

```
WARNING:

THIS OPERATION RESULTS IN FUEL
AND FUEL VAPOUR BEING
PRESENT WHICH IS EXTREMELY FLAMMABLE,
GREAT CARE MUST BE TAKEN WHEN
WORK IS BEING CARRIED OUT ON
THE FUEL SYSTEM AND
THE FOLLOWING PRECAUTIONS MUST
BE STRICTLY ADHERED TO:

SMOKING MUST NOT BE ALLOWED NEAR THE
AREA. 'NO SMOKING' WARNING SIGNS MUST BE
POSTED ROUND THE AREA.

A CO² FIRE EXTINGUISHER MUST BE CLOSE AT
HAND.

DRY SAND MUST BE AVAILABLE TO SOAK UP
ANY SPILLAGE.

ENSURE THE FUEL IS EMPTIED FROM THE
TANK USING SUITABLE FIREPROOF EQUIPMENT.

FUEL MUST NOT BE EMPTIED INTO A PIT.

THE WORKING AREA MUST BE WELL VENTILATED.

ENSURE THE FUEL IS EMPTIED INTO AN AUTHO-
RISED EXPLOSION PROOF CONTAINER.

THE DISCARDED TANK MUST NOT BE
DISPOSED OF UNTIL IT IS RENDERED SAFE
FROM EXPLOSION.
```

Open the luggage compartment and displace the right-hand front side liner.

Displace the fuel pump relay (silver relay on black / yellow base) from mounting at rear of hockey stick panel and remove the relay (Fig. 11).

Crank the engine to depressurise the fuel system.
Disconnect and remove the battery.
Remove the spare wheel.
Empty the fuel tank into a explosion proof container using

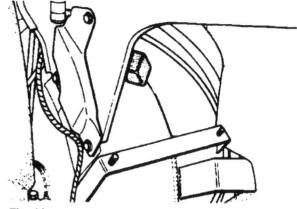

Fig. 11

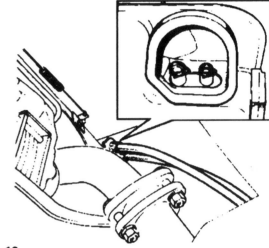

Fig. 12

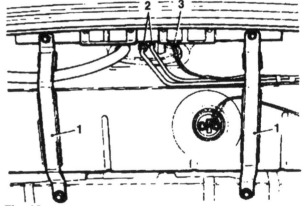

Fig. 13

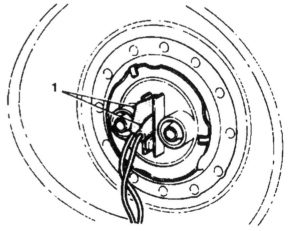

Fig. 14

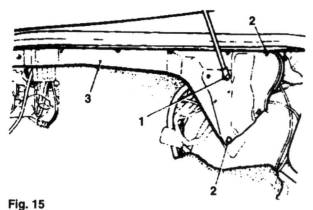

Fig. 15

Fig. 16

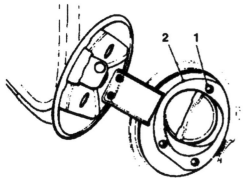

Fig. 17

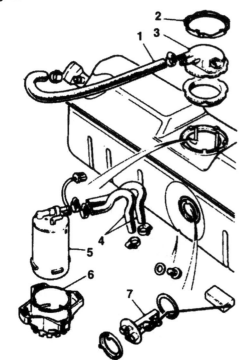

Fig. 18

suitable fireproof equipment.
Raise the vehicle on a ramp.
Remove the clips securing the fuel supply and return hoses (Fig. 12).
Disconnect the hoses from the tank.
Fit protective plugs to the hoses and the tank.
Lower the vehicle.
Displace and remove left-hand side liner.
Remove compartment floor carpet.
Remove the boot seal from the front flange and remove the front liner to flange trim.
Displace and remove the spare wheel securing stud.

Release securing nuts and bolts and remove the fuel tank retaining straps (1 Fig. 13).

Disconnect gauge unit harness connections (1 Fig. 14) and displace harness.
Disconnect pump harness multi-plug and displace harness (3 Fig. 13).

Release P clips securing purge and overflow pipes.
Note and disconnect plastic purge and overflow pipes from the evaporative loss flange (2 Fig. 13).
Position a suitable support to prop the boot lid open.
Displace the left-hand and right-hand boot lid gas strut lower ball mountings (1 Fig. 15).
Remove securing nuts and bolts (2 Fig 15) and displace the left-hand and right-hand hockey stick panels (3 Fig. 15).
Slacken the fuel filler neck to tank hose securing clips (1 Fig. 16).
Remove the fuel filler neck retaining bracket (2 Fig. 16).

Open the fuel filler flap and remove the filler neck seal securing screws (1 Fig. 17).
Displace and remove the retaining plate (2 Fig. 17), seal and filler neck.
Remove fuel filler neck hose (3 Fig. 16) and securing clips.
Disconnect drain tubes (4 Fig. 16) at both sides from the boot aperture drain stub pipes (5 Fig. 16).
Withdraw and remove tank assembly from the vehicle, place on bench.
Remove seal from tank at feed and return stub pipes.
Cut insulation pad from tank.

Release clip and disconnect breather hose (1 Fig. 18) from evaporative loss flange.

Note: Operations within fuel tank assembly must only be performed with use of screwdriver JD 175.

Using tool JD 174, displace and remove the evaporative loss flange retaining ring (2 Fig. 18).
Reposition the flange (3 Fig. 18) for access to the fuel pump harness multi-plug, disconnect the multi-plug.
Using screwdriver JD 175 release the internal hose (4 Fig. 18) securing clips and disconnect hoses from the pump assembly.
Displace and remove the pump assembly (5 Fig. 18).
Displace and remove the pump rubber mounting (6 Fig. 18).
Using tool 18G 1001 displace and remove the fuel gauge unit securing ring.
Remove the gauge unit and seal (7 Fig. 18).
Fitting a replacement tank is the reversal of the removal procedure.
Refit the components from the original tank.
Ensure new 'O' rings and seals are fitted.
Ensure all electrical connections are clean and secure.
Check all flange, hose and pipe connections for security and signs of leakage.

Service Tools

Evaporative Loss Flange Lock Ring Wrench	JD 174
Hose Clip to Fuel Pump Screwdriver	JD 175
Tank Sender Unit Wrench	18G 1001

Torque Figures

Tank strap securing nut	40-50 Nm
Fuel sender unit locking ring	5.5-13 Nm
Internal hose securing clips	1.5-2 Nm
Fuel filler neck hose clips	3.5-4.5 Nm
Tank breather hose clips	1.5-2 Nm
Fuel filler neck retaining bolts	3.5-4.5 Nm

FUEL PUMP, RENEW

Remove the fuel tank assembly as described in the operation (and noting the 'Warnings' in) 'Fuel Tank, Renew'.

Note: Operations within fuel tank assembly must only be performed with use of screwdriver JD 175.

Release clip and disconnect breather hose from the evaporative loss flange.
Using tool JD 174, displace and remove the evaporative loss flange retaining ring.
Reposition the flange for access to the fuel pump harness multi-plug, disconnect the multi-plug.

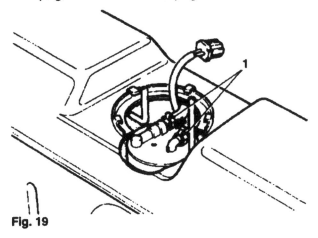

Fig. 19

Using screwdriver JD 175 release the internal hose securing clips (1 Fig. 19) and disconnect hoses from the pump assembly.
Release tie wrap from around the mounting rubber and using a twisting action displace and remove the pump assembly.
Fit and fully seat the replacement fuel pump assembly to the mounting, engaging the master lug.
Secure with new tie wrap.
Reposition hoses to the pump and tighten securing clips.
Fit and seat new evaporative flange seal.
Connect evaporative loss flange to pump multi-plug.
Seat evaporative loss flange to seal.
Using tool JD 174 secure flange with retaining ring.
Connect and secure breather hose to flange stub pipe.
Reposition the tank assembly to the vehicle.
Refit the tank by reversing the removal operations. Ensure new 'O' rings and seals are fitted.
Ensure all electrical connections are clean and secure. Check all flange, hose and pipe connections for security and signs of leakage.
Service Tools and Torque Figures are as listed under the operation 'Fuel Tank, Renew'.

FUEL PUMP MOUNTING RUBBER, RENEW

Remove the fuel tank assembly as described in the operation (and noting the 'Warnings' in) 'Fuel Tank, Renew'.

Note: Operations within fuel tank assembly must only be performed with use of screwdriver JD 175.

Release clip and disconnect breather hose from the evaporative loss flange.
Using tool JD 174, displace and remove the evaporative loss flange retaining ring.
Reposition the flange for access to the fuel pump harness multi-plug, disconnect the multi-plug.
Using screwdriver JD 175 release the internal hose securing clips (1 Fig. 19) and disconnect hoses from the pump assembly.
Release tie wrap from around the mounting rubber and using a twisting action displace and remove the pump assembly.
Displace and remove the pump mounting rubber.
Fit and align the new mounting rubber to the internal brackets. Note, the mounting rubber has a master lug so can only be fitted in one position.
Fit and fully seat the fuel pump assembly to the mounting, engaging the master lug. Secure with new tie wrap.
Reposition hoses to the pump and tighten securing clips.
Fit and seat new evaporative flange seal.
Connect evaporative loss flange to pump multi-plug.
Seat evaporative loss flange to seal. Using tool JD 174 secure flange with retaining ring.
Connect and secure breather hose to flange stub pipe.
Reposition the tank assembly to the vehicle.
Refit the tank by reversing the removal operations. Ensure new 'O' rings and seals are fitted.
Ensure all electrical connections are clean and secure. Check all flange, hose and pipe connections for security and signs of leakage.
Service Tools and Torque Figures are as listed under the operation 'Fuel Tank, Renew'.

FUEL TANK EVAPORATIVE LOSS FLANGE, RENEW

Remove the fuel tank assembly as described in the operation (and noting the 'Warnings' in) 'Fuel Tank, Renew'.

Note: Operations within fuel tank assembly must only be performed with use of screwdriver JD 175.

Release clip and disconnect breather hose from the evaporative loss flange (Fig. 20).

Using tool JD 174, displace and remove the evaporative loss flange retaining ring.

Fig. 20

Reposition the flange for access to the fuel pump harness multi-plug, disconnect the multi-plug.

Fit and seat new evaporative flange seal.

Connect evaporative loss flange to pump multi-plug.

Seat evaporative loss flange to seal.

Using tool JD 174 secure flange with retaining ring.

Connect and secure breather hose to flange stub pipe.

Reposition the tank assembly to the vehicle.

Refit the tank by reversing the removal operations.

Ensure new 'O' rings and seals are fitted.

Ensure all electrical connections are clean and secure.

Check all flange, hose and pipe connections for security and signs of leakage.

Service Tools and Torque Figures are as listed under the operation 'Fuel Tank, Renew'.

INTERNAL FUEL FEED & RETURN HOSES TO PUMP MODULE, RENEW

Remove the fuel tank assembly as described in the operation (and noting the 'Warnings' in) 'Fuel Tank, Renew'.

Caution:

Operations within fuel tank assembly must
only be performed with use of screwdriver JD 175.

Release clip and disconnect breather hose from the evaporative loss flange.

Using tool JD 174, displace and remove the evaporative loss flange retaining ring.

Reposition the flange for access to the fuel pump harness multi-plug, disconnect the multi-plug.

Using screwdriver JD 175 release the internal feed and return hose securing clips and disconnect hoses from the pump assembly.

Using a twisting action displace and remove the pump assembly.

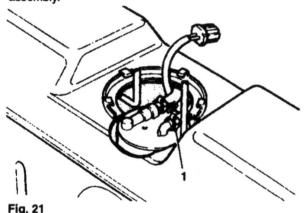

Fig. 21

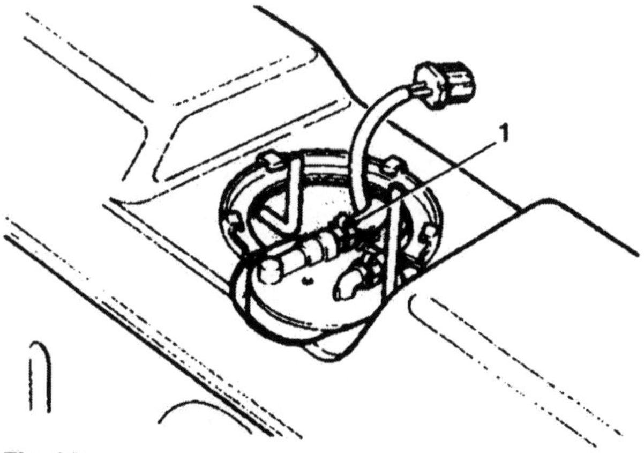

Fig. 22

Slacken the hose to internal stub pipe securing clips (1 Fig. 21 & 1 Fig. 22), displace clips and remove the hoses.

Position and secure new hoses to the internal stub pipes.

Fit and fully seat the fuel pump assembly to the mounting, engaging the master lug.

Reposition hoses to the pump and tighten securing clips.

Fit and seat new evaporative flange seal.

Connect evaporative loss flange to pump multi-plug.

Seat evaporative loss flange to seal.

Using tool JD 174 secure flange with retaining ring.

Connect and secure breather hose to flange stub pipe.

Reposition the tank assembly to the vehicle.

Refit the tank by reversing the removal operations.

Ensure new 'O' rings and seals are fitted.

Ensure all electrical connections are clean and secure.

Check all flange, hose and pipe connections for security and signs of leakage.

Service Tools and Torque Figures are as listed under the operation 'Fuel Tank, Renew'.

FUEL RAIL, RENEW

Open the luggage compartment and displace the right-hand front side liner.

Displace the fuel pump relay (silver relay on black / yellow base) from mounting at rear of hockey stick panel and remove the relay.

Crank the engine to depressurise the fuel system.

Reposition the relay and trim liner.

Disconnect the vehicle battery.

Open the bonnet and fit wing protection.

Position suitable cloth to catch possible fuel residue.

Release the union nut and disconnect the fuel feed hose from the fuel rail (2 Fig. 23).

Release the hose retaining 'P' clips.

Fit blanking plugs to the fuel feed hose and fuel rail.

Undo the fuel rail to pressure regulator union nut (3 Fig. 23).

Slacken the nut securing the pressure regulator to the mounting bracket (4 Fig. 23).

Reposition the regulator from the rail mounting.

Fit blanking plugs to the regulator and the rail.

Disconnect the harness multi-plugs from fuel injectors.

Remove the four bolts (5 Fig. 23) securing the fuel rail.

Remove the fuel rail and injector assembly.

Fit blanking plugs to the intake manifold.

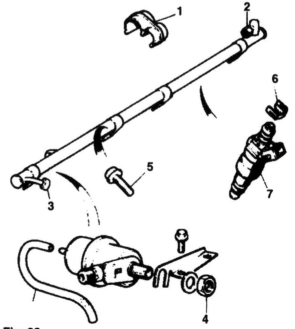

Fig. 23

Withdraw the clips (6 Fig. 23) securing the injectors (7 Fig. 23) and remove the injectors from the rail.
Remove and discard injector 'O' rings.
Fit and lubricate new 'O' ring seals and reverse the above procedure when refitting the fuel rail assembly.

Torque Figures

Fuel rail union nut	10-12 Nm
Rail securing bolts	7-10 Nm

PRESSURE REGULATOR VALVE, RENEW

Open the luggage compartment and displace the right-hand front side liner.
Displace the fuel pump relay (silver relay on black / yellow base) from mounting at rear of hockey stick panel and remove the relay.
Crank the engine to depressurise the fuel system.
Reposition the relay and trim liner.
Disconnect the vehicle battery.
Open the bonnet and fit wing protection.
Disconnect the vacuum hose (1 Fig. 24) from the pressure regulator valve.

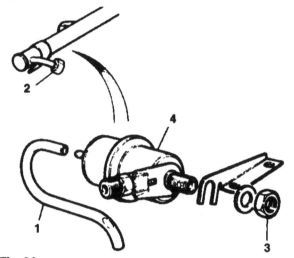

Fig. 24

Position suitable cloth to catch possible fuel residue.
Undo and release the fuel return hose from the regulator.
Fit blanking plugs to the regulator and hose.
Undo the fuel rail to regulator union nut (2 Fig. 24).
Remove the nut securing the unit to the mounting bracket (3 Fig. 24) and remove the regulator (4 Fig. 24).
Fit blanking plugs to the regulator and fuel rail.
Reverse the above procedure when replacing the regulator.

Torque Figures

Fuel rail union nut	10-12 Nm
Regulator to mounting bracket	7-10 Nm

MAIN FUEL FILTER, RENEW

Open the luggage compartment and displace the right-hand front side liner.
Displace the fuel pump relay (silver relay on black / yellow base) from mounting at rear of hockey stick panel and remove the relay.
Crank the engine to depressurise the fuel system.
Reposition the relay and disconnect the battery earth lead.
Empty the fuel tank into an explosive proof container using suitable fireproof equipment.
Raise the vehicle on a ramp.
Filter assembly is located underbelly, LHS, forward of the axle carrier.

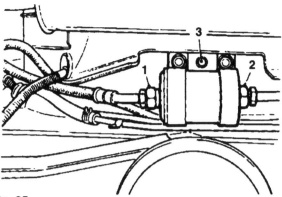

Fig. 25

Undo and remove the filter to fuel tank feed pipe union (1 Fig. 25), remove and discard 'O' ring seal.
Fit blanking plugs to feed pipe and filter.
Undo and remove the fuel feed pipe to filter union (2 Fig. 25), remove and discard 'O' ring seal.
Fit blanking plugs to feed pipe and filter.
Remove the filter mounting clamp bolt (3 Fig. 25) and remove the assembly.
Reverse the previous operations to fit the replacement filter assembly.
Use new 'O' ring seals and check pipe connections for signs of leakage.
Lower the vehicle, reposition the trim panel and reconnect the battery.
Refill the fuel tank.

Torque Figures

Filter clamp bolt	6-7 Nm
Fuel feed hose to filter unit	27-33 Nm
Filter to engine feed pipe union nut	27-33 Nm

AIR CLEANER ASSEMBLY, RENEW

Open the bonnet and fit wing protection.
Remove the intake elbow mounting nut for access.

Release the filter assembly to air mass meter retaining clips (1 Fig. 26).
Slacken the filter assembly to intake tube securing clip (2 Fig. 26).
Remove the upper wing nut (3 Fig. 26) from the filter assembly mounting.

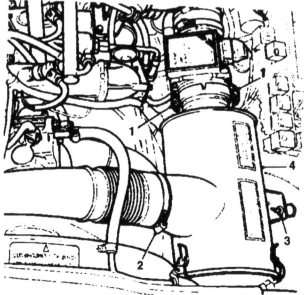

Fig. 26

Slacken but do not remove the lower wing nut.
Displace and remove the air box assembly (4 Fig. 26).
Remove and discard the filter to air mass meter 'O' ring seal.
Release the end cover retaining clips.
Displace the cover and remove the element retaining plate.
Remove the element.
Use a new 'O' ring seal and reverse the above procedure when replacing the air cleaner assembly.

AIR CLEANER ELEMENT, RENEW

Open the bonnet and fit wing protection.
Remove the intake elbow mounting nut for access.
Release the filter assembly to air mass meter retaining clips (1 Fig. 26).
Slacken the filter assembly to intake tube securing clip (2 Fig. 26).
Remove the upper wing nut (3 Fig. 26) from the filter assembly mounting.
Slacken but do not remove the lower wing nut.
Displace and remove the air box assembly (4 Fig. 26).
Remove and discard the filter to air mass meter 'O' ring seal.
Release the end cover retaining clips.
Displace the cover and remove the element retaining plate.
Remove the element.
Clean the filter box, end cover and element retaining plate.
Fit and fully seat new filter to the air box.
Fit and secure retaining plate and reposition end cover.
Use a new 'O' ring seal and reverse the above procedure when replacing the air cleaner assembly.

THROTTLE PEDAL, RENEW

LHD Vehicles

Displace the tunnel carpet from around the throttle pedal.
Remove the split pin, withdraw the clevis pin and displace the inner cable from the pedal (1 Fig. 27).

Remove the assembly mounting bracket (2 Fig. 27) upper securing nut and forward securing screw.
Remove the assembly from the vehicle.
Withdraw 'C' clips, rotate and withdraw pivot bushes (3 Fig. 27).
Remove the pedal from the mounting bracket and remove the spring.
Withdraw stop rubber from the pedal.
Refitting a new pedal is the reversal of the removal procedure.

RHD Vehicles

Displace the tunnel carpet from around the throttle pedal.

Remove the split pin from the cable retaining sleeve (1 Fig. 28).

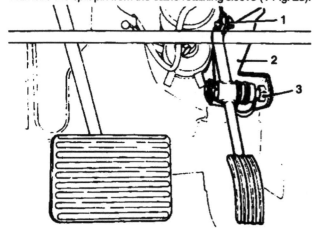

Fig. 27

Slide the sleeve from position and disconnect the inner cable from the pedal.
Remove the three pedal mounting bracket securing screws and remove the assembly from the vehicle.
Disengage the return spring.
Withdraw 'C' clips (2 Fig. 28), rotate and withdraw pivot bushes (3 Fig. 28).
Remove the pedal from the mounting bracket.
Withdraw stop rubber from the mounting bracket.
Refitting a new pedal is the reversal of the removal procedure.

THROTTLE PEDAL RETURN SPRING, RENEW

LHD Vehicles

Remove and dismantle the pedal assembly in accordance with the method detailed previously.
Remove and replace the pedal return spring.
Reassemble and refit the assembly to the vehicle.

RHD Vehicles

In the drivers footwell; unhook and displace the pedal return spring from the pedal assembly.
Engage the replacement spring onto pedal and pedal mounting bracket.

RAM TUBE, RENEW

Open the bonnet and fit wing protection.
Slacken the clip securing the ram tube to the air cleaner assembly (1 Fig. 29).

Displace the ram tube from the bell mouth location (2 Fig. 29).
Reposition the ram tube to disengage from the bell mouth.
Remove the ram tube (3 Fig. 29) from the air cleaner and remove from the vehicle.
Withdraw the bell mouth from the cross member.
Replacing the ram tube is the reversal of the above operations.

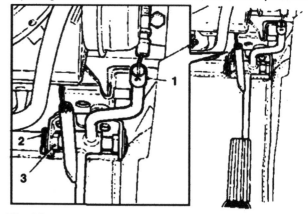

Fig. 28

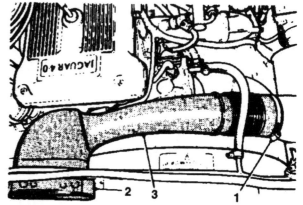

Fig. 29

INERTIA SWITCH, RENEW

Disconnect battery earth lead.
Remove the inertia switch trim cover (1 Fig. 30).

Release the switch block connector (2 Fig. 30) retaining catch and withdraw the connector.
Remove the passenger side dash liner.
Remove the two screws securing the switch and remove the switch (3 Fig. 30).
On refitting a new switch ensure it is correctly reset.

THROTTLE PEDAL BUSHES, RENEW

Remove the throttle pedal assembly.
Withdraw 'C' clips, rotate and withdraw pivot bushes.
Lubricate pedal pivot shaft and assemble to mounting.
Insert bushes and rotate quarter turn to lock.
Reposition 'C' clips.
Refit the pedal assembly to the vehicle.

THROTTLE VALVE, CHECK AND ADJUST

Open the bonnet and fit wing protection.
Slacken the clips securing the air intake elbow hoses to the breather and stepper motor and disconnect hoses.
Slacken the clip securing the elbow hose to the air flow meter.
Slacken the clip securing the intake hose to the throttle housing.
Remove the securing nut from the elbow mounting bracket and displace the elbow assembly from the air flow meter.
Disconnect sensor multi-plugs from the elbow assembly and remove the assembly from the throttle housing.
Slacken the locknut securing the throttle cable to the manifold bracket.
Ensure the throttle cable does not restrict the linkage adjustment.
Slacken the throttle stop locknut.

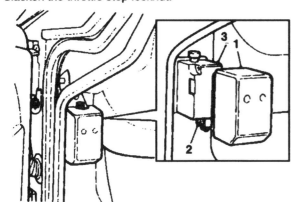

Fig. 30

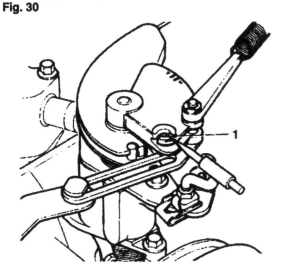

Fig. 31

Using a feeler gauge adjust the throttle stop screw to obtain an air gap of 0.05mm (0.002in) between the butterfly valve and the throttle housing.
Adjust the stop and tighten the throttle stop locknut.
Recheck the butterfly valve setting.
Fit and align the full throttle setting adjustment tool JD 131 to the throttle stop plate.
Ensure the pointer aligns with the idle mark 'I' on the throttle linkage (1 Fig. 31).
Slacken the full throttle stop locknut.

Fully open the throttle and adjust the full throttle adjusting screw to align the pointer of the special tool to mark 'A' for automatic vehicles or 'M' for manual vehicles (1 Fig. 32).

Release the throttle linkage and tighten the full throttle stop locknut.
Remove the full throttle setting tool.
Tighten the throttle cable locknut.
Check and adjust the throttle potentiometer using JDS.
Fit and align the elbow and hose assembly to the throttle housing, position and tighten the securing clip.
Connect the elbow to the air flow meter and tighten securing clip.
Seat the elbow to the rubber mounting and fit and tighten the securing nut.
Reconnect the breather and stepper motor hoses and secure.
Run the engine until normal operating temperature is reached and switch off.

Disconnect the coolant thermistor multi-plug connector at the thermostat housing (Fig 33).
Connect a 100 ohm 1/4 watt resistor across the multi-plug connections.
Ensure the leads of the resistor do not touch any metal parts of the vehicle.

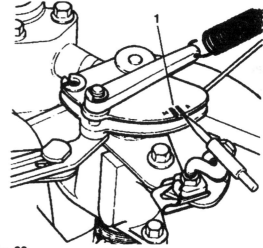

Fig. 32

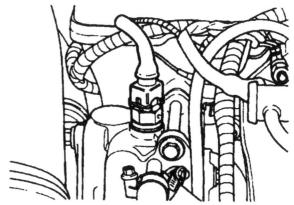

Fig. 33

Switch the ignition 'ON' and wait 10 seconds.
The idle speed stepper motor should now be fully wound in, verify by holding the motor and feeling for vibration.
With the ignition still 'ON' disconnect the idle speed stepper motor harness multi-plug connector.
Remove the resistor from the thermistor harness connector and reconnect the connector to the thermistor.
Start the engine and if necessary adjust the idle speed actuator air bypass screw with a 7/32in. Allen key to 550 to 600 rev/min.
Stop the engine.
Reconnect the idle speed stepper motor connector.
Close the bonnet.

Service Tools

Full throttle adjustment tool   JD 131

Torque Figures

Elbow to mounting securing nut	12.5-15 Nm
Cable locknut	14-18 Nm
Throttle stop locknut	3.5-4.5 Nm

THROTTLE CABLE, RENEW

LHD Vehicles

Displace and remove the throttle cable to linkage retaining clip.
Disconnect the inner cable from the throttle linkage.

Undo the cable to bracket locknut and reposition the cable from the mounting bracket (Fig 34).
Displace the tunnel carpet from around the throttle pedal.
Remove the split pin, withdraw the clevis pin and displace the inner cable from the pedal.

Displace the plenum chamber drain tube for access (1 Fig. 35).
Remove the bulkhead plate securing bolts (lower LH position has retaining nut) (2 Fig. 35).
Displace the cable and plate assembly (3 Fig. 35) from the bulkhead.
Remove the split pin and displace the cable from the lever arm.
Withdraw the cable from the plate.
Fit the replacement cable to the plate, connect to the lever arm and insert new split pin.
Clean the gasket faces and apply new gasket sealant.
Fit and align assembly to the bulkhead ensuring connecting rod passes into footwell.
Secure plate to bulkhead.
Reconnect rod to the pedal assembly.
Position the cable to the throttle linkage.
Connect and secure the inner cable to the linkage with retaining clip.
Adjust to remove excessive free play and tighten locknut.

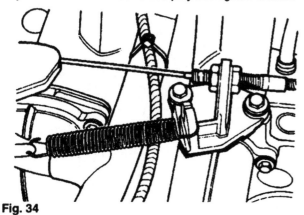

Fig. 34

Fig. 36

RHD Vehicles

Displace and remove the throttle cable to linkage retaining clip.
Disconnect the inner cable from the throttle linkage.
Undo the cable to bracket locknut and reposition the cable from the mounting bracket.
Displace the tunnel carpet from around the throttle pedal.
Remove the split pin from the cable retaining sleeve. Slide the sleeve from position and disconnect the inner cable from the pedal.
Withdraw the outer cable from the bracket adjacent to the pedal.

Withdraw the cable (1 Fig. 36) through the bulkhead grommet and into the engine compartment.
Fit the replacement cable in position.
Connect to the pedal assembly.
Position the cable to the throttle linkage.
Connect and secure the inner cable to the linkage with retaining clip.
Adjust to remove excessive free play and tighten locknut.
Seal the bulkhead grommet with silicone sealant.

SPEED CONTROL SYSTEM, DESCRIPTION

The speed control system maintains the vehicle at a driver's selected speed over normal driving conditions, without the need for constant surveillance of the speedometer and continuous driver use of the throttle pedal.
Once engaged and set the speed control system will control the position of the throttle over a wide range of movement from idle to full throttle to maintain the speed set under different conditions.

SYSTEM COMPONENTS

Throttle actuator

A vacuum operated device containing bellows attached to the throttle by a mechanical linkage.

Electronic Control Unit (ECU)

A device containing a microprocessor and associated electronic circuits which evaluates the signal provided by the drivers controls, the pedal switches and road speed sensors. It then transmits signals to a vacuum control unit in order to control the vehicle speed. It also retains in memory the speed setting.

Vacuum pump and regulator valve

An assembly which increases and decreases the vacuum system pressure depending on signals received from the ECU.

Dump valve

A device which vents the vacuum system pressure to atmosphere when braking or the throttle returns to idle.

Operating switches

The system is controlled by a centre console mounted master switch and a resume button on the end of the combination switch stalk. A brake switch disengages the system.

THROTTLE ACTUATOR, RENEW

Fig. 37

Open the bonnet and fit wing protection.
Remove the securing nut from the operating rod to actuator pivot (1 Fig. 37).
Remove the operating rod pivot pin and bush.
Reposition the rod from the actuator.
Disconnect the vacuum hose (2 Fig. 37) from the front of the throttle actuator.
Remove the actuator to mounting bracket securing nut (3 Fig. 37).
Displace and remove the actuator (4 Fig. 37).
Fit and align the replacement actuator assembly to the mounting bracket.
Fit and tighten securing nut.
Connect vacuum hose to the front of the actuator.
Reconnect the actuator linkage, insert and secure the pivot pin.
Remove wing protection and close bonnet.

Torque Figures

Actuator mounting bracket nut	3.5-4.5 Nm
Link rod securing nut	3.5-4.5 Nm

THROTTLE ACTUATOR LINKAGE ROD, RENEW

Open the bonnet and fit wing protection.

Remove the actuator to link rod securing nut (1 Fig. 38).
Remove the link rod pivot pin and bush.
Displace and remove the actuator link to throttle linkage securing clip (2 Fig. 38).
Remove the actuator link rod (3 Fig. 38).
Remove the actuator link retaining pin.
Fitting a replacement actuator linkage rod is the reversal of the above procedure. Ensure the pivot pin is adjusted to remove any free play in the link rod before tightening the actuator pivot pin securing nut.
Remove wing protection and close bonnet.

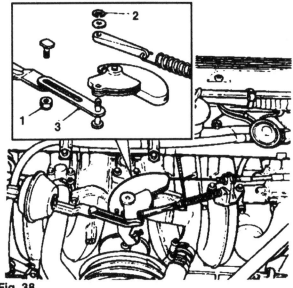

Fig. 38

Torque Figures

Link rod securing nut	3.5-4.5 Nm

VACUUM DUMP VALVE, RENEW

Open the bonnet and fit wing protection.
Remove the relay mounting bracket upper and lower securing bolts situated left-hand inner wing.

Reposition the relay mounting bracket for access (1 Fig. 39).
Release the dump valve harness tie wrap.
Disconnect the vacuum hose (2 Fig. 39) from the dump valve.
Disconnect the harness multi-plug (3 Fig. 39).
Remove the valve securing nuts and remove the valve assembly (4 Fig. 39).
Position and secure the replacement valve assembly with nuts.
Connect the harness multi-plug and refit the vacuum hose.
Secure harness with tie wrap.
Reposition and secure relay mounting bracket.
Remove wing covers and close bonnet.

VACUUM SUPPLY HOSE, RENEW

Open the bonnet and fit wing protection.
Remove the relay mounting bracket securing bolts situated left-hand inner wing.
Reposition the relay mounting bracket for access.
Disconnect the hose from the vacuum dump valve.
Disconnect the hose from the vacuum pump.
Disconnect the vacuum hose from the actuator.
Remove the hose assembly and disconnect hoses from 'T' piece.

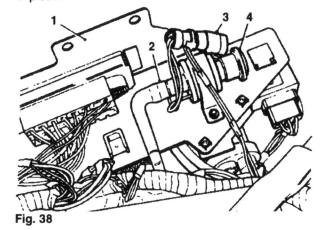

Fig. 38

Connect new hoses to the 'T' piece and fit the assembly, routing the longest hose to the actuator.
Remove wing covers and close bonnet.

SPEED CONTROL ECU, RENEW

Reposition the passenger footwell carpet for access to the ECU panel securing screws.
Undo and remove the floor panel screws.
Displace the panel and disconnect the ECU harness multi-plug.
Release the retaining clip and displace and remove the unit.
Align the new ECU, secure with retaining clip and connect the harness multi-plug.
Reposition and secure the panel, reposition carpet.

VACUUM PUMP, RENEW

Open the bonnet and fit wing protection.

Disconnect the vacuum pump harness multi-plug (1 Fig. 40).
Disconnect the vacuum hose (2 Fig. 40) from the vacuum pump.

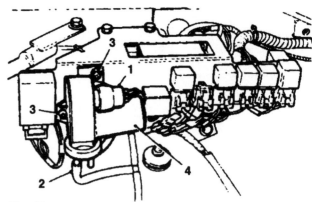

Fig. 40

Remove the vacuum pump mounting screws (3 Fig. 40).
Displace and remove the vacuum pump (4 Fig. 40).
Fit and align the replacement vacuum pump to mounting position.
Fit and tighten vacuum pump securing screws.
Connect vacuum pump harness multi-plug.
Connect vacuum hose.
Remove wing covers and close bonnet.

Chapter 5

COOLING SYSTEM - 4.0 LITRE ENGINE

RADIATOR COOLING FAN, DESCRIPTION

A new cooling fan is used. The fan has improved cooling performance and full radio interference suppression. It is mounted on the rear of the radiator cowl (Fig. 1).

ENGINE OIL COOLER, DESCRIPTION

European models have a new 28 mm full width oil cooler (2 Fig. 2). The cooler is mounted on the bonnet hinge panel and the pipework routed under the radiator (Fig. 2).
The condenser is three-quarter depth and 'O' ring connections are now standard to reduce the possibility of leaks.
USA models retain the smaller type oil cooler (1 Fig. 2). All other features are as European models.

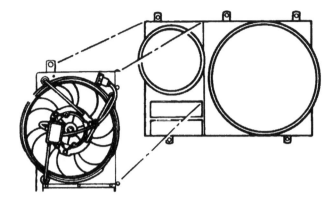

Fig. 1

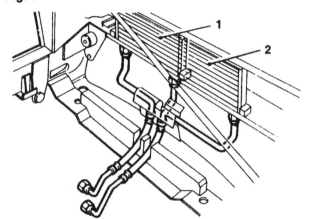

Fig. 2

Chapter 6

EXHAUST SYSTEM - 4.0 LITRE

SYSTEM WITH LOW LOSS CATALYTIC CONVERTERS, REMOVE AND REFIT PROCEDURES

INTRODUCTION

The following instructions detail the method for removing the exhaust system as two Vehicle Sets. The Vehicle Sets comprise:

Each may be removed as a set and dismantled on a bench, or each component part may be removed individually.

FRONT CATALYTIC CONVERTER SECTION, RENEW

Removal

Drive the vehicle onto the ramp.
Release the bonnet.
Open the bonnet and fit a wing protection cover to the right hand wing.
Remove the cable of the exhaust oxygen sensor from the clips situated along the right hand wing flange.
Disconnect the connector of the oxygen sensor cable.
Raise the ramp.

Remove the oxygen sensor from the downpipe catalyst assembly (1 Fig. 2).
Loosen the four nuts which secure the downpipe flanges to the exhaust manifold. Remove three of the nuts and leave the fourth to support the system (2 Fig. 2).
Remove three nuts and bolts from one of the rear flange type joints at the rear of the underfloor catalytic converters (3 Fig. 3).

Remove two nuts and bolts from the other flange type joint at the rear of the underfloor catalytic converters; loosen the third nut and bolt.

With the help of an assistant to support one end of the system, remove the two remaining fixings.
Carefully separate the assembly from the manifold flanges and rear section flange joints. Remove the assembly from the vehicle.
Remove the sealing rings from the manifold flange joints and the olives from the rear flange joints.

Disassembly

Downpipe Catalytic Converters:

Remove the nut, bolt and washer (1 Fig. 4) from the clamping ring (downpipe catalytic converters to intermediate pipe). Open the clamp if necessary and slide it onto the intermediate pipe.
Carefully separate the downpipe converters assembly from the intermediate pipe (2 Fig. 4).
Remove the clamp from the intermediate pipe and remove the olive (3 Fig. 4).

Underfloor Catalytic Converters:

Remove the nut, bolt and washers from the right hand side clamping ring (1 Fig. 5) (underfloor catalytic converters to intermediate pipe).
Open the clamp (2 Fig. 5) if necessary and slide it onto the intermediate pipe. Carefully separate the converter (3 Fig. 5) from the intermediate pipe.
Remove the clamp from the intermediate pipe.
Repeat the above instructions for the left hand side underfloor converter.

Cleaning and Inspection

For parts which are intended for re-use, clean the sealing faces, clamps and olives; inspect for cracks and deep pitting. Renew as necessary.

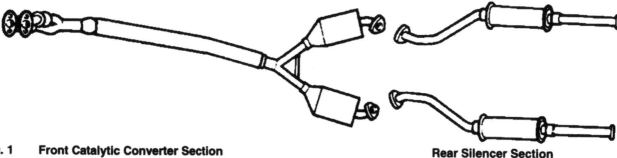

Fig. 1   Front Catalytic Converter Section

Rear Silencer Section

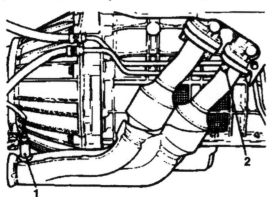

Fig. 2

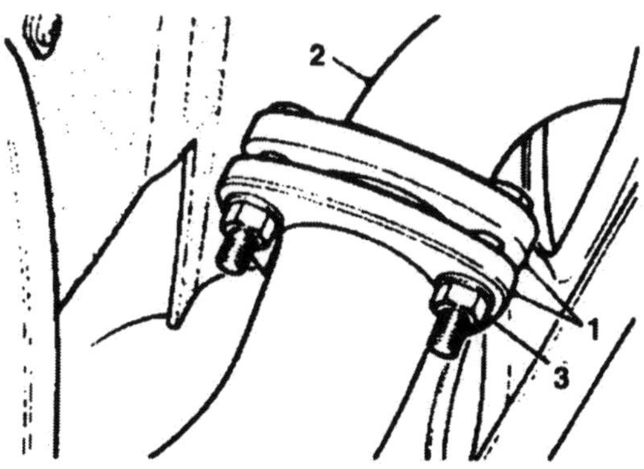

Fig. 3

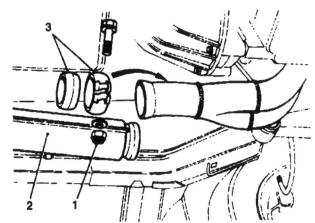

Fig. 4

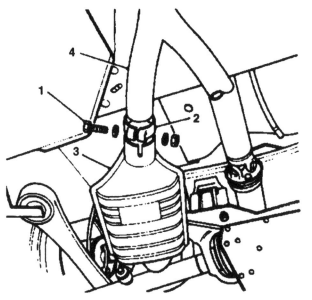

Fig. 5

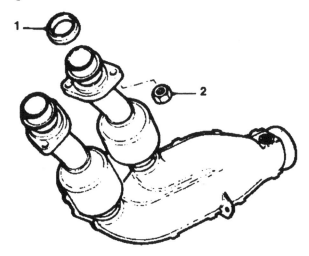

Fig. 6

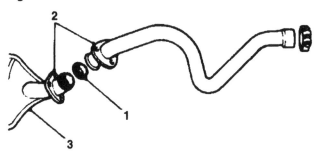

Fig. 7

Obtain new locking nuts prior to assembly, as follows:

Helicoil 3/8 in.	- 4 off -	Manifold flange
Helicoil 5/16 in.	- 1 off -	Centre flange joint
Helicoil 5/16 in.	- 6 off -	Rear flange joints
Steel locknuts 1/4 in.	- 2 off -	Rear clamps

Reassembly

Downpipe Catalytic Converters:

Check for and remove any debris from inside the downpipe and intermediate pipe assemblies.
Fit the clamping ring over the front section of the intermediate pipe.
Apply exhaust sealant to the outer surface of the olive.
Fit the olive and assemble the downpipe catalytic converter to the intermediate pipe.
Position the clamping ring over the joint and fit the retaining bolt, washers and Helicoil locking nut.
Tighten the clamp bolt sufficiently to hold the assembly together.

Underfloor Catalytic Converters:

Check for and remove any debris from inside the underfloor catalytic converter pipes and intermediate pipe assemblies.
Fit the clamping ring over the right hand rear section of the intermediate pipe.
Apply exhaust sealant sparingly to approximately 1 inch of the outer surface at the end of the intermediate pipe. Do not apply sealant to the inner surface of the catalytic converter pipe.
Fit the catalytic converter to the intermediate pipe.
Position the clamping ring over the joint and fit the retaining bolt, washers and steel locking nut.
Tighten the clamp bolt sufficiently to hold the assembly together.
Repeat the above instructions for the left hand underfloor converter.

Refitting

Locate the sealing rings onto the downpipe flange joints (1 Fig. 6)
Apply exhaust sealant sparingly to the outer surface of the rear olives and locate the olives into the rear converter joints (1 Fig. 7).

With the help of an assistant to support one end, locate the system to the vehicle.
Fit one nut to the manifold flange (2 Fig. 6) and one bolt, washer and Helicoil locking nut to each rear joint (2 Fig. 7), to support the system.
Fit and partly tighten the remaining four bolts, washers and Helicoil locking nuts to the rear joints.
Fit and partly tighten the Helicoil locking nuts to the manifold flange.
Check for adequate clearance around the downpipe catalytic converters and tighten the nuts to the correct torque, ensuring that the gap between the pipe flanges is even.
Carefully reposition the intermediate pipe (3 Fig. 7) and the underfloor converters as necessary.
Set the position of the clamping rings to ensure that the bolts and clamps will not be damaged when driving over uneven ground.
To provide the maximum clearance between the exhaust and suspension arm, set the position of each rear flange joint clamping ring (2 Fig. 7) so that a straight section of the flange is adjacent to the suspension arm (Fig. 8).

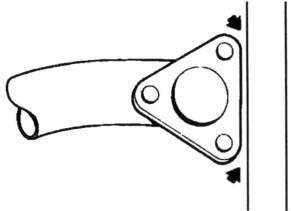

Fig. 8

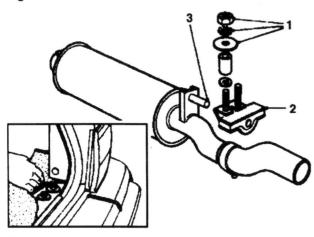

Fig. 9

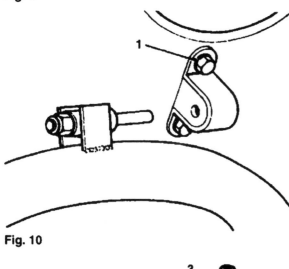

Fig. 10

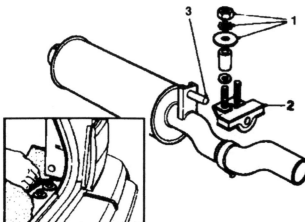

Fig. 11

Fit the oxygen sensor to the downpipe.

Lower the ramp.

Connect the oxygen sensor cable to its mating connector and fit the cable into the clips located along the right hand inner wing.

Remove the wing protection sheet.

Close the bonnet.

Torque Figures

Downpipe to manifold flange		30-35 Nm
Oxygen sensor		50-60 Nm
Clamps:	Downpipe to intermediate pipe.	15-18 Nm
	Intermediate pipe to U/F cats.	15-18 Nm
	U/F cats to over-axle pipe.	15-18 Nm

Oils / Lubricants / Sealants

Exhaust sealer   TIVOLI KAY ADHESIVES NO 5696

REAR SILENCER SECTION, RENEW

Drive the vehicle onto the ramp.

Mounting Rubber(s) - Rear

Open the boot.

Working in the boot, ease back the side trim and carpet each side to give access to the nuts which retain the exhaust rear mountings; remove the nuts and washers (1 Fig. 9). Raise the ramp.

Loosen the bolts which secure the front mounting to the boot floor.

Carefully ease down the tail pipe and remove the rear mounting(s) (2 Fig. 9) and spacers from the silencer mounting rod(s) (3 Fig. 9).

Fit the spacers onto the studs of the mounting.

Fit the rear mounting onto the mounting rod.

Lift the tail pipe so that the studs of the mounting locate into their locating holes on the boot floor.

Tighten the bolts which retain the front mounting(s).

Lower the ramp.

Fit the nuts, washers and shake-proof washers to the rear mounting studs and tighten to the correct torque. Fit the boot carpet and side trim.

Close the boot.

Mounting Rubber(s) - Front:

Raise the ramp.

Remove the two bolts (1 Fig. 10) which secure the exhaust front mounting(s) to the boot floor and remove the mounting(s) from the mounting rod(s).

Fit the mounting(s) onto the mounting rod(s) and align the fixing holes with the locating holes in the boot floor. Fit the retaining bolts and tighten to the correct torque.

Lower the ramp.

Removal

Rear Silencer Section:

Open the boot.

Working in the boot, ease back the side trim and carpet each side to give access to the nuts which retain the exhaust rear mountings; remove the nuts and washers (1 Fig. 11).

Raise the ramp.

Loosen the bolts which retain the front mounting(s) to the boot floor.

Remove the rear mounting(s) from the silencer mounting rod(s) (2 Fig. 11).

Remove the three nuts, bolts and washers from the right hand rear flange type joints (Underfloor catalytic converter to over-axle pipe).

Carefully separate the rear flange type joint.

Remove the bolts from the front mounting and remove the rear silencer section.

Remove the olive from the flange type joint and cover the output pipe of the underfloor catalytic converter to prevent ingress of contaminants.

Repeat for the left hand rear silencer.

Disassembly

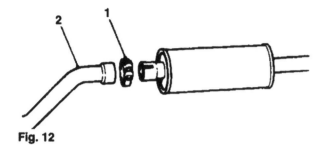

Fig. 12

Over-axle Pipe:

Remove the nut, bolt and washers (1 Fig. 12) from the clamping ring (over-axle pipe to silencer).

Open the clamp if necessary and slide it onto the over-axle pipe (2 Fig. 12).

Carefully separate the silencer and pipe.

Tail Pipe:

Loosen the socket head grub screw of the tail pipe clamp.
Remove the tail pipe and the clamp.

Front Mounting Rod:

Remove the Nyloc nut (1 Fig. 13) which secures the mounting rod to the over-axle pipe (2 Fig. 13).

Remove the mounting rod and insulating collars.

Cleaning and Inspection

For parts which are intended for re-use, clean the sealing faces, clamps and olives; inspect for cracks and deep pitting. Renew as necessary.

Obtain new locking nuts prior to assembly, as follows:

Helicoil 5/16 in.	- 6 off -	Front flange joints
Nyloc nut 3/8 in.	- 2 off -	Front mounting rods
Steel locknut 1/4 in.	- 2 off -	Silencer clamps

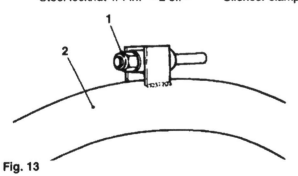

Fig. 13

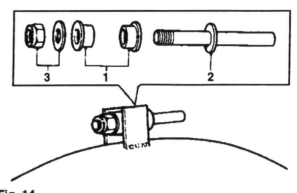

Fig. 14

Reassembly

Front Mounting Rod:

Assemble the insulation collar (1 Fig. 14), flanged end first onto the threaded end of the mounting rod (2 Fig. 14). Fit the mounting rod to the over-axle pipe, threaded section towards the front of the vehicle.

Fit an insulation collar to the threaded end of the rod and locate it into the mounting tube of the over-axle pipe.

Fit a new Nyloc nut and washer (3 Fig. 14).

Over-axle pipe to Silencer:

Check for and remove any debris from inside the silencer pipe and over-axle pipe assemblies.

Fit the clamping ring over the rear section of the over-axle pipe.

Apply exhaust sealant sparingly to approximately 1 inch of the outer surface at the end of the over-axle pipe.

Fit the silencer to the over-axle pipe.

Position the clamping ring over the joint and fit the retaining bolt, washers and steel locking nut.

Tighten the clamp bolt sufficiently to hold the assembly together.

Tail Pipe:

Apply exhaust sealant sparingly to the outer surface of the silencer rear pipe.

Slide the clamp (2 Fig. 15) over the tail pipe.

Fit the tailpipe onto the silencer output pipe.

Position the clamp and tighten the socket head grub screw (1 Fig. 15).

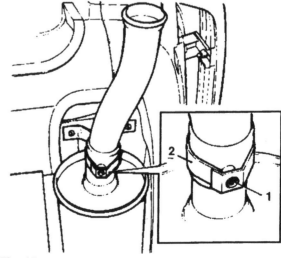

Fig. 15

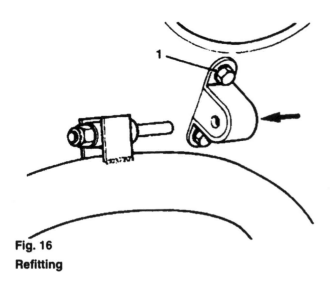

Fig. 16

Refitting

Rear Silencer Section:

Fit the front mounting assembly (Fig. 16) onto the mounting rod of the right hand silencer section.

Remove the protective cover from the underfloor catalytic converter output pipe.

Apply exhaust sealant sparingly to the outer surface of the olive and locate the olive into the underfloor converter joint. Locate the system to the vehicle.

Fit the bolts (1 Fig. 16) to the front mounting but tighten only sufficiently to support the system.

Align the front flange joint and fit the three Helicoil nuts, bolts and washers.

Tighten sufficiently to clamp the components, but allow movement of the joint.

Fit the rear mounting rubber assembly onto the mounting rod of the right hand silencer section.

Lift the tail pipe so that the studs of the mounting locate into their locating holes on the boot floor.

Use a small jack or axle stand to lightly retain the exhaust in this position.

Tighten the forward mounting bolts to the correct torque.

To provide the maximum clearance between the exhaust and suspension arm, set the position of the rear flange joint clamping ring so that a straight section of the flange is adjacent to the suspension arm (Fig. 8).

Tighten the flange bolts to the correct torque, ensuring that the flanges are pulled together evenly.

Carefully rotate the silencer and tail pipe to make final adjustment for clearances and tighten the clamp to the correct torque.

Repeat the above instruction for the left hand rear silencer section.

Lower the ramp.

Working in the boot, fit the nuts and washers to the exhaust rear mounting studs and tighten to the correct torque.

Fit the boot carpet and side trim.

Remove the jack that was supporting the rear of the exhaust.

Close the boot.

Torque Figures

Rear mounting to floor		11-15 Nm
Front mounting to floor		11-15 Nm
Mounting rod to silencer		13-17 Nm
Clamps:	U/F catalyst to over-axle pipe	15-18 Nm
	Over-axle pipe to silencer	15-18 Nm
	Tail pipe trim	2-3 Nm

Oils / Lubricants / Sealants

Exhaust sealant   TIVOLI KAY ADHESIVES No 5696

EXHAUST MANIFOLD GASKETS, RENEW

Removal

Open the bonnet and fit a wing protection cover to the right hand wing.

Jack up and support the vehicle on axle stands.

Remove the cable of the exhaust oxygen sensor from the clips situated along the right hand wing flange.

Disconnect the connectors of the oxygen sensor cable.

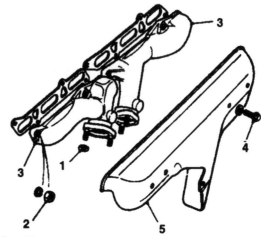

Fig. 17

Remove the five bolts (4 Fig. 17) from the manifold heatshield; remove the heatshield (5 Fig. 17).

Remove the four nuts from the downpipe flange (1 Fig. 17).

Carefully lower the downpipe from the manifold studs.

Remove the nuts from the front manifold (2 Fig. 17)

Remove the front manifold (3 Fig. 17).

Remove and discard the gasket.

Remove the nuts from the rear manifold (2 Fig. 17).

Remove the manifold (3 Fig. 17).

Remove and discard the gasket.

Cleaning and Inspection

Clean the sealing faces of the manifold(s) and inspect for cracks and deep pitting. Renew as necessary.

Obtain new locking nuts prior to assembly, as follows:

      Helicoil 3/8 in. - 4 off - Manifold flange Refitting

Fit a new gasket to the front manifold studs.

Fit the front manifold, nuts and washers.

Tighten the nuts sufficiently to clamp the manifold but allow.

Fit the rear manifold, nuts and washers.

Tighten the nuts sufficiently to clamp the manifold but allow to position the manifold(s) may be necessary.

Fit the old Helicoil nuts or ordinary nuts and tighten sufficiently to clamp the flange joint.

Tighten the manifold nuts to the specified torque.

Remove the old nuts from the downpipe flange and carefully lower the downpipe.

Fit new sealing rings to the downpipe flanges.

Fit the downpipe to the manifold flanges and secure with new Helicoil nuts (1 Fig. 18); check that the flanges clamp together evenly.

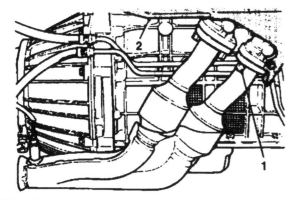

Fig. 18

Position the heatshield (2 Fig. 18) over the manifold and fit the five bolts.

Connect the oxygen sensor cable to its mating connectors and fit the cable into the clips located along the right hand inner wing.

Lower the vehicle from the axle stands.

Remove the wing protection sheet.

Close the bonnet.

Torque Figures

Manifold to cylinder head	49-54 Nm
Manifold to downpipe	30-35 Nm

INDUCTION MANIFOLD GASKET, RENEW

Removal

Open the bonnet and fit a wing protection cover to the left hand wing.

Depressurise the fuel system.

Remove the bolt which secures the induction elbow to the left hand wing (1 Fig. 19).

Loosen the clip (2 Fig. 19) which retains the air intake rubber hose to the throttle body.

Carefully ease the induction elbow and hose assembly away from the throttle body (3 Fig. 19).

Disconnect the hoses from the auxiliary air valve.

Disconnect the block connector of the throttle potentiometer.

Disconnect the pressure regulator return hose.

Fit blanking plugs to the open ends of the hose and the regulator.

Disconnect the fuel feed hose union nut and three retaining clips from the fuel rail. Fit blanking plugs to the hose and fuel rail union.

Remove the throttle return spring.

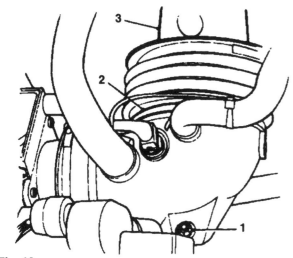

Fig. 19

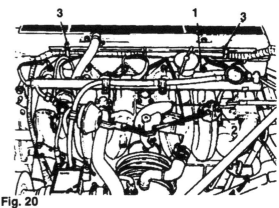

Fig. 20

Disconnect the throttle cable.

Remove the bolts from the throttle cable bracket and remove the bracket (2 Fig. 20).

Disconnect the hoses from the part throttle orifice control housing and fit blanking plugs to the hoses.

Disconnect the air conditioning vacuum hose from the manifold.

Disconnect the air temperature sensor and injector block connectors.

Disconnect the idle adjustment stepper motor connector.

Disconnect the breather hose from the cam cover.

Disconnect the hoses from the induction elbow and orifice control housing.

Remove the bolt from the oil filler tube and remove the tube from the lower housing.

Remove the bolt (4 Fig. 20) which secures the dipstick tube.

Disconnect the re-usable harness straps from the support brackets at the upper front and rear of the manifold (3 Fig. 20).

Remove the manifold nuts, bolts and washers (1 Fig. 20).

Remove the harness support brackets from the manifold studs.

Remove the filler tube bracket.

Remove the earth leads from the front retaining stud of the manifold.

Remove the manifold assembly, whilst at the same time carefully displacing the harnesses through the spaces in the manifold.

Remove and discard the manifold gasket and the oil filler tube seal.

Cleaning and Inspection

Clean the manifold and cylinder head mating faces. Inspect all mating faces and bolt holes for cracks.

Refitting

Fit a new seal (2 Fig. 21) to the oil filler tube lower housing (3 Fig. 21).

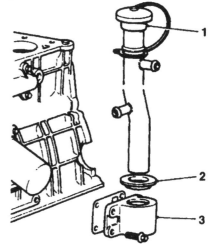

Fig. 21

Fit a new manifold gasket onto the studs.
Place the manifold assembly adjacent to its mounting studs and position the hoses and cables through the spaces in the manifold. Locate the manifold onto the mounting studs, ensuring that no cables are trapped between the faces.
Fit the harness support brackets and the oil filler tube bracket to the manifold studs.
Fit the earth leads to the manifold front stud.

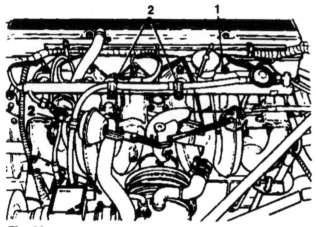

Fig. 22

Fit the nuts, bolts and washers to secure the manifold, tightening in stages to ensure that the gasket is clamped evenly between the manifold faces (1 Fig. 22).
Fit the securing bolt to the dipstick tube.
Fit the oil filler tube and retaining bolt.
Connect the breather hose to the cam cover.
Refit the hoses between the induction elbow, oil filler tube and the orifice control housing.
Connect the electrical connections to the injectors and air sensor block.
Connect the air conditioning hose to the inlet manifold.
Remove the blanking plugs from the control housing hoses and connect the hoses.
Tighten the clips.
Fit the throttle cable bracket to the manifold and tighten the securing bolts.
Connect the throttle cable and the return spring.
Remove the blanking plugs from the fuel rail and the feed hose. Connect the feed hose to the fuel rail.
Fit the three retaining clips to the feed hose (2 Fig. 22).
Remove the blanking plugs from the pressure regulator and feed hose. Connect the hose to the pressure regulator.
Connect the throttle potentiometer block connector.
Connect the idle adjustment stepper motor connector.
Connect the hoses to the auxiliary air valve.
Secure the harnesses to the support mounting brackets using the Ty-wraps.
Connect the air intake hose to the throttle body, align and tighten the hose clip.
Fit the bolt which secures the induction elbow to the left side inner wing.
Start and run the engine to check for leaks.
Stop the engine.
Close the bonnet.

Torque Figures

Manifold nuts	12-16 Nm
Manifold bolts	23-27 Nm
Throttle cable bracket	12-16 Nm
Fuel feed hose union	13-15 Nm

INTAKE TUBE GASKETS, RENEW

Removal

Open the bonnet and fit a wing protection cover to the left hand wing.
Intake Tubes 1 and 2:
Disconnect the harness Ty-wrap from the bracket which is bolted to the intake tubes.

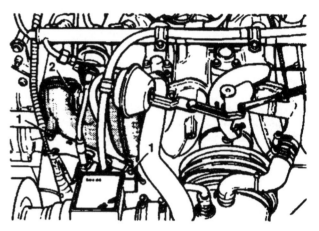

Fig. 23

Remove the two bolts which secure the bracket to the intake tubes (1 Fig. 23).
Slide the bracket down the two fuel pipes (fuel input to gallery and pressure regulator) complete with the connector for the idle speed stepper motor cable.
Remove the four bolts and remove intake tube 1.
Remove the four bolts and remove intake tube 2 (2 Fig. 23).

Intake Tubes 3 to 6:

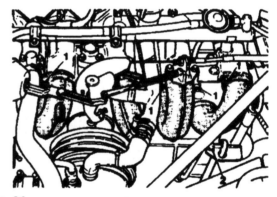

Fig. 24

On each tube, remove the four bolts (1 Fig. 24) and remove the tube. When removing intake tube 6, the dipstick tube is very close and consequently, in some instances, the upper retaining bolt may need to be removed to provide clearance (2 Fig. 24).

Cleaning and Inspection

Clean all gasket faces on the intake tubes and on the inlet manifold.
Check all gasket faces for evidence of cracking; renew as necessary.

Refitting

Intake Tubes 3 to 6:

Apply Hylosil 102 to each gasket face. Position each tube

onto the manifold and secure with the four bolts; tightening in stages to ensure that the faces are clamped evenly.
Fit the dipstick tube upper retaining bolt, if removed whilst dismantling.

Intake Tubes 1 and 2:

Apply Hylosil 102 to each gasket face.

Position each tube onto the manifold and secure with the four bolts, tightening in stages to ensure that the faces are clamped evenly.
Slide the bracket and clips along the fuel pipes and when in position, fit the two bolts securing it to the intake tubes.
Align the harness to the bracket and secure with a Ty-wrap.
Start and run the engine to check for induction leaks.
Remove the wing protection cover and close the bonnet.

Torque Figures

Intake tube bolts	10-12Nm
Bracket retaining bolts	10-13Nm

Oils / Lubricants / Sealants

Intake tube to manifold	HYLOSIL 102

LOW LOSS CATALYTIC CONVERTER SYSTEM - VARIATIONS FOR JAPAN, DESCRIPTION

In terms of removal and refitting of components, there is only one difference in the Japanese version. This is the inclusion of an exhaust gas temperature sensor which is located on the downpipe adjacent to the oxygen sensor.

Remove The Temperature Sensor:

The sensor may be removed as part of another job on the exhaust system or as a job in its own right.
Drive the vehicle onto the ramp and open the bonnet.
Fit a wing protection cover to the right hand wing.
Disconnect the temperature sensor cable connector and remove the cable from the retaining clips.
Raise the ramp.
Remove the temperature sensor from the downpipe.

Refit The Temperature Sensor:

Fit the temperature sensor to the downpipe.
Lower the ramp.
Connect the cable connector of the temperature sensor and fit the cable into the retaining clips.
Remove the wing protection cover and close the bonnet.
For the removal and refitting of other components, refer to the relevant instructions within this section for Remove And Refit Vehicle Set.

NON-CATALYTIC EXHAUST SYSTEM, DESCRIPTION

The non-catalytic exhaust system differs from the low loss catalytic converter exhaust system as follows:

> The downpipe converters are replaced by a conventional twin branch downpipe.
> Consequently, no oxygen sensor or exhaust gas temperature sensors are fitted.
> The underfloor converters are replaced by conventional silencers.
> For the removal and refitting of components, refer to the relevant instructions within this section.

EXHAUST SYSTEM 1993.5 ON, DESCRIPTION

From 1993.5 MY variations to the exhaust system are as follows:

From the front of the system:

> All joints with the exception of the downpipe are replaced with integral 'Torca' clamps. These reduce the possibility of leaks, rattles and knocks.
> Underfloor catalysts have changed and are now closer to the engine. This reduces the warm-up time before the catalysts become fully operational.
> There is an additional mounting toward the rear of the propshaft tunnel. This gives additional support, aids assembly and also helps locate the system square in the vehicle.
> Larger diameter over-axle pipes are routed to give good clearance through the rear suspension cradle.
> Rear silencers: Body to silencer clearance is improved.
> Tailpipe trims now have a square profile to match the bumper aperture.

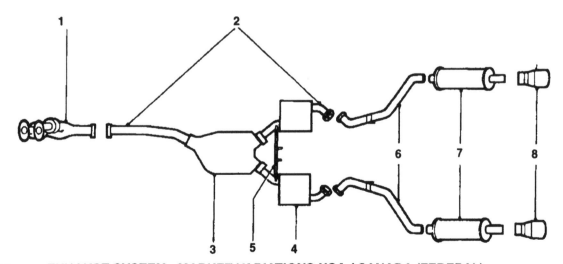

Fig. 25   EXHAUST SYSTEM - MARKET VARIATIONS USA / CANADA (FEDERAL)

Key to Fig. 25

1. *Downpipe assembly*
2. *Intermediate pipe assembly*
3. *Underfloor catalyst*

4. *Intermediate muffler*
5. *Additional mounting point*
6. *Over-axle pipe*
7. *Rear muffler*
8. *Tailpipe trim*

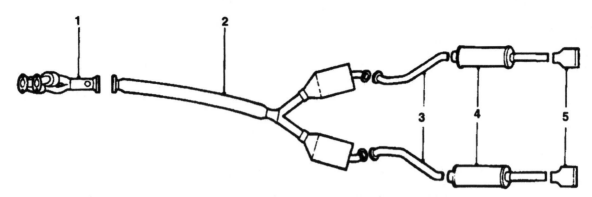

**Fig. 26   EXHAUST SYSTEM - MARKET VARIATIONS USA / CANADA (FEDERAL)
E.E.C. / EUROPEAN R.O.W. (Rest of the World), JAPAN**

Key to Fig. 26

1.   *Downpipe assembly*
2.   *Intermediate pipe assembly*

3.   *Over-axle pipe*
4.   *Rear silencer*
5.   *Tailpipe trim*

Note: The exhaust system for the above markets is similar. The following schematic illustration covers those
markets. E.E.C. / European differs from R.O.W. by
having a different intermediate pipe assembly (catalyst / silencer). Japan differs from R.O.W. by the
addition of grass shields and over-heat sensors to
the downpipe assembly.

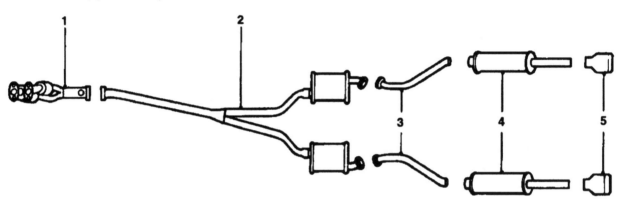

Fig. 27   NON - CATALYST

Key to Fig. 27

1.   *Downpipe assembly*
2.   *Intermediate pipe assembly*

3.   *Over-axle pipe*
4.   *Rear silencer*
5.   *Tailpipe trim*

CLUTCH - 4.0 LITRE

CLUTCH, DESCRIPTION

The clutch fitted to the AJ6 / AJ16 4.0 Litre engined vehicle is a single plate, diaphragm type, and is operated hydraulically by a slave cylinder mounted to the gearbox bell housing. This is in turn operated by a clutch master cylinder through a series of hydraulic pipes and a damper. The piston in the master cylinder is moved by a push rod by the clutch pedal, which is mounted to the bulkhead and is positioned to the left of the brake pedal.

The engine uses a twin-mass flywheel configuration with built-in springs to provide a smooth take-up of drive. The flywheel is extremely heavy and great care must be taken when lifting it from the vehicle.

CAUTION:

The hydraulic fluid used in the clutch hydraulic system is injurious to paintwork.
Utmost precautions MUST at all times be taken to prevent spillage of fluid.
Should fluid be accidentally spilled onto the paintwork, wipe fluid off immediately with a cloth moistened with denatured alcohol (methylated spirits).

CLUTCH ASSEMBLY, RENEW

Remove the rear inlet manifold securing nuts and the injector harness mounting bracket.
Fit lifting eye 18G 1465 to rear inlet manifold studs and tighten securing nuts.

Fit Service Tool MS 53B (Fig. 2) and take the weight of the engine with the hook.
Remove the air filter ram tube.
Remove fan to torquatrol unit.
Disconnect the exhaust system from the front pipe and remove the sealing olive.
Ease the exhaust system down for access and remove heatshield.
Remove the propshaft assembly and fit a blanking plug to the rear of the gearbox.
Remove the exhaust front pipe assembly.
Remove the slave cylinder nuts from the bell housing and remove the push rod.
Disconnect the gearbox switch connectors.

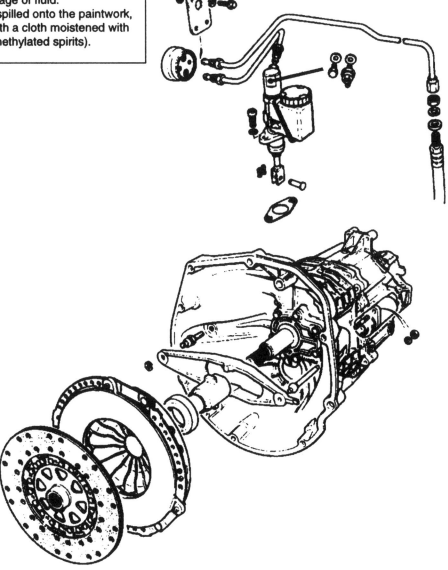

Fig. 1   Clutch Layout

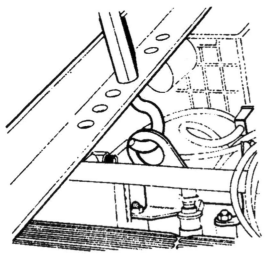

Fig. 2

Using a jack / support channel, take the weight of the gearbox and remove the bolts securing the rear mounting to the body. Lower the jack and remove the rear mounting assembly.
Using MS 53B, undo the tool hook nut to lower the rear of the gearbox.
Disconnect the gear selector shaft universal joint securing nut / bolt. Remove the wavy washer from the selector and remove the gear selector remote control.
Remove the mounting rubbers and washers.

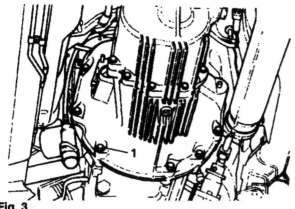

Fig. 3

Remove the gearbox to engine adaptor plate securing bolts (1 Fig. 3).

Note: Leave two opposing bolts in position for safety.

Fit a suitable hoist, securely to the gearbox.
Take the weight of the gearbox and remove the remaining two securing bolts.
Remove the gearbox from the rear of the engine.
Lower the gearbox and remove from the rear of the engine.
Hold the flywheel in one position and remove the bolts securing the clutch cover to the flywheel.

CAUTION:

Make a note of the position of any balance weights relative to the clutch cover.

Remove the balance weights.
Remove the clutch cover / drive plate assembly.

Remove the eight securing bolts (1 Fig. 4) and remove the flywheel (2 Fig. 4).

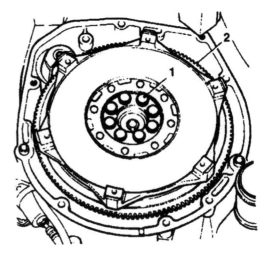

Fig. 4

WARNING:

THE TWIN-MASS FLYWHEEL IS EXTREMELY HEAVY.
ASSISTANCE MAY BE REQUIRED
DURING REMOVAL.

Clean the face of the flywheel and dowels.
Check the flywheel face for scoring, should this be excessive the flywheel may be skimmed by a maximum of 1 mm.

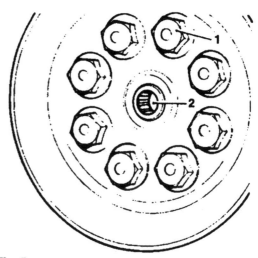

Fig. 5

Fit a new spigot bearing to the flywheel (2 Fig. 5).
Refit the flywheel to the crankshaft and torque tighten the securing bolts (1 Fig. 5).
Fit the clutch assembly to the flywheel ensuring that the larger, rounded boss faces the flywheel.
Align the clutch with an input shaft.
Fit the balance weights to the clutch cover and torque tighten the securing bolts.
Remove the input shaft.
Remove the circlip securing the clutch release arm pivot and remove the assembly.
Slacken and remove the release arm pivot pin.

Remove the bearing (1 Fig. 6) from the release arm assembly (2 Fig. 6), grease the bearing seat, fit the new bearing to the release arm assembly.
Re-assemble the release arm ensuring that the pivots are greased with Molykote FB 180.
Select third forward gear.
Move the gearbox under the ramp and raise it into position.

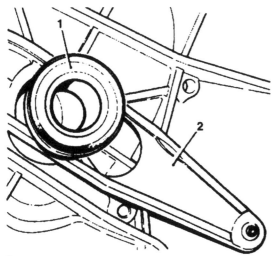

Fig. 6

Reposition the gearbox into the clutch.

Seat against the adaptor plate and secure with the bolts.

Remove the securing chain from the unit lift and fit front clamp.

Refit the gearbox switch and position the multi-plug into the securing clip.

Refit the slave cylinder.

Fit the mounting rubbers and spacers, reconnect the remote control.

Refit the selector to the gear lever.

Lift the gearbox into position, fit and torque tighten the bolts.

Refit the exhaust front pipes to the manifold.

Remove the blanking plug from the rear of the gearbox.

Refit the propeller shaft.

Refit the heat shield and reconnect the exhaust system.

Refit the fan to the torquatrol unit.

Refit the air filter ram tube to the air filter and body.

Lower the ramp. Release and remove the service tool MS 53B.

Remove the engine lifting eye 18G 1465.

Refit the injector harness mounting bracket and tighten the rear inlet manifold securing nuts.

Service Tools

Engine support bracket	MS 53B
Engine lifting bracket	18G 1465

Torque Figures

Bell housing to adaptor plate	49-54 Nm
Clutch cover to flywheel	23-27 Nm
Flywheel to crankshaft	95-105 Nm
Propeller shaft to gearbox	95-105 Nm
Slave cylinder to bell housing	23-27 Nm

Oils / Sealants / Lubricants

Withdrawal arm pivots	Molykote FB 180 grease

MASTER CYLINDER, RENEW

Open the driving side door.

Remove the master cylinder to clutch pedal clevis pin securing clip (1 Fig. 7).

Remove the clevis pin from the clutch pedal (2 Fig. 7).

Open the bonnet.

Remove the master cylinder securing nuts (3 Fig. 7).

Disconnect the hydraulic pipe from the master cylinder.

Fit blanking plugs to the pipe and master cylinder.

Remove the master cylinder assembly (4 Fig. 7) and the

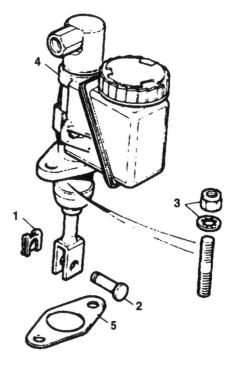

Fig. 7

gasket (5 Fig. 7).

Fit the new master cylinder to the bench vice.

Remove the new reservoir top and fill up the reservoir.

Refit and tighten the reservoir top.

Remove the blanking plug from the master cylinder.

Prime the master cylinder by 'working' the master cylinder until fluid appears at the hydraulic pipe outlet.

Remove the master cylinder from the vice.

Position a new gasket over the studs.

Fit and align the master cylinder assembly to the studs.

Fit and tighten the securing nuts.

Remove the blanking plug from the hydraulic pipe.

Ensure that the union is clean and no foreign matter enters the system.

Reconnect the pipe to the cylinder.

Tighten the union nut.

Align the push rod to the pedal.

Clean and grease the clevis pin.

Fit and align the clevis pin to the pedal and push rod.

Fit the securing clip.

Bleed the clutch hydraulic system.

Close the door.

Close the bonnet.

Torque Figures

Master cylinder to bulkhead	15-17.5 Nm
Hydraulic pipe to master cylinder	20-24 Nm
Bleed nipple to slave cylinder	16-19 Nm

Oils / Greases / Fluids

Castrol/Girling Universal Brake and Clutch Fluid

CLUTCH MASTER CYLINDER, OVERHAUL

> **WARNING:**
>
> USE ONLY CLEAN BRAKE FLUID OR DENATURED ALCOHOL (METHYLATED SPIRITS) FOR CLEANING. ALL TRACES OF CLEANING FLUID MUST BE REMOVED BEFORE REASSEMBLY. ALL COMPONENTS MUST BE LUBRICATED WITH CLEAN BRAKE FLUID AND ASSEMBLED USING THE FINGERS ONLY.

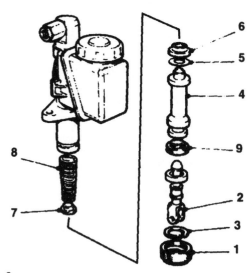

Fig. 8

Dismantling:

Remove the master cylinder.

Detach the rubber boot (1 Fig. 8) from the end of the barrel and move the boot along the push rod (2 Fig. 8).
Press the push rod, remove the circlip (3 Fig. 8).
Withdraw the push rod, piston (4 Fig. 8), piston washer (5 Fig. 8), main cup, (6 Fig. 8) spring retainer (7 Fig. 8) and spring (8 Fig. 8).
Remove the secondary cup (9 Fig. 8) from the piston.

Inspection:

Examine the cylinder bore for scoring.
Thoroughly wash out the reservoir and ensure the by-pass hole in the cylinder bore is clear. Dry using compressed air or a lint-free cloth.
Lubricate the new seals with clean brake fluid.

Reassembling:

If necessary, fit end plug on the new gasket.
Fit the spring retainer to the small end of the spring, bend over the retainer ears to secure.
Insert the spring, large end leading into the cylinder bore, follow with the main cup, lip foremost. Ensure the lip is not damaged on the circlip groove.
Using the fingers only, stretch the secondary cup on the piston with the small end towards the drilled end and groove engaging the ridge.
Gently work round the cup with the fingers to ensure correct bedding.
Insert the piston washer into the bore, curved edge towards main cup.
Insert the piston, drilled end foremost.
Fit rubber boot to the push rod.
Offer the push rod to the piston and press into the bore until the circlip can be fitted behind the push rod stop ring.

CAUTION:

It is important to ensure that the
circlip is fitted
correctly into the groove.
Locate the rubber boot in the groove.

Torque Figures

Master cylinder to housing	22-28 Nm
Hydraulic pipe to master cylinder	16-19 Nm
Bleed nipple to slave cylinder	16-19 Nm

Oils / Greases / Fluids

Jaguar Brake Fluid or Castrol Girling Universal/Disc

CLUTCH SLAVE CYLINDER, RENEW

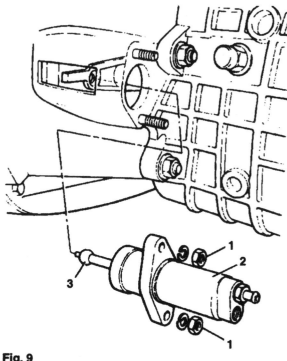

Fig. 9

Disconnect the pipe from the slave cylinder, plug or tape the pipe to prevent the ingress of any dirt.
Remove the nuts (1 Fig. 9) and spring washers securing the slave cylinder to the gearbox.
Slide the slave cylinder (2 Fig. 9) off the mounting studs, slide the rubber boot along the push rod, withdraw the cylinder from the push rod (3 Fig. 9).
To reassemble, reverse the removal operations and bleed the clutch hydraulic system.

Torque Figures

Slave cylinder to bell housing	23-27 Nm
Hydraulic pipe to slave cylinder	16-19 Nm
Bleed nipple to slave cylinder	16-19 Nm

Oils / Greases / Fluids

Jaguar Brake Fluid or Castrol Girling Universal/Disc

SLAVE CYLINDER, OVERHAUL

Remove the slave cylinder.
Dismantle the cylinder.
The new parts in the kit will indicate which used parts should be discarded.
Clean the remaining parts and the cylinder thoroughly with unused brake fluid of the recommended type and place the cleaned parts on to a clean sheet of paper.
Examine the cylinder bore and the pistons for signs of corrosion, ridges or score marks. Provided the working surfaces are in perfect condition, new seals from the kit can be fitted, but if there is any doubt as to the condition of the

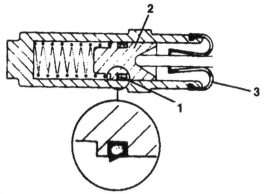

Fig. 10

parts then a new cylinder must be fitted.

Fit the new seal (1 Fig. 10) to the piston (2 Fig. 10) with the flat back of the seal against the shoulder.

Lubricate the seal and the cylinder bore with unused brake fluid of the recommended type and reassemble the cylinder.

Before fitting the dust cover (2 Fig. 10), smear the sealing areas with rubber grease.

Squeeze the remainder of the grease from the sachet into the cover to help protect the internal parts.

Refit the slave cylinder to the vehicle and bleed the system.

Torque Figures

Slave cylinder to bell housing	23-27 Nm
Hydraulic pipe to slave cylinder	34-40 Nm
Bleed nipple to slave cylinder	16-19 Nm

Oils / Greases / Fluids

Jaguar Brake Fluid or Castrol Girling Universal/Disc

CLUTCH HYDRAULIC PIPE DAMPER, RENEW

Disconnect the hydraulic pipes (1 Fig. 11).

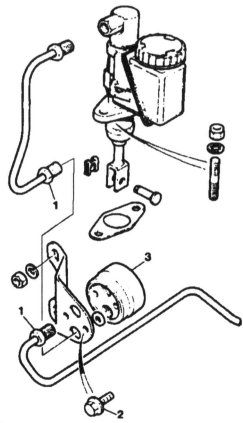

Fig. 11

Remove the securing bolts (2 Fig. 11) and remove the damper (3 Fig. 11). Clean the pipe unions, ensure that you do not contaminate the hydraulic system.

Refit the new damper and secure with the bolts, reconnect the hydraulic pipes and bleed the system.

Torque Figures

Bleed nipple to slave cylinder	16-19 Nm
Clutch damper to mounting bracket	7-10 Nm
Hydraulic pipes to clutch damper	16-19 Nm

HYDRAULIC SYSTEM, BLEED

Remove reservoir filler cap. Top up the reservoir to the correct level with hydraulic fluid.

Attach one end of a bleed tube (1 Fig. 12) to the slave cylinder bleed nipple (2 Fig. 12).

Partially fill a clean container (3 Fig. 12) with hydraulic

CAUTION:

Only use Castrol-Girling brake fluid in the hydraulic system.

---

fluid. Immerse the other end of the bleed tube in the fluid.

Slacken the slave cylinder bleed nipple.

Pump the clutch pedal slowly up and down, pausing between each stroke.

Top up the reservoir with fresh hydraulic fluid after every

CAUTION:

Do not use fluid bled from the system for topping up purposes as this will contain air.
If the fluid has been in use for some time it should be discarded. Fresh fluid bled from the system may be used after it has stood for a few hours allowing all the air bubbles to disperse.

---

three pedal strokes.

Pump the clutch pedal until the pedal becomes firm, tighten the bleed nipple to 16 to 19 Nm.

Top up the reservoir, refit the filler cap.

Apply the working pressure to the clutch pedal for two to three minutes and examine the system for leaks.

Oils / Greases / Fluids

Clutch Master Cylinder Castrol Girling Universal / Disc

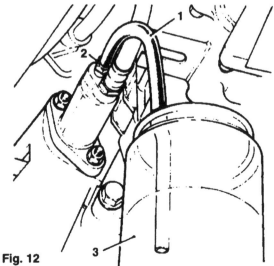

Fig. 12

Chapter 8

MANUAL GEARBOX - 4.0 LITRE

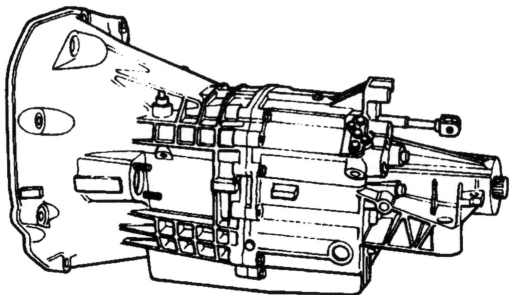

Fig. 1

GEARBOX, DESCRIPTION

The Getrag 290 5-speed transmission fitted to this vehicle incorporates synchromesh on all gears.
Externally it is most easily recognised by the unified bell-housing/gear case construction. Earlier versions have a removable bell-housing.

Gear selection is by a centrally mounted lever, connected to the transmission selector shaft via a pivoting joint.
All the gears are engaged by a single selector shaft operating three rods which move the selector forks.
The drive pinion is supported at the rear by a duplex ball bearing situated in the front casing and at the front, a spigot engages in a needle roller bearing in the flywheel.
The mainshaft is supported at the front by a caged roller bearing situated in the drive pinion counterbore: in the centre by a roller bearing supported by the intermediate casing: at the rear by a duplex bearing in the transmission rear casing.
Each of the forward speed mainshaft gears incorporates an integral synchromesh mechanism, with the clutch hubs splined to the mainshaft and situated between each pair of gears.
The countershaft is supported at the front by a roller bearing situated in the front casing: in the centre by a roller bearing in the intermediate casing: at the rear by a roller bearing situated in the tail housing.
The reverse idler gear is supported by two caged roller bearings, is in constant mesh and is situated on a stationary shaft.
Longitudinal location of the idler gear is controlled by a spacer abutting on the shaft.

GEAR RATIOS

1	3.55 :1
2	2.04 :1
3	1.40 :1
4	1.00 :1
5	0.76 :1
R	3.55 :1

GEARBOX ASSEMBLY, RENEW

Disconnect the battery.
Undo and remove inlet manifold rear securing nuts.
Remove injector harness mounting bracket.
Fit dummy lifting eye.
Fit and tighten lifting eye securing nuts.

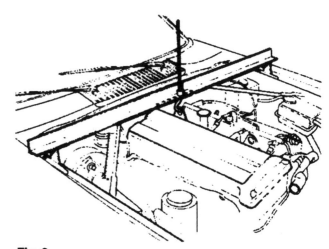

Fig. 2

Fit Service Tool MS 53B (Fig. 2) across the wing channels and align to rear lifting eye.
Fit and engage retaining hook.
Tighten hook nut to take weight of engine.
Displace and remove air filter ram tube.
Remove bolts securing fan to torquatrol unit.
Displace fan rearwards.
Disconnect lambda sensor block connectors.
Raise the vehicle on a ramp.
Disconnect exhaust system from front pipe.
Remove sealing olive.
Carefully ease exhaust system down for access.
Displace and remove heatshield.
Undo and remove propshaft to differential flange securing nuts / bolts.

Push vehicle back / forward to position flange for access to propshaft securing nuts / bolts.
Displace and remove propshaft assembly.
Fit blanking plug to rear of gearbox.
Displace and remove front pipe assembly.
Displace slave cylinder from bell housing and secure clear.
Remove slave cylinder push rod.
Displace gearbox switch block connector from securing clip.
Reposition and disconnect gearbox switch block connector.
Place jack channel to position.
Take weight of mounting on jack.
Undo and remove rear mounting to body securing bolts.
Lower jack and remove rear mounting assembly.
Remove jack and jacking channel.
Remove mounting spring and spring mounting rubber.
From above: Undo retaining tool hook nut to lower rear of gearbox (do not allow engine to foul rack).

Note: During the above operation ensure fan does not foul the fan cowl.

From below: Undo and remove selector shaft U/J securing nut / bolt.
Reposition selector from lower gear lever and remove wavy washer from selector.
Undo and remove gear selector remote control securing

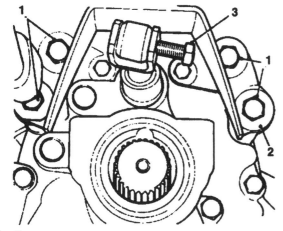

Fig. 3

bolts (1 Fig. 3).
Remove mounting rubbers and washers (2 Fig. 3).
Reposition remote control assembly for access.
Undo and remove gearbox to engine adaptor plate securing bolts (1 Fig. 4).

Note: Leave two opposing bolts in for safety.

Remove front clamp from unit lift.
Raise / lower unit with jack, no stands.

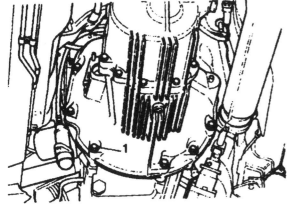

Fig. 4

Traverse lift under ramp and take weight of gearbox.
Adjust tilt angles to suit gearbox.
Adjust side and rear clamps to suit gearbox.
Tighten clamp wing nuts.
Fit safety chain assembly to left hand arm of lift.
Fit securing peg.
Pass safety chain over gearbox and engage in front arm of lift.
Tighten safety chain adjuster.
Undo and remove remaining gearbox to adaptor plate securing bolts.
Displace gearbox from engine and carefully lower unit lift.
Remove gearbox from ramp area.
Displace and remove clutch release bearing assembly from release lever.
Displace and remove clutch release lever retaining clip.
Displace and remove lever.
Undo and remove release lever pivot pin.
Undo and remove gearbox switch.
Undo and remove spring retainer securing nut.
Remove rear mounting spring retainer.
Undo and remove gearbox mounting pin.
Reposition selector shaft pin cover.
Displace and remove selector shaft to U/J pin.
Remove U/J body.
Undo and remove slave cylinder securing studs.
Slacken chain adaptor.
Release chain from front arm.
Slacken side wing nuts.
Reposition clamps aside.
Place gearbox to drain area.
Undo and remove gearbox drain plug and allow to drain.
Replace and tighten drain plug.
Fit new gearbox to unit lift.
Reposition clamps to side of gearbox.
Tighten clamp wing nuts.
Reposition chain to front arm.
Tighten chain adjuster.
Fill gearbox with oil and replace level plug.
Clean components and mating faces.
Fit and tighten slave cylinder securing nuts.
Lubricate selector output shaft.
Fit U/J assembly to selector shaft.
Fit and seat U/J pin.
Reposition retaining cover over U/J pin.
Fit and tighten gearbox mounting pin.
Fit and tighten spring retainer securing nut.
Fit and tighten gearbox switch.
Fit and tighten release lever pivot pin.
Lubricate pivot pin.
Lubricate clutch release lever.
Fit and align lever to gearbox, engage onto pivot pin.
Fit and fully seat lever to pivot retaining clip.
Lubricate release bearing housing.
Fit and fully seat bearing assembly to lever.
Select third forward gear.
Move gearbox under ramp and raise gearbox into position.
Raise / lower unit with jack, no stands.
Reposition gearbox into clutch and fully seat against adaptor plate.
During above operation adjust tilt angles.
Fit and tighten gearbox to adaptor plate securing bolts.
Slacken chain adjuster.
Release securing peg from adjuster.
Displace securing chain from unit lift.
Slacken clamp from wing nuts.
Reposition clamps from gearbox.
Lower lift and traverse aside.
Fit front clamp to unit lift.
Connect gearbox switch and position multi plug into securing clip.

Clean clutch slave cylinder push rod.
Lubricate push rod.
Fit push rod to slave cylinder.
Reposition slave cylinder to mounting studs.
Fit and tighten slave cylinder to mounting studs.
Fit and tighten slave cylinder securing nuts.
Fit and seat remote control mounting rubbers and spacers.
Align remote control to gearbox.
Fit mounting rubber backing washers.
Fit and tighten remote control mounting securing bolts.
Fit and align wavy washer to lower gear lever.
Position selector shaft to lower gear lever.
Apply lubricant to selector / lever assembly.
Fit and tighten selector to gear lever securing nut / bolt.
From above: Tighten MS 53B hook nut to raise gearbox into position.
From below: Place jacking channel to position.
Fit spring to rear mounting assembly.
Fit upper rubber to mounting spring.
Raise / lower unit with jack, no stands.
Using a jack, fit and seat mounting assembly to body / gearbox.
Fit but do not fully tighten mounting securing bolts.
Lower and remove jack.
Remove jack channel.
From above: Fully undo MS 53B hook nut.
From below: Final align mounting to gearbox / body.
Final tighten mounting bolts.
Place front pipe to vice and tighten vice.
Remove and discard front pipe to manifold sealing rings.
Clean faces.
Fit and fully seat new rings to pipe.
Fit and align front pipe to manifold.
Align retaining rings to studs.
Fit and tighten securing nuts.
Remove blanking plug from rear of gearbox.
Lubricate spline and oil seal.
Clean propshaft and mating faces.
Fit and align propshaft to gearbox and differential flange.
Fit and tighten nuts / bolts securing propshaft to differential flange.
Push vehicle back / forward to reposition flange for access to propshaft securing nuts / bolts.
Fit rear heatshield.
Clean exhaust sealing olive.
Smear sealing olive with sealant.
Fit sealing olive to intermediate pipe.
Connect exhaust system to front pipe.
Fit and tighten clamp securing nut / bolt.
Reposition lambda sensor harness to allow connection from above.
Lower vehicle on ramp.
Reconnect lambda sensor block connectors.
Reposition fan to mounting.
Fit fan to torquatrol unit.
Fit and align air filter ram tube.
Remove MS 53B hook and retaining tool.
Undo and remove dummy lifting eye securing nuts.
Remove dummy lifting eye.
Fit injector harness mounting bracket.
Fit and tighten manifold securing nuts.
Secure injector harness to mounting bracket.

Service Tools

Engine lifting brackets	18G 1465.
Engine support bracket	MS 53B.

Torque Figures

Bell housing to cylinder block	49-54 Nm
Front pipe to intermediate pipe	15-18 Nm
Front pipe to manifold	30-35 Nm
Gear lever housing to gearbox	23-27 Nm

Oils / Lubricants / Sealants

Exhaust sealer	TIVOLI KAY ADHESIVES No 5696

Gearbox oil:

Automatic Transmission Fluid	DEXRON IID

FRONT OIL SEAL, RENEW

Drive the vehicle on to a ramp.
Open the bonnet.
Raise the vehicle on the ramp.
Remove the gearbox assembly.
Place the gearbox on a bench.
Place the drain tray in position.
Undo and remove the front securing bolts (1 Fig. 5).
Displace and remove the front cover.
Note and remove the shims.
Fit the assembly to a vice.

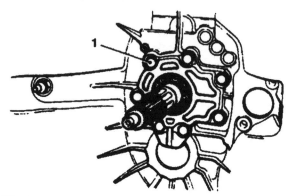

Fig. 5

Displace and remove the oil seal from the assembly.
Clean the front cover and gasket faces.
Clean the shims.
Lubricate the seal face.
Fit and seat the seal to the assembly.
Fit the shims to the cover.
Apply sealant to the front cover.
Lubricate the seal lip.
Fit and seat the front cover to the gearbox.
Fit and tighten the cover securing bolts.
Undo and remove the gearbox filler plug.
Fill the gearbox to the correct level.
Fit and tighten the filler plug.
Remove the drain tray.
Fit the gearbox using a unit lift.
Refit the gearbox assembly.
Lower the vehicle on the ramp.

Service Tools

Engine lifting brackets	18G 1465.
Engine support bracket	MS 53B.

Torque Figures

Front cover to gearbox	23- 27 Nm
Front pipe to intermediate pipe	15-18 Nm

Oils / Lubricants / Sealants

Exhaust sealer	TIVOLI KAY ADHESIVES No 5696
Front cover to gearbox	LOCTITE 573
Front cover securing bolts	LOCTITE 573
Gearbox oil Automatic Transmission Fluid	DEXRON IID

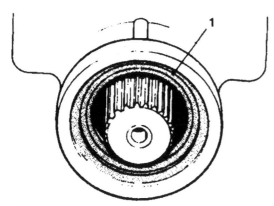

Fig. 6

REAR OIL SEAL, RENEW

Raise the vehicle on a ramp.
Disconnect the propeller shaft, noting the propeller shaft / differential flange relationship.

Using a 'Snap On' oil seal remover, displace and remove the rear oil seal (1 Fig. 6).
Clean the seal mounting face.
Apply the sealant to the seal outer face.
Fit and seat the seal to the gearbox.
Refit the propshaft.
Lower the vehicle onto the ramp.

Service Tools

'SNAP ON' Oil seal remover.

Oils / Lubricants / Sealants

Gearbox Oil Automatic Transmission Fluid   DEXRON IID

GEAR LEVER KNOB, RENEW

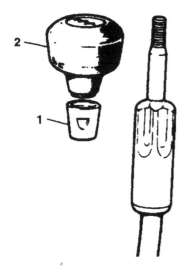

Fig. 7

Open the door.
Displace the gear lever gaiter for access and slacken the gear lever knob lock nut (1 Fig. 7).
Unscrew the gear lever knob (2 Fig. 7).
Remove the lock nut.
Refit but do not tighten the lock nut.
Fit the new gear lever knob.
Rotate the knob to its final position.
Tighten the lock nut.
Reposition the gear lever gaiter.

GEAR LEVER DRAUGHT EXCLUDER, RENEW

Open the door.
Displace the gear lever gaiter for access and slacken the gear lever knob lock nut (1 Fig. 7).

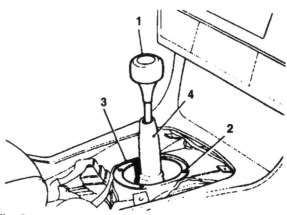

Fig. 8

Unscrew the gear lever knob (2 Fig. 8).
Remove the lock nut.
Open the centre console storage compartment.
Remove the centre console securing screws.
Displace the centre console, disconnect the block connectors and remove the console.
Remove the foam sealing ring.
Remove the draught excluder securing screws (2 Fig. 8) and ring (3 Fig. 8).
Remove the draught excluder (4 Fig. 8).
Fit the new draught excluder over the gear lever.
Fit the securing ring and screws.
Refit the foam sealing ring.
Refit the centre console.
Reconnect the block connectors.
Fit and tighten the console securing screws.
Close the centre console storage compartment.
Fit the gear lever knob / lock nut.
Rotate the knob to its final position.
Tighten the lock nut.

GEAR LEVER / REMOTE CONTROL ASSEMBLY, OVERHAUL

Drive the vehicle on to a ramp.
Open the bonnet.

Fit Service Tool MS 53B (Fig. 9) across the wing channels.
Align the hook to the rear lifting eye and take the weight of the engine.
Undo and remove fan to torquatrol unit securing bolts.
Displace fan rearwards.
Select third gear. Remove the gear lever knob.

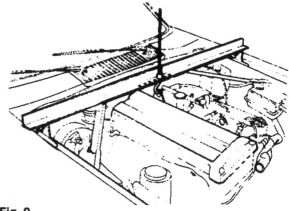

Fig. 9

Raise the ramp.

Disconnect exhaust system from front pipe.

Remove sealing olive.

Take the weight of the mounting using a jack / jack channel and suitable block.

Remove the rear mounting securing bolts.

Lower and remove the mounting assembly.

Remove the rear heat shield securing screws.

Carefully ease the exhaust system down for access and remove the heat shield.

Undo and remove propshaft to differential flange securing nuts / bolts.

Push vehicle back / forward to position flange for access to propshaft securing nuts / bolts.

Displace and remove propshaft assembly.

Slacken off MS 53B hook nut, lowering the rear of the gearbox.

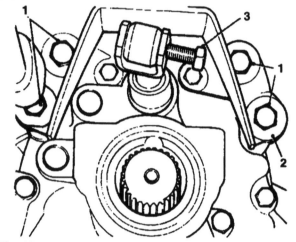

Fig. 10

Slacken off but do not remove the gear lever to rear shaft securing bolt (3 Fig. 10).

Remove the control mounting securing bolts (1 Fig. 10).

Reposition the remote control assembly for access.

Finally remove the gear lever to rear shaft securing bolt (3 Fig. 10).

Disconnect the gear lever from the yoke.

Remove the remote control / gear lever assembly.

Note: To aid removal, invert the remote control / gear lever assembly, i.e. gear lever pointing downwards.

Ensure that the draught excluder is not displaced from its position.

Remove the mounting rubbers and washers.

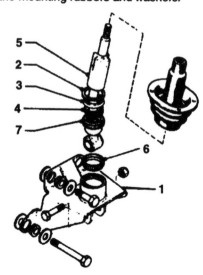

Fig. 11

Fit the remote assembly (1 Fig. 11) to a vice.

Remove the snap ring (2 Fig. 11).

Remove the gear lever (5 Fig. 11).

Remove the collar (3 Fig. 11) and spring (4 Fig. 11) from the gear lever.

Displace and remove the gear lever lower nylon cup (6 Fig. 11) and upper nylon spacer (7 Fig. 11).

Remove the remote control from the vice.

Clean all components thoroughly and examine all component parts for wear and damage. Replace worn or damaged components as necessary.

Ensure mating faces are clean and grease-free.

Fit the remote control to the vice.

Fit and seat the lower nylon cup.

Grease the gear lever ball.

Fit the spring and collar to the lever.

Fit and seat the gear lever assembly.

Compress the spring. Fit and seat the snap ring.

Remove the assembly from the vice.

Lubricate the gear lever to ease fitment through the draught excluder.

Lubricate the lower spacer.

Position the assembly to the mounting location, enter the lever into the gaiter.

Reposition the lever into the yoke.

Fit but do not tighten the securing bolt (3 Fig. 10).

Fit the lower LH mounting rubber and washer.

Fit the remaining mounting rubbers / washers (2 Fig. 10) and securing bolts (1 Fig. 10).

Give final tightening to the securing bolts.

Give final tightening to the gear lever yoke securing bolt (3 Fig. 10).

Tighten MS 53B hook nut to raise the gearbox into position.

Using a jack, fit and seat the rear mounting to the gearbox / body, ensuring that the spring is seated correctly in the spring pan.

Fit but do not fully tighten the mounting securing bolts.

Remove the jack / channel.

Give final tightening to the mounting securing bolts.

Slacken off MS 53B hook nut.

Refit the propeller shaft.

Refit the rear heat shield.

Refit the exhaust.

Fit and tighten bolts securing fan to torquatrol unit.

Lower the ramp. Remove Service Tool MS 53B.

Close the bonnet.

Refit the gear lever knob.

Service Tools

Engine support tool	MS 53B

Torque Figures

Gear lever pivot bolt	23-27 Nm
Remote control to gearbox	23-27 Nm

Oils / Lubricants / Sealants

Exhaust sealer	TIVOLI KAY ADHESIVES No 5696

GEAR SELECTOR SHAFT - REAR, OVERHAUL

Drive the vehicle on to a ramp.

Open the bonnet.

Fit Service Tool MS 53B across the wing channels (Fig. 9).

Align the hook to the rear lifting eye and take the weight of the engine.

Undo and remove fan to torquatrol unit securing bolts.

Displace fan rearwards.

Select third gear. Remove the gear lever knob.

Raise the ramp.

Disconnect exhaust system from front pipe.

Remove sealing olive.

Take the weight of the mounting using a jack / jack channel and suitable block.

Remove the rear mounting securing bolts.

Lower and remove the mounting assembly.

Remove the rear heat shield securing screws.

Carefully ease the exhaust system down for access and remove the heat shield.

Undo and remove propshaft to differential flange securing nuts / bolts.

Push vehicle back / forward to position flange for access to propshaft securing nuts / bolts.

Displace and remove propshaft assembly.

Slacken off MS 53B hook nut, lowering the rear of the gearbox.

Slacken off, but do not remove the gear lever to rear shaft securing bolt (3 Fig. 10).

Remove the control mounting securing bolts (1 Fig. 10).

Reposition the remote control assembly for access.

Finally remove the gear lever to rear shaft securing bolt.

Disconnect the gear lever from the yoke.

Remove the remote control / gear lever assembly.

Note: To aid removal, invert the remote control/gear lever assembly, i.e. gear lever pointing downwards.

Ensure that the draught excluder is not displaced from its position.

Remove the mounting rubbers and washers.

Displace the gear selector shaft universal joint cover (1 Fig. 12).

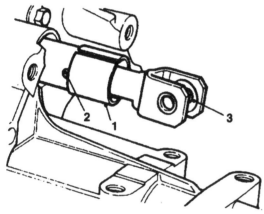

Fig. 12

Remove the selector shaft retaining pin (2 Fig. 12), and remove the shaft (3 Fig. 12).

Remove the universal joint retaining pin and remove the joint.

Clean all components thoroughly and examine all component parts for wear and damage.

Replace worn or damaged components as necessary.

Lubricate the universal joint.

Fit the joint to the selector shaft. Fit the retaining pin.

Fit the assembly to the gearbox. Fit the retaining pin.

Reposition the universal joint cover.

Lubricate the gear lever to ease fitment through the draught excluder.

Lubricate the lower spacer.

Position the remote control assembly to the mounting location, enter the lever into the gaiter.

Reposition the lever into the yoke.

Fit but do not tighten the securing bolt.

Fit the lower LH mounting rubber and washer.

Fit but do not tighten the securing bolt.

Fit the remaining mounting rubbers / washers and securing bolts.

Give final tightening to the securing bolts.

Give final tightening to the gear lever securing bolt.

Tighten MS 53B hook nut to raise the gearbox into position.

Using a jack, fit and seat the rear mounting to the gearbox / body, ensuring that the spring is seated correctly in the spring pan.

Fit but do not fully tighten the mounting securing bolts.

Remove the jack / channel.

Give final tightening to the mounting securing bolts.

Slacken off MS 53B hook nut.

Fit and seat the rear mounting centre spacers.

Refit the intermediate heat shield.

Refit the propeller shaft.

Refit the rear heat shield.

Refit the exhaust.

Lower the ramp.

Remove Service Tool MS 53B.

Close the bonnet.

Refit the gear lever knob.

Service Tools

Engine support tool   MS 53B

Torque Figures

Gear lever pivot bolt   23-27 Nm
Remote control to gearbox   23-27 Nm

Oils / Greases / Fluids

Gear lever ball   MOLYKOTE FB 180 grease.

REVERSE LAMP SWITCH, RENEW

Raise the vehicle on a ramp.

Note and disconnect reverse light switch wires.

Slacken off and remove the switch.

Clean the mating face.

Fit and tighten a new switch.

Reconnect the feed wires.

Lower the ramp.

LAYSHAFT FRONT SEAL, RENEW

Raise the vehicle on a ramp.

Remove gearbox assembly.

Remove front oil seal.

Using 'Snap On' oil seal remover tool remove and discard layshaft plug seal.

Clean gearbox face.

Place new seal to front.

Apply suitable oil sealant to seal face.

Fit and fully seat seal to casing.

Refit gearbox front seal.

Refit gearbox.

Lower ramp.

Chapter 9

AUTOMATIC TRANSMISSION - 4.0 LITRE

AUTOMATIC TRANSMISSION, DESCRIPTION

The ZF 4-speed automatic transmission utilizes a micro-processor equipped Transmission Control Unit to control system functions by means of solenoid actuated valves.

TRANSMISSION CONTROL UNIT

The TCU continuously monitors the gear selected, the speed of the output shaft and the throttle position. This information plus input from the Engine Control Unit of engine speed and load in conjunction with a pre programmed control map, enables the most suitable gear for the prevailing conditions to be selected.

The TCU controls the gear shift speed with 3 solenoid valves and gear shift quality with a solenoid operated pressure control valve. Information is fed to the TCU from sensors and if any electronic component fails, the basic functions will be performed by the hydro-mechanical system i.e. Park, Reverse, Neutral, D3 or D4. The TCU is located behind the right hand side rear quarter trim panel and is supplied with battery voltage when the ignition is switched on; there is also a separate uninterrupted battery voltage supply for the diagnostic memory. Gear change quality on upshifts is improved by the TCU retarding the ignition momentarily to reduce the torque input as the gear change takes place.

ROTARY TRANSMISSION SWITCH

The transmission control unit is informed of selector position by the rotary transmission switch which is attached to the gearbox side by a bracket and operated by a square extension of the selector shaft.

SPORT MODE SWITCH

The Sport Mode Switch is located on the gear change lever surround and provides two alternative shift speed patterns.

'Normal Mode' - designed for everyday use. Torque converter lock up occurs in fourth gear.
'Sport Mode' - gear changes take place at higher road speeds in order to enhance performance.

Torque converter lock up occurs in fourth gear.

SPEED SENSOR

An electro-magnetic sensor detects output shaft revolutions by means of a toothed wheel attached to the shaft.

SOLENOID VALVES

Gear selection and gear shift speeds are controlled by the manually operated selector valve, a solenoid operated pressure regulator and three solenoid valves. On receipt of signals from the TCU, the three solenoid valves SV1, SV2 and SV3, in various combinations with the selector valve determine the appropriate gear range. The TCU on receipt of information of engine state and road speed determines the shift speed. The pressure regulator is factory calibrated and is not adjustable in service.

Note: If reverse is selected when the vehicle is travelling at more than 5 m.p.h., solenoid valve SV2 will be energised to prevent engagement of reverse gear.

KICK-DOWN SWITCH

The kick down switch is located beneath the accelerator pedal. The switch, actuated by the final travel of the pedal, signals to the TCU that a downward change is required. The switch is adjustable for height in order to obtain the precise kickdown speed.

GENERAL DATA

Fluid capacity	10.2 litres/18 pints
Fluid type	Dexron II D
Weight (inc. fluid & converter)	82.1 kg. (181 lb.) Transmission ratios
First	2.48 : 1
Second	1.48 : 1
Third	1.00 : 1
Fourth	0.73 : 1
Reverse	2.09 : 1
Torque converter	
Nominal size (dia.)	280 mm
Torque	322 Nm at 2000 rev/min
Stall speed	2000 rev/min (+150 rev/min)
Ratio	2.12:1 @ Stall
Gear-train end-float	0.2 mm.- 0.4 mm. (0.008 in.- 0.016 in.)

ELECTRICAL DATA

Solenoid valves SV1 / SV2 / SV3	28 - 60 ohms (33.5 ohms @ 20ºC)
Pressure regulator solenoid valve	5 - 7 ohms
Output shaft sensor	300 ohms (+ 10%)
Sensor air gap	0.5 - 1 mm (0.020 in.-0.040 in.)
Throttle potentiometer	
Track resistance	5K ohms
Idle	0.6 volt
90% throttle	4.1 (+1) volts
100% throttle (kickdown)	4.5 (+1) volts

OPERATION

'P' Park   Transmission is mechanically locked.

CAUTION:
Select only when vehicle is stationary.

'R' Reverse   Reverse gear is engaged, no upshift is possible.

'N' Neutral   Engine is disconnected from driveline.

'D' Drive   Transmission will upshift and downshift automatically between the four ratios.

                Note: Positions 1 and 2 can be selected at any road speed. Safety valves prevent premature downshifts and over speeding of the engine.

'3'   Transmission will upshift to 3rd but not engage fourth and can upshift and downshift automatically in the lower three ratios.

'2'   Transmission will upshift to 2nd but will not engage 3rd and will upshift and downshift automatically in the lower two ratios.

'1'   Transmission is fixed in 1st gear and will not upshift to 2nd gear. Maximum engine braking is available.

Kickdown   Depress the accelerator pedal beyond the normal full throttle position. The transmission will then downshift to a lower ratio.

'Sport' Mode   Sport mode, engaged by actuation of the mode switch, ensures all upshifts and downshifts are at higher road and engine speeds.

FAULT DIAGNOSIS

The following tables are intended as a guide to diagnosis of possible faults in the transmission unit.

Where the fault concerns a leak it is recommended that it is located by the use of a crack detection fluid such as for example Met L Chek. This product is available on the market in spray form and permits the leak to be located after a short test drive.

INITIAL CHECKS

Note: Before road testing ensure the following settings are checked.

Kickdown switch adjustment: Refer to adjustment procedure.

Engine tune

Transmission oil level

    Ensure the vehicle is on level ground.
Run the engine.
Withdraw the dipstick and clean with a lint free cloth.
Replace the dipstick and withdraw it noting the oil level.
Stop the engine.
Top up as required.
Run the engine and recheck the level when the transmission is hot.

Selector cable adjustment

    Check gear selection in all selector positions.
If in doubt select N, detach the cable at the gearbox selector lever, check the gearbox lever is in N position (third detent from the rear) and refit the selector cable ball pin to the lever.
Adjust as necessary.

Stall test

CAUTION:
This test must not last more than 5 seconds. Always allow the engine to idle for at least two minutes between tests to allow the transmission fluid to cool down. Do not carry out more than three tests in any half hour.

    Ensure the transmission is at normal running temperature.
Fully apply the handbrake.
Start the engine.
Fully depress the footbrake.
Select 'D' drive.
Fully depress the accelerator (kickdown switch fully depressed).
Note the tachometer reading.
Compare the tachometer reading to the specification.

Road test

    Observe transmission general behaviour, noises, leaks etc. and consult the following Fault Finding Chart.

FAULT FINDING CHART

Note: The following abbreviations are used in this chart.

TCU	Transmission Control Unit
SV1	Solenoid Valve 1
SV2	Solenoid Valve 2
SV3	Solenoid Valve 3
PRSV Valve	Pressure Regulating Solenoid
RS	Rotary Switch
SS	Speed Sensor

PROBLEM	POSSIBLE CAUSE	REMEDY
Converter		
Stall speed too low	Stator freewheel faulty allowing stator to revolve	Renew torque converter
Stall speed too high	Transmission slip	Check oil level Check mechanical failure Renew gear unit
Acceleration below specification	Stator freewheel faulty allowing stator to revolve	Renew torque converter
Top speed below specification	Stator freewheel seized	Renew torque converter
'P' Park		
Does not engage	Selector cable adjustment. Parking pawl mechanism sticking	Rectify
Does not hold (ratcheting)	Selector cable adjustment. Parking pawl mechanism damaged	Rectify
Engine does not start	RS adjustment. RS faulty	Rectify
Reverse		
Reverse gear inoperative	Selector cable adjustment	Rectify
	SV2 faulty	Renew SV2
	Wire to SV2 earthed	Renew harness
	Oil filter dirty	Clean / replace filter
	Damper B faulty	Renew valve block
	Clutch B faulty (also no 3rd gear)	Renew gear unit
	Brake D faulty (also no engine braking in 1st gear, '1' selected)	Renew gear unit
	Clutch E faulty (also no engine braking in 2nd and 3rd gears plus no engine braking in 1st gear '1' selected)	Renew gear unit
	Reverse gear safety device remains activated	Check TCU Renew valve block
No reverse or forward gear	Main pressure control valve seized	Renew valve block
Slipping and juddering	Clutch B, Clutch E or Brake D faulty	Renew gear unit
Harsh engagement P to R change, N to R change	Damper B faulty (also no 2 - 3 shift)	Renew valve block
	Modulation pressure too high	Check TCU
Distinct harsh double engagement P to R or N to R below 1500 rev/min	Clutch F sealing rings faulty	Renew gear unit
Reverse light does not illuminate	Rotary Switch setting Rotary Switch faulty	Rectify Renew
'N' Neutral		
Starter will not operate	Rotary Switch setting Rotary Switch faulty	Rectify Renew
Vehicle moves	Selector cable adjustment Clutch A seized	Renew Rectify
'D' Drive		
Drive not transmitted (neutral condition)	Selector cable adjustment Oil filter dirty Main pressure control valve seized Clutch A faulty	Rectify Clean / replace filter Renew valve block Renew gear unit
	1st gear freewheel slips	Renew gear unit

Slipping and juddering at start	Clutch A faulty	Renew gear unit
Harsh engagement N to D change under 1500 rev/min	Clutch A faulty Damper A faulty	Renew gear unit
No upshifts or downshifts (hot or cold)	SS faulty	Renew SS
1-2 upshift and 2-1 downshift at kickdown speeds	Kickdown switch faulty	Renew kickdown switch
No 1-2 upshift or 2-1 downshift	SV1 wire earthed (Remains in 2)	Renew harness
	SV1 faulty	Renew SV1
	Shift valve 1-2 seized	Renew valve block
	Pressure reducing valve 1 seized	Renew valve block
No upshift 1-2	Brakes C1 and C	Renew gear unit
No upshift 2-3 and downshift 3-2	SV2 wire earthed (remains in 2nd gear)	Renew harness
	SV2 faulty	Renew SV2
No upshift 2-3	Shift valve 2-3 seized	Renew valve block
	Clutch B faulty	Renew gear unit
No upshift 3-4 and downshift 4-3	Shift valve 3-4 seized	Renew valve block
	SV1 wire earthed	Renew harness
	SV1 faulty	Renew SV1
No upshift 3-4	Shift valve 3-4 seized	Renew valve block
	Brake F faulty	Renew gear unit
	SS faulty	Renew SS
Downshift 4-3 too hard	Drain orifice F partially blocked	Clean / renew valve block
Manual gearshift D4-3 too hard	Damper E faulty	Renew valve block
	Drain orifice F partially blocked	Clean valve block
Manual gearshift D3-2 too hard	Damper C1 faulty	Renew valve block
	Drain orifice C1 partially blocked	Clean valve block
No 1st gear; 2nd gear start only	SS faulty	Renew SS
	SV1 faulty	Renew SV1
	SV1 wire earthed	Renew harness
	Shift valve 1-2 seized	Renew valve block
No 1st or 2nd gear; 3rd gear start only	SS faulty	Renew SS
	SV1 or SV2 faulty	Renew SV1 or SV2
	Shift valves 1-2 and 2-3 seized	Renew valve block
No 2nd gear; transmission shifts 1st to 3rd gear.	SS faulty	Renew SS
	Shift valve 2-3 seized	Renew valve block
No load gear change speeds not to specification	TCU faulty	Check TCU
Full load gear change speeds not to specification	TCU faulty	Check TCU
No kickdown change	Kickdown switch faulty	Renew kickdown switch
No-load change speeds only	TCU faulty	Check TCU
Kickdown change speeds only	Kickdown switch faulty	Renew kickdown switch
No load gear change too hard	TCU faulty	Check TCU
	Modulation pressure too high	Renew valve block
	Damper faulty	Renew valve block
	Clutch plates faulty	Renew gear unit
1-2, 2-3 and 3-4 changes too long	Damper faulty	Renew valve block
	PRSV faulty	Renew PRSV
	PRSV wire earthed	Renew harness
	Modulation valve seized	Renew valve block
	Pressure regulator valves 1 and 2 seized	Renew valve block
Full load and kickdown changes too long	TCU faulty	Check TCU
	Modulation pressure too low	Renew valve block
	Clutch plates faulty	Renew gear unit
Full load and kickdown changes too hard	TCU faulty	Check TCU
	Modulation pressure too high	Renew valve block
	Damper faulty	Renew valve block
'3' 3rd gear		
No engine braking	Clutch E faulty	Renew gear unit
'2' 2nd gear		
No engine braking	Brake C1 or Clutch E faulty	Renew gear unit
Manual gearchange '3' to '2' inoperative	SS faulty	Renew SS
	SV2 faulty	Renew SV2

'1' 1st gear

No engine braking	Brake D or Clutch E faulty	Renew gear unit
Manual gearchange '2' to '1' inoperative	SS faulty	Renew SS
	SV1 faulty	Renew SV1
	Damper D faulty	Renew valve body
	PRSV faulty	Renew PRSV
	Modulator valve seized	Renew valve body

Converter Clutch

Clutch engagement speed not to specification	TCU faulty	Check TCU
Engagement too hard	Converter clutch damper faulty	Renew valve block
	Converter unit faulty	Renew converter unit
No clutch engagement	TCU faulty	Check TCU
	SV3 faulty	Renew SV3
	Converter clutch valve seized	Renew valve block
	Converter faulty	Renew converter
	Pressure reducing valve 1 seized	Renew valve block
Converter clutch permanently engaged (engine will not idle in 'D')	TCU faulty	Check TCU
	SV3 faulty	Renew SV3
	SV3 wire earthed	Renew harness
	SV3 faulty	Renew SV3
	SV3 wire earthed	Renew harness

General

Noise and associated interruption of power	Oil filter dirty	Clean / replace filter. If clutch debris found, renew gear unit and clean valve block
Loud noise with no drive forward or reverse	Drive plate to converter connection damaged	Renew drive plate and converter as necessary
	Pump drive damaged	Renew gear unit
High pitched noise in all selector positions especially when cold. Sucking noise from pump	Oil level too low	Rectify
	Valve block leaking	Renew valve block
Loud noise when converter clutch engages	Torsion damper defective	Renew converter
Engine vibrations when converter clutch engaged	Change point too low	Check TCU

Leaks

Oil in bell housing	Oil pump seal leaking	Renew oil pump seal
	Pump housing leaking	Renew oil pump
	Converter seam leaking	Renew converter
Leakage between gearbox and sump	Loose fastening screws	Tighten to specified torque figure
	Sump gasket faulty	Renew sump gasket
Leakage between intermediate plate and gearbox	Bell housing to gearbox fastening screws loose	Tighten screws to specified torque figure
Leakage at harness plug connector	Connector O ring faulty	Renew O ring
Leakage at drive flange	Rear oil seal faulty	Renew oil seal
Leakage at breather	Oil level too high	Rectify
	Incorrect oil specification	Change oil and if necessary flush system
	Breather cover missing	Replace breather cover
	Breather cover O ring faulty	Renew O ring
	Breather cover loose	Renew lock washer
Leakage from oil cooler circuit	Pipe lock nuts loose	Tighten locknuts to specified torque figure
	Pipe line faulty	Renew pipes as necessary
	Oil cooler leaking	Renew oil cooler
Leakage at intermediate plate	Intermediate plate plug leaking	Tighten plug to specified torque figure Renew sealing washer
Leakage between gearbox and extension housing	Fastening screws loose	Tighten screws to specified torque figure
	Gasket faulty	Renew gasket

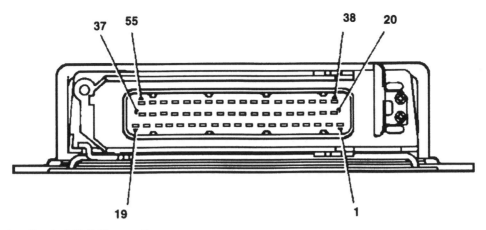

Fig. 1 Transmission Control Unit Connections

(Key to Fig. 1)

1.	Power supply	27.	-
2.	Speed sensor positive	28.	Pin code 2
3.	Engine speed	29.	-
4.	Solenoid valve 1	30.	-
	-	31.	-
6.	Pressure regulator	32.	Torque control (ignition retard)
7.	Digital ground	33.	Position code Z
8.	Throttle potentiometer input	34.	-
9.	Pin code 1 - Catalyst	35.	-
10.	Pin code 3	36.	-
11.	-	37.	-
12.	-	38.	Speed sensor - minus
13.	-	39.	Permanent battery supply diagnostic memory
14.	Position code Y	40.	-
15.	Diagnostic L line	41.	Kickdown switch
16.	Fault indicator	42.	Solenoid valve 3
17.	-	43.	Sport mode switch
18.	-	44.	Throttle potentiometer ground
19.	Solenoid valves supply	45.	Throttle potentiometer supply
20.	Vehicle speed sensor screen	46.	-
21.	Injector time (load signal)	47.	-
22.	-	48.	-
23.	Plus ignition - limp home	49.	-
24.	Solenoid valve 2	50.	Position code X
25.	Torque control UP / DOWN	51.	Diagnostic K line
26.	Power ground		

Rotary Switch Harness

(Key to Fig. 2)

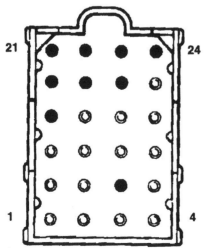

Fig. 2 Main harness connector

Pin	Cable colour	Function
1.	BK	Switch common
2.	LGB	Code X output
3.	LGP	Code Y output
4.	LGW	Code Z output
5.	WP	Start inhibit ground
6.	RW	Start inhibit output
8.	SG	Reverse lights output
9.	RY	Solenoid supply
10.	YB	Solenoid Valve 1
11.	YP	Solenoid Valve 2
12.	YU	Solenoid Valve 3
14.	OG	Pressure Regulator Solenoid Valve
15.	R	Speed Sensor
16.	U	Speed Sensor
20.	W	Ground (screen)

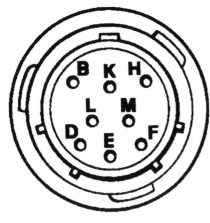

Fig. 3 Gearbox connector

Pin	Cable colour	Function
B	OG	Pressure Regulator Solenoid Valve
D	-	-
E	U	Speed Sensor
F	R	Speed Sensor
H	YB	Solenoid Valve 1
K	YP	Solenoid Valve 2
L	YU	Solenoid Valve 3
M	RY	Solenoid supply

Colour code					
N	Brown	G	Green	S	Slate
B	Black	R	Red	L	Light
W	White	Y	Yellow	P	Purple
K	Pink	O	Orange	U	Blue

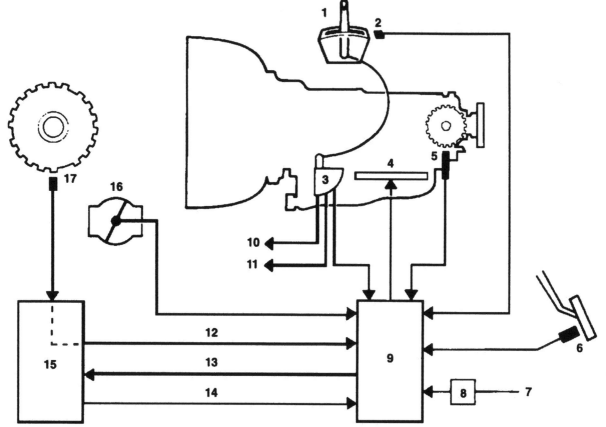

Fig. 4 Control Layout

Key to Fig. 4

1.	Selector	10.	Reverse Lights
2.	Mode Switch	11.	Start Inhibitor
3.	Rotary Switch	12.	Engine Speed
4.	Valve Block	13.	Ignition Retard
5.	Output Shaft Speed Sensor	14.	Engine Load
6.	Kickdown Switch	15.	Engine Control
7.	Battery	16.	Throttle Potentiometer
8.	Transmission Relay	17.	Engine Speed Sensor
9.	Transmission Control Unit		

TRANSMISSION UNIT, RENEW

Disconnect the battery.
Remove the dipstick tube to manifold securing bolt.
Remove the manifold to cylinder head rear securing nuts.
Remove the rear support bracket and strap assembly from the manifold studs.

Fit Service Tool 18G 1465 Engine Lifting Eye to the manifold rear studs and secure.
Fit Service Tool MS 53B Engine Support across the wing channels with the hook connected to the lifting eye (Fig. 5).
Adjust the support hook nut to take the weight of the engine.
Raise the vehicle on a hoist.

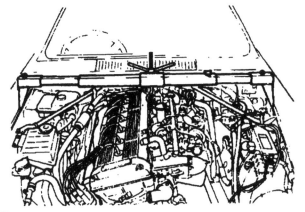

Fig. 5

Note: The vehicle should be in 'N' neutral gear selection, with the handbrake off, to assist the removal and refitting of the propshaft.

Drain the transmission.
Refit the gearbox drain plug.
Remove the front exhaust pipe.
Ensure that the exhaust system is held aside.
Support the transmission and remove the rear mounting securing bolts with the bracket and spring assembly.
Remove the propshaft to transmission drive flange securing bolts and separate the flanges.
Remove the rotary switch guard securing bolts and the switch guard.
Remove the rotary switch securing nuts and detach the switch from the selector shaft and the mounting studs.
Remove the selector cable mounting securing nuts.
Remove the selector cable ball joint securing nut and withdraw the ball pin from the lever.
Withdraw the selector cable mounting from the mounting studs.
Rotate the bayonet collar to withdraw the rotary switch harness plug from the gearbox case socket and ensure that the selector cable, rotary switch and harness are held aside.
Remove the dipstick.
Re position the oil container beneath the dipstick tube union, disconnect the union and drain the remaining fluid.
Remove the dipstick tube assembly.
Fit a blanking plug to the sump.
Remove the transmission cooler pipes clamp screw and clamp.
Remove the intermediate housing to cooler pipe banjo bolts and seal both the intermediate housing and the pipes with blanking plugs.

Remove the converter housing stone guard (2 Fig. 6) and the lower rubber blanking plug (1 Fig. 6) to expose the drive plate and torque converter.
Turn the crank to align each of the six drive plate to torque converter securing bolts with the housing access hole,

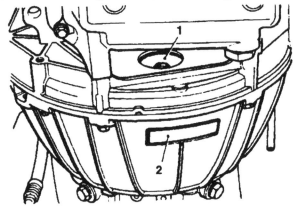

Fig. 6

remove them and separate the converter from the drive plate.
Remove the fan cowl securing clips and release the fan cowl from the radiator.
Lower the rear of the transmission by releasing the suspension nut on MS 53B Engine Support.

Note: Do not allow the engine to foul the steering rack.

Remove the foam insulation pad. Disconnect the engine earth lead.
Remove eight converter housing to intermediate plate securing bolts, (1 Fig. 7) leaving the two upper bolts to

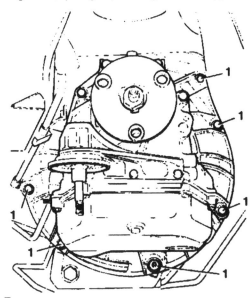

Fig. 7

retain the transmission in position.
Utilizing an Epco V 1000 unit lift:
Remove the front and rear clamps.

Traverse the lift (1 Fig. 8) under the transmission unit.
Take the weight of the transmission on the lift.
Adjust the tilt angle and side clamps (2 Fig. 8).
Fit the chain assembly to the right-hand arm, pass the chain (3 Fig. 8) over the transmission into the front arm and tighten the chain adjuster (4 Fig. 8).
Remove the remaining two securing bolts and remove the transmission from the engine.
Lower the transmission and remove from below the vehicle hoist (Fig. 9).

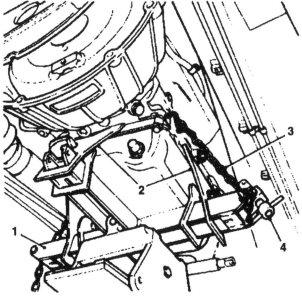

Fig. 8

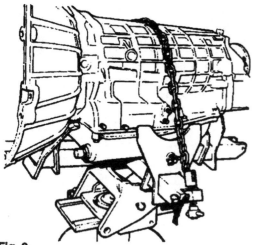

Fig. 9

Transfer the torque converter retaining strap from the replacement gearbox to the newly detached unit.

Remove the transmission from the hoist.

Remove the rear mounting spring locating cup securing nut, mounting cup and stud.

Remove the selector lever securing nut and the selector lever.

Remove the rotary switch mounting bracket securing bolt and studs together with the mounting bracket.

Flush the transmission cooling system to eliminate any contaminated fluid.

Clean all mating faces on the replacement transmission.

Refit the rotary switch mounting plate, tighten the studs and bolts to the recommended torque figure.

Refit the selector lever and tighten the locknut to the recommended torque figure.

Fit the rear mounting stud and tighten to the recommended torque figure.

Fit the rear mounting spring locating cup to the stud and secure with the nut.

Tighten the nut to the recommended torque figure.

Check that the selector lever is in the 'N' neutral position.

Fit the replacement transmission securely to the hoist.

Move the hoist into position, elevate the transmission and carefully position the torque converter spigot bearing into the crankshaft location.

Fit the engine earth lead.

Fit four converter housing securing bolts.

Remove the transmission hoist.

Fit the remaining converter housing securing bolts and tighten them all to the recommended torque figure.

Turn the crank and drive plate to align the drive plate to torque converter securing bolt holes then fit the securing bolts through the access aperture.

Fit all six bolts, then tighten them to the recommended torque figure.

Refit the blanking plug to the intermediate plate.

Refit the stone guard to the converter housing.

Remove the plugs from the transmission cooler ports in the intermediate housing and refit the oil cooler supply and return pipes.

Tighten the banjo bolts to the recommended torque figure.

Secure the cooler pipes to the engine sump bracket and tighten the securing screw to the recommended torque figure.

Remove the plug from the dipstick tube union on the sump pan and refit the dipstick tube.

Tighten the dipstick tube union nut to the recommended torque figure.

Refit the foam insulation pad.

Raise the engine and gearbox unit using the support hook nut on Service Tool MS 53B.

Check that the gear selector is in 'N' neutral position.

Release the selector cable and gearbox harness then reconnect the multi pin plug to the gearbox connector.

Refit the selector cable ball pin to the selector lever, checking that the ball pin shank is a free fit in the selector lever, then secure with the locknut and tighten to the recommended torque figure.

Fit the outer selector cable mounting to the gearbox studs with the nuts and tighten to the recommended torque figure.

Refit and adjust the rotary switch.

Refit the rotary switch guard and tighten the securing bolts to the recommended torque figure.

Refit the gearbox drive flange to the propshaft with the securing bolts and nuts and tighten to the recommended torque figure.

Lift the rear mounting assembly into position and locate with the securing bolts.

Fully release the engine lifting tool MS 53B.

Tighten the rear mounting assembly securing bolts to the recommended torque figure.

Release the exhaust system and refit the front pipe.

Lower the hoist.

Remove the engine support MS 53B.

Remove the engine lifting eye 18G 1465.

Refit the rear support bracket and strap assembly to the manifold studs, secure with the nuts and tighten to the recommended torque figure.

Secure the injector harness to the inlet manifold with a ratchet strap.

Fit the dipstick tube to the inlet manifold bracket and secure with the bolt, then tighten the bolt to the recommended torque figure.

Refit the radiator cowl into position and fit the securing clips.

Reconnect the battery.

Refill the transmission.

Service Tools

Engine support bracket	MS 53B
Engine lifting bracket	18G 1465

Torque Figures

Dipstick tube nut	20 Nm
Down pipe to intermediate pipe	15-18 Nm
Drain plug	15 Nm
Drive plate to torque converter	49-54 Nm
Intermediate pipe to silencer	15-18 Nm
Oil cooler pipe banjo bolt	32-36 Nm
Drive flange to propeller shaft	95-105 Nm
Rear engine mounting bracket to body	20-25 Nm
Torque converter housing to block	49-54 Nm

Oils / Sealants / Lubricants

Exhaust sealer	TIVOLI KAY ADHESIVES No 5696
Transmission fluid	DEXRON IID

TORQUE CONVERTER, RENEW

Drain the transmission unit.
Remove the transmission unit.

Fit Service Tool JD 105 lifting handles (1 Fig. 10) to the torque converter.

Remove the converter (2 Fig. 11) from the gearbox.
Remove the lifting handles.
Clean and lubricate the replacement torque converter drive tang and pump seal area.
Fit the lifting handles to the replacement torque converter.
Refit the torque converter to the gearbox taking care not to damage the pump oil seal.
Refit the transmission unit.
Refill the transmission unit.

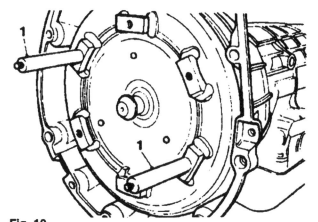

Fig. 10

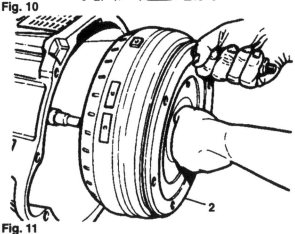

Fig. 11

Service Tools

Converter lifting handles   JD 105

EXTENSION HOUSING GASKET, RENEW

Disconnect the battery.
Remove the dipstick tube to manifold securing bolt.
Remove the manifold to cylinder head rear securing nuts.
Remove the rear support bracket and strap assembly from the manifold studs.
Fit Service Tool 18G 1465 Engine Lifting Eye to the manifold rear studs and secure with the nuts.
Fit Service Tool MS 53B Engine Support across the wing channels with the hook engaging the lifting eye (Fig. 5).
Adjust the support hook nut to take the weight of the engine.
Raise the vehicle on a hoist.

Note: The vehicle should be in 'N' neutral gear selection, with the handbrake off, to assist the removal and refitting of the propshaft.

Drain the transmission.
Refit the gearbox drain plug.
Remove the front exhaust pipe.
Ensure that the exhaust system is held aside.
Support the transmission and remove the rear mounting securing bolts with the bracket and spring assembly.
Remove the propshaft to transmission drive flange securing bolts and separate the flanges.
Remove the fan cowl securing clips and release the fan cowl from the radiator.
Lower the rear of the transmission by releasing the suspension nut on MS 53B Engine Support.
Note: Do not allow the engine to foul the steering rack.
Select 'P' park.
Remove the gearbox drive flange locking washer.
Remove the flange locking nut and the drive flange.
Remove the selector cable mounting bracket clamp securing bolt.

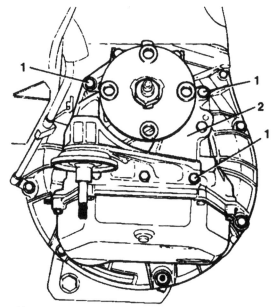

Fig. 12

Remove the rear housing securing bolts (1 Fig. 12) and then withdraw the rear housing assembly. (2 Fig. 12)
Clean both the gearbox and extension housing sealing faces.
Lubricate the rear bearing.
Fit the new gasket to the gearbox rear face, refit the rear housing assembly with the securing bolts and tighten to the recommended torque figure.
Lubricate the oil seal and apply 'Hylomar' to the output shaft splines.
Refit the drive flange to the output shaft and tighten the securing nut to the recommended torque figure.
Fit a new securing nut locking washer and using a suitable metal punch, punch it into position.
Raise the rear of the transmission by tightening the hook nut on MS 53B Engine Support.
Refit the propshaft flange to the drive flange and tighten the securing nuts to the recommended torque figure.
Refit the selector cable clamp and tighten the securing bolt to the recommended torque figure.
Refit and align the rear mounting assembly with the securing bolts, slacken MS 53B Engine Support hook nut and tighten the bolts to the recommended torque figure.
Clean the exhaust sealing olive and apply sealant.
Refit the front pipe to the exhaust system with the sealing olive and clamp, then tighten the clamp securing nut and bolt to the recommended torque figure.
Lower the hoist.
Remove the engine support hook.
Remove MS 53B Engine Support.
Remove 18G 1465 Engine Lifting Eye, refit the rear support bracket and strap assembly with the manifold securing nuts and tighten them to the recommended torque figure.
Refit the dipstick tube to the manifold and tighten the bolt to the recommended torque figure.
Reconnect the battery.
Refill the transmission.

Service Tools

Engine support bracket	MS 53B
Engine lifting bracket	JD 124

Torque Figures

Down pipe to intermediate pipe	15-18 Nm
Drain plug	15 Nm
Drive flange to propeller shaft	95-105 Nm
Extension housing to gearbox case	23 Nm

Oils / Sealants / Lubricants

Exhaust sealer   TIVOLI KAY ADHESIVES No 5696.
Output shaft splines   HYLOMAR
Transmission fluid   DEXRON IID

EXTENSION HOUSING, RENEW

Remove the rear extension housing gasket.
Remove the bolts securing the selector cable mounting bracket and remove the bracket.
Remove the rear mounting spring locating cup securing nut, mounting cup and stud.
Remove the breather cap (1 Fig. 13), '0' ring (2 Fig. 13) and circlip.

Fit a new bearing to the replacement extension housing using Service Tool JD 104 (1 Fig. 14).
Fit a new oil seal to the housing using Service Tool JD 102.
Fit a new 'O' ring to the breather and fit the assembly to the extension housing.
Refit the selector mounting bracket to the housing and tighten the securing bolts to the recommended torque figure.
Refit the spring locating cup stud and tighten to the recommended torque figure.
Refit the locating cup and tighten the locknut to the recommended torque figure.
Refit the extension housing to the gearbox .

Service Tools

Rear extension oil seal replacer   JD 102
Rear extension bearing replacer   JD 104

Torque Figures

Down pipe to intermediate pipe	15-18 Nm
Drain plug	15 Nm
Drive flange to propeller shaft	95-105 Nm
Extension housing to gearbox case	23 Nm
Spring locating cup stud	125-145 Nm
Locating cup locknut	75-80 Nm

Oils / Sealants / Lubricants

Exhaust sealer   TIVOLI KAY ADHESIVES No 5696.
Output shaft splines   HYLOMAR
Transmission fluid   DEXRON IID

FLUID, RENEW

Raise the vehicle on a hoist.

Remove the drain plug (1 Fig. 15) and drain the transmission fluid into a suitable container.
Disconnect the dipstick tube union nut (2 Fig. 15) and the tube from the transmission sump. Drain the remaining fluid into a container. Allow the transmission to drain for at least 15 minutes (approximately 4 litres / 7.0 pints of fluid will drain).

Note: Do not re-use the fluid.

If debris are found in the fluid thoroughly check the gear train for damage and the valve block for debris then renew the torque converter and the oil cooler.
Renew the sealing washer, refit the drain plug and tighten to the recommended torque figure.

Reconnect the union nut (2 Fig. 15)
Lower the hoist.

Remove the transmission dipstick (Fig. 16).

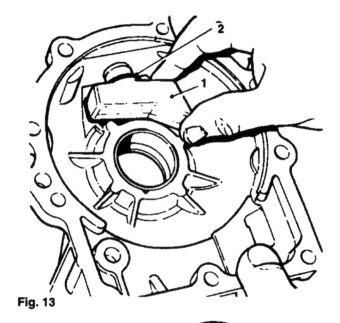

Fig. 13

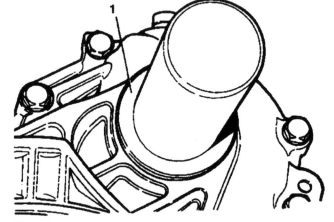

Fig. 14

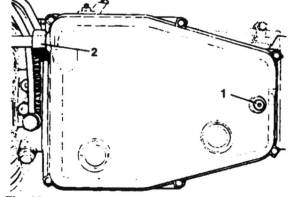

Fig. 15

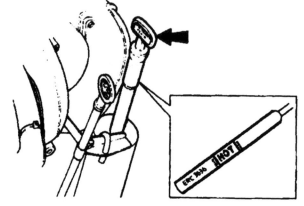

Fig. 16

Fill the transmission with the correct amount of recommended fluid (approximately 4 litres / 7.0 pints).

Note: It will not be possible to fully fill the gearbox and converter until the engine has been run for the pump to prime the system.

Stop the engine, top up the fluid level as required with the recommended fluid and refit the dipstick.

Torque Figures

Dipstick tube union nut	81 Nm
Drain plug	15 Nm

Oils / Sealants / Lubricants

Transmission fluid	DEXRON IID

FLUID FILTER, RENEW

Raise the vehicle on a hoist.
Drain the transmission fluid.
Remove the sump pan.
Remove the filter securing screws and the filter.
Remove and discard the filter 'O' ring.
Fit a new 'O' ring to the new filter.
Fit the filter to the valve block and secure with the screws.
Tighten the filter securing screws to the recommended torque figure.
Refit the sump pan.
Refill the transmission with fluid and check the level.

Torque Figures

Fluid filter screws	8 Nm

KICKDOWN SWITCH, ADJUSTMENT

Check for the correct settings of maximum and minimum throttle.
Locate the throttle potentiometer harness connector.
Identify the potentiometer to Transmission Control Unit signal wire (green / blue).
Connect a suitable pointed probe to the Jaguar Diagnostic System or a multi meter and set the meter to the DC voltage scale.

Insert the probe through the No. 5 hole (Fig. 17) in the rear of the connector and ease it alongside the green / blue wire, through the rubber backing washer, to contact the rear of no. 5 pin.
Switch the ignition on.

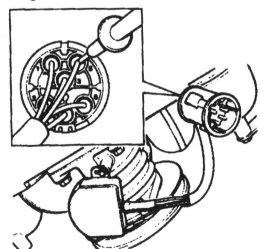

Fig. 17

With the accelerator pedal released i.e. idle position, the reading on the voltmeter should be 0.6 volt (+ 0.02 volt).
With the accelerator pedal fully depressed i.e. full kickdown, the reading on the voltmeter should be 4.5 volts (+ 0.1 volt).
Reset the accelerator pedal to approximately 90% of the total pedal travel, note the voltmeter reading and adjust the pedal position to achieve a voltmeter reading of 4.1 volts (+ 0.1 volt).
This position is the kickdown point.

With the accelerator pedal held at the kickdown point adjust the kickdown switch to lightly contact the underside of the accelerator pedal (Fig. 18).
This position can be checked by inserting a thin piece of paper between the underside of the accelerator pedal and the kickdown switch. When the paper can be withdrawn with a slight resistance the setting is correct.
Road test to check the kickdown switch setting.

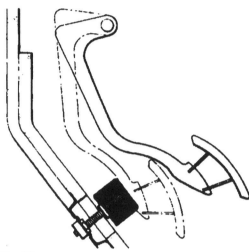

Fig. 18

KICKDOWN SWITCH, RENEW

Displace the draught welting from the lower 'A' post area.
Remove the 'A' post lower trim pad securing screws and remove the pad.
From inside the car move the footwell carpet from the kickdown switch area.
Cut and remove the ratchet strap securing the kickdown switch harness to the side harness.
Disconnect the kickdown switch block connector.
Unscrew and remove the kickdown switch (making a note of its approximate position on the stud) from the adjustment stud.
Slacken off and remove the adjustment stud, remove the floor pan retainer (dished pressing).
Fit the floor pan retainer to the under carpet.
Fit and tighten the adjustment stud to the floor pan.
Fit the new switch to its approximate original position on the adjustment stud.
Reconnect the switch connector and secure the harness with a new ratchet strap.
Set kickdown switch adjustment.
Refit the footwell carpet and trim pad.
Attach the harness to the valve block with the securing clip.
Refit the sump pan.
Refill the transmission.
Lower the hoist.

ROTARY SWITCH, ADJUSTMENT

Raise the vehicle on a hoist.
Select 'N' Neutral.

Remove the rotary switch protection cover securing bolts and remove the cover (1 Fig. 19).

Slacken off the rotary switch to mounting bracket securing nuts.

Displace the rubber blanking plug from the switch centre boss (2 Fig. 19).

The switch is now free to rotate around the selector shaft within the limits of the slotted holes.

Fit Service Tool JD 161 (Fig. 20) to the square section bore

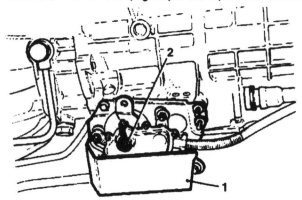

Fig. 19

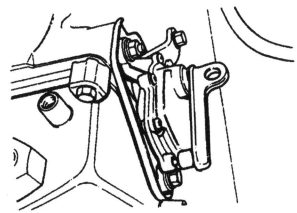

Fig. 20

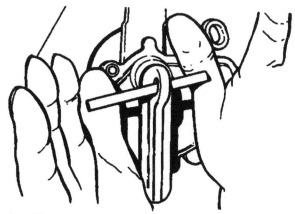

Fig. 21

in the switch shaft and the locating hole in the switch body, this will correctly align the switch to the selector shaft.

Lightly tighten the securing nuts and check that the tool can be withdrawn easily with the aid of a bar (Fig. 21).

If it is difficult to withdraw the tool, the switch is not aligned and must be re adjusted.

When the adjustment is correct, tighten the nuts to the recommended torque figure.

Remove the service tool and refit the rubber blanking plug.

Refit the switch guard and tighten the bolts to the recommended torque figure.

Lower the hoist.

Service Tools

Rotary switch setting tool	JD 161

Torque Figures

Centre console securing screws	8-12 Nm
Switch guard to sump pan bolts	15-18 Nm
Switch to mounting bracket nuts	15-18 Nm

ROTARY SWITCH, RENEW

Place the vehicle on a hoist.

Disconnect the battery.

Remove the passenger's side dash liner.

Chock the road wheels and apply the handbrake.

Select 'N' neutral position.

Carefully displace the console carpet from the side casing area for access to the rotary switch harness.

Disconnect the rotary switch harness block connector.

Displace the harness to transmission tunnel grommet.

Reposition the harness through the transmission tunnel aperture.

Raise the hoist.

Remove the rotary switch guard securing bolts and the guard (Fig. 22 and 1 Fig. 23).

Remove the rotary switch securing nuts.

Remove the 'P' clip securing screw from the selector cable mounting bracket and remove the clip from the harness.

Rotate the bayonet collar and withdraw the rotary switch harness plug from the gearbox case socket.

Withdraw the rotary switch from the selector shaft and the mounting studs.

Withdraw the harness and block connector through the tunnel aperture.

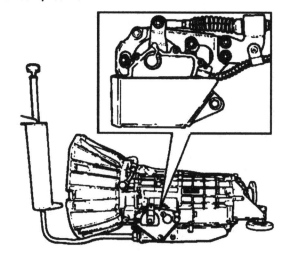

Fig. 22

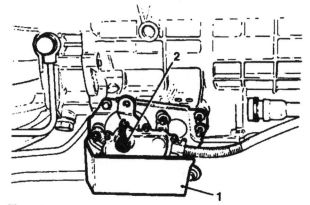

Fig. 23

Insert the replacement harness and block connector through the tunnel aperture.

Fit the rotary switch harness plug to the gearbox case socket and secure with the bayonet collar.

Remove the rubber plug from the rotary switch centre boss (2 Fig. 23).

Fit Service Tool JD 161 to the rotary switch to align the switch rotor to the switch body.

Check that the selector and the gearbox selector shaft are in the 'N' neutral position.

Fit the rotary switch with the tool to the selector shaft and the switch mounting studs (Fig. 24).

Fit the securing nuts to the studs and adjust to touch.

Check that the tool can be withdrawn easily before tightening the nuts to the recommended torque figure.

The tool is provided with a hole to aid withdrawal (Fig. 21).

Tighten the nuts to the recommended torque figure and refit the rubber plug to the centre boss.

Fit the P clip to the harness and secure to the selector cable mounting bracket with the screw.

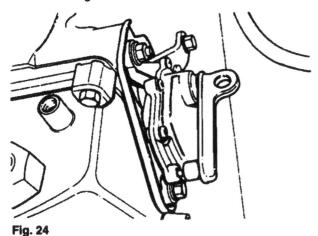

Fig. 24

Note: Ensure that the harness is positioned to clear the propshaft.

Fit the rotary switch guard and tighten the securing bolts to the recommended torque figure.

Lower the hoist.

Route the rotary switch harness to the main harness multiplug.

Fully seat the grommet to the transmission tunnel.

Connect the rotary switch block connector.

Reposition the carpet to cover the rotary switch harness tunnel grommet.

Refit the passenger's side dash liner.

Reconnect the battery.

Service Tools

Rotary switch setting tool   JD 161

Torque Figures

Centre console securing screws   8-12 Nm
Switch guard to sump pan bolts   15-18 Nm
Rotary switch to mounting bracket nuts   15-18 Nm

SELECTOR CABLE, ADJUSTMENT

WARNING:
DO NOT ATTEMPT TO DRIVE THE VEHICLE WITH THE GEAR CHANGE LINKAGE DISCONNECTED OR THE SELECTOR GATE DETACHED.

Disconnect the battery.

Unscrew the gear selector knob.

Remove the screws securing the rear of the gear control lever escutcheon.

Using a thin blade, carefully prise out the speed control switch and the cigar lighter.

Note the positions of the terminals and disconnect.

Prise out the window switch pack, note the terminals and disconnect.

Working through the speed control and cigar lighter panel holes, slacken (do not remove) the two bolts securing the escutcheon.

Pull the escutcheon slightly rearwards and upwards to remove.

Slacken the locknuts (1 Fig. 25) at the abutment bracket to allow free movement of the outer cable.

Select 'N' Neutral (3 Fig. 25).

Ensure that the gearbox selector lever (2 Fig. 25) is also in

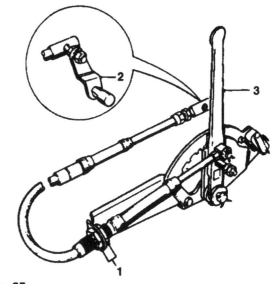

Fig. 25

'N' Neutral position.

Tighten the locknuts to achieve the maximum free play of the selector in the neutral position.

Refit the escutcheon and secure with the bolts.

Refit the switches and cigar lighter.

Refit the gear selector knob.

Reconnect the battery.

Torque Figures

Cable to abutment bracket locknuts   18.5 - 24 Nm.

SELECTOR CABLE, RENEW

WARNING:
DO NOT ATTEMPT TO DRIVE THE VEHICLE WITH THE GEAR CHANGE LINKAGE DISCONNECTED OR THE SELECTOR GATE DETACHED.

Place the vehicle on a hoist.

Disconnect the battery.

Unscrew the gear selector knob.

Remove the screws securing the rear of the gear control lever escutcheon.

Using a thin blade, carefully prise out the speed control switch and the cigar lighter. Note the positions of the terminals and disconnect.

Prise out the window switch pack, note the terminals and disconnect.

Working through the speed control and cigar lighter panel holes, slacken (do not remove) the two bolts securing the escutcheon.
Pull the escutcheon slightly rearwards and upwards to remove.

Slacken and remove the locknuts (1 Fig. 25), disconnect the inner selector cable from the selector lever (3 Fig. 25) and withdraw the cable.
Raise the hoist.

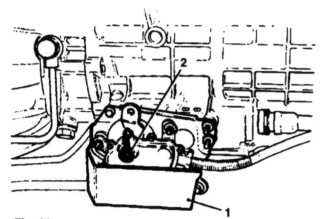

Fig. 26

Remove the switch guard (1 Fig. 26).
Remove the switch securing nuts.
Remove the rotary switch from the selector shaft and mounting studs.
Remove the locknut securing the selector cable ball pin to the selector lever (2 Fig. 25 and Fig. 27).

Withdraw the ball pin shank from the selector lever.
Remove the outer selector cable clamp bolts from the mounting bracket.
Withdraw the cable and grommet.

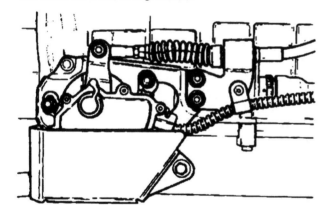

Fig. 27

Note: New cable pre set:- Eyelet to ball pin centre line measurement = 1147 mm (with ball pin trunnion 10 - 11 mm thread engagement).
Fit the grommet to the replacement cable.
Insert the cable through the body aperture and refit the grommet.
Refit the ball pin shank to the selector lever and tighten the locknut to the recommended torque figure.
Refit the outer cable clamp bolts and tighten to the recommended torque figure.
Refit the rotary switch to the selector shaft and the mounting studs.
Fit and adjust the rotary switch.
Lower the hoist.
Adjust the selector cable.

Refit the gear lever knob.
Reconnect the battery earth lead.
Check that the starter will operate only in 'P' Park and 'N' Neutral gear selections.

Torque Figures

Cable bracket to transmission unit	7-10 Nm
Ball pin locknut to selector lever	8-12 Nm
Control cable to ball pin joint	40-50 Nm

SUMP PAN AND SEAL, RENEW

Raise the vehicle on a hoist.
Drain the transmission fluid.
Remove the dipstick.
Disconnect the dipstick tube from the fluid pan.
Remove the rotary switch guard to sump pan securing bolts.
Remove rotary switch guard.
Remove sump pan securing bolts and clamps (2 Fig. 28).
Remove the sump pan and sealing rubber (3 Fig. 28).

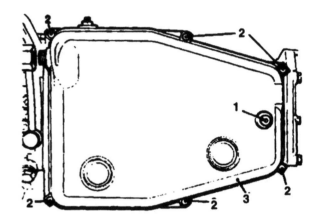

Fig. 28

Remove the magnets.
Discard the sump and the sealing rubber.
Thoroughly clean the replacement sump pan and the magnets.
Refit the magnets.
Fit a replacement sealing rubber to the sump pan.
Lift the sump pan and sealing rubber into position and secure with the clamps and bolts.
Tighten the sump pan securing bolts to the recommended torque figure.
Refit the rotary switch guard.
Tighten the switch guard securing bolts to the recommended torque figure.
Reconnect the dipstick tube to the sump pan.
Tighten the dipstick tube to sump pan nut to the recommended torque figure.
Refit the sump pan drain plug (1 Fig. 28) with a new sealing washer and tighten to the recommended torque figure.
Refill the transmission with fluid and check the level.

Torque Figures

Dipstick tube nut	20 Nm
Drain plug	15 Nm
Sump pan securing bolts	8 Nm

Chapter 10

PROPELLER SHAFT AND DRIVE SHAFTS - 4.0 LITRE

PROPELLER SHAFT AND DRIVE SHAFTS, DESCRIPTION

DRIVE SHAFTS

The drive shafts transmit drive from the final drive unit to the hubs and also serve as upper transverse members to locate the rear wheels.

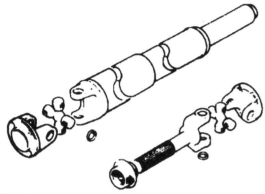

Fig. 1

PROPELLER SHAFT - Automatic transmission (Fig. 1)

The propeller shaft fitted to 4.0 litre automatic transmission models is a two universal joint type, which is of the flange-fitting type at both ends. The front flange bolts directly to the corresponding output flange of the automatic transmission unit, while the rear flange mounts in a similar fashion to the input flange of the final drive unit. The shaft is split by a splined sliding joint, which is protected by a flexible rubber gaiter.

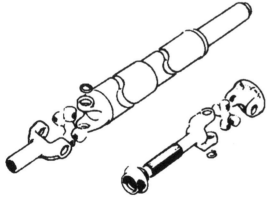

Fig. 2

PROPELLER SHAFT - Manual transmission (Fig. 2)

The propeller shaft fitted to the 4.0 litre manual transmission XJS is also of the two universal joint type but with a reverse spline fitting coupled to the gearbox. The rear coupling is of a similar flanged type to the automatic transmission types. The shaft is split by a splined sliding joint, which is protected by a flexible rubber gaiter.

UNIVERSAL JOINTS

When fitting a propeller shaft it is essential to ensure that

the universal joints operate freely; any stiffness, even in a single joint, will initiate propeller shaft vibration.
Universal joints must always be fitted as a set.

DRIVE SHAFT, RENEW

Note: This procedure is the same for left the or right hand drive shaft.

Jack up the rear of vehicle and support with axle stands.
Remove the road wheel.
Slacken off the inner shroud clip (1 Fig. 3) and slide the shroud away from flange.

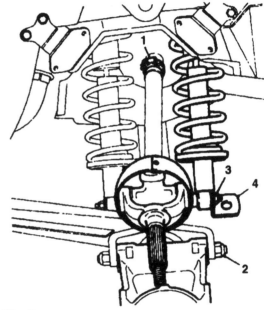

Fig. 3

Remove the flange securing nuts (1 Fig. 4).
Withdraw the split pin and remove the castellated hub nut and plain washer.
Discard the split pin.
Remove the grease nipple from the hub carrier.

Fit hub puller JD 1D (1 Fig. 5) to the hub (2 Fig. 5).
Fit and tighten the securing bolts / nuts.
Withdraw the hub assembly from the splined shaft and

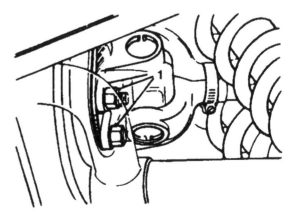

Fig. 4

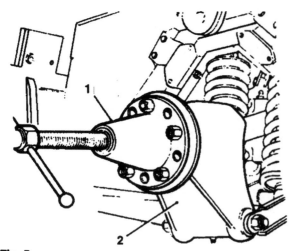

Fig. 5

remove the hub puller.
Pivot the hub assembly on the fulcrum shaft (2 Fig 3).
Remove the damper mounting nut (3 Fig. 3), detach the lash down bracket (4 Fig. 3), drift the pin forward and release the lower end of the damper.
Remove the drive shaft assembly; collect the flange / brake disc shims.
Remove the phosphorus bronze spacer and inner oil seal track from the splined shaft.
Drill out the pop rivets from the shrouds, remove the clips and collect the shroud halves.*
(The above steps also apply to the next procedure [Overhaul]).

Place the shroud halves on to the replacement shaft, line up holes and pop rivet the halves together.
Seal the shroud joints with underseal, refit the clips, place the outer shroud in position and tighten the clip.
Fit the inner oil seal track (note that the chamfer clears the radius on the drive shaft) and spacer to the splined shaft.
Clean and refit the camber shims, place the flanged end of the shaft in position and refit the securing nuts, but do not tighten at this stage.
Refit the damper to lower mounting pin, refit the lash down bracket.
Fit and tighten the securing nut.
Apply Loctite to the splines, check that the spacer and oil seal track are installed, raise the hub carrier and enter the splined shaft into the hub.
Tap the hub home on to the splines, replace the washer, fit and tighten the hub securing nut.
Fit a new split pin.
Check the hub bearing end float.
Refit the fulcrum grease nipple into the hub carrier.
Final tighten of the drive shaft flange nuts.
Slide the inner shroud into position and tighten the clip.
Refit the road wheel.
Remove the axle stands and lower the vehicle.
Check / adjust the camber angle of the rear wheel.

Service Tools

Hub remover   JD 1D

Torque Figures

Drive shaft to final drive unit	66.5-74.5 Nm
Drive shaft to hub carrier	136-163 Nm
Lower damper mounting	41-48 Nm
Outer pivot pin	131-144 Nm

Oils / Sealants / Lubricants

Drive shaft splines   LOCTITE

DRIVE SHAFT, OVERHAUL

Note: This procedure is the same for left or right hand drive shaft

Follow the procedure under 'DRIVE SHAFT, OVERHAUL' as far as the asterisk(*).

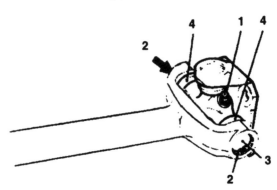

Fig. 6

Remove the grease nipples (1 Fig. 6).
Secure the shaft in vice and remove two opposed circlips from one universal joint (2 Fig. 6).

Note: Tap the bearings slightly inwards to assist removal of the circlips.

Tap one bearing inwards to displace the opposite bearing.
Remove the shaft assembly from the vice, and trap the displaced bearing in the vice.
Tap the shaft away from the bearing.
Retrieve the bearing. Refit the shaft to the vice.
Displace the second bearing by tapping the joint spider along the length of the shaft yoke and extract the second bearing.
Remove the two grease seals from the spider (4 Fig. 6).
Detach the spider, with the end section of the shaft, from the centre section of the shaft.
Place the end section of the shaft in the vice and repeat the removal procedure for the remaining two bearings.
Remove the spider from the end section of the shaft.
Repeat the removal procedure for the universal joint at the opposite end of the drive shaft.
Wash all component parts in petrol.
Check the splined yoke for wear of the splines.
Examine the bearing races and spider journals for signs of looseness, load markings, scoring or distortion. It is essential that bearing races are a light drive fit in the yoke trunnion.

Note: Spider or bearings should not be renewed separately, as this will cause premature failure of the replacement.

Remove the bearing assemblies from one replacement spider; if necessary, retain the rollers in their respective housings with petroleum jelly.
Leave the grease shields in position.
Fit the spider to one end section of the shaft.
Fit two bearings and circlips into the end section trunnions.
Use a soft round drift against the bearing housings.
Insert the spider into the trunnions of the centre section of the shaft.
Fit two bearings and circlips into the centre section trunnions.
Fit the grease nipple to the spider.
Repeat the procedure for the opposite end universal joint.
Grease the joints with a hand grease gun.

Refit the shroud halves on to the drive shaft, line up the holes and pop rivet the halves together.

Seal the shroud joints with underseal, refit the clips, place the outer shroud in position and tighten the clip.

Fit the inner oil seal track (note that the chamfer clears the radius on the drive shaft) and spacer to the splined shaft.

Clean and refit the camber shims, place the flanged end of the shaft in position and refit the securing nuts but do not tighten at this stage.

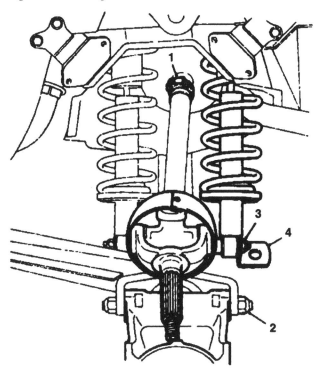

Fig. 7

Refit the damper to lower mounting pin, refit the lash down bracket (4 Fig. 7). Fit and tighten the securing nut (3 Fig. 7). Apply Loctite to the splines, check that the spacer and oil seal track are installed, raise the hub carrier and enter the splined shaft into the hub.

Tap the hub home on to the splines, replace the washer, fit and tighten the hub securing nut.

Fig. 8

Fit a new split pin.
Check the hub bearing end float.
Refit the fulcrum grease nipple into the hub carrier.

Final tighten the drive shaft flange nuts (1 Fig. 8).
Slide the inner shroud into position and tighten the clip.
Refit the road wheel.
Remove the axle stands and lower the vehicle.
Check / adjust the camber angle of the rear wheel.

Service Tools

Hub remover	JD 1D

Torque Figures

Drive shaft to final drive unit	66.5-74.5 Nm
Drive shaft to hub carrier	136-163 Nm
Lower damper mounting	41-48 Nm
Outer pivot pin	31-144 Nm

Oils / Sealants / Lubricants

Drive shaft splines	LOCTITE

PROPELLER SHAFT, RENEW

Drive the vehicle on to a ramp. Raise the ramp.
Slacken off the front to intermediate exhaust pipe clamp bolt and separate the two sections of the system to permit removal of the heat shield.

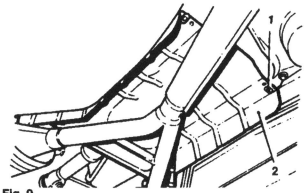

Fig. 9

Remove the four heat shield securing screws and washers (1 Fig. 9).

Pull the heat shield (2 Fig. 9) forward and down and remove from the vehicle.

Mark the propshaft flange relative to the differential flange, remove the four flange securing bolts / nuts, and remove the propshaft from the vehicle.

Fit the new propshaft to the vehicle, noting the flange relationship.

Fit and tighten the securing bolts / nuts.
Refit the heat shield.
Refit the exhaust.
Lower the ramp.

Torque Figures

Propeller shaft flange bolts	35-43 Nm
Front pipe to intermediate pipe	15-18 Nm

PROPELLER SHAFT, OVERHAUL

Note: The following is the procedure for automatic transmission-equipped vehicles. On manual gearbox-equipped vehicles the procedure is fundamentally the same although the shaft differs visually in having a splined coupling at the gearbox end.

Remove the propeller shaft assembly.
Thoroughly clean the shaft.

Secure the shaft in a vice and remove two opposed circlips from one universal joint (1 Fig. 10).

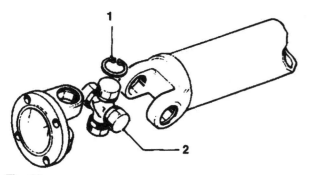

Fig. 10

Note: Tap the bearings slightly inwards to assist removal of the circlips.

Tap one bearing inwards to displace the opposite bearing.
Remove the shaft assembly from the vice and trap the displaced bearing in the vice. Tap the shaft away from the bearing.
Retrieve the bearing (2 Fig. 10).
Refit the shaft to the vice.
Displace the second bearing by tapping the joint spider along the length of the shaft yoke and extract the second bearing.
Remove the two grease seals from the spider.

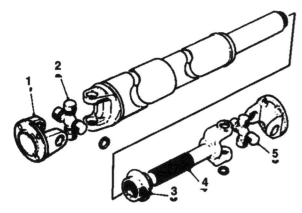

Fig. 11

Detach the spider, with the end section of the shaft (1 Fig. 11), from the centre section of the shaft.
Place the end section of the shaft in the vice and repeat the removal procedure for the remaining two bearings (2 Fig. 11).
Remove the spider from the end section of the shaft.
Repeat the removal procedure for the universal joint at the opposite end of the propeller shaft (5 Fig. 11).
Slide the gaiter along the shaft (3 Fig. 11) and mark the sleeve and shaft for reassembly.

Separate the two components of the shaft.
Wash all parts in petrol, taking care not to transpose the bearings and cups from mating journals.
Examine all bearing races and spider journals for signs of looseness, load markings, scoring or distortion.

Note: Spiders or bearings may only be renewed as a complete set. Bearing cups must be a light drive fit into the trunnions.

Examine the spline (4 Fig. 11) for wear.
Check the circumferential play. If this exceeds 0.004 in. (0,1 mm) measured on the spline outside diameter, the complete propeller shaft must be renewed.
Remove the bearing assemblies from one replacement spider and ensure that they are properly greased. Fit the spider to one end section of the shaft.
Fit two bearings and circlips into the end section trunnions. Use a soft round drift against the bearing housings.
Insert the spider into the trunnions of the centre section of the shaft.
Fit two bearings and circlips into the centre section trunnions.
Repeat the procedure for the opposite end universal joint.
Check the stiffness of each joint, which should be between 0,035 and 0,092 kgf.m (3 and 8 lbf. in.) about either axis.
Exercise the joints if necessary to reduce stiffness.
Fit a new gaiter to the sleeve yoke.
Grease the splines.

Note: If the shaft is not to be fitted immediately, retain in a fully compressed position with 18 swg. (1,2 mm) wire between the flange yokes.

Service Tools

Engine support bracket	MS 53B

Torque Figures

Propeller shaft flange bolts	35-43 Nm
Exhaust pipe flange	15-17.5 Nm
Rear mounting front bolts	19-24 Nm
Rear mounting rear bolts	36-43 Nm
Rear mounting spigot	34-41 Nm

Oils / Sealants / Lubricants

Exhaust down pipe	TIVOLI KAY ADHESIVES 5696

Chapter 11

FRONT SUSPENSION (93.5 MODEL YEAR)

SUSPENSION UNIT, RENEW

Place the vehicle on a ramp.
Disconnect the battery.
Remove the left and right hand air cleaner elements.
Remove the left and right hand, upper engine mounting nuts.
Using a syringe, clear all fluid from the power steering reservoir.
Take the hooks from the Engine Support Tool, Service Tool MS 53A, and position them in the engine front lifting eyes.
Place the tool MS 53A in position over the hooks and locate in the wing channels (Fig. 1).
Remove the left hand shock absorber thread protector.
Remove the shock absorber securing nuts (1 Fig. 2).
Remove the washer, micron buffer and cup washer (2, 3 and 4 Fig. 2).
Repeat the procedure on the right hand shock absorber.

Jack up the front of the vehicle and support on two stands.
Remove the wheel and tyre assemblies.

From the wheel arch:
Unscrew the caliper brake pipe to brake hose securing union nut (1 Fig. 3). Fit a blanking plug to the pipe.

Remove the brake hose lock nut (3 Fig. 3) at the caliper abutment bracket.
Reposition the hose (4 Fig. 3) from the abutment bracket.
Remove the left hand anti-roll bar to anti-roll bar link securing nut (1 Fig. 4).

Remove the cup washer and mounting rubber.

From the engine bay:
Disconnect the left hand ABS sensor harness connector.
Remove the sensor harness 'P' clip securing bolt.
Remove the 'P' clip.
Remove the harness grommet from position in the inner wing.
From the wheel arch:
Displace the sensor harness from the retaining bracket.
Repeat the procedure on the right hand sensor harness, brake pipe / hose and anti-roll bar.
Raise the vehicle on the ramp.

Note: On convertible models only, remove the front cruci-form bracing assembly.

Displace the anti-roll bar from the anti-roll bar links and position aside.
Unscrew the securing bolts and remove the right hand steering rack heat-shield.
Remove the right hand down-pipe catalytic converter.
Turn the steering to allow access to the lower column to steering rack pinion securing bolt (1 Fig. 5).

Remove the securing bolt.
Lower the ramp.

From inside the vehicle:
Remove the lower steering column to upper universal joint securing bolt (1 Fig. 6).

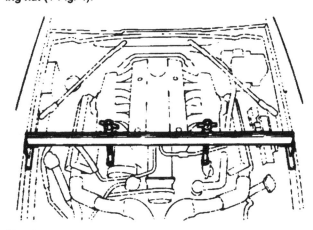

Fig. 1

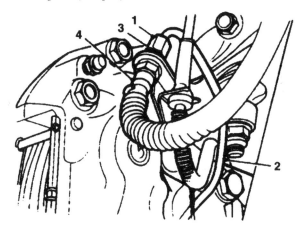

Fig. 3

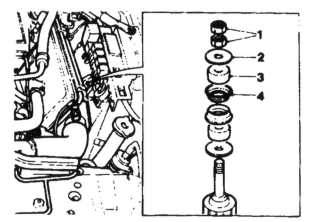

Fig. 2

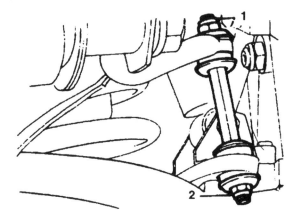

Fig. 4

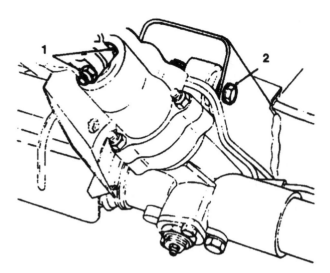

Fig. 5

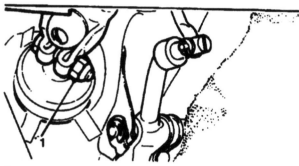

Fig. 6

Rotate the steering to give access to, and remove, the remaining universal joint securing bolt.

Ease the upper universal column clear from the bottom of the upper column.

Pull the lower column up into the vehicle and clear of the pinion.

Raise the vehicle on the ramp.

Remove all the screws securing the under-tray and remove the under-tray.

Remove the power steering feed hose union nut from the steering pump.

Reposition the hose.

Slacken the power steering return hose to pump securing clip.

Reposition the return hose and remove the hose clip.

Ensure the disconnected hoses are fitted with blanking plugs.

Remove the engine and chassis earth strap from the front beam.

Align a jack to the suspension unit and jack up the unit.

With the jack in place, remove the suspension forward mounting securing nuts/bolts (3 Fig. 7).

Note: Fig. 7 shows the under-tray removed.

Remove the spacers between the mounting and body.

Remove the rear beam mounting securing nuts. Carefully lower the jack with suspension unit from the vehicle.

Position a pulley block over the suspension unit and attach a lifting chain from the unit to the block.

Raise the lifting chain to clear the suspension unit from the jack.

Swing the pulley block clear of the jack and lower the suspension unit safely to one side.

Remove the lifting chain from the suspension unit and pulley block.

Fitting a new suspension unit is a reversal of the removal procedure.

Upon completion, refill the power steering reservoir to the correct level.

Reconnect the battery.

Bleed the ABS hydraulic system.

Remove the vehicle from the ramp.

Special Tools

Engine support tool MS 53A

Oils / Greases / Sealants

Brake fluid to a minimum DOT 4 specification

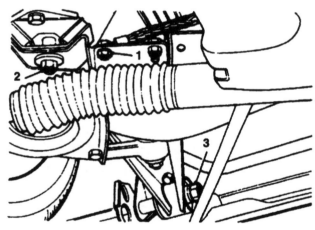

Fig. 7

Chapter 12

REAR SUSPENSION (93.5 MODEL YEAR)

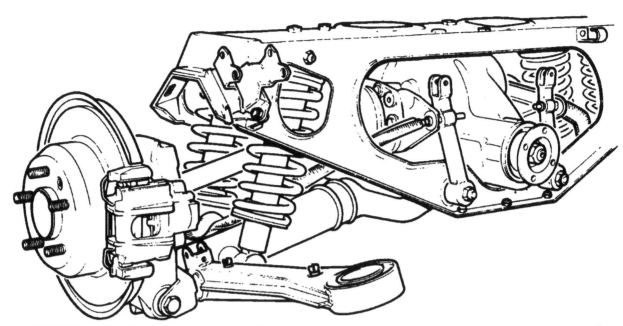

Fig. 1   1993.5 Model Year Rear Axle Arrangement

REAR HUB BEARING, RENEW

REAR HUB OIL SEAL, RENEW

REAR HUB CARRIER FULCRUM BEARING, RENEW

These three Renew Operations are covered in full within the **BRAKES** Section of this Supplement.

Chapter 13

ANTI-LOCK BRAKING SYSTEM

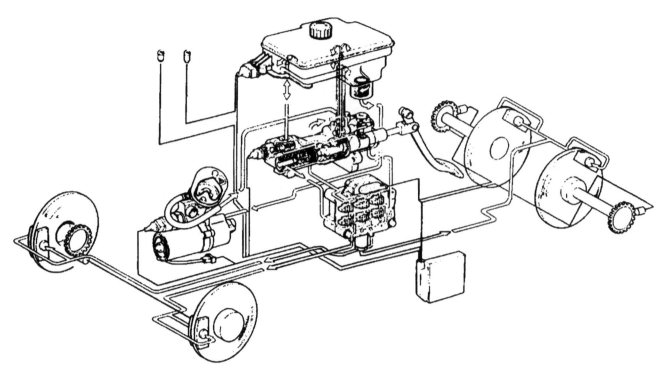

DESCRIPTION

The hydraulic components necessary for the anti-lock control are integrated with a servo-hydraulic booster and its operating components; brake fluid is used as the single working fluid. An energy source consisting of an electric motor, a pump, a pressure switch and an accumulator provides high pressure fluid to operate the booster during normal braking as well as providing a means of increasing brake line pressure when required during anti-lock braking (ABS). Also contained in this assembly are the master cylinder the solenoid valves and the reservoir.

During installation the front hydraulic circuit is operated conventionally by the master cylinder, assisted by the booster. The rear circuit is operated directly by the controlled, pressurized fluid in the booster. The front circuit is static and the rear circuit dynamic .

However when the anti-lock control is required, the front also becomes dynamic. Then inlet and outlet solenoid valves operate in each of the three circuits to control the pressure as required to prevent wheel-lock.

Sensors are installed at each wheel. Their wheel speed related signals are processed by an electronic control module (ECM) which operates the solenoid valves in the hydraulic system. The front wheels are controlled individually and the rear wheels, which are on a single hydraulic circuit, are controlled together on the select low principle. Therefore a tendency for one wheel to lock results in control of both wheels according to the need of the 'locking' wheel.

The state of the anti-lock system is continuously monitored by the ECM, which automatically switches off the system if a failure is identified, illuminating a warning lamp and leaving full, boosted braking to all wheels. Warning lamps also indicate low accumulator pressure or low fluid level in the reservoir.

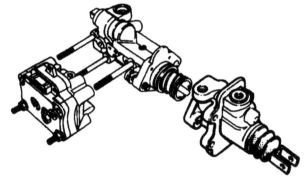

Fig. 1

If the front hydraulic circuit fails, pedal TRAVEL will increase. If the rear circuit fails, pedal EFFORT will increase.

Hydraulic Brake Booster

One important aspect of the ABS system is the hydraulic booster (Fig. 1) which boosts the pedal force by means of hydraulic pressure. The dynamic circuit of the rear brakes is supplied from the hydraulic accumulator via a control valve in the booster.

The pressure in the booster and the rear brake circuit is proportional to the pedal force i.e.:

Low pedal force- low pressure
High pedal force- high pressure

The booster comprises an actuating piston (1 Fig. 2) and a booster piston (2 Fig. 2). The movable mechanical connection between the control valve (3 Fig. 2) and the two pistons is made by means of a scissor-lever mechanism (4 Fig. 2). The control valve (3 Fig. 2) opens the unpressurized booster chamber to the reservoir (9 Fig. 2); simultaneously the channel from the hydraulic accumulator is closed. The accumulator is constantly maintained in an operating pressure range between 140 to 180 bar.

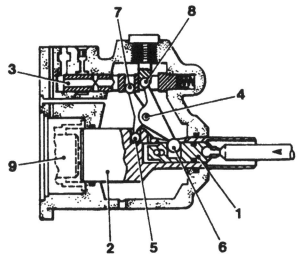

Fig. 2

As force is applied to the brake pedal, the actuating piston (1 Fig. 2) with the scissor (4 Fig. 2) moves forward. The two lower articulated balls (5 & 6 Fig. 2) move towards one another while the upper balls (7 & 8 Fig. 2) move apart. Due to this movement, the control valve (3 Fig. 2) opens the intake channel from the accumulator just after it closes the return flow. In the brake booster, a pressure is built up which is transmitted to the rear brakes and which acting simultaneously on the booster piston (2 Fig. 2) boosting the actuating force on the master cylinder piston. At the same time the pressure acts between the booster piston (2 Fig. 2) and the actuating piston (1 Fig. 2) separating the two parts. The lower articulated balls (5 & 6 Fig. 2) move apart whilst the upper balls move towards one another, this movement closes the intake by means of the control valve; the return flow remains closed.

The control valve (3 Fig. 2) is closed when the pressure acting on the actuating plate (1 Fig. 2) causes a force which is equal to the preset pedal force, i.e. when there is a balance of forces. The pressure acting on the annular circle of the booster piston (2 Fig. 2) increases the pedal force. The pedal force is increased in the ratio 1: 4, the booster ratio depends on the ratio of actuating piston area (1 Fig. 2) to booster piston area (2 Fig. 2).

The pressure in the booster is proportional to the pedal force. In maximum braking position, the control valve is open completely, the entire accumulator pressure of 180 bar acts on the booster piston. The maximum possible brake boosting is utilized. The brake pressure to the front wheel brakes can only be increased when the pedal force is increased. The brake pressure in rear wheel brakes cannot exceed 180 bar even in the case of the pedal force increase.

Electronic Control Module (ECM)

The ECM processes the signals from the four wheel sensors, converts their frequency information into values which correspond to wheel speed and then, with the data received, controls the solenoid valves during braking under ABS control.

The ECM checks the input and output signals in order to indicate any ABS disturbances. A self-monitoring function is integrated in the ECM which, in the case of a system failure, illuminates the ABS warning lamp.
During ABS control the respective solenoid valves in the wheel circuit concerned are controlled. The main valve is controlled when a front wheel is under ABS control. In order to enable the wheel to transmit the optimum brake

force under all road conditions, the control of the valves must be operated very rapidly, up to 6 times a second.

When the ignition is switched on, battery voltage is supplied to pin 2 of the ECM which is fed internally to pin 8. From pin 8, the main relay coil is activated closing the relay contacts thereby allowing Battery Voltage to be applied to pins 3 and 20. which switches on the ECM thus starting the test routines. The module is protected by a 30A fuse.

During the test routines, the ABS warning lamp is illuminated by being switched to earth via pin 1. The time the lamp is illuminated depends on the charging of the hydraulic accumulator. If the system test routines prove satisfactory the electronic control module opens the earth circuit between pins 1 and 27, the warning light is then extinguished. The warning lamp is supplied with battery voltage from the ignition switch when the ignition is switched on.

ABS and Brake Warning Lamp

The ABS warning lamp warns the driver in the case of a malfunction in the ABS system i.e. loss of pressure and indicates that the ECM has switched off the ABS system. The consequence is the loss of the anti-lock function, the conventional brake with brake boosting is, however, maintained.

In the case of a failure due to faulty plug connections, broken cables or defective components, such as sensors and solenoid valves, the voltage supply to the main relay is switched off by the ECM via terminal 8. The relay de-energises disconnecting the supply to pins 3 and 20.

Voltage (Refer to circuit Fig. 3) is available to the ABS warning lamp from the ignition switch. The earth circuit for the warning lamp is via the diode and contacts 30/87a of the main relay.
If the ECM is not connected the warning lamp will illuminate because it is connected to earth via the main relay.

Should the ABS and the brake warning lamp illuminate simultaneously brake pedal feels normal but the fluid is low.

The brake warning is illuminated because an earth connection for the warning lamp is made via the switch contacts located in the reservoir.

The ABS warning lamp is illuminated because fluid level switch is open and the circuit between terminals 9/10 of the ECM is broken. During the system self test, the ECM will

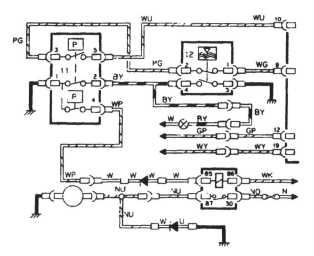

Fig. 3

419

recognize this failure and switch terminal 27 from the warning lamp to earth, at the same time, the system will be partially inhibited.

Should the pedal feel hard (after some brake applications) i.e. in the case of accumulator pressure below 85 bar there is no boost braking.

The brake warning lamp illuminates because the pressure is below 105 bar. The warning lamp switch contacts in the combined pressure switch close so a circuit is made to earth via terminals 1 and 2.

The ABS warning lamp illuminates simultaneously because the warning switch contacts open breaking the circuit between terminals 9 and 10 of the ECM. During the system self test, the ECM recognizes this failure and switches terminal 27 from the warning lamp to earth, at the same time, the system will be partially inhibited.

Pump Motor Operation

With the pressure in the hydraulic accumulator below 140 bar the motor will switch on.

When the ignition is switched on battery power is supplied to the relay coil, if the pressure in the hydraulic accumulator is below 140 bar, the coil will be earthed.

Battery power is then supplied to the pump motor via the closed contacts of the relay and a 30 amp fuse.

The motor operates until a pressure of 180 bar is achieved. Having reached a pressure of 180 bar the pressure switch contacts open de-activating the relay thus switching off the pump motor. When the pressure drops to 140 bar the pressure switch closes, switching on the pump motor so that the system pressure is maintained at between 140 to 180 bar.

Motor Pump Unit

With the motor pump unit, an independent energy supply is obtained by means of a motor and a pump which generates hydraulic energy, and accumulates the pressure energy in a hydraulic accumulator. From the hydraulic accumulator the pressure supply for the dynamic circuit of the rear wheel brakes, the hydraulic brake booster and the static circuit of the front wheel brakes is provided during a braking with ABS control.

The motor (1 Fig. 4) drives via a coupling (2 Fig. 4) a rotor (3 Fig. 4) which includes two pistons (4 Fig. 4) and two balls (5 Fig. 4) which move in an eccentric ring (6 Fig. 4). Brake fluid is drawn via the suction channel (7 Fig. 4), a filter (8 Fig. 4), through the control shaft (9 Fig. 4) on the upper side of the lower piston. The rotation causes a reduction of space due to the eccentric ring and ball. The piston is moved towards the control shaft and a pressure is generated. The pressure opens a check valve (10 Fig. 4),is transmitted to the accumulator and to the annular chamber of the control piston in the booster. Simultaneously the pressure acts on the tappet (11 Fig. 4) of the combined pressure warning switch (12 Fig. 4) and moves it till the system pressure of 180 bar is achieved, the pressure switch in the combined pressure switch switches the motor off.

The hydraulic system is protected against damage by the pressure control valve which releases pressure at 210 bar.

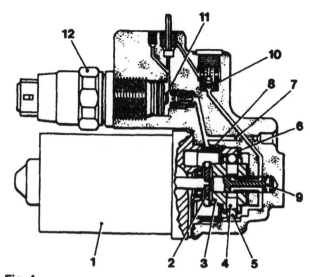

Fig. 4

Hydraulic Accumulator

The hydraulic accumulator (Fig. 5) accumulates the hydraulic pressure and makes it available to the hydraulic system for the booster and rear wheel brakes.

Owing to the fact that fluids are almost incompressible they cannot be used for an accumulation of energy. The additional use of compressible nitrogen gas in the hydraulic accumulator allows an energy accumulation. The gas and fluid must be separated by means of a membrane (1 Fig. 5).

The hydraulic accumulator has a reservoir which is divided into two chambers by a membrane (1 Fig. 5). The upper chamber (2 Fig. 5) is filled with nitrogen gas to an initial pressure of 84 bar. The lower chamber (3 Fig. 5) is filled with brake fluid supplied by the pump.

With an increasing amount of brake fluid the pressure in the system also increases, the nitrogen gas is compressed and the gas pressure is increased i.e. the nitrogen volume becomes smaller. Equally the accumulator volume for the brake fluid increases up to the cut-off pressure of 180 bar. The pressure in the accumulator is maintained by means of a check valve which is integrated in the motor pump unit and is available up to the annular chamber of the control piston.

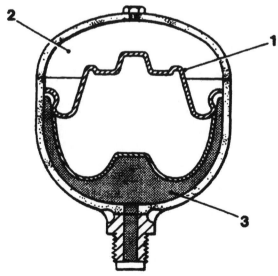

Fig. 5

Combined Pressure Warning Switch

The pump generates a pressure in the system which acts on the tappet (1 Fig. 6). The tappet moves against a spring (2 Fig. 6). At a pressure of 130 bar the warning switch (3 Fig. 6) is actuated.

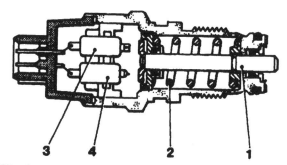

Fig. 6

First, the brake warning light will be extinguished followed a short time after by the ABS warning light. The tappet (1 Fig. 6) is moved further by pressure and opens the pressure switch (4 Fig. 6) at a pressure of 180 bar to open the pump relay coil circuit so stopping the pump motor.

If, after a number of brake applications or normal leak down, the system pressure falls below 140 bar, the pressure switch (5 Fig. 6) closes and the pump will operate until the cut-off pressure of 180 bar is reached. In this way, the system pressure is maintained within the range of 140 to 180 bar. If due to a hydraulic failure or a electrical failure, a pressure drop below 105 bar occurs, the brake warning switch (4 Fig. 6) contacts close. The brake warning lamp lights, simultaneously the ABS warning switch (3 Fig. 6) contacts open and the circuit between terminals 9 and 10 of the ECM is broken. The ECM is partially inhibited, ABS control for the front wheel brakes is switched off and the ABS warning light is illuminated.

Wheel Sensors

At each road wheel a speed sensor and a toothed rotor is fitted. By means of a magnetic core (1 Fig. 7) in a coil (2 Fig. 7) a sinusoidal alternating voltage signal is generated by a toothed rotor (3 Fig. 7) breaking the magnetic field and inducing voltage in the coil. The frequency of which is dependent on the wheel speed. The voltage is transmitted to the ECM via screened cables. The gap between the

sensor and the toothed wheel is very important i.e. a large gap will generate a low voltage, and a small gap a high voltage.

Valve Block

The valve block contains three pairs of solenoid valves, one pair for each of the front brakes and one pair for the rear brakes. Each pair contain an outlet and an inlet valve. The valves are electronically operated by signals from the ECM. During braking with ABS control, the ECM provides a voltage to the inlet and outlet solenoid valves, which influence the hydraulic pressure to the brakes. The control to the rear brakes is determined by the wheel which first shows a tendency to lock. The brake pressure at both rear wheels is thus determined by the wheel having the lowest friction coefficient.

This ensures that in the case of braking on a surface with low friction coefficient neither of the rear wheels will lock.

All the valves with the exemption of the main valve have a common earth point which is connected to pin 11 of the ECM.

The earth connection pin 11 is known as the reference earth, via this connection the ECM receives test pulses for the valves.

During normal braking the anti-lock system will not be activated. However, if the braking force applied is sufficient to overcome tyre/road adhesion the anti-lock system will automatically be activated preventing the road wheel from locking.

With no current flowing through the inlet solenoids the valves are open so that during braking the brake pressure can be applied direct to the wheel brakes.

With no current flowing through the outlet solenoids the valves are closed and disconnect the wheel brakes from the reservoir.

To maintain pressure the outlet valve (1 Fig. 8) stays closed, and the inlet valve (2 Fig. 8) to a wheel with a tendency to lock' closes, ensuring the brake pressure to that wheel cannot be increased. To decrease pressure the inlet

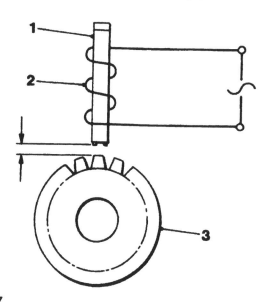

Fig. 7

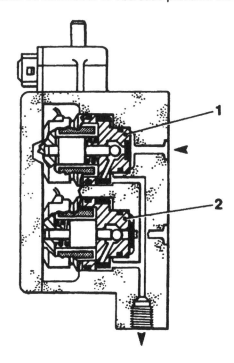

Fig. 8

valve closes and the outlet valve opens. The brake pressure to the wheel is decreased.

To increase the pressure to the wheel the inlet valve opens and the outlet valve closes and the brake pressure to the wheel is again increased almost up to the locking pressure limit.

These phases are repeated up to six times a second until the tendency for the wheel to lock is eliminated.

Main valve

The main valve is a solenoid operated valve which during ABS controls supplies the front wheel brakes with pressure and causes a push back of the pedal by applying a pressure to the positioning sleeve. In a braking position without ABS control the solenoid valve (1 Fig. 9) is de-energized and connects the reservoir (2 Fig. 9) with the master cylinder (3 Fig. 9).
The connection between the booster, the master cylinder and the return flow is closed.
At the beginning of ABS control the main solenoid valve is energized and connects the dynamic circuit of the booster with the static circuit of the master cylinder.

The connection between the master cylinder and the reservoir is closed.

Master cylinder

The master cylinder (Fig. 10) acts exclusively on the front wheel brakes, the static circuit. The pressure in the master cylinder is generated by means of force acting from the booster piston to static fluid column.

In addition, the positioning sleeve and the main valve are integrated in the master cylinder.

The brake pressure to the rear wheel brakes is supplied by the hydraulic accumulator through the control valve.

As pressure is applied to the brake pedal, the master cylinder piston (1 Fig. 10) is moved forward by the booster piston (5 Fig. 10). The control valve (2 Fig. 10) closes and a pressure is built up in the front wheel brakes. Simultaneously to the forward movement of the booster piston, the positioning sleeve (3 Fig. 10) is pulled to the left. The main valve (4 Fig. 10) is in an inoperative position. The connection between the master cylinder and the reservoir is open, the connection between the booster and the master cylinder is closed.
If there is a wheel locking tendency during braking at one or more front wheels, the main valve (4 Fig. 10) is controlled by the ECM. The main valve closes the connection between the static circuit of the master cylinder and the reservoir.

Simultaneously, the connection between the dynamic circuit in the rear wheel brakes, the booster, and the static circuit in the front wheel brakes is made.

The dynamic pressure is applied to the positioning sleeve (3 Fig. 10) which is moved to its stopping point. This movement causes the push back of the booster piston and of the pedal. At the same time, the dynamic flow-in over the primary seal (6 Fig. 10) of the master cylinder piston (1 Fig. 10) and thus the direct pressure supply in the wheels occurs.

The pressure on the right side of the primary seal is higher than on the left side and due to this the primary seal is pushed forward.

Due to the pressure compensation, the master cylinder piston (1 Fig. 10) is pushed to the booster piston (5 Fig. 10).

The direct flow into the front wheel brakes and the push back of the pedal, avoids a pedal pulsation during braking under ABS control and guarantees at the same time that sufficient reserve is available in case of rear circuit failure.

Positioning sleeve

The positioning sleeve is a safety feature which ensures that in the case of a circuit failure in the rear wheel brakes there is sufficient master cylinder stroke left during ABS controlled braking. In its inoperative position the sleeve is at its stopping point. The chamber left of the positioning sleeve is connected with the reservoir via the main valve. The chamber right of the sleeve is connected direct with the reservoir.

Braking without ABS control, pressure is applied to the brake pedal, the booster piston moves to the left pushing the positioning sleeve to the left against the spring tension.

Braking on surfaces with a high friction coefficient under ABS control, pressure is applied from the dynamic circuit by opening the main valve. The positioning sleeve moves to the right to its stopping point and pushes back the booster piston as well as the brake pedal.

At the beginning of ABS control the brake pedal has a hard pedal feel.

During braking on surfaces with a low friction coefficient the booster piston and the brake pedal moves gradually to the stopping point of the positioning sleeve. Therefore, no push back movement of the pedal can be felt.

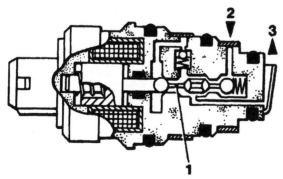

Fig. 9

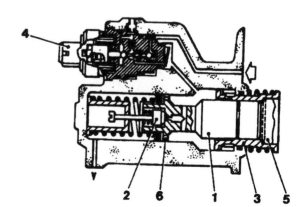

Fig. 10

Reservoir

The reservoir is made up of separate chambers. The most important chambers deliver supply to:

1. The front wheel brakes, (1 Fig. 11) i.e. for the master cylinder via the main valve.
2. The motor pump unit, i.e. for the booster and rear wheel brakes (2 Fig. 11).
3. For the return flow from the booster (3 Fig. 11), the rear brakes (2 Fig. 11), the valve block and the position sleeve (4 Fig. 11).

If due to a leakage, e.g. in the front wheel brakes, brake fluid is lost, the necessary amount of fluid for operating the rear wheel brakes is still available. The ECM is partially inhibited, the ABS for the front wheel brakes is switched off and the ABS warning light is illuminated.

In the case of a leakage in the rear wheel brakes, a residual amount of fluid is available for operating the front wheel brakes to decelerate the vehicle. The ECM is partially inhibited, the ABS for the front wheel brakes is switched off and the ABS warning light is illuminated.

The fluid level is monitored by two reed contacts on different levels inside the reservoir. The contacts are switched by contact plates (1 Fig. 12) on a float stem.

With fluid loss, the lower reed contact (2 Fig. 12) switches on the brake warning lamp, the same as a conventional brake system.

With further loss of brake fluid the upper reed contact (3 Fig. 12) is opened. The circuit between terminals 9 and 10 is broken. The ECM senses the open circuit during its continuous tests and is partially inhibited. The ABS for the

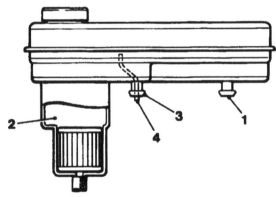

Fig. 11

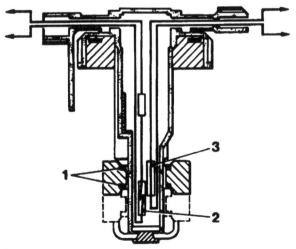

Fig. 12

front wheel brakes is switched off and the ABS warning lamp is illuminated.

WARNING LAMP INDICATIONS WITHOUT ERROR CODE OUTPUT

The on-board diagnosis can only monitor errors that generate electrical signals. The error code information is triggered by the diagnosis trigger input, and displayed by the warning lamp.

TEST CYCLE FOR WARNING SWITCH PATH:

After the ignition is switched 'ON' (providing the brake pressure warning light is 'OFF'), the warning lamp (WL) remains 'ON' for approximately 1.7 seconds. Then it flickers for approximately one second to test the reservoir and pressure switch path. If the warning lamp flickers continuously, this path is open or short circuited to ground potential.

IMPROPER INSTALLATION:

If the main connector is not installed in the ECM (or if the connector is loose), and the ignition is switched 'ON' (position 2), the main relay remains 'resting' and the warning lamp is switched 'ON' by the 'resting' contact of the main relay.

FAILURES OF THE ECM

FAILURES DETECTED BY INTERNAL TIME-OUT:

Certain hardware faults cause the ECM to be switched off by internal time-out. Any hardware fault will cause the warning lamp to light continuously and, since main power is cut off, the ECM is no longer capable of storing/outputting failure codes.

SHORT CIRCUIT AT THE DIAGNOSIS TRIGGER INPUT:

If the diagnosis trigger input is shorted to ground potential, the ECM goes into the diagnosis output mode when the ignition is switched on and if a failure is stored in the continuous memory. As the car accelerates and reaches 8 km/hr (5 mph), the short to ground on the diagnosis trigger unit still exists, the ECM is switched off, and the warning lamp lights continuously.

WARNING LAMP PATH FAULTS:

Short circuiting the warning lamp wire to ground potential will activate the warning lamp, but will not affect the anti-lock braking facility. The ECM cannot recognise this short circuit.

DEFECTIVE WARNING LAMP DRIVER:

If there is a defect in the warning lamp driver inside the ECM, either the warning lamp will remain continuously 'OFF', or will stay continuously 'ON', depending on the internal failure cause.

MISCELLANEOUS WARNING LAMP DISPLAYS/ CONDITIONS:

In the case of intermittent V defective contacts or leads in the warning lamp driving path, the warning lamp may flicker 'ON/OFF' for undefined periods.
If the warning lamp is 'blown-out' or otherwise damaged/destroyed, no information about the status of the ECM is possible.
Note: The driver will realise that the ABS warning lamp circuitry is faulty, because the lamp will not illuminate on the ignition cycle.

FAULT DIAGNOSIS

FAILURE	EFFECT	RESULT	INDICATOR
Brake fluid low	Requires topping up	Brake fluid low at level 1	Brake warning lamp on
Broken sensor			No ABS
Partial intermittent failure on front axle		No ABS on rears only	ABS warning lamp on
Partial intermittent failure on front axle above 40 km/hr (25 mph)		No ABS	ABS warning lamp on
Partial intermittent failure on front axle above 20 km/hr (12 mph)		No ABS	ABS warning lamp on
Partial intermittent failure on front axle above 20 km/hr (12 mph)		No ABS	ABS warning lamp on
Pressure Switch connection broken	Accumulator will not charge	Loss of power assistance Unboosted front brakes only. No ABS	ABS warning lamp on
30 A main fuse blown (pump motor)	Accumulator will not charge	Loss of power assistance. Unboosted front brakes only No ABS	ABS and brake warning lamps on
30 A ABS fuse blown		No ABS	ABS warning lamp on
Pump connection broken	Accumulator will not charge	Loss of power assistance. Unboosted front brakes only. No ABS	ABS and brake warning lamps on when pressure drops.
Brake fluid low at level 2		Boosted brakes. ABS on rear only	ABS and brake warning lamps on
Failed front hydraulic circuit	Loss of fluid to level 2	Boosted rears with ABS only.	ABS and brake warning lamps on
Failed rear hydraulic circuit*	Loss of fluid	Unboosted front. brakes only. No ABS	ABS and brake warning lamps on

*Note: If the front hydraulic circuit fails, pedal TRAVEL will increase. If the rear circuit fails, pedal EFFORT will increase.

BRAKE WARNING LIGHT SWITCH

During warning light switch renewal/adjustment, the operator must ensure that the brake pedal is fully returned against its stop PRIOR TO SETTING THE SWITCH. Failure to do this may result in a no-warning lights condition.

BRAKE FLUID LEVEL

Correct brake fluid level is essential for the efficient operation of the brake system.

There are two 'MAX' marks on the reservoir. The brake fluid level must be at the highest 'MAX' level on the reservoir (Fig. 13).

Note: In some cases the fluid may be above the 'MAX' mark, this is dependent upon the charged state of the hydraulic unit. Therefore the following procedure for checking or topping up the hydraulic brake fluid level must be followed.

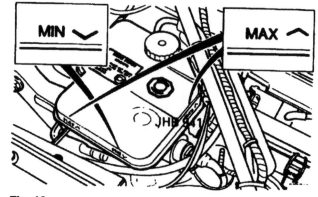

Fig. 13

424

CAUTION:

Fluid must not be allowed to contact the vehicle paintwork. Remove any spilt fluid from the paintwork by rinsing away with running water.

1.   Ensure the vehicle is on a level surface.
2.   With the ignition switched 'OFF' pump the brake pedal at least 20 times, or until pedal travel becomes hard.
3.   Switch the ignition 'ON'.
4.   Wait for the pump to stop running.
5.   Check brake fluid level. Fluid level must be at the highest 'MAX' level on the reservoir (Fig. 13). Top up using the recommended brake fluid.

The efficiency of the brakes may be impaired if fluid is used which does not meet JAGUAR specifications. Use ONLY Jaguar Brake and Clutch Fluid or Castrol Girling Universal to/or exceeding DOT 4 specification.

Also do not use brake fluid that has been exposed to atmosphere for any length of time. Moisture absorbed from the atmosphere dilutes the fluid and impairs its efficiency.

BRAKE HYDRAULIC SYSTEM - ABS, BLEEDING

CAUTION:

Fluid must not be allowed to contact the vehicle paintwork. Remove any spilt fluid from the paintwork by rinsing away with running water. Methylated spirit (Denatured alcohol) MUST NOT be used to clean the contaminated area.
Use only Castrol/Girling brake cleaning fluid.

CAUTION:

Never use methylated spirit (Denatured alcohol) for component cleaning purposes.

CAUTION:

Throughout the following maintenance/service operations, absolute cleanliness must be observed to prevent grit or other foreign matter contaminating the brake system.

CAUTION:

The following notes MUST BE read before commencing bleed procedure.

Bleeding the ABS brake system is not a routine maintenance operation and should only be necessary when either, part of the system has been disconnected or, the fluid is contaminated.

During the bleeding procedure, it is important that the level of fluid in the reservoir is maintained at approximately 2 mm below the bottom of the filler neck.

Note: The motor/pump unit cannot charge the accumulator if air, and not fluid, is standing on the low pressure side of the pump.
Therefore, if the motor/pump unit, the fluid intake hose, or the hydraulic unit of the ABS system have been renewed/disconnected, the FLUID INTAKE HOSE bleeding procedure must be used.

For all other maintenance work, use the FRONT BRAKES/REAR BRAKES bleeding procedure(s).

CAUTION:

DO NOT allow the pump motor to run for more than two minutes. If the motor does exceed the two minute time limit, immediately switch off the ignition (POSITION 'O'). Allow the motor to cool for at least ten minutes before continuing with the bleed procedure.

BLEED PROCEDURE (FRONT BRAKES) - FOR ALL MAINTENANCE WORK, EXCLUDING: THE MOTOR/PUMP UNIT, FLUID INTAKE HOSE, OR ACTUATION UNIT.

Ensure that the vehicle is standing level. Switch the ignition off. Discharge the hydraulic accumulator by operating the brake pedal (approximately 20 times) until the pedal travel goes hard. Top up the reservoir to approximately 2 mm below the bottom of the filler neck (Fig. 14).

Bleed the caliper furthest away from the actuation unit first, i.e., the left hand caliper on RHD cars, the right hand caliper on LHD cars.
Note: The following procedure must be done at a rate not faster than 3 seconds per cycle.

Open the bleed nipple (1 Fig. 15), fully depress the brake pedal, and hold down. Close the bleed nipple after a minimal one second. Release the pedal slowly.
Repeat until the fluid flows air free.

Note: If more than 20 cycles are required, the reservoir must be topped up.

Bleed the remaining caliper using the same procedure.

BLEED PROCEDURE (REAR BRAKES) - FOR ALL MAINTENANCE WORK, EXCLUDING: THE MOTOR/PUMP UNIT, FLUID INTAKE HOSE, OR ACTUATION UNIT.

Ensure that the vehicle is standing level. Switch the ignition off. Discharge the hydraulic accumulator by operating the brake pedal (approximately 20 times) until the pedal travel goes hard. Top up the reservoir to approximately 2 mm below the bottom of the filler neck (Fig. 14).
Open one bleed nipple (Fig. 16), fully depress the brake pedal, and hold down. Switch on the ignition and wait (A MINIMUM OF 15 SECONDS) until the fluid flows air free. Close the bleed nipple. Release the pedal slowly. Switch off the ignition. Bleed the remaining caliper using the same procedure.

Note: Ensure that the fluid level does not drop more than 10 mm below the maximum mark during the above procedure.

BLEED PROCEDURE (FLUID INTAKE HOSE) - FOR ALL MAINTENANCE WORK INVOLVING: THE MOTOR/PUMP UNIT, FLUID INTAKE HOSE, OR ACTUATION UNIT.

Ensure that the vehicle is standing level.
Switch the ignition off. Top up the reservoir to approximately 2 mm below the bottom of the filler neck. Disconnect the

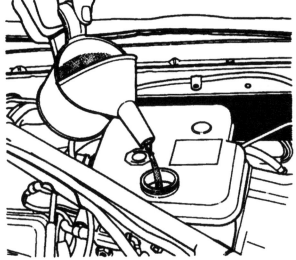

Fig. 14

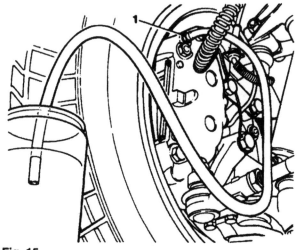

Fig. 15

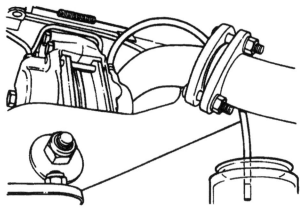

Fig. 16

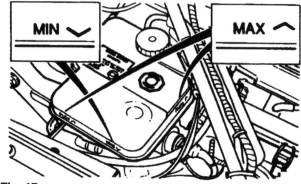

Fig. 17

fluid intake hose at the pump and allow the fluid to flow into a container until it is air free. Check that the plastic elbow '0' ring is not damaged, and reconnect the hose WHILE FLUID IS FLOWING.

Switch the ignition on and operate the brake pedal several times. If the motor/pump unit is charging, the fluid level in the reservoir will decrease. The upper cut-out point should be reached in less than 60 seconds.

ON COMPLETION OF THE BLEED PROCESS: Switch on the ignition.
Wait until the accumulator is charged to the upper cut-out point, and top up the reservoir to the maximum level (Fig. 17).

Note: If the fluid level is too high, any excess fluid must be drawn off. This can be done by using the procedure for bleeding the rear brakes.

Tighten the bleed screw to 7-13 Nm

FLUID RESERVOIR, RENEW

Note: Read the notes relating to the ABS system bleed procedure before commencing this operation.

Discharge the hydraulic accumulator by operating the brake pedal (approximately 20 times) until the pedal travel goes hard.
Open the bonnet.
Place suitable absorbent material around the reservoir area to catch any spilt fluid.

Remove the low pressure hose to pump securing clip (1 Fig. 18), disconnect the hose, remove and discard the '0' ring.
Route the end of the hose into a container. Drain the fluid. Disconnect the reservoir multi plug.

Remove the reservoir retaining clip bolt (1 Fig. 19).

426

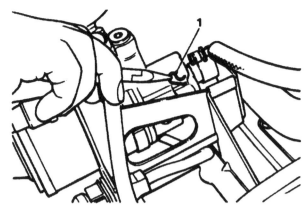

Fig. 18

Fig. 19

Displace the reservoir from position, disconnect the low pressure hose, and remove the reservoir taking care not to spill any fluid.

Remove the retaining clip, remove and discard the '0' rings/sleeve from the actuation unit.

Fit the retaining clip to the new reservoir.

Fit new '0' rings/sleeve to the actuation unit.

Place the reservoir in position. Connect the low pressure hose.

Fit the reservoir to the actuation unit ensuring correct location to the sleeve and '0' rings.

Fit and tighten the retaining clip bolt (1 Fig. 19).

Reconnect the multi plug.

Connect the low pressure hose to the pump; fit the retaining clip (1 Fig. 18).

Bleed the ABS hydraulic system.

Switch on the ignition.

Wait until the accumulator is charged to the upper cut-out point, and top up the reservoir to the maximum level (Fig. 17)

ACTUATION UNIT, RENEW

Note: Read the notes relating to the ABS system bleed procedure before commencing this operation.

Discharge the hydraulic accumulator by operating the brake pedal (approximately 20 times) until the pedal travel goes hard.

Open the bonnet.

Remove the harness to RH wing stay ratchet straps, displace the harness and remove the wing stay.

Place suitable absorbent material around the reservoir area to catch any spilt fluid.

Remove the low pressure hose to pump securing clip (1 Fig. 18), disconnect the hose and remove and discard the '0' ring.

Route the end of the hose into a container. Drain the fluid.

Disconnect the reservoir multi plug.

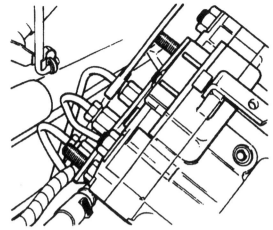

Fig. 20

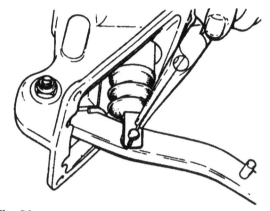

Fig. 21

Remove the reservoir retaining clip bolt (Fig. 19).

Displace the reservoir, disconnect the low pressure hose, and remove the reservoir taking care not to spill any fluid.

Disconnect the actuation unit multi plug. Disconnect the earth lead.

Place suitable absorbent material below the valve block unions to catch any spilt fluid.

Disconnect the pipes from the valve block (Fig. 20).

Disconnect the actuation unit harness.

Disconnect the high pressure pipe.

Remove the pedal box to bulkhead securing bolts, displace and remove the actuation unit/pedal box assembly.

Remove the actuation unit to pedal split pin and clevis pin (Fig. 21). Remove the actuation unit securing bolts, displace and remove the unit.

Remove the high pressure pipe union. Fit blanking plugs to all ports to prevent fluid leakage.

Remove all blanking plugs from the new actuation unit.

Fit the high pressure pipe union to the unit.

Fit the actuation unit to the pedal box, fit and tighten the securing bolts.

Connect the brake pedal, fit the clevis and split pins.

Fit the assembly to the vehicle, position the valve block pipes, fully seat the assembly.

Fit and tighten the valve block pipe unions.

Fit and tighten the pedal box securing bolts.

Reconnect the high pressure pipe.

Reconnect the actuation unit harness.

Reconnect the earth lead and multi plug.

Fit new '0' rings and sleeve (Fig. 22) to the actuation unit.

Place the reservoir in position. Connect the low pressure hose.

Fit the reservoir to the actuation unit, ensuring correct location to the sleeve and '0' rings. Fit and tighten the retaining clip bolt.

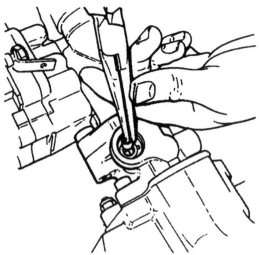

Fig. 22

Reconnect the multi plug.
Connect the low pressure hose to the pump and fit the retaining clip.
Refit the wing stay, securing the harness with new ratchet straps.
Bleed the ABS hydraulic system.
Switch on the ignition.
Wait until the accumulator is charged to the upper cut-out point, and top up the reservoir to the maximum level (Fig. 17).

Torque Figures

Bleed screw to caliper	7 - 13 Nm
High pressure pipe union	7 - 12 Nm
Pedal box to body (5/16 UNF bolt)	7 - 18 Nm
Pedal box to body (5/16 UNF nut)	1 - 18 Nm
Reservoir retaining clip bolt	6 - 8 Nm

ACCUMULATOR - ABS, RENEW

```
WARNING:

THE ACCUMULATOR CONTAINS PRESSURISED
NITROGEN GAS. DO NOT PUNCTURE
OR INCINERATE.
```

Note: Read the notes relating to the ABS system bleed procedure before commencing this operation .

Discharge the hydraulic accumulator by operating the brake pedal (approximately 20 times) until the pedal travel goes hard.

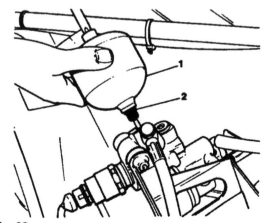

Fig. 23

Place suitable absorbent material around the accumulator area to catch any spilt fluid. Slacken off and remove the accumulator (1 Fig. 23).
Remove and discard the '0' ring seal (2 Fig. 23). Clean the surrounding area.
Fit a new '0' ring to the new accumulator.
Fit and tighten the accumulator.
Switch on the ignition.
Wait until the accumulator is charged to the upper cut-out point, and the ABS warning light is extinguished.
Switch off the ignition.

Torque Figures

Accumulator to pump	40 - 46 Nm

MOTOR AND PUMP UNIT, RENEW

Note: Read the notes relating to the ABS system bleed procedure before commencing this operation.

Discharge the hydraulic accumulator by operating the brake pedal (approximately 20 times) until the pedal travel goes hard.
Open the bonnet. Place suitable absorbent material around the reservoir area to catch any spilt fluid.
Remove the low pressure hose to pump securing clip (1 Fig. 18), disconnect the hose, remove and discard the '0' ring.
Route the end of the hose into a container.
Drain the fluid.

Disconnect the pressure warning switch and motor multi plugs (Fig. 24).

Disconnect the bundy pipe (1 Fig. 25), remove and discard the '0' rings.
Remove the motor/pump special mounting bolt.
Displace and remove the motor/pump unit.
Slacken off and remove the accumulator.
Remove and discard the '0' ring seal.
Remove the pressure warning switch. Clean the surrounding area.

Fit a new '0' ring to the pressure warning switch.
Fit the pressure warning switch to the new motor/pump unit.
Fit a new '0' ring to the accumulator. Fit and tighten the accumulator.
Fit the assembly to the vehicle, routing the harness through the mounting bracket.
Fit and tighten a new special bolt.
Fit a new '0' ring to the low pressure hose; fit the hose and retaining clip.
Fit new '0' rings to the bundy pipe and fit the pipe.

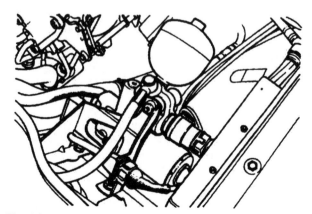

Fig. 24

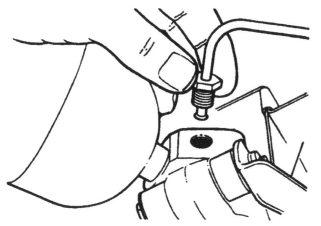

Fig. 25

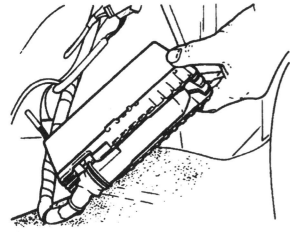

Fig. 26

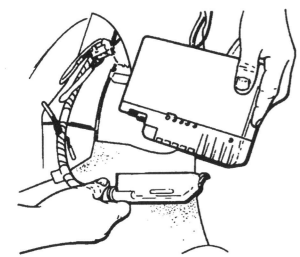

Fig. 27

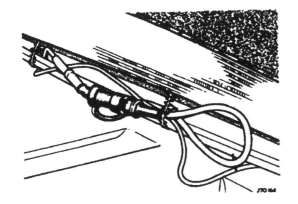

Fig. 28

Reconnect the multi plugs.
Bleed the ABS hydraulic system.
Switch on the ignition.
Wait until the accumulator is charged to the upper cut-out point, and top up the reservoir to the maximum level. Close the bonnet.

Torque Figures

Accumulator to pump	4 - 46 Nm
Bleed screw to caliper	7 - 13 Nm
High pressure pipe union	7 - 12 Nm
Pump/motor mounting bolt	7 - 9 Nm

ELECTRONIC CONTROL MODULE (ECM), RENEW

Remove the left hand side boot liner.

Release the ECM retaining strap and displace the ECM from the carrier (Fig. 26).

Disconnect the multi plug (Fig. 27) and remove the ECM.
Offer the new ECM up to position.
Connect the multi plug.
Fit the ECM to the carrier and fit the retaining strap.
Refit the boot side liner.

WHEEL SPEED SENSOR REAR, RENEW

Move the front seats fully forward. Where fitted, open the stowage compartment lid and remove the carpet.
(RH SENSOR RENEWAL ONLY). Where fitted, release the hood motor cover fasteners and remove the cover.

Displace the stowage compartment rear carpet for access to the sensor multi plug (Fig. 28).
Disconnect the multi plug.
Attach a drawstring to the sensor harness, and remove the inner grommet.
Locate a trolley jack below the outer fork of the rear wishbone.

CAUTION:

To prevent wishbone component damage,
use a suitably shaped block of wood
between the jack and wishbone.

Jack up the vehicle. Place an axle stand below the jacking spigot. Lower the car/jacking spigot on to the axle stand.
Leave the jack in position as safety measure.
Remove one road wheel nut, and mark the stud-to-wheel relationship.
Remove the remaining nuts, and remove the road wheel.

Remove the sensor securing bolt and displace the sensor from the hub (Fig. 29).
Cut and remove the harness ratchet clips.

Displace the harness outer body grommet (1 Fig. 30).
From below, carefully withdraw the harness and drawstring (2 Fig. 30).
Transfer the drawstring to the new sensor harness.
Carefully draw the new harness into position and seat the outer body grommet.

CAUTION:

Ensure that two washers (one green, one black)
are fitted to the body of the new sensor.

429

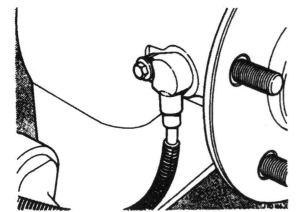

Fig. 29

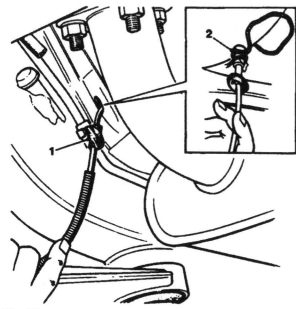

Fig. 30

Fig. 31

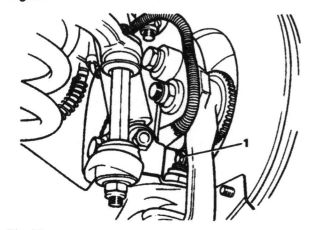

Fig. 32

Route the harness to the hub, fit the sensor to the hub.
Fit and tighten the sensor securing bolt.
Secure the harness with new ratchet straps.
Lift the road wheel up to the hub, align the wheel with the marked stud and fit the wheel.

Fit and tighten the wheel nuts to 88 to 102 Nm in the sequence shown in Fig 31.

Note: It will not be possible to fully torque tighten the wheel nuts until the car has been lowered.

Take the weight of the car with the trolley jack, remove the axle stand and lower the vehicle.
Fully seat the inner body grommet.
Route the harness, remove the drawstring and connect the multi plug.
Refit the cover, carpets etc.

WHEEL SPEED SENSOR FRONT, RENEW

Drive the vehicle on to a ramp.
Open the bonnet. Disconnect the sensor multi plug.
Displace the body grommet from the inner wing valance.
Raise the ramp.
Remove the brake disc cover plate.
Remove the retaining bracket.
Remove the upper cover securing bolt and remove the cover.

Remove the sensor securing bolt (1 Fig. 32) and displace the sensor from the hub.

Displace the harness carrier grommets (1 Fig. 33).
Displace and remove the sensor and harness.
Fit the new sensor in position, route through the inner wing and seat the harness to the carrier grommets.
Fit the sensor to the hub, and fit and tighten the securing bolt to 8 - 11 Nm.
Fit the body grommet to the inner wing (2 Fig. 33).
Refit the brake disc cover plate.
Refit the retaining bracket.
Lower the ramp.
Route the harness to the connector and connect the multi plug.
Secure the harness with new ratchet straps.
Close the bonnet.

ANTI-LOCKING BRAKING SYSTEM, PRESSURE TEST

Notes:
1.   Read the notes relating to the ABS system bleed procedure before commencing this operation
2.   Use only the Jaguar approved ABS pressure test equipment.
3.   Under ideal conditions this test should be carried out at 20 degrees Celsius (room temperature).

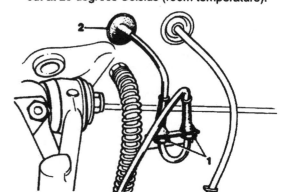

Fig. 33

Drive the car on to a ramp.
Switch the ignition to '0'.
Discharge the hydraulic accumulator by operating the brake pedal (approximately 20 times) until the pedal travel goes hard.
Open the bonnet.

Disconnect the bundy pipe from the motor/pump unit (Fig. 34). Ensure that the '0' rings and mating faces are clean.

Fit the hydraulic test point No. 07 to the high pressure outlet (Fig. 35).

Raise the lift. Remove the bleed nipples and fit the correct test points (M10 x 1) No. 6 to both front brake calipers (Fig. 36).

Remove the bleed nipple and fit the correct test point (3/8 x 24 UNF) No. 08 to one rear brake caliper (Fig. 37).
Lower the lift.
Connect the flexible pipe (with yellow protective caps) to the 250 bar gauge.
Connect the other end to the test point on the accumulator block noting that different threads are used.

CAUTION:

Always connect the flexible pipe to the gauge first, then to the test point. This is because the adaptor has a one way valve fitted to it and incorrect fitment could result in loss of fluid at high pressure.

Switch the ignition on (position 2), noting the immediate reading up to approximately 80 bar.
Note the elapsed time when the brake warning and the ABS lights extinguish - should be up to a maximum of 40 seconds with an approximate 1 second differential between the two lights.
Note the reading on the pressure gauge - should be 125 bar.

FAULT DIAGNOSIS:

1. **If the immediate reading shows less than 80 bar, there is an internal leakage within the accumulator. Change the accumulator.**
2. **If the pressure gauge does not read 125 bar when the warning lights extinguish change the pressure switch.**
3. **If the pressure gauge reads more than 190 bar change the pressure switch.**

Note the elapsed time when the pump cuts out - up to a maximum of 60 seconds.
Note the reading on the pressure gauge - should be no more than 190 bar.

FAULT DIAGNOSIS:

If the gauge does not read 190 bar or the elapsed time is greater than 60 seconds:

1. **Check the pump/motor voltage (motor multi plug pin 2). If it is above 10 volts renew the pump unit. If it is below 10 volts check for voltage drop at the pump/motor line.**
2. **Check for corroded contacts.**
3. **Check the reservoir filter: remove the hose from the pump and check for free flow through the filter hose. If free flow is evident reconnect the hose to the pump and repeat the test. If elapsed time is greater than 60 seconds renew the pump.**

Ensure that the hydraulic circuit is pressurised (pump not running) and the ignition is on (position 2).
Press the brake pedal continuously.
Note the reading on the pressure gauge when the pump is activated - should be approximately 140 bar.
Stop braking.

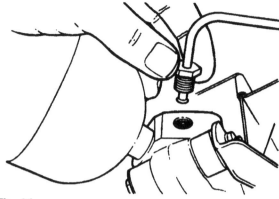

Fig. 34

Fig. 35

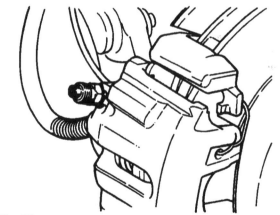

Fig. 36

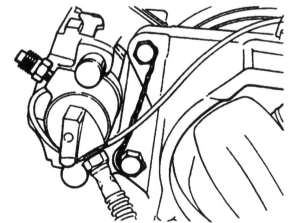

Fig. 37

431

The pressure should then be restored to a maximum of 190 bar.

Switch the ignition off (position 0).

FAULT DIAGNOSIS:

1. **If the gauge does not read approximately 140 bar when the pump is activated renew the pressure warning switch.**
2. **If maximum pressure is not reached (approximately 190 bar) renew the pressure warning switch.**

Remove the pump/motor fuse.

Switch the ignition on (position 2).

Press the brake pedal hard (approximately 7 times).

Note the reading on the pressure gauge when the warning light comes on - should be approximately 105 bar.

Continue pressing the pedal (approximately 20 times) until the pressure gauge reads 0 (zero) bar. Switch the ignition off (position 0).

Refit the pump/motor fuse.

FAULT DIAGNOSIS:

If the warning light does not come on at approximately 150 bar renew the pressure switch.

Switch the ignition on (position 2), and wait until maximum pressure is attained -190 bar.

Switch the ignition off (position 0), and wait 3 minutes for the pressure to stabilise.

Set the 'tell-tale' needle on the pressure gauge.

After 5 minutes has elapsed, check that the pressure leak is no more than 10 bar.

FAULT DIAGNOSIS:

If the pressure loss is more than 10 bar check for external leakage.
If no external leakage is evident renew the hydraulic actuation unit.

Discharge the system by operating the brake pedal 20 times.

Remove the gauge and flexible pipe from the motor/pump unit.
Raise the lift.

Fig. 38

Connect the second flexible pipe (with yellow protective caps) to the second pressure gauge.

Connect both flexible pipes to the front caliper test points.

Lower the lift.

Switch the ignition on (position 2) and wait until maximum pressure is attained (up to 60 seconds).

Switch the ignition off (position 0).

Fit the brake pedal operating tool (Fig. 38) from the brake pedal to the floor crossmember beneath the seat and adjust the knurled nut until both pressure gauges read 100 bar.

Wait 3 minutes for the pressure to stabilise.

Reset the 'tell-tale' needle on both pressure gauges.

After 5 minutes has elapsed, check that the pressure leak is no more than 5 bar.

Remove the brake pedal operating tool.

FAULT DIAGNOSIS:

If the pressure loss is more than 10 bar check for external leakage. If no external leakage is evident renew the hydraulic actuation unit.

Leaving one front gauge still in circuit, transfer a flexible pipe and pressure gauge to the rear adaptor.

Switch the ignition on (position 2), and wait (up to a maximum of 60 seconds) until maximum pressure is attained - 180 bar.

Switch the ignition off (position 0).

Fit the brake pedal operating tool (Fig. 38) and adjust the knurled nut until the rear pressure gauge reads 100 bar.

THE READINGS ON THE FRONT AND REAR SYSTEMS SHOULD DIFFER BY NO MORE THAN 7.5 BAR (HIGHER ON THE FRONT SYSTEM).

Wait 3 minutes for the pressure to stabilise.

Reset the 'tell-tale' needle on both pressure gauges.

After 5 minutes has elapsed, check that the pressure leak is no more than 5 bar.

FAULT DIAGNOSIS:

If the pressure loss is more than 10 bar check for external leakage.
If no external leakage is evident renew the hydraulic actuation unit.

Remove the brake pedal operating tool (Fig. 38).

Discharge the system.

Remove all test points from the circuit.

Refit the bleed nipples and high pressure hose, ensuring that new '0' rings are fitted to the high pressure hose.

Bleed the ABS hydraulic system.

Examine the system for external leaks.

Torque Figures

Bleed screw to caliper	7 - 13 Nm
High pressure pipe union	7 - 12 Nm

BRAKES, 1993.5 MODEL YEAR ON

INTRODUCTION

For 1993.5 Model Year, the front brake assemblies and service procedures remain unchanged.

The rear brakes and handbrake, however, change to the outboard type as fitted to 93 MY saloon models.

The handbrake cable layout/linkage varies and the method of setting and adjustment differs from previous XJS models.

Service details and operations for rear brakes and some rear suspension components are as follows:

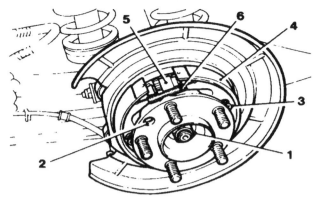

Fig. 39

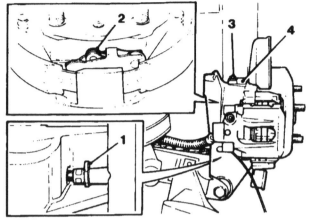

Fig. 40

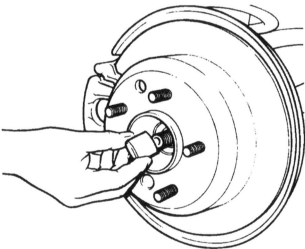

Fig. 41

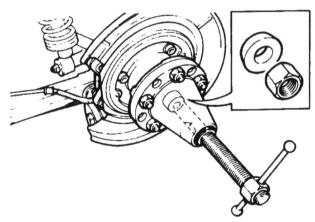

Fig. 42

REAR HUB BEARING, RENEW

Jack up the rear of the vehicle and place on stands.
Remove the roadwheel and tyre assembly.

Unscrew, but do not remove the drive shaft to hub assembly securing nut (1 Fig. 39).
Remove the rear disc.
Position the hub access hole (2 Fig. 39) to expose the front handbrake shoe retaining clip (3 Fig. 39).
Remove the retaining clip assembly.
Displace and reposition the forward shoe (4 Fig. 39) to allow removal of the upper adjuster (5 Fig. 39).
Position the hub access hole (2 Fig. 39) to expose the rear handbrake shoe retaining clip. Remove the retaining clip assembly.
Displace the front and rear shoes from the handbrake cable link.
Displace and remove the shoe/lower spring assembly from the brake backplate.
Unhook the upper and lower springs (6 Fig. 39) from the shoes and place the shoes and springs to one side.

Remove the handbrake cable securing clip (1 Fig. 40).
Reposition the handbrake link and remove the cable clevis pin (2 Fig. 40).
Cut and remove the tie strap securing the ABS harness to the hub.
Remove the ABS sensor securing bolt (3 Fig. 40).
Displace and reposition the sensor (4 Fig. 40).
Fully remove the drive shaft to hub securing nut (1 Fig. 40) and displace and remove the cone washer.

Fit a thread protector to the drive shaft (Fig. 41).
Unscrew the centre bolt of tool JD1D, hub removal tool.

Fit and tighten the tool to the hub (Fig. 42).
Tighten the tool centre bolt to withdraw the hub from the drive shaft.
Unscrew the tool securing nuts and remove the tool from the hub.
Remove the thread protector from the drive shaft.
Reposition the handbrake cable from the hub carrier.
Remove the fulcrum shaft securing nut. Remove the fulcrum shaft.
Remove the wishbone/damper tie bracket spacer washers.
Carefully displace the hub assembly and remove the shims from between the hub and wishbone. Place the hub assembly on a clean work bench.
Clean the drive shaft spline and faces.
Position hub tool JD 132-1 on a press.

Place the hub assembly to the tool/press. Ensure it is aligned correctly with the handbrake cable housing located into the tool cut out (Fig. 43).

Fit and align the tool button JD 132-2 to the hub (1 Fig. 44).
Press and remove the hub from the carrier.
Remove the hub carrier, ABS rotor and inner bearing as an assembly.
Remove the tool button JD 132-2.
Remove the ABS rotor from the hub carrier.
Remove the hub and tool assembly from the press and remove tool JD 132-1 from the hub. Place tool to one side.
Using a suitable drift, remove the outer hub seal and bearing cone.
Remove the bearing spacer and shim(s).
Reposition the hub carrier on the bench.
Using a drift remove the inner bearing cone and seal .

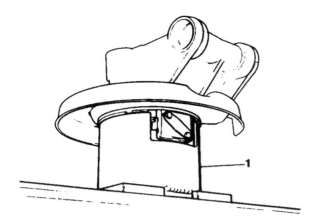

Fig. 43

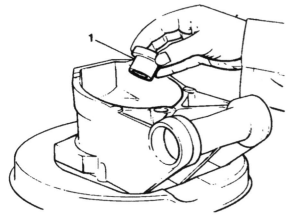

Fig. 44

Note: Always keep bearing cups and cones together in original pairs. Never mix cups and cones.

Position the hub carrier on a suitable block of wood.
Using a drift remove the outer bearing cup.
Reposition the hub carrier on the wooden block and remove the remaining, inner, bearing cup.
Thoroughly clean all components to be reused prior to reassembly.

Position the hub carrier on the press with the inner bearing side uppermost.
Position the new inner bearing cup in the hub carrier. Fit and align the bearing replacer tool JD550-4/2 to the inner bearing cup.
Position press tool 18G 134 on the bearing replacer tool.
Using the press, fully seat the new inner bearing cup to the hub carrier.
Reposition the hub carrier on the press with the outer bearing side uppermost.
Position the new outer bearing cup in the hub carrier. Fit and align the bearing replacer tool JD550-4/1 on the outer bearing cup.
Position press tool 18G 134 on the bearing replacer tool.
Using the press, fully seat the new outer bearing cup on the hub carrier.
Remove the tool assembly. Remove tool 18G 134 from the replacer tool. Place the hub carrier on the workbench.

Fit and align the new outer bearing cone to the hub carrier.
Fully seat the hub to the carrier/bearing cone assembly.
Reposition the assembly on the bench.
Fit the bearing spacer to the hub followed by the largest shim, (0.031 in.) (0.001 inch = 0.0254mm for conversion purposes).
Fit and align the new inner bearing cone to the carrier / hub assembly.

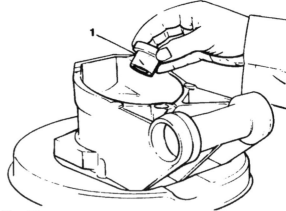

Fig. 45

Position a suitable block of wood on the press.
Fit the carrier/hub assembly on the press/wood and ensure, when using the press, the load is not put upon the wheel studs.
Fit the ABS rotor to the inner hub.
Using the press, fully seat the ABS rotor to the hub assembly.
Fit the hub securely in a vice.
Fit hub end float tool JD 15 to the hub.
Fit and tighten the hub end float dial gauge JD 13A to the hub carrier.
Using two suitable levers, insert between the hub and carrier and prise lightly back and forth to check the end float.
Make a note of the end float reading.

Note: Care must be taken not to use excess leverage as this could move the spacer and prevent a correct reading being obtained.

Remove the dial gauge JD 13A and the end float tool JD 15.
Position hub tool JD 132-1 on the press.
Align the hub assembly and fit the button JD 132-2 to the hub (Fig. 45).
Using the press, remove the hub from the carrier.
Remove the hub carrier, ABS rotor and inner bearing as an assembly.
Remove the button, tool JD 132-2.
Remove the hub and tool assembly from the press and remove tool JD132-1 from the hub. Place tool to one side.
Remove the ABS rotor and bearing cone from the hub carrier.
Remove the shim, spacer and bearing cone from the hub.
Lubricate the bearings with the correct amount of grease (see DATA).
Fit and align the new outer bearing cone to the hub carrier.
Fit and fully seat a new seal to the hub carrier. Fully seat the hub to the carrier/bearing cone assembly.

Reposition the hub/carrier assembly on the bench and fit the spacer.
Place the large shim (0.031 in.) aside and calculate the shim(s) required to give a 0.003 in. pre-load.
Fit the shim(s) required to the hub.
Fit and align the new inner bearing cone to the hub.
Fully seat a new inner seal to the hub / carrier assembly.
Position a suitable block of wood on the press.
Fit the carrier/hub assembly on the press / wood and ensure, when using the press, the load is not put upon the wheel studs.
Fit the ABS rotor to the inner hub.
Using the press, fully seat the ABS rotor to the hub assembly.

Position the hub/carrier assembly on the wishbone.
Fit the shims, as removed, between the wishbone and hub assembly.

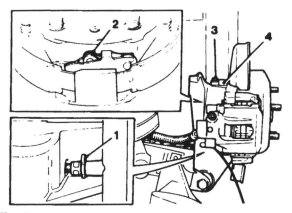

Fig. 46

Align the hub assembly to the lower wishbone.
Fit the wishbone/damper tie bracket spacer washers.
Fit and fully seat the fulcrum shaft to the wishbone / hub assembly.
Fit, but do not tighten, the fulcrum shaft securing nut.
Apply 'Loctite' to the drive shaft spline.
Pivot the hub and position the handbrake cable through the cable housing.
Position the drive shaft to the hub and pivot the hub to fully seat the shaft.
Fit the cone washer.
Fit, but do not fully tighten, the drive shaft securing nut.
Fit the handbrake link to the cable.
Fit the cable to link clevis pin.
Reposition the cable link.
Fit and seat the cable securing clip.
Position the ABS sensor to the hub / carrier assembly.
Fit and tighten the sensor securing bolt.
Fit a new tie strap to secure the sensor harness.

Lightly lubricate the adjuster with Copperslip grease.
Position the shoes on the backplate.
Ensure the forward shoe is correctly seated in the handbrake link.
Fit the lower return spring.
Align the rear shoe to the backplate.
Align the hub access hole to the rear shoe retaining clip position.
Fit and fully seat the rear shoe retaining clip.
Fit the rear shoe to handbrake link.
Fit the upper return spring. Pull both shoes outward to allow the adjuster to be fitted.
Fit and align the adjuster to the shoes.
Align the front shoe to the backplate.
Align the hub access hole to the front shoe retaining clip position.
Fit and fully seat the front shoe retaining clip.
Refit the discs.
Refit the roadwheels .
Ensure fixings/fastenings are torque tightened to the specified tolerances.
Lower the vehicle from the axle stands.

Oils / Greases / Sealants

Hub bearings Shell Retinax 'A'	11.5 cc inner bearing cone
	9.0 cc outer bearing cone
Brake shoe adjuster	Copperslip Grease
Drive shaft splines	Loctite 270

Service Tools

Hub remover JD 1D
Hub press tool JD 132-1
Hub press tool button JD 132-2

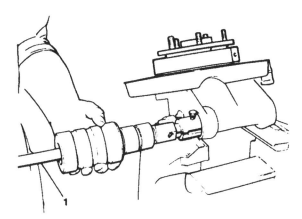

Fig. 47

Press tool handle 18G 134
Bearing replacer JD 550-4 / 1
Bearing replacer JD 550-4 / 2
Dial gauge JD 13A
End float measurement base JD 15

REAR HUB OIL SEAL, RENEW

Follow the instructions given under the preceding operation, omitting to renew the cups and cones unless they are found to be damaged.

REAR HUB CARRIER FULCRUM BEARING, RENEW

Raise the vehicle on a ramp.
Using suitable levers reposition the handbrake ratio levers inward.
Disconnect the main handbrake cable from the right hand ratio lever. Position the lever outwards toward the abutment bracket.
Remove the rear cable retaining collet and clip.
Reposition rear cable from the abutment bracket.
Lower the ramp to a suitable working position to carry out the following procedures.
Raise the rear end of the vehicle and support on stands.
Remove the roadwheel and tyre assembly.
Unscrew, but do not remove the drive shaft to hub assembly securing nut (1 Fig. 39).
Remove the rear disc.
Cut and remove the tie strap securing the ABS harness to the hub.
Remove the ABS sensor securing bolt (3 Fig. 46).
Displace and reposition the sensor (4 Fig. 46).
Remove the fulcrum shaft securing nut.
Fully remove the drive shaft to hub securing nut (1 Fig. 39) and displace and remove the cone washer.
Fit a thread protector to the drive shaft (Fig. 41).
Unscrew the centre bolt of tool JD1D, hub removal tool.
Fit and tighten the tool to the hub (Fig. 42).
Tighten the tool centre bolt to withdraw the hub from the drive shaft.
Unscrew the tool securing nuts and remove the tool from the hub.
Remove the thread protector from the drive shaft.
Remove the fulcrum shaft.
Remove the wishbone/damper tie bracket spacer washers.
Carefully displace the hub assembly and remove the shims from between the hub and wishbone.

Carefully secure the hub assembly in a vice.
Using a suitable drift, remove the fulcrum bearing tube.
Remove the shim and bearing cone from the tube.
Remove the remaining bearing cone and shim from the hub carrier.

Fit extractor tool 18G 284 AAH (1 Fig. 47) to the fulcrum bearing cup.
Tighten the tool cross bolt and ensure the tool legs locate behind the bearing cup.
Fit the slide hammer 18G 284 to the tool.
Using the slide hammer, remove the bearing cup.
Unscrew, but do not remove, the tool cross bolt.
Place the bearing cup aside.
Repeat the procedure to remove the remaining bearing cup. Place the bearing cup aside. Remove the slide hammer 18G 284 from tool 18G 284 AAH.
Remove the hub/carrier assembly from the vice.
Clean the hub carrier and fulcrum bearing faces.
Clean the cross shaft and tube.
Position the hub/carrier assembly on a press.
Fit the tool handle 18G 134 to the bearing replacer tool JD 550-6.
Fit a new bearing cup to the replacer tool.
Align the tool/bearing cup to the hub/carrier assembly.
Using the press, fully seat the new bearing cup to the hub carrier.
Remove the replacer tool.
Reposition the hub/carrier assembly on the press.
Fit the remaining new bearing cup.
Remove the replacer tool.
Remove the tool handle 18G 134 from the bearing replacer tool JD 550-6.
Move the assembly from the press and secure in position in a vice.
Apply grease to the new bearings.
Fit a new bearing cone to the fulcrum tube.
Fit the minimum thickness of shim to the fulcrum tube/bearing cone.
Fit the tube/bearing cone assembly to the hub carrier.
Fit the remaining bearing cone to the tube/carrier assembly.
Fit the fulcrum shaft.
Fit and fully seat the remaining shim to the tube.
Fit a suitable piece of tube to seat onto the shim.
Fit and tighten the cross shaft securing nut to fully seat the shims onto the cross tube. Remove the shaft securing nut.
Remove the tube.
Fit a suitable flat washer to the shaft and tighten the shaft securing nut.
Using feelers, measure and note the gap between the shim and the washer.
Remove the shaft securing nut.
Remove the washer and displace the cross shaft.
Using a suitable drift, move the cross tube from the shims.
Retrieve the shims, measure, then place aside.

Note: Calculate equal size shims to be used to give a 0.003 in. (0.0762 mm) pre-load.

Fit the new shims to the cross tube.
Fit the cross shaft.
Align a suitable tube then tighten the shaft securing nut to fully seat the shims on to the cross tube.
Unscrew and remove the nut.
Remove the tube.
Carefully move the cross shaft from the tube.
Remove the hub/carrier assembly from the vice.

Position the hub/carrier assembly on the lower wishbone.
Route the handbrake cable into position.
Fit the shims, as removed, between the wishbone and hub assembly.
Align the hub assembly to the lower wishbone.
Fit the wishbone/damper tie bracket spacer washers.

Fit and fully seat the fulcrum shaft to the wishbone/hub assembly.
Fit, but do not tighten, the fulcrum shaft securing nut.
Apply 'Loctite' to the drive shaft spline.
Pivot the hub to engage on the drive shaft.
Fit the cone washer.
Fit, but do not fully tighten, the drive shaft securing nut.

Position the ABS sensor to the hub / carrier assembly.
Fit and tighten the sensor securing bolt.
Fit a new tie strap to secure the sensor harness.
Fully tighten the drive shaft and fulcrum shaft securing nuts.
Refit the disc.
Raise the ramp.
Position the rear handbrake cable through the abutment bracket.
Fit the cable securing clip.
Position the ratio lever outwards.
Position the cable through the ratio lever.
Fit the cable retaining collet.
Move both ratio levers inward and fit the main handbrake cable to the right hand ratio lever.
Lower the ramp.
Refit the roadwheel.
Ensure fixings/fastenings are torque tightened to the specified tolerances.
Lower the vehicle from the axle stands.

Oils / Greases / Sealants

Fulcrum shaft and bearings - LM Multipurpose Grease
Drive shaft splines - Loctite 270

Service Tools

Hub remover JD 1D
Bearing cup remover 18G 284 AAH
Bearing cup replacer JD 550 6
Press tool handle 18G 134
Slide hammer 18G 284

REAR DISC, RENEW

Raise the rear of the vehicle and support on stands.
Remove the roadwheel and tyre.
Carefully remove the caliper retaining spring (1 Fig. 48).
Remove the caliper to carrier securing bolt covers (2 Fig. 48).
Remove the socket head securing bolts (3 Fig. 48).
Displace the caliper from the carrier.
Remove the brake pads (4 Fig. 48) from the caliper.
Position the caliper to one side to allow access to the carrier/hub assembly.

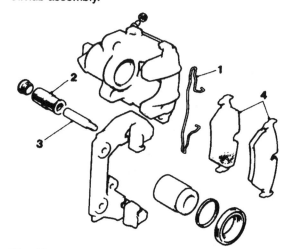

Fig. 48

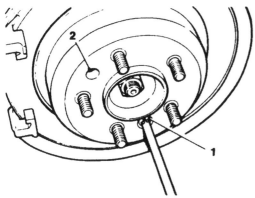

Fig. 49

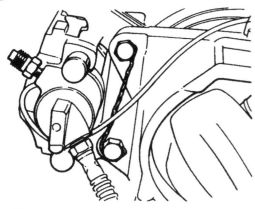

Fig. 50

Note: Do not allow the caliper to hang with the weight on the feed hose. Ensure it is placed safely and securely aside.

Cut and remove the lock wire securing the carrier to hub assembly securing bolts. Unscrew and remove the securing bolts. Remove the carrier.

Unscrew and remove the disc securing screw (1 Fig. 49). Remove the disc.
Carefully remove any brake dust / dirt from the caliper and handbrake shoe areas.

Fit a new disc.
Fit and tighten the disc securing screw (1 Fig. 49).

Operate the handbrake lever to centralize the shoes.
Align the hub access hole (2 Fig. 49) to the handbrake shoe adjuster.
Turn the adjuster anti-clockwise until the shoes contact the disc.
Turn the adjuster clockwise until the disc can be turned quite freely.
Fit the carrier to the hub assembly and tighten the carrier securing bolts.

WARNING:

BRAKE PAD/LINING DUST CAN CONTAIN ASBESTOS WHICH IF INHALED, CAN DAMAGE YOUR HEALTH. ALWAYS USE A VACUUM BRUSH TO REMOVE DRY BRAKE PAD/LINING DUST. NEVER USE AN AIR LINE.

Secure the carrier bolts with new lockwire (Fig. 50). Ensure the caliper piston is fully retracted and refit the brake pads.

Note: If the pads are worn to the minimum thickness, new pads must be fitted.

Position the caliper/pad assembly to the carrier.
Lubricate, then fit and tighten the caliper securing bolts.
Fit the caliper securing bolt covers.
Fit the caliper retaining spring.
Refit the roadwheel.
Lower the vehicle from the axle stands.
Check the brake fluid and top up as necessary.
Ensure fixings/fastenings are torque tightened to the specified tolerances.

Note: It may be necessary to run the engine to give power assistance to the brake pedal.

WARNING:

APPLICATION OF THE BRAKE PEDAL MUST BE CARRIED OUT, AS THE BRAKE WILL NOT OPERATE EFFICIENTLY UNTIL THE PADS ARE CORRECTLY POSITIONED.

Oils / Greases / Sealants

Brake fluid to a minimum DOT 4 specification.
Caliper securing bolts Molykote 111.

REAR HANDBRAKE CABLE ASSEMBLY, LH (and RH), RENEW

Raise the vehicle on a ramp.
Using suitable levers reposition the handbrake ratio levers inward.
Disconnect the main handbrake cable from the right hand ratio lever.
Position the lever outward toward the abutment bracket.
Remove the rear cable retaining collet and clip.

Reposition the rear cable from the abutment bracket.
Lower the ramp to a suitable working position to carry out the following procedures.
Raise the rear end of the vehicle and support on stands.
Remove the roadwheel and tyre assembly.

Unscrew, but do not remove the drive shaft to hub assembly securing nut (1 Fig. 51).
Remove the rear disc.
Position the hub access hole (2 Fig. 51) to expose the front handbrake shoe retaining clip (3 Fig. 51).
Remove the retaining clip assembly.
Displace and reposition the forward shoe (4 Fig. 51) to allow removal of the upper adjuster (5 Fig. 51).

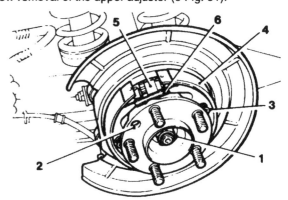

Fig. 51

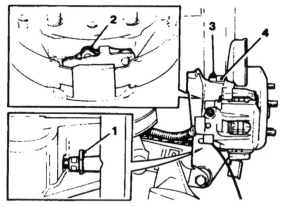

Fig. 52

Position the hub access hole (2 Fig. 51) to expose the rear handbrake shoe retaining clip. Remove the retaining clip assembly.
Displace the front and rear shoes from the handbrake cable link.
Displace and remove the shoe/lower spring assembly from the brake backplate.
Unhook the upper and lower springs (6 Fig. 51) from the shoes and place the shoes and springs to one side.

Remove the handbrake cable securing clip (1 Fig. 52).
Reposition the handbrake link and remove the cable clevis pin (2 Fig. 52).
Remove the cable from the hub assembly.

Fit the new handbrake cable to the cable housing.
Fit the handbrake link to the cable.
Fit the cable to link clevis pin.
Reposition the cable link.
Fit and seat the cable securing clip.
Lightly lubricate the adjuster.
Position the shoes on the backplate. Ensure the forward shoe is correctly seated in the handbrake link.
Fit the lower return spring.
Align the rear shoe to the backplate.
Align the hub access hole to the rear shoe retaining clip position.
Fit and fully seat the rear shoe retaining clip.
Fit the rear shoe to handbrake link.
Fit the upper return spring.
Pull both shoes outward to allow the adjuster to be fitted.
Fit and align the adjuster to the shoes.
Align the front shoe to the backplate.
Align the hub access hole to the front shoe retaining clip position.
Fit and fully seat the front shoe retaining clip.

Refit the disc.
Refit the roadwheel.
Raise the ramp.
Position the rear handbrake cable through the abutment bracket.
Fit the cable securing clip.
Position the ratio lever outwards.
Position the cable through the ratio lever.
Fit the cable retaining collet.
Move both ratio levers inward and fit the main handbrake cable to the right hand ratio lever. Lower the ramp.

Open the driver's door.
Switch on the ignition.
Power the driver's seat fully forward.
Position the seat squab fully forward.
Switch off the ignition.
Unscrew the handbrake lever cover securing screw.

Remove the cover.
Position the carpet for access to the seat belt slider rail cover.
Remove the cover.
Remove the slider rail securing bolt.
Move the slider rail and sill carpet to allow access to remove the slider rail spacer.
Unscrew the handbrake cable adjuster locknuts.
Adjust the cable until upon application of the handbrake, only three to four notches of the ratchet are felt/heard.
Tighten the locknuts.
Refit the slider rail spacer.
Reposition the sill carpet.
Refit the slider rail to position and fit and tighten the securing bolt.
Fit the securing bolt cover and reposition the floor carpet.
Fit the handbrake lever cover.
Fit and tighten the cover securing screw.
Reposition the seat squab back to the original position.
Switch on the ignition.
Power the seat back to the original position.
Switch off the ignition.
Close the door.
Ensure fixings/fastenings are torque tightened to the specified tolerances.
Lower the vehicle from the stands.

To renew the rear handbrake cable assembly, right hand, carry out the operation on the opposite rear side of the vehicle.

Oils / Greases / Sealants

Brake shoe adjuster - Copperslip Grease

HANDBRAKE, SETTING PROCEDURE

Open the driver's door.
Switch on the ignition.
Power the driver's seat fully forward.
Position the seat squab fully forward.
Switch off the ignition.
Unscrew the handbrake lever cover securing screw.
Remove the cover.
Position the carpet for access to the seat belt slider rail cover.
Remove the cover. Remove the slider rail securing bolt.
Move the slider rail and sill carpet to allow access to remove the slider rail spacer.
Unscrew the handbrake cable adjuster locknuts.
Raise the vehicle on a ramp.
Reposition the main handbrake cable rubber seal.
Move the handbrake ratio levers inward.
Disconnect the main handbrake cable from the ratio lever cable abutments and reposition the cable.
Unscrew, but do not remove the fulcrum shaft/ratio lever securing nuts.
Position the ratio levers to the abutment brackets to give a 1 mm to 3 mm (0.039 to 0.118 in.) gap.
Tighten the fulcrum shaft/ratio lever securing nuts.

Note: Ensure the ratio levers do not move whilst tightening the fulcrum shaft/ratio lever securing nuts.

Offer the handbrake cable up to the ratio lever. Fit and fully seat the cable in the abutment. Position the rubber cable seal.
Position the ratio levers inward. Fit and fully seat the cable to the opposite ratio lever abutment.
Lower the ramp to a suitable working position to carry out the following procedures.
Raise the rear end of the vehicle and support on stands.
Remove the roadwheel and tyre assembly.

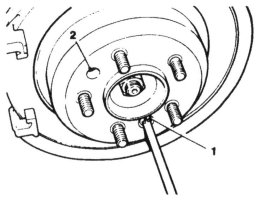

Fig. 53

Turn the hub assembly and align the disc/hub access hole (2 Fig. 53) to the handbrake shoe adjuster.
Turn the adjuster anti-clockwise until the shoes contact the disc.
Turn the adjuster clockwise until the disc can be turned quite freely.
Repeat the procedure to adjust the opposite side rear handbrake shoes.
Refit the roadwheel.
Lower the ramp.
From inside the vehicle:
Adjust the cable until upon application of the handbrake, only three to four notches of the ratchet are felt/heard.
Tighten the locknuts.
Refit the slider rail spacer.
Reposition the sill carpet.
Refit the slider rail to position and fit and tighten the securing bolt.
Fit the securing bolt cover and reposition the floor carpet.
Fit the handbrake lever cover. Fit and tighten the cover securing screw.
Reposition the seat squab back to the original position.
Switch on the ignition. Power the seat back to the original position. Switch off the ignition. Close the door.
Lower the vehicle from the stands.

REAR BRAKE PADS, RENEW

Raise the rear of the vehicle and support on stands.
Remove the roadwheel and tyre assembly.

Carefully remove the caliper retaining spring (1 Fig. 54).
Remove the caliper to carrier securing bolt covers (2 Fig. 54).
Remove the socket head securing bolts (3 Fig. 54).
Displace the caliper from the carrier.
Remove the brake pads (4 Fig. 54) from the caliper.
Safely discard the worn brake pads.
Position the caliper to one side.

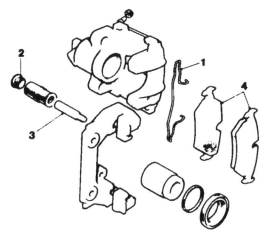

Fig. 54

Note: Do not allow the caliper to hang with the weight on the feed hose. Ensure it is placed safely and securely aside.

WARNING:

BRAKE PAD/LINING DUST CAN CONTAIN ASBESTOS WHICH IF INHALED, CAN DAMAGE YOUR HEALTH. ALWAYS USE A VACUUM BRUSH TO REMOVE DRY BRAKE PA /LINING DUST. NEVER USE AN AIR LINE.

Carefully remove any brake dust/dirt from the caliper area.

Ensure the caliper piston is fully retracted and fit new brake pads.

Note: The pad with the spring clip attached locates into the caliper piston.

Offer up the caliper/pad assembly to the carrier.
Lubricate, then fit and tighten the caliper securing bolts.
Fit the caliper securing bolt covers.
Fit the caliper retaining spring.
Repeat the procedure on the opposite rear side of the vehicle. Refit the roadwheel.
Lower the vehicle from the axle stands.
Check the brake fluid and top up as necessary.

WARNING:

IT IS ESSENTIAL THAT THE BRAKE PEDAL IS REPEATEDLY APPLIED TO ENSURE THAT THE PADS ARE FULLY IN CONTACT WITH THE DISC.

Note: It may be necessary to run the engine to give power assistance to the brake pedal.

Oils / Greases / Sealants

Brake fluid to a minimum DOT 4 specification.
Caliper securing bolts - Molykote 111.

HANDBRAKE SHOES, RENEW

Jack up the rear of the vehicle and place on stands.
Remove the roadwheel and tyre assembly.
Remove the rear disc.
Position the hub access hole (2 Fig. 55) to expose the front handbrake shoe retaining clip (3 Fig. 55). Remove the retaining clip assembly.
Displace and reposition the forward shoe (4 Fig. 55) to allow removal of the upper adjuster (5 Fig. 55).
Position the hub access hole (2 Fig. 55) to expose the rear handbrake shoe retaining clip. Remove the retaining clip assembly.
Displace the front and rear shoes from the handbrake cable link.
Displace and remove the shoe/lower spring assembly from the brake backplate.
Unhook the upper and lower springs (6 Fig. 55) from the shoes and place the shoes and springs to one side.
Carefully remove brake dust/dirt from the caliper and shoe area.

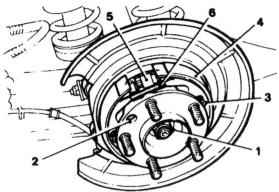

Fig. 55

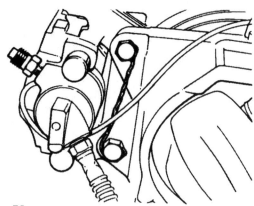

Fig. 56

Clean the springs, retaining pins and adjuster.
Lightly lubricate the adjuster.
Offer the shoes up to the backplate.
Ensure the forward shoe is correctly seated in the hand-brake link.
Fit the lower return spring.
Align the rear shoe to the backplate. Align the hub access hole to the rear shoe retaining clip position.
Fit and fully seat the rear shoe retaining clip.
Fit the rear shoe to handbrake link.
Fit the upper return spring.
Pull both shoes outward to allow the adjuster to be fitted.
Fit and align the adjuster to the shoes.
Align the front shoe to the backplate.
Align the hub access hole to the front shoe retaining clip position.
Fit and fully seat the front shoe retaining clip.
Repeat the procedure for the opposite rear side of the vehicle.

Refit the discs.
Refit the roadwheels.
Lower the vehicle from the axle stands.

Oils / Greases / Sealants

Brake shoe adjuster - Copperslip Grease

REAR BRAKE CALIPER, RENEW

Open the bonnet and place a cloth around the brake reservoir filler cap/neck to absorb any possible spillage.
Jack up the rear of the vehicle and place on stands.
Remove the roadwheel and tyre.
Carefully remove the caliper retaining spring (1 Fig. 54).
Partially unscrew the brake pipe to feed hose union nut.
Remove the caliper to carrier securing bolt covers (2 Fig. 54).
Remove the socket head securing bolts (3 Fig. 54).
Displace the caliper from the carrier.
Remove the brake pads (4 Fig. 54) from the caliper.
Unscrew and remove the caliper from the feed hose.
Fit suitable plugs to the caliper and hose.
Cut and remove the lockwire securing the carrier to hub assembly securing bolts.

Unscrew and remove the securing bolts.
Remove the carrier.
Carefully remove any brake dust/dirt from the carrier and disc.

Fit the carrier to the hub assembly and tighten the carrier securing bolts.
Secure the carrier bolts with new lockwire (Fig. 56).
Remove the plugs from the new caliper and the feed hose.
Fit, but do not fully tighten the caliper to the feed hose.
Ensure the caliper piston on the new caliper is fully retracted and refit the brake pads.
Note: If the pads are worn to the minimum thickness, new pads must be fitted.

Position the caliper/pad assembly to the carrier.
Lubricate, then fit and tighten the caliper securing bolts. Fit the caliper securing bolt covers.

Fully tighten the feed hose to the caliper.
Fully tighten the brake pipe to feed hose union nut.
Fit the caliper retaining spring.
Refit the roadwheel and tyre assembly.
Lower the vehicle from the axle stands.
Bleed the brakes.

Oils / Greases / Sealants

Brake fluid to a minimum DOT 4 specification.

Torque Figures (for 1993.5 MY vehicles)

(All figures in Newton Metres)

Brake disc to hub	11 - 16
Caliper to carrier	25 - 30
Caliper carrier to hub carriers	54 - 66
Fulcrum shaft nut	70 - 80
Bleed screw/nipple to caliper	7 - 13
Brake hydraulic hose to caliper	10 - 12
ABS sensor securing bolt	8,5 - 11
Handbrake lever to body	15 - 21
Handbrake cable locknuts	19 - 25
Handbrake lever cover screw	0.6 - 1.0
Wheel nut to wheel stud	88 - 102

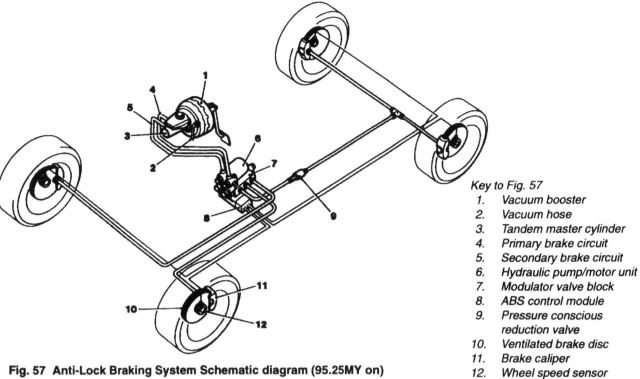

Key to Fig. 57
1. Vacuum booster
2. Vacuum hose
3. Tandem master cylinder
4. Primary brake circuit
5. Secondary brake circuit
6. Hydraulic pump/motor unit
7. Modulator valve block
8. ABS control module
9. Pressure conscious reduction valve
10. Ventilated brake disc
11. Brake caliper
12. Wheel speed sensor

Fig. 57 Anti-Lock Braking System Schematic diagram (95.25MY on)

BRAKES, 1995.25 MODEL YEAR ON

DESCRIPTION

The anti-lock braking system (ABS) components are combined with a hydraulic booster, tandem master cylinder (TMC) and ventilated disc brakes on all four wheels, to provide a two-circuit braking system. The front brakes are fitted with four-piston calipers; the rear brakes have single-piston calipers and drums for the cable operated handbrake. The anti-lock braking system comprises the following:

- Hydraulic control module comprising electric motor driven pump, two low-pressure accumulators, modulator valve block and ABS control module (ABS CM)
- Four inductive wheel speed sensors, hub end mounted
- ABS warning indicator, mounted on the instrument panel
- Auxiliary inputs providing information to the ABS CM
- Diagnostic ISO communication BUS input/output link

The valve block houses solenoid operated valves which are activated by voltage signals from the control module. The signals are generated using wheel speed information received from the wheel speed sensors.

The valves regulate the supply of pressure individually to the front wheels and collectively to the rear wheels, as necessary, to prevent wheel locking during braking.

When the ignition is switched on, an ABS self test is initiated. During this test, the ABS warning indicator is lit for approximately 1.7 seconds and then extinguished. After this time delay the control module is ready to process signals provided from the various input sources and, using the software defined algorithm, control the electrical and hydraulic circuits. A fault is indicated if the warning indicator remains lit or comes on whilst the vehicle is being driven. Under fault conditions the system is inhibited or disabled, although conventional braking is unaffected.

The fluid level indicator lamp, mounted on the instrument panel, is lit when the brake fluid falls below the MIN mark on the brake fluid reservoir.

Fault conditions detected by the ABS CM disable the ABS until the fault is rectified. The system will be disabled when any of the following conditions occur:

- Valve failure
- Sensor failure
- Main driver failure (internal ABS CM fault)
- Redundancy error (internal ABS CM fault)
- Over-voltage/under-voltage
- Pump motor failure
- Under-voltage condition.

The input frequency from each wheel speed sensor signal is translated by the ABS CM into a comparable wheel speed. The ABS CM continually monitors the system. False wheel speed information, such as sudden speed changes or excessive speeds, is detected as a sensor malfunction. The ABS CM reacts to fault conditions in the following ways:

Inhibit - ABS is inhibited until the sensed speed returns to within an acceptable limit, whereupon ABS is restored. Conventional braking is unaffected. Depending on vehicle speed, the ABS warning indicator may come on.

Disable - ABS is disabled (switched off) and the ABS warning indicator comes on. The system will not be restored until the engine is switched off and restarted or the fault has been rectified. After the system has been disabled, the warning indicator remains on until the vehicle has reached a speed of 20 km / h (12.5 mile / h) during the first ignition cycle after fault rectification.

Vacuum booster

The vacuum booster (1 Fig. 58) is mounted on the brake pedal box and secured by three bolts. The tandem master cylinder (TMC) (2 Fig. 58) locates on two studs on the vacuum booster. Two lugs locate the fluid reservoir (3 Fig. 58) on the TMC which is secured by a split pin.

The vacuum is drawn from the inlet manifold. At the vacuum booster, the vacuum hose is connected to the vacuum chamber via an elbow connector. At the inlet manifold, the vacuum hose connector is of the push-on quick-release type.

Applied pedal force is increased by the vacuum booster which actuates the intermediate piston of the TMC.

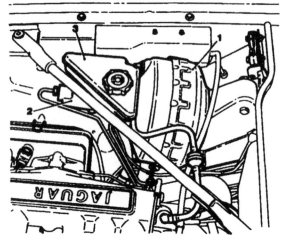

Fig. 58

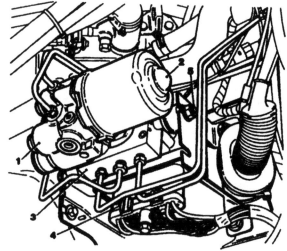

Fig. 59

The boost ratio supplied by the vacuum booster is 6.5 : 1.

Note: The vacuum booster and the TMC are supplied as a unit but are individually serviceable.

The brake fluid reservoir is fitted with a fluid level switch, which opens when fluid level is low and lights the low fluid level indicator.

Hydraulic control module

The hydraulic control module is located under the bonnet adjacent to the engine compartment bulkhead. It is secured within a steel mounting bracket at three securing points.

The hydraulic pump (1 Fig. 59) is a reciprocating two-circuit pump in which one brake circuit is assigned to each pump circuit. The pump supplies adequate pressure and volume supply to the brake circuits under anti-lock braking conditions. The pump housing incorporates a low pressure accumulator and damping chamber for each brake circuit. The pump is driven by an electric motor (2 Fig. 59) which draws 32A current at peak operation and has an internal resistance of 0.8 ohms.

The modulator valve block (3 Fig. 59) houses six solenoid valves: three normally open (NO) inlet valves and three normally closed (NC) outlet valves. This provides three outlet ports, one for each front brake and one for both rear brakes.

All electronic and power connections are made through one cable loom connector to the ABS control module (4 Fig. 59) located below the valve block.

The pump, motor and valve block are non-serviceable. If a fault occurs in any of these components, the whole hydraulic control module must be replaced. The ABS CM control module can be replaced separately.

ABS control module

The control module (CM) (4 Fig. 59), located beneath the modulator valve block (3 Fig. 59), is the system controller and processes all the information supplied from the external sensors and probes. Refer to the Control Module Connection Diagram. The signals from the four wheel speed sensors are independently processed by the CM, calculating numerical values which correspond directly to

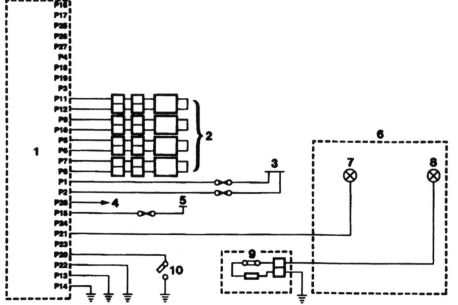

Key to Fig 60
1. *ABS control module*
2. *Wheel speed sensors*
3. *Battery voltage inputs*
4. *Diagnostic communication bus*
5. *Ignition voltage input*
6. *Instrument pack*
7. *ABS warning indicator*
8. *Low fluid level indicator*
9. *Fluid level switch*
10. *Brake pedal switch*

Fig. 60 Control Module Connection Diagram

the wheel speed. These values are converted into control signals for pressure modulation during ABS control.

The CM continuously monitors ABS operation, lighting the ABS warning indicator and inhibiting or disabling the system when faults are detected. In a fault condition, conventional braking is unaffected. The CM is self testing and cannot be fault diagnosed beyond 'black box' level, i.e. a faulty module. The CM houses the solenoids which operate the inlet and outlet valves of the modulator valve block. There is no electrical connection between the CM and the modulator valve block, but there is an electrical connection from the CM to the pump motor.

The CM functions include the following:

- Providing control signals for the operation of ABS solenoid valves
- Calculating wheel speed from voltage signals transmitted by the wheel speed sensors
- Monitoring of all electrical components
- On Board Diagnostics (OBD): storage of fault codes in a non-volatile memory

The fault codes generated by the CM are stored in a non-volatile memory which can be read via the OBD link. The ABS warning indicator is lit if the CM connector is loose or not fitted.

Control module connections

Control module connections, numbered 1 to 28, provide the necessary input / output signals to enable the module to control and monitor ABS operation.

Connections are as follows:

1	Battery positive feed (via fuse F22)
2	Battery positive feed (via fuse F11)
3	Not used
4	Not used
5/6	Wheel sensor, left-hand front
7/8	Wheel sensor, right-hand front
9/10	Wheel sensor, left-hand rear
11/12	Wheel sensor, right-hand rear
13	Ground
14	Ground
15	Ignition feed (via fuse F10)
16	Not used
17	Not used
18	Not used
19	Not used
20	Brake pedal switch
21	ABS warning indicator
22	ABS ground
23	Not used
24	Not used
25	Not used
26	Not used
27	Not used
28	Diagnostic ISO communication bus

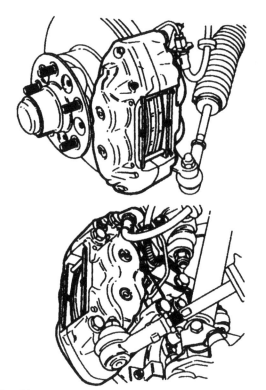

Fig. 61

Brake calipers

The front brakes are fitted with four-piston calipers acting upon 24 mm (15/16 in) thick ventilated brake discs (Fig. 61). The caliper carrier is secured by two bolts to the suspension vertical link.

The rear brakes are fitted with single-piston calipers acting upon 20 mm (25/32 in) thick ventilated brake discs (Fig. 62). The caliper carrier is secured by two bolts (wire locked) to the hub carrier.

The brake discs must be renewed when the minimum thicknesses specified below are reached:

Front brake disc - 22.9 mm (29/32 in)
Rear brake disc - 18.5 mm (47/64 in)
The rear brake caliper (1 Fig. 63) is mounted on the carrier (2 Fig. 63) by means of two guiding pins (3 Fig. 63) and a caliper retaining clip (5 Fig. 63). The guiding pins slide in bushes (4 Fig. 63) fitted to the caliper.

Note: The guiding pins are fitted with dust caps which must be fitted when reassembling the caliper.

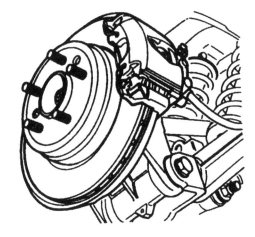

Fig. 62

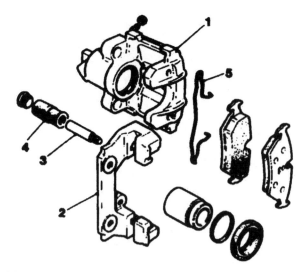

Fig. 63

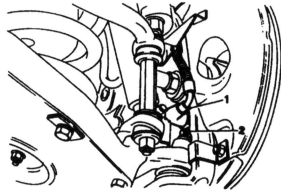

Fig. 64

Fig. 65

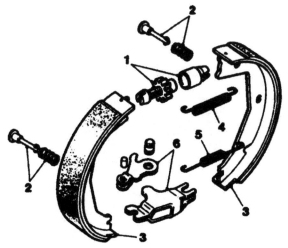

Fig. 66

Wheel speed sensors

Speed sensors are provided for each road wheel. The front sensors (1 Fig. 64) are mounted on the vertical link, while the rear sensors (1 Fig. 65) are mounted on the hub carrier. A toothed wheel, which turns with the road wheel, induces an a.c. voltage signal in the sensor. The frequency and amplitude of the a.c. voltage varies directly in relation to wheel speed, providing the control module with wheel speed information to give a comparison between the speed of each individual wheel, controlling braking as necessary.

Each sensor is monitored for open and short circuit failure, causing ABS control to be disabled on detection of a fault condition. ABS is also disabled should any sensed speed in excess of 330 km/h (205 mile/h) be detected. Similarly, ABS control is inhibited (switched off until fault condition is cleared) at speeds up to 40 km/h (25 mile/h) when frequency fluctuations are detected that are inconsistent with wheel rotation. At speeds above 40 km/h (25 mile/h), ABS control is disabled when inconsistencies are detected.

The sensor coil has a resistance value of 1100 ohms and has a voltage of 2.5V present on each connecting pin when the vehicle is stationary.

Handbrake

The handbrake comprises drum brakes on each rear wheel, cable operated by the handbrake lever which is located between the driver's seat and the inner sill.

When the handbrake lever is operated, the cable system applies equal force to both RH and LH brake shoe expander assemblies. The brake shoes expand and press against the hub assembly, locking the rear wheels.

The handbrake switch latches when the lever is operated and lights the handbrake warning indicator mounted on the instrument panel.

The drum brakes are of the duo-servo type. The expander assembly (6 Fig. 66) is mounted on the backplate mounting lug. The brake shoes locate on the expander assembly and the adjuster (1 Fig. 66). These are held in position by the upper and lower return springs (4 and 5 Fig. 66) and the hold-down springs (2 Fig. 66). The adjuster allows manual adjustment of the brake shoes.

OPERATION

The wheel speed sensors, fitted to all four road wheels, transmit wheel speed information to the control module. The module uses this information to modulate brake pressure during anti-lock braking.

Brake pedal force is increased by the vacuum booster which activates the Tandem Master Cylinder (TMC) intermediate piston. Brake fluid is supplied to the pump inlet ports on two separate circuits. The primary circuit supplies the front brakes whilst the secondary circuit supplies the rear brakes.

The rear wheels are controlled collectively on a 'select-low' principle during ABS operation. This means that if locking in either rear wheel is detected, controlled brake pressure is applied to both wheels.

A pressure conscious reduction valve (PCRV) is fitted

FAULT DIAGNOSIS

Trouble	Cause	Remedy
Long brake pedal	Brake caliper piston(s) or caliper guide pins (rear only) sticking Worn/damaged brake pads	Service or renew caliper or caliper guide pins (rear only) Renew brake pads
Vibration during braking	Worn/damaged brake pads Loose caliper mounting bolts Insufficient grease on sliding parts Foreign material or scratches on brake disc contact surface Damaged brake disc contact surface	Renew brake pads Tighten caliper mounting bolts Apply grease where necessary Clean brake disc contact surface Renew brake disc
Poor braking performance	Leak in hydraulic system Air in hydraulic system Worn/damaged brake pads Foreign material on brake pads Brake caliper piston malfunction Tandem master cylinder malfunction Vacuum booster fault Disconnected or damaged vacuum hose Low brake fluid level	Repair leak. Check all pipework connections. Refill and bleed the system Check for leaks and bleed the system Renew brake pads Examine brake pads and clean or renew as necessary Renew faulty brake caliper piston Service or renew tandem master cylinder Renew vacuum booster Renew vacuum hose Check for leaks, refill and bleed the system
Brakes pull to one side	Worn/damaged brake pads Foreign material on brake pad Failing valves in ABS valve block Abnormal wear or distortion on brake disc Incorrect tyre pressure	Renew brake pads Examine brake pads and clean or renew as necessary Renew hydraulic control module Examine front brake disc and service or renew as necessary Inflate tyre to correct pressure
Brakes do not release	No brake pedal free play Vacuum booster binding Tandem master cylinder return port faulty Faulty valve in ABS valve block	Adjust brake pedal free play Renew vacuum booster Clean return port on tandem master cylinder Renew hydraulic control module
Excessive pedal travel	Leak in hydraulic system Air in hydraulic system Worn tandem master cylinder piston seals or scored cylinder bore 'Knock back'. Excessive brake disc run-out or loose wheel bearings	Repair leak. Check all pipework connections. Refill and bleed the system Check for leaks and bleed the system Renew tandem master cylinder Check brake disc run-out and renew as necessary. Adjust wheel bearing
Brakes grab	Brake pads contaminated by grease or brake fluid Brake pads distorted, cracked or loose Loose caliper mounting bolts or guide pins (rear only)	Renew brake pads. Check pipework for leaks Renew brake pads Check caliper and repair/renew as necessary
Brakes drag	Seized or incorrectly adjusted handbrake or cable Broken or weak handbrake return springs Caliper piston(s) seized Brake pedal binding at pivot points Vacuum booster binding Tandem master cylinder faulty	Examine handbrake and repair/renew as necessary Renew handbrake return springs Examine calipers and repair/renew as necessary Examine brake pedal bushings and repair/renew as necessary Renew vacuum booster Examine tandem master cylinder and repair/renew as necessary
Hard brake pedal when pressed	Lack of vacuum at the vacuum booster Tandem master cylinder push-rod binding Frozen tandem master cylinder piston Brake caliper piston or caliper guide pins (rear only) seized	Check vacuum hose. Repair or renew as necessary Renew tandem master cylinder Renew tandem master cylinder Examine caliper and renew/repair as necessary
Excessive brake noise	Worn brake pads Bent or cracked handbrake shoes Foreign objects in brake pads or handbrake shoes Broken/loose handbrake hold-down springs or return springs Loose caliper mounting bolts	Renew brake pads Renew handbrake shoes Examine brake pads and handbrake shoes. Clean or renew as necessary Examine handbrake assembly. Repair or renew as necessary Torque-tighten caliper mounting bolt

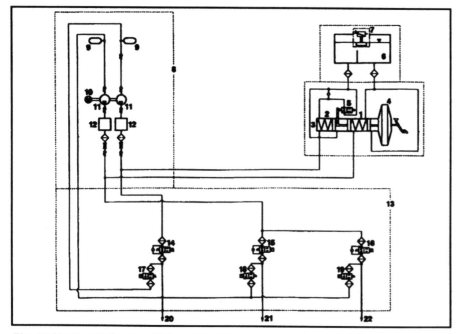

Fig. 67 Hydraulic System Schematic Diagram

between the outlet of the valve block and the rear brake circuit. The valve is fitted to prevent over-braking at the rear wheels. Up to a threshold of 25 bar (363 lbf/in²), brake pressure to the front and rear brakes is equal. Above this threshold, the PCRV reduces pressure to the rear brakes to provide a closer balance between front and rear brakes and optimize road adhesion.

Hydraulic operation

Referring to the Hydraulic System Schematic Diagram (Fig. 67), the TMC primary circuit (item 1) applies brake pressure to the front brakes. Individual control of the front wheels is provided by solenoid valves. One pair of valves (items 15 and 18) controls the front left brake circuit (item 21) and another pair of valves (items 16 and 19) controls the front right brake circuit (item 22). The TMC secondary circuit (item 2) applies brake pressure to the rear brake circuit (item 20) via valves (items 14 and 17), on a 'select low' principle.

Should the ABS be initiated by a locking tendency of any wheel during braking, the pump unit (item 8) is started and the appropriate NO inlet valve (item 14, 15 or 16) closes in response to signals from the control module. This action prevents further increase of brake pressure by blocking the supply of brake fluid from the TMC (item 3). If excessive deceleration continues, the appropriate NC outlet valve (item 17, 18 or 19) opens, releasing brake pressure to the low pressure accumulators (item 9) until the wheel accelerates again.

From the low pressure accumulators, volume is pumped back into the TMC, forcing the brake pedal back. To optimize the friction coefficient between tyre and road surface, brake pressure is increased in small steps by closing the outlet valve and opening the inlet valve and recharging brake pressure.

During the pressure build-up phase, the volume required for replenishment is supplied by the TMC and additionally by the pump from the low pressure accumulators. Since the delivered flow is generally greater than volume flow drained from the brake circuits, the low pressure accumulators serve as intermediate accumulators to compensate for temporary volume flow peaks.

The TMC piston positions, and therefore the brake pedal, vary with the fluid displacement in the brake caliper. As controlled pressure in the brake caliper decreases and increases during ABS, the brake pedal 'cycles', informing the driver that controlled braking is in progress.

Actuation of the brake pedal, causes the central valve (item 5) in the TMC to close. This action prevents damage to the TMC piston seals.

At the end of a brake application, volume is restored to the TMC, at low pressure from the fluid reservoir (item 6).

SYSTEM FAULT INDICATIONS

Fault indication:

ABS warning indicator still illuminated after ignition switch on and ABS CM self test.

Possible causes

Fuses blown
Faulty wheel speed sensor or harness
Faulty wiring
Faulty ABS CM.

Fault diagnosis

Note: After a fault has been successfully diagnosed and corrected, the ABS warning indicator will remain lit until the fault codes have been cleared from the non-volatile memory. This can be done by using Jaguar Diagnostic Equipment or, alternatively, by driving the vehicle to a speed above 20 km/h (12.5 mile/h). If the indicator remains on after this procedure, repeat fault diagnosis.

1.   Check the fuses (F11 & F22) in the battery feed lines and fuse (F10) in the ignition line. The fuses are located in the left and right-hand scuttle fuse boxes

2.   Unbolt the 28-way multi-plug connector from the ABS CM

3.   Measure the resistance across each wheel speed sensor. The value should be 1100R ± 50%. If not, unplug the sensor flying lead and measure again. If the value is now within range, inspect the harness

between ABS CM and sensor, otherwise renew the sensor

4. Check continuity to ground from ABS CM harness connections 13 and 14. If the value is much greater than 0.1R, renew the harness
5. Measure the voltage between ABS CM harness connection 14 and connections 1 and 2 respectively. If the value is not approximately equal to battery voltage, renew the harness
6. With the ignition switch ON, measure the voltage between ABS CM harness connections 14 and 15. If the value is not approximately equal to battery voltage, renew the harness.

Renew the ABS CM if the fault has not been located after carrying out the above procedures.

Fault indication:

ABS warning indicator illuminates at 20 km/h (12.5 mile/h).

Possible causes

Fuses blown
Faulty hydraulic control module pump/motor unit or circuitry
Faulty ABS CM.

Fault diagnosis

1. Check fuses
2. Disconnect the pump / motor unit and measure the resistance across the two pin connector. The measured value should be in the region of 0.8R. Renew the complete hydraulic control module if the measured value indicates excessive resistance or a short circuit
3. Unbolt the 28-way connector from the ABS CM and measure the voltage between harness connections 1 and 14. If the value is not approximately equal to battery voltage, renew the harness.

Renew the ABS CM if the fault has not been located after carrying out the above procedures.

Fault indication:

ABS warning indicator illuminates on 'pull-away' or during driving.

Possible causes

Faulty sensor or wiring
Faulty brake disc or wheel bearing installation giving inconsistent signals to the ABS CM.

Fault diagnosis

Check sensor installation for:

1. Security of sensor lead fixing bolt
2. Damage to sensor lead
3. Possible damage to brake disc
4. Excessive play in wheel bearing
5. Intermittent faults caused by poor harness connection or damage.

Pin-point tests

Wheel sensor

Testing between pins of the 28-way multi-plug connector, check that the resistance of each sensor coil is 1100 ohms ± 50%.

Code	Fault	Comment
5242h	Outlet valve, rear	
5250h	Inlet valve, rear	
5120h	Outlet valve, front right	
5214h	Inlet valve, front right	
5194h	Outlet valve, front left	
5198h	Inlet valve, front left	
5168h	Sensor, rear right	Sensor failure recognised by monitoring of d.c. voltage with vehicle stationary
5178h	Sensor, rear left	
5148h	Sensor front right	
5158h	Sensor, front left	
5165h	Sensor, rear right	Sensor failure recognised by monitoring of wheel speed continuity
5175h	Sensor, rear left	
5145h	Sensor, front right	
5155h	Sensor, front left	
5260h	Sensor, rear right	Sensor failure recognised by wheel speed comparison
5261h	Sensor, rear left	
5259h	Sensor, front right	
5258h	Sensor, front left	
5235h	Sensor, rear right	Sensor failure recognised by long term detection of missing sensor signal
5236h	Sensor, rear left	
5234h	Sensor, front right	
5233h	Sensor, front left	
9317h	Over-voltage	
9342h	CPU failure	
5095h	Pump motor	
5267h	Disturbance detection	

Note: See Control Module Connections for the ABS CM wheel sensor connector references.

Hydraulic pump motor

Disconnect the pump motor bi-pin connector and check that the resistance of the motor winding is approximately 0.8 ohms.

Stored fault codes

The following fault codes are stored automatically within the ABS CM and may be accessed, as an aid to fault diagnosis, using Jaguar Diagnostic Equipment.

BRAKE HYDRAULIC SYSTEM

Fluid level

> **WARNING:**
>
> AVOID SKIN/EYE CONTACT OR INGESTION OF BRAKE FLUID.IF SKIN OR EYES ARE ACCIDENTALLY SPLASHED WITH BRAKE FLUID, RINSE THE AFFECTED AREA IMMEDIATELY WITH PLENTY OF WATER AND SEEK MEDICAL ATTENTION IMMEDIATELY.

Checking the fluid level

Correct brake fluid level is essential for the efficient operation of the brake system. Check that the fluid level is between the MAX and MIN marks on the fluid reservoir (Fig. 68). Top up if necessary with recommended brake fluid.

Note: The efficiency of the brakes may be impaired if fluid is used which does not meet specifications. Use ONLY brake and clutch fluid that conforms to a minimum DOT 4 specification. Do not use brake fluid that has been exposed to atmosphere for any length of time. Moisture absorbed from the atmosphere impairs the efficiency of the brake fluid.

Fig. 68

BRAKE HYDRAULIC SYSTEM, BLEED

General instructions

Use a brake bleeder bottle with a clear bleeder tube. Also recommended is a filler unit with a fill pressure of 1.0 bar (14.5 lbf/in²). If a filler unit is not used, ensure that there is sufficient brake fluid in the reservoir throughout the bleeding procedure.

Note: Always bleed the caliper furthest away from the actuation unit first. On right-hand drive vehicles, bleed in the following order: front left (FL), front right (FR), rear left (RL) and rear right (RR). On left-hand drive vehicles bleed in the following order: FR, FL, RR and RL.

System Bleeding After Brake Fluid Renewal

Ensure that the vehicle is standing level. Switch the ignition off. Check that the fluid level in the reservoir is between the MIN and MAX marks.
Connect the bleeder bottle tube to the relevant front caliper bleeder screw (Fig. 69) and open the screw (see preceding 'Note').
Bleed until new, clear, bubble free fluid is observed in the tube and then close the bleeder screw.
Repeat this procedure at each remaining caliper.
With the engine running, check brake pedal travel. If excessive, check for leaks and repeat the bleed procedure.
Fill the reservoir to the MAX level.
System bleeding after tandem master cylinder renewal
Ensure that the vehicle is standing level. Switch the ignition off. Check that the fluid level in the reservoir is between the MIN and MAX marks.
Connect the bleeder bottle tube to the relevant front caliper bleeder screw (Fig. 69) and open the screw (see Note on previous page).
Actuate the brake pedal to the floor; hold for approximately two seconds and then release the pedal. Wait another two

Fig. 69

seconds and actuate the brake pedal again for a further two seconds. Repeat this action 20 to 30 times until clear, bubble free brake fluid streams out.
With the brake pedal actuated, close the bleeder screw. Build up fluid pressure by pumping the pedal and then open the bleeder screw. Repeat this action three to five times.

Note: If a filler unit is not used, observe the fluid level in the reservoir and top up if necessary.

Repeat this procedure for the remaining three calipers: (Fig. 70) shows the bleeder screw of the rear left caliper.

With the engine running, check brake pedal travel. If excessive, check for leaks and repeat the bleed procedure. Fill the reservoir to the MAX level.

System bleeding after hydraulic control module renewal

Hydraulic control modules are supplied pre-filled to enable the brake system to be bled in the conventional way.
Ensure that the vehicle is standing level. Switch the ignition off.
Check that the fluid level in the reservoir is between the MIN and MAX marks.
Connect the bleeder bottle tube to the relevant front caliper bleeder screw (Fig. 69) and open the screw (see Note on previous page).
Actuate the brake pedal full stroke, wait a moment and then release. Wait two to three seconds and then actuate the brake pedal full stroke again. This allows the TMC to be completely re-filled with fluid.
Repeat 20 to 30 times until the fluid in the bleeder tube is clear and bubble free.
With the brake pedal actuated, close the bleeder screw. Build up fluid pressure by pumping the pedal and then open the bleeder screw. Repeat this action three times.

Repeat this procedure for the remaining three calipers.
With the engine running, check brake pedal travel. If pedal travel is excessive, check the system for leaks and repeat the bleed procedure.
Fill the reservoir to the MAX level.

Bleeding after renewal of caliper

Follow the procedure above but only at the affected caliper.

BRAKE CALIPERS, INSPECTION AND CLEANING

WARNING:

BRAKE LINING DUST CAN, IF INHALED, DAMAGE YOUR HEALTH. ALWAYS USE A VACUUM BRUSH TO REMOVE DRY BRAKE LINING DUST. NEVER USE AN AIR LINE.

When fitting new brake pads always take necessary precautions and remove the brake dust from around the caliper area. After renewal, pump the brake pedal several times to centralize the new brake pads.

.CAUTION:

When cleaning brake components only use a proprietary fluid. Never use petrol. Use of petrol, paraffin or other mineral based fluids can prove dangerous.

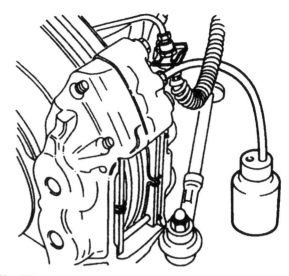

Fig. 70

Remove all brake dust from the caliper, carrier and brake disc. Thoroughly clean the pad abutment areas. Avoid damaging the piston and dust cover.
Examine all the components for signs of wear, damage and corrosion. Pay particular attention to the piston and piston bore.
Remove caliper body corrosion with a wire brush or wire wool. No attempt should be made to clean a badly corroded or scored piston bore. The caliper must be renewed.

CAUTION:

No attempt should be made to clean corroded bolts.

Inspect the caliper guide pins. Ensure that they are not corroded or seized and that the caliper moves freely. If they are difficult to remove or corroded in any way, they must be replaced together with new dust covers.
When reassembling always renew piston seals. Lubricate the new piston seal and fit carefully to the inner groove of the piston bore.

CAUTION:

Ensure that working surfaces and hands are clean. Use only brake fluid of the correct specification to lubricate the new seals when fitting.

WHEEL SPEED SENSOR, REAR, RENEW

Open the door. Move the front seats fully forward.
Remove the rear seat cushion and squab.
Displace the insulation material and release the ratchet strap securing the sensor multi-plug (1 Fig. 71) to the transmission tunnel.

Disconnect the multi-plug. Attach a drawstring to the sensor harness.
Locate a trolley jack below the outer fork of the rear wishbone.

CAUTION:

To prevent wishbone component damage, use a suitably shaped block of wood between the jack and wishbone.

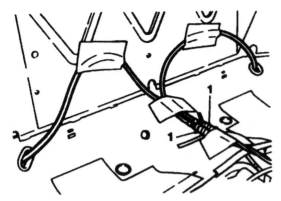

Fig. 71

Fig. 72

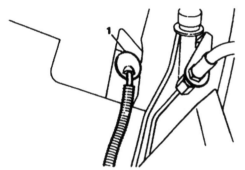

Fig. 73

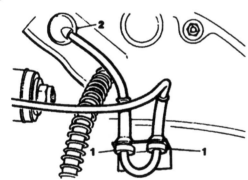

Fig. 74

Fig. 75

Jack up the vehicle. Place an axle stand below the jacking spigot. Lower the car/jacking spigot on to the axle stand.
Leave the jack in position as a safety measure.
Remove the road wheel.
Remove the sensor securing bolt (2 Fig. 72) and displace the sensor (1 Fig. 72) from the hub.
Cut and remove the harness ratchet strap.
Displace the harness outer body grommet (1 Fig. 73). From below, carefully withdraw the harness and drawstring. Transfer the drawstring to the new sensor harness.

Carefully draw the new harness into position and seat the outer body grommet (1 Fig. 73).
Route the harness to the hub and fit the sensor to the hub (1 Fig. 72).
Fit and tighten the sensor securing bolt (2 Fig. 72).
Secure the harness with a new ratchet strap.
Fit the road wheel.
Route the harness, remove the drawstring and connect the multi-plug (1 Fig. 71).
Fasten the ratchet strap to secure the sensor multi-plug to the transmission tunnel.
Replace the rear seat cushion and squab.
Reposition the front seats and close the door.

Torque Figures

Wheel nut to wheel stud	88 - 102 Nm
Wheel sensor bolt	8 - 10 Nm

WHEEL SPEED SENSOR, FRONT, RENEW

Drive the vehicle on to a ramp.
Open the bonnet. Disconnect the sensor multi-plug. Cut and remove the ratchet strap if required.
Displace the body grommet (2 Fig. 74) from the inner wing valance.
Raise the ramp.
Remove the brake disc cover plate. Remove the retaining bracket.
Remove the upper cover securing bolt and remove the cover.
Remove the sensor securing bolt (2 Fig. 75) and displace the sensor (1 Fig. 75) from the hub.

Displace the harness carrier grommets from the inner-wing and caliper brackets. The inner-wing bracket is shown (1 Fig. 74). Cut and remove the harness ratchet strap.
Displace and remove the sensor and harness.
Route the new sensor harness through the inner wing, seat the harness to the carrier grommets and fit a new ratchet strap.
Fit the sensor to the hub. Fit and tighten the securing bolt.
Fit the body grommet to the inner wing (2 Fig. 74).
Fit the brake disc cover plate and the retaining bracket.
Lower the ramp. Route the harness to the connector and connect the multi-plug. Secure the harness with a new ratchet strap if required.
Close the bonnet.

Torque Figures

Wheel sensor bolt	8 - 10 Nm

HYDRAULIC CONTROL MODULE, RENEW

Note: Refer to Brake Hydraulic System, Bleed before carrying out this procedure. Pay particular attention to the warnings and cautions relating to brake fluid, cleanliness and cleaning materials.

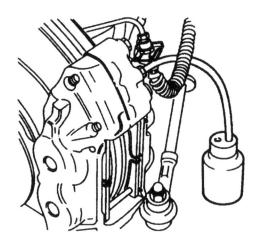

Fig. 76

Raise the vehicle.

Remove the bleed screw dust cap from the front LH caliper. Connect a bleeder tube and bottle to the bleed screw (Fig. 76) and open the bleed screw.

Fit a brake pedal hold-down tool (JDS-9013) between the brake pedal and the steering wheel. Adjust the tool to operate the brake pedal 60 mm down. This operation is necessary to prevent fluid loss from the reservoir through disconnected brake pipes.

Tighten the front LH caliper bleeder screw. Disconnect the bleed tube from the bleed screw and remove the tube and bottle. Fit the bleed screw dust cap.

Undo the securing bolt of the multi-plug connector (4 Fig. 77). The bolt will remain captive.

Disconnect the multi-plug connector and reposition safely.

Place absorbent material underneath the hydraulic control module to absorb any spillages.

Undo the tandem master cylinder (TMC) brake pipe unions at the pump unit (1 Fig. 77) and disconnect the brake pipes.

Fit plugs immediately to the brake pipes and the pump unit to prevent fluid loss.

Undo the front and rear brake pipe unions at the valve block (3 Fig. 77).

Remove the brake pipes.

Fit plugs immediately to the brake pipes and the valve block to prevent fluid loss.

Undo and remove the three securing nuts (5 Fig. 77) and remove the hydraulic control module.

Remove the absorbent material and clean the mounting bracket and surrounding area.

Fit and align a new hydraulic control module to the mounting bracket. Ensure that the mounting cup tangs fully engage the bracket slots.

Fit and tighten the securing nuts (5 Fig. 77).

Place absorbent material underneath the hydraulic control module to absorb any spillages.

Connect the front and rear brake pipes to the valve block (3 Fig. 77), removing the plugs immediately prior to connection. Tighten the pipe unions.

Connect the TMC brake pipes to the pump unit (1 Fig. 77), removing the plugs immediately prior to connection. Tighten the pipe unions.

Remove the absorbent material and clean the surrounding area.

Reposition and connect the multi-plug connector (4 Fig. 77). Tighten the securing bolt.

Ensure that all fixings are torque-tightened to specified tolerances.

Release the brake pedal hold-down tool and remove.

Bleed the system. See Brake Hydraulic System, Bleed.

Examine the hydraulic control module for leaks.

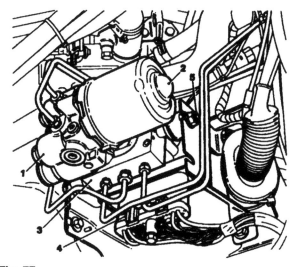

Fig. 77

Torque Figures

Hydraulic module to module bracket	18 - 26 Nm
28-way connector to control module	4 - 5.6 Nm
M12 pipe connectors	15 - 19 Nm
M10 pipe connectors	12 - 16 Nm

Oils / Greases

Jaguar Brake and Clutch Fluid or Castrol/Girling Universal to/or exceeding DOT 4 specification.

PRESSURE CONSCIOUS REDUCING VALVE, RENEW

Note: Refer to Brake Hydraulic System, Bleed before carrying out this procedure. Pay particular attention to the warnings and cautions relating to brake fluid, cleanliness and cleaning materials.

Remove the air cleaner cover and element.

Remove the bleeder screw dust cap from the front LH caliper. Connect a bleeder tube and bottle to the bleeder screw (Fig. 76) and open the bleeder screw.

Fit a brake pedal hold-down tool (JDS-9013) between the brake pedal and the steering wheel. Adjust the tool to operate the brake pedal 60 mm down. This operation is necessary to prevent fluid loss from the reservoir through disconnected brake pipes.

Tighten the front LH caliper bleeder screw. Disconnect the bleeder tube from the bleeder screw and remove the tube and bottle. Fit the bleeder screw dust cap.

Place suitable absorbent material around the pressure conscious reducing valve (PCRV) area to catch any spilt fluid.

Supporting the PCRV unions, disconnect the pipe unions (1 Fig. 78) from the PCRV (2 Fig. 78). Fit plugs immediately to the brake pipes and PCRV to prevent fluid loss.

Remove the PCRV from the spring clip.

Fit the new PCRV into the spring clip. While supporting the PCRV unions, connect the pipe unions (1 Fig. 78) to the PCRV (2 Fig. 78), removing the plugs immediately prior to connection.

Tighten the pipe unions.

Remove the absorbent material and clean the surrounding area.

Fit the air cleaner element and cover.

Bleed the system. See Brake Hydraulic System, Bleed.

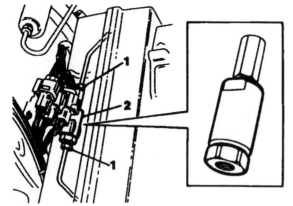

Fig. 78

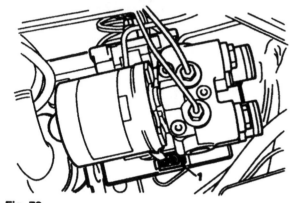

Fig. 79

Torque Figures

M12 pipe connectors	15 - 19 Nm
M10 pipe connectors	12 - 16 Nm

Oils / Greases

Jaguar Brake and Clutch Fluid or Castrol/Girling Universal to/or exceeding DOT 4 specification.

ABS Control Module, renew

Note: Refer to Brake Hydraulic System, Bleed before carrying out this procedure. Pay particular attention to the warnings and cautions relating to brake fluid, cleanliness and cleaning materials.

Remove the hydraulic control module.
Disconnect the pump electric motor to ABS CM multi-plug (1 Fig 79).
Undo and remove two securing screws and remove the ABS CM.
Clean the mating faces of the hydraulic control module and the new ABS CM.
Fit and tighten the two securing screws.
Connect the pump electric motor to ABS CM multi-plug (1 Fig 79).
Fit the hydraulic control module.

Torque Figures

Hydraulic module to module bracket	18 - 26 Nm
Control module to hydraulic module	1.8 - 2.8 Nm
28-way connector to control module	4 - 5.6 Nm
M12 pipe connectors	15 - 19 Nm
M10 pipe connectors	12 - 16 Nm

Oils / Greases

Jaguar Brake and Clutch Fluid or Castrol/Girling Universal to/or exceeding DOT 4 specification.

CHAPTER 14

BODY - 1992 MODEL YEAR ONWARDS (INCLUDING SRS SYSTEMS)

Introduction

Many improvements were made to all variants of XJS at 1992 and 1993/93.5 Model Years.

Of most note were the improvements to Body-in-White construction methods and improvements to the glasshouse at 1992, and the new bumper systems and associated changes at the later point.

Also covered in this Section are the Supplementary Restraint Systems (SRS) which have been fitted at various points from 1989 on (limited market production). Later versions benefit from airbags on both sides of the vehicle, together with 'tear-loop' type seatbelts.

Sealants

In production, the vehicle is given an under body protection. Stone chip primer is also used on the sill panels, while hot wax injection is applied to all box sections and closed members. Recommended sealants are listed in the Jaguar Body Sealing and Preservation Manual (AKM 9137).

PLENUM CHAMBER FINISHER, RENEW

Open the bonnet and remove the wiper arms.
Use a suitable tool to carefully remove the wiper pinion plastic trim covers (2 Fig. 1).

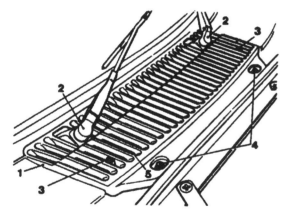

Fig. 1

Displace the washer jets (3 Fig. 1) down through the plenum chamber finisher (1 Fig. 1).
Remove the plenum chamber finisher (1 Fig. 1) by undoing the screws (4 Fig. 1) which secure the plenum chamber finisher to the body.
Remove the plenum finisher gauze from the clips (5 Fig. 1) which hold it in position.

Fit the gauze into position in the plenum chamber finisher, using the clips (5 Fig. 1) to hold it in place. Place the plenum chamber finisher in the position where it is to be mounted, and locate the washer jets (3 Fig. 1) through the finisher and into their positions.
Fix the plenum chamber finisher firmly into position using the securing screws (4 Fig. 1).
Mount the wiper pinion plastic trim covers (2 Fig. 1) firmly on to their seatings.
Fit the wiper arms and close the bonnet.

STEERING COLUMN SWITCH HOUSING, RENEW

Note: The air bag module should first be removed before proceeding to remove the steering column switch housing.

Remove the steering wheel.
Remove the steering column lower cowl.
Remove the steering column upper cowl.

Removing the steering column switch housing.
Remove the fibre optic illumination bulb cover (1 Fig. 2).
Displace both fibre optic switch leads (2 Fig. 2) from the housing.
Release both stalk switches from their mountings by squeezing their securing lugs (1 Fig. 3).

Displace them from their mountings by carefully sliding them out.
Disconnect the horn contact wire (2 Fig. 3).
Undo and remove the housing to column securing bolts (3 Fig. 3) and displace the housing (4 Fig. 3) from the steering column.
Displace the illumination bulb holder (5 Fig. 3) from the switch housing.
Remove the switch housing (4 Fig. 3) from the steering column.

Place the switch housing (4 Fig. 3) by the position where it is to be mounted and fully seat the switch illumination bulb holder (5 Fig. 3) into the switch housing.

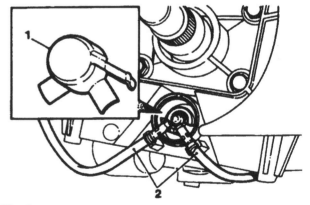

Fig. 2

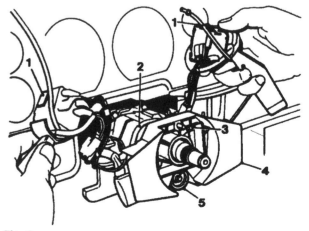

Fig. 3

Locate the switch housing on to the steering column and when it is fully seated, fix it in position by firmly tightening the securing bolts (3 Fig. 3).

Connect the horn contact wire (2 Fig. 3).

Place each stalk switch (1 Fig. 3) in position and fully seat them into their locations in the switch housing to fix them.

Locate the fibre optic switch leads into their positions in the switch housing (2 Fig. 2).

Firmly locate the fibre optic illumination bulb cover (1 Fig. 2) into its position on the switch housing.

Fit the steering column upper cowl.
Fit the steering column lower cowl.
Fit the steering wheel.

Note: The air bag module should be fitted on to the steering wheel as the final stage of fitting the steering wheel.

DOOR GLASS, RENEW

Remove the door trim pad.
Remove the front speaker.
Remove and discard the tape securing the rear upper edge

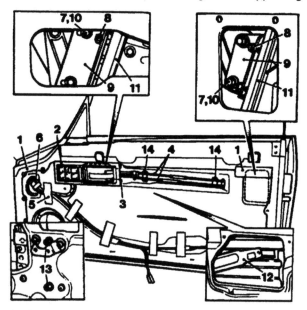

Fig. 4

of the water shedder (1 Fig. 4) to the door panel.
Remove the electric seat switch surround (2 Fig. 4).
Remove the inner door release handle (3 Fig. 4) securing screws and displace the inner door handle operating rods (4 Fig. 4) from the guide retaining clips (14 Fig. 4).
Reposition the inner door release handle to allow access to the seat switch.
Displace and remove the clips (14 Fig. 4) which retain the inner door handle operating rods (4 Fig. 4). Remove and discard the seat switch harness (5 Fig. 4) securing tape.
Undo and remove the remaining seat switch securing screw.
Disconnect the seat switch harness multi plug and remove the seat switch assembly (6 Fig. 4).
Remove and discard the tape securing the rear upper edge of the water shedder (1 Fig. 4) to the door, and reposition the shedder to give access to the door glass securing screws (7 Fig. 4).
Switch the ignition on, motor the glass down to obtain access to the glass securing bolts, and switch the ignition off.
Mark the position where the glass securing bolts (8 Fig. 4) are fixed to the glass carrier brackets (9 Fig. 4).
Take the weight of the door glass and release this from the carrier by undoing the securing bolts (8 Fig. 4).

Note: In performing this operation, care must be taken to avoid dislodging the securing nuts (10 Fig. 4) mounted on the glass.

Carefully remove the door glass.
After lubricating the runner (11 Fig. 4), fit the door glass to the door and engage the runner to the regulator arm (12 Fig. 4) when the door glass is aligned.
Align the glass carrier brackets (9 Fig. 4) with the glass and fit the glass to carrier securing bolts (8 Fig. 4) without fully tightening them.
Line up the door glass with the position marked when the glass was removed. When correctly positioned, tighten two of the glass to carrier securing bolts, one in the front and the other in the rear carrier. Switch the ignition on, motor the door glass fully up and switch the ignition off.

Note: The ignition needs to be turned on whenever the glass is to be raised or lowered and off when adjustments are being made to the glass.

Carefully close the door and note how the door glass is positioned against the aperture frame and seals.
Open the door and lower the glass down to give access to the securing bolts.
Loosen the glass to carrier bracket securing bolts (8 Fig. 4) to allow the position of the door glass to be adjusted.
Make any necessary adjustment and hold the glass in position by tightening two glass to carrier securing bolts.
Fully raise the door glass and carefully close the door to check the position of the door glass to aperture frame and seals, repeating the procedure if further adjustment is still required.
Lower the door glass and adjust in turn the upper and lower regulator stop adjusters (13 Fig. 4), slackening the stop adjuster lock nuts to enable the adjustment to be made, and tightening them after making the adjustment.
Close the door and check the glass height, finally tightening all the glass to carrier securing bolts when the glass is satisfactorily positioned.
Reposition the water shedder (1 Fig. 4) to the door and secure the front and rear edges with suitable tape.
Fit and align the electric seat switch assembly (2 Fig. 4), holding this in position with the front securing screw which should not be fully tightened. Fit and align the operating rod retaining clips (14 Fig. 4) and fully seat the door handle operating rods (4 Fig. 4) into these.
Position the inner handle (3 Fig. 4) to its location and firmly tighten the screws which secure it in position.
Tighten the seat switch securing screw.
Connect the seat switch harness (5 Fig. 4) multi plug and reposition the harness, securing it with suitable tape.

Fit the front speaker.
Fit the door trim pad.

DOOR GLASS, ADJUST

Open the door and remove the door trim pad.

If the door glass merely requires an adjustment in an upwards or downwards direction, carry out a stage 1 adjust.

Stage 1 adjust:

Partially lower the glass.
Displace the water shedder.
Slacken the regulator stop locknut (1 Fig. 5). Adjust the regulator stop stud (2 Fig. 5).

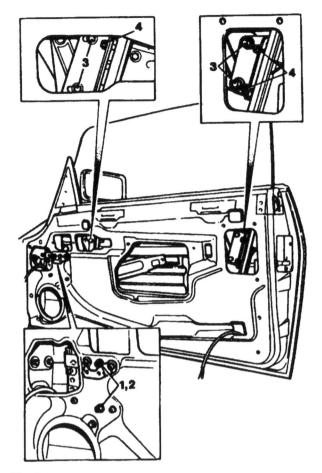

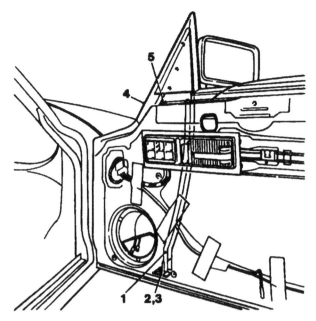

Fig. 5

Tighten the locknut (1 Fig. 5).

Raise the window glass and close the door.

Check the alignment of the glass and, if necessary, repeat the procedure until satisfied that the glass is correctly aligned.

Replace the door trim pad and close the door.

If the glass needs to be adjusted in its carrier and the runners need to be aligned with the glass, carry out a stage 2 adjust.

Stage 2 adjust:

Open the door and remove the door trim pad.

Partially lower the glass and displace the water shedder to gain access to the screws securing the glass (3 Fig. 5).

Slacken these screws and move the glass in the desired direction.

Tighten the glass securing screws (3 Fig. 5).

Fully raise the window glass and close the door.

Check the alignment of the glass and if necessary, repeat the procedure to readjust the glass.

Partially lower the glass to gain access to the glass runner securing bolts (4 Fig. 5).

Slacken these and adjust the glass.

Tighten the bolts securing the glass runner.

Complete the operation with a stage 1 adjust, as described above.

Fit the door trim pad and close the door.

DOOR CHEATER FRAME, RENEW

Switch the ignition on to lower the glass and then switch the ignition off.

Remove the door trim pad.

Remove the front speaker.

Fig. 6

Undo and remove the screws securing the cheater frame channel (1 Fig. 6) to the lower bracket (2 Fig. 6), using the speaker aperture to gain access to these screws.

Undo the bolt fixing the bracket to the door (3 Fig. 6) but do not remove this.

Remove the cheater frame inner and outer plates.

Remove the fir-tree retaining clip (4 Fig. 6) which attaches the seal to the door.

Pull the seal away from the door to release the adhesion.

Undo and remove the screws securing the cheater frame to the upper door (5 Fig. 6).

Remove the cheater frame and seal assembly from the door and pull the seal away from the frame.

Position the seal against the door cheater frame and fit these together.

Place the cheater frame and seal assembly into position in the door and hold in place with the upper and lower door securing screws (5 Fig. 6), but do not tighten these until the position of the cheater frame assembly has been adjusted.

Once the cheater frame assembly has been satisfactorily positioned, the screws securing the cheater frame to the upper and lower door (5 Fig. 6), and the bolt securing the lower bracket to the door (3 Fig. 6) should all be tightened.

Apply a suitable adhesive to the seal and fully seat this seal to the door panel.

Fit the fir-tree clip which retains the seal to the door, and fully seat this.

Fit the cheater frame outer and inner plates.

Fit and tighten the screws securing the cheater frame channel (1 Fig. 6) to the lower bracket (2 Fig. 6).

Fit the front speaker.

Fit the door trim pad.

Switch on the ignition, raise the glass, and then switch the ignition off.

DOOR TRIM PAD, RENEW

Open the door.

Using a suitable implement, ease the arm rest escutchion plate from position to gain access to the armrest upper securing screw (1 Fig. 7).

Undo and remove the armrest upper securing screw.

Undo and remove the door pocket lower securing screws (2 Fig. 7).

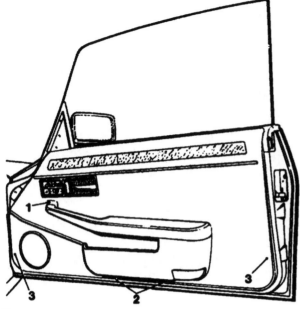

Fig. 7

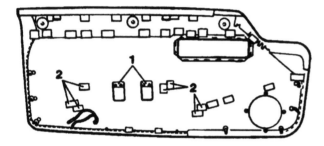

Fig. 8

Displace the trim pad to door securing clips (3 Fig. 7).
Lift the trim pad to disengage from the locating tangs and displace the trim pad from the door.
Note and disconnect the door guard lamp wires. Remove the trim pad.

Place the trim pad face downwards on a bench.
Undo and remove the veneer panel securing screws.
Displace and remove the veneer panel.
Undo and remove the screws securing the retaining brackets and remove the retaining brackets (1 Fig. 8).

Remove the remaining door pocket securing screw.
Release the door pocket retaining tangs (2 Fig. 8). Turn the trim pad over and displace and remove the door pocket from the trim pad.
Displace and remove the radio speaker grille by first removing the grille spire clips.
Displace and remove the inner waist draught seal from the trim pad.
Displace and remove the trim pad retaining clips. Put the dismantled trim pad aside.

Place the trim pad on a bench.
Fit the trim pad to door securing clips (3 Fig. 7). Fit and fully seat the inner waist draught seal.
Fit and align the speaker grill and secure this in position with spire clips.
Position the door pocket assembly.
Secure the pocket retaining tangs and fit and fully seat the new lower securing clips.
Secure the door trim pad clip retaining tangs.
Position the door trim pad retaining brackets. Fit and tighten the bracket securing screws.

Fit and align the veneer panel and tighten the screws which secure this.

Place the door trim pad in position against the door.
Connect the door guard lamp wires.
Locate the door trim pad securing clips against the retaining tangs and seat the trim pad fully home.
Fit and tighten the door pocket lower securing screws (2 Fig. 7).
Fit and tighten the armrest upper securing screw (1 Fig. 7).
Fully reseat the armrest escutcheon plate (1 Fig. 7).

DOOR TRIM PAD CHROME FINISHERS, RENEW

Remove the door trim pad.
Place the trim pad face downwards on a bench.
Remove and discard the tape from securing clip tangs of whichever of the two finishers is to be replaced.
Displace the centre finisher retaining tangs (1 Fig. 9) or lower finisher retaining tangs (2 Fig. 9) as appropriate.

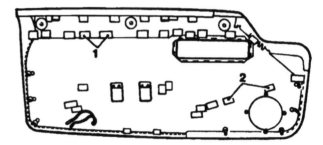

Fig. 9

Separate the finisher assembly from the trim pad and remove the finisher retaining clips.
Place the finisher to one side.

Take the centre or lower chrome trim pad finisher as appropriate and fit retaining clips to the finisher.
Fit the finisher to the door trim pad and hold it securely in position with retaining tangs.
Cover the exposed tangs with suitable adhesive tape.

Fit the door trim pad.

DOOR LOCK MECHANISM, RENEW

Remove the door trim pad.

Disconnect the multiplug connecting the harness to the door lock motor leads (1 Fig. 10) and remove and discard the tape which holds these leads to the door.
Reposition the water shedder (2 Fig. 10) to gain access to the door lock operating rods (3 Fig. 10).
Displace the door lock operating rods from the guide clips (4 Fig. 10).
Disconnect the rod which operates the outer lock mechanism from this mechanism and reposition the rod.
Disconnect the clip (5 Fig. 10) which connects the rod which operates the outer release handle to the handle and reposition this rod.
Disconnect the clip (6 Fig. 10) which connects the rod which operates the door lock to the door lock barrel and reposition this rod.
Disconnect the clip (7 Fig. 10) which connects the rod which operates the door release to the door release mechanism and reposition this rod.
Undo the screws (8 Fig. 10) securing the outer lock to the door and remove the outer lock assembly.

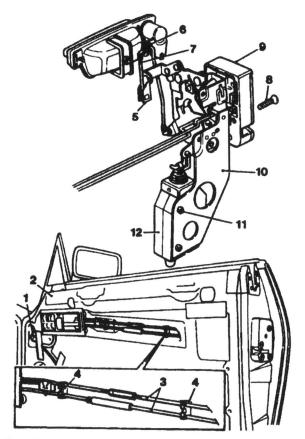

Fig. 10

Remove the motor/lock mechanism assembly (10 Fig. 10). Undo the screws (11 Fig. 10) securing the motor to the door lock mechanism, remove the motor assembly (12 Fig. 10), and place the mechanism to one side.

Fit the motor (12 Fig. 10) to the door lock mechanism, positioning it so that the screw holes line up, and firmly fix them together with the securing screws (11 Fig. 10).
Lubricate the linkage between the motor and the door lock mechanism.
Position the motor assembly (10 Fig. 10) to the door, ensuring that no link rods or harnesses become trapped.
Fit the outer lock assembly (9 Fig. 10) to the door, fixing it firmly in position with the securing screws (8 Fig. 10).
Reposition the door lock operating rods (3 Fig. 10) and secure them with the retaining clips (5, 6 & 7 Fig. 10).
Route the leads (1 Fig. 10) to the door lock motor harness through the door panel.
Adjust the rod connected to the release lever and secure it with a retaining clip (7 Fig. 10).
Reposition the rod connected to the lock barrel and secure it with a retaining clip (6 Fig. 10).
Reposition the harness through the weather strip (2 Fig. 10) and position the weather strip against the door, securing the top edge with suitable adhesive tape.
Fit the operating rod guide clips (4 Fig. 10) to the door, fully seating them into their locations
Fit the door lock operating rods (3 Fig. 10) into the guide clips, fully seating them into position.
Connect the door lock motor harness multiple (1 Fig. 10) and secure the leads to the door lock with suitable adhesive tape.
Fit the door trim pad.

FRONT SEAT RUNNER, RENEW

Switch on the ignition and motor the seat fully forwards.
Undo the bolts securing the back of the seat runner to the floor panel (1 Fig. 11).

Fig. 11

Motor the seat fully backwards and switch the ignition off.
Undo the bolts securing the front of the seat runner to the floor panel (2 Fig. 11).
Reposition the footwell carpet to give access to the ECU cover panel.

Undo the screws which secure the ECU cover panel and reposition this panel to gain access to the seat harness multiplugs (3 Fig. 11).
Undo the screws which secure the seat harness plastic clamp (1 Fig. 12) and remove this.

Cut and remove the seat harness ratchet straps (4 Fig. 11).
Disconnect the seat harness multiplugs (3 Fig. 11) including the harness to ECU multiplug (5 Fig. 11). Reposition the seat to gain access to the seat belt anchorage bolt (2 Fig. 12).
Remove the seat belt anchorage bolt trim cap and undo the bolt (2 Fig. 12).
Reposition the seat belt.
Remove the seat and runner assembly and place it on a bench.

Undo the screws securing the motor drive shaft retaining plate assemblies (3 Fig. 12).
Remove the motor drive shaft retaining plate assemblies.
Cut and remove the ratchet straps (4 Fig. 12) holding the seat belt buckle harness and disconnect the seat belt buckle harness multiplug (5 Fig. 12).
Reposition the runners to give access to the bolts which secure the front of the seat frame (6 Fig. 12) and undo these bolts.
Reposition the runners to give access to the bolts which secure the rear of the seat frame (7 Fig. 12) and undo these bolts.
Reposition the runner assembly to gain access to the harness
Displace the harness multiplugs (8 Fig. 12) from the retaining clips and disconnect the multiplugs.
Remove the runner assembly.

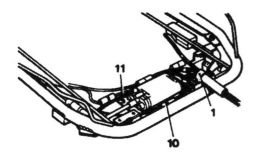

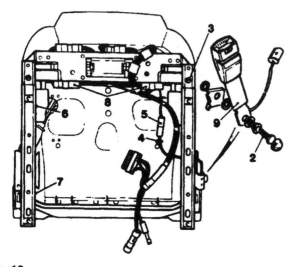

Fig. 12

Remove the trim cap from the bolt which secures the seat belt buckle and undo the bolt (2 Fig. 12).

Remove the seat belt buckle.

Remove the belt buckle bracket (9 Fig. 12).

Undo the screws which secure the seat harness plastic clamp plate (1 Fig. 12) and remove this plate. Displace the motor harness multiplug (10 Fig. 12) from its retaining clip.

Undo, without fully removing, the screws (11 Fig. 12) which secure the motor to the runner.

Remove the motor/drive and harness assembly.

Remove the clips retaining the harness.

Place the runner assembly to one side.

Fit the motor/drive and harness assembly to the runner, lining up the screws with the holes on the runner and taking care not to trap the harness.

Fix the motor/drive and harness assembly to the runner by tightening the securing screws (11 Fig. 12).

Position the harness to the retaining bracket and place the seat harness plastic clamp plate (1 Fig. 12) by the bracket so that the screw holes line up.

Firmly screw the clamp plate into position.

Fit and align the seat belt buckle bracket (9 Fig. 12).

Position the seat belt buckle against the bracket and fix with the securing bolt (2 Fig. 12) which should be tightened and covered with a trim cap.

Place the runner assembly to the seat and connect the harness multiplugs (8 Fig. 12).

Position the multiplugs into their retaining clips.

Align the runner assembly with the seat frame and fit and tighten the rear bolts (7 Fig. 12) which secure the runner to the seat frame.

Reposition the seat runners to allow access to the front securing bolts (6 Fig. 12) which should be fitted and tightened.

Connect the seat belt buckle multiplug (5 Fig. 12) and position this to the seat diaphragm where it should be secured with ratchet straps (4 Fig. 12).

Reposition the seat runners fully forward and fit and fully seat the motor drive shaft assemblies, manoeuvring the runners to engage in the driveshaft couplings.

Fit and tighten the screws which secure the motor drive shaft retaining plate (3 Fig. 12).

Place the seat and runner assembly in position in the vehicle. Position the seat belt to its anchorage point on the seat and fix it in position with the anchorage bolt (2 Fig. 12) which should then be covered with a trim cap.

Reposition the seat to its mounting.

Connect the seat harness multiplugs (3 Fig. 11).

Position the harness to the bracket where it is to be clamped and secure it with the plastic clamp (1 Fig. 12) which should be lined up with its screw holes and firmly screwed into position.

Fit the bolts (2 Fig. 11) which secure the front of the seat runner to the floor and tighten these.

Switch on the ignition and motor the seat fully forwards to enable the bolts (1 Fig. 11) which secure the rear of the seat runner to the floor to be fitted and tightened.

Motor the seat fully forwards and backwards to enable the ECU to read the potentiometer switch.

Switch off the ignition.

Fully seat the ECU cover panel into position and firmly fix it in place with the securing screws.

DASH LINER - PASSENGER'S SIDE, RENEW

Open the door. Remove the fuse indicator panel from the liner. Remove the dash liner screws (1 Fig. 13) Remove the liner (2 Fig. 13).

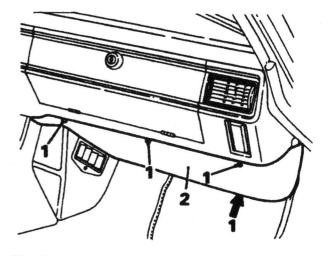

Fig. 13

Place the new liner in position.

Align the screw holes and fit and tighten the liner securing screws.

Fit the fuse indicator panel to the liner.

DASH LINER - DRIVER'S SIDE, RENEW

Open the door.

Remove the dash liner fastener (1 Fig. 14).

Pivot the liner down from the fascia underframe and remove the liner 'P' clip securing nut. Displace the 'P' clip from the underframe.

Remove the liner and place it onto a bench.

Remove the liner mounting clips securing nuts and remove the mounting clips.

Remove the 'P' clips.

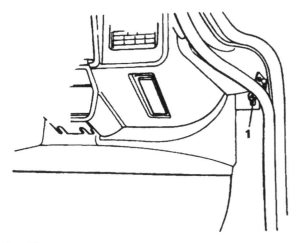

Fig. 14

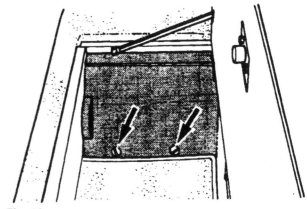

Fig. 16

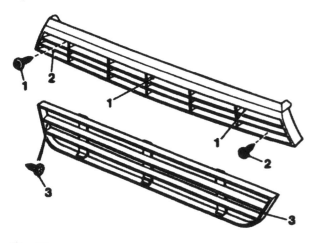

Fig. 15

Fit the 'P' clip to the new liner mounting stud and fit the mounting clips to the liner.

Fit but do not fully tighten the clips securing nuts.

Fit the liner assembly to the underframe and align the liner to the remaining trim.

Pivot the liner down from the fascia underframe and fully tighten the clips securing nuts.

Position the 'P' clip to the underframe and tighten the securing nut.

Pivot the liner to engage the underframe and secure the fastener.

Close the door.

RADIATOR GRILLE, RENEW

Undo and remove the three screws which secure the grille to the bonnet (1 Fig. 15).

Undo and remove the two screws which secure the grille to the hinge bracket (2 Fig. 15).
Remove the grille.

Fit the grille to the vehicle and firmly fix into position with the securing screws.

RADIATOR LOWER GRILLE PANEL, RENEW

Drive the car on to a ramp and raise the ramp.
Undo the two screws (3 Fig. 15) which secure the grille to the panel.

Position the grille.
Fit and tighten the screws which secure it in position.
Lower the ramp.

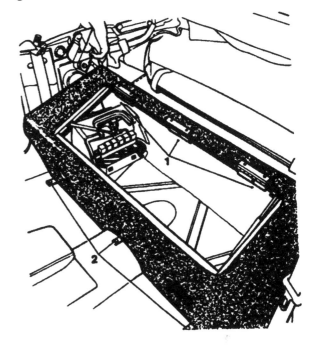

Fig. 17

REAR STOWAGE COMPARTMENT CONVERTIBLE, REMOVE AND REFIT

Switch on the ignition.
Release the hood locking catches and lower the convertible hood.
Move the seats forward.
Switch off the ignition.
Remove the rear quarter trim panels for access.
Open the stowage compartment lid and release the retaining clips (Fig. 16) on the cover of the hood motor.

Remove the cover.
Remove the carpets from the stowage compartment floor and rear panel.
Disconnect the Ty-wraps which secure the harness and hydraulic pipes to the lower edge of the internal rear vertical panel in the stowage compartment.
Carefully ease back the Velcro from the internal rear vertical panel to provide access to the three securing bolts (1 Fig. 17).
Remove the three securing bolts. Lift the footwell carpets.
Remove the screws (2 Fig. 17) which secure the stowage compartment to the floor brackets.
Disconnect the speaker wires.
Close the lid and remove the stowage compartment.

Fit the stowage compartment to the vehicle.
Fit the upper securing bolts, but do not tighten.

Fit the lower securing screws, but do not tighten. Fully tighten the securing screws and bolts.
Reposition the carpet to the footwells.
Fit the harness and hydraulic pipes to the Ty-wraps.
Connect the speaker wires.
Glue the Velcro in place, as required, over the bolt heads.
Fit the carpets to the stowage compartment. Fit the cover to the hood motor.
Close the lid of the stowage compartment. Fit the rear quarter trim panels.
Switch on the ignition.
Reposition the seats.
Raise the hood and secure the hood locking catches.
Switch off the ignition.

CONVERTIBLE HOOD, DESCRIPTION

Operation

When the OPEN (hood down) position is selected, the hood and rear quarter windows move in unison. The rear quarter windows open fully in approximately two seconds and the hood within twelve seconds.
With hood CLOSED (hood up) selected, the control module initially energizes the hydraulic pump and delays the rear quarter windows closure by approximately eight seconds.
Irrespective of how long the switch is held down, the control module will only run one cycle (twelve seconds), thus all relays will be de-energized after twelve seconds.

Safety

1. The hood cannot be raised or lowered electrically unless the handbrake is on and the gear selector in either the 'P' or 'N' position.

2. Should the hood switch be released before completion of a cycle, or if the ignition/auxiliary switch is turned off during a cycle, the system will stop immediately.

3. After interruption;

a) If the original direction is reselected, the sequence will continue to completion from where it left off (ONLY if the ignition has not been switched off).

b) If the opposite direction is selected, the timing sequence will commence at the beginning of the newly selected direction.

c) If the ignition has been switched off, the timing will commence at the beginning of its cycle.

Component Location

Control Module & Relays:
    En bloc with the pump/motor package.

Hood Switch:
    Centre console.

Hydraulic Pump Assembly:
    Convertible: Located within the stowage compartment on the right hand side concealed by a quick release trim cover.
    2 + 2: Mounted on the right hand side of the luggage compartment above the battery and concealed by a quick release trim cover.

Hydraulic Rams:
    Located underneath the rear quarter trim panels and connected to the pump by flexible nylon tubes.

HOOD HYDRAULIC CONTROL SYSTEM, RENEW

Ensure that the handbrake is on and the gearshift is in park or neutral.
Move the seats foward for access.
Release the hood handles and catches.
Turn on the ignition.
Depress the 'hood down' switch until the hood is fully lowered.
Remove the rear quarter lower trim pads and remove the rear stowage compartment lid.
Remove the rear stowage compartment (Fig. 17).
Undo and remove the bolts securing the pump mounting plate and displace the mounting plate assembly.

Undo and remove the screws securing the pump cover and remove the pump cover.
Undo and remove the cylinder (ram) upper pivot bolt and remove the spacer washers.
Undo and remove the bolts securing the cylinder pivot bracket and displace the cylinder from position (Fig. 18) - do not kink the pipes.

Repeat the operation for the opposite side cylinder.
Displace the floor carpets from the pipes. Release the pipe securing clips.
Release the hydraulic pump mounting rubbers from the plate.
Disconnect the multi-plug connector and displace and remove the complete system (Fig. 19).

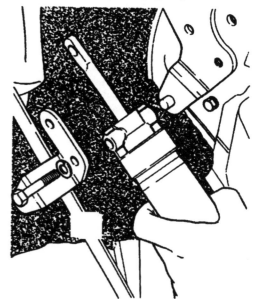

Fig. 18

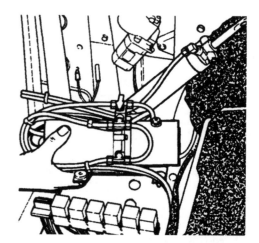

Fig. 19

Place the hydraulic system in position, route the pipes and fully seat the mounting rubbers to the plate.
Locate the cylinders to the correct position.
Reconnect the multi-plug connector.
Fit the pump cover and fit and tighten the screws which secure this cover.
Fit the cylinder pivot bracket and fit and tighten the bolts which secure this.
Position the cylinder to the hood, fit the spacers, and fit and tighten the securing bolt. Repeat the operation for the opposite side cylinder.
Position the pump mounting plate by aligning the holes and fit and tighten the bolts which secure this in position.
Secure the pipes with clips.
Apply glue to the carpet and position the carpet to the floor.
Refit the rear stowage compartment.
Refit the rear quarter trim pads.
Reposition the seats.
Depress the 'hood up' switch to close the hood.
Lock the hood with the catches and handles.
Switch off the ignition.

CONVERTIBLE HOOD HYDRAULIC SYSTEM, BLEED

Remove the Hood Hydraulic Control System from the vehicle.
Disconnect the **Left Hand** side cylinder pipes from the pump (1 Fig. 20).

Submerge the two tubes in hydraulic fluid (Univis J13) and fully extend the ram to prime the cylinder with fluid.
Invert the cylinder, fully compress the ram and reconnect the pipes.
Remove the reservoir filler plug (2 Fig. 20) and fill to highest level.
Refit the filler plug.
Select the ELECTRIC position on the manual/electric valve (3 Fig. 20).
Disconnect the Left Hand side cylinder pipes from the cylinder.
Submerge the two tubes in hydraulic fluid (Univis J13) and fully extend the Right Hand side ram.
Invert the cylinder, fully compress the ram, and reconnect the pipes to the Left Hand cylinder.
Remove the reservoir filler plug and fill to the MAX level mark, refit the filler plug.
Refit the system to the vehicle.
Power the hood several times over its full cycle ending with it fully down.
Recheck the fluid level and correct as required.

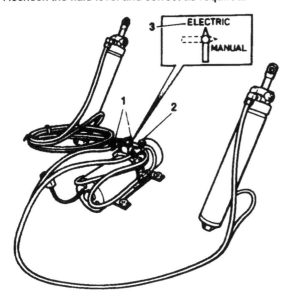

Fig. 20

CONVERTIBLE HOOD - HEADLINING, RENEW

Note: This procedure describes replacement of the original FIXED headlining with a DETACHABLE type.

Removal

With the gearshift in PARK or NEUTRAL, power the hood to the closed position. As required, disconnect the vehicle battery ground lead. Protect the vehicle against dirt or damage.

Note: Selection of MANUAL on the hydraulic pump assembly will allow the hood to be moved by hand.
Remove; Rear quarter trim pads and the rear seats (2 + 2 only).
Pull the rear shelf carpet clear at the sides to reveal backlight frame pivot brackets (1 Fig. 21) and remove the fixings.

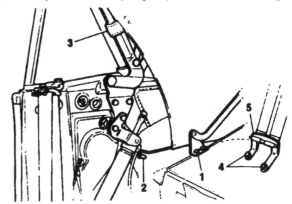

Fig. 21

Remove headlining side curtain stiffener rod from retention clip (2 Fig. 21) and disengage rod from its forward location.
Release headlining side curtain press studs at the main column (3 Fig. 21).
At the extreme rear body sides RH and LH, between the headlining and canopy, release webbing retainer fixings (4 Fig. 21).
Using a trimmers knife, cut the headlining free from the area behind the webbing (5 Fig. 21).

Position the hood to gain access to the HEADER and remove:

Headlining retainer strip (1 Fig. 22),
Seal end plates (2 Fig. 22) and
Header location pins (3 Fig. 22).

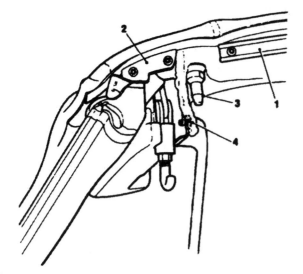

Fig. 22

Drill out rivet - headlining to link (4 Fig. 22).
Release the headlining from the HEADER and peel material back to the FIRST roof bow.
Cut the material from the FIRST bow and fold to the SECOND. At the SECOND bow, chisel (there is no access for a drill) strap retaining rivets (1 Fig. 23), cut the lining from the bow and fold back to the MAIN bow.

At the MAIN bow remove the inner screws (2 Fig. 23) and cut lining material to leave a strip approximately 40 mm wide attached to the bow.
Remove the headlining from the vehicle.

Replacement

Apply adhesive to the MAIN bow and the 40 mm strip of the original headlining - secure the strip to the bow by folding forward and over.
Apply adhesive to the outer rear webbing (MAIN bow to rear body) and velcro strips and secure velcro to webbing (1 Fig. 24).

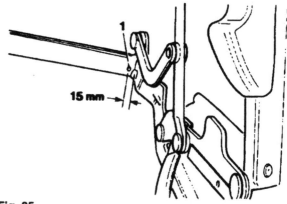

Fig. 25

Position listing rails to their respective roof bows, place centrally, mark and drill through into roof bow. Use a drill suitable for the provided self tapping screws.
The listing rails are colour coded: RED = 1st bow, YELLOW = 2nd bow, BLUE = main bow.
On the underside of the SECOND roof bow, drill one hole LH and RH, 15 mm inboard (1 Fig. 25).
These holes will provide location for fixing of the tapes which were released by chiselling from the upper surface in the removal procedure.

Feed the FIRST and SECOND roof bow listing rails through respective sleeves in the headlining taking care not to 'snag' the material.
Apply double sided adhesive tape to the MAIN bow, position the headlining 'fly' as illustrated (1 Fig. 26).

Secure headlining and listing rail to the MAIN bow (2 Fig. 26). Thread tapes through loops on MAIN bow (1 Fig. 27), headlining (2 Fig. 27) and secure at previously drilled inboard holes (3 Fig. 27).

Please note that, in the interest of clarity the outer canopy is NOT shown.
Secure the listing rails and thus headlining at the SECOND and FIRST bow (pull the material to cover the rail ends before fixing the final screw).

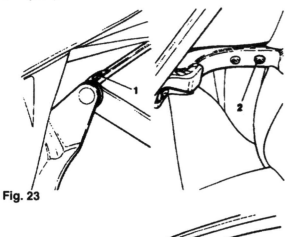

Fig. 23

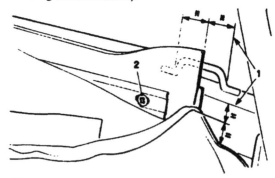

Fig. 26

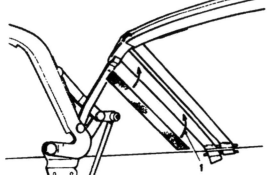

Fig. 24

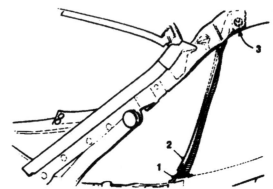

Fig. 27

Apply double sided adhesive tape at the HEADER across the retaining strip fixing holes.

Using a garnish awl or similar; pierce the headlining and fix at the side inner face adjacent to header latches.

Carefully align the headlining centre 'notch', evenly tension the material and locate to the adhesive tape.

Check tension by fully closing the hood and if acceptable, fix the headlining with the original retaining strip.

Cut off excess material above the retaining strip, if any, and replace header location pins.

At the rear, wrap the side curtains around the back of the webbing straps and secure webbing brackets to the rear body panel.

Ensuring that the rear side curtains are not creased, engage the velcro to retain in position.

Refit and tighten to the specified torque, backlight frame pivot brackets and replace the rear shelf carpet.

Engage side curtain press studs, refit stiffener rod and clips.

Verify the fit of the headlining and rectify if required.

CONVERTIBLE HOOD - CANOPY, RENEW

> **CAUTION:**
>
> It must be stressed that the following procedure should only be carried out by fully skilled personnel who have been trained in this type of work.

Removal

With the gearshift in PARK or NEUTRAL, power the hood to the closed position.

As required, disconnect the vehicle battery ground lead.

Protect the vehicle against dirt or damage.

Note: Selection of MANUAL on the hydraulic pump assembly will allow the hood to be moved by hand.

Remove the rear quarter trim pads.

Remove rear seats (2 + 2 only).

Pull the rear shelf carpet clear at the sides to reveal backlight frame pivot brackets and remove fixings.

Disconnect heated backlight feed wires.

Remove canopy fixing (1 Fig. 28), finisher (2 Fig. 28) and end cap (3 Fig. 28) at the 'B' / 'C' post buttress.

Note: Do not allow the end cap to fall into the body cavity.

Position the hood to gain access to the HEADER and remove the retaining strip (1 Fig. 29), seal end plates (2 Fig. 29) and header location pins (3 Fig. 29).

Release headlining adhesive at the leading edge and peel the lining back.

Note the position of, and remove drop glass seals from the carriers (upper only).

Remove seal carriers taking care to note position and shims if any.

Remove canopy rubber sealing strip (4 Fig. 29) and peel canopy from HEADER leading and side edges (5 Fig. 29).

Release canopy side tension wire from HEADER on both sides (6 Fig. 29).

Release canopy side tension wire from spring at main column.

Drill out rivet securing canopy side tie to side link on RH and LH (1 Fig. 30).

Fold the canopy back clear of the frame.

With a wax crayon or similar, mark the position of the tonneau loops on the rear panel - for future reference.

Pull the shelf carpet aside to reveal canopy tensioning cable adjusters and remove M6 nuts (1 Fig. 31), taking

care to prevent rotation of the cable.

Withdraw canopy and cable from the body and fold forwards and over the MAIN roof bow.

Release webbing from backlight upper loop by pushing webbing slack through frame and pulling the plastic retainer pin clear.

Remove the canopy/backlight assembly from the vehicle.

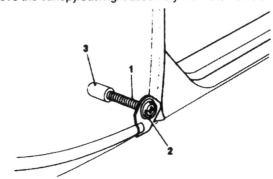

Fig. 28

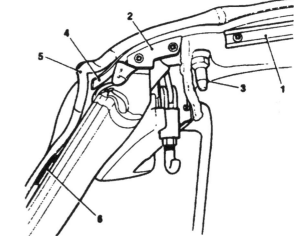

Fig. 29

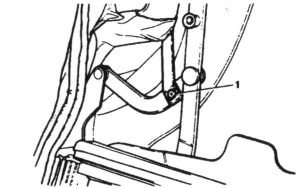

Fig. 30

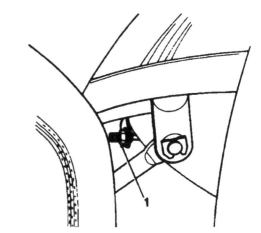

Fig. 31

Preparation

With the canopy on a suitable workbench, withdraw the side tension wires and cut the backlight free.

Remove interior and exterior glass seals.

At one corner of the backlight, carefully cut a hole in the bonding material through which a 'cheese wire' may be passed.

Using caution not to damage frame or glass, release the glass to frame bond.

Remove staples and original material from backlight frame.

Clean bonding material from glass and frame.

Tools

It is advisable to manufacture special tools:

A   (Fig. 32) used to seat the canopy material into the rear panel 'U' section and close the canopy to piping gap.
     Dimension A = 135 mm; B = 32 mm; C = 16 mm; D = 65 mm; E = 15 mm

B   (Fig. 33) used to lever the cable over the canopy material and thus under the 'U' section.
     Dimension A = 270 mm; B = 190 mm; C = 6 mm; D = 25 mm

Fig. 32

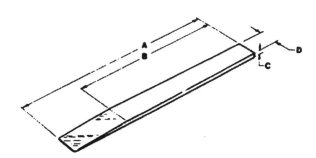

Fig. 33

Replacement

Canopy to body and frame

Lay the replacement canopy onto a clean workbench and fit the side tension wires (previously removed) large eyelet to the rear.

Position the canopy to the frame and secure side tension wires at the header and spring (main column).

Rivet side tie to link RH and LH (see Fig. 30).

Fold canopy material over MAIN roof bow to gain access to backlight webbing.

Secure backlight frame pivot brackets to the body in the most FORWARD position and attach the webbing to the frame.

Pull rear of canopy over the rear body 'U' section and fit tensioning cables through body and assemble M6 nuts by 2 or 3 threads only.

Note: The natural tendency is for the cable to stay inside the body, therefore, lever the cable over the leading edge of the 'U' section to lie outside and across the rear panel.

Position tonneau loops to previously marked locations and using special tool 'B' lever the cable over the canopy and thus into the 'U' section. It is necessary that a second person holds the cable in position on the opposite side to the lever to prevent the cable pulling out of the 'U' section (Fig. 34).

Reposition the tonneau loops if required.

Pull canopy sides evenly to the 'B'/'C' post buttress, pierce the material with a garnish awl (Fig. 35) and secure each

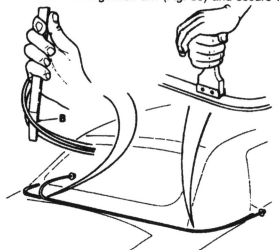

Fig. 34

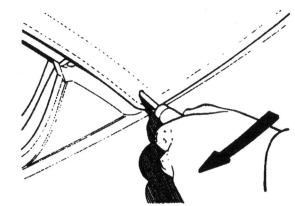

Fig. 35

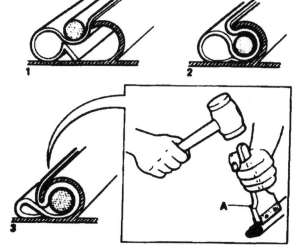

Fig. 36

side using previously removed fixings.
Using special tools 'A' and 'B' with a soft faced mallet, 'chase' the canopy material and retaining cable into the 'U' section to achieve the illustrated condition:

(1 Fig. 36) Initial fitting
(2 Fig. 36) Canopy fully seated
(3 Fig. 36) 'U' section and canopy gap fully closed.

Tighten the tensioning cable nuts evenly on each side (prevent rotation of the cable) as the previous operation is being carried out to achieve a crease-free canopy appearance.

Note: Ensure that the tonneau loops remain in the correct location during the final fitting operations.

The correct canopy fit and cable tension should result in the canopy material having even contact with the body all round.

CAUTION:

To avoid future canopy damage, do not omit the screw end caps.

Reposition the carpet over retaining cable adjusters.

With the hood raised for access - and to provide material tension:
Apply adhesive to the main column seal carrier flanges and canopy material; secure the canopy ensuring that the tuck under is symmetrical and the 'edge sew line' is parallel to the seal flange edge.
Apply adhesive to HEADER and canopy material; secure the canopy firstly by folding the sides under, followed by the leading edge material.

Note: Whilst lining up the material, ensure that the rain channels are straight and even side to side. Fully close the hood and check for correct material ten-

sion and appearance - rectify as required.
Fitting of the remaining items is the reversal of previous procedures, noting that special attention should be paid to seal/carrier alignment.

CAUTION:

Be absolutely sure that ALL aspects of the canopy fitting are acceptable before proceeding with the following operation.

Backlight Aperture

Fully latch the frame at the HEADER.
Centralize the backlight frame ensuring that an equal gap by measurement exists between the outer seams and the frame (1 Fig. 37).

Note: For 2 + 2 only, the backlight frame must be held away from the main roof bow by either inserting a distance piece between the frame and the bow or by pulling the frame away from the main roof bow toward the rear shelf. Special attention therefore must be paid to keep the frame parallel across car. These actions are required because this model has elasticated backlight frame upper attachments.

With two people working equally from both sides, staple the material to the inner vertical edges of the backlight frame (2 Fig. 37) whilst gradually relieving material tension within the frame by cutting (3 Fig. 37). **DO NOT CUT RIGHT ACROSS.**
Continue to staple for upper and lower edges, cutting to relieve tension as you go.
When the material is fixed all round, trim off excess.
Refit backlight.

All other fitting is the reversal of previous procedures.

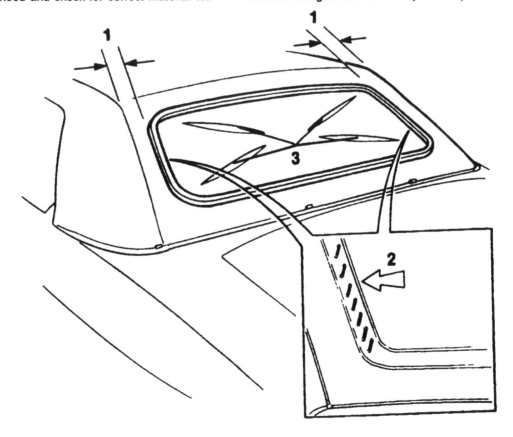

Fig. 37

CONVERTIBLE HOOD 'UP' RELAY, RENEW

Open the storage compartment lid.
Twist and release the pump cover release catches.
Displace and remove the cover.
Remove the relay (4 Fig. 38).
Fit the relay.

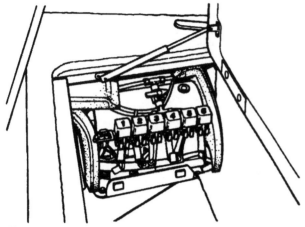

Fig. 38

Refit the cover and reposition the carpet.
Close the lid.

CONVERTIBLE HOOD 'DOWN' RELAY, RENEW

Open the storage compartment lid.
Twist and release the pump cover release catches.
Displace and remove the cover.
Remove the relay (3 Fig. 38).

Fit the relay.
Refit the cover and reposition the carpet.
Close the lid.

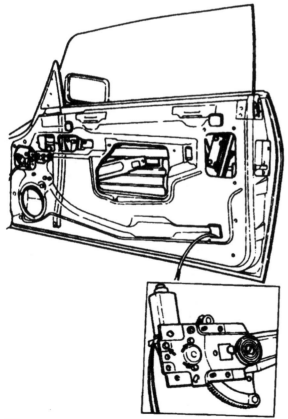

Fig. 39

DOOR - GLASS LIFT MOTOR, RENEW

As required, disconnect vehicle battery ground lead.

Remove:
Door trim pad
Door inner handle
Door glass regulator

Remove the motor (Fig. 39) from the regulator assembly

Reassembly and fitting is the reversal of this procedure noting that the glass height setting procedure must be observed.

DOOR GLASS - ADJUST

The door glass may be adjusted in two planes:

Vertically -   to set or correct height and front to rear inclination.
Horizontally -   to set or correct angle and pressure against seals.

These adjustments are critical to the prevention of wind noise and water ingress.
As required, disconnect vehicle battery earth lead.
Remove door trim pad, slacken and move the cheater forward and remove the water curtain aside.
To take maximum advantage of the setting procedure, slacken the regulator assembly and move it to the TOP of its mounting holes.
Two people are needed to successfully carry out the glass adjustments; one inside the vehicle to operate the regulator, adjust and secure the stops and the other outside to position and hold the glass.
With the glass partially lowered slacken the 'UP' stop stabilizer and glass mounting bracket fixings front (1 Fig. 40) and rear (2 Fig. 40).

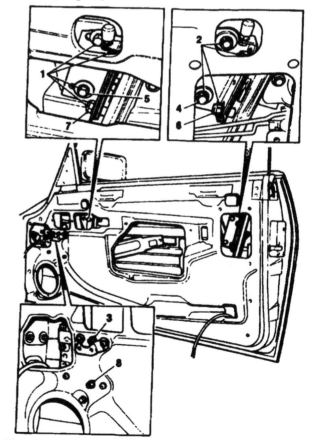

Fig. 40

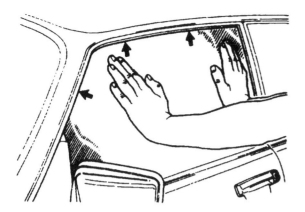

Fig. 41

Slacken the 'UP' cam quadrant stop locknut and rotate the cam anticlockwise (3 Fig. 40) to allow maximum upward glass travel.
Power the drop glass to maximum upward travel and position to contact the cantrail and 'A' post seals (Fig. 41).

Secure in order, lower rear (4 Fig. 40) and lower front (5 Fig. 40) mounting bracket fixings.
Hold the stabilizer bracket buffer against the door panel and secure in order, lower rear (6 Fig. 40) and lower front (7 Fig. 40) fixings.
Lower the drop glass to access and secure the front and rear stabilizer and glass mounting bracket upper fixings.
Verify seal contact and rectify as required.
Power the drop glass to maximum upward travel.
Reset the "UP" cam quadrant stop by rotating the cam clockwise until resistance is felt and secure the locknut.
Check the glass fully lowered position which should be with the glass upper edge at, or just below, seal lip height.
To rectify an incorrect glass to seal lower relationship, power the glass upward, release the 'DOWN' cam quadrant locknut and rotate the cam anticlockwise (8 Fig. 40).
Lower the glass to the correct level, rotate the cam quadrant clockwise until resistance is felt and secure the locknut.
Cycle the glass two or three times and verify the set position.
Power the convertible hood closed and re-check seal engagement.

Note: It is important to carry out these last two checks, the action of the hood may alter your initial settings.

Fitting is the reversal of this procedure.

REAR QUARTER GLASS - LIFT MOTOR, (CONVERTIBLE), RENEW

With the gearshift in the park or neutral position, power the hood to the fully stowed position.
As required, disconnect vehicle battery ground lead.

Remove:
    Stowage compartment
    Rear quarter trim pad
    As required, hood operating cylinder - RH ONLY
    (NOT necessary for 2 + 2)

Disconnect motor multiplug and release the regulator assembly fixings.
Disengage the regulator from the glass slide and remove the assembly from the vehicle (Fig. 42).

Remove the motor from the regulator assembly.
Reassembly and fitting is the reversal of this procedure noting that the glass height setting procedure must be observed.

Fig. 42

REAR QUARTER GLASS (CONVERTIBLE), ADJUST

The door glass may be adjusted in two planes:

Vertically -   to set or correct height and front to rear inclination.
Horizontally -   to set or correct angle and pressure against seals.

These adjustments are critical to the prevention of wind noise and water ingress.
As required, disconnect vehicle battery earth lead.
Remove rear quarter trim.
Two people are needed to successfully carry out the glass adjustments; one inside the vehicle to operate the regulator, adjust and secure the stops and the other outside to position and hold the glass.

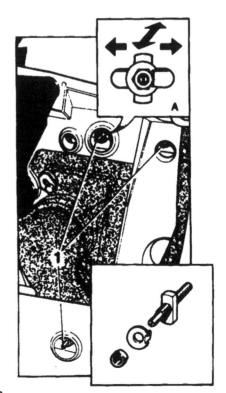

Fig. 43

Lower convertible hood and move the hydraulic selector to MANUAL (in order that that glass may be electrically raised or lowered independent of the hood).
Slacken glass adjuster locknuts (1 Fig. 43).

Power the quarter glass upwards and position using the adjuster screws relative to the correctly set door drop glass:

CRITERIA	SET AT
Height (Fig. 44)	+ 1mm
Parallel trailing edge gap (A Fig. 45)	2mm
Profile	As door

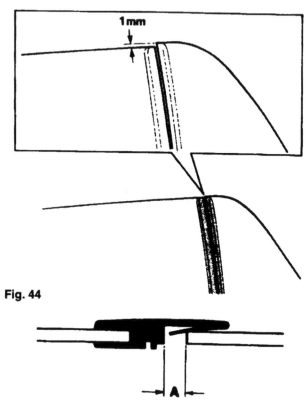

Fig. 44

Fig. 45

Secure the adjuster locknuts.
Cycle the glass two or three times and verify the set position. Power the convertible hood closed and recheck seal engagement and door drop glass relationship.

Note: It is important to carry out these last two checks, the action of the hood may alter your initial settings.

Fitting is the reversal of this procedure.

SUPPLEMENTARY RESTRAINT SYSTEM INTRODUCTION

General

All 1994 Model Year-on vehicles are fitted as standard with both driver and passenger side 'Supplementary Restraint System' (SRS), a system introduced at 1993.5 model year. North American and Canadian vehicles are fitted with a tear loop seat belt buckle and knee bolster on the driver side to complement those units currently fitted on the passenger side.
The passenger airbag is fitted in the area normally occupied by the glove box and when activated, exits through veneer faced deployment doors on the fascia.

> **WARNING:**
>
> THE VEHICLE MAY NOT COMPLY WITH LEGISLATIVE SAFETY STANDARDS IF ANY 'SRS' COMPONENT DOES NOT CONFORM TO THE ORIGINAL MANUFACTURERS SPECIFICATION FOR THAT MARKET.
> THE NORTH AMERICA /CANADIAN STEERING WHEEL AND PASSENGER AIRBAG MODULES ARE NOT INTERCHANGABLE WITH THOSE OF OTHER MARKETS.

Note: To ensure operator safety during removal and handling of airbags, a mechanism is built into each module to allow it to be armed and disarmed.

The airbag modules CANNOT be removed from either the steering wheel or fascia unless that module has been disarmed.

Tear Loop Seat Belt Buckle

This unit is designed to control the rate of forward travel of the occupant towards the deployed airbag.

Knee Bolster

Both the driver and passenger side knee bolsters replace the underscuttle pads found on other market vehicles. Although similar in appearance to the non 'SRS' underscuttle pad, the knee bolster is specifically designed to work as part of the occupant restraint system and comply with USA and Canadian market legislation.

System Recognition

The following features will allow easy identification of an 'SRS' equipped vehicle.

> Unique four spoke steering wheel for airbag application only.
> 'SRS' logo on the steering wheel centre pad.
> 'SRS' logo on the passenger side fascia veneer (Fig. 46).
> Inclusion of an airbag symbol on the VIN plate, located at the lower left hand corner of the windscreen (Fig. 47).
> Warning labels located in the vehicle interior.

Fig. 46

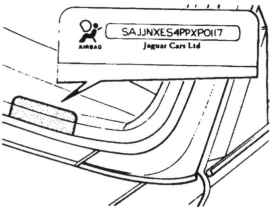

Fig. 47

WORKING PRACTICES: AIRBAG

General

Be aware of, and comply with all health and safety requirements, whether they be legislative or common sense. This applies to conditions set both for the operator and workshop. Before commencing any repair or service procedure, disconnect the vehicle battery ground connection and protect the vehicle where appropriate, from dirt or damage. Wherever possible, disarm the airbag module when working in the vicinity of it.

Use only the correct tools and equipment as described in the working text. Do not transfer an airbag module to another vehicle.

Fig. 48

North American/Canadian specification units are NOT compatible with those of other markets and cannot be interchanged.

Handling the Airbag - Undeployed

Always wear eye and ear protection and impervious rubber gloves. **THE MODULE IS NON-SERVICEABLE; DO NOT TAMPER WITH IT.**
Do not subject the airbag module to excessive movement, sharp blows, electricity and heat.

Never carry the module against your body, hold it as shown (Fig. 48).

Note: Driver airbag shown. The passenger module should be held in a similar manner, with the deployment apperture facing either to the front or rear and at the side of the body.

The module should be stored, deployment apperture uppermost, in a secure cabinet: **NEVER** store face down, against a vertical surface, or stacked.

Handling the Airbag - Deployed

Wear eye, nose and mouth protection and impervious rubber gloves at all times. Should the materials from a deployed airbag come into contact with your eyes or skin. Wash the affected area with cool water and seek medical advice. Do not attempt to treat yourself.
Inhalation of airbag propellant residue may cause irritation to your respiratory system.
Seal the deployed module in a plastic bag in preparation for disposal.

Disposal

Contact your importer or Jaguar Service for instruction in the disposal of an undeployed module where:

Service life has been exceeded.
The module has been removed from a damaged vehicle.
There is any doubt concerning the condition of the arming mechanism.

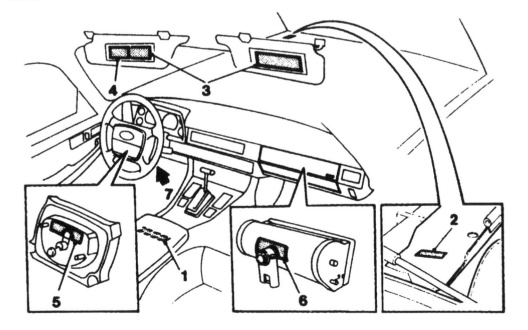

Fig. 49

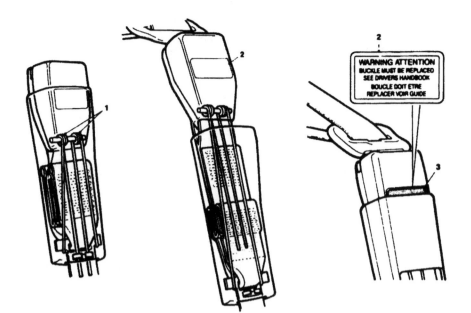

Fig. 50

If a vehicle is to be scrapped, a deployed module(s) may be disposed of with it and therefore need not be removed from the vehicle.

A deployed module which is to be disposed of separately, should be done so with regard to current local legislative requirements. If in any doubt, contact your local Environmental Agency.

Special Notes

Airbag modules for the North American/Canadian markets differ in calibration to those of other markets and therefore cannot be interchanged.

To make identification easy, an airbag module calibrated for full FMVS 208 operation (North American/Canadian) is coloured GOLD, as opposed to BLACK for all other markets. Further, with respect to the passenger side airbag, the mounting brackets are colour coded to match the module.

Should an attempt be made to ignore the colour coding, FMVS 208 modules also have unique fixing dimensions and brackets, thus further inhibiting incorrect fitting.

Warning Labels. To emphasize the need for caution and to convey maintenance information, warning labels are located at various points in the vehicle interior and in the case of label 2, under the bonnet (Fig. 49).

Label 1	Service life data, Passenger and Driver airbag (North American/Canada only).
Label 2.	Service life data, Passenger and Driver airbag (All other markets).
Label 3.	Seat belt/airbag warning, Passenger and Driver airbag only.
Label 4.	Seat belt/airbag warning, Driver airbag only.
Label 5.	Warning of misuse, Driver airbag.
Label 6.	Warning of misuse, Passenger airbag.
Label 7.	Warning, steering column removal.

TEAR LOOP SEAT BELT BUCKLE, DESCRIPTION

The mechanism within the buckle assembly is designed to release additional webbing when the stitching, which retains the webbing loops, breaks under a predetermined load.

The wires within the buckle (1 Fig. 50) have the following functions:

Protect the stitching from 'normal' loads such as heavy braking or cornering.
Control the rate of deployment.
Support the buckle assembly.

When the unit has been activated the buckle will extend from the shroud and reveal a warning label (2 Fig. 50). The extent of deployment will depend upon the severity of the load.

WARNING:

IF THE LABEL IS VISIBLE AT ALL (3 Fig. 50), THE COMPLETE ASSEMBLY MUST BE RENEWED, AS MUST ANY SEAT BELT WHICH HAS BEEN WORN IN AN ACCIDENT.

AIRBAG MODULE, DRIVER SIDE, RENEW

WARNING:

PLEASE READ THE SECTION ENTITLED 'SUPPLEMENTARY RESTRAINT SYSTEM', BEFORE PROCEEDING WITH ANY AIRBAG RELATED OPERATIONS.

Disconnect vehicle battery ground lead.
Tilt the steering wheel fully downwards.
Rotate steering wheel 90 degrees from the straight ahead to remove airbag nut cover (where fitted) and nut. Repeat for opposite side (Fig. 51).

Rotate steering wheel 180 degrees from the straight ahead to open the cover for the arming screw and third module fixing (1 Fig. 52).

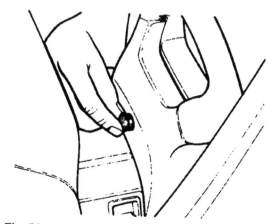

Fig. 51

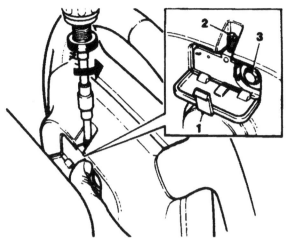

Fig. 52

Using special tool JD 159, rotate the arming screw anti-clockwise (2 Fig. 52) approximately 12 turns, or until resistance is felt.

Note: This action will also release the slide interlock for access to the third module fixing.

Do not rotate the steering wheel with the cover open.
Release the third module fixing (3 Fig. 52).
Remove the module from the vehicle and observe all safety considerations.

```
WARNING:

ENSURE THAT THE MODULE ARMING PIN
IS IN THE DISARMED POSITION (B FIG. 53)
IMMEDIATELY THAT THE ASSEMBLY IS
REMOVED FROM THE VEHICLE. IF IT IS IN THE
ARMED POSITION (A FIG. 53),
CAREFULLY PLACE THE MODULE IN A
SAFE PLACE AND CONTACT YOUR
IMPORTER OR JAGUAR SERVICE.
```

Fitting is the reversal of this procedure noting that the arming screw and all fixings must be tightened to the specified torque.

Torque Figures

Airbag to steering wheel - Nut	9.5 - 12.5 Nm
Airbag to steering wheel - Screw	9.5 - 12.5 Nm
Arming screw	1 - 2 Nm

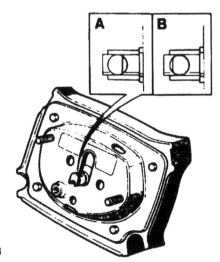

Fig. 53

AIRBAG MODULE, PASSENGER SIDE, RENEW

```
WARNING:

PLEASE READ THE SECTION ENTITLED:
SUPPLEMENTARY RESTRAINT SYSTEM,
BEFORE PROCEEDING WITH ANY
AIRBAG RELATED OPERATIONS.
```

Remove fascia assembly.
Slacken airbag upper M10 fixings (1 Fig. 54) and catch plate nuts M6 (2 Fig. 54).

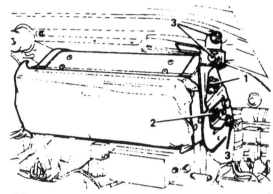

Fig. 54

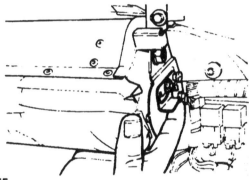

Fig. 55

Lift catch plates RH and LH (Fig. 55) and allow the airbag assembly to pivot downwards to the DISARMED position.

Remove fixings outer bracket to dash rail and crossbeam assembly (3 Fig. 54).

```
CAUTION:

As the catches are released the arming
mechanism will apply considerable force.
Do not allow the airbag assembly to 'snap'
down; fully support it with both hands and
ease to the disarmed position.
```

Remove previously slackened fixings, M10 and M6, airbag to inner bracket and remove the airbag/outer bracket assembly from the vehicle.
Check that the arming mechanism slide is fully down in the DISARMED position (1 Fig. 56).

Should the slide NOT be in the disarmed position, carefully place the airbag on a suitable work surface so that a SIDE face is towards your body and the deployment apperture is NOT facing downwards. Pull the slide downwards by finger pressure only, if this cannot be achieved, store the unit and

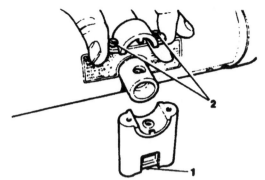

Fig. 56

contact your importer or Jaguar Service.
Only on a disarmed assembly, release split cap fixings (2 Fig. 56) and carefully remove the arming mechanism from the airbag module.

> **WARNING:**
>
> ENSURE THAT THE MODULE ARMING PIN IS IN THE DISARMED POSITION (B Fig. 57) IMMEDIATELY THAT THE ARMING MECHANISM IS DISENGAGED FROM THE MODULE. IF IT IS IN THE ARMED POSITION (A Fig. 57), CAREFULLY PLACE THE MODULE IN A SAFE PLACE AND CONTACT YOUR IMPORTER OR JAGUAR SERVICE. DO NOT TAMPER WITH THE MODULE.

Remove outer bracket and anti tamper bracket.
Reassembly and fitting is the reversal of this procedure ensuring that:
Upon assembly, the arming mechanism spigot is fully engaged onto the module.
When located in the vehicle and pivoted back to the armed position, that resistance is felt from the arming mechanism.
The anti tamper bracket fully obscures the upper outer fixing.
All fixings must be tightened to the specified torque.

Torque Figures

Airbag inner mounting bracket to body	8.5 - 10.5 Nm
Crossbeam to upper dash	23 - 31 Nm
Crossbeam to inner and outer mounting bracket.	23 - 31 Nm
Arming mechanism to module	1.7 - 2.3 Nm
Airbag outer mounting bracket to body (nut)	7.5 - 10.5 Nm
Airbag outer mounting bracket to body (bolt)	23 - 31 Nm
Airbag module to inner and outer bracket (M10)	13.5 - 18.5 Nm

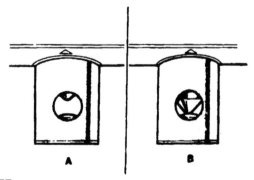

Fig. 57

UNDERFLOOR CROSS STRUT, FRONT, RENEW

> **CAUTION:**
>
> Do not lift the vehicle under the cross strut.

Remove screws, cross strut to front crossmember.
Remove jacking studs, cross strut to front floor and place the strut assembly aside.
Fitting is the reversal of this procedure noting the position of washers and spacers (Fig. 58).

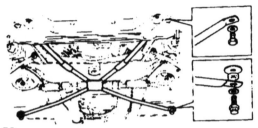

Fig. 58

Tighten fixings to the specified torque.

Torque Figures

Screw - front and rear	55-75 Nm
Jacking stud	41-55 Nm

UNDERFLOOR CROSS STRUT, REAR RENEW

> **CAUTION:**
>
> Do not lift the vehicle under the cross strut.

Remove screws, cross strut to rear longitudinal.
Remove screws, cross strut to rear floor and place the strut assembly aside.
Fitting is the reversal of this procedure, ensuring that the strut is aligned before tightening the fixings to the specified torque (Fig. 59).

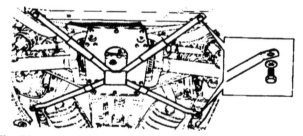

Fig. 59

Torque Figures

Screw - front and rear	55-75 Nm

WHEELARCH BAFFLE, FRONT, RENEW

Disconnect vehicle battery ground lead.
Raise front of vehicle and remove road wheel.
Release securing studs baffle to brake duct (1 Fig. 60).
Remove fixings, baffle to undertray (2 Fig. 60), inner valance (3 Fig. 60) and wing extension (4 Fig. 60).
Fitting is the reversal of this procedure.

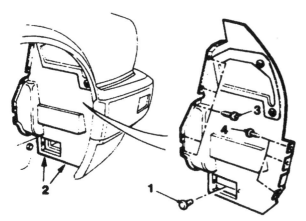

Fig. 60

CROSSMEMBER - FRONT, RENEW

Disconnect vehicle battery ground lead.
Open bonnet and protect paintwork.
Using a syringe, remove fluid from the power steering reservoir.
Remove RH and LH air cleaner element.
Remove RH and LH upper shock absorber fixings and bushings.
Attach MS 53 engine support bracket assembly to the front lifting eye and take the weight of the engine.
Remove RH and LH engine mounting upper securing nuts.
Raise the front of the vehicle and position suitable supports.
Remove front wheels noting wheel to stud relationship.
Convertible only: Remove underfloor cross strut.
All: Disconnect anti roll bar from vertical links RH and LH.
From inside engine bay, disconnect ABS harness connectors and feed cable through inner wing panel.
Release locking wire and bolts and remove front brake calipers.

Note: It is not necessary to break into the hydraulic system, but the calipers must be supported and NOT allowed to hang on the flexible hoses.

Release the fixings on the spoiler undertray at wheelarch baffle, crossmember and bumper cover and remove the undertray.

RHD only: Remove front right hand catalytic converter.
LHD only: Remove oil filter.
All: Release steering column lower shaft from steering rack pinion and lower column, and pull the shaft clear of the rack pinion.
Disconnect the power steering feed and return pipes at the

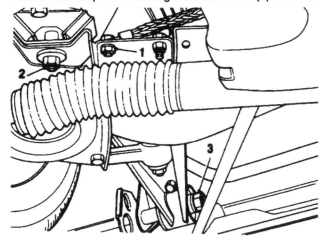

Fig. 61

pump, drain remaining fluid into a suitable container.
Fit blanking plugs to the pipes and pump.
Release earth leads from crossmember fixing (1 Fig. 61).
From below, support the crossmember assembly.
Remove crossmember to mounting rear nuts RH and LH (2 Fig. 61).
Remove crossmember forward fixings and tie down brackets (3 Fig. 61).

WARNING:

ENSURE THAT THE ENGINE IS FULLY
SUPPORTED BY THE ENGINE
SUPPORT BRACKET MS 53.

Lower the crossmember assembly from the vehicle ensuring that the brake calipers, pipes and steering rack are clear.
Place the crossmember assembly onto a suitable work surface.

Remove:

Front mounting bush RH and LH.
Road spring (using spring compressor JD 6G).
Lower wishbone fulcrum shaft RH & LH.
Upper wishbone fulcrum to tower bolts RH &LH (Support the hub/swivel assembly).
Steering rack and steering arm assembly.

Note: Note position of upper wishbone fulcrum camber shims.

Remove engine mounting RH & LH.
Reassembly of the crossmember and fitting to the vehicle is the reversal of this procedure with special attention to the following:

- Renew all self locking nuts.
- Renew all bolts that were originally fitted with thread locking adhesive.
- Renew all locking wire.
- Renew all split pins (cotter pins).
- Tighten all fixings to the specified torque.
- Check steering geometry and adjust as required.
- Check operation of brakes.

BUMPER, FRONT, RENEW

Introduction

The following instructions describe the sequential procedures for the removal, strip, assembly and setting of the front bumper (all of its separate subsystems; select the appropriate procedure for the task).

CAUTION:

IT IS ESSENTIAL THAT THE ORDER OF THE
SETTING PROCEDURES AND COMPONENT
TIGHTENING IS ADHERED TO WHEN
REPLACING COMPONENTS WHICH AFFECT
BUMPER ALIGNMENT.

For all operations, disconnect the vehicle battery ground, open the bonnet, fit wing protection, raise the front of the vehicle and, as required, remove the front wheels.

Removal And Disassembly Procedures

Blade Assembly

From inside the front wheel arches RH and LH, remove the headlamp access panel. Remove quarter blade side fixings (1 Fig. 62). Release fixings, centre blade to front panel (2 Fig. 62) and pull the blade assembly forwards.

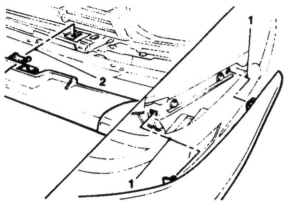

Fig. 62

Where fitted, disconnect the multiplugs and hoses to the headlamp power wash jets.
Remove the front blade assembly from the vehicle and place on a suitable workbench.

> **CAUTION:**
>
> Ensure that the blades are adequately supported and that paint damage cannot occur during removal. Remove the fixings on the reverse side of the blades to separate the quarters from the centre section and flip seal.

Bumper Cover

Disconnect direction indicator and foglamp multiplugs RH and LH.
Remove foglamps and brackets.
Remove fixings, bumper cover to body panel extensions (1 Fig. 63).

Release the fixings on the spoiler undertray at wheelarch baffle, crossmember and bumper cover (2 Fig. 63) and remove the undertray.
Remove fixings at the grille and crossmember to remove the lower air duct (Fig. 64).

Remove vehicle licence/registration plate and plinth.

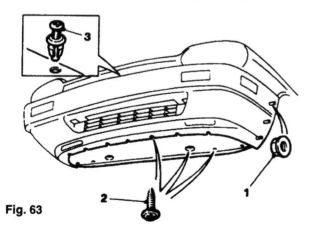

Fig. 63

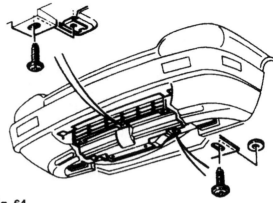

Fig. 64

Release the fixings securing bumper cover to beam (3 Fig. 63), carefully remove the bumper cover from the vehicle and place on a suitable workbench.
Remove direction indicators and spire clips. Release edge clips at the lower rear edge of the stoneguard grille and cover aperture, disengage forward lugs and pull the grille clear of the bumper cover.

Bumper Beam

Remove bolts (1 Fig. 65) securing the beam to the mounting struts noting the spacers and place the beam aside.

Fig. 65

Note: Specified washer fitted under the bolt head: Oval (2 Fig. 65) for 'Energy Absorbing' strut equipped vehicles, and tanged (3 Fig. 65) for 'Fixed' strut types. One washer per bumper only.

Bumper Mounting Strut

Remove the retaining nut and washer (1 Fig. 66) whilst preventing the threaded adjuster (2 Fig. 66) from rotating and withdraw the strut assembly from the chassis tube.

Note: It may be necessary to drive the strut from its rubber bush by rotating the threaded adjuster anti-clockwise. In this case, ensure that the length of thread (3 Fig. 66), is known prior to disassembly.

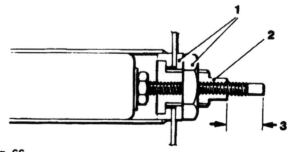

Fig. 66

With the strut assembly on a suitable workbench, remove the threaded adjuster and beam adjuster (vertical). Note the positions of the adjusters relative to the strut for initial setting of new components.

Assembly And Setting Procedures

Bumper Mounting Strut

Assembly and fitting is the reversal of the above procedure with adjusters nominally set. Insert the strut, with the beam adjuster slotted face downwards, and loosely assemble plain washer and nut (1 Fig. 66) but do not tighten.

Bumper Beam

Assemble the beam to the struts and retain with bolts and appropriate washer, do not fit spacers or nuts.

Note: Ensure correct location of bolt head and washer in beam slot.

Bumper Cover

Reassemble the cover, as required, prior to fitting and apply protection around the towing eye aperture.
Position the cover to the beam and align centre hole in the cover with the centre hole in the beam and secure with a scrivet.

Note: It is recommended that two people fit the bumper cover in order that it may be sprung over the side markers.

Working from the outside to the centre, fit the remaining scrivets.

Bumper Setting, Horizontal

Hold the strut threaded adjuster and rotate the large nut CLOCKWISE to draw the bumper assembly rearwards until the studs on the cover sides locate in the body panel extensions. Loosely assemble the cover side fixings.
The bumper cover and wheelarch should now be aligned.
Rotate the threaded adjuster ANTI-CLOCKWISE until fully seated on the inner face of the body tube, but does not begin to jack the bumper forward. Tighten the large nut to the specified torque.

Note: If sufficient rearward adjustment cannot be attained, slacken the large nut and turn the threaded adjuster CLOCKWISE 2 or 3 turns. Repeat the operations Bumper Setting Horizontal to obtain the desired condition.

Blade Assembly

Reassemble the blades as required and refit to the vehicle. Ensure that the flip seal is located against the body correctly.

Bumper Setting, Vertical

Rotate the beam adjusters, RH and LH to adjust the vertical height of the bumper assembly. The bumper cover should just contact the blade mounted buffers (Fig. 67) to give an all round 3.0mm gap.

Note: Ensure that the cover does not distort the blade vertically.

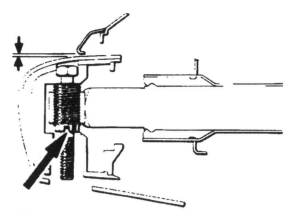

Fig. 67
Secure Fixings

Secure the cover side fixings to the specified torque.
Assemble spacers and nyloc nuts to the bumper beam bolts and tighten to the specified torque.
All other reassembly and fitting procedures are the reversal of disassembly.

BUMPER, REAR, RENEW

Introduction

The principals of fixing and adjustment are common to front and rear bumpers and many repair operations are similar.
However, the rear bumper cover, when fitted, masks the beam to strut fixings. Therefore, height adjustment may only be carried out with the cover removed.
It should also be noted that with Non Energy Absorbing struts, the beam to strut bolt head, is not captive in the beam.
For all operations open the luggage compartment, disconnect the vehicle battery ground, raise the rear of the vehicle and, as required, remove the rear wheels.
See front bumper procedures for all aspects of operations which are not detailed on this or the next page.

REMOVAL AND DISASSEMBLY PROCEDURES

Blade Assembly

Remove luggage compartment side liner RH, LH and rear protection plate to gain access to blade fixings.

Note: The blade assembly is dowelled together unlike the front which is thread secured.

Bumper Cover

See procedure for front cover noting that the lower edge is clipped to the floor pan.

Bumper Beam

See procedure for front beam noting washer types and location.

CAUTION:

Having unclipped the cover, do not attempt to gain access to the beam fixings by forcing the cover downwards.

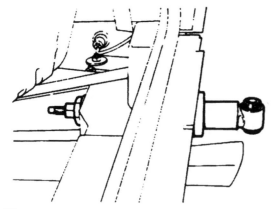

Fig. 68

Bumper Mounting Strut

See procedure for front strut, noting that access to the large nut and threaded adjuster is from inside the luggage compartment (Fig. 68).

Assembly And Setting Procedures

Bumper Mounting Strut and Initial Vertical Bumper Setting

See procedure for front strut, noting that the initial height of the bumper beam, controlled by the beam adjuster, must be set at this point.

With the strut initially located, washer and large nut engaged, set the distance from the beam adjuster top face to the rear panel flange with a simple tool, as shown, or set dimension 'A' at 59.0mm by rule measurement (Fig. 69).

Bumper Beam

As front, but the spacer and nut should be fitted and secured to the specified torque prior to fitting the cover.

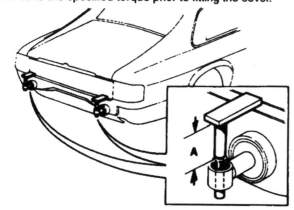

Fig. 69

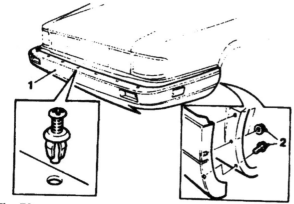

Fig. 70

Bumper Cover (1 Fig. 70)

See procedure for front cover.

Bumper Setting Horizontal

See procedure for front bumper, but secure the cover side fixings (2 Fig. 70) to the specified torque IMMEDIATELY after horizontal setting.

Blade Assembly

See procedure for front blade. To achieve the 3.0mm gap condition, apply a light downward pressure on the blade as the fixings are secured.

Note: Should the gap condition be unattainable, remove the cover and reset the beam height accordingly.

Torque figures

Bumper beam to bumper strut (All)	39-51 Nm
Bumper blade - centre to quarter	1.5-2.5 Nm
Bumper blade - centre to body	5-7 Nm
Bumper blade - quarter to body	5-7 Nm
Bumper cover to wing extension (nut)	5-7 Nm
Bumper cover to body (screw)	5-7 Nm
Bumper cover to undertray	3.5-4.5 Nm
Bumper strut to body	39-51 Nm
Fog lamp bracket to beam	5-7 Nm
Undertray to crossmember	3.5-4.5 Nm
Undertray to wheelarch baffle	1.5-2.5 Nm

Fascia board, Renew

Disconnect the vehicle battery ground lead and position the gear control lever fully rearward.

Remove:

Driver side airbag.
Steering wheel.
Upper and lower steering column cowls.
Underscuttle pad or knee bolster, driver and passenger side.
Instrument module.
Centre fascia veneer panel.
Centre fascia vent outlet.
Glove box liner (where applicable).

Through the instrument aperture, remove the upper fascia retaining nut (1 Fig. 71).

Remove RH and LH side fixing, brace to 'A' post (2 Fig. 71). Through the centre vent aperture, remove the upper fascia retaining tube nut (3 Fig. 71) (Passenger airbag only), or, through the glove box aperture, remove the upper fascia retaining nut (4 Fig. 71) (non passenger airbag vehicles). Remove A/C control panel knobs and retaining collars. Pull the A/C control panel clear to release centre console to fascia fixings.

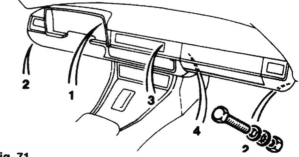

Fig. 71

Remove the RH and LH side 'A' post to fascia trims and
fascia mounted interior lamps.

Remove the hazard warning/heated backlight switch and
exterior mirror consoles (for access).

Release the switch and trip computer harness from the
clips, and pull clear.

Disconnect the fog lamp, trip computer and interior lamp
multiplugs.

Remove the fascia assembly from the vehicle and place on
a suitably protected work bench.

Remove the trip computer and switch/veneer assembly.

Remove the RH and LH side air vent veneer and vent
assembly.

Remove the dimmer control.

Remove the airbag deployment door and fascia brackets
(passenger airbag only).

Remove the glove box lid (non passenger airbag).

Remove clips and retainers as required.

Reassembly and fitting is the reversal of this procedure.

REAR QUARTER GLASS, RENEW

Remove the following components:

The left and right hand rear seat cushions.
Left and right hand rear seat squabs.
Rear seat cushion centre panel.
Rear quarter trim pad veneer panel.
Rear seat squab surround trim panel.
Right hand rear quarter upper trim pad.

Displace and remove the cantrail trim and the 'B' post seat
belt adjuster knob.

Undo and remove the 'B' post trim upper and lower securing screws.

Displace and remove the door waist plastic capping, the 'B'
post upper trim and the seat belt adjuster trim cover.

Displace the door seal from the area adjacent to the 'B' post.

Undo and remove the seal to the 'B' post channel retention
securing screws.

Displace and remove the channel, the 'B' post finisher and
quarter light lower waist finisher. Unscrew and remove the
quarter light securing nuts, noting that the rear window
studs carry larger washers to span the aperture flange rear
'V' slots.

Carefully remove the quarter light assembly.

Remove and discard the old 'memory' foam from the quarter light aperture flange.

Fit the finishing brightwork ('banana') into the rear of the
body aperture.

Fit and align new 'memory' foam to the aperture flange (B
Fig. 72).

Fit and align foam strip to the new glass bottom edge and
replace the large washers onto the rear studs with the nuts
just started.

Slide the lower finisher on to the bottom of the encapsula-
tion and offer the whole assembly into the body aperture.

The rear of the glass may need to be offered at an angle to
engage the studs into the 'V' slots at first, then the whole is
pressed into position by locating all the studs through the
holes in the flange and tightening the nuts in the sequence
shown (A Fig.72).

Access to the rear stud nuts is achieved via 'rat holes'
through the inner bodywork (1 & 2 Fig. 73).

The remainder of the work is completed by carrying out the
previous operations in reverse order.

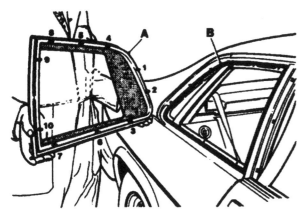

Fig. 72

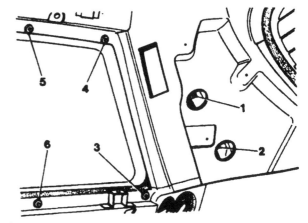

Fig. 73

Chapter 15

AIR CONDITIONING SYSTEM - 1994 MODEL YEAR ON

INTRODUCTION

This Section is primarily concerned with the 1994 Model Year introduction of HFC 134A (Hydro fluorocarbon) refrigerant. However, the opportunity has also been taken to update certain critical test and fault finding operations which are more appropriate to the new refrigerant.

Working practices and safety related procedures are relevant to all refrigerant types, but certain aspects are more critical to HFC 134A than R12; take special note of the section 'Handling refrigerant' and notes concerning moisture contamination.

GENERAL SPECIFICATIONS

REFRIGERANT		
Designation	Charge weight	Manufacturer and Type
HFC 134A	950 g + / - 50 g	ICI Klea or equivalent

COMPRESSOR		
Type & model	Configuration	Manufacturer
SD-7H15	7 Cylinder 155 cm³ per revolution	Sanden

COMPRESSOR LUBRICATION		
Designation	System Capacity	Manufacturer and Type
Polyalkylene glycol (PAG)	120 - 150ml	Sanden

STANDARD FOR RECOVERY / RECYCLE / RECHARGE EQUIPMENT.

Feature	Requirement
Recovery rate	0.014 - 0.062 m³ / min. (1.36 kg in 20 minutes)
Cleaning capability	15 parts per million (ppm) moisture; 4000 ppm oil; 330 ppm non condensable gases in air
Oil separator	With hermetic compressor and automatic oil return
Moisture indicator	Sight glass type, sensitive to 15 ppm minimum
Vacuum pump	2 stage 0.07 - 0.127 m³ / min.
Filter	Replaceable with moisture indicator
Charge	Selectable charge weight and automatic delivery
Hoses	Dedicated HFC 134A port connections.
Charge pressure	Heating element to increase pressure

COMPRESSION BELT TENSION

Burroughs method	New belt 600 N. If tension is below 230 N, reset at 400 N
Clavis method	New belt 103 to 107 Hz. If tension is below 70 Hz, reset at 83 to 87 Hz

Note: Tension measured midway between compressor and air pump/idler pulleys.

For new belt, rotate engine 3 revolutions minimum and recheck tension.

CLIMATE CONTROL SYSTEM, GENERAL

The climate control system fitted to 1994-on XJS has many features which make it unique. These differences demand changes to current system maintenance and rectification working practices.

Feature:
1. Refrigerant HFC 134A (Hydro fluorocarbon), non ozone depletory.
2. PAG (polyalkylene glycol) synthetic compressor lubricating oil.
3. Dedicated and improved compressor for HFC 134A refrigerant.
4. Quick fit/release self sealing charge and discharge ports.
5. Dual pressure switch to control the compressor (incorporated into the liquid line).
6. Clamp retained '0' ring seals at the expansion valve and evaporator.
7. All aluminium evaporator matrix and pipework.
8. Aluminium receiver/dryer (without sight glass) and HFC dedicated desiccant.
9. Parallel flow extended height condenser.
10. Single muffler situated in the suction hose.
11. Improved electrical system connectors.
12. Improved system control panel.

System Recognition

The following features will facilitate easy identification of the R134A system:

1. Aluminium pipes.
2. Large diameter, quick release charge and recovery ports.
3. HFC 134A labelling on compressor. No sight glass.

CAUTION:

The system refrigerant HFC 134A, is NOT compatible with any other previously fitted XJS system. The PAG compressor lubrication oil is NOT compatible with previously-used mineral based oils and must be treated exactly as detailed in the following sections.

WORKING PRACTICES

General

Be aware of, and comply with all health and safety requirements, whether they be legislative or common sense. This applies to conditions set both for the operator and workshop.
Before commencing any repair or service procedure, disconnect the vehicle battery ground connection and protect the vehicle where appropriate, from dirt or damage.
Work in a well ventilated, clean and tidy area.
Recovery and charge equipment must comply with, or exceed the standard detailed in General Specifications.

Handling Refrigerant

Wear eye protection at all times.
Use gloves, keep skin that may come into contact with HFC 134A covered.
Should refrigerant come into contact with your eyes or skin; wash the affected area with cool water and seek medical advice, do not attempt to treat yourself.
Avoid breathing refrigerant vapour, it may cause irritation to your respiratory system.
Never use high pressure compressed air to flush out a system. Under certain circumstances HFC 134A + compressed air + a source of combustion (welding and brazing operations in the vicinity), may result in an explosion and the release of potentially toxic compounds.
HFC 134A and CFC 12 must never come into contact with each other, they will form an inseparable mixture which can only be disposed of by incineration.
Do not vent refrigerant directly to atmosphere, always use Jaguar approved recovery equipment. Remember, HFC 134A is costly but recyclable.
Because HFC 134A is fully recyclable it may be 'cleaned' by the recovery equipment and reused following removal from a system.
Leak tests should only be carried out with an electronic analyser which is dedicated to HFC 134A. Never use a CFC 12 analyser or naked flame type.
Do not attempt to 'guess' the amount of refrigerant in a system, always recover and recharge with the correct charge weight. In this context do not depress the charge or discharge port valves to check for the presence of refrigerant.

Handling Lubricating Oil

Avoid breathing lubricant mist, it may cause irritation to your respiratory system.

Always decant fresh oil from a sealed container and do not leave oil exposed to the atmosphere for any reason other than to fill or empty a system. PAG oil is very hygroscopic (absorbs water) and will rapidly become contaminated by atmospheric moisture.
PAG oil is NOT compatible with previously used mineral based oils and must NEVER be mixed.
Do not reuse oil when it has been separated from refrigerant following a recovery cycle and dispose of the oil safely.

System Maintenance

When depressurizing a system do not vent refrigerant directly to atmosphere, always use Jaguar approved recovery equipment. Remember, HFC 134A is costly but recyclable.
Always decant compressor oil from a sealed container and do not leave oil exposed to the atmosphere for any reason other than to fill or empty a system. PAG oil is very hygroscopic and will rapidly become contaminated by atmospheric moisture.
Plug pipes and units immediately after disconnection and only unplug immediately prior to connection. Do not leave the system open to atmosphere.
It is not necessary to renew the receiver dryer whenever system has been 'opened' as previously advised. However, if a unit or part of the system is left open for more than five minutes, it may be advisable to renew the receiver dryer. This guidance is based on U.K. average humidity levels; therefore, locations with lower humidity will be less critical to moisture contamination of the unit. It must be stressed that there is not a 'safe' period for work to be carried out in: ALWAYS plug pipes and units immediately after disconnection and only remove plugs immediately prior to connection.
If replacement parts are supplied without transit plugs and seals do not use the parts. Return them to your supplier.
Diagnostic equipment for pressure, mass and volume should be calibrated regularly and certified by a third party organisation.
Use extreme care when handling and securing aluminium fittings, always use a backing spanner and take special care when handling the evaporator.
Use only the correct or recommended tools for the job and apply the manufacturer's torque specifications.
Keep the working area, all components and tools clean.

SYSTEM

Air Conditioning Control Module

The electronic control module (A/CCM) is located on the right hand side of the heater unit.
A digital micro-processor within the A/CCM receives data signals from operator controlled switches. Comparison of these signals with those returned from system temperature sensors and feedback devices results in the appropriate output voltage changes needed to vary: Blower motor speed, Flap position and those Solenoids which respond to operator selected temperature demand.
The A/CCM is a non-serviceable item but may be interrogated for system test. Care must be exercised when connecting test equipment, the A/CCM may be irreparably damaged should any of the test pins be shorted or bent.

Air Conditioning Control Panel

The Control Panel (Fig. 2) contains: Fan speed/defrost rotary switch, Manual and Demist mode buttons, Temperature differential slider, Temperature rotary control, A/C On and Recirculation mode buttons. The control panel relays information to the electronic control module.

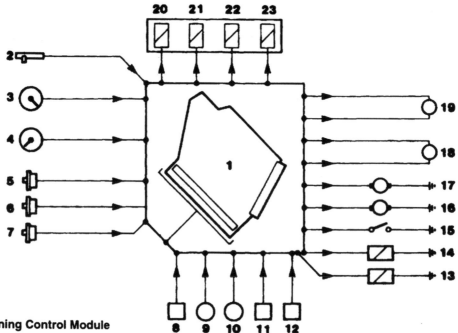

Fig. 1 Air Conditioning Control Module

key to Fig. 1

1.	A/C control module (A/CCM)
2.	Differential temperature control
3.	Temperature control
4.	Blower motor switch
5.	Ambient temperature sensor
6.	Motorized in-car aspirator
7.	Evaporator temperature sensor
8.	Coolant temperature switch
9.	Lower flap feedback potentiometer
10.	Upper flap feedback potentiometer
11.	LH Blower motor feedback
12.	RH blower motor feedback

13.	High speed relay
14.	High speed relay
15.	Compressor clutch
16.	Blower motor
17.	Blower motor
18.	Lower flap servo motor
19.	Upper flap servo motor
20.	Defrost vacuum solenoid
21.	Auto recirculation vacuum solenoid
22.	Centre vent vacuum solenoid
23.	Water valve vacuum solenoid

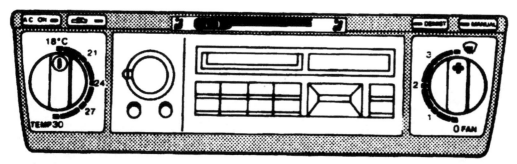

Fig. 2 Air Conditioning Control Panel

The FAN SPEED CONTROL rotary switch controls airflow from the blower motors. The switch has five positions: 0, 1, 2, 3 and DEFROST. In the 0 position the system is not operational, however, a residual signal to the control module (A/CCM) ensures that the blower flaps are closed, thus preventing outside air from entering the system.

Information regarding selection by the control switch of 1, 2 and 3 fan speeds is relayed to the A/CCM. Signals are also relayed to the A/CCM from the temperature selector feedback circuits and various sensors. Fan speed is steplessly controlled by the A/CCM, within the ranges 1, 2 and 3.

When DEFROST is selected the fans operate at maximum speed, front screen vents open fully, lower flaps close fully and maximum output is directed to the windscreen (there may be a delay of up to 30 seconds from selection to execution of this function).

The DEMIST mode button, when pressed, causes increased air flow to the front screen.

The FACE LEVEL TEMPERATURE DIFFERENTIAL sliding control is used to vary the temperature of face vent distributed air to that of footwell delivered air.

The TEMPERATURE ROTARY CONTROL is used to pre set the in car heat level in either 'automatic' or 'manual' mode. There are three temperature sensors located in the system - Exterior ambient, In-car and Evaporator.

An input voltage is supplied to the sensors from AC4-13 of the control module. The temperature sensing signal from the sensors is transmitted to the control module via AC4-4 and AC2-4 respectively. The sensors are semi-conductor devices which provide a voltage output proportional to the sensed temperature.

When pressed, the A/C ON button, causes the system to engage the Air conditioning compressor via its electromagnetic clutch.

When engaged, with indicator lamp lit, the MANUAL MODE facility provides operator selection of fan speed and in car temperature. In car temperature will not be thermostatically corrected to a predetermined level via the system sensors.

Automatic temperature control is resumed when the button is pressed and the indicator lamp is unlit.

RECIRCULATION mode closes the blower flaps and circulates only that air which is in the vehicle. When the ignition is turned off the blower flaps revert to the fresh air position.

REFRIGERATION CYCLE:

The Compressor draws low pressure refrigerant from the evaporator and by compression, raises refrigerant temperature and pressure. High pressure, hot vaporized refrigerant enters the Condenser where it is cooled by the flow of ambient air. A change of state occurs as the refrigerant cools in the condenser and it becomes a reduced temperature high pressure liquid.

From the condenser the liquid passes into the Receiver/Drier which has three functions,

- a) Storage vessel for varying system refrigerant demand.
- b) Filter to remove system contaminants.
- c) Moisture removal via the desiccant.

With the passage through the receiver/drier completed the still high pressure liquid refrigerant, enters the Expansion Valve where it is metered through a controlled orifice which has the effect of reducing the pressure and temperature.

The refrigerant, now in a cold atomized state, flows into the Evaporator and cools the air which is passing through the matrix.

As heat is absorbed by the refrigerant it once again changes state, into a vapour, and returns to the compressor for the cycle to be repeated (Fig. 3).

There is an automatic safety valve incorporated in the compressor which will operate should the system pressure be in excess of 41 bar. The valve will re-seat when the pressure drops below 35 bar.

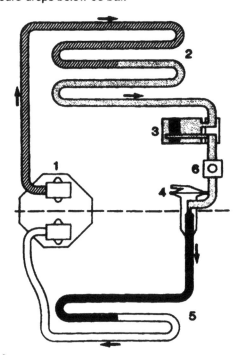

Fig. 3

key to Fig. 3

1.	*Compressor*	*4.*	*Expansion Valve*
2.	*Condenser*	*5.*	*Evaporator*
3.	*Receiver/Drier*	*6.*	*Dual pressure switch*

Note: The division of HIGH and LOW side is simply the system pressure differential created by the compressor discharge (pressure), suction (inlet) ports and the relative inlet and outlet ports of the expansion valve. This differential is critical to system fault diagnosis and efficiency checks.

System protection

The Dual pressure switch, located in the liquid line, cuts electrical power to the compressor clutch if the system pressure is outside of the range of 2 Bar (1st Function) to 27 Bar (2nd Function).

SYSTEM TROUBLE-SHOOTING

There are five basic symptoms associated with air conditioning fault diagnosis. A slightly different approach to problem solving will be necessary since the deletion of the sight glass. It is very important to positively identify the area of concern before starting a rectification procedure. A little time spent with your customer on problem identification, and use of the following trouble shooting guides will be beneficial.

The following conditions are not in order of priority.

NO COOLING

- Is the electrical circuit to the compressor clutch functional?
- Is the electrical circuit to the blower motor/s functional?
- Slack or broken compressor drive belt.
- Compressor partially or completely seized.
- Compressor shaft seal leak.
- Compressor valve or piston damage (may be indicated by small variation between HIGH & LOW side pressures relative to engine speed).
- Broken refrigerant pipe (causing total loss of refrigerant).
- Leak in system (causing total loss of refrigerant).
- Blocked filter in the receiver drier.
- Evaporator sensor disconnected?
- Dual pressure switch faulty?

Note: Should a leak or low refrigerant be established as the cause, follow the procedures as detailed under 'Recovery/Recycle/Recharge', and observe all refrigerant and oil handling instructions.

INSUFFICIENT COOLING

- Blower motor/s sluggish.
- Restricted blower inlet or outlet passage
- Blocked or partially restricted condenser matrix or fins.
- Blocked or partially restricted evaporator matrix.
- Blocked or partially restricted filter in the receiver drier.
- Blocked or partially restricted expansion valve.
- Partially collapsed flexible pipe.
- Expansion valve temperature sensor faulty (this sensor is integral with valve and is not serviceable).
- Excessive moisture in the system.
- Air in the system.
- Low refrigerant charge.
- Compressor clutch slipping.
- Blower flaps or distribution vents closed or partially seized.
- Water valve not closed.
- Evaporator sensor detached from evaporator.

INTERMITTENT COOLING

- Is the electrical circuit to the compressor clutch consistent?
- Is the electrical circuit to the blower motor/s consistent?
- Compressor clutch slipping.
- Faulty air distribution flap potentiometer or motor.
- Motorized in-car aspirator or evaporator temperature sensor faulty, causing temperature variations.
- Blocked or partially restricted evaporator or condenser.

NOISY SYSTEM

- Loose or damaged compressor drive belt.
- Loose or damaged compressor mountings.
- Compressor oil level low, look for evidence of leakage.
- Compressor damage caused by low oil level or internal debris.
- Blower/s motor/s noisy.
- Excessive refrigerant charge, witnessed by vibration and 'thumping' in the high pressure line (may be indicated by high HIGH & high LOW side pressures).
- Low refrigerant charge causing 'hissing' at the expansion valve (may be indicated by low HIGH side pressure).
- Excessive moisture in the system causing expansion valve noise.

Note: Electrical faults may be more rapidly traced using JDS or PDU.

INSUFFICIENT HEATING

- Water valve stuck in the closed position.
- Motorized in-car aspirator seized.
- Blend flaps stuck or seized.
- Blocked or restricted blower inlet or outlet.
- Low coolant level.
- Blower fan speed low.
- Coolant thermostat faulty or seized open.

MANIFOLD GAUGE SET

The manifold gauge set is a most important tool for fault diagnosis and system efficiency assessment. The relationship to each other of HIGH and LOW pressures and their correlation to AMBIENT and EVAPORATOR temperatures

must be compared to determine system status (see Figs 5 & 6, Pressure/Temperature Graphs).

Because of the heavy reliance upon this piece of equipment for service diagnosis, ensure that the gauges are calibrated regularly and the equipment is treated with care.

The gauge set (Fig. 4) consists of a manifold fitted with:

- Low side service hose - BLUE.
- Low side hand valve - BLUE.
- Low pressure compound gauge - BLUE.
- High pressure gauge - RED.
- High side hand valve - RED.
- High side service hose - RED.
- System service hose - NEUTRAL COLOUR (commonly yellow).

Manifold

The manifold is designed to control refrigerant flow. When connected into the system, pressure is registered on both gauges at all times. During system tests both the high and low side hand valves should be closed (rotate clockwise to seat the valves). The hand valves isolate the low and the high sides from the centre (service) hose.

Low Side Pressure Gauge

This compound gauge, is designed to register positive and negative pressure and may be typically calibrated - Full Scale Deflection, 0 to 10 bar (0 to 150 lbf/in²) pressure in a clockwise direction; 0 to 1000 mbar (0 to 30 in Hg) FSD negative pressure in a counter clockwise direction.

High Side Pressure Gauge

This pressure gauge may be typically calibrated from 0 to 30 bar (0 to 500 lbf/in²) FSD in a clockwise direction. Depending on the manufacturer, this gauge may also be of the compound type.

SYSTEM CHECKING WITH THE MANIFOLD GAUGE SET

Evacuating the Manifold Gauge Set

Attach the centre (service) hose to a vacuum pump and start the pump. Open fully both high and low valves and allow the vacuum to remove air and moisture from the manifold set for at least five minutes.
Turn the vacuum pump off and isolate it from the centre service hose but do not open the hose to atmosphere.
Observe the manufacturer's recommendation with regard to vacuum pump oil changes.

Connecting the Manifold Gauge Set

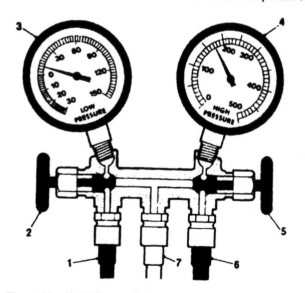

Fig. 4 Manifold Gauge Set

CAUTION:

It is imperative that the vacuum pump is not subjected to a positive pressure of any degree. Therefore the pump must be fitted with an isolation valve at the centre (service hose) connection and this valve must be closed before the pump is switched off. This operation replaces the purge' procedure used on previous systems. Observe the manufacturer's recommendation with regard to vacuum pump oil changes.

Attachment of the hose quick release connectors to the
high and low side system ports is straightforward, provided
that the high and low valves are closed and the system is
NOT operational.

Assessment of system operating efficiency and fault diagnosis may be achieved by using the facilities on your
Recovery/Recharging/Recycling station, follow the manufacturers instructions implicitly and observe all safety considerations.

Stabilizing the System

Accurate test gauge data will only be attained if the system
temperatures and pressures are stabilized.

Ensure that equipment and hoses cannot come into contact with engine moving parts or sources of heat.

It is recommended that a free standing air mover is placed
in front of the vehicle to provide mass air flow through the
condenser/cooling system see illustration below.

Start the engine, allow it to attain normal working temperature and set at fast idle (typically 1200 to 1500 rpm).

Select full air conditioning performance.

With all temperatures and pressures stable or displaying
symptoms of faults, begin relevant test procedures.

Pressure/Temperature Graphs
(To obtain Bar, multiply the lbf/in^2 figure by 0.069.
To obtain kgf/cm^2 multiply the lbf/in^2 figure by 0.070)

Note: The system controls will prevent the evaporator
temperature from falling below 0° C. The graph is
typical of HFC 134A

SYSTEM FAULT DIAGNOSIS

Probable causes of faults may be found by comparing
actual system pressures, registered on your manifold
gauge set or recovery / recharge / recycle station, and the
pressure to temperature relationship graphs found on the
previous page. The chart below shows the interpretation
that may be made by this difference. The 'Normal' condition is that which is relevant to the prevailing ambient and
evaporator temperatures.

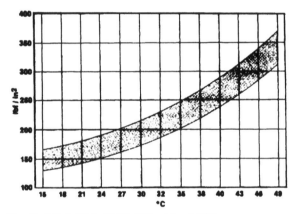

Fig. 5 High Side (lbf/in^2)/Ambient (°C)

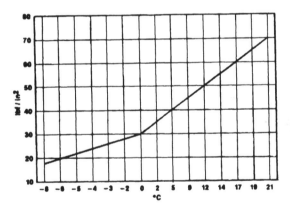

Fig. 6 Low Side (lbf/in^2)/Evaporator (°C)

Low Side Gauge	High Side Gauge	Symptom	Diagnosis
Normal	Normal	Discharge air initially cool then warms up	Moisture in system
Normal to low	Normal	As above	As above
Low	Low	Discharge air slightly cool	HFC 134A charge low
Low	Low	Discharge air warm	HFC 134A charge very low
Low	Low	Discharge air slightly cool or frost build up at expansion valve	Expansion valve stuck closed
Low	Normal to high	Discharge air slightly cool	Restriction in High side of system
High	Low	Compressor noisy	Defective reed valve
High	High	Discharge air warm and high side pipes hot	HFC 134A charge high or condenser malfunction
High	High	Discharge air warm Sweating or frost at evaporator	Expansion valve stuck open

Note: If erratic or unusual gauge movements are experienced, check the equipment against a known manifold gauge set.

GENERAL SYSTEM PROCEDURES

Leak Test

Faults associated with low refrigerant charge weight and low pressure may be caused by leakage. Leaks traced to mechanical connections may be caused by torque relaxation or joint face contamination. Evidence of oil around such areas is an indicator of leakage. When checking for non visible leaks use only a dedicated HFC 134A electronic analyser and apply the probe all round the joint / connection. Should a leak be traced to a joint, check that the fixing is secured to the correct tightening torque before any other action is taken.

Do not forget to check the compressor shaft seal and evaporator.

Note: Never use a dedicated CFC 12 or naked flame type analyser.

Charge Recovery (System depressurization)

The process of HFC 134A recovery will depend on the basic characteristics of your chosen recovery / recycle / recharge equipment, therefore, follow the manufacturers instructions carefully.

Remember that compressor oil may be drawn out of the system by this process, take note of the quantity recovered so that it may be replaced.

CAUTION:

Observe all relevant safety requirements.
Do not vent refrigerant directly to atmosphere
and always use Jaguar approved
recovery/recycle/recharge equipment.

Wear suitable eye and skin protection.
Do not mix HFC 134A with CFC 12.
Take note of the amount of recovered refrigerant, it will indicate the state of the system and thus the magnitude of any problem.

Evacuating the System

This process, the removal of unwanted air and moisture, is critical to the correct operation of the air conditioning system. The specific procedures will vary depending on the individual characteristics of your chosen recovery/recycle /recharge equipment and must be carried out exactly in accordance with the manufacturers instructions.

Moisture can be highly destructive and may cause internal blockages due to freezing, but more importantly, water suspended in the PAG oil will damage the compressor. Once the system has been opened for repairs, or the refrigerant charge recovered, all traces of moisture MUST be removed before recharging with new or recycled HFC 134A.

Adding Compressor Lubricating Oil

Oil may be added by three methods, two of which are direct into the system - 1) via the recovery/recycle/ recharge station, 2) by proprietary oil injector.

Equipment manufacturer's instructions must be adhered to when using direct oil introduction.

The third method may be required because of rectification work to the existing compressor, or the need to fit a new compressor.

From an existing compressor, drain the oil into a measur-

ing cylinder and record the amount. Flush the unit out with fresh PAG oil and drain thoroughly. Replenish the compressor with the same amount of PAG oil that was originally drained out and immediately plug all orifices ready for refitting to the vehicle. The transit lubricating oil must be drained and discarded from a new compressor before it may be fitted. An adjustment should be made to the system oil level by taking into account, a) the quantity found in the original compressor, and b) the quantity deposited in the recovery equipment oil separator from the charge recovery operation. Typically, 80 ml may be drained from the original compressor and 30 ml found in the oil separator; if these quantities are added together - 80 + 30 = 110 ml, then this is the amount of fresh PAG oil that must be put into the new compressor prior to fitting.

Please note that the discrepancy between this figure and the nominal capacity of 135 ml is caused by normally unrecoverable oil being trapped in components such as the receiver/drier or evaporator. The previous statements are only valid if there is NO evidence of an oil leak from the system. If oil has been lost and the fault attended to, then the compressor, whether original or replacement, should be filled with the specified quantity.

CAUTION:

Always decant fresh oil from a sealed container
and do not leave oil exposed to the atmosphere.
PAG oil is very hygroscopic (absorbs water)
and will rapidly attract atmospheric moisture.

PAG oil must NEVER be mixed with mineral based oils.
Do not reuse oil following a recovery cycle, dispose of it safely.

Depending on the state of the air conditioning system immediately prior to charge recovery and the rate of recovery, an amount of oil will be drawn out with the refrigerant. The quantity will be approximately 30 to 40 ml; this may vary, and the figure is given only for guidance. It is most important that the oil separator vessel in the recovery equipment is clean and empty at the start of the process so that the amount drawn out may be accurately measured.

Adding Refrigerant

In order that the air conditioning system may operate efficiently it must contain a full refrigerant charge. The indications of some system defects, and the results of certain tests, will show that a low charge is the most probable cause of the fault. In such cases the charge should be recovered from the system, the weight noted, and the correct amount installed.

CAUTION:

If oil was drawn out during the recovery process,
the correct amount may be added directly from
your recovery / recycle /recharge station
(if so equipped) prior to the 'charging process'.

Note: Never attempt to 'guess' the amount of refrigerant in a system.
Always recover and recharge with the correct charge weight, this is the only accurate method.

It must be stressed that the need to protect compressor oil from moisture is vital, observe the procedures in **HANDLING LUBRICATING OIL.**

CHAPTER 16

ELECTRICAL SYSTEM (1992 MODEL YEAR ON)

ALTERNATOR DRIVE BELT, RENEW AND ADJUST

Loosen the pivot bolts securing the air conditioning compressor.
Loosen the adjusting link securing bolt and trunnion block bolt.
Loosen the adjusting link locknut and adjust the compressor towards the engine until compressor drive belt can be removed.
Loosen the alternator pivot nut and bolt.
Loosen the adjusting link pivot bolt and the trunnion block bolt.
Loosen the adjusting link locknut and adjust the alternator towards the engine by means of the adjusting nut.
Remove the trunnion block bolt and push the alternator towards the engine until the drive belt can be removed from the pulleys.
On fitting the new belt ensure that the drive belt is adjusted to the correct tension.
A load of 1.5 kg must give a total belt deflection of 4.5 mm when applied at the mid point of the belt.

Operate the glass motor switch. If the glass operates as normal, the circuit breaker is faulty and should be renewed.

2. With the ignition switched on, battery voltage should be obtained at the brown/blue lead terminal of the left hand switch. Operate the switch. Battery voltage should be obtained at the red/blue lead terminal when the switch is operated in one direction, and the green/blue lead terminal when the switch is operated in the opposite direction.
Should a zero reading be obtained at either test point, renew the switch.

Note: The same test applies to the right hand switch. The switch cable colours are red/green for one direction and green/red for the opposite direction.

If the above tests prove satisfactory, check the lift motor for continuity.

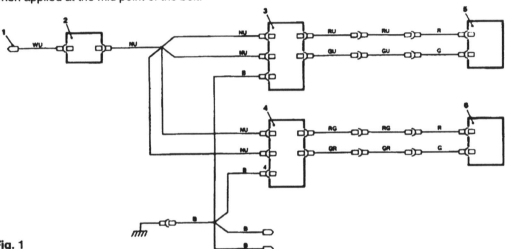

Fig. 1

Key to Fig. 1
1. To auxiliary controlled relay
2. Thermal circuit breaker
3. LH drop glass switch
4. RH drop glass switch
5. LH motor
6. RH motor

ELECTRICALLY OPERATED DOOR GLASS, DESCRIPTION

Power is supplied via auxiliary controlled load relay contacts from the main battery supply through a thermal circuit breaker. The auxiliary controlled load relay is energised when the ignition switch is closed and this energises the circuit to the window lift motor. When a control switch is operated to lower a window, current flows via contacts within the switch to the motor (circuit to earth is also via the switch). When the switch is operated to raise the window, current flows in the opposite direction through the switch and motor.

TESTING (Refer to Fig. 1)

If the drop glass fails to operate check the fuse and all connectors and ensure that all connections are clean and tight.
1. Check the thermal circuit breaker:
Connect the white/blue and brown/blue leads together.
Switch on the ignition.

If the wiring continuity proves satisfactory, remove the lift motor for bench testing.

CIRCUIT BREAKER, RENEW

Disconnect the battery earth lead.
Remove the passenger side dash liner.
Remove nuts and shakeproof washers (1 Fig. 2) securing mounting plate to fan motor.
Ease the mounting plate from studs.

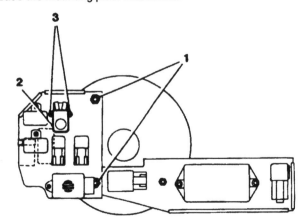

Fig. 2

Note position of cables and disconnect at lucars on relevant circuit breaker.
Remove the two screws (3 Fig. 2) securing the unit and remove the unit (2 Fig. 2).
Fitting a new circuit breaker is a reversal of the removal procedure.

GLASS LIFT SWITCHES, RENEW

Disconnect the battery earth lead.
Carefully displace the lift switchpack (Fig. 3) from the console veneer panel.

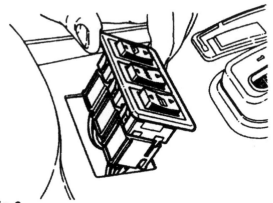

Fig. 3

Disconnect the harness multi-plug from the window lift switch.
Displace and remove the switch from the panel.
The refitting procedure is a reversal of the removal procedure.

KIEKERT CENTRAL LOCKING SYSTEM

DESCRIPTION

The system comprises an actuator in both doors and the boot lid, and is controlled by an ECM located in the passenger side 'A' post.
The doors will lock and unlock simultaneously. The boot will either open in unison with the doors or remain locked dependent on the door lock position, as follows:

Boot lock positions

1.   Lock turned fully clockwise. In this position the lid locks/unlocks in unison with the door locks.
2.   Lock turned fully anti-clockwise. In this position the boot lid is permanently locked and cannot be unlocked using the central locking system.
3.   Lock turned to the central position. With the lock in this position the key cannot be removed.

The boot lid can be opened irrespective of the central door locking mode (locked or unlocked). This enables access to the boot if the car is centrally locked.

FRONT AND REAR PARKING LAMPS, CIRCUIT DESCRIPTION

With the master lamp switch in the parking lamp on position, current flows to the side lamp relay to energise circuit to side lamps and bulb failure units. The current flowing through the bulb failure units will cause the bulb failure warning lamp to glow for 15 to 30 seconds. If the warning lamp fails to go out then there is a bulb failure or a circuit fault in the front parking lamp, rear lamps or number plate lamps.

Fault Finding

Check the fuses and all connections, ensuring the earth connections are clean and tight.
With the master light switch in the parking lamp on position, battery voltage should be obtained at terminals 85 and 87 of the side lamp relay 2. This in turn makes the circuit to the lamp failure units 6, 11, 17 and 23.
If battery voltage is obtained at the B terminal of a lamp failure unit but a zero reading at the L terminal, renew the bulb failure unit.

HEADLAMP, ALIGNMENT

ADJUSTMENT

Headlamp beam setting should only be carried out with approved beam setting apparatus.

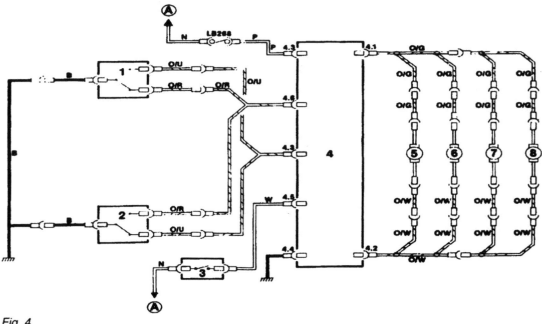

Fig. 4

Key to Fig. 4

1. RH door lock switch	3. Ignition switch	6. LH door lock motor
2. LH door lock switch	4. Door lock control module	7. Boot lock motor
	5. RH door lock motor	8. Fuel filler flap lock (coupe only)

Vertical and horizontal adjustment of the headlamp beam is made with two adjusting screws set in a yellow plastic moulding sited above either front wheel arch inside the bonnet. The silver screw adjusts the horizontal alignment and the black screw adjusts the vertical alignment. For horizontal beam adjustment of the right hand headlamp turn the screw clockwise to move the beam to the right. Turn the screw anti-clockwise to move the beam to the left.

Turn in the opposite directions for adjustment of the left hand headlamp. Where a headlamp levelling motor is fitted (i.e. interposed in the drive cable), setting the beam vertically is exactly the same as for the non-headlamp beam levelling lamp. Three positions are however available for setting the levelling motor to cater for differing rear end loads.

HEADLAMP ASSEMBLY, RENEW

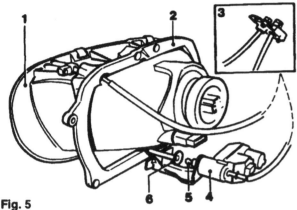

Fig. 5

Key to Fig. 5

1. Headlamp
2. Headlamp carrier
3. Manual adjuster cable
4. Motor
5. Motor drive ball
6. Headlamp carrier socket

Disconnect the battery earth lead.
Remove headlamp rim finisher.
Displace manual adjuster retaining block from inner wing.
Displace adjuster cables from block.
Remove adjuster block and place aside.
Reposition adjuster cables through inner wing.
Undo and remove headlamp carrier to body securing screws.
Displace headlamp assembly from aperture.
Disconnect headlamp multi-plug.
Remove headlamp assembly.

The refitting procedure is a reversal of the removal procedure.
Take care not to touch the bulb glass envelope (1 Fig. 6).

HEADLAMP BULB, RENEW

Disconnect the battery earth lead.
Reposition the front wheels to gain access to the plate located inside the wheel arch.
Turn fastener anti-clockwise and remove the plate.
Disconnect the multi-connector socket from the bulb and remove the rubber cover.
Disconnect the wire clip securing bulb to the headlamp unit and remove the bulb 1 Fig. 6 (not applicable to USA).
The refitting procedure is a reversal of the removal procedure. Take care not to touch the bulb glass envelope.

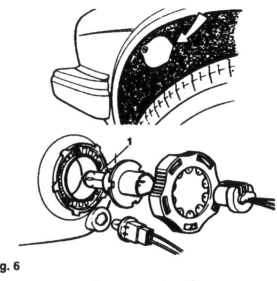

Fig. 6

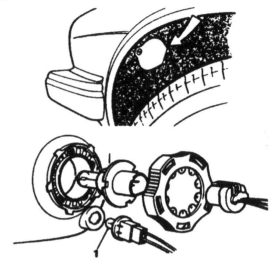

Fig. 7

FRONT PARKING LAMP BULB, RENEW

Disconnect the battery earth lead.
Reposition the front wheels to gain access to the plate located inside the wheel arch.
Turn fastener anti-clockwise and remove the plate.
Rotate the bulb holder anti-clockwise and remove from the headlamp.
Pull the capless bulb (1 Fig. 7) from the holder and replace with one of the correct type.

FOG AND REVERSE LAMP ASSEMBLY, RENEW

Disconnect the battery earth lead.

Undo and remove screws (1 Fig. 8) securing lamp assembly cover to boot lid.
Disconnect bulb wires from fog and reverse bulbs. Undo and remove nuts (2 Fig. 8) securing lamp assembly to boot lid.
Remove lamp assembly (3 Fig. 8) from boot lid.
To fit a new lamp assembly is the reversal of the removal procedure.

TAIL, STOP AND FLASHER LAMP ASSEMBLY, RENEW

Disconnect the battery earth lead.
Displace the boot rear side trim for access.
Disconnect wires from stop/tail and flasher bulbs.
Undo and remove lamp assembly to body securing nuts (1 Fig. 9).

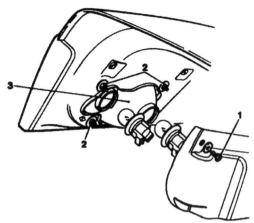

Fig. 8

Remove lamp assembly from rear wing housing.
The refitting procedure is a reversal of the removal procedure.

TAIL, STOP AND FLASHER LAMP BULB, RENEW

Displace boot rear side trim for access.
Displace bulb holder from lamp assembly (red-stop/tail; blue-flasher).
Remove bulb from holder (2 Fig. 9).

Fitting a new bulb is a reversal of the removal procedure.

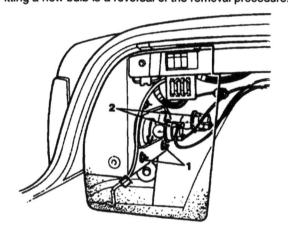

Fig. 9

FRONT FLASHER LAMP BULB, RENEW

Rotate the bulb holder anticlockwise by hand and withdraw from behind the front bumper. Remove the bulb from the holder and replace with one of the correct type. Refit the bulb holder and turn clockwise.

INSTRUMENT ILLUMINATION BULBS, RENEW

SPEEDOMETER ILLUMINATION BULB
TACHOMETER ILLUMINATION BULB
OIL AND TEMPERATURE GAUGE ILLUMINATION BULB
BATTERY AND FUEL GAUGE ILLUMINATION BULB

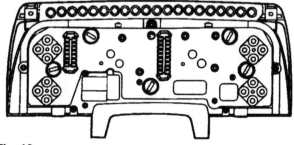

Fig. 10

Refer to Fig. 10
Remove the central finisher.
Displace and remove the side finishers.
Displace and remove the instrument panel to fascia securing screw cover plates.
Undo and remove screws securing fascia to panel.
Carefully displace the instrument panel for access.
Displace and remove the illumination bulb from the instrument panel.

Fitting a new bulb is a reversal of the removal procedure.

INSTRUMENT PACK WARNING BULBS, RENEW

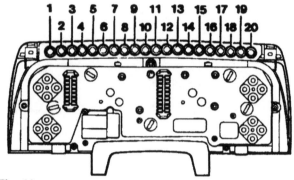

Fig. 11

Key to Fig. 11

Low washer bottle warning bulb (1)
Automatic transmission failure warning indicator bulb (3)
Flasher bulb (5 & 16)
Ignition bulb (7)
Headlamp high beam bulb (15)
Oil pressure bulb (11)
Handbrake bulb (10)
Brake warning bulb (12)
Fuel warning bulb (9)
Low coolant warning bulb (13)
Seat belt warning bulb (6)
Fog lamp warning bulb (17 & 20)
Bulb failure warning bulb (19)
Caravan warning bulb (4)
Exhaust temperature warning bulb (2)
Sport mode bulb (18)
Check engine warning bulb (8)
Anti lock brake warning bulb (14)

Disconnect the battery earth lead.
Position tilt steering column to lowest position.
Undo and remove instrument panel finishers securing screws.
Remove central finisher.
Displace and remove side finishers.
Displace and remove instrument panel to fascia securing screw cover plates.
Undo and remove panel to fascia securing screws.
Carefully displace instrument panel for access.
Displace and remove indicator bulb from instrument panel (refer to Fig. 11 for locations).

Fitting a new bulb is the reversal of the removal procedure.

VARIATIONS FOR 1993.5 MODEL YEAR - ON VEHICLES

HARNESSES

New harnesses are fitted to suit the new emissions hardware and fuel tank system.

The harnesses now have ultrasonic splices which are superior for electrical transmission. This type of splicing cannot be modified and is weakened when peeled apart. For this reason the heat-shrink sleeving should never be removed.

A new type of diagnostic connector is fitted in the boot adjacent to the 12-way fuse box.

CLIMATE CONTROL SYSTEM

An ELMOS control unit is fitted which is more reliable due to an improved design and the use of fewer components. The case is coloured brown on the XJS thereby differentiating it from the black item fitted to the saloons.

The solar sensor/alarm is modified, but functions exactly as before.

The electrical harness incorporates changes to accommodate manual recirculation, the MAX cooling function and the omission of the humidity control.

COOLING SYSTEM

An electric fan is fitted to enhance the cooling performance.

This is fitted with air flow flaps on the fan mounting assembly.

FUEL SYSTEM

The fuel tank, on vehicles fitted with the 6.0 Litre engine, has twin fuel pumps and a modified pump control module. One pump operates at engine speeds up to 2840 RPM, at which point the second pump switches in.

New in-tank electrical connecting leads are specified for interconnection with the new four-way header connector.

BATTERY

The battery is a 72 Ah item produced by Varta.

STARTER MOTOR

A Magneti Marelli 1.8 kW starter motor is fitted to give better cold start cranking.

GENERATOR

A NipponDenso 120 Amp generator is fitted to improve the balance of electrical loads and to provide a higher output at lower engine speed.

Electrical connection to the generator is by one plug and one eyelet.

The load dump module is now no longer necessary and has been omitted from the specification.

ENGINE MANAGEMENT SYSTEM

A revised Lucas Marelli system with new calibration and modifications to the hardware and harnesses has been fitted.

TRANSMISSION

The software on the 4L80E automatic transmission is modified to match the characteristics of the 6.0 litre engine. A mode switch is fitted to all vehicles.

SUN VISOR MOUNTED ILLUMINATED VANITY MIRRORS

Sun visors, which incorporate an illuminated vanity mirror, are fitted to the driver's side and passenger's side. The sun-visors also incorporate interior courtesy lights.

SECURITY SYSTEM

A security system, which currently complies with most European and overseas legislation, is offered as a factory fit option. The remote arming/disarming control unit is a radio transmitter or infra-red controller depending on market legislation. The features offered by the new system are:

Radio frequency arming/disarming which also controls the central door locking.
Entry sensing on doors and boot. Ignition anti-tamper.
Start inhibit.
Headlamp convenience (illumination for twenty-five seconds).
Starter disable.
Arm and disarm audible and visual indication.
Warn away potential thieves (reduced sound alarm).
Remote panic alarm.
Escalating siren response.
Headlamp warning flash.
Error tone.
Security OFF (Valet) switch.

There are options available which may be initialized on the system, by the dealer, using JDS.

IN-CAR ENTERTAINMENT

An additional 3 inch co-axial, loudspeaker is added to the top of each rear quarter trim panel on the two-seater convertible. This improves the quality of the mid-range sound and provides a better sound field for the front seat occupants.

Changes to the body manufacture now provide the option to supply the Compact Disc Autochanger as a production line fitted unit or as a dealer fitted unit.

The link lead which was fitted to the rear of the radio has been omitted. A telephone muting facility has been added to the radio which operates automatically when the radio telephone is in use.

Head cleaning tapes, type Allsop-3, are now accepted by the tape player.

Changes to the Radio Data System (RDS) software are made as follows:

When the RDS data is lost and no alternative frequency is available, the radio will remain on-station and display the station frequency only; RDS will remain on in readiness for the signal to recover.

When TP is selected, the radio will check the current station for TP. If it is not a TP station the radio will search for a TP station, and while doing so, the display will show TRAFFIC, flashing on and off, during the search.

The function of the RDS and TP switches has been reversed compared to the previous radio. The RDS switch is now a momentary on/off switch and the TP switch operates after being pressed for two seconds, with a beep to confirm acceptance of on/off. The volume minimum pre-set level for a traffic announcement has been reduced because the original volume level was considered to be too loud.

MULTI-FUNCTION UNIT

The multi-function unit is located under the fascia, mounted on a bracket behind the glovebox. The unit is new and combines the functions previously performed by six separate modules, plus two new functions.

The separate modules which have been replaced by the multi-function unit are:

Seat belt warning module
Lights-on buffer module
Over-speed module (Saudi Arabia only)
Interior lamp delay module
Heated rear window timer
Bulb check unit

The two new functions performed by the multi-function unit are:

Ignition key-in audible warning, and
Security system interface.

A description of each function follows:

SEAT BELT WARNING

With the ignition switched on, the seat belt visual warning located in the instrument pack will be activated for approximately six seconds. With the ignition switched on and the driver's seat belt unlatched, an audible warning will sound for approximately six seconds; there is no audible warning if the seat belt is latched before the ignition is switched on. Both warning devices will reset (and cancel) when the ignition is switched off.

LIGHTS-ON WARNING

An audible warning will sound if the following three conditions exist - the light switch is in the side lamp or headlamp position, the ignition key is removed and the driver's door is opened. The alarm can be cancelled by moving the light switch to the off position or closing the driver's door or inserting the key in the ignition switch. It should be noted that when the driver's door is open and the key is in the ignition switch, then the key-in alarm will also sound.

OVER-SPEED WARNING (SAUDI ARABIA ONLY)

An audible warning will be given when the road speed is between 120 and 130 km / h.

INTERIOR LAMP DELAY

The interior lamps will operate immediately if any door is opened. When leaving the vehicle parked, the interior lamps will remain on for approximately ten seconds after the last door has been closed. On entering the vehicle, and then closing both doors, the lamps will turn off either after approximately ten seconds or when the ignition is switched on. If the door(s) are left open with the ignition key removed, the lamps will go off after approximately two minutes (to prevent draining the battery) and will reset only after the doors have been closed for approximately ten seconds.

HEATED REAR WINDOW TIMER

The heated rear window is activated, by the multi-function unit and a relay, for approximately ten minutes when the heated rear screen switch is operated. It can be switched off before the time-out period if required, by operating the switch again. The timing function will also switch off and reset if the ignition is switched off.

BULB CHECK

The following bulbs; park brake, brake fluid and catalyst overheat (Japan only) are activated for approximately 2.5 seconds when the ignition is switched on, to indicate that the bulbs are functional.

IGNITION KEY-IN WARNING

When the ignition is in the off position but the key is left in the ignition switch, the alarm will sound when the driver's door is opened. The warning can be cancelled by removing the ignition key or closing the driver's door or switching on the ignition.

SECURITY SYSTEM INTERFACE

An output from the multi-function unit to the security system is provided which reflects the state (on / off) of the interior lamp door switches.

MULTI-FUNCTION UNIT SELF DIAGNOSTICS

The Multi-Function Unit has its own diagnostic mode which is capable of identifying faults within the module, open circuit and short circuit faults in the wiring and faults in the components of the vehicle systems which supply inputs to the module and which receive signals from the module.

INPUT DIAGNOSTICS

Before starting the diagnostic mode, sit in the front of the car with the doors closed and the seat belt unlatched. Under these conditions, the audible chime on the multi-function unit should not operate in the diagnostic mode. The input diagnostic mode is activated as follows:

1. Ensure that the sidelights are switched off.
2. Press and hold in the heated rear window switch.
3. Switch on the ignition switch whilst still holding in the heated rear screen switch.
4. Release the heated rear window switch.

If the alarm chime does not sound, this indicates that there is no fault.
The following checks should now be performed:

Latch the seat belt; the chime should sound. Unlatch the seat belt and the chime should be silenced.

Open the driver's door; the chime should sound. Close the driver's door and the chime should be silenced.

Open the passenger door; the chime should sound. Close the passenger door and the chime should be silenced.

Note that the vehicle road speed sensor is also being tested and could be causing a fault indication. Where this is suspected, disconnect it and look for a change in the above tests. Due to the method involved in selecting the diagnostics mode, the integrity of the following function/components is tested by default:

key-in ignition switch,
ignition input to the multi-function unit and
the heated rear window switch.

OUTPUT DIAGNOSTICS

If not already in the input diagnostic mode, repeat steps 1 to 4 described above.
Switch on the sidelights to select the output diagnosis mode; all inputs will now be inhibited. As each output is selected, it should operate its circuit load and the chime will remain silent. If it fails to operate its particular circuit due to a short circuit load, open circuit etc., then the chime will sound.
Monitoring of the output functions is achieved by cycling through each, using the heated rear window switch:

All outputs inactive, but output diagnosis mode enabled. Press and release the heated rear window (HRW) switch to select the heated rear window circuit. Press and release the HRW switch to de-select the heated rear window circuit.

Press and release the HRW switch to select the interior lamp circuit. Press and release the HRW switch to de-select the interior lamp circuit.

Press and release the HRW switch to select the seat belt visual warning circuit. Press and release the HRW switch to de-select the seat belt visual warning circuit.

Press and release the HRW switch to select the bulb check circuit. This will test the circuits and warning lamps associated with low wash, park brake, brake fluid and catalyst overheat (Japan). Press and release the HRW switch to de-select the bulb check circuit.

Switch off the sidelights to de-select the output diagnostics mode. Switch off the ignition.

The multi-function unit will reset and operate normally next time the ignition is switched on.

LAMPS

INTRODUCTION

New lamps and reflectors are specified, styled to match the new design of the front and rear bumper assemblies.
Access to the front and rear side marker lamps/reflectors and rear reflectors involves the removal of parts of the bumper. Removal of the bumper components is described in the Body section.

FRONT DIRECTION INDICATOR LAMPS

These are retained by two screws and are similar to those on the 1992 model year vehicles.

FRONT DIRECTION INDICATOR LAMP, BULB CHANGE

Remove the two screws which secure the lamp to the bumper cover. Remove the lamp. Release the bulb holder and remove the bulb.

Refit the bulb and the bulb holder to the lamp. Position the lamp to the bumper cover and fit the retaining screws. Check the operation of the lamp.

FRONT DIRECTION INDICATOR LAMP, RENEW

Remove the two screws which secure the lamp to the bumper cover. Remove the lamp. Disconnect the electrical connector from the bulb holder.

Connect the electrical connector to the bulb holder. Position the lamp to the bumper cover and fit the retaining screws. Check the operation of the lamp.

SIDE MARKER LAMPS / REFLECTORS (FRONT)

These lamps/reflectors mount directly to the vehicle body and not to the bumpers. However, the side marker lamp bulbs may be reached through an access panel in the front wheelarch liner and the rear side marker lamp bulbs may be reached from inside the luggage compartment.

FRONT SIDE MARKER LAMP ASSEMBLY, RENEW

Support the front of the vehicle and remove the front road wheel(s).
Referring to the Body section remove the (plated) front bumper blade assembly.
Remove the plastic 'scrivets' which secure the top of the bumper cover to the body. Remove the three body-to-bumper cover fixings from the relevant side of the car, and carefully reposition the bumper cover to give access to the marker lamp (if both lamps are to be renewed simultaneously, the bumper cover will have to be completely removed).

Reaching through the wheelarch access panel, disconnect the harness connector. Release the tangs which secure the lamp to the body, and remove the lamp.

Fit the new lamp, ensuring that the locking tangs have fully located. Reaching through the wheel arch access panel, connect the harness connector. Check that the lamp operates.

Reposition the bumper cover, and referring to the Body section, secure the cover and refit the blade, using new scrivets.

Fit the front road wheels and lower the vehicle.

REAR SIDE MARKER LAMP / REFLECTOR ASSEMBLY, RENEW

Support the rear of the vehicle and remove the rear road wheel(s).

Referring to the Body section, remove the (plated) rear quarter bumper blade assembly from the relevant side of the vehicle. Remove the three body-to-bumper cover fixings at the rear of the wheel arch.

Reposition the bumper cover to give access to the marker lamp.

From inside the boot, move the trim panel and disconnect the harness connector. Release the tangs which secure the lamp to the body, and remove the lamp from beneath the bumper cover.

Fit the new lamp to the body, ensuring that the locking tangs have fully located. From inside the boot, connect the harness connector and reposition the trim panel. Check that the lamp operates.

Referring to the Body section, secure the bumper cover and refit the rear quarter bumper blade assembly.

Fit the rear road wheel(s) and lower the vehicle.

REAR REFLECTOR LENS - RENEW

Support the rear of the vehicle and remove the rear road wheels.

Referring to the Body section, remove the bumper cover/blade assembly.

Release the spire clips which secure the reflector(s) to the bumper cover. Remove the reflector(s).

Fit the reflector(s) to the bumper cover and secure with new spire clips.

Referring to the Body section, refit the bumper cover/blade assembly.

Fit the rear road wheels and lower the vehicle.

CHAPTER 17

INSTRUMENTS - 1992 MODEL YEAR ON

DESCRIPTION

The instrument pack (Fig. 1) features the traditional layout of two large main dials with four small supplementary gauges. These are conventional analogue gauges comprising 90 degree movements for the four minor gauges and 270 degree movements for the two major gauges.

The trip meter is integrated into the tachometer and the odometer is included in the speedometer. The pulse signal required to operate the speedometer is controlled by a speed sensor situated in the differential unit. The engine speed signal received by the tachometer is derived from the ignition coil negative terminal. The voltage wave form at this point can reach as much as 400 volts when a spark is generated and it is desirable to suppress this voltage before allowing it into the wiring harness.

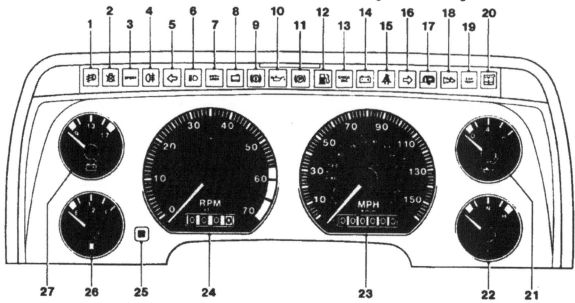

Fig. 1

Key to Fig. 1
(not all features are fitted to all models/markets)

1. Front fog lamps
2. Bulb failure
3. Sport mode
4. Rear fog guard
5. Left direction indicator
6. Headlamp main beam
7. Anti-lock braking system
8. Low coolant level
9. Low brake fluid/ABS low hydraulic pressure
10. Low oil pressure
11. Handbrake
12. Low fuel
13. Check engine
14. Ignition
15. Seat belt
16. Right direction indicator
17. Caravan DI indicator
18. Automatic transmission failure
19. Exhaust temperature
20. Low windscreen wash reservoir level
21. Oil pressure
22. Coolant temperature
23. Speedometer/odometer
24. Tachometer/trip meter
25. Trip reset
26. Fuel level
27. Battery condition

There is an array of 'secret-till-lit' warning lights which are situated in a row at the top of the instrument pack (Fig. 1).

INSTRUMENT PACK, TESTING

Fuel Gauge Tank Unit, Convertible, Calibration Limits

LEVEL	RESISTANCE (Ohms)	FUEL REMAINING (Litres)
Empty	240-250	6
Warning Light	185-215	12
Quarter Full	102-104	24.5
Half Full	68-70	43
Three Quarters Full	44-46	61.5
Full	16-18	80

Fuel Gauge Tank Unit, Coupe, Calibration Limits

LEVEL	RESISTANCE (Ohms)	FUEL REMAINING (Litres)
Empty	240-250	6
Warning Light	185-215	12
Quarter Full	102-104	26
Half Full	68-70	46
Three Quarters Full	44-46	66
Full	16-18	86

INSTRUMENT PACK CONNECTIONS (FIG. 2)

SOCKET A		SOCKET B	
PIN	CIRCUIT INPUT	PIN	CIRCUIT INPUT
01	Brake systems	01	Panel lamps (5 off), ground
02	Main beam	02	Panel lamps (5 off), positive
03	LH turn	03	Oil pressure gauge
04	Rear fog	04	Not used
05	Auto transmission sports mode	05	Supply for exhaust temperature/speed warning, oil pressure gauge, washer level
06	Bulb failure	06	Washer level
07	Front fog	07	Exhaust temperature (Japan)/Speed warning (Saudi Arabia)
08	Tachometer	08	Gear box fail
09	Fuel gauge	09	Caravan/trailer
10	Speedometer	10	RH turn
11	Check engine	11	Seat belt
12	Low fuel tell-tale	12	Ignition
13	Low fuel tell-tale	13	Supply for analogue instruments & A14 to A17, B11, B12
14	Park brake	14	Temperature gauge
15	Oil pressure tell-tale		
16	Brake systems		
17	Low coolant		
18	Anti-lock brakes		

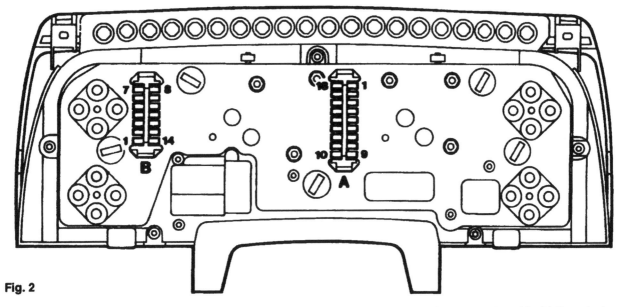

Fig. 2

TRIP COMPUTER, DESCRIPTION

When the ignition is switched ON, the LCD back lights illuminate and the trip computer defaults to the time of day. The time button is dual function; first press gives the time of day, a subsequent press causes the elapsed time since reset to be displayed, with leading zero suppressed. After five seconds the elapsed time reverts to the time of day.

The trip computer provides information on vehicle speed, fuel usage and distance travelled, all of which are calculated by a microprocessor. It computes fuel consumption, both average and 'at the moment' usage, fuel used on a journey or period; distance travelled, average speed and time elapsed since the start of the journey or over a period. The information may be displayed in either litres and kilometres or in miles and gallons.

The unit also provides a warning of fuel failure. In the event of an engine management fault occurring, the 'check engine' warning light and the words 'check engine' are permanently displayed on the trip computer until the engine is switched off. When the fault has been signalled, the likely area of malfunction can be indicated when the vehicle is stationary. Switch off the engine, wait at least five seconds turn the ignition switch to position II (do not start the engine). The relevant failure code FF11 to FF99 is displayed.

INSTRUMENT PANEL MODULE, RENEW

Disconnect the battery earth cable.
Position the tilt steering column to the lowest position.
Remove the instrument panel finishers' securing screws and remove the central and side finishers (1 Fig. 3).

493

Fig. 3

Remove the instrument panel to fascia securing screw cover plates and remove the screws securing the instrument panel to the fascia (2 Fig. 3).
Ease the instrument panel forward for access and disconnect the multi-plugs.
Remove the instrument panel.

Note: To minimise the risk of damage and contamination, all repairs conducted on the Instrument Pack should be performed in a non-static dust free environment.

Position the new instrument panel adjacent to the mounting position and connect the multi-plugs.
Fully seat the instrument panel.
Fit and tighten the instrument panel to the fascia and secure the screws.
Fit the securing screw finishers.
Position and fit the side and central finishers and secure the screws.
Reposition the steering column to the original position.
Reconnect the battery earth cable.

INSTRUMENT PRINTED CIRCUIT, RENEW

Disconnect the battery earth lead. Position the tilt steering column to the lowest position. Remove the instrument panel finishers' securing screws and remove the central and side finishers (1 Fig. 3). Remove the instrument panel to fascia securing screw cover plates and remove the screws securing the instrument panel to the fascia (2 Fig. 3).

Note: To minimise the risk of damage and contamination, all repairs conducted on the Instrument Pack should be performed in a non-static dust free environment.

Place a protective cover on a workbench and position the instrument panel down onto the cover.
Remove the panel illumination bulbs (1 Fig. 4).

Remove the printed circuit securing nut rubber covers.
Remove the printed circuit to instrument panel securing nuts and remove the printed circuit from the instrument panel (2 Fig. 4).
Carefully fit the new printed circuit to the instrument panel and secure the nuts.
Refit the printed circuit securing nut rubber covers.

Note: Ensure the printed circuits are not torn or deformed.

Fit the panel illumination bulbs.

Refit the instrument panel assembly to the vehicle.
Reconnect the battery earth lead.

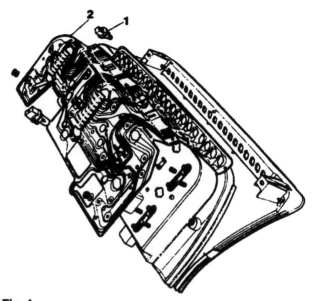

Fig. 4

SPEEDOMETER, RENEW

Disconnect the battery earth cable.
Position the tilt steering column to the lowest position.
Remove the instrument panel finishers' securing screws and remove the central and side finishers (1 Fig. 3).
Remove the instrument panel to fascia securing screw cover plates and remove the screws securing the instrument panel to the fascia (2 Fig. 3).

Note: To minimise the risk of damage and contamination, all repairs conducted on the Instrument Pack should be performed in a non-static dust free environment.

Place a protective cover on a workbench and position the instrument panel lens down onto the cover.
Remove the gauge illumination bulbs (1 Fig. 5).

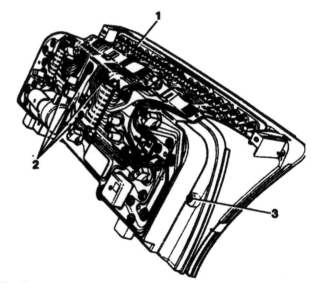

Fig. 5

Remove the printed circuit to instrument panel securing nuts (2 Fig. 5) and remove the printed circuit from the locating lugs.
Remove the screws securing the rear cover to the lens/veneer panel (3 Fig. 5).
Invert the assembly and remove complete with the printed circuit from the rear cover.
Remove the tachometer from the rear cover housing (1 Fig. 6).

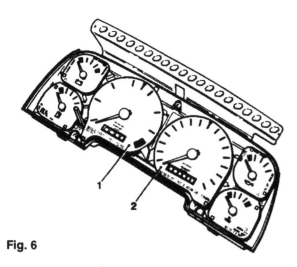

Fig. 6

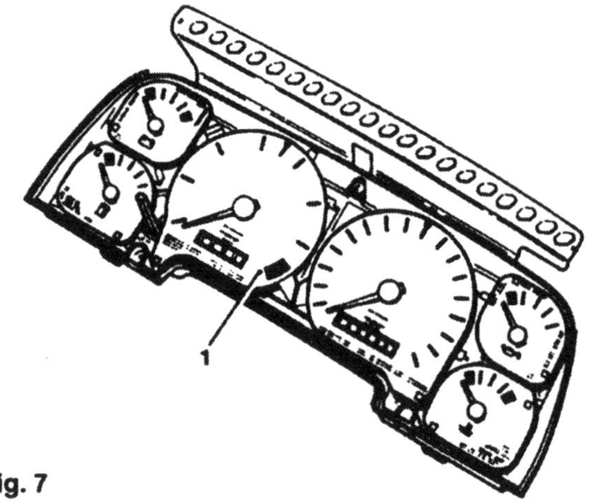

Fig. 7

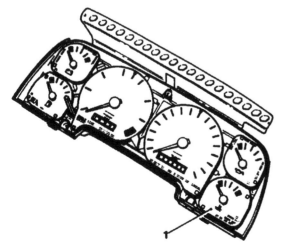

Fig. 8

Fig. 9

Remove the speedometer and odometer assembly from the rear cover housing (2 Fig. 6).
Disconnect the odometer to speedometer link lead and place the speedometer to one side.

Place the new speedometer to the front and connect the odometer to speedometer link lead connector plug.
Fit the speedometer and odometer assembly to the rear cover housing.
Fit the new tachometer to the rear cover housing.
Reposition the printed circuit around the rear cover and fit the lens/assembly to the rear cover housing.
Fit the rear cover to the lens/veneer panel and secure the screws.
Fit the printed circuit to the rear housing and secure the nuts.
Fit the gauge illumination bulbs.
Refit the instrument panel.
Reconnect the battery earth cable.

TACHOMETER, RENEW

The procedure is basically as for Speedometer. Refer to Fig. 7

BATTERY CONDITION INDICATOR, RENEW

The procedure is basically as for Speedometer. Refer to Fig. 8

COOLANT TEMPERATURE GAUGE, RENEW

The procedure is basically as for Speedometer. Refer to Fig. 9

FUEL GAUGE, RENEW

The procedure is basically as for Speedometer. Refer to Fig. 10

OIL PRESSURE GAUGE, RENEW

The procedure is basically as for Speedometer. Refer to Fig. 11

COOLANT TEMPERATURE TRANSMITTER, RENEW

WARNING:

DO NOT REMOVE THE CAP AT THE REMOTE
HEADER TANK UNLESS THE ENGINE IS COLD.

Open the bonnet. Disconnect the battery earth lead.

Remove the radiator pressure cap to depressurize the system.
Disconnect the transmitter feed wire.
Unscrew and remove the transmitter (1 Fig. 12).
Remove and discard the seal/washer.

Fit a new seal/washer to the replacement transmitter.
Fit and tighten the transmitter.
Reconnect the transmitter feed wire.
Check/top up the coolant. Refit the pressure cap.

Note: Always top-up with the recommended strength of antifreeze, never with water only.

Reconnect the battery earth lead.
Close the bonnet.

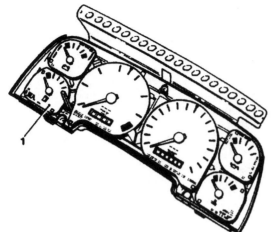

Fig. 10

Fig. 11

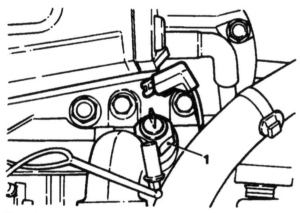

Fig. 12

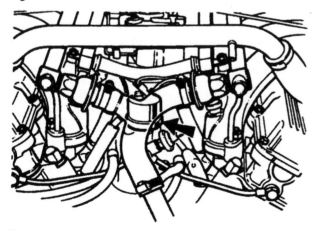

Fig. 13

Torque Figures

Transmitter to engine   49 - 54 Nm

Oils/Sealants/Lubricants

'JAGUAR UNIVERSAL' or a PHOSPHATE FREE type to
B.S. 6580 antifreeze. 50% down to -36°C (-33°f); 55%
down to -40°C (-40°F); 33% down to -19°C (-2.2°F).

OIL PRESSURE TRANSMITTER, RENEW

Open the bonnet.
Disconnect the battery earth lead.
Disconnect the rubber boot and the transmitter feed wire.
Unscrew and remove the transmitter (Fig 13).

Fit and tighten the new transmitter.
Reconnect the transmitter feed wire and the rubber boot.
Reconnect the battery earth lead.
Close the bonnet.

Torque Figures

Transmitter to engine   20 - 27 Nm

OIL PRESSURE WARNING LIGHT SWITCH, RENEW

Open the bonnet.
Disconnect the battery earth lead.
Disconnect the switch feed wire.
Unscrew and remove the switch (Fig 14).

Fit and tighten the new switch. Reconnect the switch feed
wire.
Reconnect the battery earth lead.
Close the bonnet.

Torque Figures

Switch to engine   20 - 27 Nm

FUEL GAUGE TANK UNIT, RENEW

> **WARNING:**
>
> FUEL IS HIGHLY FLAMMABLE AND GREAT CARE
> MUST BE TAKEN WHEN DRAINING THE FUEL TANK.
> NO SMOKING SIGNS MUST BE DISPLAYED NEAR
> THE WORKING AREA. DISCONNECT THE BATTERY
> LEADS BEFORE DRAINING THE TANK.
> KEEP NEARBY A CARBON DIOXIDE FIRE
> EXTINGUISHER AND DRY SAND TO SOAK
> UP ANY SPILLAGE.
> ENSURE THE AREA IS WELL VENTILATED.
> THE FUEL MUST BE DRAINED INTO AN
> AUTHORISED EXPLOSION PROOF CONTAINER.

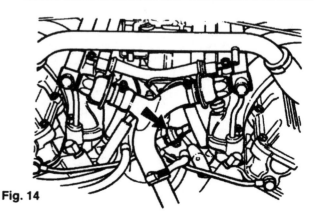

Fig. 14

Open the boot and disconnect the battery.
Drain the fuel from the tank using the approved equipment.
Remove the boot right and left hand side liners.
Remove the spare wheel trim cover and remove the spare wheel.
Remove the boot seal from the front body flange and remove the boot front liner.
Disconnect the harness to the tank unit wires.
Using the service tool 18G 1001 (Fig. 15), remove the tank unit securing ring then the tank unit assembly.

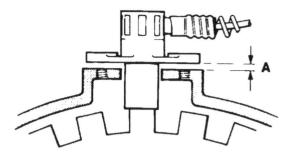

Fig. 17

backplate (without the 'O' ring), ensuring that the sensor bottoms onto the rotor tooth.
Measure and note the gap (A Fig. 17) between the sensor securing flange and the backplate.

Remove the sensor.
Fit a new 'O' ring.
Fit and align the correct size/number of shims to the sensor to give 0.010 - 0.020 in. clearance between the sensor and the rotor teeth.
Fit the sensor assembly to the backplate.
Fit and tighten the securing bolts.
Lower the ramp.

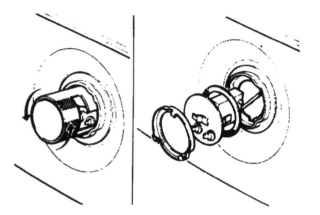

Fig. 15

Remove and discard the 'O' ring seal and clean the tank seal face.

Fit a new seal to the tank and carefully fit the tank unit, secure with the retaining ring.
Reconnect the harness to the tank unit wires.
Fit the front liner and reposition the trim over the front flange. Fully seat the boot seal to the flange. Refit the spare wheel, secure the nut and replace the wheel cover.
Fit the left and right hand front side liners.
Refill the fuel tank; check for leaks.
Close the boot.

Service Tools

18G 1001 Locking ring spanner

SPEED SENSOR AIR GAP, CHECK/ADJUST

Drive the vehicle on to a ramp. Raise the ramp.
Remove the sensor (Fig. 16) from the final drive backplate.

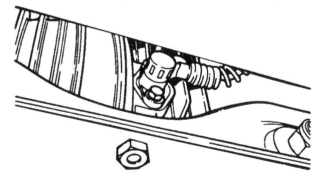

Fig. 16

Displace and remove the shims from the sensor.
Remove and discard the 'O' ring seal.
Clean the backplate face.

Position the car on the ramp to align one rotor tooth directly in front of the sensor mounting hole. Fit the sensor to the

SPEED SENSOR, RENEW

Drive the vehicle on to a ramp. Open the boot.
Remove the spare wheel for access.
Remove the LH boot liner.
Disconnect the speed sensor multi plug, located under the LH rear wing (black PMHD).
Displace the harness and body grommet.
Reposition the harness from the boot into the axle area.
Raise the ramp.
Remove the sensor from the final drive backplate (Fig. 16).
Displace and remove the sensor/harness assembly.
Displace and remove the shims from the sensor. Clean the backplate face.

Remove the protection cap from the new sensor. Displace and remove the 'O' ring.
Position the car on the ramp to align one rotor tooth directly in front of the sensor mounting hole.
Fit the sensor to the backplate, ensuring that the sensor bottoms onto the rotor tooth.
Measure and note the gap (A Fig. 17) between the sensor securing flange and the backplate.
Remove the sensor. Fit a new O ring.
Fit and align the correct size/number of shims to the sensor to give 0.010 - 0.020 in. clearance between the sensor and the rotor teeth.

Fit the sensor assembly to the backplate.
Fit and tighten the securing bolts.
Reposition the harness into the boot area.
Seat the body grommet.
Lower the ramp.
Position the harness and reconnect the multi plug.
Fit the liner and spare wheel.
Close the boot.

TRIP COMPUTER, RENEW

Disconnect the battery earth lead.
Remove the combined interior light switch.
Remove the combined heated backlight and hazard warning switch.
Remove the computer to fascia securing bolts (1 Fig. 18) and carefully displace the computer from the veneer panel.

Fig. 18

Disconnect the computer from the harness multi-plug, then connect the new computer to the multi-plug.

Fit the new computer to the veneer panel and tighten the fascia securing bolts.
Refit the combined heated backlight and hazard warning switch assembly.
Refit the combined interior light switch assembly.
Reconnect the battery earth lead.

CIRCUITS - 1990 MODEL YEAR ON

WIRING COLOUR CODE

N	BROWN	Y	YELLOW
B	BLACK	O	ORANGE
W	WHITE	S	SLATE
K	PINK	L	LIGHT
G	GREEN	U	BLUE
R	RED	P	PURPLE

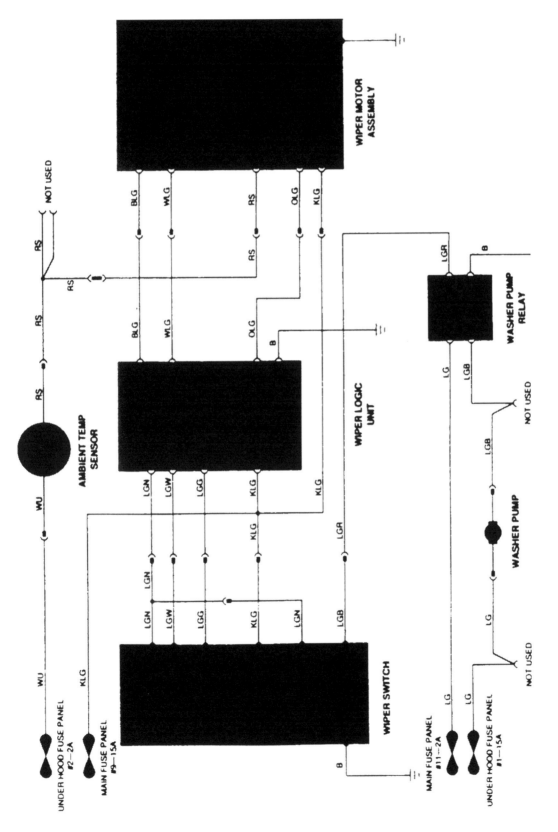

WINSHIELD WIPER AND WASHERS

499

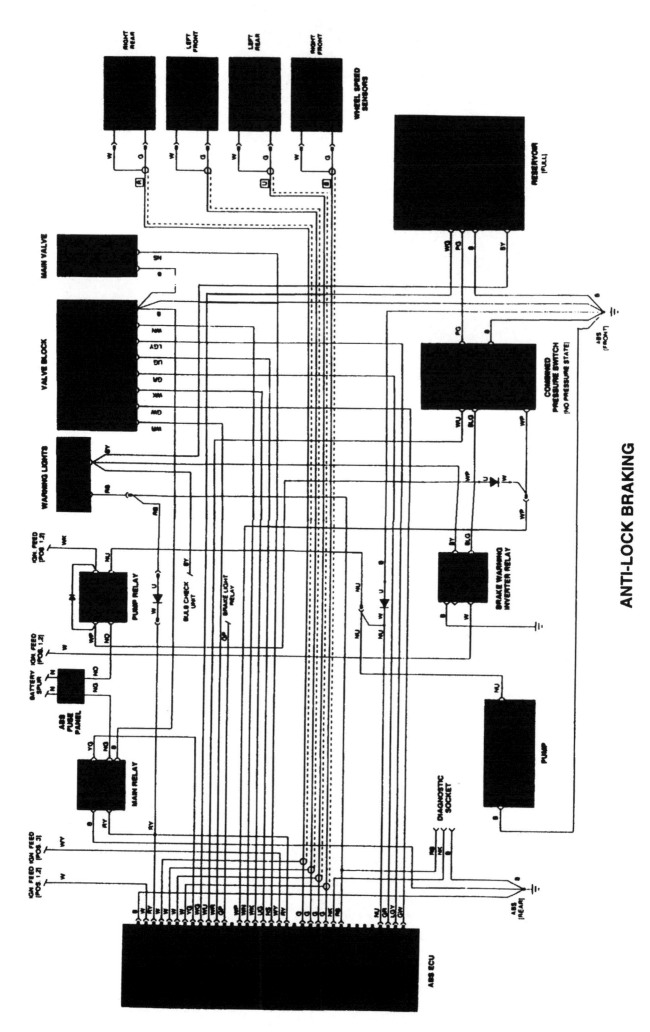

ANTI-LOCK BRAKING

500

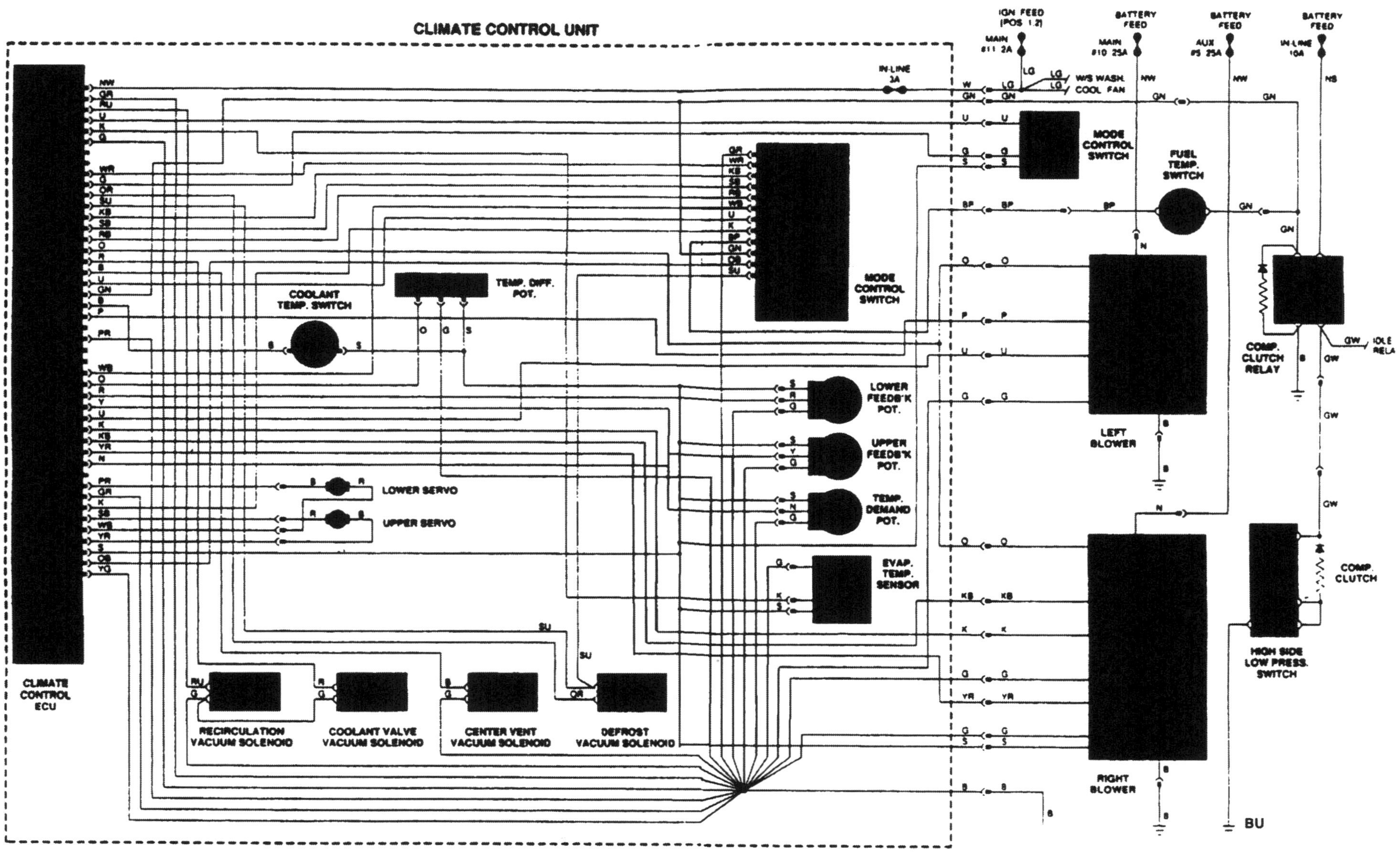

CLIMATE CONTROL SYSTEM

CLIMATE CONTROL UNIT

501

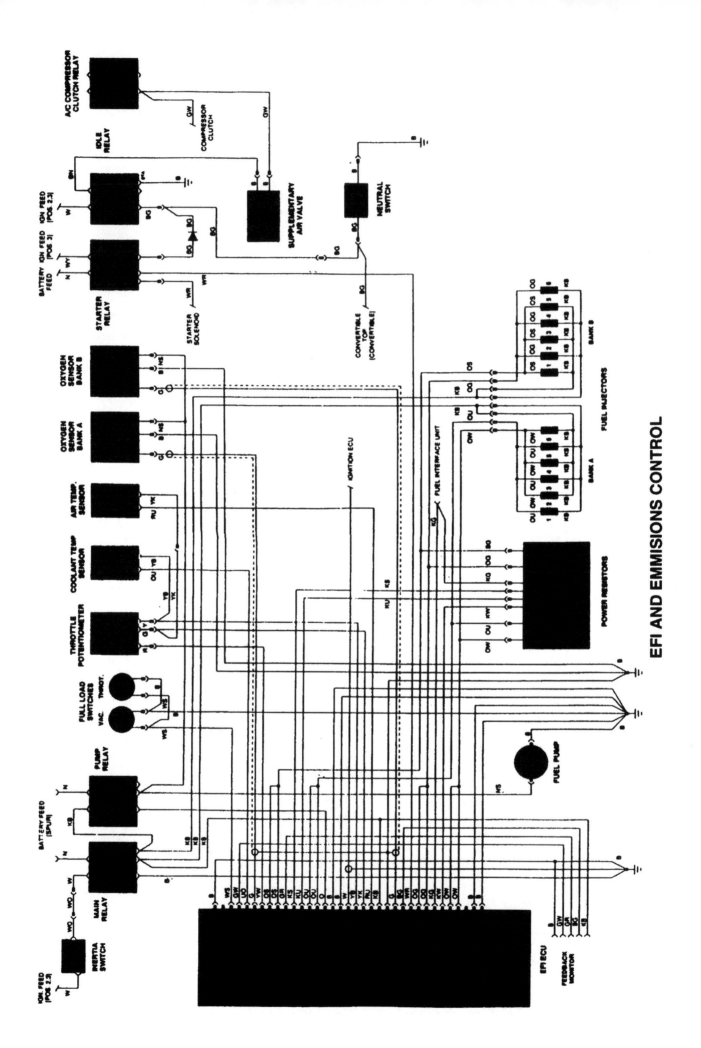

EFI AND EMMISIONS CONTROL

CIRCUITS - 1992 MODEL YEAR ON

Please note, that due to the ever increasing sophistication of these vehicles, it has been necessary to cover Electrical Circuits from 1992 on in a separate publication. A limited selection of circuits are printed here, but full coverage will only be available by reference to the 1992 Model Year-on Service Manual (Part Number JJM 10 04 06/02) and the relevant Model Year's 'Electrical Guide'.

CIRCUIT DIAGRAMS AND HARNESSES

ENGINE MANAGEMENT/PI – HIGH POWER BOARD (4.0 LITRE)

KEY TO CIRCUIT COMPONENTS

1. Inertia switch LB212
2. PI main relay LI95
3. Connector LI92 (part) to High Power ECU board
4. High Power ECU board
5. Connector LI92 (part) to High Power ECU board
6. Ignition amplifier LI113
7. Fuel pump relay RH173
8. Battery spur RH172
9. Idle speed motor LI100
10. Canister purge valve (CAT only) LI142
11. Ignition coil LI114
12. Distributor
13. Resistor PIR1
14. Fuel pump and RFI module RH104
15. Injector LI116
16. Injector LI117
17. Injector LI118
18. Injector LI119
19. Injector LI120
20. Injector LI121
21. Lambda (oxygen) heater LI122

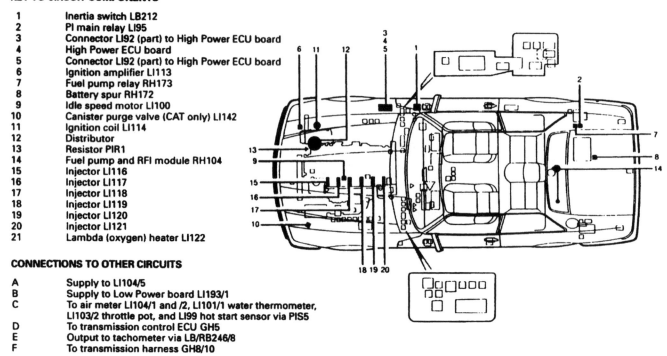

CONNECTIONS TO OTHER CIRCUITS

A. Supply to LI104/5
B. Supply to Low Power board LI193/1
C. To air meter LI104/1 and /2, LI101/1 water thermometer, LI103/2 throttle pot, and LI99 hot start sensor via PIS5
D. To transmission control ECU GH5
E. Output to tachometer via LB/RB246/8
F. To transmission harness GH8/10

November 1995 ISSUE 2

ENGINE MANAGEMENT/PI – HIGH POWER BOARD (4.0 LITRE)

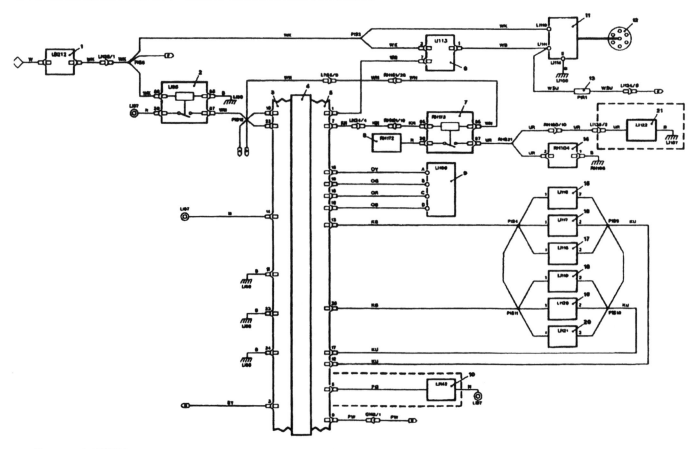

November 1995 ISSUE 2

90.5.7 – 2

ENGINE MANAGEMENT/PI – LOW POWER BOARD (4.0 LITRE)

KEY TO CIRCUIT COMPONENTS

1	Air meter LI104
2	Water thermostat LI101
3	Hot start sensor LI99
4	Throttle pot LI103
5	Crankshaft sensor LI109
6	Lambda sensor (CAT only) LI123
7	Connector LI93 (part) to low power ECM board
8	Low power ECM board
9	Connector LI93 (part) to low power ECM board
10	Breather heater relay (CAT only) LI131
11	Serial output connector LI132
12	Fuel failure code reset LI91
13	Breather heater (CAT only) LI98

CONNECTIONS TO OTHER CIRCUITS

A	Supply from high power board LI192/3
B	From electronic gearbox circuit
C	From electronic gearbox circuit
D	From electronic gearbox circuit
E	From ambient sensor via BHS39
F	Supply from PIS12 (high power board circuit)
G	To electronic gearbox circuit
H	To binnacle BN5/20
I	To air conditioning circuit
J	To instrument pack LB246/8
K	To BHS28
L	To BHS45
M	To electronic transmission GH5
N	To gear selector GH7/6
P	To bulkhead harness GH9/6

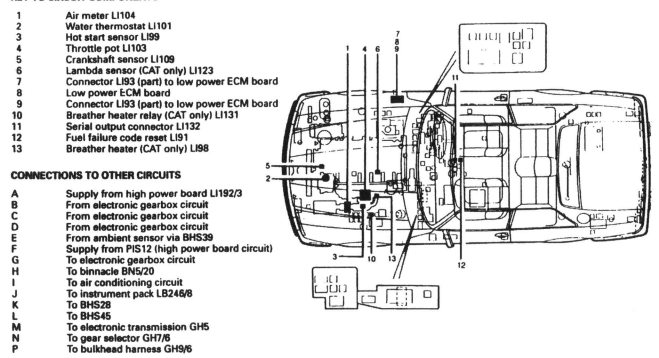

November 1996 ISSUE 2

ENGINE MANAGEMENT/PI – LOW POWER BOARD (4.0 LITRE)

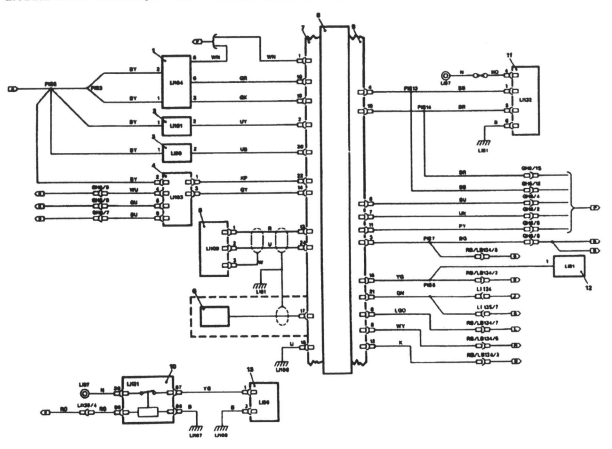

November 1995 ISSUE 2   90.5.7 – 1

CRUISE CONTROL

KEY TO CIRCUIT COMPONENTS

1	Ignition switch LB/RB202
2	Fuse 2 (RH fusebox – LH drive/ LH fusebox – RH drive)
3	Brake pedal switches LB/RB167
4	Stop light relay RH113
5	Cruise control inhibit switch (automatic transmission only) or Link (manual transmission only)
6	Electronic control unit CC2
7	Speed interface unit RH114
8	Fuse 16 (LH fusebox – LH drive/ RH fusebox RH drive)
9	PI harness connector LI134
10	Speed warning circuit/exhaust temperature connector LB/RB211 (Saudi Arabia) / XT7 (Japan)
11	Cruise control pump LF16
12	Speed control dump valve LF13
13	Cruise control switch CC1
14	Lighting switch LL30
15	Clutch switch (left forward harness–LH drive only)
16	Clutch switch (right forward harness–manual transmission and RH drive vehicles only)

CONNECTIONS TO OTHER CIRCUITS

A	To instrument pack LB/RB246/10
B	To trip computer BN3/4
C	To reversing light switch (manual gearbox only)
D	To trip computer BN3/6

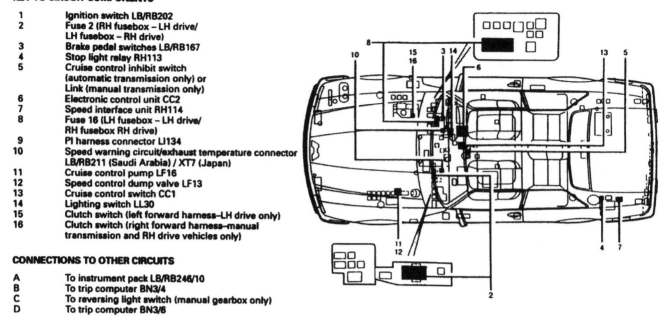

November 1995 ISSUE 2

CRUISE CONTROL

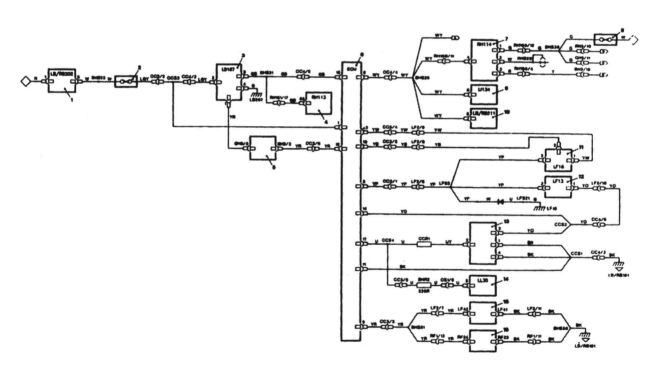

November 1995 ISSUE 2

90.5.9 – 1

CIRCUIT DIAGRAMS AND HARNESSES

ANTI–LOCK BRAKE SYSTEM

KEY TO CIRCUIT COMPONENTS

1   Fuse 11 (RH fusebox – LH drive/
    LH fusebox – RH drive)
2   ABS main relay (ABS16)
3   Diagnostic connector (ABS2)
4   Electronic control unit – ECU (ABS1)
5   Main valve (ABS9)
6   Valve block (ABS8)
7   Fuse 22 (RH fusebox – LH drive/
    LH fusebox – RH drive)
8   ABS pump relay (LB–RB144)
9   ABS pump motor (ABS14)
10 Ignition switch (LB–RB202)
11 Battery post LB–RB192)
12 Brake warning lamp (instrument pack
    LB–RB246/16)
13 ABS reservoir (ABS7)
14 ABS accumulator (ABS15)
15 LH front brake pad sensor ABS6
16 RH front brake pad sensor ABS11
17 LH rear brake pad sensor ABS5
18 RH rear brake pad sensor ABS4

CONNECTIONS TO OTHER CIRCUITS

A   To bulb check unit LB174
B   To instrument pack (brake light) LB246/1
C   To instrument pack (anti lock failure lamp) LB247/18
D   To stop lights circuit
E   To bulb fail module

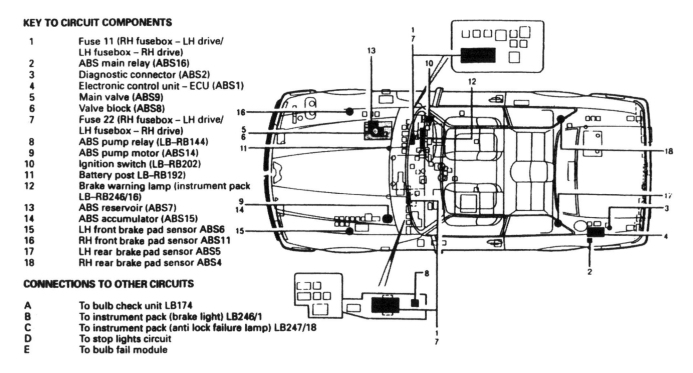

November 1995 ISSUE 2

ANTI-LOCK BRAKE SYSTEM

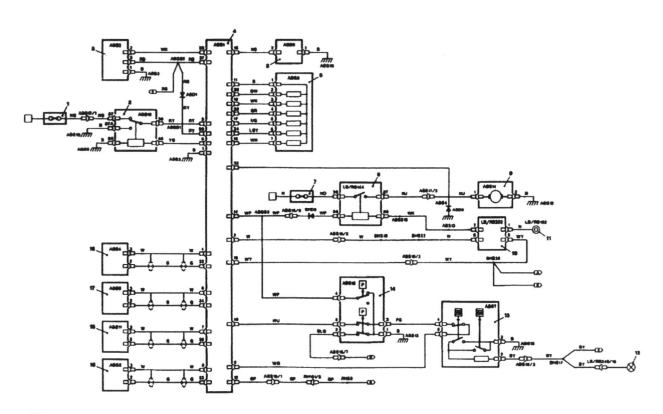

J91 370

November 1995 ISSUE 2   90.5.10 – 1

507

CIRCUIT DIAGRAMS AND HARNESSES

EXTERNAL LIGHTING — SIDE, TAIL, STOP AND REVERSE LAMPS

KEY TO CIRCUIT COMPONENTS

1	Side light relay LF10
2	Fuse 8 – forward fusebox
3	Dim dip relay DM2
4	Relay DR5
5	Fuse 18 (LH fusebox – LH drive / RH fusebox – RH drive)
6	Lights on warning device LB286
7	RH front bulb failure sensor LB124, LB125 and LB126
8	Fuse 14 – RH fusebox
9	RH front side light LF22
10	RH front side marker LF26
11	LH front bulb failure sensor LB120, LB121 and LB122
12	Fuse 13 – RH fusebox
13	LH front side light LF29
14	LH front side marker LF30
15	Lighting switch CS2
16	LH tail and number plate bulb failure sensor (RH120, RH121 and RH122)
16.1	RH tail and number plate bulb failure sensor (RH98, RH99 and RH100)
17	LH tail light RH119
18	LH rear side marker RH116
19	RH number plate light
20	RH tail light RH96
21	RH rear side marker RH95
22	LH number plate light
23	Stop lamp failure warning switch RH129
24	Handbrake switch RH133, RH134 and RH135
25	RH stop light RH96
26	LH stop light RH119
27	Fuse 19 RH fusebox – LH drive / fuse 18 LH fusebox – RH drive
28	Stop light relay RH113
29	High level stop lights HL2
30	Brake pedal and stop light switches LB167
31	Reversing lights (automatic transmission)
31.1	RH luggage compartment fuse 3R, (RH97)
31.2	Reversing lights relay GH3
31.3	Transmission control module (TCM) – GH5
31.4	Reversing light switch GH7
31.5	LH reverse light
31.6	RH reverse light
32	Reversing lights (manual gearbox)
32.1	Fuse 16 (RH fusebox – RHD, LH fusebox – LHD)
32.2	Reversing light switch GH10
32.3	LH reverse light
32.4	RH reverse light

CONNECTIONS TO OTHER CIRCUITS

A	To dimmer module RB/LB170/5
B	To trip computer BN3/4
C	To interior lighting circuits
D	To security system
E	To dimmer module RB/LB170/7
F	To headlight relay LF9, wiper delay module and front foglights shorting plug
G	To cruise control CC3/6
H	To caravan / trailer connector RH117/2
J	To caravan / trailer connector RH117/5
K	To instrument pack LB246/14 and bulb check unit LB
L	To caravan / trailer connector RH117/7
M	To ABS connector AB18/1
N	To cruise control CC4/5
P	To cruise control inhibit switch via GH9/5
Q	To cruise control CC4/2
R	To gearbox power relay GH14/87
S	To mode switch GH6/6
T	To drivers seat ECM via DS1/10
U	To trip computer via BN3/4
V	To speed interface unit RH114 via RH160/12
W	To drivers seat ECM via DS1/10

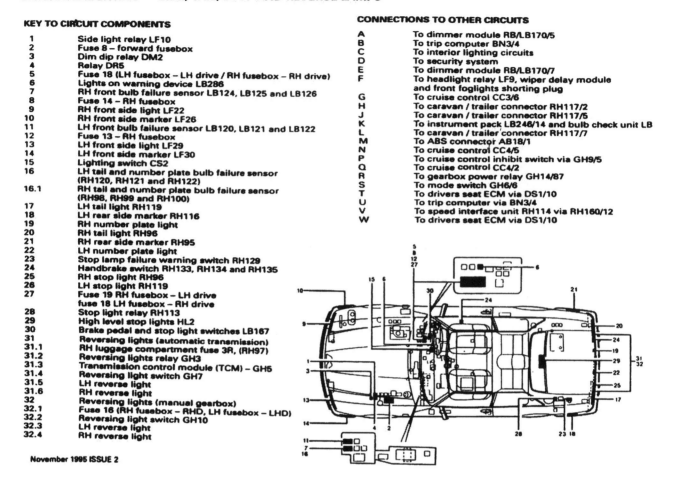

November 1995 ISSUE 2

EXTERNAL LIGHTING — SIDE, TAIL, STOP AND REVERSE LAMPS — Continued

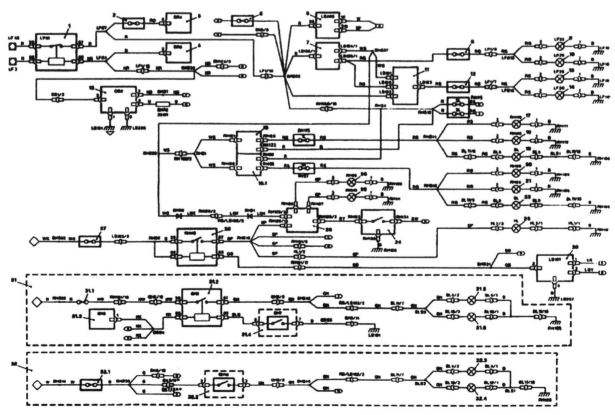

FOGLIGHTS (LHD — USA AND CANADA)

KEY TO CIRCUIT COMPONENTS

1	Battery post LB191
2	Fuse 20 (LH fusebox)
3	Front foglight relay LB259
4	Front foglights ON indicator
5	Front LH foglight LF33
6	Front RH foglight LF25
7	Front and rear foglight switches
7.1	Front foglight switch
7.2	Rear foglight switch
7.3	Foglights ON indicators
7.4	Foglights switch panel lamps
8	Shorting plug LB189 (USA / Canada configuration)
9	Lighting switch LL30
10	Battery post LF46
11	Headlights relay LF9
12	Battery post LB287
13	Fuse 17 (LH fusebox)
14	Rear foglights relay LB260
15	Rear foglights ON indicator
16	Caravan connector link (pins 9 and 10)
17	Rear RH foglight RH96
18	Rear LH foglight RH119

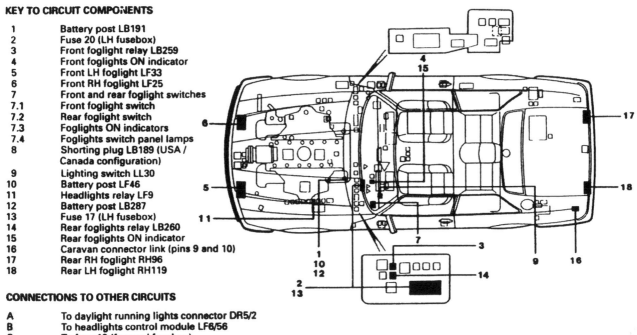

CONNECTIONS TO OTHER CIRCUITS

A	To daylight running lights connector DR5/2
B	To headlights control module LF6/56
C	To fuse 12 (forward fusebox)
D	To wiper delay module LB148/12
E	To daylight running lights connector DR4/2
F	To headlights control module LF6/56b
G	To fuse 10 (forward fusebox)
H	To fuse 9 (forward fusebox)
J	To illumination control module LB170/6
K	To radio connector LB160/11
L	To daylight running lights connector DR4/3

November 1995 ISSUE 2

FOGLIGHTS (LHD — USA AND CANADA)

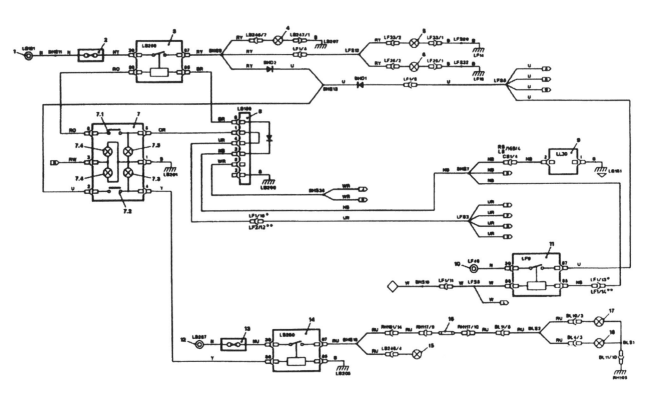

FOGLIGHTS (RHD)

KEY TO CIRCUIT COMPONENTS

1	Battery post RB192
2	Fuse 20 (RH fusebox)
3	Front foglight relay RB259
4	Front LH foglight LF33
5	Front RH foglight LF25
6	Front foglights ON indicator
7	Front and rear foglight switches
7.1	Front foglight switch
7.2	Rear foglight switch
7.3	Foglights ON indicators
7.4	Foglights switch panel lamps
8	Battery post RB191
9	Fuse 17 (RH fusebox)
10	Rear foglights relay RB260
11	Caravan connector link (pins 9 and 10)
12	Rear RH foglight RH96
13	Rear LH foglight RH119
14	Rear foglights ON indicator

CONNECTIONS TO OTHER CIRCUITS

A	To radio connector RB160/11
B	To illumination control module RB170/6
C	To headlamp relay and headlamp control module
D	To security system connector SS2/9
E	From lighting supply

November 1995 ISSUE 2

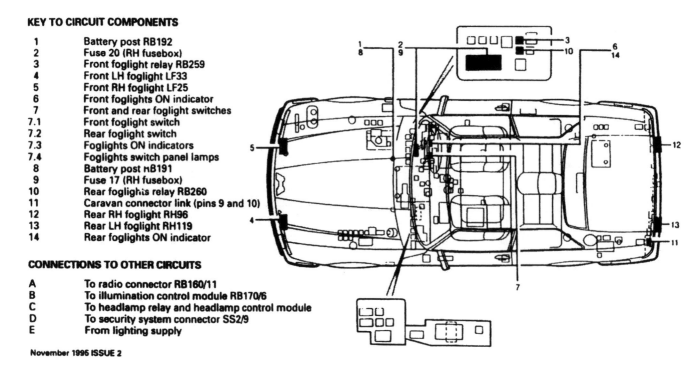

FOGLIGHTS (RHD)

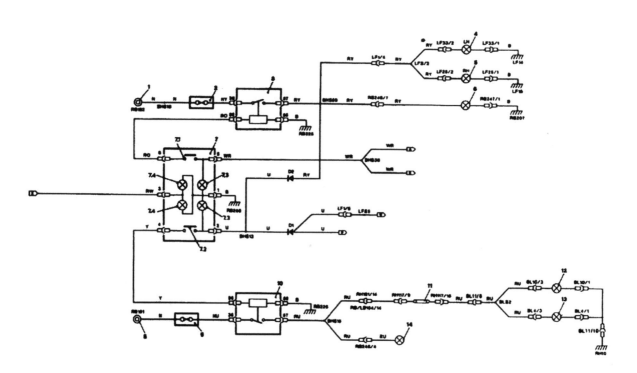

November 1995 ISSUE 2

90.5.13 – 3

CIRCUIT DIAGRAMS AND HARNESSES

FOGLIGHTS (LHD — NOT USA OR CANADA)

KEY TO CIRCUIT COMPONENTS

1	Battery post LB191
2	Fuse 20 (LH fusebox)
3	Front foglight relay LB259
4	Front foglights ON indicator
5	LH front foglight LF33
6	RH front foglight LF25
7	Front and rear foglight switches
7.1	Front foglight switch
7.2	Rear foglight switch
7.3	Foglights ON indicators
7.4	Foglights switch panel lamps
8	Shorting plug LB189
	(ROW – Rest of World – configuration)
9	Battery post LB287
10	Fuse 17 (LH fusebox)
11	Rear foglights relay LB260
12	Caravan connector link (pins 9 and 10)
13	Rear foglights ON indicator
14	Rear RH foglight RH96
15	Rear LH foglight RH119

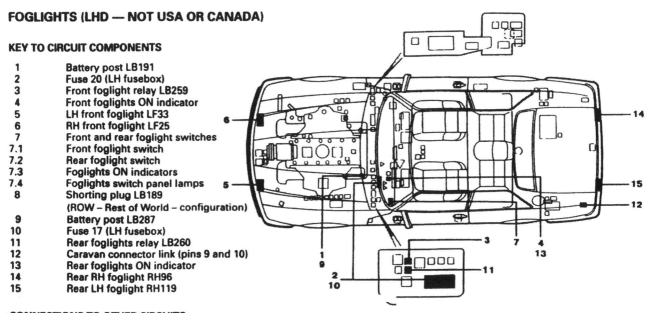

CONNECTIONS TO OTHER CIRCUITS

A	To headlight control module LF6/56
B	To fuse 12 (forward fusebox)
C	To Canada davlight running lamps connector DR5/2
D	To headlight relay LF9/87
E	From illumination control module LB170/6
F	From interior lighting circuit

November 1995 ISSUE 2

FOGLIGHTS (LHD — NOT USA OR CANADA)

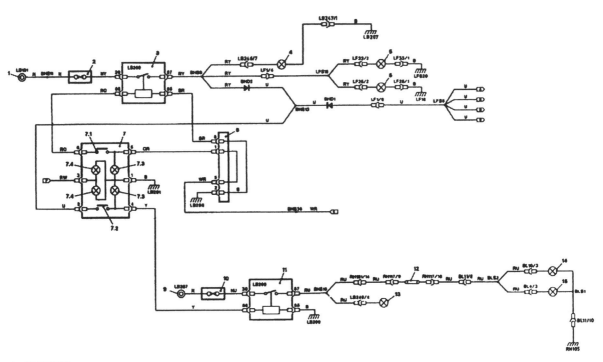

PVI 375

CIRCUIT DIAGRAMS AND HARNESSES

HEADLAMPS — DIM/DIP

KEY TO CIRCUIT COMPONENTS

1	Fuse 8 (forward fusebox)
2	Dim/dip relay DM2
3	Dim/dip resistor DM1
4	Battery post LF46
5	Headlamp relay LF9
6	Lighting switch LL30
7	Dip switch CS2
8	Headlamp control module LF6
9	Battery post LF3
10	Fuse 9 (forward fusebox)
11	LH dipped headlamp LF29
12	Fuse 5 (forward fusebox)
13	LH main beam LF29
14	Fuse 10 (forward fusebox)
15	RH dipped headlamp LF22
16	Fuse 6 (forward fusebox)
17	RH main beam LF22

CONNECTIONS TO OTHER CIRCUITS

A	To sidelamp relay LF10
B	To bulb failure sensors, trip computer and illumination control module
C	To daylight running lights, fuse 12 – forward fusebox, and front / rear foglight switch via LF1/6
D	To illumination control module LB170/7,
E	To front foglight shorting plug LB189/4 – LHD only
F	To windscreen wiper delay module LB148/12
G	To front fog shorting plug LB189/3

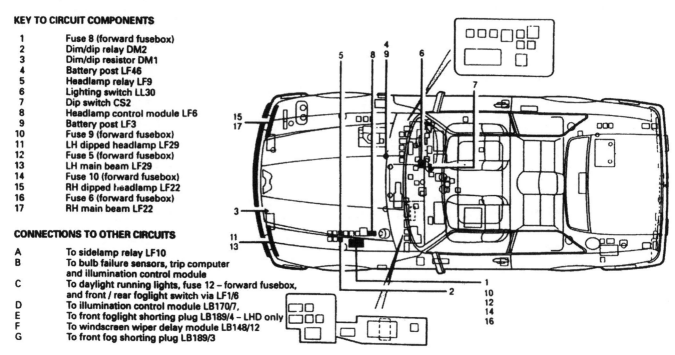

November 1995 ISSUE 2

HEADLAMPS — DIM/DIP

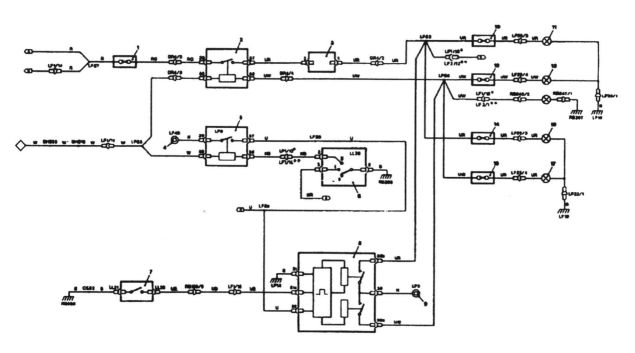

November 1995 ISSUE 2

90.5.12 – 3

CIRCUIT DIAGRAMS AND HARNESSES

HEADLAMPS — DAYLIGHT RUNNING AND HEADLAMP LEVELLING

KEY TO CIRCUIT COMPONENTS

1	Lighting switch LL30
2	Battery post LF46
3	Daylight running circuit (Canada only)
3.1	Relay DR1 (ignition on)
3.2	Relay DR3 (dipped beam)
3.3	Relay DR2 (main beam)
4	Fuse 8 (forward fusebox)
5	Lights on warning LB286
6	Headlamp relay LF9
7	Headlamp control module LF6
7.1	Pulse width modulator
7.2	Dipped beam relay
7.3	Main beam relay
8	Battery post LF3
9	Headlamp change–over (dip) switch
10	Front/rear fog lamp switches LB176
11	Headlamp levelling circuit (Germany only)
11.1	Fuse 12 (forward fusebox)
11.2	Headlamp levelling switch LL13
11.3	Headlamp levelling motor LF28 (LH)
11.4	Headlamp levelling motor LF23 (RH)
11.5	Headlamp levelling diode module
12	Fuse 9 (forward fusebox)
13	LH dipped beam
14	Fuse 5 (forward fusebox)
15	LH main beam
16	Fuse 10 (forward fusebox)
17	RH dipped beam
18	Fuse 6 (forward fusebox)
19	RH main beam
20	Main beam indicator (instrument pack)

CONNECTIONS TO OTHER CIRCUITS

A	To security system RH174/5
B	To illumination control module LB170/7
C	To sidelight relay LF10/85 and DR5/3
D	To sidelight relay LF10
E	To bulb fail sensors, trip computer, internal lighting, and fuse 18 (LH fuse box – RHD, RH fusebox – LHD)
F	To internal lights and security system
G	To front fog shorting plug
H	To wiper delay module
J	To front fog shorting plug
K	To ignition switched loads
L	To LH foglamp LF33
M	To RH foglamp LF25
N	To front foglamp relay LB259/87
P	To instrument pack LB246/7

HEADLAMPS — DAYLIGHT RUNNING AND HEADLAMP LEVELLING — Continued

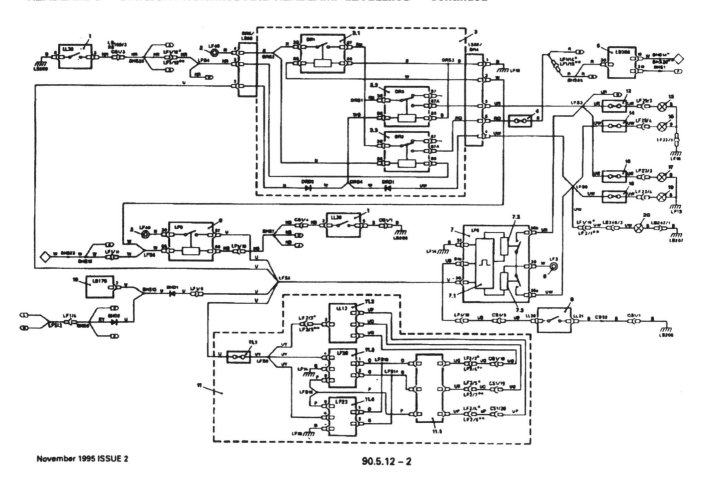

November 1995 ISSUE 2

90.5.12 – 2

INTERNAL LIGHTING

KEY TO CIRCUIT COMPONENTS

1 Battery post LB/RB145
2 Fuse 4 (RH fusebox – LH drive, LH fusebox – RH drive)
3 Driver's courtesy light
4 Driver's door lamp
5 Roof light
6 Pasenger's door lamp
7 Passenger's map light
8 Interior light LH
9 Interior light RH
10 Interior lighting switch BN4
11 Interior lights delay unit LB/RB143
12 Key switch LB/RB203
13 Door switch LB/RB188
14 Driver's seat control unit DS2
15 Lights ON warning LB/RB286
16 Passenger's seat control unit
17 Door switch LB/RB140
18 Seat belt logic unit LB/RB130
19 Battery post LB/RB191
20 Fuse 11 (LH fusebox – LH drive, RH fusebox – RH drive)
21 Luggage compartment lights
22 Luggage compartment lights switch

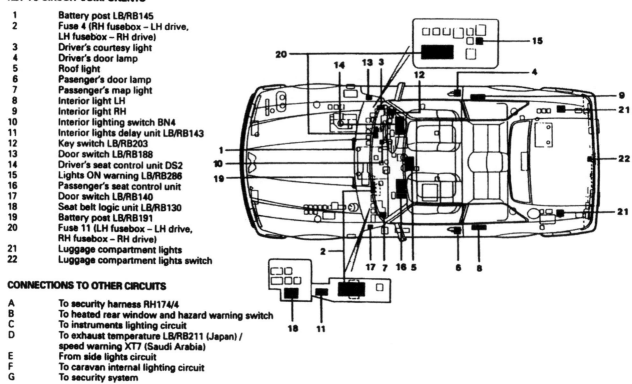

CONNECTIONS TO OTHER CIRCUITS

A To security harness RH174/4
B To heated rear window and hazard warning switch
C To instruments lighting circuit
D To exhaust temperature LB/RB211 (Japan) / speed warning XT7 (Saudi Arabia)
E From side lights circuit
F To caravan internal lighting circuit
G To security system

November 1995 ISSUE 2

INTERNAL LIGHTING

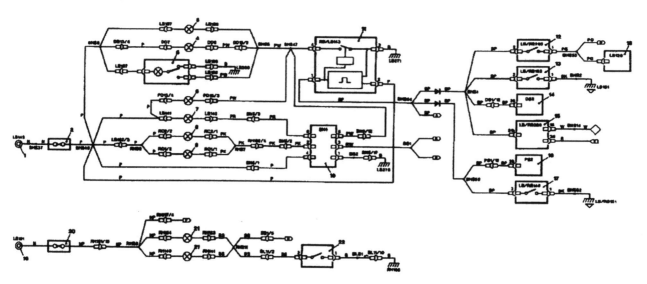

November 1995 ISSUE 2

90.5.15 – 1

CIRCUIT DIAGRAMS AND HARNESSES

DIRECTION AND HAZARD WARNING LAMPS

KEY TO CIRCUIT COMPONENTS

1	Battery post LB / RB191
2	Fuse 19 (LH fusebox–LHD / RH fusebox RHD)
3	Flasher unit LB / RB172
4	Trailer / caravan indicator (instrument pack)
5	Fuse 5 (RH fusebox–LHD / LH fusebox RHD)
6	Hazard warning switch
7	Direction indicator switch
7.1	Turn left
7.2	Turn right
8	Rear LH direction indicator (vehicle)
9	Rear LH direction indicator (caravan)
10	Front LH direction indicator
11	Front LH direction indicator repeater
12	Turn left indicator (instrument pack)
13	Front RH direction indicator
14	Front RH direction indicator repeater
15	Rear RH direction indicator (vehicle)
16	Rear RH direction indicator (caravan)
17	Turn right indicator (instrument pack)

CONNECTIONS TO OTHER CIRCUITS

A	To seat belt logic unit
B	To internal lighting circuit

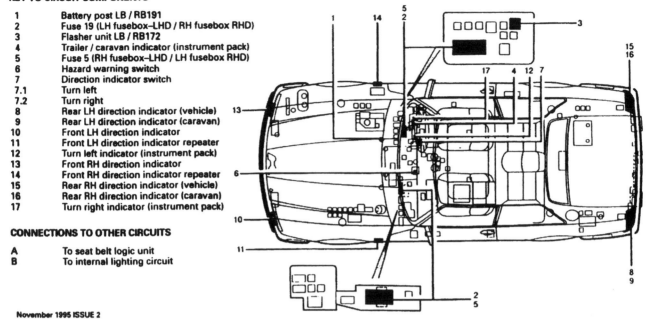

November 1995 ISSUE 2

DIRECTION AND HAZARD WARNING LAMPS

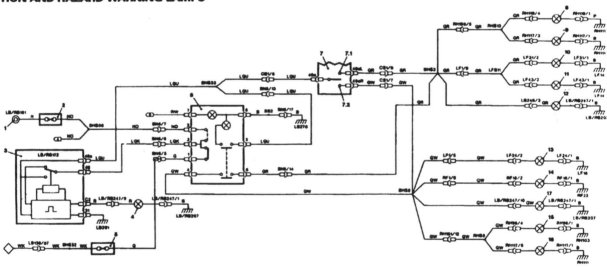

November 1995 ISSUE 2

90.5.14 – 1

J91 184

ECU and Relays Location

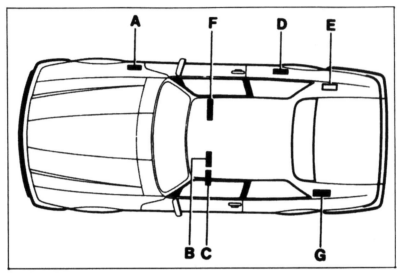

A. Engine management
   ECU (4.0L)
   Marelli ignition ECU (V12)
B. Seat drive non-memory
   ECU
C. Speed control ECU
D. Gearbox ECU
   (4.0L Automatic)

E. Fuel injection ECU
   (V12 only)
F. Seat memory ECU
   (handed – always on)
   driver's side
G. ABS ECU

ECU locations

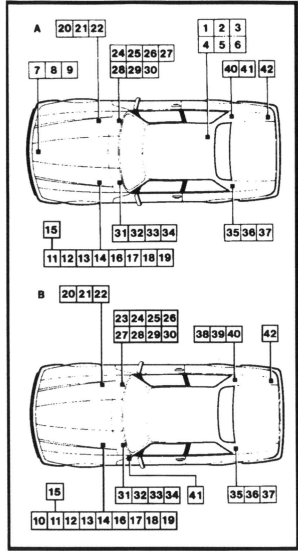

A – V12 RHD   B – 4 litre RHD

Relay Locations

Component	Case-Base Colour
1. Hood up relay	Violet-Green
2. Hood down relay	Violet-Blue
3. Quarters up RH relay	Violet-Red
4. Quarters down RH relay	Violet-Black
5. Quarters up LH relay	Violet-Yellow
6. Quarters down LH relay	Violet-Natural
7. Extra air valve 1 relay (V12 non-catalyst)	Violet-Blue
8. Extra air valve 2 relay (V12 non-catalyst)	Blue-Black
9. Anti-stall relay (V12)	Violet-Blue
10. Breather heater relay (4.0L catalyst)	Blue-Black
11. Daylight running 1 relay	Blue-Blue
12. Daylight running 2 relay	Violet-Violet
13. Daylight running 3 relay	Violet-Violet
14. Radiator fan relay	Blue-Blue
15. Dim-dip control relay	Blue-Blue
16. Sidelights relay	Blue-Blue
17. Headlight relay	Blue-Blue
18. Fan run on diode unit	Blue-Natural
19. Headlamp levelling diode module	Red-Black with red top
20. Air conditioning relay	Blue-Green
21. Horn relay	Blue-Black
22. Starter solenoid relay	Blue-Natural
23. Gearbox warning relay (4.0L)	Violet-Black
24. Ignition continuity load relay	Silver-Natural
25. Front fog lamp relay	Blue-Violet
26. Rear fog lamp relay	Blue-Brown
27. Lights-on warning unit	Black-Natural
28. Dimmer module	Red-Black with red top
29. Bulb check unit	Black-Green
30. Heated rear screen timer	Yellow-Blue
31. ABS pump relay	Yellow-Yellow
32. Auxiliary continuity load relay	Silver-Black
33. Seat heater relay	Silver-Natural
34. Interior light delay unit	Red-Red
35. ABS main relay	White-Black
36. Stop lamp relay	Blue-Yellow
37. Anti-slosh module	Black-Black
38. Reverse lights relay (4.0L)	Blue-Black
39. Gearbox power relay (4.0L)	Blue-Blue
40. Fuel pump relay	Silver-Yellow-Black
41. Main PI relay	Silver with red stripe-Red
42. Radio aerial delay relay	Green-Black

Note: Items 1 to 6 are for the convertible only. Items 11 to 13 are for Canadian market vehicles, item 15 is for UK market vehicles, item 19 is for German market vehicles. Either item 11 or item 15 occupies the one location depending on market requirement.

Fuses

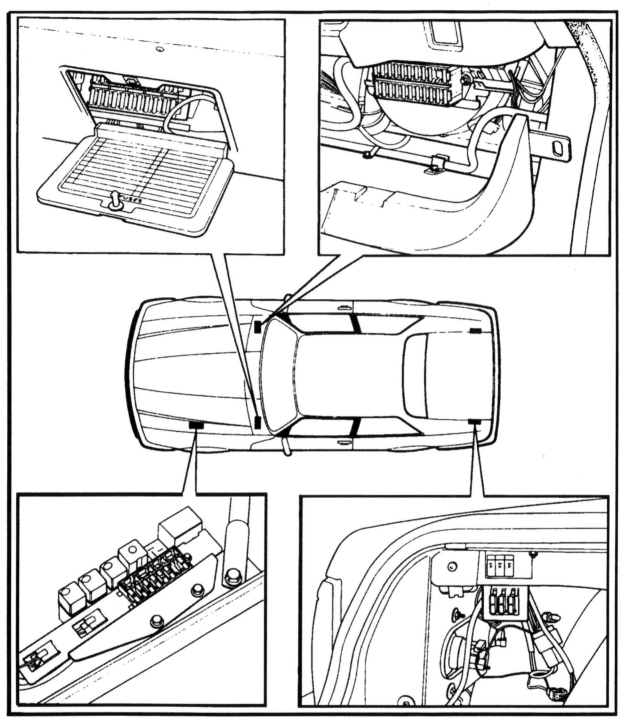

Fuse locations
Key – RHD shown, read clockwise from top left:

Left-hand side fuse box
Right-hand side fuse box
Left-hand side rearward fuse box,
(RHS rearward fuse box shown opposite)
Left foward fuse box

Left Foward Fuse Box Chart

Fuse No	Fuse Colour Code	Value	Circuit
1 to 4			Unused
5	Red	10A	Left-hand main beam headlamp
6	Red	10A	Right-hand main beam headlamp
7	Red	10A	Radiator fan
8	Light blue	15A	Dim/dip headlamps (United Kingdom only)
9	Brown	7.5A	Left-hand dipped beam headlamp
10	Brown	7.5A	Right-hand dipped beam headlamp
11			Unused
12	Violet	3A	Headlamp levelling (German market only)

Left-Hand Side Rearward Fuse Box Chart

Fuse No	Fuse Colour Code	Value	Circuit
1	Violet	3A	Left-hand tail and left-hand number plate lamps
2	Violet	3A	Right-hand caravan/trailer tail lamp
3	Violet	3A	Left-hand caravan/trailer tail lamp

Right-Hand Side Rearward Fuse Box Chart

Fuse No	Fuse Colour Code	Value	Circuit
1	Violet	3A	Right-Hand tail & right-hand number plate lamps
2	Red	10A	Radio aerial
3	Tan	5A	Reverse lights (XJ–S 4.0 automatic)

R.H. Steering – Driver's Side Fuse Box Chart – Memory Seats

Fuse No	Fuse Colour Code	Value	Circuit
1	Light blue	15 A	Driver's seat control power – fore-aft-lumbar
2	Violet	3A	Kickdown (XJ–S V12 only)
3	Yellow	20 A	R.H. air conditioning fan
4	Pink	4 A	Radio telephone ignition feed
5	Light blue	15 A	Horns
6	Tan	5 A	Radio memory
7	Red	10 A	Radio power
8	Red	10 A	Windscreen washer pump
9	Red	10 A	Driver's seat heater
10	Black	1 A	45 second timer (XJ–S V12 only)
11	Red	10 A	Luggage compartment lamps (caravan power)

R.H. Steering – Driver's Side Fuse Box Chart – Non-Memory Seats

Fuse No	Fuse Colour Code	Value	Circuit
1	Yellow	20 A	Driver's seat control power – fore-aft-recline
2	Violet	3 A	Kickdown (XJ–S V12 only)
3	Yellow	20 A	R.H. air conditioning fan
4	Pink	4 A	Radio telephone ignition feed
5	Light blue	15 A	Horns
6	Tan	5 A	Radio memory
7	Red	10 A	Radio power
8	Red	10 A	Windscreen washer pump
9	Red	10 A	Driver's seat heater
10			
11	Red	10 A	Luggage compartment lamps (caravan power)

R.H. Steering – Passenger's Side Fuse Box Chart – XJ–S 4.0

Fuse No	Fuse Colour Code	Value	Circuit
1	Yellow	20 A	Passenger seat control power – fore-aft-recline
2	Violet	3 A	Speed (cruise) control
3	Yellow	20 A	L.H. air conditioning fan
4	Tan	5 A	Interior lights
5	Brown	7.5 A	Direction indicators
6	Red	10 A	Central door locking
7	Red	10 A	Cigar lighter
8	Light blue	15 A	Windscreen wiper system
9	Light green	30 A	Headlamp power wash
10	Violet	3 A	Diagnostic memory power supply unit – auto transmission
11	Light green	30 A	ABS ECU

L.H. Steering – Driver's Side Fuse Box Chart – Memory Seats

Fuse No	Fuse Colour Code	Value	Circuit
1	Light blue	15 A	Driver's seat control power – fore-aft-lumbar
2	Violet	3A	Kickdown (XJ–S V12 only)
3	Yellow	20 A	L.H. air conditioning fan
4	Pink	4 A	Radio telephone ignition feed
5	Tan	5 A	Radio telephone power feed (where fitted)
6	Tan	5 A	Radio memory
7	Red	10 A	Radio power
8	Red	10 A	Windscreen washer pump
9	Red	10 A	Driver's seat heater
10	Black	1 A	45 second timer (XJ–S V12 only)
11	Red	10 A	Luggage compartment lamps (caravan power)

L.H. Steering – Driver's Side Fuse Box Chart – Non–Memory Seats

Fuse No	Fuse Colour Code	Value	Circuit
1	Yellow	20 A	Driver's seat control power – fore-aft-recline
2	Violet	3 A	Kickdown (XJ–S V12 only)
3	Yellow	20 A	L.H. air conditioning fan
4	Pink	4 A	Radio telephone ignition feed
5	Tan	5 A	Radio telephone power feed (where fitted)
6	Tan	5 A	Radio memory
7	Red	10 A	Radio power
8	Red	10 A	Windscreen washer pump
9	Red	10 A	Driver's seat heater
10			
11	Red	10 A	Luggage compartment lamps (caravan power)

L.H. Steering – Passenger's Side Fuse Box Chart – XJ–S 4.0

Fuse No	Fuse Colour Code	Value	Circuit
1	Yellow	20 A	Passenger seat control power – fore-aft-recline
2	Violet	3 A	Speed (cruise) control
3	Yellow	20 A	L.H. air conditioning fan
4	Tan	5 A	Interior lights
5	Brown	7.5 A	Direction indicators
6	Red	10 A	Central door locking
7	Red	10 A	Cigar lighter
8	Light blue	15 A	Windscreen wiper system
9	Light green	30 A	Headlamp power wash
10	Violet	3 A	Diagnostic memory power supply unit – auto transmission
11	Light green	30 A	ABS ECU

Convertible only – Power Operated Hood Fuse

Fuse Colour Code	Value	Circuit	Location
Light green	30A	Power operated hood pump motor	Mounted on a bracket adjacent to the pump motor in rear stowage compartment

In-Line Fuses

Fuse Colour Code	Value	Circuit	Location
Tan	5A	Exhaust temperature warning (Japan only)	Behind the centre console adjacent to the R. H. foot-well
Grey	2A	Overspeed warning (Saudi Arabia only)	Behind the centre console adjacent to the R. H. foot-well
Tan	5A	P.I. engine harness diagnostic circuit	Adjacent to the diagnostic socket behind the centre console trim (R.H. side – XJS 5.3; passenger side – XJS 4.0)

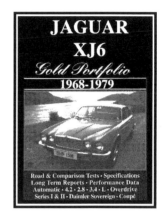
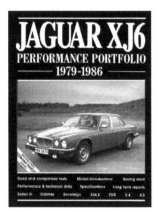

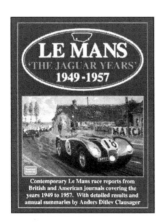

OFFICIAL TECHNICAL BOOKS

Brooklands Technical Books has been formed to supply owners,
restorers and professional
repairers with official factory literature.

Workshop Manuals

Jaguar Service Manual 1946-1948		9781855207844
Jaguar XK 120 140 150 150S & Mk 7, 8 & 9		9781870642279
Jaguar Mk 2 (2.4 3.4 3.8 240 340)	E121/7	9781870642958
Jaguar Mk 10 (3.8 & 4.2) & 420G	E136/2	9781855200814
Jaguar 'S' Type 3.4 & 3.8	E133/3	9781870642095
Jaguar E-Type 3.8 & 4.2 Series 1 & 2		
	E123/8, E123 B/3 & E156/1	9781855200203
Jaguar E-Type V12 Series 3	E165/3	9781855200012
Jaguar 420	E143/2	9781855201712
Jaguar XJ6 2.8 & 4.2 Series 1		9781855200562
Jaguar XJ6 3.4 & 4.2 Series	E188/4	9781855200302
Jaguar XJ12 Series 1		9781783180417
Jaguar XJ12 Series 2 / DD6 Series 2	E190/4	9781855201408
Jaguar XJ6 & XJ12 Series 3	AKM9006	9781855204010
Jaguar XJ6 OWM (XJ40) 1986-94		9781855207851
Jaguar XJS V12 5.3 & 6.0 Litre	AKM3455	9781855202627
Jaguar XJS 6 Cylinder 3.6 & 4.0 Litre	AKM9063	9781855204638

Owners Workshop Manuals

Jaguar E-Type V12 1971-1974	9781783181162
Jaguar XJ, Sovereign 1968-1982	9781783811179
Jaguar XJ6 Workshop Manual 1986-1994	9781855207851
Jaguar XJ12, XJ5.3 Double Six 1972-1979	9781783181186

Parts Catalogues

Jaguar Mk 2 3.4	J20	9781855201569
Jaguar Mk 2 (3.4, 3.8 & 340)	J34	9781855209084
Jaguar Series 3 12 Cyl. Saloons		9781783180592
Jaguar E-Type 3.8	J30	9781869826314
Jaguar E-Type 4.2 Series 1	J37	9781870642118
Jaguar E-Type Series 2	J37 & J38	9781855201705
Jaguar E-Type V12 Ser. 3 Open 2 Seater	RTC9014	9781869826840
Jaguar XJ6 Series 1		9781855200043
Jaguar XJ6 & Daimler Sovereign Ser. 2	RTC9883CA	9781855200579
Jaguar XJ6 & Daimler Sovereign Ser. 3	RTC9885CF	9781855202771
Jaguar XJ12 Series 2 / DD6 Series 2		9781783180585
Jaguar 2.9 & 3.6 Litre Saloons 1986-89	RTC9893CB	9781855202993
Jaguar XJ-S 3.6 & 5.3 Jan 1987 on	RTC9900CA	9781855204003

Owners Handbooks

Jaguar XK120		9781855200432
Jaguar XK140	E101/2	9781855200401
Jaguar XK150	E111/2	9781855200395
Jaguar Mk 2 (3.4)	E116/10	9781855201682
Jaguar Mk 2 (3.8)	E115/10	9781869826765
Jaguar E-Type (Tuning & prep. for competition)		9781855207905
Jaguar E-Type 3.8 Series 1	E122/7	9781870642927
Jaguar E-Type 4.2 2+2 Series 1	E131/6	9781869826383
Jaguar E-Type 4.2 Series	E154/5	9781869826499
Jaguar E-Type V12 Series 3	E160/2	9781855200029
Jaguar E-Type V12 Series 3 (US)	A181/2	9781855200036
Jaguar XJ (3.4 & 4.2) Series 2	E200/8	9781855201200
Jaguar XJ6C Series 2	E184/1	9781855207875
Jaguar XJ12 Series 3	AKM4181	9781855207868

Carburetters

SU Carburetters Tuning Tips & Techniques	9781855202559
Solex Carburetters Tuning Tips & Techniques	9781855209770
Weber Carburettors Tuning Tips and Techniques	9781855207592

Jaguar - Road Test Books

Jaguar and SS Gold Portfolio 1931-1951	9781855200630
Jaguar XK120 XK140 XK150 Gold Port. 1948-60	9781870642415
Jaguar Mk 7, 8, 9, 10 & 420G	9781855208674
Jaguar Mk 1 & Mk 2 1955-1969	9781855208599
Jaguar E-Type	9781855208360
Jaguar XJ6 1968-79 (Series 1 & 2)	9781855202641
Jaguar XJ12 XJ5.3 V12 Gold Portfolio 1972-1990	9781855200838
Jaguar XJS Gold Portfolio 1975-1988	9781855202719
Jaguar XJ-S V12 1988-1996	9781855204249
Jaguar XK8 & XKR 1996-2005	9781855207578
Road & Track on Jaguar 1950-1960	9780946489695
Road & Track on Jaguar 1968-1974	9780946489374
Road & Track On Jaguar XJ-S-XK8-XK	9781855206298

Brooklands Books Ltd., PO Box 146, Cobham,
Surrey KT11 1LG, England
E-mail: sales@brooklands-books.com   www.brooklandsbooks.com

ISBN: 9781855204638   Part No. AKM 9063 & Supplement AKM 9063BB   Ref: J84WH   2210/01T4

www.brooklandsbooks.com

Printed in Great Britain
by Amazon

46566353R00291